UNDERGRADUATE INSTRUMENTAL ANALYSIS

UNDERGRADUATE INSTRUMENTAL ANALYSIS

Fifth Edition,
Revised and Expanded

James W. Robinson
Department of Chemistry
Louisiana State University
Baton Rouge, Louisiana

Marcel Dekker, Inc. New York • Basel • Hong Kong

Library of Congress Cataloging-in-Publication Data

Robinson, James W.
Undergraduate instrumental analysis / James W. Robinson. — 5th ed.
p. cm.
Includes bibliographical references and index.
ISBN 0-8247-9215-7
1. Instrumental analysis. I. Title.
QD79.I5R6 1994
543—dc20 94-21455
CIP

The publisher offers discounts on this book when ordered in bulk quantities. For more information, write to Special Sales/Professional Marketing at the address below.

This book is printed on acid-free paper.

MARCEL DEKKER, INC.
270 Madison Avenue, New York, New York 10016

Current printing (last digit):
10 9 8 7 6 5 4 3 2 1

PRINTED IN THE UNITED STATES OF AMERICA

Preface to the Fifth Edition

The field of analytical chemistry is expanding at an explosive rate. Automation and new instrumental techniques have increased scientific knowledge by leaps and bounds while greatly reducing operator error. The undivided attention to tedious details and the almost artistic skills needed to get accurate results are no longer a cornerstone of the field.

Within living memory, the determination of components at the 0.01% concentration level was a challenge, usually accomplished by taking a large sample and concentrating the analyte by any means possible "à la Madame Curie." Today, numerous analytical techniques need no sample preparation but are capable of detecting part per billion and part per trillion concentrations. In other advances, huge protein molecules can now be fragmented and characterized, and these techniques are providing the basis for cloning and gene modification. The information gained from these powerful new techniques can be invaluable in understanding chemistry but destructive when applied blindly to ethical or legal matters.

These days, analytical chemistry is concerned with applying a multitude of instruments to answer questions we did not even know how to ask until these very techniques revealed them to us. Such questions are: What are the health effects of various toxic materials at the part per million or part per billion levels of concentration? Why are they toxic? How can we prevent deleterious health effects? How do we analyze semiconductors for impurities when the impurities are at the 10^{-8} g/g level? How do we separate optically active drugs and investigate their different physiological effects?

All these everyday problems involve using complex equipment from HPLC to ICP–MS, and the analytical chemist should know the best procedure to use to solve the problem, the limitations of the methods, and other analytical techniques that will give corroborating information.

We have documented the basic chemistry and physics of these innovations and presented them in terms understandable to the chemist and the nonchemist alike. Also, we have provided examples of the applications and shortcomings of each technique and illustrated how to report the data so that the results are meaningful to all concerned.

Each chapter is written in sufficient depth to be used for an advanced-level instrumental analysis course. The range of techniques covered allows the instructor a wide choice of topics. For analytical chemistry at less advanced levels, the book can be used by chemists and nonchemists alike.

For courses such as organic chemistry lab, the fundamentals presented in Chapters 1–6, 13, and 15 are suggested. These chapters cover qualitative and quantitative molecular analysis. To this end, sections on interpretation of IR, NMR, and MS spectra have been added to the interpretation of UV spectra. Chapter 7–10 and 16 may be added for the traditional sophomore quantitative analytical course.

To help in understanding the subject matter and for out-of-classroom assignments, numerous problems (and answers in a solutions manual, which instructors may order from the publisher) have been added for each chapter.

Keeping pace with the advances in analytical chemistry is imperative for all scientists in order to operate at maximum efficiency and to stay abreast of current research in all scientific fields utilizing analytical chemistry. We hope this text serves this function for all such scientists.

James W. Robinson

Preface to the Fourth Edition

The book covers the more common instrumental methods used in analytical chemistry. It is written descriptively for a terminal course in analytical chemistry for the nonchemistry major and chemistry major alike.

It is not anticipated that the student will become an expert in any of the fields covered after reading this book. However, he or she should become familiar with the fundamentals of the methods, such as the bases of the techniques, the information provided, the difficulties involved in obtaining this information, and what information cannot be obtained from the various methods described. If all the techniques are understood, the student should be able to select the most useful analytical approach for characterizing typical samples. He or she should also be able to suggest alternative methods for confirmational purposes. The nonanalytical chemist should find this book helpful in deciding on the analytical methods to use to obtain the needed information and the methods' reliabilities.

The various spectroscopic methods are presented in order of decreasing wavelength of the radiation involved. Consequently, the first topic discussed in this field is nuclear magnetic resonance; the last topic is x-ray spectroscopy. A particular analytical course may not cover all the chapters presented, so each chapter has been made as self-explanatory as possible. It is not, therefore, necessary to read a major part of the book in order to understand any particular chapter. To achieve this goal some aspects of the presentation have been somewhat scattered and may seem repetitious; however, it was felt that the advantages of this approach far outweigh the disadvantages.

It was also felt that, in view of the many well-written and reliable books available on volumetric and gravimetric analysis, it would be superfluous to discuss these topics here. Consequently, no detailed discussion has been devoted to volumetric or gravimetric analysis or to qualitative analysis using chemical schemes. This is not to say that these methods are not important; they continue to have their place in analytical chemistry and will do so for many years.

James W. Robinson

What Is Analytical Chemistry?

Perhaps the most apt definition of analytical chemistry is "the qualitative and quantitative characterization of a material or materials." This broad definition includes many functions carried out by analytical chemists but does not differentiate these functions from those of many other scientists. But, an analytical chemist can be described as someone who publishes in analytical journals and attends analytical meetings.

One of the prime functions of the analytical chemist is the qualitative and quantitative determination of the elements present in chemical compounds. From these determinations an empirical formula can be derived for newly synthesized compounds, natural products derived from living tissue, newly developed drugs, new plastics, and so on.

Other functions of the analytical chemist include the elucidation of the molecular structure of compounds such as proteins, the determination of the distribution of isomers in a mixture, and the determination of impurities in products as diverse as drinking water, motor oil, sea water, cigarette smoke, steel, blood, and plants. An important issue today is the effect of impurities in our environment. Are some of them toxic or carcinogenic? Analytical chemistry plays a major role in determining these effects by establishing the quality and quantity of trace materials present in foodstuffs, toxins, pollutants, drugs, cosmetics, and so on.

Physical properties must sometimes be measured. For example, crystallinity has a profound effect on the strength of polymers. A manufacturer must know if a polymer will tear when it is being used (it must tear easily if it is the wrapping on a package of cigarettes, but it must not tear if it is the protective wrapping on a scientific instrument). It may also be necessary to determine how long it will take a polymer to decompose if it is left out in the sun or rain. If it is to be used for roofing, we want it to last forever; if it is to be used as a garbage bag and then thrown away, we want it to decompose—otherwise it will become a problem in solid waste disposal. The same questions may be raised about varnishes and paints. Usually, these are the questions that must be answered by the analytical chemist.

Another problem tackled by the analytical chemist is the determination of the molecular weights of various compounds. A single compound does not present a very difficult problem, because it has only one molecular weight, but a mixture of compounds will encompass a whole range of molecular weights. This is particularly true with polymers and naturally occurring products, such as those found in plant and animal life, and it is then necessary to determine the molecular weight range and distribution of the compounds present.

The analytical chemist is also called upon to determine the compositions of mixtures. If two or three components are present in a mixture, it is often important to know how much of each is present. This information is necessary to such people as metallurgists and engineers who prepare alloys and need to know the actual composition of the alloys they have formed.

Companies that manufacture soaps or detergents need to know how much of their product is biodegradable. Federal law requires that all detergents and soaps be biodegradable. When soaps used in the household or laundry are washed down the drain and into our rivers, they must not persist as soap. Bacteria should be able to attack the soap and degrade it into other forms of matter. If the soap cannot be degraded, it will create foaming problems in rivers and streams; contamination of drinking water is possible, and under severe conditions the fish and vegetation in these rivers will be killed. The analytical chemist can determine how fast a detergent will be degraded in rivers or sewage.

In medicine, we are beginning to understand the effects of trace metals on the different parts of the body, and we are now trying to correlate the presence or absence of these metals with the occurrence of certain diseases. The analytical chemist in medical research is asked to provide methods of identifying and determining these trace metals.

For many years, analytical chemists were concerned with the chemical properties of the materials analyzed. It was common practice to dissolve the sample and then carry out some chemical reaction on the chemical component of interest. The extent of the reaction was then measured, and the concentration of the chemical component present in the sample was calculated. This system served analytical chemists well in the past. Schemes of qualitative and quantitative analysis were developed based on chemical reactions. Such schemes were valuable both to teach reaction chemistry and to carry out qualitative and quantitative analysis. The most important analytical fields involved were volumetric and gravimetric analysis. Unfortunately, these procedures demanded a high degree of skill and loving care on the part of the analytical chemist to produce reliable results. The analytical chemist who worked successfully in these fields was an artist. Care and patience were of prime importance in getting correct results every time an analysis was run.

Because modern industry has grown so quickly and the continuous control of manufactured products has become such an important and vast problem, the few skilled analytical wet chemists cannot provide the information needed. Masses of data are demanded every day by people in many walks of life. In hospitals, diagnosis of illness is dependent to a large degree on the routine determination of sodium, potassium, calcium, and magnesium in blood. In industry, many stages of manufacturing require quality control by analysis to make consistently good products. In this competitive world, we cannot afford to make substandard products.

Analytical chemists are also called upon to distinguish between master artworks and fakes. In the process of carrying out their examinations, they must not destroy or blemish the work of art. The methods used must be *nondestructive*, a factor that has become increasingly important to many areas of analytical chemistry in recent years.

Hiring more people to carry out analyses based on chemical reactions is not the answer. Frequently the newly hired analytical chemist has neither the interest nor the skill to perform like the artists of old. Furthermore, the time involved in running volumetric and gravimetric analyses is very long, and industry cannot afford to delay decisions until the analytical chemist can obtain his or her results.

As a response to these demands analytical chemists have had to turn to new ways by which to carry out analyses. These methods are generally based on the physical properties of the material. Extensive studies are being made on measuring and understanding the interactions of radiation and matter. By measuring the radiation effects, we are able to determine not only what is present in the sample, but also how much. Based on these interactions, instrumentation has been developed to carry out analyses on a rapid and even continuous basis.

A large number of important fields of physics and chemistry have been united to provide the answers required by analytical chemists. The procedures developed range in complexity. Some methods are completely automatic and require very little attention from an operator; on the other hand, some instruments have been developed that require a high degree of skill to operate. Furthermore, some equipment provides results that take years of training to interpret; however, the information obtained by using this equipment cannot be obtained by other methods.

What is the function of the analytical chemist? The first requirement is to provide data to other scientists. These scientists include other chemists, biochemists, biologists, physicians and medical technicians, agricultural scientists, metallurgists, engineers, health scientists, nutrition scientists, and environmentalists, among many others. To satisfy them, the analytical chemist must be cognizant of all the analytical methods presently available. When asked to characterize a sample, he or she should know what analytical procedure can be used to best advantage to obtain the information.

What can other scientists gain from analytical chemistry? In order to make optimal use of the techniques that have been developed in analytical chemistry, other scientists should know what information is available to them through analytical chemistry and, almost as important, what is not available to them. If scientists understand the principles of the various fields of analytical chemistry, they will know if they can obtain the information they desire, and they will also know the best way to obtain it; otherwise they may waste much time trying to get the information by some indirect means. Knowledge of analytical chemistry is highly desirable in all people who use analytical data to arrive at a decision in their work.

Scientists seeking information from analytical chemists must provide reliable samples that will give accurate results when analyzed. They should also communicate as much information concerning the samples as possible before the analytical chemist starts to work. This enables the analytical chemist to make the right choice in selecting a technique to solve the problem at hand.

Research analytical chemists have somewhat different functions. They should be aware of and contribute to new developments in chemistry and physics. Their most important role is to harness these developments for analytical chemistry so as to provide either sources of new information or more rapid and more sensitive methods of obtaining information.

ALL THOSE "ILLIONS"

Nowadays, determinations of concentrations at the part per million, part per billion, and part per trillion levels are commonplace. The mind has difficulty distinguishing these different "illions" and boggles at trying to grasp their meaning. One way to do this is as follows:

A million seconds is about 12 days (actually 11.57). So a *part per million* (ppm) is as significant as 1 second in 12 days. Similarly, a billion seconds is 32 years (actually 31.7). So a *part per billion* is as significant as 1 second in about 32 years. Similarly, a trillion seconds is 32,000 years. So, a *part per trillion* is as significant as 1 second in 32,000 years.

Christ was born 2,000 years ago, the earliest recorded human history was 5,000 years ago, the last ice age was 17,000 years ago. Does that help to get these "illions" into perspective?*

---

*It is interesting to note that the national debt is $4 trillion. If we paid $240,000 per minute, paying no interest on the principal, it would take 32 years to pay it off.

Numerous analytical techniques are capable of detecting part per billion and part per trillion concentrations. Some, like GC–electron capture, atomic absorption spectroscopy, and ICP–MS can operate at 10^{-14} concentration levels. This information can be useful in understanding body chemistry or can be destructive when blindly applied to legal matters.

Clearly, analytical chemistry is capable of answering questions we didn't even know to ask a few years ago. It is opening up new fields of endeavor at a prodigious rate. Keeping pace with the advances in analytical chemistry is imperative for all chemists in order to operate at maximum efficiency and to stay abreast of current research in all related fields.

Contents

UNDERGRADUATE INSTRUMENTAL ANALYSIS

1
Concepts of Analytical Chemistry

The prime concern of analytical chemistry is the qualitative and quantitative characterization of matter. The word "characterization" is used in a very broad sense in this instance. It may mean the qualitative identification of impurities present in a sample or the quantitative determination of a given compound in another sample; yet again, it may include the determination of the degree to which a polymer is crystalline as opposed to the degree which it is amorphous, or how fast a paint degrades in the sunlight, or how much cholesterol is in a heart patient's blood.

The skilled analytical chemist should know the best way to carry out these determinations. Other scientists who need analytical results should know what information can be obtained directly and the reliability of the results reported.

Some of the more important concepts of analytical chemistry are indicated in this chapter.

A. QUALITATIVE ANALYSIS

Qualitative analysis is the branch of analytical chemistry that is concerned with the question, "What compounds or elements are in this sample?" Numerous ways have been devised to provide the answer. The first of the many techniques available is elemental analysis. If the sample is a metal, the results of elemental analysis disclose whether it is an alloy or a pure metal. If the method of elemental analysis is sufficiently sensitive, one can determine the elements in the impurities present in the sample. Methods of detection sensitive to 1 part per billion (ppb) have been developed and used over the years. The most useful methods of qualitative elemental analysis include inductively coupled photometry (ICP) and inductively coupled photometry–mass spectrometry (ICP–MS), plasma emission, activation analysis, x-ray fluorescence, flame tests, and the application of wet chemical qualitative schemes.

If the sample is organic, qualitative analysis can be carried out at several levels. For example, if the sample were fructose, elemental analysis would reveal the empirical formula CH_2O. Molecular weight determination would indicate 180—which corresponds to the molecular formula $C_6H_{12}O_6$. Structural formula determination by infrared (IR) spectroscopy, nuclear magnetic resonance (NMR), or mass spectrometry (MS) would result in two formulas, which can be distinguished from each other by their polarizing effect on polarized light.

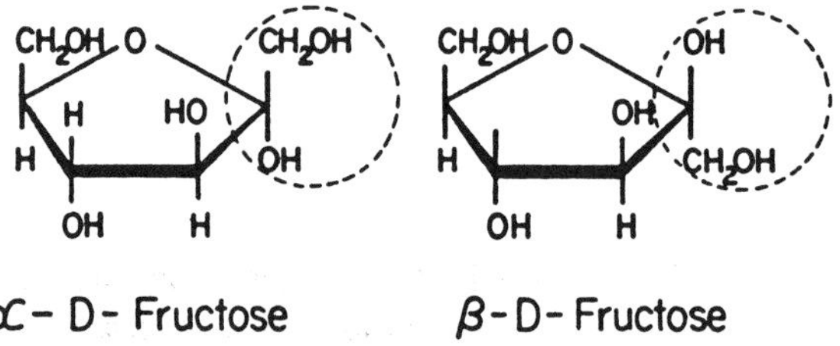

α- D- Fructose β- D- Fructose

These molecules are optically active, and this is biologically important because the body absorbs only dextrorotatory D compounds. The presence of certain functional groups may be identified by IR. Such groups include alcohols, ketones, aldehydes, esters, acids, ethers, mercaptans, thio ethers, ammines, and so on. Their positions relative to each other in the molecule can be deduced by NMR, and their molecular weight from MS.

Mixtures of compounds can be separated by gas–liquid or liquid–solid chromatography, depending on molecular weight, and each separate compound can be characterized by techniques such as those listed above.

Qualitative analysis has become increasingly important to organic chemists, biochemists, biologists, agricultural scientists, environmental health scientists, and medical research workers. The most useful methods of qualitative molecular analysis include NMR, IR, ultraviolet (UV) absorption spectroscopy, x-ray fluorescence, x-ray diffraction, and MS. Each of these fields is discussed later in this book.

B. QUANTITATIVE ANALYSIS

When qualitative analysis is completed, the next question is "How much of each or any component is present?" The determination of "how much" is called *quantitative analysis*. The first quantitative analytical fields to be developed were for *quantitative elemental analysis*, which revealed how much of each element was present in a sample. Using chemical reactions, it was possible to determine the compositions of simple salts. It was found, for example, that dry sodium chloride (NaCl) always contained 39.33% Na and 60.67% Cl. The atomic theory was founded on early quantitative results such as these, as was the concept of valency and, finally, the determination of atomic weights.

In similar fashion, quantitative elemental analysis enabled the inorganic chemist to determine the empirical formulas of new compounds. An *Empirical formula* is the simple ratio of the number of elements that make up a molecule. For any given compound, the empirical formula may or may not coincide with the molecular formula. A *molecular formula* indicates the different elements present and the total number of atoms of each element in a single molecule. For example, the empirical formula of fructose is CH_2O, but its molecular formula is $C_6H_{12}O_6$.

For many years, organic chemists could obtain only the empirical formulas of newly synthesized or isolated compounds. If the results coincided with the expected result, it was strong supporting evidence in the identification of the compound. For example, a chemist might believe that he had isolate butane, whose molecular formula is C_4H_{10} and whose analysis is 82.76% C and 17.24% H. If, however, he had actually isolated pentane, with molecular formula C_5H_{12} and analysis 83.33% C, 16.66% H, it would require great skill to distinguish between these two very similar compounds by elemental analysis. The whole field of microanalysis was devoted to this end.

Unfortunately, quantitative elemental analysis could not differentiate between isomers of the same compounds. For example,

$$CH_3\text{—}CH_2\text{—}CH_2\text{—}CH_3 \quad \text{and} \quad H_3C\text{—}\underset{\substack{|\\ CH_3}}{\overset{\substack{H\\ |}}{C}}\text{—}CH_3$$

n-butane and *i*-butane

have the same molecular formula and elemental analysis. The inability to distinguish between isomers is a handicap to the chemist. As the fields of organic chemistry and biochemistry have become increasingly complicated, quantitative molecular analysis has been developed to provide better and more useful information. The results of such analyses make it possible to determine (1) how much of a particular compound is in a mixture, (2) what functional groups are in the molecule, (3) how the functional groups are arranged in the molecule, (4) the three-dimensional configuration of the molecule, and (5) other information leading to the complete characterization of the sample.

The most important methods of quantitative molecular analysis include MS, NMR, IR, UV, x-ray diffraction, and thermal analysis. The basis and application of these methods are discussed in the relevant chapters. A summary of the major analytical fields and their applications is given in Tables 1.1 and 1.2.

Table 1.1 Principal Applications of Analytical Techniques

Molecular Analysis

Nuclear magnetic resonance

a. Qualitative analysis: Nuclear magnetic resonance primarily identifies type of hydrogen and carbon in organic molecules (aromatic, aliphatic, alcohols, aldehydes, etc.) Most importantly, it also reveals the positions of the functional groups in the molecule relative to each other. For example, it will distinguish between CH_3—CH_2—CH_2OH and CH_3—CHOH—CH_3. It does not indicate the molecular weight of the compound. It gives some indication of hetero elements such as sulfur, nitrogen, and halides.

b. Quantitative analysis: The method is useful at the percent concentration level, but not at trace levels, e.g. PPM. It is not generally considered a highly accurate method of quantitative analysis at these concentration levels.

Infrared spectroscopy

a. Qualitative analysis: Infrared readily identifies functional groups present in the molecule and gives some indication of the presence of hetero elements—O, S, N, Si, halides—and includes their environment. Infrared spectroscopy distinguishes between aliphatic and aromatic compounds. Except in special cases, it does not give the molecular weight of the compound or the positions of the functional groups relative to each other.

b. Quantitative analysis: Infrared spectroscopy is used routinely for the quantitative analysis of organic compounds, particularly at high concentration levels. The method requires skill in order to get good results. It is used mostly for liquid samples and has less application to solid sample analysis. The related field of Raman spectroscopy has been used for the characterization of polymers, as well as more convenient organic compounds. Application of IR spectroscopy to gas samples is limited because of lack of sensitivity.

Ultraviolet absorption

a. Qualitative absorption: Ultraviolet absorption is particularly useful for identifying functional groups and the structures of molecules containing unsaturated bonds (π electrons), such as

```
H       H  H      H
 \      |  |     /
  C  =  C = C = C
 /                \
H                  H
```

Table 1.1 Continued

and aromatics and lone-pair electrons, such as those in pyridine:

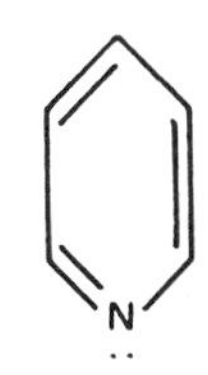

lone pair

It does not indicate molecular weight or give useful information on saturated bonds (σ bonds).

b. *Quantitative analysis*: Ultraviolet absorption is used routinely for the quantitative determination of unsaturated compounds such as those found in natural products. It is particularly useful at the percent level and the 1/10 of a percent level. Using special reagents, it is not generally useful for trace quantitative analysis. The method is subject to spectral overlap and therefore interference from other compounds in the sample, but techniques are available to overcome these problems.

Ultraviolet fluorescence

a. *Qualitative analysis*: Ultraviolet fluorescence is used for the determination of unsaturated compounds, particularly aromatics. It does not indicate molecular weight, but gives some indication of the functional groups present. It is much more sensitive than UV absorption.

b. *Quantitative analysis*: Ultraviolet fluorescence is a very sensitive method of analysis (10^{-8} g/g or 10 ppb), but it is subject to many kinds of interference, both from quenching effects and from spectral overlap from other compounds.

Spectrophotometric analysis (Visible)

a. *Qualitative analysis*: Organic or inorganic reagents are used for specific tests for many elements or compounds forming a compound that absorbs at specific wavelengths. The products may or may not be colored. If the compounds are colored, analysis may be carried out visually (colorimetric analysis).

b. *Quantitative analysis*: Sensitive and selective methods have been developed for certain elements and groups. It is used extensively in routine analysis.

X-ray diffraction

a. *Qualitative analysis*: X-ray diffraction is used for the identification of crystal lattice dimensions and to identify the structure and composition of all types of crystals and related materials such as soils, polymers, and natural products.

b. *Quantitative analysis*: X-ray diffraction is used for the determination of percent crystallinity in polymers and mixtures and mixed crystals, soils, and natural products.

Table 1.1 Continued

X-ray absorption

Qualitative analysis: X-ray absorption reveals the contours and location of high atomic weight elements in the presence of low atomic weight matrixes or voids in solid samples. Examples are bone locations in the human body, the contents of suitcases, old paintings hidden under new painting on a canvas, bubbles in welded joints, or the distribution of high atomic weight elements in various samples.

Radiotracer techniques

By radiolabeling compounds, one can trace their pathways in various systems. The systems studied include biological systems, plant systems, rivers, streams, and sewage plants.

Mass spectrometry

a. Qualitative analysis: Mass spectrometry can be used to identify very high (100,000 Da) molecular weights of organic and inorganic compounds. It is used to identify functional groups and the structures of molecules. It cannot easily be used to identify the positions of functional groups relative to each other. It confirms the presence of hetero elements in molecules.

b. Quantitative analysis: Mass spectrometry is used extensively for the quantitative determination of the components of liquid and gas samples. Solid samples have been analyzed with difficulty.

Thermal analysis

a. Qualitative analysis: Thermal analysis is used to identify inorganic and some organic compounds. It is also used to identify phase changes, chemical changes on heating, thermal diagrams, heats of fusion, melting points, boiling points, drying processes, decomposition processes, and the purity of compounds.

b. Quantitative analysis: Thermal analysis can be used for the quantitative determination of the components of an inorganic sample, particularly at high concentration levels. It is not very useful for trace quantities because of its lack of sensitivity.

Gas chromatography

a. Qualitative analysis: Gas chromatography (GC) can be used to identify the components of a gas or low-boiling liquid based on retention time. It does not reveal molecular structure or molecular weight directly, except by comparison with standards.

b. Quantitative analysis: Gas chromatography is an accurate method for quantitative analysis based on the area of the peak and comparison with standards. It is used extensively in organic and inorganic analysis.

Table 1.1 Continued

Liquid chromatography

a. Qualitative analysis: Liquid chromatography (LC) is used for the identification of components of liquid mixtures, including high molecular weight components and thermally unstable compounds. Identification is based on retention time and comparisons with standards. Liquid chromatography does not give molecular structure or molecular weight, except by comparison with standards.

b. Quantitative analysis: Liquid chromatography is used for the quantitative determination of components of high molecular weight or thermally unstable compounds. It is particularly useful for separating complicated mixtures such as natural products derived from plants or animals.

Electron microscopy

Qualitative analysis: Electron microscopy reveals the shapes and dimensions of crystals and natural products such as plants, cells, body tissues, blood cells, and dust particles.

Elemental Analysis

Emission spectrography

a. Qualitative analysis: Emission spectrography, including plasma emission, is one of the best methods for qualitative elemental analysis, particularly for metals and metalloids, but it is not usually useful for the nonmetals. Its sensitivity range is great, varying from fractional parts per million to percent levels.

b. Quantitative analysis: Emission spectrography and ICP are used extensively for the quantitative determination of metals in concentrations down to fractional parts per million. ICP-MS is used for simultaneous multielement analysis for metals and nonmetals alike. It has a wide linear range from ppb to percent levels. It also provides the isotope distribution of the elements.

Flame photometry

a. Qualitative analysis: Flame photometry is particularly useful for the determination of alkali metals and alkaline-earth metals. It provides the basis for flame tests used in qualitative analysis schemes.

b. Quantitative analysis: Flame photometry is used for the quantitative determination of alkaline metals and alkaline-earth metals plus some transition elements. It provides much simpler spectra than those found in emission spectrometry, but its sensitivity, particularly for the transition metals, is much reduced.

Atomic absorption spectroscopy

a. Qualitative analysis: Atomic absorption spectroscopy is not used for qualitative analysis, except under special circumstances since it is only possible to test for one element at a time.

Table 1.1 Continued

---

b. Quantitative analysis: Atomic absorption spectroscopy is perhaps the most accurate and sensitive methods available for the quantitative determination of metals and metalloids down to absolute amounts as low as 10^{-14} g. It cannot be used directly for the determination of nonmetals. It is more accurate than most other routine methods of analysis.

X-ray fluorescence

a. Qualitative analysis: X-ray fluorescence is useful for elements with atomic numbers greater than 10, including metals and nonmetals. The method requires elaborate procedures for qualitative determination, except under special circumstances.

b. Quantitative analysis: X-ray fluorescence is used extensively for quantitative determinations, particularly of elements with high atomic weights. It is very sensitive and produces reliable results.

Electron spectroscopy for chemical analysis

a. Qualitative analysis: Electron spectroscopy for chemical analysis (ESCA) is used for surface analysis and for the examination of sample surfaces. It is highly sensitive in terms of the number of atoms detected, but not in terms of the percentages of the elements on the surface being examined. By bombarding the surface with electrons, one can uncover successive atomic surface layers and examine their composition in situ, thus obtaining a three-dimensional structure diagram of surfaces.

b. Quantitative analysis: Electron spectroscopy for chemical analysis is useful only in a semiquantitative way, but it does reveal the distribution of the studied elements on the surface.

C. RELIABILITY OF RESULTS

The quantitative analysis of any particular sample should generate results that are reproducible, reliable, and truly representative of the sample. Unfortunately, it is virtually impossible to characterize a sample exactly, because some degree of error is always involved in the results. Under the worst conditions, this error may be gross. Not only may the results be scientifically worthless, but they may also lead to serious errors of judgment. For example, if the sample were a product from a commercial plant stream, incorrect modifications to the plant equipment might be made by the manufacturing company and the quality of the product might be degraded as a result. The overall cost to the company might be

Table 1.2 Methods of Analytical Analysis

	Qualitative		Quantitative	
Method	Elemental	Molecular	Elemental	Molecular
Activation analysis	Yes	No	Yes	No
Atomic absorption	No	No	Yes	No
Electrochemistry	Yes	Yes	Yes	Yes
Electron spectroscopy	No	Yes	No	Yes
Emission spectrography	Yes	No	Yes	No
ESCA	Yes	No	Yes	No
Flame photometry	Yes	No	Yes	No
Gas chromatography	No	Yes	No	Yes
Gravimetric analysis	No	No	Yes	Yes
ICP	Yes	No	Yes	No
ICPMS	Yes	No	Yes	No
Infrared	No	Yes	No	Yes
Liquid chromatography	No	Yes	No	Yes
Mass spectrometry	Yes	Yes	Yes	Yes
Microchemistry	No	No	Yes	No
Nuclear magnetic resonance	No	Yes	No	Yes
Nuclear science	No	Yes	No	Yes
Spectrophotometry	Yes	Yes	Yes	Yes
Thermal analysis	No	Yes	No	Yes
UV absorption	No	Yes	No	Yes
UV fluorescence	No	Yes	No	Yes
Volumetric analysis	No	No	Yes	Yes
X-ray absorption	Yes	No	Yes	No
X-ray diffraction	No	Yes	No	Yes
X-ray fluorescence	Yes	No	Yes	No

very high. On the other hand, under the best conditions, the data would lead to correct modification and an improved product.

For analytical results to be most useful, it is important to be aware of the reliability of the results. To do this it is necessary to understand the sources of error and to be able to recognize when they can be eliminated and when they cannot. The error is the difference between the true answer and the observed answer. There are numerous sources of error, some of which are described below.

1. Types of Errors

There are two principal types of errors: determinate errors and indeterminate errors. Broadly speaking, *determinate errors* are caused by faults in the analytical procedure or the equipment used in the analysis. A particular determinate error may cause the analytical results produced by the method to be always too high; another may render all results too low. Sometimes the inaccuracy produced is constant (e.g., all answers are 10% too high). For example, the true answers for the three samples were 25, 20, and 30%, but the answers obtained were 35, 30, and 40%, i.e., 10% too high. This is called a *constant error*. Sometimes the inaccuracy is proportional to the true answer, giving rise to *proportional errors* such as 10% of the answer. e.g., for the same samples as above the answers obtained were 27.5, 22.0, and 33.0%, which is 10% too high. Consequently, although the results produced by repetitive analysis of a single sample may all be too high (and inaccurate), they may nevertheless agree closely with each other, indicating high precision. It can be seen that close agreement between results (high precision) may not be an indication of the accuracy of the results. In this case, there is no determinate error. However, the small differences between each observation indicate indeterminate error.

Indeterminate errors are not constant or biased. They cause varying results, some of which may be too high and some too low. Indeterminate errors usually arise from sources that cannot be corrected. For example, a four-place balance may give the weight of the sample as 1.032 g, when the actual weight is 1.03215 g. The 0.00015 g cannot be measured on this balance and is uncorrectable. This is an indeterminate error. Suppose, for example, that two machines are used to measure how long it takes a bullet to reach a target. Unknown to the operator, machine A has a faulty relay switch, so that all the answers obtained with it are incorrect. Machine B, however, is in good working order. The results obtained from each machine for five shots are shown in Table 1.3. The true time it takes for the bullet to reach the target is 0.123 sec. The average of the results form machine B is 0.123 sec. Each individual reading varies slightly from this reading, but the errors average each other out. The error is indeterminate. The results from machine A vary slightly from one to another (indeterminate error), but all the results are too high. This indicates a determinate error in addition to the indeterminate error.

a. Determinate or Systematic Errors

Determinate errors are errors that arise because of some faulty step in the analytical process. The faulty step is repeated every time the determination is performed. Whether a sample is analyzed 5 times, or 50 times, the results may all agree with each other but differ widely from the true answer. An example is

Table 1.3 Observed Time (sec) for a Bullet to Reach a Target

Machine A (determinate and indeterminate error)	Machine B (indeterminate error)
0.130	0.120
0.133	0.122
0.132	0.123
0.136	0.124
0.134	0.126
Average 0.133	0.123

given in Table 1.4. An analyst or doctor examining the analytical results might be deceived into believing that the close agreement among the answers indicates high accuracy and that the results are close to the true answer. It is the analyst's responsibility to recognize and correct for *systematic errors* that cause results to be consistently wrong. Two methods are commonly used to do this. One is to analyze the sample by a completely different analytical procedure that is known to involve no systematic errors. The second is to run several analyses of a standard solution of known concentration. The difference between the known (true) concentration and that obtained by analysis should reveal the error. If the results of numerous analyses are consistently high (or consistently low), then a determinate error is involved in the method. This error must be identified and a correction made before the procedure is capable of giving accurate results. Errors can arise from faulty balances, volumetric flasks, electrical dials, bent recording pens, impure chemicals, incorrect analytical procedures or techniques, operator error, low line voltage on instruments, and so on.

Table 1.4 Percentage of Potassium in a Patient's Blood

Observed analytical result (%)	True answer (%)
0.53	0.40
0.55	
0.54	
0.52	
0.51	
Average 0.53	0.40

Table 1.5 Percentage of Potassium in Patients' Blood

Patient	Analytical result (%)	True answer (%)
A	0.53	0.40
B	0.75	0.62
C	0.64	0.51
D	0.43	0.30
E	0.48	0.35
Constant error of −0.13%		

In the example above, the true answer was 0.40% and the observed result was 0.53%. If the doctor had repeated this procedure for numerous patients, he might have obtained the results shown in Table 1.5. It can be seen that in all cases the error was −0.13%. This indicates a constant determinate error.

On the other hand, a second doctor, working at a different hospital with different equipment and patients, might obtain the results shown in Table 1.6. Examination of these analytical results show they are all 25% greater than the true answer. The error is *proportional* to the answer. Such information is useful in the diagnosis of the source of the determinate error.

b. Indeterminate Errors

After all the determinate errors of an analytical procedure have been detected and eliminated, the method is still not capable of giving absolutely accurate answers. Numerous small undetectable errors may be made at each step of the procedure. Each error may be positive or negative, and the total error may be slightly too high or too low. The net error involved is an indeterminate error. All analytical procedures are subject to indeterminate error. The extent of this error can be calculated, and, when known, the reliability of the method can be stated.

Indeterminate errors may arise from many sources. For example, a balance may be accurate to within 0.001 g, in which case it would not discriminate be-

Table 1.6 Percentage of Potassium in Patients' Blood

Patient	Analytical result (%)	True answer[a] (%)
A	0.60	0.48
B	0.45	0.36
C	0.90	0.72
D	1.00	0.80
E	0.75	0.60

[a]Results indicate a positive proportional error of 20% of the true answer.

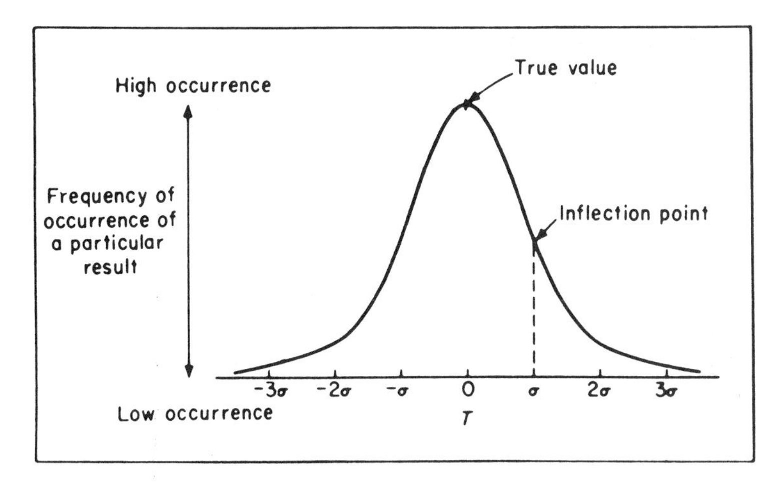

Figure 1.1 Distribution of results with indeterminate error. The range of the actual error is from low (−) to high (+); *T* stands for true value.

tween two samples that weigh 1.0151 and 1.0149 g. Other sources of error include the limit of accuracy of volumetric equipment and the limit of readout of electrical dials or recording instruments. The significance of these effects is that a small error is always involved in each step of an analytical determination.

Let us suppose that an analytical procedure has been developed in which there are no determinate errors. If an infinite number of analyses of a single sample were carried out using this procedure, the distribution of results would be shaped like a symmetrical bell (Fig. 1.1). This shape is called *Gaussian*. The frequency of occurrence of any given result would be represented graphically by Fig. 1.1.

If only indeterminate errors were involved, the most frequently occurring result would be the true result (i.e., the result at the maximum of the curve). Unfortunately, in practice it is not possible to make an infinite number of analyses of a single sample. At best, only a few analyses can be carried out, and frequently only one analysis of a particular sample is possible. We can, however, use our knowledge of statistics to determine how reliable these results are. The basis of statistical calculations is outlined below.

2. Definitions

True value T: the true or real value.
Observed value V: the value observed by experiment.

Error E: the difference between the true value T and the observed value V (it may be positive or negative),

$$E = |V - T|$$

Mean M: the arithmetic mean of the observations, that is,

$$M = \frac{\Sigma V}{N}$$

where N is the number of observations and ΣV is the sum of all the values V.

Absolute deviation d: the difference between the observed value V and the arithmetic mean M,

$$d = V - M$$

Deviation has no algebraic sign (+ or −); all differences are counted as positive.

Relative deviation D: The absolute deviation d divided by the mean M,

$$D = \frac{d}{M}$$

Percentage deviation: The deviation times 100 divided by the mean,

$$d(\%) = \frac{d \times 100}{M}$$

Relative error: The sum of the errors ΣE divided by N and the result divided by M,

$$E_{\text{rel}} = \frac{\Sigma E/N}{M}$$

Average deviation:

$$d_{\text{av}} = \frac{\Sigma d}{N}$$

This is numerically related to the standard deviation but has no real statistical significance.

Standard deviation σ:

$$\sigma = \sqrt{\frac{\Sigma d^2}{N - 1}}$$

[This may be expressed as the percent relative standard deviation (%RSD) = $(\sigma/M) \times 100$.]

Example 1.1 illustrates some of the statistical values obtained from the treatment of three observed values.

Example 1.1

V	Mean	Error	Deviation	Average deviation d
103	309/3	0	0	
101		−2	2	
105		+2	2	4/3
—	—	—	—	—
309	103	0	4	1.33

In Example 1.1, $E_{rel} = \Sigma\,(E/N)/M = (4/3)/103 = 4/309$; $d(\%) = [4/(3 \times 103)] \times 100\%$; $E = 0 - 2 + 2 = 0$. It can be seen that there are no determinate errors remaining because $E = 0$. The indeterminate errors, however, give rise to deviation of random differences from the true answer.

a. Standard Deviation σ

The most commonly used method of presenting the reliability of results is in terms of sigma (σ). In Fig. 1.1, the standard deviation coincides with the point of inflection of the curve; 68% of the results fall within ± σ of the true answer and 95% of the results obtained fall within ±2σ of the true answer. These facts enable us to give some meaning to quantitative analytical results.

For example, let us suppose that we know by previous testing that the standard deviation of a given analytical procedure for the determination of silicon (Si) is 0.1%. Also, when we analyze a particular sample using this method, we obtain a result of 28.6% Si. We can now report with 68% confidence that the analysis is 28.6% ± 0.1%. We could further report with 95% confidence that the true result is 28.6% ± 0.2% (where 0.2% = 2σ). Hence the report that an analysis of a sample indicated 28.6% Si and that 2σ for the method is 0.2% means that we are 95% confident that the true answer is 28.6% ± 0.2% Si.

Such information concerning the reliability of the method is called *precision data*. Analytical results published without such data lose much of their meaning. They indicate only the result obtained, and not the reliability of the answer or how close the data are to the true answer.

b. Variance

The square of the standard deviation (σ^2) can be used to detect the introduction of error in an analytical method. For example, after a given analytical procedure has been used on numerous occasions to analyze a certain sample, it is possible to obtain the standard deviation for the method.

We may suspect that at some time after the analytical method was developed, an error had crept into the procedure. Perhaps a new batch of chemicals used in the process was impure and gave rise to an error in the answer. Such an error would cause a shift in the level of results obtained. A check on the new results can be made by measuring the *variance* of the suspected results and comparing it with the variance of the results obtained previously. In this case, the variance σ_2^2 of the suspected results is calculated and compared with the variance σ_1^2 of earlier results. The ratio of the variances of these two sets of numbers is called the *F function*:

$$F = \frac{\sigma_1^2}{\sigma_2^2}$$

Tables of F values indicate the likelihood of two groups of numbers belonging to the same set or being from two different sets. If the F function shows that the two groups of analytical answers belong to two different sets of numbers, an error has crept into the method and should be eliminated. If the answers all belong to the same set of numbers, no new error has been introduced into the method and it may be used safely as before.

It should be remembered that the method indicates only the *probability* of the two sets of numbers belonging to one set. In all cases, a judgment has to be made. Of course, when more data are available, the judgment can be reviewed and a decision may be arrived at with more confidence.

3. Precision and Accuracy

It is very important to recognize the difference between precision and accuracy. *Accuracy* is a measure of how close a determination is to the true answer. *Precision* is a measure of how close a set of results are to each other. The difference is illustrated in Table 1.7. A superficial examination of the results provided by

Table 1.7 Results Obtained[a] (%)

	Analyst 1[b]	Analyst 2[c]	Analyst 3[d]
	10.0	8.1	13.0
	10.2	8.0	9.2
	10.0	8.3	10.3
	10.2	8.2	11.1
	10.1	8.0	13.1
	10.1	8.0	9.3
Average	10.1	8.1	11.9
Error	0.0	2.0	1.8

[a]True answer is 10.1% (obtained independently).
[b]Results are precise and accurate.
[c]Results are precise but inaccurate.
[d]Results are imprecise and inaccurate.

analyst 2 could be misleading. It is very easy to be deceived by the closeness of the answers into believing that the results are accurate. The closeness, however, shows that the results of analyst 2 are *precise*, and not that the analysis will result in our obtaining the true answer. The latter must be discovered by an independent method, such as having analyst 2 analyze a solution of known composition (e.g., a solution of pure chemical) and checking the answer against the known composition of the chemical. The U.S. National Institute of Standards and Technology (formerly the National Bureau of Standards) in Washington, D.C., has a number of such samples available for analysis.

It is important to realize that the inability to obtain the correct answer does not necessarily mean that the analyst uses poor laboratory technique or is a poor chemist. Many causes contribute to poor analyses, and only by being honest in recording the results can the causes of error be recognized and eliminated.

4. Error Analysis

When a new analytical procedure is developed or when an analytical procedure already in use is put into operation for the first time in a particular analytical laboratory, it is necessary to determine what errors are involved in the method. First, it is necessary to detect and eliminate any determinate error. The results obtained by the new method are then compared with those of an established method or those obtained by analyzing a reproducible and reliable standard. If they agree, the new method is free form determinate error; if they do not agree, an error is involved in the new method.

Each of these methods works well when the error involved is significant and obvious. However, when the difference between the results is small, the following question arises: Is the error caused by a small determinate error or an indeterminate error? This difficulty can be resolved by carrying out a number of analyses by the new method. Using a statistical approach, we can determine if the error is significant and therefore involves a determinate error, or if it is merely a part of the distribution of results encountered with indeterminate errors. If a determinate error is involved, it must be eliminated.

Having detected the presence of a determinate error, the next step is to find its source. Practical experience of the analytical method or first-hand observation of the laboratory operator using the procedure is invaluable. Much time can be wasted in an office guessing at the source of the trouble. Unexpected errors can be discovered only in the laboratory. A little data is worth a lot of discussion (ROBINSON'S LAW).

a. Common Determinate Errors

Operator Error. Operator errors are caused by the analyst performing the analysis. They may be the result of inexperience; that is, the operator may use the equipment incorrectly—for example, by placing the sample in the instrument incorrectly each time or setting the instrument to the wrong condition for analysis. Some operator-related errors are (1) *carelessness*, which is not as common as is generally believed, and (2) *poor sampling*, which is a common error. If the sample taken to the laboratory is not representative of the complete sample, no amount of good analysis can produce the correct result. Special attention should be paid to taking a representative of the complete sample for analysis. For example, it is usually fatal to send an untrained assistant to take a sample. It is equally dangerous to take the "first cupful" as a good sample.

Sample handling is also important. After a representative sample is received at the laboratory, it should be stored properly. Incorrect storage (e.g., in an open container) can cause contamination by impurities in the air or evaporation of the volatile components of the sample. After a while, the store material is not representative of the original sample. Another storage problem may be temperature: a sample that is kept too warm may decompose. Remember that even good wine goes sour if stored incorrectly. Each of the foregoing errors can be eliminated by proper operator attention.

Reagents and Equipment. Determinate errors can be caused by contaminated or decomposed *reagents*. Impurities in the reagents may interfere with the method, especially at the ppm level or below. The reagents may also be improperly labeled. the suspect reagent may be tested for purity by using a different set of reagents to recheck the work and by checking to see that the procedure used was the same as that recommended.

Numerous errors involving equipment are possible, including incorrect instrument alignment, incorrect wavelength settings, incorrect reading of values, and incorrect settings of the readout scale (i.e., zero signal should read zero, and for many samples the maximum single should be set at 100 or some other suitable number). Any variation in these settings can lead to repeated errors. These problems can be removed by a systematic check of the equipment.

Analytical Method. The analytical method proposed may be unreliable. It is possible that the original author obtained good results by a compensation of errors; that is, although he appeared to obtain accurate results, his method may have involved errors that balanced each other out. When the method is used in another laboratory, the errors may differ and not compensate for each other. A net error in the procedure may result. Errors involved include (1) incomplete reaction for chemical methods, (2) unexpected interferences from the solvent or impurities, (3) a high contribution from the blank (the *blank* is the result obtained by going through the analytical procedure, adding all reagents, and taking all steps, but either adding no sample or using pure solvent as the sample), (4) incorrect choice of wavelength for measurement of spectra, (5) an error in calculation based on incorrect assumptions in the procedure (errors can evolve from assignment of an incorrect formula or molecular weight to the sample), (6) matrix interferences, and (7) unexpected interferences. Regarding the last, most authors check all the compounds likely to be present to see if they interfere with the method; unlikely interferences may not have been checked. Checking the original publication should clear up this point.

There are other sources of error, such as the use of contaminated distilled water in a chemical method or a change in the line voltage used to operate the instruments. The later is particularly likely in industrial areas where heavy demands on local power may be made, or released, suddenly. This problem may be acute when the work shift changes or when plants begin operating in the morning or close down for the night. At such times the switching on or off of heavy machinery may significantly change the line voltage available in the laboratory.

By careful checking, all of the foregoing sources of error can be detected and eliminated.

b. Detection of Indeterminate Error

Indeterminate errors are always present. The extent of the errors is determined by statistics as described earlier. The standard deviation can be determined accurately only by carrying out an infinite number of tests, but this is not a practical proposition. However, an *estimate* of σ can be made from a finite number of observations by using Eq. (1.1). This "estimate of σ" is always greater than "true σ" to allow for the uncertainty of *estimation* of σ from a finite number of tests compared to the *determination* of σs from an infinite number of tests. An increase in the number of observations brings about a decrease in the error in the estimate, so that the calculated value approaches true σ.

The value obtained for σ is an estimate of the precision of the method. If an analyst setting up new procedure carries out 20 determinations of a standard sample, she can obtain the short-term precision of the method. This is the optimum value of σ because it was obtained using the same operator, the same equipment, and the same chemicals. In practice the data may be too optimistic. Routine analysis may be carried out for many years in a lab, such as blood analysis for Na.K.Ca.Mg determinations. Different operators, different equipment, and different chemicals may be used. The analysis of a standard sample should be carried out on a regular basis (daily, weekly, etc.) and collected at the end of the year. At that time, the long-term precision of the method can be calculated. This is a more realistic measure of the reliability of the analytical results obtained on a continuing basis from that lab.

A reduction in σ can sometimes be brought about inadvertently when eliminating determinate error. A skillful operator might reduce operator-determinate error to a small value and simultaneously reduce operator-indeterminate error. This is why one operator can carry out an analysis using a particular method and obtain a low standard deviation, whereas another, using the same method and sample, might get results with a higher standard deviation.

Table 1.8 Confidence Limits for Different Values of σ

Deviation	Confidence limit
σ	66.7
2σ	95.0
3σ	99.0
4σ	99.99

In practice, the standard deviation for a particular method used for a particular sample may be low when determination is carried out by a single operator or by the author of the method. If a method is used by 10 different operators over an extended period of time, however, the standard deviation will be significantly higher, even if the same sample is used each time. These two sets of precision data are called *short-term precision* (obtained by one operator at one time) and *long-term precision* (obtained by several operators at several different times, or even one operator over a long period of time).

Precision can be improved by paying closer attention to the details of the procedure, operating the equipment at peak performance conditions, using correct sampling procedures, and storing the sample properly.

Figure 1.1 illustrates the distribution of results obtained with normal indeterminate errors. From this curve it can be calculated that 66.7% of the results are within $\pm\sigma$, 95% are within $\pm 2\sigma$, 99% are within $\pm 3\sigma$, and 99.99% are within $\pm 4\sigma$ of the true answer. These relationships are listed in Table 1.8. These numbers represent the confidence limits the analyst has in his results.

An illustration of how these confidence limits are used is as follows. If σ for a procedure is 0.1% and the result obtained is 20.0%, the analyst reports that the result is $20.0 \pm 0.1\%$, and she is 66.7% confident of that, or it may be reported as 20.0 ± 0.2, and she is 95% confident of that. Most analysts prefer to report results with 95% confidence, i.e., within $\pm 2\sigma$.

c. Rejection of Results

Another source of imprecision is the occasional result that is obviously in error. This result may have been caused by incorrect weighing or measuring, spillage, faulty calculation of results, or carelessness. In any case, the result that includes such a gross error should be rejected and not used in any compilation of results. An acceptable rule to follow is that if the error is greater than 4σ, it should be rejected. When calculating the standard deviation for a new procedure, a suspected result should *not* be included in the calculation of σ. After the calculation, the suspected result should be examined to see if it is more than 4σ from the true value. If it is outside this limit, it should be ignored; if it is within this limit, the value for σ should be recalculated with this result included in the calculation. It is not permissible to reject more than one result on this basis. A suspected result should not be included in calculating σ. If it is included, it will automatically fall within 4σ because such a calculation includes this number.

d. Significant Figures

Analytical results, among many other things, are reported in numbers, such as 50.1%, 10 parts per million (ppm), or 25 ml/liter. The numbers should be such that there is uncertainty only in the last figure of the number. For example, the number 50.1% means that the percentage is closer to 50.1 than to 50.2 or 50.0, but it does not mean that the percentage is exactly 50.1. In short, we are sure of the "50" part of the number, but there is some doubt about the last figure. If we were to analyze two samples containing 50.08% and 50.12% of a component by using an instrument accurate to 0.1%, we would not be able to distinguish the difference in the compositions of the samples, but would report them both as 50.1%.

The number 50.1 has three significant figures (5, 0, 1). Since there is some doubt about the last significant figure, there is no point in reporting any more figures (even thought they might be mathematically obtainable), because they would be meaningless. For example, in the sample discussed above, experimentally there was shown to be 50.1% of component A. It was known from other sources that component B was 25% of component A. There should therefore be (50.1 × 25)%/100 of component B, or 12.525%. The number 12.525 was derived from a number with three significant figures, the last of which was in the tenths of a percent. The calculated value cannot contain more than three significant numbers; therefore we should report that component B was present at 12.5%.

The reporting of figures implies that all the numbers are significant and only the last number is in doubt, even if that number is zero. For example, 1.21×10^6 implies that the number is closer to 1.21×10^6 than to 1.22×10^6 or 1.20×10^6. But 1,210,000 implies that the number is closer to 1,210,000 than to 1,210,001 or 1,209,999. Furthermore, the number 50.10 implies 10 times greater accuracy than 50.1.

The following are some rules that should be observed when reporting results.

1. In enumerating data, report all significant figures, such that only the last figure is uncertain.
2. Reject all other figures, rounding off in the process. That is, if a number such as 1.325178 must be reported to four significant figures, the first five figures should be considered. If the fifth figure is greater than 5, increase the fourth figure by 1 and drop the fifth figure. If the fifth figure is less than 5, the fourth figure remains unchanged and the fifth number is dropped. If the fifth figure is 5 and the fourth figure is even, it is not increased when the 5 is dropped. Table 1.9 shows some examples of rounding to four significant figures.

Table 1.9 Rounding Off to Four Significant Figures

Number	Four significant figures
1.37285	1.373
1.27245	1.372
1.3735	1.374
1.3725	1.372
1.37251[a]	1.373

[a]The number 0.00051 is greater than 0.0005, even though the last figure is not significant; hence the fourth figure is increased by one.

3. In reporting results obtained by addition and subtraction, the figures in each number are significant only as far as the first doubtful figure of any one of the numbers to be added or subtracted. For example, the numbers in the set 21.1, 3.216, and 0.052 are reliable to the first decimal point; that is 21.1 is reported as 21.1, 3.216 is rounded off to 3.2, and 0.052 to 0.1. The sum of the rounded-off numbers (21.1 + 3.2 + 0.1) is 24.4. In the first number, uncertainty arises at the tenths place. All other numbers are rounded off to the tenths place prior to addition.

4. For multiplication and division, the number of significant figures in each term and in the answer should be no greater than that of the term with the least number of significant figures. For example, $1.236 \times 3.1 \times 0.18721 \times 2.36$ is rounded off to $1.2 \times 3.1 \times 0.2 \times 2.4$. The answer is rounded off to 1.7. In this case, the term 3.1 contains only two significant figures. It is meaningless to give the other terms or the answer to more than two significant figures.

5. The characteristic of a *logarithm* indicates an order of magnitude; it is not a significant number. The mantissa should contain no more significant numbers than are in the original number. For example, the number 12.7 has a log of 1.1038. The log is rounded off to three significant figures and is reported as 1.104.

6. If several analyses have been obtained for a particular sample (*replicate analysis*), it should be noted at what point there is doubt in the significant numbers of the result. The final answer should be reported accordingly. For example, given the triplicate answer 11.32, 11.35, 11.32, there was no doubt about 11.3, but there was doubt about the fourth figure. The average should be reported as 11.33 [i.e., $(11.32 + 11.35 + 11.32) \div 3$]. In the triplicate set of answers 11.42, 11.35, 11.22, there was no doubt about 11, but there was doubt about the next figure. The average should be reported as 11.3[$11.4 + 11.3 + 11.2) \div 3$].

D. SIGNAL AND NOISE (SENSITIVITY)

If a recording is made of a signal versus wavelength or a signal versus time, the ideal tracing would be a smooth line and a peak, as shown in figure 1.2a. In practice, however, the recorded trace is seldom smooth, but includes random signals called *noise*. Noise can originate from small fluctuations in the light source, the detector, stray light, the recorder, the monochromator system, and so on. Provided that the signal is significantly greater than the noise, it is not difficult to make a reliable measurement of it.

In Fig. 1.2b the noise level is increased and is superimposed on the signal. The value of the signal is less certain. The results are less precise, but clearly discernible. In Fig. 1.2c the noise level is as great as the signal, and it is virtually impossible to make a meaningful measurement of the latter.

The noise can be reduced by optimizing the equipment settings (i.e., source, monochromator slit, detector, amplifier, etc.). It may further be improved by "*smoothing*" out the signal with a condenser. Unfortunately, this has the effect of slowing the response of the instrument. So a trade-off is usually made by the operator, who uses as much smoothing as possible but preserves a response time fast enough to be useful.

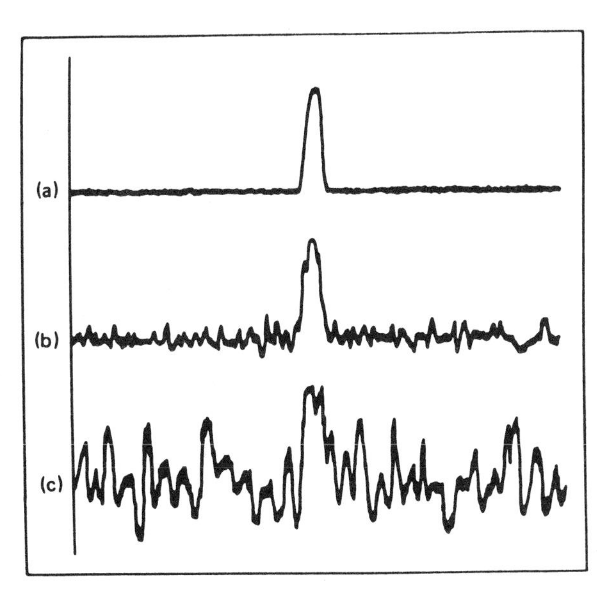

Figure 1.2 Signal versus wavelength with different noise levels: (a) no detectable noise, (b) moderate noise, and (c) high noise.

The sensitivity of an analytical procedure is ultimately limited by the noise level. When the signal can no longer be distinguished from the noise, the sensitivity limit of the system has been reached.

It is conventional to state the *signal-to-noise ratio* when defining the sensitivity of a method. This is an arbitrary decision that varies from one worker to another. Generally, a ratio of 1:1 is used (i.e., signal = noise), but other ratios are also commonly used.

A method that has improved the sensitivity limit is *time averaging*. In this case, a signal, such as that illustrated by Fig. 1.2c, is recorded 100 or more times very rapidly and overlapped and stored in a computer. Since the noise level is random, it averages out to a curve like Fig 1.2a. The signal itself, however, is not random, but additive. Each run results in an accumulated signal that increases with the number of scans. Using this technique, signals that would normally be lost in the noise can be detected and measured. The technique has been used in NMR, UV, IR, and nuclear science methods.

1. Signal-to-Noise Ratio

The signal-to-noise ratio (S/N) is a limiting factor in making absorption or emission measurements. When the signal is not discernible from the noise level, it is impossible to make any meaningful measurements. In the Fourier transform system, there is an immediate advantage in the fact that the signal intensity falling on the detector is greater. The detector operates under optimum conditions and therefore there is a reduction in noise level.

Furthermore, the S/N level is proportional to the square root of T, where T is the time over which the signal and noise are accumulated. If an observation time of 1 sec is necessary to give an S/N of 1:1, this must be increased to 4 sec to give an S/N of 2:1. It can be seen that there is some advantage to this system, but logistics prevent very long time measurements form being made.

A second way to improve S/N is to use repetitive scans. In this instance advantage is again taken of the fact that noise is random but the signal is additive. With the Fourier transform system, a complete scan can be taken using a single burst of radiation from the light source, permitting rapid scanning and recording on computer tape over a reasonable time period after a laser source is used. If a scan is taken many, many times and each instance is recorded on the same piece of computer tape, then the signal will increase, whereas the noise level will tend to average out. If the scans are of a short enough duration, it is possible to take hundreds of scans in a relatively short time. The relationship between S/N and the number of scans is shown in Fig. 1.3.

A third method to improve the signal noise level is to add a condenser between the detector output and the amplifier. The net effect is to average out the noise level, but the signal accumulates. The capacity of the condenser used de-

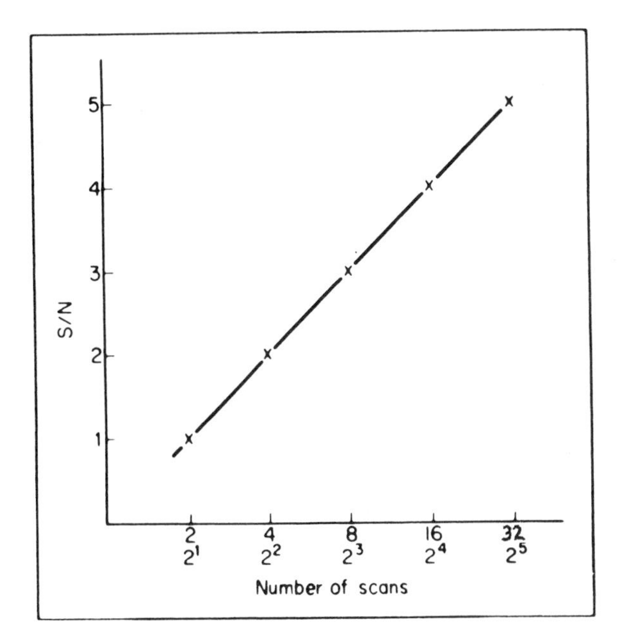

Figure 1.3 The signal-to-noise ratio increases with the square of the number of observations or scans.

pends on the required reduction of the noise level. The greater the capacity, the better the smoothing of the signal. Unfortunately, the system is limited, because as the capacity of the condenser increases, the response time of the system also increases. Transient signals are often not detected at all. In practice, the operator must select a compromise capacity so that the optimum response time and S/N are obtained.

The decision to improve S/N is always an arbitrary one. It is theoretically possible to measure any signal in any noise level, provided that the measurements are taken over a long enough period of time and repeated enough times to accumulate the signal. In practice, there is usually a time limitation on such observations and the operator must make this decision based on the circumstances under which he is working.

The Fourier transform system has found many applications, not only in IR spectroscopy, but also in NMR, UV, and so on. The disadvantage of this system is that the readout from the detector does not in any way resemble an IR spectrum. It is composed primarily of an infinite number of interference patterns, and it is necessary to use a computer to unscramble these in order to reveal the true IR spectrum. Programs are available to do this, however, so it is no longer a practical problem.

2. The Wheatstone Bridge and the Null Point Method

The circuitry of the Wheatstone bridge is used extensively is spectroscopic instrumentation. It is a very convenient method for converting a radiation signal into an electrical signal that can be measured and recorded. The principal is as follows.

Suppose we have a uniform wire *AB* that is 1 m long. The electrical resistance of the wire is proportional to the length of the wire. Furthermore, let us suppose that the ends of the wire are connected to a steady dc voltage source such as that supplied by a battery. This system can be seen in Fig. 1.4a. If the connections from the battery to the wire have zero resistance, then the potential difference between the ends of the wire is the same as the voltage of the battery. Let us suppose that this equals 1 V. The voltage between *A* and *B* is therefore 1

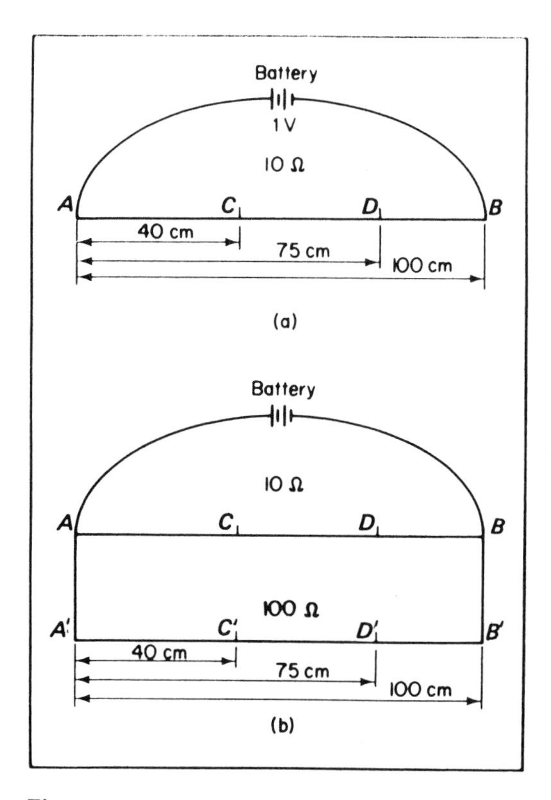

Figure 1.4 (a) The potential drop at *C* and *D* depends on the lengths of wires *AC* and *AD*, provided that the resistance is proportional to length. (b) The same potential drop, but the resistance of *A'B'* is different from that of *AB*.

V. The voltage drop across the resistance wire is proportional to its resistance. For example, if the total resistance of the wire *AB* is 10 Ω, every portion of the wire that generates 1 Ω resistance will experience a voltage drop of 0.1 V. We could therefore say that in wire *AB*, which is 1 m in length and has a uniform resistance drop of a total of 10 Ω, we can calculate the voltage at any particular point on the wire by measuring the relative lengths of the wire from the ends. For example, at point *C*, which is 40 cm from point *A* and therefore 60 cm from point *B*, there should be a voltage drop of 40/100 of the total voltage drop at point *C*, or 0.4 V. Therefore the actual voltage is 1 V − 0.4 V = 0.6 V. Similarly, at point *D* the voltage drop is 75/100 or 0.75 V. The actual voltage is therefore 1 V − 0.75 V = 0.25 V.

In a similar fashion, we can replace wire *AB* with a second wire *A'B'*, also 1 m long, but with a resistance of 100 Ω. This wire is in all respects similar to the previous wire, except that the resistance is different. As before, the voltage drop at any particular point can be predicted from the voltage drop to the total voltage. Thus at point *C'*, 40 cm from *A'*, the voltage drop is 40/100 = 0.4 V and the actual voltage is 1 V − 0.4 = 0.6 V. This can be seen in Fig. 1.4b.

If we now connect points *A* and *A'* and *B* and *B'*, leaving the ends connected to the battery as before, we have circuits joined together in parallel.

If we connect *C* and *C'*, there will be no current flowing in the *C–C'* circuit because the voltages at *C* and *C'* are equal to each other. Similarly, if we join points *D* and *D'*, no voltage will flow because both *D* and *D'* are at the same potential.

From this basic experiment we can deduce the conditions for the Wheatstone bridge. If the ratio of the resistance of the wires *A*, *C* and *C*, *B* are equal to *A'C'* and *C'B'*, then the two points *C* and *C'* are at equal potential and the bridge is balanced; that is, no current flows of *C* and *C'* are connected.

A more common way to express the Wheatstone bridge arrangement is shown in Fig. 1.5. In this system, if the ratios

$$\frac{R_1}{R_2} = \frac{R_3}{R_4}$$

then no current flows between the points *C* and *C'* and the bridge is said to be balanced. This system has been used extensively in analytical spectroscopic instrumentation by using fixed resistances R_2, R_3, R_4, and using R_1 as the detector in the system. When the radiation falls on the detector, there is frequently a change in current or the electrical resistance of the detector, so an imbalance in the Wheatstone bridge occurs and the current flows through the galvanometer in a circuit joining *C* and *C'*. This current is a direct measure of an imbalance of the system and is therefore a measure of the change in resistance in R_1, the detector used in the instrument.

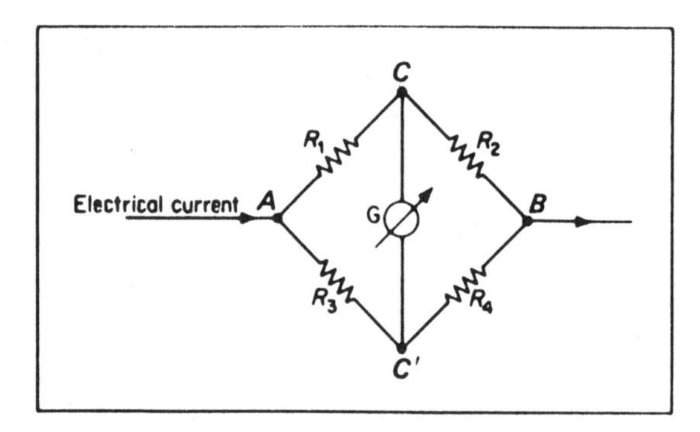

Figure 1.5 A Wheatstone bridge.

There are basically two methods of utilizing this imbalance. The first is to record the current flowing through CC' directly and to use this as a measure of the radiation intensity. Inasmuch as this current is very small, an amplification system is commonly used before the signal is relayed to a recorder.

An alternative procedure to the direct measurement of resistance change is to use the *null point system*. In this system, resistance R_1 again is the detector used in the instrument, but R_2 is a variable resistance as shown in Fig. 1.6. When radiation falls on R_1, its resistance changes and generates an imbalance in the Wheatstone bridge, so current flows through circuit CC'. This current drives a feedback mechanism that changes the resistance of R_2 until no current flows through circuit CC' and the bridge once again is in balance. A measure of R_2 allows R_1 to be calculated, since $R_1/R_2 = R_3/R_4$ and R_3 and R_4 are constants.

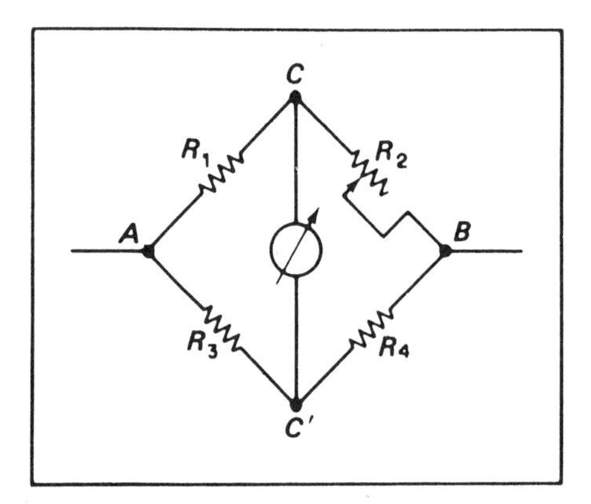

Figure 1.6 A null point Wheatstone bridge, where R_1 is the detection device, R_2 is varied by feedback from the galvanometer until balance is achieved, and $R_1/R_2 = R_3/R_4$.

In this system, the current flowing through the circuits CC' is reduced to zero and the circuit once again has reached the "null point." A measure is made of R_2 or the change in R_2 in order to calculate R_1 or the change in R_1. This signal is then amplified and relayed to the readout system. The advantage of the null point system is that it is not necessary to measure small currents directly, something that is always a difficult task. Such small currents are frequently encountered when low-intensity signals fall on R_1 and only a small imbalance is generated in the circuit. The null point principle successfully overcomes this problem, although it too is ultimately limited by the fact that small imbalances in the ratio R_1/R_2 are insufficient to drive the feedback mechanism.

E. SAMPLING

The most important single step in analysis is taking a sample of the material to be analyzed. If a nonrepresentative sample is taken, no matter how excellent the analytical procedure may be or how expert the analyst, the result obtained will be incorrect. There are two requirements for correct sampling. First, a sample must be taken that is sufficient for all the analyses to be carried out comfortably with enough for at least duplicate analyses if necessary. Of course, if only a small quantity of sample is available, the analyst must do his best with what is provided. Second, the sample must be representative of the material being sampled. If the sample does not represent the material, the analysis cannot provide a reliable characterization of the material and the result is in error regardless of the accuracy of the analytical method.

1. Sampling of Different Sample Types

a. Gas Samples

Two types of gas samples can be taken. First, single spot samples can be taken in a balloon or syringe after the sample has been well stirred with a fan or paddle. Several composite samples of gas can be taken by bleeding gas slowly into a suitable container and collecting the gas over a period of time. If necessary, this can be done over several hours and an average analysis of the gas stream obtained.

b. Liquid Samples

It is important to stir liquid samples adequately and to take samples remote from sources of contamination. For example, a liquid brought to the laboratory in a container should be sampled so as to avoid floating froth or sludge on the bottom of the container. Spot or composite samples can be taken in a manner similar to that used for gas samples. River or sea samples should be taken at points where contamination is likely to be minimal. With river samples it is important to avoid river bank contamination, floating froth contamination, or concentrated discharge from manufacturing processes. River and sea water samples should be taken at several depths and distances from the shore. Continuous analysis of commercial products on stream should be taken by placing the analysis probe in a position where the sample is representative of the product.

c. Solid Samples

Solid samples are the most difficult to handle because they cannot be conveniently "stirred up." Moreover, unlike the situation with fluids, there are no diffusion or convection currents in solids to ensure some mixing. Numerous methods have been devised to reduce this problem. The recommended method should be used for each particular type of sample, such as polymers, metals, soils, cements, foods, cloth, plant materials, and biological specimens. Frequently, this involves taking portions from different parts of the sample, mixing them, and making a representative part.

d. Bulk Samples

The process of sampling bulk materials, such as coal, metal ore, and soil samples, requires several steps. First, a gross representative sample is gathered. The size of the representative sample should be at least 500 times as big as the largest particle in the bulk sample. Numerous portions of the sample should be taken from various locations. The combination of these portions should be representative of the whole sample. With soil samples from a field, drillings are taken from different areas of the field. Unique areas should be represented but not overemphasized.

The gross representative sample may be too large to transport or to handle in the laboratory, so it must be reduced in size. Large particles (as in coal) should be broken up and mixed into the main sample. A smaller representative sample may be obtained by several methods. In the *long pile and alternate shovel method*, the sample is formed into a long pile. It is then separated into two piles by shoveling alternate shovelfuls into first one pile and then the other. One pile is discarded. If necessary, the second pile may the be further reduced by additional piling. The *cone and quarter method* is widely used. In this method, the sample is made into a circular pile and mixed well. It is then separated into imaginary quadrants. A second pile is made up from two opposite quarters of the first, the remainder of which is discarded. This process is shown in Fig. 1.7. As the total size of the sample is reduced, it should be broken down to successively smaller pieces. The final sample should be representative of the sample and large enough to provide sufficient material for all the necessary analyses.

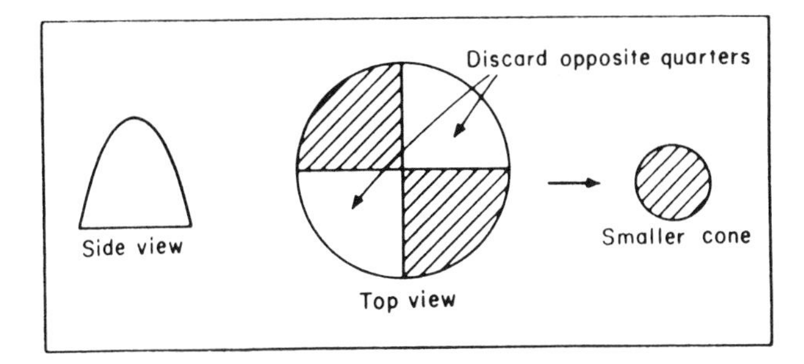

Figure 1.7 The cone and quarter method of sampling bulk materials.

Metal samples need special care. When a molten metal solidifies in a crucible or other container, the first solid metal formed tends to be purer than the remainder. The last metal to solidify is a the most impure and is generally located in the center, or *core*, of the solidified metal. Consequently, most of the impurities are found in the core. It is important to bear this in mind when sampling casting or pressings. If convenient, a sample should be ground from a representative cross section of the sample. Otherwise, a hole may be drilled across the sample at a suitable place and the drillings mixed and used as the sample.

2. Storage of Samples

When samples cannot be analyzed immediately, they must be stored. The storage container must be clean and airtight. Liquid or gas samples can be stored in plastic containers. Corrections must be made for any absorption effect on the walls of the container, particularly in trace metal analysis of fluids. Solid samples should also be sealed. All samples should be stored in rooms that are not too hot or too humid. Storage for long periods of time should be avoided, because decomposition or evaporation of the sample may take place. Special care should be taken with samples containing trace quantities (parts per million) of materials that must be determined subsequently. Trace components may plate out on the glass or plastic vessel upon standing or may be contaminated by leaching impurities from the wall of the container.

It has been observed that trace metals tend to plate out along strain lines in glass. Such strain lines are not reproducible from one bottle to another, therefore, the loss of trace metals cannot be estimated for one container by measuring the loss in a similar but different container. It is best to avoid using glass bottles when trace analyses are involved.

Many of the problems can be avoided by freezing the samples. In the field the samples may be quick-frozen before transportation back to the lab. In any case, if a delay between sampling and analyzing is expected, freeze the samples and keep them frozen until ready for analysis for best results.

It is most important that storage be carried out properly and not casually, because a contaminated sample—or one that has plated out—will give enormous analysis error, no matter how much care is taken during analysis.

BIBLIOGRAPHY

Bauer, E. L., *A Statistical Manual for Chemists*, 2d ed., Academic, New York, 1971.
Drago, R. S., *Physical Methods in Chemistry*, Saunders, Philadelphia, 1977.
Harris, D. C., *Quantitative Chemical Analysis*, W. H. Freeman, New York, 1992.
Kask, U., and Rawn, J. D., *General Chemistry*, W. C. Brown, Dubuque, IA, 1993.
Pietrzyk, D. J., and Frank, C. W., *Analytical Chemistry*, Academic, New York, 1974.
Noble, D., *Anal Chem*, (1994), *66*(4), 251A.

PROBLEMS

1.1 (a) What is a determinate error? (b) In a volumetric analysis, the burette used was inaccurate. Would this cause determinate or indeterminate errors?

1.2 (a) What is precision? (b) Do determinate errors affect precision?

1.3 (a) What is accuracy? (b) How can the accuracy of an analytical procedure be determined?

1.4 (a) What is the statistical definition of sigma (σ)? (b) What percentage of answers should fall within 2σ of the true answer?

1.5 What is the standard deviation of the following set of numbers?
3.1 3.2 3.1 3.3 3.2 3.1 3.2 3.4 3.1 3.4 3.1

1.6 What is the range of true data of (a) 0.1 g weighed to within 0.001 g, (b) 0.1 ml measured to within 0.002 ml, (c) 10.3°C measured to within 0.01°C, and (d) 1.0 hr measured to within 1 sec?

1.7 How many significant numbers are there in each of the following numbers?
3.216 32.1 30 3×10^6 3.21×10^6 321,000

1.8 Round off the following additions according to the rules of significant figures:

	3.2	1.9632	1.0×10^6
	0.135	0.0013	1.321×10^6
	3.12	1.0	1.13216×10^6
	0.028	0.0234	4.32×10^6
Total	6.463	2.9879	7.77316×10^6

1.9 What are (a) the arithmetic mean, (b) the standard deviation, and (c) the 95% confidence limits of the following set of numbers?
2.13 2.51 2.15 2.17 2.09 2.12 2.17 2.09 2.11 2.12
(d) Which numbers are outside 4σ and should be ignored in the calculation?

1.10 (a) What is the importance of good sampling? (b) What precautions should be taken in sample storing?

1.11 (a) Illustrate the difference between precision and accuracy. (b) Do indeterminate errors affect precision or accuracy?

1.12 The results in Problem 1.9 were obtained for the nickel content (in parts per million) of a patient's liver. The doctor knew that he must operate if the nickel content was greater than 2.17 ppm. (a) Are the results greater than 2.17 ppm with 95% confidence? (b) If it were found that under somewhat different clinical conditions the doctor should operate if the nickel content was greater than 2.00 ppm with 95% confidence, should the doctor operate if the results obtained were those shown in Problem 1.9? If it were your liver, would you let him operate?

1.13 The result of a single copper determination on a patient's kidney was 10.3 ppm. The standard deviation for the method was 0.9 ppm. If the copper content was greater than 8.8 ppm, the treatment for the patient should be altered. Based on these results, would the doctor change her treatment? If she were unsure of the significance of the analytical result, how would she obtain further information?

1.14 What is the standard deviation of the following sets of numbers?
10.1, 10.0, 10.2, 10.0, 9.8, 9.9, 10.1, 11.5, 10.1, 9.9, 10.0, 9.9.

1.15 With what confidence can an analytical chemist report data using σ as the degree of uncertainty.?

1.16 What is the standard deviation of the following set of numbers?
10.1, 10.0, 10.2, 10.0, 9.8, 9.9, 10.1, 11.5, 11.6, 10.1, 9.9, 9.9, 10.0.

1.17 An analysis was reported as 10.0 with $\sigma = 0.1$. What is the probability of a result occuring within (a) ± 0.4 or (b) ± 0.2 of 10?

1.18 (a) If the signal is 20 units and the noise level 80 units, what is the S/N level? (b) If the noise is reduced by 25%, what is the new S/N level?

1.19 If the signal is 10 units and the noise 10 units, what percentage must the noise be reduced to give an S/N ratio of 2/1?

1.20 If the data in Problem 1.19 giving an S/N ratio of one were obtained from the average of four scans, how many scans would be necessary to obtain an S/N of 2?

1.21 (a) Describe a Wheatstone bridge. (b) What is the advantage of a null point Wheatstone bridge? (c) When is this system a disadvantage?

1.22 Assuming that the results of many analyses of the same sample present a Gaussian distribution, what part of the curve defines the standard deviation σ?

1.23 A liquid sample is stored in a bottle for the determination of trace metals at the ppm level. What factors can cause the results to be (a) too low or (b) too high?

1.24 Round off the following numbers according to the rules of significant figures: (a) $10.2 \div 3$, (b) $10.0 \div 3$, (c) $11 \div 4$, (d) $11.0 \div 4$, and (e) $11.00 \div 4$.

1.25 Which methods are used for elemental analysis?

1.26 (a) What methods are best for elemental qualitative analysis? (b) What methods are best for quantitative analysis?

1.27 (a) Which methods are best for molecular functional group analysis? (b) What methods best indicate which groups are next to each other in a molecule?

1.28 What methods are best for quantitative analysis of (a) complex mixtures, (b) simple mixtures, or (c) pure compounds?

1.29 What methods are best for measurements of molecular weights?

1.30 Which methods indicate the distribution of high atomic weight elements in a sample?

1.31 If the sensitivity as defined by the S/N ratio of one run is 1 ppm, how many accumulated repetitive scans would be needed to increase the sensitivity (a) an order of magnitude, i.e., 10-fold, (b) two orders of magnitude, i.e, twofold?

2
Introduction to Spectroscopy

A. THE INTERACTION BETWEEN RADIATION AND MATTER

If a beam of white light is passed through a beaker of water, it remains white. If potassium permanganate is added to the water, the white light appears purple after it passes through the solution. The permanganate solution allows the blue and red light to pass through but absorbs the other colors from the original beam. This is an example of the interaction between radiant energy and matter. In this case the radiant energy is visible light, and we can see the effect of absorption with our eyes. However, absorption of radiation can take place over a wide range of radiant energy, most of which cannot be seen. Such absorption effects are not visible to the human eye, but they occur nevertheless and can be measured with suitable instruments.

The absorption of energy is not haphazard, but follows well-documented rules with respect to both the wavelengths of light absorbed and the extent of absorption. The subject of spectroscopy is concerned with the interaction of radiant energy and matter.

1. What Is Radiant Energy?

The nature of radiant energy, including visible light, has baffled scientists for many years. At times it appears to behave like a wave; at other times it behaves as though it were composed of small particles. Despite our limited understanding of its nature, we have come to recognize certain well-defined properties of radiation. We know, for example, that it is a form of energy.

The wavelike character of light is supported by the fact that we can measure its wavelength λ, the crest-to-crest length of its waves, quite easily and accurately. The wavelength of light is related to its frequency ν by the equation

$$c = \nu\lambda \tag{2.1}$$

where

c = speed of light (cm/sec)
ν = frequency (cycles per second, or cps)
λ = wavelength (cm)

We also know that there is a relationship between the frequency of light and its energy. This relationship is given by the important formula

$$E = h\nu \tag{2.2}$$

where

E = energy (ergs)
h = Planck's constant (6.6256×10^{-27} erg sec)
ν = frequency of the light in cycles per second (cps)

From Eqs. (2.1) and (2.2), we can deduce that

$$E = \frac{hc}{\lambda} \tag{2.3}$$

These relationships show that as the wavelength of light decreases, its energy increases. Put another way, as the frequency increases, the energy increases.

2. What Is Matter?

Matter is composed of molecules or atoms. These atoms or molecules may assume various physical forms. For example, they may be rather widely separated from each other, as they are in gases, or they may exist in agglomerates as liquids or solids. The molecules of a solid may be arranged as crystals, as they are in inorganic salts, or may be randomly distributed, as they are in plastics. Whatever their arrangement, the molecules are in constant motion. Many types of motion are involved, such as rotational, vibrational, and translational (i.e., uniform straight line) movement. In the case of individual atoms, the orbiting electrons can move from one orbit to another. A change in any of these forms of motion involves a change in energy.

Both the chemical structure of molecules and the way these molecules are arranged relative to each other affect the way in which any given material interacts with energy.

3. How Does Radiant Energy Interact with Matter?

If we pass white light through blue glass, the emerging light is blue, because the red and yellow light have been absorbed by the glass. We can confirm this absorption by shining a red or yellow light through the blue glass. If the glass absorption is strong enough, all the red or yellow light is absorbed; no light emerges and when one looks through the glass, the light source appears black.

Absorption is not a chance occurrence, but conforms to well-established physical laws. The motions of the molecules and atoms in matter are all associated with energy. For example, a vibrating molecule is said to have vibrational energy. If the molecule acquires more vibrational energy, it vibrates faster and is said to have moved to a state of higher vibrational energy. As shown in Fig. 2.1, this energy may be acquired by absorbing radiant energy. The physical law governing the relationship between the radiant energy and the vibrational energy is expressed mathematically as $E = h\nu$ [Eq. (2.2)] or $\nu = E/h$.

The energy difference E between the two vibrational states E_A and E_B must equal the energy of the absorbed radiation $h\nu$. Furthermore, the molecule can *only* absorb radiation with energy equal to the difference between the energy states. Consequently, a given molecule can *only* absorb radiation of characteristic energies (and therefore frequencies) that correspond to differences in the energy levels of the molecule. A given molecule can only absorb radiation of certain definite frequencies.

In everyday terms, if a glass passes blue light, but not red or yellow, it has energy states such that

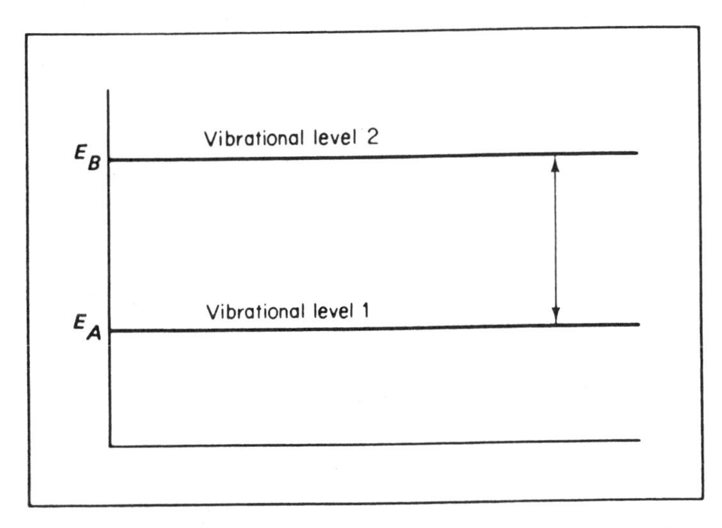

Figure 2.1 Absorption of radiant energy. $E_A - E_B = E$; $E = h\nu$.

$$E_1 = h\nu_1$$

where $\nu_1 = 0.6 \times 10^{15}$ cps (a frequency characteristic of red radiation) and E_1 is the difference between two energy states E_A and E_B in which the glass can exist.

The glass also has energy states E_A and E_C, the difference between them being E_2,

$$E_A - E_C = E_2 = h\nu_2$$

where $\nu_1 = 0.66 \times 10^{15}$ cps (a frequency of yellow radiation). It does not, however, have energy states to satisfy the relationship

$$E_A - E_D = E_3 = h\nu_3$$

where $\nu_3 = 0.77 \times 10^{15}$ cps (a frequency of blue radiation).

As a result, the glass absorbs red and yellow but not blue radiation. When white light passes through the glass, the light's red and yellow components are absorbed, but its blue component passes through and the glass appears to be blue. Of course, "red," "blue," and "yellow" light span a range of wavelengths, and not just a single wavelength or frequency. The laws still apply within these ranges.

For many years it was considered that a given molecule (e.g., hydrogen) could exist in any energy level varying over wide energy ranges. It was believed that the energy difference could have continuous values and that the molecule could absorb over a continuous frequency range. It has been shown that this is not so. Molecules can exist only in certain well-defined energy states. This is expressed scientifically by saying that the energy levels are not continuous, but "*quantized*" (see Sec. C.1). As a direct consequence, the energy difference between two defined energy levels is fixed; therefore only radiation of well-defined energy levels is absorbed, and therefore, only radiation of well-defined frequencies can be absorbed by a particular molecule or atom. These absorption frequencies are the same for all molecules of the same chemical species. The same rule applies to atoms. The uniqueness of the frequencies at which a given molecular species absorbs radiation is the basis of absorption spectroscopy. The experimental basis of these very important rules is illustrated below.

B. THE ABSORPTION OF ENERGY BY ATOMS

1. The Franck–Hertz Experiment

When two bodies collide, an exchange of energy usually takes place. After an inelastic collision, one body usually travels faster and the other slower than before the collision. By contrast, in an *elastic* collision there is no exchange of energy and both bodies travel at the same speed after as before the collision.

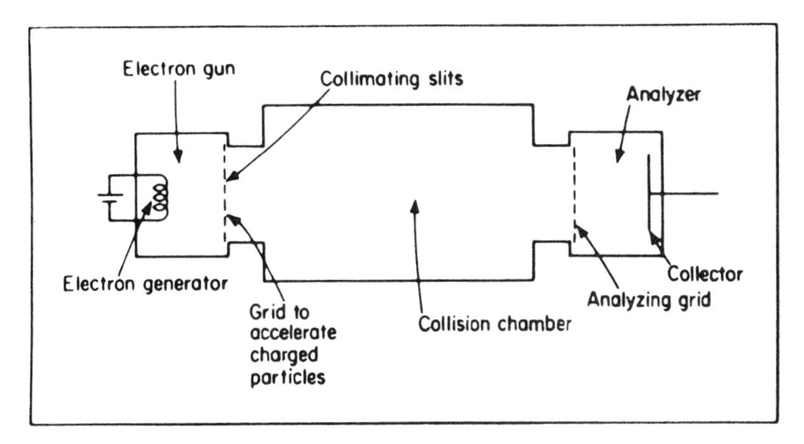

Figure 2.2 Schematic diagram of the equipment used by James Franck and Gustav Hertz to study collision processes between atoms and electrons.

If we measure the speed of one body before and after an inelastic collision, we can calculate how much energy was transferred in the process. In 1914, James Franck and Gustav Hertz used this principle to study the collision processes between atoms and electrons.* Their equipment (see Fig. 2.2) was composed of three main components. The first was an electron gun, a device containing a hot filament similar to that in an electric light bulb. The electron gun released large numbers of electrons with very low energy. A positively charged electrode, in this case a grid, attracted and accelerated the electrons. Perforations in the positive electrode allowed the electrons to pass through and enter the collimating slits, through which the electrons also passed. Only those electrons traveling in the desired direction were admitted by the collimating slits to the second unit of the Franck–Hertz apparatus. This unit was a collision chamber containing a monatomic gas. The pressure in the chamber was reduced so that an electron would not generally experience more than one collision while passing through the chamber. The third major component was a detector that measured the energy of the electrons after they had traversed the collision chamber. A simple detector, such as a photocell or collector, was placed behind an electrode (once again, a grid was used) set at a negative potential. Electrons with low energy were repelled by the negatively charged grid and never reached the photocell. Only electrons whose speed (i.e., translational energy) was sufficient to propel them past the negative electrode reached the detector. This setup enabled discrimination between electrons with energy greater than that re-

*See, for instance, U. Fano and L. Fano, *Basic Physics of Atoms and Molecules*, Wiley, New York, 1959.

quired to penetrate the negative grid and those electrons with insufficient energy to do so. By varying the grid voltage, the energy of the electrons could be measured.

2. Energy Levels of Atoms

The Franck–Hertz experiment showed that those electrons that were accelerated only to low kinetic energies by the positive grid experienced only elastic collisions in the collision chamber. As the electrons' energy increased, their collisions with atoms in the collision chamber remained elastic until the energy reached a critical value. Above this critical value the collisions become inelastic. This value, called the *threshold value*, is constant for all atoms of the same element, but different for atoms of different elements. Examples are 1.6 electron volts (eV) for cesium, 4.9 eV for mercury, and 16.6 eV for neon.

An example of the Franck–Hertz experiment is as follows. A collision chamber containing many atoms of mercury was bombarded with electrons whose energy was 2 eV. The energy of the emerging electrons remained 2 eV, showing that the collisions between the atoms and electrons in the chamber were elastic. In a second set of conditions, the energy of the bombarding electrons was increased to 6.0 eV. Many emerging electrons had an unchanged energy of 6.0 eV, but it was observed that the energy of some of the emerging electrons was 1.1 eV. This result indicated that inelastic collisions, which removed 4.9 eV (i.e., 6.0 − 1.1 eV) from the electrons, had occurred in the chamber.

During the passage of the electrons through the collision chamber, an atom was struck that was in a steady energy state E_0 before the collision. During the collision, 4.9 eV of energy were transferred to it, and it entered a different steady energy state E_1. The difference in the energy states E_0 and E_1 was 4.9 eV. The experiment told us that for the atom there were two energy states E_0 and E_1, and that the difference in energy between these states was 4.9 eV.

By increasing the energy of the bombarding electrons, we can identify numerous higher-energy states E_3, E_4, and so on, in the atom. The energy absorbed by the atom to reach these states is $E_3 - E_0$, and so on. The double transition from E_0 to E_1 and then from E_1 to E_2 is possible but unlikely, because in the collision chamber only very few atoms exist at any one time in state E_1; therefore very few collisions occur between electrons and atoms in state E_1. Because there are so few of these collisions, we do not observe many transitions from E_1 to E_2; hence this transition is unlikely. In this manner, we are able to derive an energy level chart (Fig. 2.3) for the particular atoms in the chamber.

When atoms attain the high energy levels E_1, E_2, E_3, . . . , they usually emit radiation and descend back to the lowest energy state E_0. The radiated energy is the origin of atomic emission spectra. In practice, an atomic emission

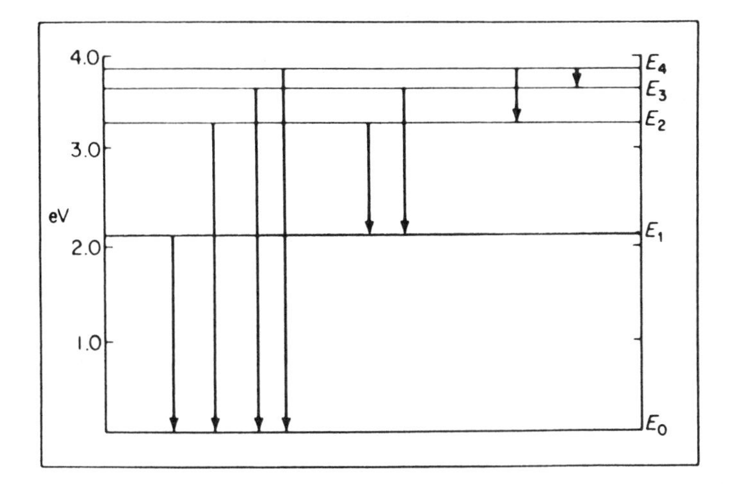

Figure 2.3 Energy levels (in electron volts) of mercury atoms.

spectrum is used to derive the energy level charts (Grotrian diagrams) of the elements.* (See Fig. 9.1 for a partial Grotrian diagram.)

As we noted earlier [Eq. (2.2)], the relationship between the energy levels and the frequency of the emitted radiation is given by $E = h\nu$, where h is Planck's constant, E ($=E_1$, $- E_0$) is the energy lost by the atom, and ν is the frequency of the emitted radiation.

The Franck–Hertz experiment clearly illustrated that an atom cannot absorb all the energy that is available to it, but only that energy that will cause it to move from one steady, well-defined energy level E_0 to another E_1. The system is said to be *quantized*; that is, only certain energy levels are found to exist. The conclusions have been extended to state that only certain energy levels are permitted in atoms of molecules—furthermore, the energy absorbed may be radiant energy.

C. THE ABSORPTION OF ENERGY BY MOLECULES

A magnetic field exerts a force on any electrical charge that passes through it. If the electrical charge is a flowing electric current, the force is perpendicular to the direction of the current's flow and to the magnetic field. If the electric current flows around a loop, the forces exerted by the magnetic field on the loop tend to cancel each other out, leaving no net force. [In such a case, however, a torque (twisting) effect is created.] But if the magnetic field is not uniform,

*See, for example, C. Chandler, *Atomic Spectra*, Van Nostrand, New York, 1964.

there is a net magnetic force on the loop. At the center of the loop, the potential magnetic energy is proportional to the magnetic field B. This energy may take the form of either a torque on the loop or a magnetic field. It is described by

$$E = \mu_{eff}B \tag{2.4}$$

where B is the magnetic field strength and μ_{eff} is a proportionality constant.

Based on this phenomenon, Otto Stern and Walther Gerlach in 1921 examined the magnetic properties of molecules. First, they built a nonuniform magnetic field using what would become known as a Stern–Gerlach magnet. This nonuniform field was necessary to produce a net magnetic effect at all points. Then they passed a beam of molecules through the magnetic field. The beam of molecules was finely collimated so that the translational energy of all the molecules was the same before entering the magnetic field. Even small deflections of the molecules were revealed on a screen placed beyond the magnet. Figure 2.4 is an illustration of their equipment.

The finely collimated molecular beam was split into a number of smaller beams by the strong nonuniform magnetic field. For example, a beam of hydrogen atoms or silver atoms was split into two beams, oxygen molecules into three beams, nitrogen molecules into four beams, and oxygen atoms into five beams; helium was undeflected. The experimenters also observed that each component beam was equally intense. When a second Stern–Gerlach magnet was put in the path of one of the emerging molecular beams, it was found that no further splitting of the beam took place, provided that the second magnet was oriented in the same direction as the first.

This experiment showed that the molecules possess a magnetic moment that causes deflection of the molecular beam upon passing through a magnetic field. Second, it showed that all the molecules do not have the same magnetic moment, but that some have higher moments than others. This disparity proved true in molecular beams and atomic beams. Third, it was shown that the beam was

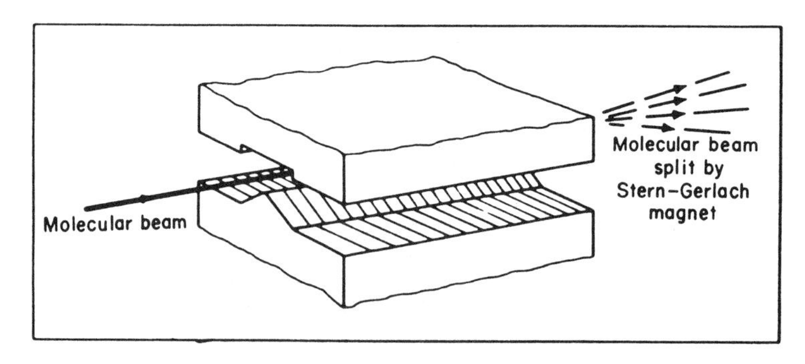

Figure 2.4 Stern–Gerlach magnet.

split into other discrete, well-defined beams that were equal to each other in intensity. This finding indicated that molecules exist only in well-defined states. These magnetic states are called *permitted states*, or *eigenstates*, and the energy levels associated with them are called *permitted energy levels*, or *eigenvalues*. The largest of these eigenvalues is indicated by the symbol μ and is called the *magnetic moment* of the atom or molecule.

The difference between adjacent eigenvalues is represented by μ/j, and the sequence of all the eigenvalues for a given species is represented by

$$\mu_{eff} = \frac{\mu m}{j} \tag{2.5}$$

where μ_{eff} is the effective magnetic moment and m and j are quantum numbers, with m being any whole number between $-m$ and $+m$. For example, if the maximum number for m is 2, then $m = 2$, $j = 1$, $\mu_{\text{eff}} = \mu_0$ (the largest value), the minimum value of μ_{eff} is $-\mu_0$, and the difference between each energy level is $\mu(1/j)$, where j is equal to μ_0 divided by the number of levels between $+\mu_0$ and $-\mu_0$ (see Fig. 2.5). This magnetic property provided the basis for nuclear magnetic resonance.

1. Quantum Theory

The Stern–Gerlach experiment illustrated a very important concept in spectroscopy, namely, the existence of permitted states, called eigenstates. These are energy states in which a molecule or an atom may exist for long or short periods of time. A molecule or atom can exist in intermediate energy states only when it is descending or ascending from one state to another, which takes only a very short time.

The energy states referred to are not confined to the magnetic moment, but also correspond to rotational energy, vibrational energy, electronic energy, and

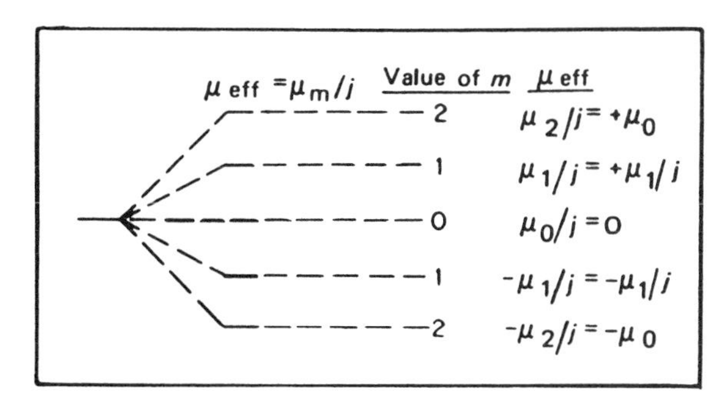

Figure 2.5 The eigenvalues of a molecular beam.

nuclear energy levels. These energy levels in turn correspond to NMR, IR, UV, x-ray, and nuclear science. The energy levels related to these forms of the motion of the molecules and atoms are governed by quantum theory. For example, the vibrational energy is not continuous, but may exist only in permitted vibrational energy states.

The transition of a molecule or atom from a lower to a higher energy state, which is accomplished by the absorption of radiant energy, is the basis of all methods of absorption spectroscopy, including infrared (IR) absorption, atomic absorption, ultraviolet (UV) absorption, x-ray absorption, and nuclear magnetic resonance (NMR).

An atom or molecule may also descend from a higher energy state, or eigenvalue, to a lower one, a transition that is usually accompanied by the emission of radiant energy. This transition is the basis of all methods of emission spectroscopy, such as molecular fluorescence, x-ray fluorescence, flame photometry, emission spectrography, and ICP.

The relationship between energy states and the energy of the radiation absorbed is given by Eq. (2.2). As described earlier, E is the *difference* in energy between the energy states in which the molecule existed before and after the transition; that is, $E = E_1 - E_0$, where E_0 and E_1 are the absolute energy levels of the eigenstates concerned (ν is the frequency of the absorbed or emitted radiation and h is Planck's constant). If the molecule absorbs energy, it passes from E_0 (lower energy state) to E_1 (higher energy state). If it emits energy, it passes from energy state E_1 to E_0.

a. The Boltzmann Distribution

In a system of atoms or molecules, each individual atom or molecule will tend to stay in the lowest energy level. At low temperatures this is the *ground state*. However, there is always a finite possibility that some of the atoms or molecules will be in upper energy states by virtue of the temperature of the system and the transfer of energy that can occur during collision, among other things.

If the temperature is raised, the number of atoms or molecules in an upper state increases and, conversely, the number in the ground state decreases. If the energy difference ΔE between the upper state and the lower state of a molecule is great, it is difficult to reach the higher state. Under these conditions, for a given temperature, the number in the upper state is low. If for a different compound the energy difference is small, it is easier to reach the upper state, and, for the same temperature, the number in the upper state is greater.

The relationship between the populations of atoms or molecules in upper and lower states is given by the Maxwell–Boltzmann equation, also known as the Boltzmann distribution. This is given by the equation

$$\frac{N_1}{N_0} = e^{-E/kT} \tag{2.6}$$

where

N_1 = number of atoms in the upper state
N_0 = number of atoms in the lower state
E = energy difference between the upper and lower states
k = Boltzmann distribution constant
T = absolute temperature

Equation (2.6) assumes that the system is in thermal equilibrium and that the energy levels are not *degenerate*, that is, they do not consist of several sublevels. More accurately, the equation is

$$\frac{N_1}{N_0} = \frac{g_1}{g_o} e^{-E/kt} \tag{2.7}$$

where g_1 and g_0 are the statistical weights of the states and allow a correction to be made for degeneracy.

2. Molecular Motion

a. Rotation of Molecules

Like all free bodies, molecules rotate in space. Associated with this rotation is rotational energy. As with the magnetic moment, the molecules may exist only in certain well-defined states of rotational energy. The absorption of energy causes transitions from one level to another and the molecule rotates faster. This gives rise to *rotational* absorption spectra. The rotational energy of a molecule depends on the angular velocity of the molecule, which is variable. It also depends on the molecule's shape and weight distribution, which change as bond angles change or the molecule flip-flops. The change in shape is restricted in dimers such as oxygen (O_2), but in molecules with numerous atoms, such as a common molecule like C_6H_{14}, there are many shapes possible and therefore many rotational energy levels possible. Furthermore, the substitution of isotopes such as ^{13}C for ^{12}C generates new sets of energy levels. Consequently, even simple molecules have many permitted rotational energy levels. The energies involved are very small, and the absorbable radiation involved is in the radiofrequency, or *microwave*, region of the spectrum. The microwave region of spectroscopy is largely unexploited in analytical chemistry because of the experimental complexities involved and the richness of the spectra, which makes their interpretation difficult.

b. Vibration of Molecules

The combined atoms that form the components of a molecule vibrate in much the same way as do larger objects. The molecules can be visualized as a set of weights (atoms) joined by springs (chemical bonds). The atoms can vibrate to-

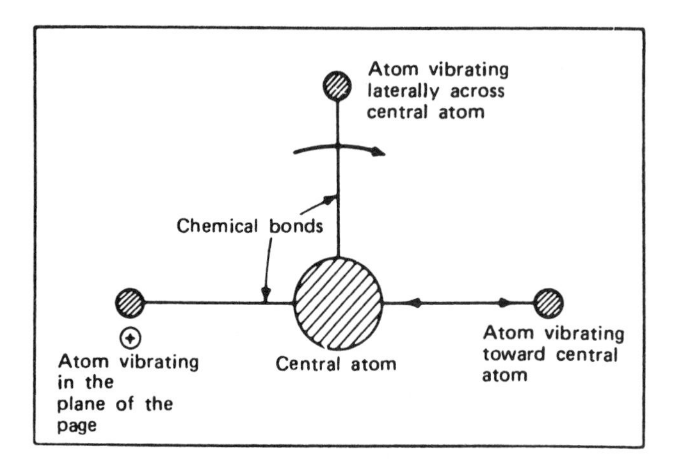

Figure 2.6 The vibration of atoms in a molecule.

ward and away from each other, or they may bend at various angles to each other, as shown in Fig. 2.6. In doing so, they exhibit vibrational energy. As with the magnetic moment, only certain energy levels, or eigenvalues, are permitted, and, as before, the radiation energy associated with transitions between these energy levels is described by $E = h\nu$.

The wavelength range of this radiant energy is in the infrared, generally considered to be between 0.8 μm and about 200 μm. Absorption of IR radiation usually causes vibration of parts of the absorbing molecule. The vibration of the molecule is also usually accompanied by increased molecular rotation; hence, in practice, the energy absorbed corresponds to a combination of vibrational and rotational energy. This combination provides the basis for IR absorption spectroscopy, one of the most important methods of spectroscopic analysis. With this technique, the shape and structure of molecules can be deduced; also, the compositions of mixtures of molecules (solutions) can be determined.

c. Absorption Caused by Electronic Transitions in Molecules

Outer electrons in molecules orbit in well-defined energy levels. Under normal conditions the molecule is *unexcited* and the electrons forming a chemical bond are in a *bonding orbital*. This orbital is the lowest energy state, or *ground state*.

If the molecule absorbs sufficient radiant energy, the electron moves from the bonding orbital to the *antibonding orbital* with associated higher energy. The molecule thereby becomes excited, and the electron enters an excited energy level. As we saw earlier, these energy levels are well defined and their values well documented. The radiant energy required to make an electron move from a bonding orbital to an antibonding orbital corresponds to that found in the visible and ultraviolet regions of the electromagnetic spectrum. This wavelength range is considered to be between 800 and 200 nm. When the electrons are part of a *free atom*, no vibrational or rotational energy changes are involved. Therefore, the wavelength of the radiation involved is very well defined and stretches over very narrow frequency bands. But when the orbiting electron is part of a *molecule*, the energy required to move the electron from the ground state to an excited state varies with the type of molecular bond in which the electron exists. For example, electrons in paraffinic bonds (σ bonds) require more energy than electrons in unsaturated bonds (π bonds), such as in aromatics. But when a molecule is electronically excited, there is often a simultaneous change in vibrational and rotational energy. As a consequence, the total energy change involved is the change in electronic energy, modified by the associated rotational and vibrational energy changes.

If we examine the absorption of UV radiation by a large population of molecules of the same compound, we find that different individual molecules require different amounts of energy brought about by a difference in the vibrational or rotational energies in the different molecules. This difference leads to the absorption of energy over wide frequency ranges, or *absorption bands*, rather than absorption lines occurring at one frequency only. But studying UV absorption bands or lines, we are able to identify chemical compounds or atoms. In molecules the energy levels of the pertinent electrons are in bonding and antibonding orbitals. The UV absorption spectra are usually simple and cover a broad wavelength range. Similar broadening is experienced in the infrared, since the vibrational energy levels are modified by the rotational energy levels.

d. Effect of the Absorption of Energy

We have seen that there are three principal types of motion in atoms and molecules, giving rise to rotational, vibrational, and electronic excitation. The vibration and rotation of all molecules are controlled by the shape of the complete molecule, the weight of its component parts, and the strength of the bonds. The most pertinent groups are the functional groups such as carboxylic acids and aldehydes, or combined elements such as nitrogen, oxygen, and hydrogen. Infrared absorption is concerned with functional group analysis and the components of molecules, particularly organic molecules. On the other hand, electronic excitation is brought about by the absorption of visible or UV light and is particularly important when the electrons involved can be easily excited. Molecules containing halides or nitrogen compounds contain electrons that are easily excited, as do unsaturated organic molecules such as aromatics and olefins. Methods of analysis based on absorption are usually nondestructive, because only the movement of the molecules is involved, and assessment of this motion does not lead to destruction of the molecules.

If the frequency of the radiation is increased beyond the UV range, the radiation becomes energetic enough to excite nonvalence electrons of the outer shell. With still further increases the radiation has enough energy to remove electrons from the inner electron shells of an atom. When this happens, electrons from outer shells drop down and fill the inner shells. This displacement generates x-rays. Atoms are able to absorb and emit x-rays. However, the wavelengths absorbed or emitted depend on the energy levels of the inner and outer electron shells. These levels are not noticeably affected by the state of chemical combination of the atom. For this reason, x-ray absorption and emission are particularly useful for elemental analysis and are almost independent of the chemical form of the elements. Diffraction of x-ray beams is also used for measuring the dimensions of crystals.

Electromagnetic radiation with energy greater than that of x-rays is emitted by some radioactive atoms. This radiation is so energetic that it can penetrate to the nuclei of atoms. If it is absorbed by the nucleus, the atom becomes a different element (i.e., it is transmuted), which may be radioactive. The absorption and emission of such very high-energy radiation are the basis of nuclear science and activation analysis. In this field only the nuclei of atoms are involved in the interaction with radiation. The radiation is therefore useful for elemental analysis. In a separate branch of nuclear science, by putting a radioactive isotope of an element, such as ^{14}C, in place of the normal isotope (i.e., ^{12}C), we can follow the path of a particular atom in a given chemical reaction. For example, if a precipitate forms and is found to be radioactive, molecules containing a radioactive isotope must be in the precipitate. If the original mixture contained only one radioactive isotope, a measurement of the radioactivity of the original solution and of the final precipitate gives a measure of the completeness of precipitation. The ease of locating radioactive atoms and studying reactions and molecular pathways are the basis of radiotracer chemistry.

D. THE EMISSION OF RADIANT ENERGY BY ATOMS AND MOLECULES; METHODS OF ELECTRONIC EXCITATION OF ATOMS

Molecules and atoms emit radiation in all regions of the spectrum, from radio waves to radioactive radiation. The most commonly exploited region of the spectrum is the UV range. This range is involved in the excitation of the outer electrons of a molecule or atom, called electronic excitation, which we introduced in Section C.2.c.

When atoms or molecules absorb radiant energy, thereby becoming excited, as described in Section C.2.c, they usually remain in the excited state for a very short time (in the neighborhood of 10^{-6} sec). The atom or molecule then emits a photon of energy and returns to the ground, or unexcited, state. Figure 2.7,

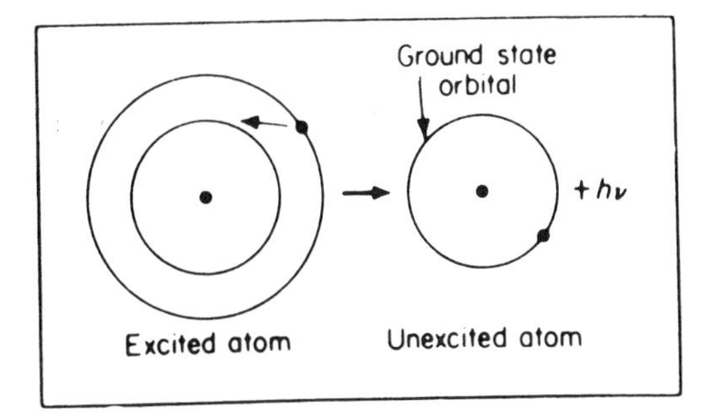

Figure 2.7 The emission of spectral energy.

which illustrates the process, does not, however, represent the paths of the electrons around the atomic nucleus. These paths are not clear to us at present. We are only able to derive the *probability* of an electron being in any one place at any particular time. Nevertheless, this probability does reflect the charge distribution that the electron contributes to the atom as a whole. The energy levels of an atom of one element are different from the energy levels of an atom of a different element. In all atoms there are numerous energy levels in which the electron may exist (as described in Sec. C.2.c). As we saw earlier, the difference in energy between these levels is equal to $E_2 - E_1$ or $E_3 - E_2$ or $E_4 - E_3$, and so on (Fig. 2.8).

The energy levels are defined by quantum theory. For example, the lowest electron shells of sodium are filled and the valence electron is the 3*s* electron. All upper orbitals are empty and constitute the upper excited electronic energy levels. In the case of sodium, E_0 is the 3*s* level, E_1 is the 3*p* level, E_2 is the 4*p* level, and so forth. All empty upper orbitals are available, but the rules concerning forbidden and permitted transitions must be obeyed.

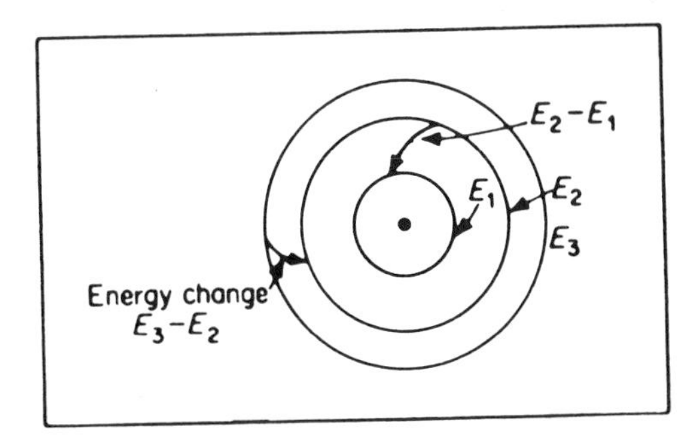

Figure 2.8 Emission of energy between different excitation levels.

For each change, the energy jump from one level to the next is different, and therefore $h\nu$ is different. Hence for each transition, radiation is emitted at a different frequency ν. This means that each element is capable of emitting radiation at many different characteristic wavelengths. These different emission lines constitute the *emission spectrum* of the element. The emission spectrum of each element is different from that of other elements. This difference provides the basis of emission spectroscopy and enables us to identify the element emitting the radiation. It provides the basis of the most important method of elemental qualitative analysis (see Chap. 10). For many elements literally hundreds of transitions can be identified from the standard tables.

The same type of energy changes take place with electrons in molecules, except that in this case the electrons involved are in molecular orbitals, not atomic orbitals. There are fewer molecular orbitals; hence molecular spectra are comparatively simple. Molecular emission gives rise to molecular fluorescence or phosphorescence. Not all molecules fluoresce, and very few phosphoresce. These phenomena can be used to detect and identify very small concentrations of certain compounds.

The analytical applications of radiofrequency radiation and far-infrared radiation are limited but are being studied. Energy changes involving inner shell electrons result in the absorption or emission of x-rays, and these will be discussed in Chapter 11. The most commonly used energy changes are those involving outer shell electrons, such as in emission spectrography, UV absorption, flame photometry, atomic absorption spectroscopy, and so on. The phenomena involving outer shell electrons are described below.

1. Excitation Involving an Electrical Discharge

There are several methods of exciting atoms. One of the earliest and most important of these is to put the sample in an electrical discharge between two electrodes (Fig. 2.9). The electrical discharge, of very high energy, breaks down the sample, which is probably in the form of molecules or ions, into excited atoms. Atoms of the different elements present emit a characteristic emission spectrum. This is the basis of emission spectroscopy, which is discussed in Chapter 10. Recently excitation using inductively coupled plasma has been developed and has significantly extended the sensitivity and use of emission spectrography in quantitative and qualitative analysis.

2. Flame Excitation

Atoms may also be excited by putting them into a flame, the thermal energy of which breaks down or reduces the molecules or ions of the sample into atoms. The atoms become excited and emit their characteristic spectra. This is the basis of flame photometry, which is discussed in Chapter 9.

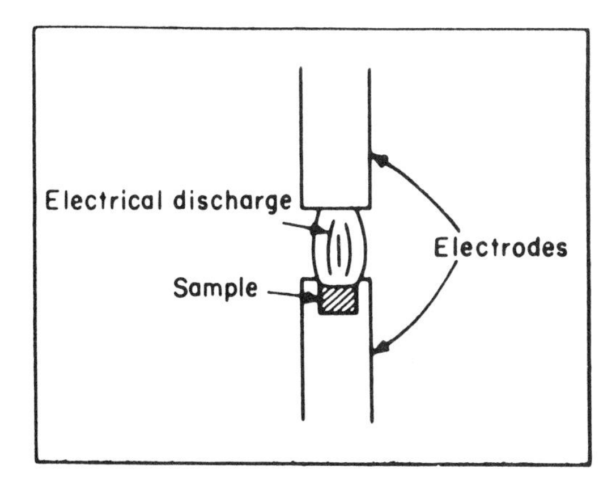

Figure 2.9 Excitation by electrical discharge.

The energy of a flame is lower than that of an electrical discharge; hence atoms can be excited only to lower states by this method. Fewer transitions are possible; therefore the spectrum consists of fewer spectral lines than in emission spectroscopy. This subject is discussed in greater detail in Chapters 9 and 10.

3. Excitation by Radiation

By irradiating atoms with light of the correct wavelength, we cause the atoms to absorb the radiation and to become excited in the process. The excited atoms reemit the absorbed radiation and become unexcited; the emission of radiation following excitation by radiant energy is called *fluorescence*. The process of fluorescence of radiation by atoms is the basis of x-ray fluorescence and atomic fluorescence. It is also the basis of molecular fluorescence, a very important analytical field. In practice, the sample is irradiated with UV light, which is absorbed by the molecules. The molecules become excited and reemit the spectral energy. The reemitted fluorescence radiates in all directions. By measuring the intensity of fluorescence at right angles to the radiation that excites the sample, we can distinguish between the beam of light causing excitation and the fluoresced radiation emitted by the sample (Fig. 2.10).

Excited molecules normally return to the unexcited state without any other change taking place in the molecular orbital. Occasionally, however, the electron involved changes its direction of spin while in the excited state and enters a triplet state. This spin change makes emission of radiation difficult, because the transitions involved are forbidden. In practice, this emission of radiation is delayed compared to fluorescence. This is the basis of *phosphorescence*, which is not an important analytical field because of the strict controls needed to make the intensity of phosphorescence a reliable measure of the concentrations of emitting molecules present.

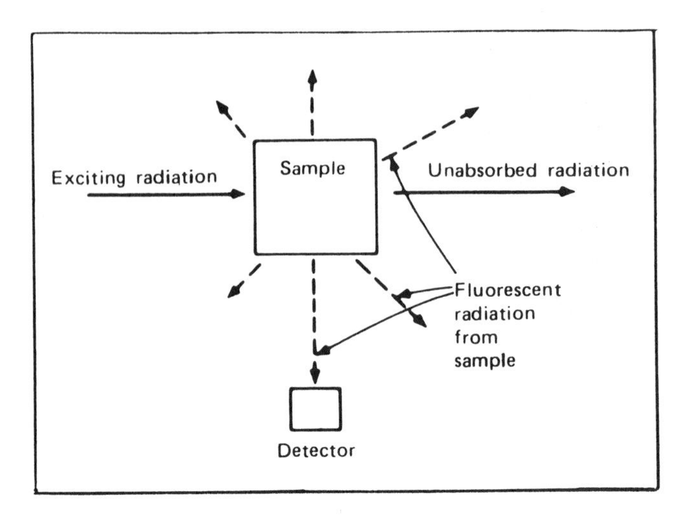

Figure 2.10 Molecular fluorescence.

E. ABSORPTION LAWS

When a beam of light passes through a solution that absorbs radiation, a certain amount of light is absorbed by the solution. The quantity of light absorbed is well defined and follows definite physical laws. For example, if the intensity of a beam of monochromatic radiation passing through a solution is I_0, then the intensity of the radiation emerging from the solution is I_1, and not I_0 (Fig. 2.11). The ratio of I_1 to I_0 is known as the *transmittance*, and the relationship is described mathematically as

$$T = \frac{I_1}{I_0} \tag{2.8}$$

where T is the transmittance, or fraction of light transmitted by the solution. To the first approximation the ratio I_1/I_0 remains constant even if I_0 changes; hence T is independent of the actual intensity I_0.

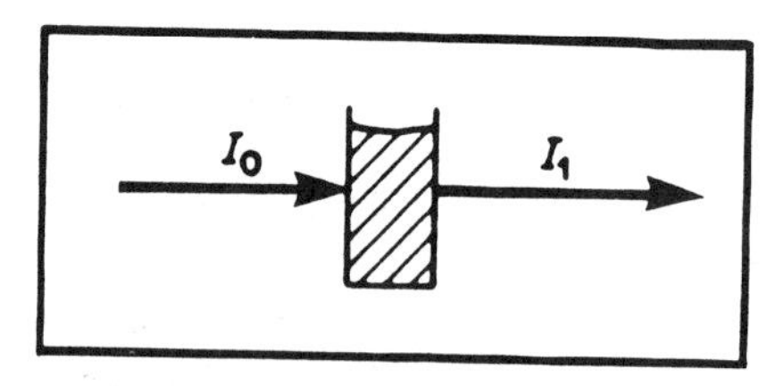

Figure 2.11 Absorption of radiation.

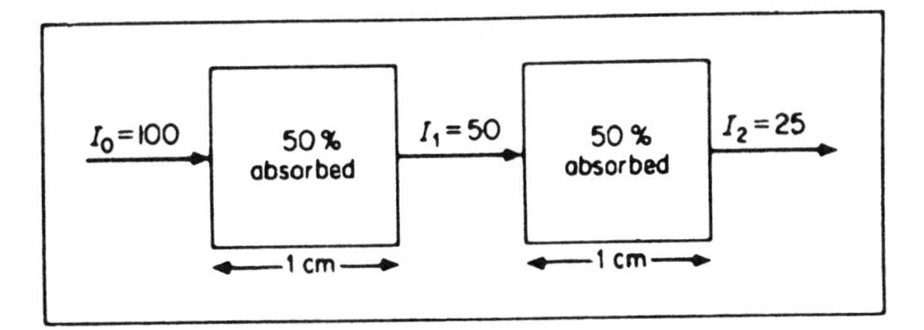

Figure 2.12 Absorption of radiation by two cells.

To illustrate, suppose that a solution of unit concentration is 1 cm thick and absorbs 50% of the radiation of a given wavelength that fall on it. Then

$$\frac{I_1}{I_0} = \frac{50}{100 \text{ units}}$$

If a second similar solution, also 1 cm thick, is placed behind the first solution, again, 50% of the radiation is absorbed. In this case, however, only 50 units of light fall on the second solution; therefore 25 units emerge (Fig. 2.12). By increasing the cell path length b systematically to $3b$, $4b$, etc., we can determine that $I_1 = 10^{-b}$. The I_0 relationship between T and b is logarithmic with a negative sign. A similar relationship exists between c, concentration, and a, absorptivity, the ability of the sample molecules to absorb at that wavelength.

If instead of examining this phenomenon in terms of the amount of light transmitted, we examine it in terms of how much light is absorbed, we arrive at a different relationship. We may now introduce the term A, absorbance, which is defined as:

$$A = -\log T$$

Therefore,

$$A = \log_{10}\left(\frac{1}{T}\right) \tag{2.9}$$

or

$$A = -\log\left(\frac{I_1}{I_0}\right) = \log_{10}\left(\frac{I_0}{I_1}\right) \tag{2.10}$$

where I_0 is the original intensity of radiation and I_1 is the intensity of radiation after absorption.

If we continue to add sample cells 1 cm long to the system, we could measure I_1 after each addition and plot T and A versus the number of sample cells added. The number of cells is also a measure of sample path length. The relationship would appear as shown in Fig. 2.13. Note that the slope of the curves relating T and the sample path length is logarithmic, but the slope relating A and the path length is linear. A linear relationship exists between A and b. Most researchers utilize A as a function of path length for this reason.

1. Lambert's Law

The relationship in Eq. (2.9) forms the basis of Lambert's law (also known as Bouguer's law). For a solution of unit concentration, the relationship can be expressed as

$$A = ab$$

where a is the absorptivity of the liquid and b is the optical path length. The interrelationship of these measurements is given by

$$A = ab = -\log T = -\log\left(\frac{I_1}{I_0}\right) = \log\left(\frac{I_0}{I_1}\right) \tag{2.11}$$

Since

$$\log\left(\frac{I_0}{I_1}\right) = ab \tag{2.12}$$

we have

$$\frac{I_0}{I_1} = 10^{ab} \tag{2.13}$$

and

$$I_1 = I_0 \times 10^{-ab} \tag{2.14}$$

or

$$I_0 = I_1 \times 10^{ab} \tag{2.15}$$

Each of these relationships is a mathematical expression of Lambert's law. They show that the amount of light absorbed depends on the absorptivity of the liquid and the length of the optical path through the solution, provided that the wavelength and the sample remain constant.

2. Beer's Law

Lambert's law shows that there is a logarithmic relationship between transmittance and the length of the optical path through the sample. A similar relationship (see Fig. 2.13) holds between transmittance and the concentration of a

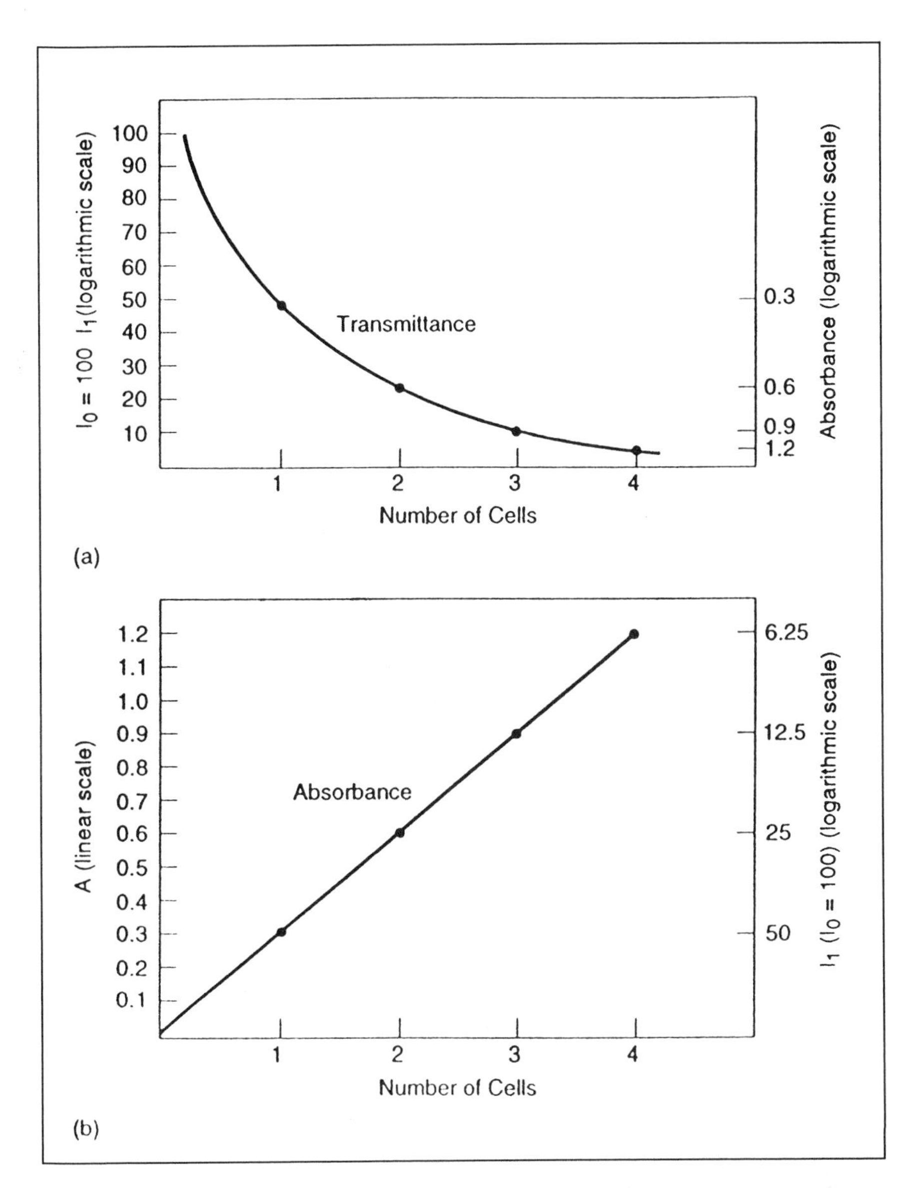

Figure 2.13 Relationship between transmittance (a) and absorbance (b) with increasing number of cells, that is, an increase in path length.

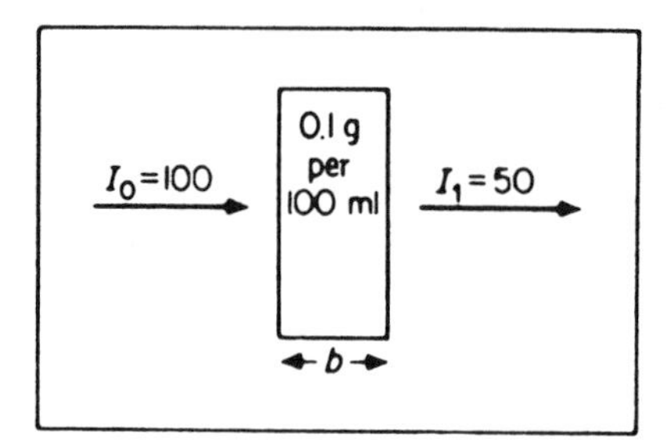

Figure 2.14 Optical arrangement illustrating Beer's law.

solution. Suppose that we have the optical arrangement shown in Fig. 2.14, where the incident radiation I_0 is 100 units, the emerging radiation I_1 is 50 units, and the concentration of the absorbing solution is 0.1 g/100 ml. If we place a second similar container in the path of the emerging light, the intensity of radiation I_1 leaving this container is 25 units (Fig. 2.15). Let us now change the conditions so that *all* the absorbing material is in the first container; that is, the concentration of the absorbing material is 0.2 g/100 ml. We measure the intensity of the emerging radiation and find that it is 25 units (Fig. 2.16).

From this illustration of the relationship between concentration and transmittance we can deduce that for a given path length

$$A = ac$$

where c is the concentration of absorbing material, A is the absorbance, and a is the absorptivity of the sample. Also, as before [Eq. (2.11)], the relationships are

$$A = ac = -\log T = -\log\left(\frac{I_1}{I_0}\right) = \log\left(\frac{I_0}{I_1}\right) \tag{2.16}$$

Moreover,

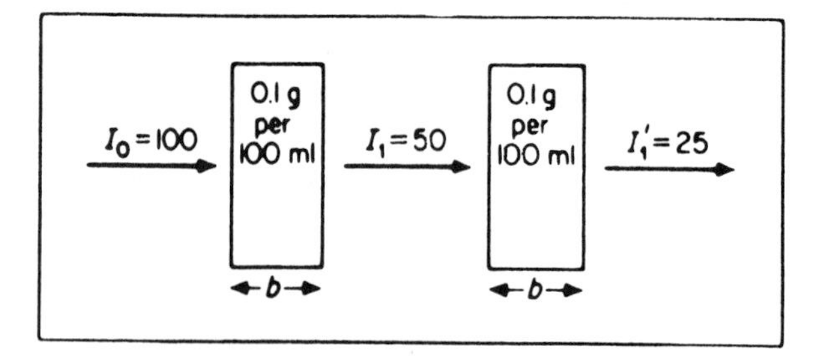

Figure 2.15 Relationship between the transmittance and the path length of a solution.

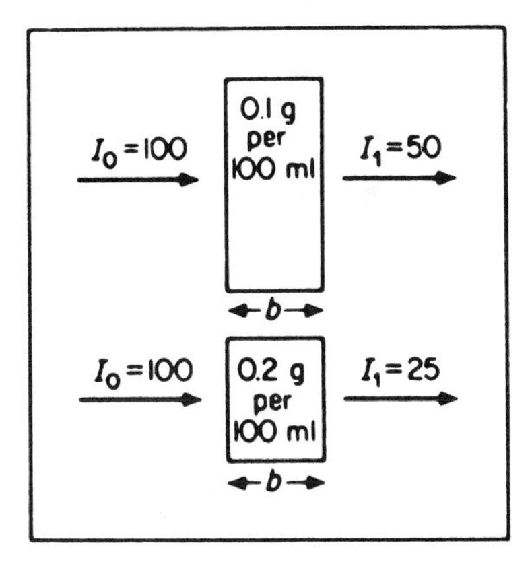

Figure 2.16 Relationship between concentration and transmittance.

$$\log\left(\frac{I_0}{I_1}\right) = ac$$

$$\frac{I_0}{I_1} = 10^{ac} \qquad (2.17)$$

$$I_1 = I_0 \times 10^{-ac}$$

or

$$I_0 = I_1 \times 10^{ac}$$

These expressions, which are all forms of Beer's law, show the relationship between the degree of light absorption and the concentration of a solution at a given wavelength and cell thickness.

The term *molar absorptivity* (ϵ) is used to designate the absorptivity when the concentration is expressed as one mole per liter and the light path b is expressed as one mole per centimeter. A *mole* of a substance is the weight of 6.02×10^{23} molecules of that substance, that is, the same number of grams as the molecular weight. A mole of acetic acid (CH_3COOH) is 60 g. A molar solution contains this weight in 1 liter of solution; for example, if the molecular weight of a solute (such as acetic acid) is 60, a molar solution of acetic acid is 60 g/liter.

3. Beer–Lambert Law

The combination of Beer's law and Lambert's law gives rise to the Beer–Lambert law, also known as the Beer–Bouguer law or simply Beer's law. The combination of the two relationships can be expressed as

$$A = abc \tag{2.18}$$

This leads to

$$A = abc = -\log T = -\log\left(\frac{I_1}{I_0}\right) = \log\left(\frac{I_0}{I_1}\right) = \log\left(\frac{I}{T}\right) \tag{2.19}$$

$$\log\left(\frac{I_0}{I_1}\right) = abc, \qquad \frac{I_0}{I_1} = 10^{-abc}$$

or

$$I_1 = I_0 \times 10^{-abc} \quad \text{and} \quad I_0 = I_1 \times 10^{abc} \tag{2.20}$$

These expressions show that there is a linear relationship between the absorbance A and concentration c of a given solution if the optical path length and the wavelength of the radiation are kept constant. The analytical implications of this relationship are very important. By measuring the ratio I_1/I_0, we can derive the absorbance A, and from this value we can calculate c, the concentration of the solution. This method of calculating solution concentrations from absorption measurements is widely used in spectroscopy.

The Beer–Lambert law is usually found to hold at low concentrations, but deviations are common at higher concentrations. Nevertheless, this law is a cornerstone of all phases of absorption spectrometry, such as spectrophometry, IR absorption spectrometry, UV absorption spectrometry, and atomic absorption.

It is of interest to note that from Eq. (2.19) $A = \log I_0/I_1$. If $A = 1$ and $I_0 = 100$, then

$$\log\left(\frac{I_0}{I_1}\right) = \frac{100}{I_1} = 1$$

Hence, $I_1 = 10$, i.e., the amount of light absorbed $I_0 - I_1 = 90\%$. Similarly, if $A = 2$, $\log 100/I_1 = 2$ and $I_1 = 1$. The amount of light absorbed is $100 - 1 = 99\%$. If $A = 3$, the amount of light absorbed is 99.9%.

In practice, any readings involving numbers greater than $A = 1$ are suspect. If $A = 2$, the error involved is probably very great (see Fig. 2.25).

4. Deviations from Beer's Law

It has been found that in practice Beer's law applies only under ideal circumstances. There is usually a deviation from a linear relationship between concentration and absorbance, as indicated in Fig. 2.17. Such deviations can arise from impurities that fluoresce or absorb at the fluorescence or absorption wavelength. This interference causes an error in the measurement of the intensity of the radiation penetrating the sample. Variations in the concentration of the impurity from one sample to another cause a variation in the error. Also, the material being measured may be in chemical equilibrium with other compounds present in the sample. At high concentrations, the equilibrium may be displaced, which

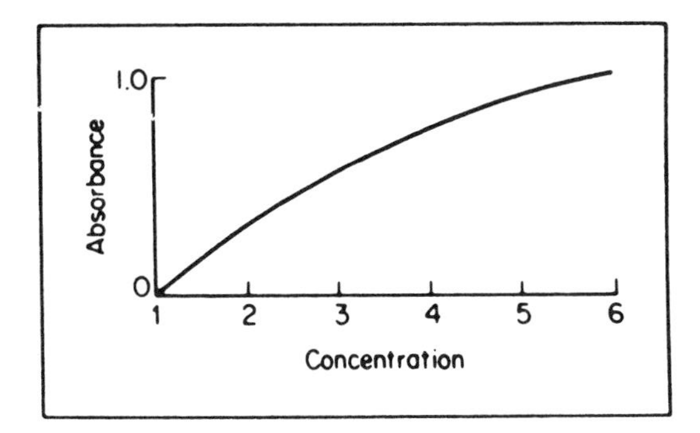

Figure 2.17 Deviation from Beer's law.

would cause an apparent decrease in the concentration of the compound being determined, and therefore a deviation in the absorbance–concentration relationship. A third source of error is the optical slit used on the instrument. If this slit is too wide, it allows unabsorbable radiation to fall on the detector, causing a change in the apparent absorbance–concentration. Also radiation with a wavelength slightly greater or slightly less than that desired would pass through the slit and reach the detector. Such radiation might not be absorbed by the sample, but might be absorbed by impurities in the sample, which would cause an apparent change in the absorbance. The error would be increased at higher concentrations. Other sources of error include dimerization of the compound and interaction between the sample compound and the solvent, causing loss of absorption by the sample.

Because of the deviations, the extrapolation of the absorbance–concentration relationships from standards with low concentrations to samples with high concentrations will cause errors. This difficulty can be overcome by preparing calibration curves that show the experimental relationship between A (and therefore I_0/I_1) and the actual sample concentration. Errors caused by absorption by interfering molecules, or variations in sample cell width, may be eliminated if they are constant in the sample being analyzed and the calibration samples. This requires extra care when making up the standard solutions.

F. METHODS OF CALIBRATION

1. Standard Calibration Curves

Calibration curves are determined experimentally by preparing a series of mixtures, each with a known concentration of the absorbing sample in a suitable solvent. The absorbance of each solution is then measured and a curve relating the experimentally determined absorbance and the concentration of the standard sample is prepared.

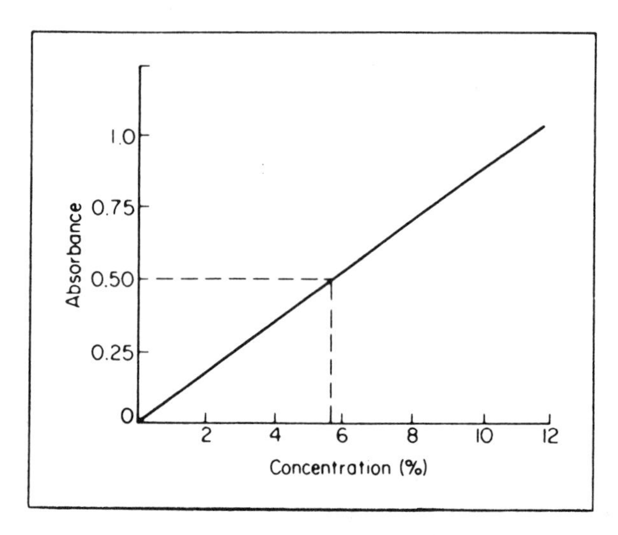

Figure 2.18 Standard calibration curve.

A typical calibration curve is shown in Fig. 2.18, prepared by plotting A (at 9.4 μm) against the concentration of the solution. When a sample of unknown concentration is to be determined, the absorbance A is measured at a fixed wavelength. From the calibration curve we may determine c, which is the concentration of hexanol. For example, if A were found to be equal to 0.50, the concentration would be 5.8% (Fig. 2.18). A typical series of absorption values and the resultant calibration curve are shown in Figs. 2.19 and 2.20 for hexanol measured at 9.4 μm.

An experimental determination of the concentration of a solution can be simplified by keeping the intensity of the incident radiation I_0 constant; then only I_1, the intensity of the emerging radiation, needs to be measured. Since [Eq. (2.19)]

$$A = -\log\left(\frac{I_1}{I_0}\right) = abc$$

then $-\log(I_1/I_0)$ is related to c by the same curve, and

$$c = -\log\left(\frac{I_1}{I_0}\right) \div ab \qquad (2.21)$$

For a given system, however, a, b, and I_0 are constant. Therefore a direct relationship exists between c and $-\log I_1$ or $\log(1/I_1)$.

When a solution of unknown concentration is to be determined, it is put into the spectrometer and $I_1(I_1/I_0)$ is measured. Using the calibration curve, we ascertain the experimental relationship between I_1 and c. Therefore, by measuring

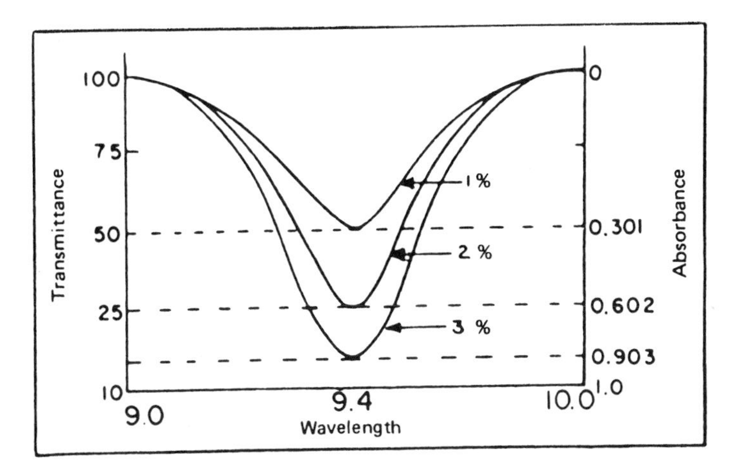

Figure 2.19 Infrared absorption curves for hexanol at 9.4 μm at different concentrations.

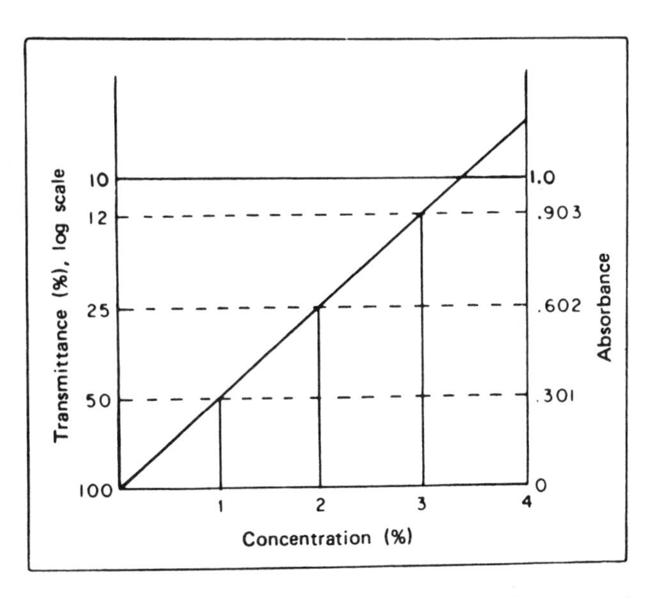

Figure 2.20 Calibration curve for hexanol at 9.4 μm derived from the absorption curve in Fig. 2.19.

I_1, we can determine the concentration of the absorbing compound in the sample. One problem that must always be checked is due to absorption by molecules of interfering compounds that may be present in the sample. This absorption may cause a shift in the absorption baselines, as indicated in Fig. 2.21, where in the normal case (Fig. 2.21a) I_0 can be measured directly from the baseline. However, when the baseline is disturbed (Fig. 2.21b), a correction must be made.

This correction can be done in two ways. In each case an estimation of the disturbed baseline is made, as in Fig. 2.21b. The ration I_1/I_0 can then be measured directly as 60/90. However, if a calibration curve has been prepared, the following approach may be more convenient. We may first measure I_1 and use the calibration curve to read off the apparent sample concentration. Then, from reading off the baseline absorption in terms of concentration, a value can be obtained for the background absorption in terms of sample concentration. This concentration is then subtracted from the apparent concentration to give the true answer. This procedure is illustrated mathematically as follows.

Case a—no interfering compound present:

$$T = \frac{66}{100} = \frac{2}{3}$$

$$A = \log\left(\frac{3}{2}\right) = 0.1761$$

concentration from calibration curve = 0.59%

If the calibration curve is linear, then a concentration of 1% will result in an absorbance of 0.1761/0.59 = 0.301.

Case b—interfering compound present:

$$T = \frac{60}{90} = \frac{2}{3}$$

$$A = \log\left(\frac{3}{2}\right) = 0.1761$$

concentration from calibration curve = 0.59%

Alternative method for case b:

$$T = \frac{60}{90}$$

$$A = 0.2201$$

apparent concentration from calibration curve = 0.74%

The error in T introduced by the concentration obtained from the false baseline is 90/100, and

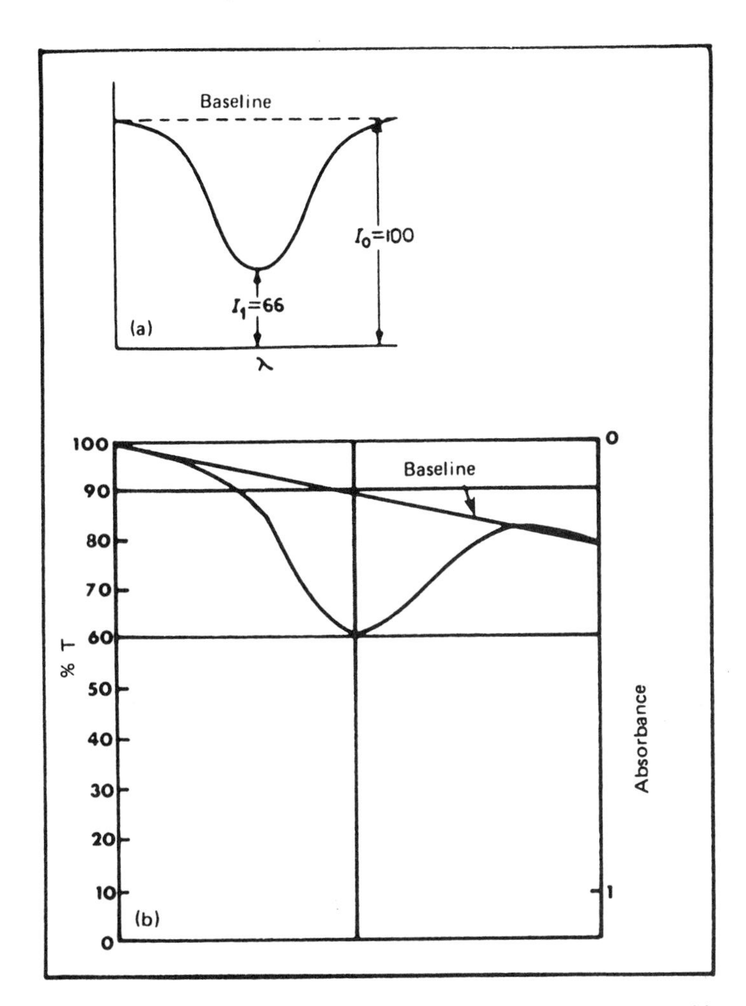

Figure 2.21 Absorption baseline shift caused by extraneous absorbing material. In the normal case (a), $T = 66/100 = 2/3$ and $A = \log(3/2)$; the concentration from the calibration curve (not shown) is 17%. The baseline in (b) shows the effect of an interfacing compound; there $T = 60/90 = 2/3$, $A = \log(3/2)$, and the concentration from the calibration curve is 17% (see text for alternative calculation.

error in $A = 0.045$

error in concentration from calibration curve = 0.15%
true result = 0.74% − 0.15% = 0.59%

If a double-beam instrument is used (see Chap. 3) and the interfering compound is introduced into the *reference* cell, a correction is made automatically

Table 2.1 Radiation Intensity

Sample	Emission intensity	Na added to sample (ppm)
1	2.9	0
2	4.2	1
3	5.5	2
4	6.8	3
5[a]	0.5	Background emission from flame

[a]Flame only; no sample added.

by the instrumentation, which measures the ratio of the absorption of the sample beam to that of the reference beam. The ratio is independent of the actual values of I_0 and I_1. The interfering absorption is common to both beams. The net result is that the common interfering substance causes no signal to be recorded by the instrument. The absorbing sample, however, causes an imbalance in the two beams, and the final signal is a correct measure of the absorption by the sample.

2. Standard Addition Method

The standard addition method may be used if no suitable calibration curves have already been prepared and it is undesirable to prepare such a curve. This situation may arise if, for example, there is no time to prepare a calibration curve, or the lack of sufficient information on the interferences in the sample makes it impossible to prepare a valid curve. In a typical example, sodium in an industrial plant stream of unknown composition may be determined by the method of standard addition. Several 100-ml aliquots of the sample are taken, and different but known quantities of the metal being determined are added to each aliquot except one, which is left untreated. The radiation intensity from each aliquot is then measured on the flame photometer. The results are shown in Table 2.1.

The data can be displayed in the form of a graph (Fig. 2.22). A measure of the background radiation can be made either from a similar sample containing none of the element being determined or by measuring the radiation intensity at a wavelength slightly different from the characteristic wavelength of the metal. It can be seen from Fig. 2.22 that, over the range examined, the relationship between emission intensity and the sodium concentration added was linear. From the difference in readings between samples 1 and 2 and samples 3 and 4, it can be seen that 1 mg of sodium per 100 ml generated an emission intensity of 1.3 units.

The emission from the sample with no sodium added was 2.9 units. But since the background was 0.5 unit, the emission from the sample was 2.9 − 0.5 = 2.4 units. But 1.3 units is equivalent to 1 mg/ml; therefore it is calculated

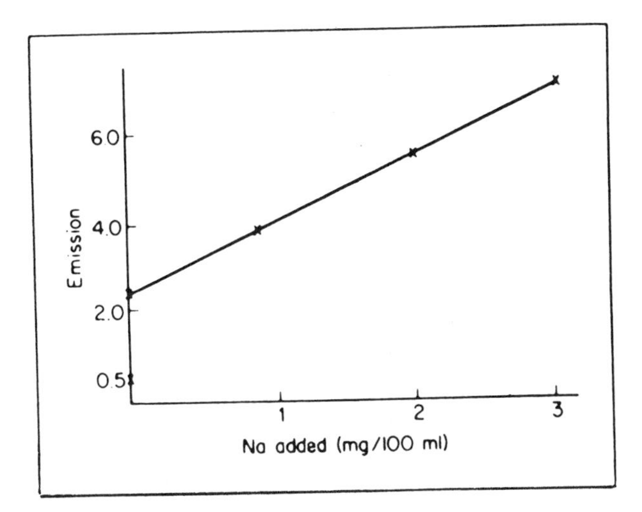

Figure 2.22 Determination of sodium concentration by the standard addition method.

as $(1 \times 2.4)/1.3 = 1.85$ mg/ml. Thus the concentration of sodium in the original sample was 1.85 mg/ml.

The concentration can be calculated from the relationship

$$\text{concentration} = \frac{R - B}{\Delta_{em}/\Delta_{conc}}$$

where R is the radiation emission from the sample, B is the background emission, and $\Delta_{em}/\Delta_{conc}$ is the rate of change of emission with change in sample concentration.

In an alternative method of calibration, the calibration curve produced by standard addition (Fig. 2.23) may be extended backward. The x intercept indicates the "concentration" in the sample, in this case, 2.23 ppm. A parallel line is drawn through the background contribution (0.5), indicating a contribution of 0.38. The corrected sample concentration is $2.23 - 0.38 = 1.85$ mg/100 ml.

As yet another alternative, a second line parallel to the y axis and passing through the background measurement (0.5) is drawn. The concentration is read at the point of intersection between the line described and the extended calibration curve as shown in Fig. 2.23.

3. Errors Associated with Beer's Law Relationships: The Ringbom Plot

Whenever a measurement is made, there is always an associated indeterminate error (see Chap. 1, Sec. C). This is as true of measurements of radiation as any other measurement. These errors may involve the light source, the monochromator, the detector system, or other parts of the system, but, because they are

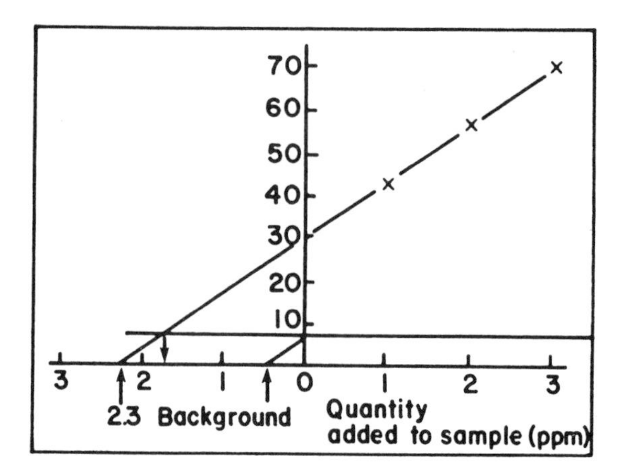

Figure 2.23 Alternative method for calculating concentration using the standard addition method. A correction is made for the background in the final answer.

indeterminate errors, they are not correctable. The errors in measurements of radiation intensity lead directly to errors in the measurement of the concentration when using calibration curves.

Let us suppose that the absolute error is a constant with respect to the radiation intensity falling on the detector and is equal to 1 unit of recording chart paper. At high concentrations of sample, the degree of absorption is high and very little radiant energy falls on the detector. If the amount of energy falling on the detector is equivalent to 2 units, then the error involved is 1 unit. This is a 50% relative error.

At the other end of the scale (see Fig. 2.24), if the concentration is very low, then the amount of radiation falling on the detector is high (e.g., 97 units), corresponding to a 3% absorption. If there is a 1-unit error in the measurement, then the radiation intensity falling on the detector may be 96 units, corresponding to a 4% absorption. The error in measuring the percent absorption is 33%, and there is a corresponding relative error of 33% in determining the concentration in the sample from the calibration curve. These two examples occur at the extreme ends of the concentration ranges used in a Beer's law relationship, and it can be seen that significant error is involved.

We can plot this relative error as a function of transmittance, and it will give us a curve such as that shown in Fig. 2.25. This is known as a *Ringbom plot*. The minimum relative error occurs at 37% transmittance, although satisfactory results can be achieved over the range of 15–65% transmittance, that is, an absorbance range of 0.82–0.19.

As a consequence of this relationship, it is always advisable to determine the concentration of samples from absorbance readings that lie within this range.

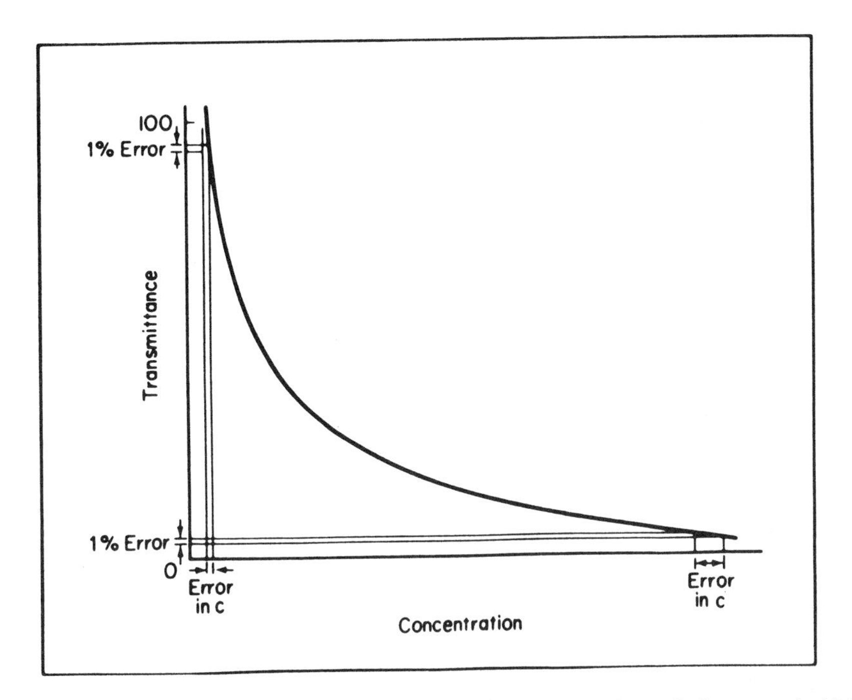

Figure 2.24 Concentration versus transmittance. Note that relative error is high at each end of the scale and low in between.

If the samples are too dilute, they may be concentrated by solvent extraction or evaporation. If they are too concentrated, they may be diluted to bring them into the desirable range. However, if it is not possible to alter the samples and they are outside the range indicated, the only alternative is to continue the analysis, knowing that the precision of the procedure will be diminished depending on the percent transmittance measured.

Recommended nomenclature and definitions for spectroscopy experiments are given in Table 2.2.

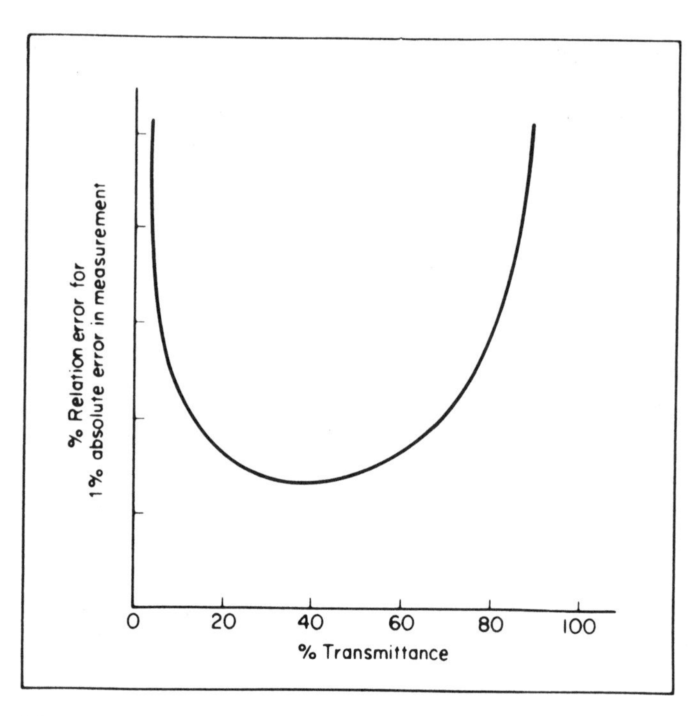

Figure 2.25 Ringbom plot relating relative error and absolute error. The minimum relative error is at about 37% transmittance. It is acceptable between 15 and 65%.

Table 2.2 Nomenclature and Definitions for Spectroscopy Experiments

Term	Symbol	Definition
Transmittance	$\frac{I_1}{I_0}$	Ratio of light intensity after passing through sample I_1 to light intensity before passing through sample I_0
Absorbance	A	$-\log T = abc$
Absorptivity	a	A/bc, where b = path length and c = concentration (g/liter)
Path length	b	Optical path length through sample
Sample concentration	c	Concentration of sample (g/liter)
Molar absorptivity	·	A/bc^x, where c^x = concentration (mol/liter)
Absorption maximum	λ_{max}	Wavelength at which highest absorption occurs
Wavelength	λ	Distance between wave crests
Frequency	ν	Number of waves per unit time
Wave number	$\bar{\nu}$	Number of waves per centimeter
Wavelength unit	Å	Angstrom, 10^{-10} m
	nm	Nanometer, 10^{-9} m or 10 Å
	μm	Micrometer, 10^{-6} m

BIBLIOGRAPHY

Ayres, G. H., *Quantitative Chemical Analysis*, 2nd ed., Harper and Row, New York, 1968.

Bauer, H. H., Christian, G. D., and O'Reilley, J. E., *Instrumental Analysis*, Allyn and Bacon, Boston, 1978.

Ewing, G. W., *Instrumental Methods of Chemical Analysis*, 4th ed., McGraw-Hill, New York, 1975.

Harris, D. C., *Quantitative Chemical Analysis*, W. H. Freeman, New York, 1992.

Koenig, J. L., Industrial problem solving with molecular spectroscopy, *Anal. Chem.*, (1994), *66*(9): 515A.

Kolthoff, I. M., Sandell, E. B., Mechan, E. J., and Bruckenstein, S., *Quantitative Chemical Analysis*, 4th ed., Macmillan, London, 1969.

Willard, H. H., Merrit, L. L., Dean, J. A., and Settle, F. A., *Instrumental Methods of Analysis*, 7th ed., Van Nostrand, New York, 1988.

SUGGESTED EXPERIMENTS

2.1 You will need a UV-visible spectrophotometer for this experiment.
(a) Prepare suitable standard solutions of (1) 0.1 g of $KMnO_4$ per liter of water, (2) 1.0 g of $K_2Cr_2O_7$ per liter of water, and (3) water-soluble red ink diluted 50% with water.
(b) Plot the absorption spectrum from 700 to 350 nm and determine the wavelength of maximum absorption for each solution.
(c) Measure the transmittance I_1/I_0 at the wavelengths of maximum absorption determined for each solution.

2.2 (a) Choose one of the standard solutions prepared in Experiment 2.1a and measure the transmittance at the wavelength where maximum absorption occurs. Take 50 ml of the standard solution (solution A) and transfer it to a 100-ml standard flask; make up to volume with water. This is solution B. Measure the transmittance of solution B. Dilute solution B by 50% to obtain solution C. Measure the transmittance of solution C. Repeat this process to produce solutions D, E, and F.
(b) Prepare a graph correlating transmittance T and the concentrations of solutions A, B, C, D, E, and F.
(c) From the graph obtained in step (b), prepare a graph correlating A, the absorbance, with the concentrations of solutions A, B, C, D, E, and F.

2.3 Add a measured quantity of standard solution of one of the other compounds suggested in Experiment 2.1 to the standard solution prepared in Experiment 2.2. Is there a change in the transmittance of the solution at the wavelength used in Experiment 2.2? Measure the transmittance of the second standard alone at this wavelength. Is the total amount of light absorbed by the separated solutions equal to the amount absorbed by the mixture? Would this change in absorbance be a source of error in practice?

2.4 Measure the absorbance of neutral $KMnO_4$ at its wavelength of maximum absorption. Leave the container open to the atmosphere for 5 min and then measure the absorbance of the solution again. Repeat measurements at 5-min intervals. Plot the measured absorbance against the time exposed to the air. The change is caused by the chemical instability of the standard (it reacts with the air). All standards are subject to this error to a greater or lesser extent, and precautions must be taken to prevent and avoid this source of trouble.

PROBLEMS

2.1 A molecule absorbs radiation of frequency 3×10^{14} cps. What is the energy difference between the molecular energy states involved?

2.2 What frequency of radiation has a wavelength of 500 nm?

2.3 Describe the Franck–Hertz experiment.

2.4 Describe the Stern–Gerlach experiment.

2.5 What are the principal modes of movement of a molecule?

2.6 What are eigenstates and eigenvalues?

2.7 What is the magnetic moment of a molecule?

2.8 What is the Boltzmann distribution?

2.9 If E, the excitation energy in the Boltzmann distribution, is doubled, how much must the temperature be increased to keep the ratio N_1/N_0 constant?

2.10 A molecular compound A absorbs at $\nu = 10^{16}$ cps and has a given ratio N_1/N_0 at temp 1000° Abs. If a second molecular compound B absorbs at 2×10^{16} cps, to what temperature must it be heated so that N_1/N_0 is the same for both molecules?

2.11 If in Problem 2.10, the frequency for a compound C was 0.5×10^{16} cps, what would be the temperature necessary to maintain the same N_1/N_0 ratio as compound A?

2.12 An atom type A absorbs at 400 nm and has a given ratio N_1/N_0 at temperature 1500°A. If an atom type B absorbs at 200 nm, what must the temperature be to have the same N_1/N_0 ratio?

2.13 Arrange the following types of radiation in order of increasing wavelength: IR, radiowaves, x-rays, UV, visible light.

2.14 Does the degree of absorption by a population of atoms or molecules depend on the number in the ground state or the excited state? Explain.

2.15 Does the intensity of emission by a population of atoms or molecules depend on the number in the ground state or the excited state? Explain.

2.16 What are the modes of vibration of a molecule?

2.17 What methods of excitation are used to cause emission useful in analytical chemistry?

2.18 What is the Beer–Lambert law?

2.19 (a) Define transmittance and absorbance.

(b) What is the relationship between concentration and (1) transmittance, (2) absorbance?

2.20 The following data were obtained in a standard series calibration for the determination of iron by measuring the transmittance, at 506 μm and 1.00 cm optical path, of solutions of iron(II) color-developed with 1,10-phenanthroline.

Fe conc. (ppm)	% *T*	Fe conc. (ppm)	% *T*
0.20	90.0	3.00	26.3
0.40	82.5	4.00	17.0
0.60	76.0	5.00	10.9
0.80	69.5	6.00	7.0
2.00	41.0	7.00	4.5

(a) On 2-cycle semilog graph paper, plot % *T* against concentration of iron. Does the system conform to Beer's law over the entire concentration range? (b) Calculate the average molar absorptivity of iron when it is determined by this method. (c) On 2-cycle semilog graph paper, plot 100 − % *T* against log concentration (Ringbom method). (1) What is the optimum concentration range and the maximum accuracy (percent relative error per 1% photometric error) in this range? (2) In what concentration range will the relative analysis error per 1% photometric error not exceed 5%?

2.21 The following data were obtained in a standard calibration for the determination of copper, as $Cu(NH_3)_4^{2+}$, by measuring the transmittance in a filter photometer.

Cu conc. (ppm)	% *T*	Cu conc. (ppm)	% *T*
0.020	96.0	0.800	27.8
0.050	90.6	1.0	23.2
0.080	84.7	1.40	17.2
0.100	81.4	2.00	12.9
0.200	66.7	3.00	9.7
0.400	47.3	4.00	8.1
0.600	35.8		

(a) On 2-cycle semilog graph paper, plot % *T* against copper concentration. Does the system, measured under these conditions, conform to Beer's law over the entire concentration range? Is any deviation from the law of small or of large magnitude? Suggest a plausible cause for any deviation. (b) Plot (100 − % T) against the log concentration (Ringbom method). (1) What is the concentration range for highest accuracy? (2) In this range, what is the percent relative error per 1% photometric error? (3) What is the concentration range within which the relative analysis error will not exceed 5%; 10%?

2.22 0.20 g of copper is dissolved in nitric acid. Excess ammonia is added to form Cu $(NH_3)_4^{2+}$, and the solution is made up to 1 liter. The following aliquots of the solution are taken and diluted to 10 ml: 10, 8, 5, 4, 3, 2, and 1 ml. The absorbances of the diluted solution were 0.5, 0.4, 0.25, 0.2, 0.15, 0.1, and 0.05, respectively. A series of samples was analyzed for copper concentration by forming the $Cu(NH_3)_4^{2+}$ complex and measuring the absorbance. The absorbances were (a) 0.45, (b) 0.3, and (c) 0.2. What were the respective concentrations in the three copper solutions? If these three samples were obtained by weighing out separately (a) 1 g, (b) 2 g, and (c) 3 g of sample, dissolving and diluting to 10 ml, what was the original concentration of copper in each sample?

2.23 What is the standard addition method for measuring concentration of an unknown?

2.24 What are the advantages of the standard addition method?

2.25 What is the Ringbom plot?

2.26 What range of % transmittance results in the best precision?

2.27 What is A if the % light absorbed is (a) 90%, (b) 99%, (c) 99.9%, (d) 99.99%.

3
Concepts of Spectroscopy

A. OPTICAL SYSTEMS USED IN SPECTROSCOPY

Analytical spectroscopy is involved primarily in measuring emitted radiation or the absorption of radiation by a sample. In each case, it is important to know (1) the wavelengths at which the sample emits or absorbs radiation and (2) the intensity of radiation (at the different wavelengths) *or* the degree of absorption involved. In this chapter we discuss the instruments that have been developed to facilitate the study of these phenomena.

The equipment for all wavelength ranges has the same basic design. For example, all instruments include a light source, a monochromator (for selecting the wavelength to be studied), a sample holder, a detector (to measure the intensity of radiation), and some means of displaying the signal from the detector. If we are measuring emitted radiation, the sample, excited by some means, is the light source. If the equipment is used to study light-absorption, a separate source of radiation is used.

The specific arrangement of the components listed above is referred to as the *optics* of the instrument. Two basic designs are commonly used-sing-beam optics and double-beam optics.

1. Components of a Single-Beam Optical System

Figure 3.1 is a schematic diagram of a single-beam instrument used for absorption measurements. The functions of the components are as follows.

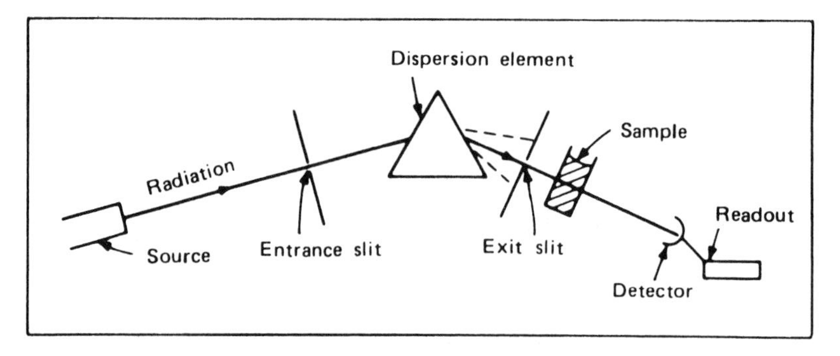

Figure 3.1 Single-beam instrument.

a. Radiation Source

A good radiation source should have the following characteristics: (1) The beam must emit radiation over the entire wavelength range to be studied. (2) The intensity of the radiation over the whole wavelength must be high enough so that extensive amplification of the detector output is not necessary. Moreover, the intensity of the radiation (3) should not vary significantly at different wavelengths, (4) should not fluctuate over long time intervals, and (5) should not flicker over short time intervals.

The design of the radiation source varies with the wavelength range for which it is to be used (e.g., x-ray, infrared, ultraviolet). In all cases a reliable, steady supply of electrical power to the source is necessary to ensure that the foregoing requirements are met. For example, the power supply should deliver a constant electrical output to the radiation source even if there is a voltage variation in the power mains. This occurs when industrial concerns in the vicinity of the laboratory start up or shut down their machinery. Devices are available to compensate for any voltage variation in the source power supply to the light source and keep it constant. If the voltage to the light source varies, usually the intensity of the light will vary. The detector measures only the intensity of light falling on it, and any variation in intensity therefore leads to a direct error.

b. Monochromator

The monochromator consists of a dispersion element and a slit system as shown in Fig. 3.1. The function of the dispersion element is to spread out, or disperse, the radiation according to wavelength. This does it in the same way that a prism splits the component colors of sunlight into the colors of the rainbow. The two most common types of dispersion elements are prisms and gratings.

Prisms. Prisms are among the most widely used monochromators. Shaped like a bar with a triangular cross section, they disperse incident radiation according to wavelengths (Fig. 3.2). The degree to which light is bent depends on the refractive index of light passing from air to glass (in a glass prism) and from glass to air. The refractive index varies with wavelength; hence different wavelengths can be separated. The prism bends longer-wavelength radiation less than it does shorter-wavelength radiation. The long-wavelength end of the spectrum is often called the *red end* and the short-wavelength end the *blue end*. This terminology is sometimes used even when these colors are not involved at all; in infrared spectroscopy, for example, both the short and long wavelengths are in the infrared region, but the short-wavelength end is still referred to as the blue end and the long-wavelength end as the red end. This prism is able to bend short-wavelength light more than long-wavelength light because its refractive index is greater for short-wavelength light than it is for long-wavelength light.

Gratings. Gratings have become increasingly popular in recent years. A grating is essentially a series of parallel straight lines cut into a plane surface (Fig. 3.3). For gratings to be used with UV light, there may be 15,000–30,000 such grooves per inch of surface, according to the quality of the grating. The more lines per inch on the grating, the shorter the wavelength of radiation that the grating can disperse and the greater the dispersive power.

Dispersion by a grating follows the laws of diffraction. (The path of the diffracted light is shown in Fig. 3.4.) The angle at which light is dispersed by a grating is controlled by the physical requirements for light reinforcement, as opposed to light destruction. Reinforcement is necessary for the very existence of the beam. Two beams reinforce each other when they are in phase with each

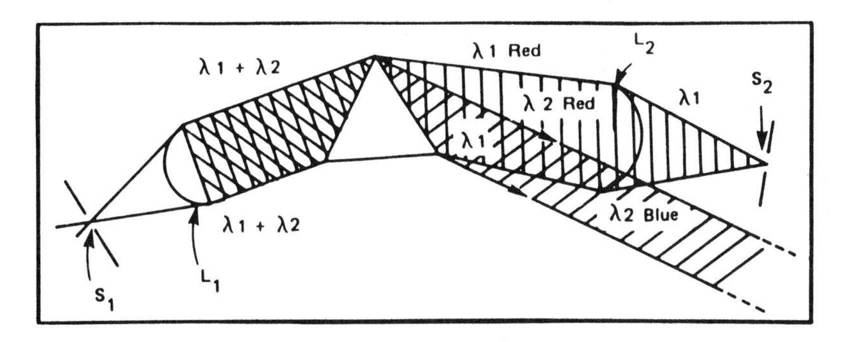

Figure 3.2 Dispersion of radiation into different wavelengths. Focusing lenses are used to control the light. Here the radiation source emits red (long wavelength) and blue (short wavelength) light. The light is focused on the entrance slit S_1 and proceeds to a collimating Lens L_1, which converts it to a parallel beam falling on one face of the prism. The face is completely filled with light. Here the light is dispersed and falls on a second lens L_2, which focuses it onto the exit slit S_2. The latter is placed so that only light of the derived wavelength proceeds to the detector. Other light is blocked out.

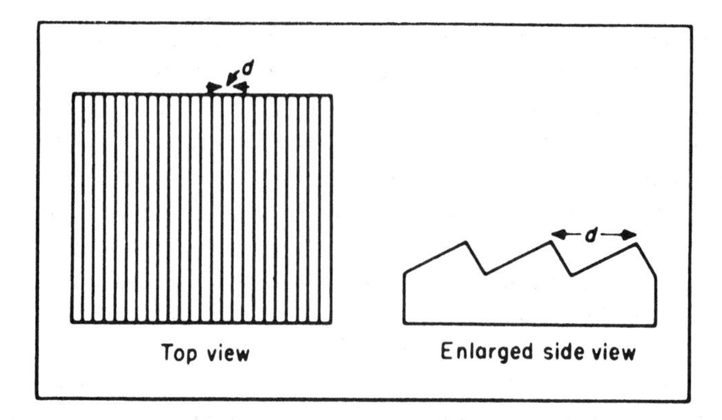

Figure 3.3 Highly magnified view of a grating monochromator.

other. This occurs when the wave fronts are parallel to each other, as shown in Fig. 3.4. However, the beams destroy each other when they are out of phase. Reinforcement of light occurs when

$$n\lambda = d(\sin i \pm \sin \theta) \tag{3.1}$$

where

n = order (a whole number)
λ = wavelength of the radiation
d = distance between grooves
i = angle of incidence of the beam of light
θ = angle of dispersion of light in a particular wavelength

The negative sign in the relationship is applied when the angle of incidence i and the angle of dispersion θ are both on the same side of the vertical. For light of

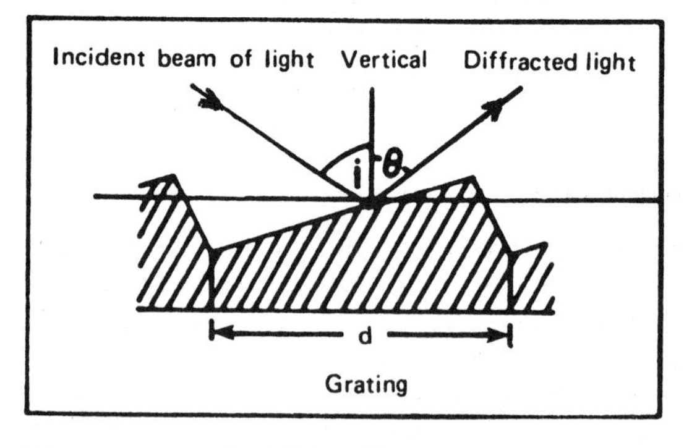

Figure 3.4 Path of light diffracted by a grating.

different wavelengths λ, the angle of dispersion θ is different. Separation of light occurs because light of different wavelengths is dispersed at different angles.

One problem with gratings is that light with several different wavelengths may leave the grating at the same angle of dispersion. For example, suppose that a beam of radiation falls on a grating at an angle i. The angle of dispersion of the radiation is described by Eq. (3.1). For a given angle of dispersion θ, the product $n\lambda$ is a constant. Any combination of n and λ that equals this constant will satisfy Eq. (3.1). If λ is 600 nm and n is 1, then the angle of dispersion is θ; furthermore, if $\lambda = 200$ nm and n is 3, it is also θ, and so on. In practice, radiation with each of these wavelengths is dispersed at an angle θ and travels down the same light path. This is illustrated in Fig. 3.5.

Equation (3.1) enables us to calculate the angle θ at which radiation of a known wavelength λ will leave a grating when we known the angle of incidence i. Wavelengths of light that are related in this way are said to be different *orders* of radiation. They are not separated by gratings. The wavelengths of radiation traveling the same path after dispersion are related by the simple number n, which may take the value of any whole number. On the best equipment, different orders are separated by using a small prism or a filter system as an *order sorter* in conjunction with the grating (Fig. 3.6 and 3.7). It is common for infrared instruments to use filter order sorters. As the grating rotates to different wavelength ranges, the filters rotate to prevent order overlap, and only one wavelength reaches the detector.

Resolution Required to Separate Two Lines of Different Wavelength. The ability to disperse radiation is called *resolving power*. Alternative designations include *dispersive power* and *resolution*. For example, in order to observe an absorption band at 599.9 nm without interference from an absorption band at 600.1 nm, we must be able to resolve, or separate, the two bands.

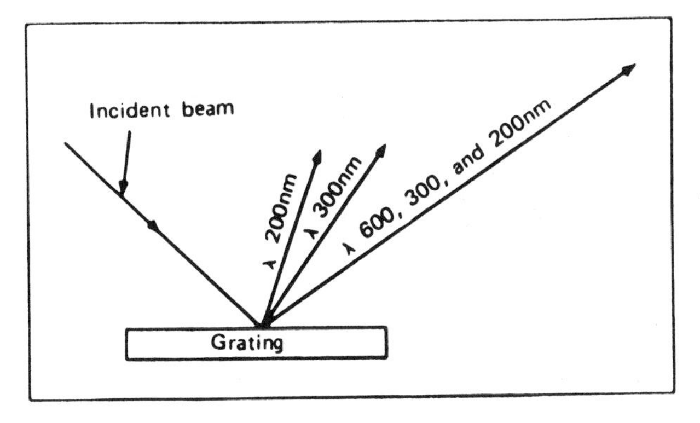

Figure 3.5 Angle of dispersion of light of different wavelengths.

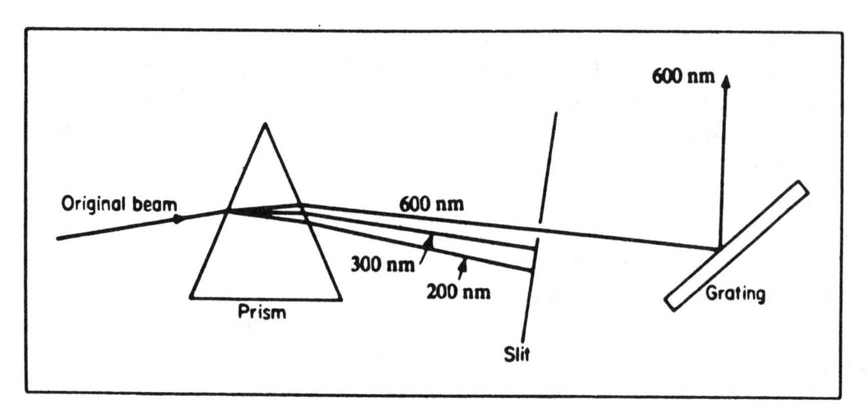

Figure 3.6 Prism used as an order sorter.

The resolving power R of a monochromator is equal to $\bar{\lambda}/\delta\lambda$, where $\bar{\lambda}$ is the average of the wavelengths of the two lines to be resolved and $\delta\lambda$ is the difference in wavelength between these lines. In the present example the required resolution is

$$R = \frac{\text{average of 599.9 and 600.1}}{\text{absolute difference between 599.9 and 600.1}} = \frac{600}{0.2} = 3000 \quad (3.2)$$

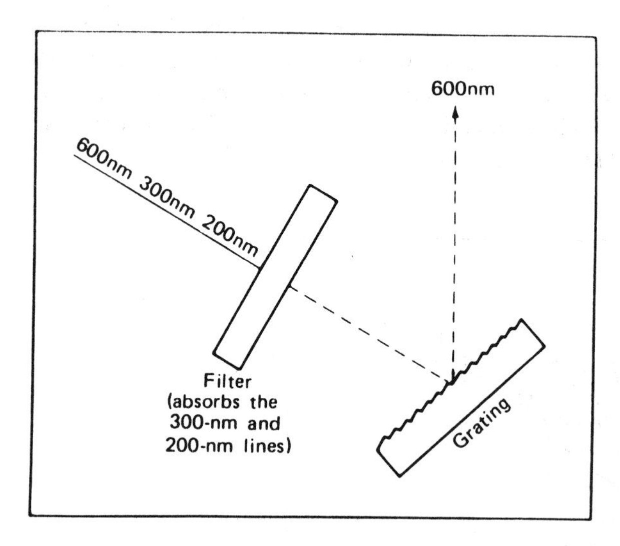

Figure 3.7 Filter order sorter. This filter prevents different orders from reaching the grating monochromator.

Resolution of a Prism. The resolving power R of a prism is given by

$$R = t\frac{d\eta}{d\lambda}$$

where t is the thickness of the base of the prism and $d\eta/d\lambda$ is the rate of change of dispersive power (or *refractive index*) η of the material of the prism with wavelength.

For the resolution of two beams at two wavelengths λ_1 and λ_2, it is necessary that the refractive index of the prism be different at these wavelengths. If it is constant, no resolution occurs. The resolving power of a prism increases with the thickness of the prism. It also increases when the refractive index of the material used is improved. The greater the resolving power of a prism, the more easily it will differentiate emission (or absorption) bands with similar wavelengths.

The greater the change in η with wavelength, the greater the dispersive power of the prism. It is found in practice that η is maximum as it approaches the wavelength at which it becomes opaque to light. A typical curve illustrating this relationship is shown in Fig. 3.8.

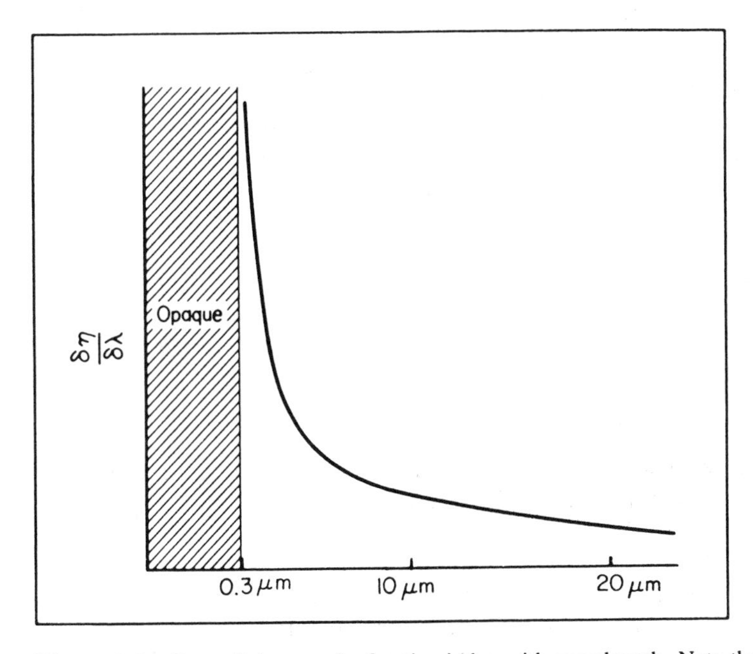

Figure 3.8 Rate of change of refractive index with wavelength. Note the high rate of change as the wavelength approaches the cutoff.

For maximum dispersion, the prism is most effective at wavelengths close to the wavelength at which it ceases to be transparent. At longer wavelengths, the resolving power decreases, resulting in decreasing usefulness of the monochromator. At best, use of the prism results in a change in resolution with wavelength and a possible need to control slit widths to compensate. At worst, the crowding of wavelengths leads to a loss of resolution and therefore less analytical or spectroscopic usefulness. This problem is particularly severe in infrared spectroscopy.

Resolution of a Grating. The resolving power of a *grating* is given by

$$R = nN \tag{3.3}$$

where n is the order and N is the total number of grooves in the grating. Suppose that we can obtain a grating with 500 lines/cm. How large a grating would be required to separate the sodium D lines at 589.5 and 589.0 nm in first order?

We know from Eq. (3.2) that the required resolution R is given by

$$R = \frac{589.25}{0.5} = 1178.5$$

The resolution of the grating must therefore be at least 1178.5. But R (for a grating) $= nN$; therefore $1178.5 = nN$. Since we stipulated first order, $n = 1$; hence N, the total number of lines, is 1178.5. But the grating contains 500 lines/cm. It must be 1178.5/500 cm long, or 2.357 cm.

In a separate example, we may ask how many lines per centimeter must be cut on a grating 3 cm long to resolve the same sodium D lines.

As before, the required resolution is 1178.5, and

$$nN = 1178.5$$

If $n = 1$, then

$$N = 1178.5$$

and

$$N/\text{cm} = 1178.5/3 = 392.83$$

But it is not possible to cut a fraction of a line; hence N must be the next whole number, 393/cm, or 1179 lines on the entire grating.

c. Optical Slits

A system of slits (Fig. 3.9) is used to select radiation from the light beam both before and after it has been dispersed by the dispersive elements. The jaws of the slit are made of metal and are usually shaped like two knife edges. They can be moved relative to each other to change the mechanical width of the slit as desired. The *mechanical slit* is the physical distance between the jaws of the slit.

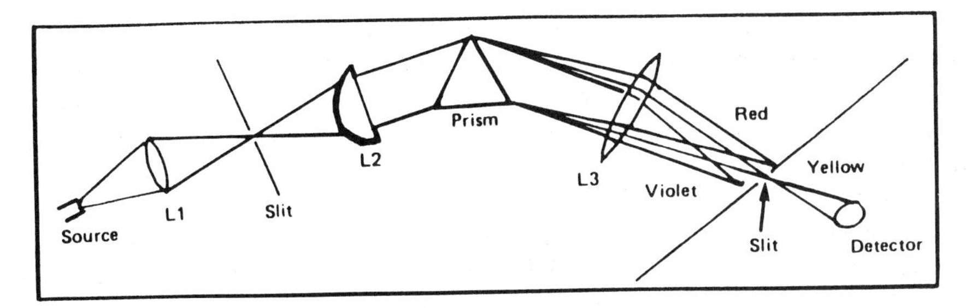

Figure 3.9 System of entrance and exit slits.

The *entrance slit* permits passage of a beam of light from the source. Radiation from the light source is collected by the gathering lens and focused on the entrance slit. Stray radiation is excluded. After passing through the entrance slit, the radiation falls on a collimating lens that transforms it into a parallel beam of light, which falls onto and completely covers one side of the prism (or grating). The prism disperses the light in different directions according to wavelength. At the setting selected, one beam is refocused by a second plane convex lens onto the exit slit. Light of other wavelengths is also focused, but not onto the exit slit. Ideally the light is an image of the entrance slit. It is redirected and focused onto the detector for intensity measurement.

It should be noted that front-faced mirrors instead of lenses are frequently used for focusing and collimating the light. In the infrared these are always more efficient and do not absorb the radiation. They are also easily scratched, since the refracting surface is on the front and not protected by glass, as is the case with conventional mirrors. Back-faced mirrors are not used because the supporting material (e.g., glass) will absorb the radiation.

The physical distance between the jaws of the slits is called the *mechanical slit width*. This can be measured directly with a ruler if necessary. Usually the instrument has a ruler or scale attached so that one can read off the mechanical slit width directly. In UV absorption spectroscopy mechanical slit widths are of the order of 0.1 mm. In IR spectroscopy slit widths between 0.1 and 2.0 cm are common.

The wavelength range of the radiation that passes through the exit slit is called the *spectral slit width*. This width can be measured by passing an emission line of very narrow bandwidth through the slits to the detector. By rotating the dispersion element we can record the wavelength range over which response occurs. After correcting for the actual width of the emission line, we can calculate the spectral slit width. For example, the emission line from cadmium appeared at 228.8–229.4 nm. This means that the cadmium line reached the detector even though the wavelength reading of the instrument varied between

228.2 and 229.4 nm. It is apparent that the wavelength range of the radiation falling on the detector was 1.2 nm wide and that this was the special linewidth reaching the detector. In this example, no correction was made for the actual width of the cadmium 228.8-nm line, which in this case is about 0.001 nm and is negligible.

If the mechanical slit width were made wider, the spectral slit width would simultaneously increase. For example, with the mechanical slit settings described above, it would not be possible to resolve an emission line at 229.0 nm from the 228.8-nm line, because both would pass through the slits. However, if the spectral slit width were reduced at 0.1 nm, it would be possible to resolve these lines. This might be done by reducing the mechanical slit to approximately 0.3 nm.

In practice, the slits are kept as narrow as possible to ensure optimum resolution; however, they must be wide enough to admit sufficient light to be measured by the detector. The final choice of slit width is determined by the operator based on the particular sample at hand. A good rule of thumb is to keep the slits as narrow as possible without impairing the functioning of the detector.

By rotating the prism or grating or by moving the exit slit across the light beam from the monochromator, the wavelength range passing through the exit slit can be changed. By continuously rotating the dispersion element from one extreme to another, the complete spectrum can be scanned.

d. Detector

The detector is used to measure the intensity of the radiation that falls on it. Normally, it does this by converting the radiation energy into electrical energy. The amount of energy produced is usually low and must be amplified. The signal from the detector must be steady and representative of the intensity of radiation falling on it. If the signal is amplified too much, it becomes erratic and unsteady; it is said to be *noisy*. The degree of random variation in the signal is called the *noise level*. Amplifying the signal from the detector increases its response. In practice, the response can be increased until the noise level of the signal becomes too great; at this point the amplification is decreased until the noise level becomes acceptable.

e. Uses of Single-Beam Optics

Single-beam optics are used for all spectroscopic emission methods. In emission procedures the sample is put where the source is located in Fig. 3.1. The method allows the emission intensity and wavelength to be measured accurately and rapidly.

In spectroscopic absorption studies the intensity of radiation before and after passing through the sample must be measured. When single-beam optics are used, any variation in the intensity of the source while measurements are being

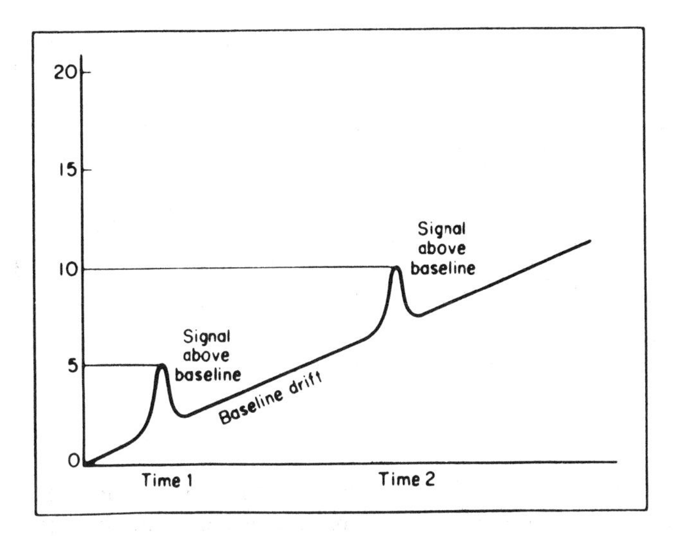

Figure 3.10 The absolute signal measured changes from 5 to 10 units as a result of baseline drift with time.

made may lead to analytical errors. Slow variation in the average signal (not noise) with time is called *drift*. Drift can cause a direct error in the results obtained and must be corrected in practice. Single-beam optics are particularly subject to errors caused by drift. This can be seen in Fig. 3.10.

There are several sources of drift. The *radiation source* intensity may change because of line voltage changes, the source warming up after being recently turned on, or the source deteriorating with time. The *monochromator* may shift position as a result of vibration or heating and cooling causing expansion and contraction. The line voltage to the *detector* may change, or the detector may deteriorate with time and cause a change in response.

Errors caused by drift lead to an error in the measurement of the emission signal or the absorption signal compared to the standards used in calibration. The problem can be reduced by constantly checking the light intensity or by using a standard emission sample. However, the effects of drift can be greatly decreased by using a double-beam system.

2. The Double-Beam System

The double-beam system is used extensively for spectroscopic absorption studies. The individual components of the system have the same function as in the single-beam system, with one very important difference. The radiation from the source is split into two beams of approximately equal intensity. One beam is

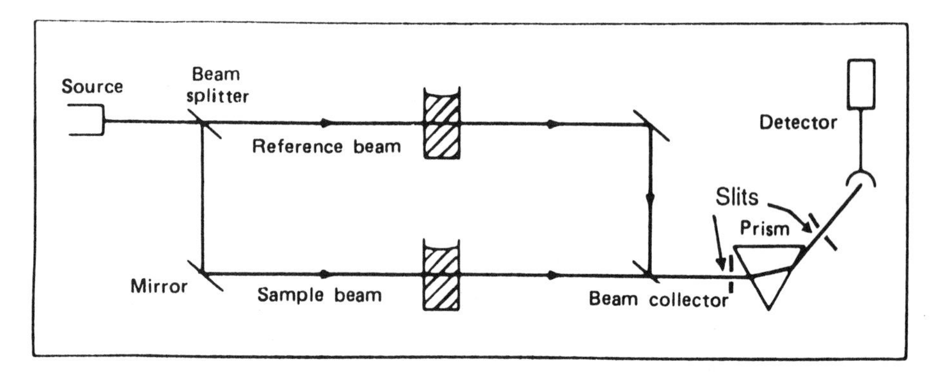

Figure 3.11 Double-beam optical system.

termed the *reference beam*; the second, which passes through the sample, is called the *sample beam*. The two beams are then recombined and pass through the monochromator and slit systems to the detector. This is illustrated in Fig. 3.11.

As shown in Fig. 3.12a, the beam splitter may be a simple mirror plate into which a number of holes are drilled. Light is reflected by the mirror plate and passes down the sample beam path. An equal portion of light passes through the holes in the plate and forms the reference beam.

Another convenient beam splitter is a disk with opposite quadrants removed (Fig. 3.12b). The disk rotates in front of the radiation beam and the mirrored surface reflects light into the sample path. The missing quadrants permit radiation to pass down the reference path. Each beam of light is intermittent and arrives at the detector in the form of an alternating signal. When no radiation is

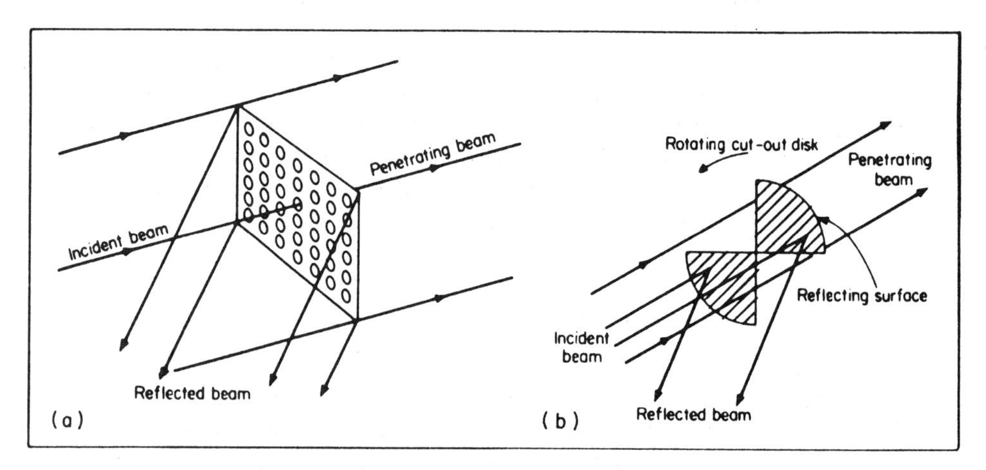

Figure 3.12 (a) Plate beam splitter and (b) disk beam splitter.

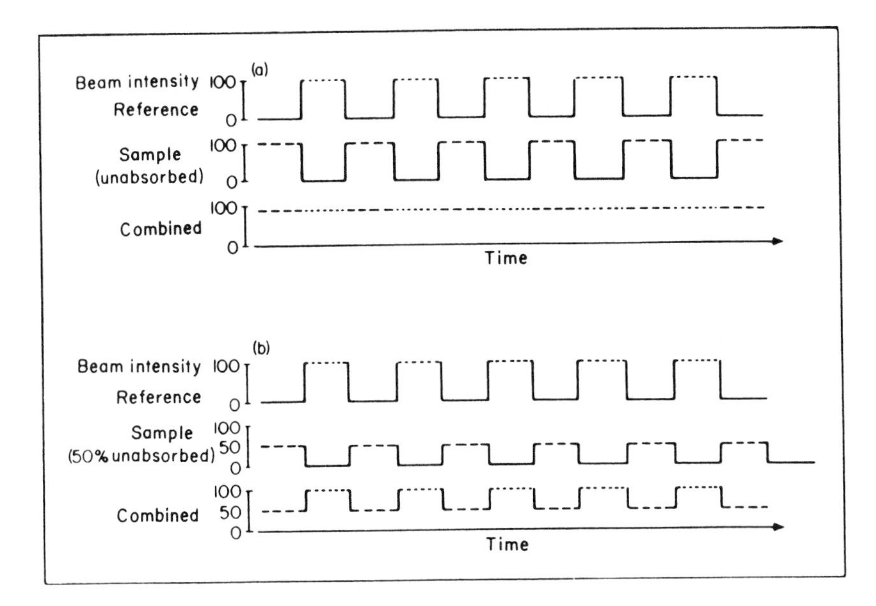

Figure 3.13 Form of radiation intensity reaching the detector using double-beam optics: (a) no absorption and (b) the two signals reaching the detector with 50% absorption by the sample.

absorbed by the sample, the two beams are equal and recombine and form a steady beam of light. However, when radiation is absorbed by the sample the two beams are not equal, and an alternating signal arrives at the detector. This is illustrated in Fig. 3.13.

Using this system, we can measure the *ratio* of the reference beam intensity to the sample beam intensity. Because the ratio is used, any variation in the intensity of radiation from the source during measurement does not introduce analytical error. This advantage revolutionized the studies of absorption spectroscopy. If there is a drift in the signal, it affects the sample and reference beams equally. The recombined beam will continue to give a straight line unless the drift is very great, in which case correction is not complete. If absorption occurs, the alternating signal is obtained by measuring the reference beam, and the sample beam is virtually independent of drift and therefore more accurate.

B. ANALYTICAL METHODS USED IN SPECTROSCOPY

There are numerous analytical fields in spectroscopy. They utilize the interaction between radiation and all parts of the atoms, molecules, and the arrangements of molecules in matter. The more important fields are discussed in subsequent chapters. They are summarized for convenience in Table 3.1.

Table 3.1 Analytical Fields in Spectroscopy

Radiant energy	Radiofrequency	Infrared	Visible	Ultraviolet	X-ray	γ-ray
Analytical field	Nuclear magnetic resonance	Absorption	Absorption Spectrophotometry	Absorption	Absorption	Activation analysis
	Microwave	Attenuated total reflectance	Nephelometry	Fluorescence	Fluorescence	
		Optical rotary dispersion	Turbidimentry	Phosphorescence	Diffraction	Radiotracer techniques
				Emission spectrography		
				Flame photometry		
				Atomic absorption		
Interaction of matter	Nuclear disintegration	Vibration and rotation of molecules	Electronic excitation of atoms or molecules		Inner electrons of atoms displaced	Nuclear disintegration
					Crystal lattice diffracts lights	Stable molecules

Each of the fields listed in Table 3.1 has an optimum concentration range over which it is useful. These ranges are illustrated in Fig. 3.14. It should be remembered that the concentration range quoted is for untreated or unconcentrated samples. The sensitivity limit of all methods can be extended by suitable concentration, extraction, or gathering techniques. Table 3.2 lists some of the most commonly used instrumental methods of analysis together with comments on their respective sensitivity and analytical characteristics.

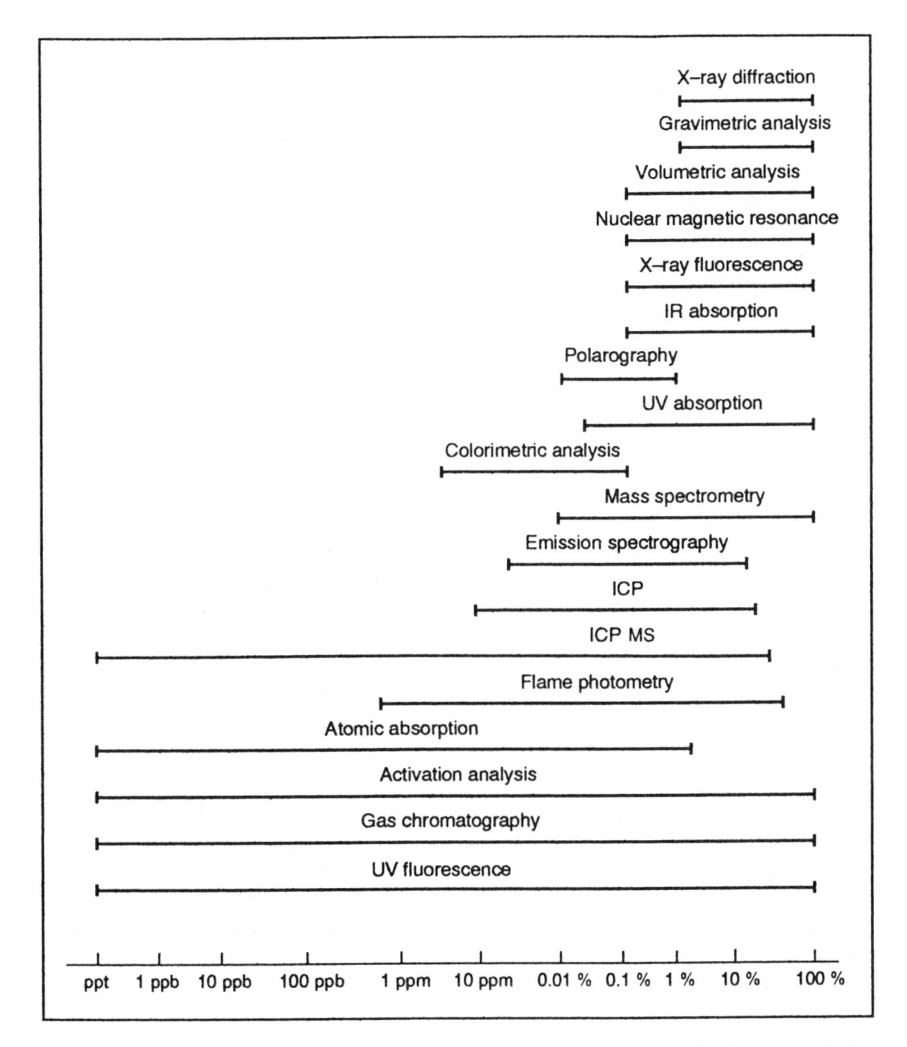

Figure 3.14 Optimum concentration ranges for the various fields of analytical spectroscopy.

Table 3.2 Instrumental Methods

Method	Sensitivity (g)	Comments
Gas chromatography	10^{-8} to 10^{-14}	Depends on detector
Thin-layer chromatography		Nondestructive
Fluorescence	10^{-9}	
Mass spectrometry	10^{-12}	Detects all elements and most volatile inorganic and organic compounds
Liquid chromatography		UV detector limited to those compounds with such absorption characteristics
Refractive index detector	10^{-6}	
Ultraviolet/visible detector	10^{-9}	
Neutron activation analysis	10^{-12}	Responds to elements with variable sensitivity
Atomic absorption spectroscopy		Detects metals and metalloids
Flame	10^{-9}	
Thermal	10^{-14}	
Atomic emission spectroscopy	10^{-9}	High sensitivity and multielement analysis via inductively coupled plasma; best for qualitative analysis
ICP	10^{-10}	Detects metals and metalloids
ICP MS	10^{-12}	Detects all elements, including nonmetals; high sensitivity; more interference than expected; measures isotope ratios
Infrared spectroscopy		Sensitivity fair; detects organic functional groups
Standard techniques (pure samples)	10^{-6}	
Fourier transform infrared	10^{-9}	
X-ray fluorescence	10^{-7}	Used for elements with atomic numbers above 11
Optical microscopy	10^{-12}	Simple, rapid method for particulate analysis
Anodic stripping voltammetry	10^{-8}	Can analyze from 10 to 20 elements; best for Cu, Pb, Zn, Cd
Surface analysis		Detects and identifies atoms in first several atomic layers of a surface; among the most sensitive methods known
ESCA	10^{-10}	
Ion scattering spectroscopy	10^{-10}	
Auger	10^{-10}	
Secondary ion mass spectrometry	10^{-15}	

Table 3.2 Continued

Method	Sensitivity (g)	Comments
Polarography		Detects most metallic elements and compounds; also organics
DC polarography	10^{-8}	
Pulsed polarography	10^{-10}	
Stripping voltammetry	10^{-11}	
Ion-selective electrodes	10^{-15}	Sensitivity shown is for detection of Cu; otherwise, sensitivities vary, depending on element and electrode
GC/UV-photoelectron spectroscopy	10^{-5}	Using a direct-coupled GC, spectra can be obtained in less than 1 min on 10^{-5}-g quantities
Wet chemical methods	10^{-6}	Many methods involving chemical reaction, colorimetric detection give ppm and ppb or finer measurements in large sample sources such as air, water
Nitrate by enzyme reagent	10^{-9}	
Hg in water	10^{-9}	
Refractive index	10^{-6}	Simple, rapid method useful only for binary mixtures whose components have a wide range of refractive indices
Proton NMR		Detects all organic and diamagnetic organometallic compounds that contain hydrogen atoms; instrument costs range from moderate to expensive; NMR offers additional ability to give structure and identity for compounds
Continuous wave (single scan)	10^{-4}	
Fourier transform (~10,000 scans)	10^{-6}	
UV absorption	10^{-7}	Best for unsaturated and aromatic samples
Combination method		Quantitative analysis based on peak area and qualitative analysis based on data obtained by instrument linked to GC; computer interfacing greatly extends potential and speed of method
GC–MS	10^{-12}	
GC–Fourier transform infrared	10^{-9}	
GC–UV	10^{-6}	

BIBLIOGRAPHY

Meehan, E. J., *Optical Methods of Analysis. Treatise an Analytical Chemistry*, 2nd ed., Wiley, New York, 1981.

PROBLEMS

3.1 An optical cell containing a solution was placed in a beam of light. The original intensity of the light was 100 units. After passing through the solution, its intensity was 80 units. A second similar cell containing more of the same solution was also placed in the light beam behind the first cell. Calculate the intensity of radiation emerging from the second cell.

3.2 The transmittance of a solution 1 cm thick of unknown concentration is 0.7. The transmittance of a standard solution of the same material is also 0.7. The concentration of the standard solution is 100 ppm; the cell length of the standard is 4 cm.

3.3 A solution contains 1 mg of $KMnO_4$ per liter. When measured in a 1-cm cell at 525 nm, the transmittance was 0.3. When measured under similar conditions at 500 nm, the transmittance was 0.35. (a) Calculate the absorbance A at each wavelength. (b) Calculate the molar absorptivity at each wavelength. (c) What would T be if the cell length were in each case 2 cm?

3.4 A series of standard ammoniacal copper solutions were prepared and the transmittance measured. The following data were obtained:

Cu concentration	Transmittance	Sample	Transmittance
0.20	0.900	1	0.840
0.40	0.825	2	0.470
0.60	0.760	3	0.710
0.80	0.695	4	0.130
1.00	0.635		
2.00	0.410		
3.00	0.263		
4.00	0.170		
5.00	0.109		
6.00	0.070		

Plot the concentration against the transmittance on semilog paper. The transmittance of solutions of copper of unknown concentrations was also measured in the same way and the data in the following table were obtained:

From the standard calibration curve preparation from the data above, calculate the concentration of each solution.

3.5 What are the components of a single-beam optical system?

3.6 Describe a monochromater.

3.7 (a) What is the resolution of a prism? (b)What is the general relationship between n and λ?

3.8 (a) What is the mechanical slit width? (b) What is the spectral slit width?
3.9 What is the effect of slit width on resolution?
3.10 What is the resolution of a grating in second order?
3.11 What resolution is required to separate two lines λ_1 and λ_2?
3.12 What resolution is required to resolve the Na D lines 589.0 and 589.5 nm?
3.13 How many lines must be on a grating to separate the D lines in second order?
3.14 A grating contains 1000 lines. Will it resolve two lines of 500.0 λ and 499.8 nm?
3.15 What are the components of double-beam optics?
3.16 How does double-beam optics correct for drift?
3.17 What are the principal types of electromagnetic radiation in order of increasing energy? (p. 80)
3.18 Match common spectroscopic analytical fields with the relevant radiation.
3.19 Which spectroscopic methods are used for (a) molecular analysis and (b) elemental analysis?

4
Nuclear Magnetic Resonance

In the past few years, the technology associated with *nuclear magnetic resonance* (NMR) has advanced dramatically. For a number of years, NMR was associated only with proton NMR in liquid samples at high concentration levels. True, research workers were involved in studying nuclei other than H and studying solid samples, but commercial equipment was not available for routine use.

That has now changed—much more powerful magnets are available, and 500 MHz NMR instruments are not common but are not rare either. Magic angle NMR is commercially available for studying solids such as polymers, and ^{13}C NMR is also commonly practiced. Imaging techniques under the name "magnetic resonance" are in widespread medical use for noninvasive diagnosis of cancer. In short, the field has broadened greatly in scope and gives every indication of continuing to advance for many years to come.

NMR involves the interaction of radio waves and the spinning of precessing nuclei of the combined atoms in a molecule. Radio waves are the lowest-energy form of electromagnetic radiation that are useful in analytical chemistry. Their frequency is on the order of 10^7 cps. The energy of *radiofrequency* (RF) radiation can therefore be calculated from Eq. (2.2) as

$$E = h\nu$$

where Planck's constant h is 6.6×10^{-27} erg sec and ν (the frequency) is between 10^7 and 10^8 cps:

$$E = 6.6 \times 10^{-27} \times 10^{7}$$
$$= 6.6 \times 10^{-20} \text{ erg sec}$$

The quantity of energy involved in RF radiation is very small. In fact, it is too small to vibrate, rotate, or electronically excite an atom or molecule. It is, however, great enough to affect the nuclear spin of the atoms that make up a molecule. As a result, spinning nuclei of atoms in a molecule can absorb RF radiation and change the direction of the spinning axis. The analytical field involved with the interaction between nuclei and RF radiation is nuclear magnetic resonance.

In analytical chemistry, NMR is a technique that enables us to study the shape and structure of molecules. In particular, it reveals the different chemical environments of the various forms of hydrogen present in a molecule, from which we can ascertain the structures of the molecules with which we are dealing. If we already know what types of compounds are present, NMR provides a means of determining how much of each is in the mixture. It is thus a method of both qualitative and quantitative analysis, particularly of organic compounds.

As we just indicated, the most important application of NMR is in the study of hydrogen atoms in organic molecules. The hydrogen atom is perhaps the easiest to understand from the point of view of its physical properties. The nuclei of other elements may not behave in exactly the same way as that of hydrogen; many do not respond to radio waves at all. Of those that do respond, the most important are the fluorine-19, phosphorus-31, and carbon-13 isotopes. If these nuclei are to be studied in depth, you should consult a more comprehensive treatise on the subject; some suggestions are listed in the Bibliography at the end of the chapter.

A. PROPERTIES OF NUCLEI

As the hydrogen nucleus ^{1}H spins about its axis, it displays two forms of energy. Because the nucleus has a mass and because that mass is in motion (it is spinning), the nucleus has spin angular momentum and therefore mechanical energy. So the first form of energy is *mechanical energy*. The formula for the mechanical energy of the hydrogen nucleus is

$$\text{spin angular momentum} = \frac{h}{2\pi}[I(I + 1)] \qquad (4.1)$$

where I is the spin number, which is 1/2 for ^{1}H.

The spin number I is a physical property of the nucleus, which in turn is made up of protons and neutrons. An odd number of protons or of neutrons produces an angular momentum. An even number of protons or of neutrons results

Table 4.1 Rules Predicting Spin Numbers of Nuclei

Mass ($P + N$) (atomic weight)	Charge (P) (atomic number)	Spin number
Odd	Odd or even	1/2, 3/2, 5/2, . .
Even	Even	0
Even	Odd	1,2,3

in zero angular momentum. A rule of thumb, therefore, is that *neutrons cancel neutrons and protons cancel protons*. For a given element, if we know the atomic weight (protons P + neutrons N) and the atomic number (protons P only), we can predict the spin number of the element. For example, ^{12}C has atomic weight 12 and atomic number 6. Hence it has 6 protons (atomic number = 6) and 6 neutrons (atomic weight − atomic number = 12 − 6). Since both these numbers are even, the net spin number is zero. Therefore the angular momentum [Eq. (4.1)] is zero. Consequently, ^{12}C does not respond in NMR.

For ^{13}C, on the other hand, the atomic weight is 13 ($P + N = 13$) and the atomic number is 6 ($P = 6$). Thus $N = 7$. The spin is 1/2. So ^{13}C can be used in NMR, and although it represents only 1% of the total C present, it has proven to be very valuable in elucidating the structure of complicated molecules.

The spin numbers of the elements can be derived from the empirical rules listed in Table 4.1. The physical properties predict whether the spin number is equal to zero, a half integer, or a whole integer, but the actual spin number—for example, 1/2 or 3/2—must be determined experimentally. It can be seen from Table 4.1 that many elements in the periodic table will not respond to NMR, notably ^{12}C and ^{16}O, which are very important biologically. Some of the more important elements and their spin numbers are shown in Table 4.2.

Table 4.2 Spin Numbers of Elements Responding to NMR

Element	Spin Number
^{13}C	1/2
^{17}O	5/2
^{1}H	1/2
^{2}H (deuterium)	1
^{3}H (tritium)	1/2
^{19}F	1/2
^{31}P	1/2
^{29}Si	1/2

The second form of nuclear energy is *magnetic*. It is attributable to the electrical charge of the nucleus. Any electrical charge in motion sets up a magnetic field. This field may interact with other local magnetic fields. The spinning of the nucleus and its electrical charge produce magnetic energy. Knowledge of the mechanical and magnetic energies of the elements enables us to interpret NMR spectra.

There are many other isotopes that respond to NMR. Table 4.1 shows that any isotope with an odd mass and/or an odd charge will respond with low intensity. Most elements have at least one isotope that responds. This is the case with ^{13}C or ^{17}O.

All combined halides respond strongly, but for an element such as Ca, $A = 40$, $Z = 20$, its principal isotope ^{40}Ca (97%) does not respond, but ^{43}Ca (0.15%) responds weakly because it is present in such small amounts.

It should be noted that elements in the ionic form do not respond, but cause significant line broadening.

B. QUANTIZATION OF ^1H NUCLEI IN A MAGNETIC FIELD

When a nucleus is placed in a very strong magnetic field B_0, the nucleus tends to become lined up in definite directions relative to the direction of the magnetic field; that is, it becomes quantized (see Chap. 2). Each relative direction of alignment is associated with an energy level. Only certain well-defined energy levels are permitted; hence the nucleus can become aligned only in well-defined directions relative to the magnetic field B_0.

The *number of orientations* is a function of the physical properties of the nuclei and is *numerically equal* to $2I + 1$:

$$\textit{number of orientations} = 2I + 1 \tag{4.2}$$

In the case of ^1H, $I = 1/2$; hence the number of orientations is $2 \times (1/2) + 1 = 2$. Consequently, for ^1H only two energy levels are permitted.

In these energy levels the hydrogen nucleus is spinning about an axis of rotation. The axis of rotation also rotates in a circular manner, as shown in Fig. 4.1. This rotation is called *precession*. The direction of precession is either *with* the applied field B_0 or *against* the applied field. These states are also depicted in Fig. 4.1.

By digressing briefly, we can now gain further insight into the involvement of nuclei in NMR by looking at Fig. 4.2. This diagram is a representation of the nuclei of a molecule of methane (CH_4). Normally the valence electrons move in bonding orbitals holding together the hydrogen nuclei and the carbon nucleus. At the same time the hydrogen nuclei and carbon nucleus are spinning about their axes. A change in the precession angle of the axes of rotation of the atomic

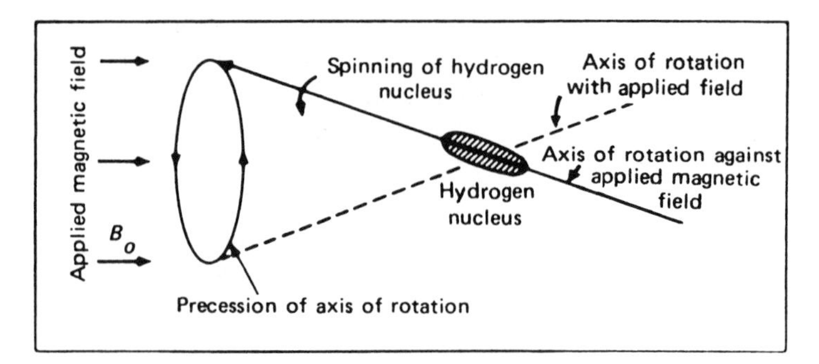

Figure 4.1 Permitted energy levels of quantized hydrogen nuclei. Shown is the energy state with the axis against the applied magnetic field B_0. There exists a similar energy state having the axis with the applied magnetic field.

nuclei is involved in NMR. In the case of hydrogen the direction of precession changes from *with* the magnetic field to *against* the magnetic field.

Returning to our quantized hydrogen nuclei (Fig. 4.1), it can be deduced that the difference in energy between the two quantum levels depends on the applied magnetic field B_0 and the magnetic moment μ of the nucleus. The energy differences are modified from the Stern–Gerlach experiment by the fact that they are dependent on the applied magnetic field B_0 and the spin number I of the nucleus. The relationship between these energy levels and the frequency ν of absorbed radiation is calculated as follows:

$$E = -M\left(\frac{\mu}{I}\frac{h}{2\pi}\right)B_0 \tag{4.3}$$

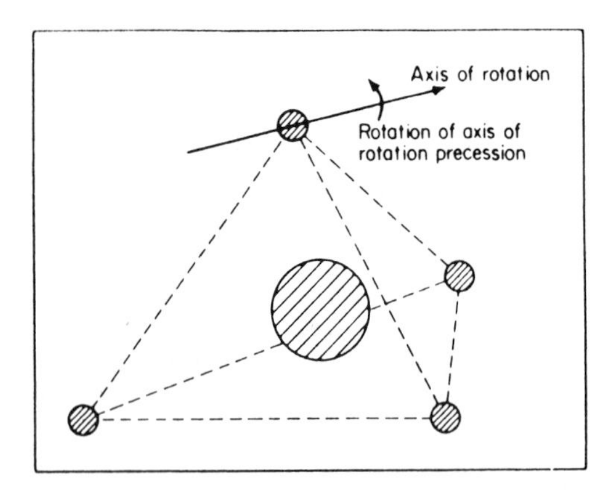

Figure 4.2 Nuclei of a molecule of methane.

where

M = magnetic quantum number
μ = nuclear magnetic spin
B_0 = applied magnetic field
I = spin angular momentum

Equation 4.3 is the general equation for all nuclei that respond NMR. However, if we confine our discussion to the H nucleus, which is by far the most important nucleus, then $I = 1/2$. Therefore, there are only two levels.

For two energy levels $M = +1/2$ and $-1/2$, then

$$\Delta E = h\nu = -\frac{+1}{2}\left(\frac{\mu}{I}\frac{h}{2\pi}\right)B_0 - \frac{-1}{2}\left(\frac{\mu}{I}\frac{h}{2\pi}\right)B_0 \qquad (4.4)$$

and

$$E = h\nu = \frac{\mu}{I}\frac{h}{2\pi}B_0 \qquad (4.5)$$

$$\nu = \frac{\mu}{I}\frac{B_0}{2\pi}$$

or

$$2\pi\nu = \frac{\mu}{I}B_0$$

This can be written as

$$\omega = \gamma B_0 \qquad (4.6)$$

where γ is the *gyromagnetic ratio*, μ/I:

$$\gamma = \frac{\text{nuclear magnetic moment}}{\text{spin angular momentum}} \qquad (4.7)$$

ω = frequency in rad/sec

But since $2\pi\nu = \omega$, we have

$$\nu = \frac{\gamma B_0}{2\pi} \qquad (4.8)$$

Since $\gamma = \mu/I$, it can be seen that Eqs. (4.8) and 4.5) are the same.

Equation (4.8) is the *Larmor equation,* which is fundamental to NMR. It indicates that for a given nucleus there is a direct relationship between the resonance frequency ν *and the applied magnetic field* B_0. This relationship is the mathematical basis of NMR.

In practice, it is found that if the applied field is 14,092 gauss, the frequency of radiation (rf) absorbed is 60 MHz. The nomenclature 60 MHz NMR indicates the rf used and also defines the strength of the applied magnetic field.

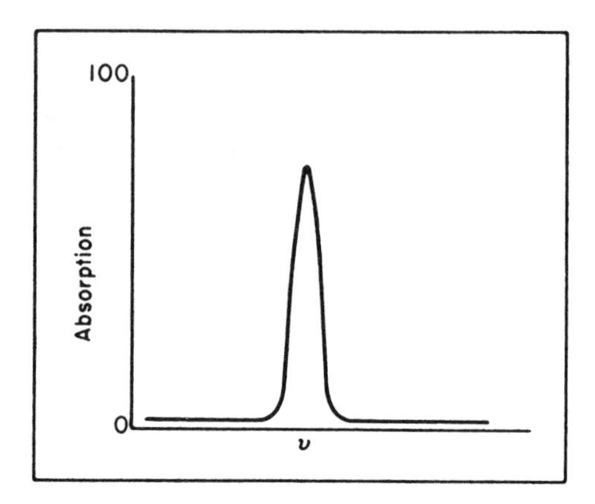

Figure 4.3 Absorption versus the frequency ν of RF radiation.

Similarly, a 100 MHz NMR uses 100 MHz rf and a magnetic field of 14,092 × 100/60 = 23,486 gauss.

A 500 MHz NMR uses a magnetic field of 23,486 × 5 = 117,330 gauss. Such intense fields cannot be achieved with a permanent ferromagnet, but use circular coils maintained at liquid N_2 temperature. They have essentially zero resistance and carry a permanent current. This system generates the intense magnetic fields required for such instruments.

The physical process occurring is that the spinning nucleus is aligned with the magnetic field. The nucleus precesses in the same manner as a falling top. When the rate of precession equals the frequency of the RF field, absorption of RF radiation takes place and the nucleus becomes aligned *opposed* to the magnetic field and is in an excited state. In practice, the samples, which contain hydrogen nuclei in the organic molecules, are first put into a magnetic field and then irradiated with RF radiation. When the frequency of the radiation satisfies Eq. (4.8), the radiant energy becomes absorbed. If the magnetic field B_0 is kept constant, we may plot the absorption against the frequency ν of the RF radiation. The resulting absorption curve should be similar to that shown in Fig. 4.3.

When a nucleus absorbs energy, it becomes excited and reaches an excited state. It then loses energy and returns to the unexcited state. Then it reabsorbs radiant energy and again enters an excited state. The nucleus alternately becomes excited and unexcited and is said to be in a state of *resonance*.

1. Saturation

The number of nuclei in the ground state is the number lined up with the magnetic field B_0. The degree of saturation is equal to the proportion of nuclei aligned with the magnetic field. There is always a portion of excited nuclei, and the remainder are unexcited as defined by the Boltzmann distribution The excess of unexcited nuclei over excited nuclei is called the *Boltzmann excess*. When no radiation falls on the sample, the Boltzmann excess is maximum, N_x. However, when radiation falls on the sample, an increased number of ground-state nuclei become excited and a reduced number remain in the ground state. If the RF field is kept constant, a new equilibrium is reached and the Boltzmann excess decreases to N_s. When $N_s = N_x$, absorption is maximum. When $N_s = 0$, absorption is zero. The ratio N_s/N_x is called Z_0, the *saturation factor*.

If the applied RF field is too intense, all the excess nuclei will be excited, $N_s \rightarrow 0$, and absorption $\rightarrow 0$. The sample is said to be saturated. The saturation factor Z_0 is

$$Z_0 = (1 + \gamma^2 B_1^2 T_1 T_2)^{-1} \tag{4.9}$$

where

γ = gyromagnetic ratio
B_1 = intensity of RF field
T_1, T_2 = longitudinal and transverse relaxation (see Sec. C.2.a)

As a consequence of this relationship, the RF field must not be very strong so as to avoid saturation. On the other hand, if a strong RF field is used, fine structure due to spin–spin splitting is collapsed and different information on the structure of the molecule is obtained (see Sec. E.2).

A system of molecules in the ground state may absorb energy and enter an excited state. A system of molecules in an excited state may emit energy and return to the ground state. If the number of molecules in the ground state is equal to the number in the excited state, the net signal observed is zero and no absorption is noted. Consequently, a signal can be seen only if there is an excess of molecules in the ground state.

The Boltzmann distribution [(Eq. (2.6)] is very pertinent to NMR because the energy difference ΔE between ground-state and excited-state nuclei is very small. Consequently, there are always almost as many nuclei in the excited state, N_1, as in the ground state, N_0. The difference between N_0 and N_1 is very small at room temperature. Typically, for every 100,000 nuclei in the excited state, there may be 100,001 in the ground state.

C. WIDTH OF ABSORPTION LINES

The resolution or separation of two absorption lines depends on how close they are to each other and the absorption linewidth. The width of the absorption line (i.e., the frequency range over which absorption takes place) is affected by a number of factors, only some of which we can control. These factors are discussed below.

1. The Homogeneous Field

The most important factor controlling the absorption linewidth is the applied magnetic field B_0. It is very important that this field be constant over all parts of the sample, which may be 1–2 in. long. If it is not, B_0 is different for different parts of the sample and therefore ν, the frequency of the absorbed radiation, will vary in different parts of the sample [see Eq. (4.8)]. This variation results in a wide absorption line. For qualitative or quantitative analysis a wide absorption line is very undesirable, since we may get overlap between neighboring peaks. In practice, the magnetic field must be constant over the entire sample to 1 part per 10 million. Fortunately, we are able to control the magnetic field well enough to eliminate this as a serious problem.

2. Relaxation Time

The second important feature that influences the absorption linewidth is the length of time that an excited nucleus stays in the excited state. The *Uncertainty Principle tells* us that

$$\Delta E \, \Delta t = \text{constant} \tag{4.10}$$

where ΔE is the uncertainty in the value of E [see Eq. (2.2)] and Δt is the length of time a nucleus spends in the excited state. Since $\Delta E \, \Delta t$ is a constant, when Δt is small, ΔE is large. But we know that $E = h\nu$ [Eq. (2.2)] and that h is a constant. Therefore any variation in E will result in a variation in ν. If E is not an exact number but varies over the range $E + \Delta E$, then ν will not be exact but will vary over the corresponding range $\nu + \Delta\nu$. Equation (2.2) can therefore be restated as

$$E + \Delta E = h(\nu + \Delta\nu) \tag{4.11}$$

We can summarize this relationship by saying that when Δt is small, ΔE is large and therefore $\Delta\nu$ is large. If $\Delta\nu$ is large, then the frequency range over which absorption takes place is wide and a wide absorption line results.

The length of time the nucleus spends in the excited state is Δt. It is controlled by the rate at which the excited nucleus loses its energy of excitation and returns to the unexcited state. The process of losing energy is called *relaxation*, and the time spent in the excited state is the *relaxation time*. There are two principal modes of relaxation: longitudinal and transverse.

a. Longitudinal Relaxation T_1

When the nucleus loses its excitation energy to the surrounding molecules, the system becomes warm as the energy is changed to heat. This process is quite

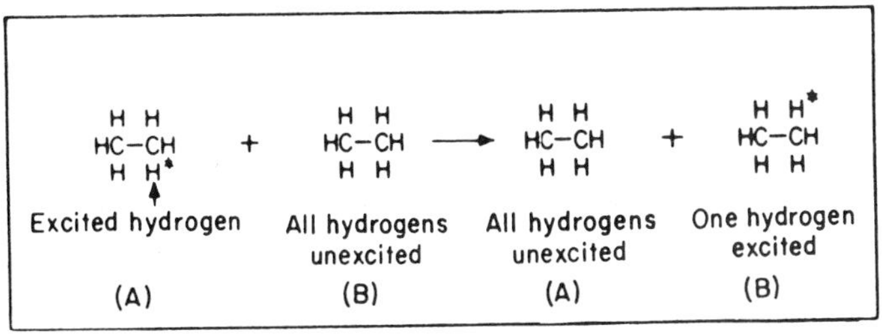

H H HC–CH H H*	+	H H HC–CH H H	⟶	H H HC–CH H H	+	H H* HC–CH H H
Excited hydrogen		All hydrogens unexcited		All hydrogens unexcited		One hydrogen excited
(A)		(B)		(A)		(B)

Figure 4.4 Transverse relaxation in ethane.

fast when the molecules are able to move quickly. This is the state of affairs in liquids. The excitation energy becomes dispersed throughout the whole system of molecules in which the sample finds itself. No radiant energy appears; no other nuclei become excited. Instead, as numerous nuclei lose their energy in this fashion, the temperature of the sample goes up. This process is called *longitudinal relaxation* T_1.

b. Transverse Relaxation T_2

An excited nucleus may transfer its energy to an unexcited nucleus of a similar molecule that is nearby. This may be octane (C_8H_{18}) with one excited hydrogen nucleus that interacts with another octane molecule. In the process, the nearby unexcited nucleus becomes excited and the previously excited nucleus becomes unexcited. There is no net change in energy of the system, but the length of time that one nucleus stays excited is shortened. This process, which is called *transverse relaxation* T_2, is illustrated for ethane in Fig. 4.4.

It is found in practice that in liquid samples the net relaxation time is comparatively long and narrow absorption lines are observed. In solid samples, however, the transverse time T_2 is very short. Consequently ΔE and therefore $\Delta \nu$ are large. For this reason solid samples generally give wide absorption lines. For most work, the absorption bands are too wide to be useful analytically. This means that solid samples generally cannot be examined directly by NMR without prior treatment.

However, methods have been developed to overcome this problem. For example, polymer samples normally give broad band spectra. But if they are "solvated," narrower lines are obtained and the spectra are useful. A sample is "solvated" by dissolving a small amount of solvent into the polymer. The latter becomes jellylike but does not lose many of its physical properties. The solvating process greatly slows down transverse relaxation and the *net* relaxation time is increased. Based on Eq. (4.10), the linewidth is decreased and resolution of the spectrum is possible.

3. Magic Angle NMR

A problem with the examination of solids is that the nuclei can be considered to be frozen in space and cannot freely line up in the magnetic field. The NMR signals generated are dependent, among other things, on the orientation of the nuclei to the magnetic field. This results in an apparent change in the chemical shift producing broad band spectra. This phenomenon is chemical *shift anisotropy* and causes such spectra to be of little analytical value.

However, it has been shown theoretically and experimentally that by spinning the sample at an angle of 54.7 degrees (the magic angle) to the magnetic field rather than 90 degrees, the chemical shift anisotropy is greatly reduced and narrow line spectra are obtained. The spinning is carried out at very high frequencies (5–15 kHz) for optimum performance. This, of course, is more useful analytically because it allows better resolution and therefore better measurement of chemical shift and spin–spin splitting. In turn, this is very informative of the functional groups and their positions relative to each other in the sample molecule.

Special probes have been developed for magic angle NMR which automatically position the sample, making the method a more practical procedure with ordinary instruments. The method has proved very useful for studying solid samples that hitherto could not be examined by NMR. Such samples include synthetic polymers, where spatial arrangement of the components greatly affects the physical properties of the polymer, such as its strength or its ability to stay strong at higher temperatures.

4. Other Sources of Line Broadening

Any process of *deactivating*, or relaxing, an excited molecule results in a decrease in the lifetime of the excited state. This in turn causes line broadening [see Eq. (4.10)]. Other causes of deactivation include (1) the presence of ions (the large local charge deactivates the nucleus), (2) paramagnetic molecules such as dissolved O_2 (the magnetic moment of electrons is about 10^3 times as great as nuclear magnetic moments; the high local field causes line broadening), and (3) nuclei with quadrupole moments such as adjacent combined N (usually, nuclei in which $I > 1/2$ have quadrupole moments, which causes electronic interactions and line broadening; an important nucleus with a quadrupole field is ^{14}N).

Other sources of line broadening included: (1) ions, (2) paramagnetic molecules, such as dissolved oxygen (remove by bubbling N_2 gas through the sample, if possible), and (3) atoms with a quadrapole instead of a dipole, such as N.

D. CHEMICAL SHIFTS

If NMR were suitable only for detecting and measuring the presence of hydrogen in organic compounds, it would be a very limited technique. There are a number of fast, inexpensive methods for detecting and measuring hydrogen in organic compounds. Fortunately, combined hydrogens in different types of molecules are found in practice to absorb at slightly different frequencies, thus extending the range of information that can be obtained by NMR. This variation in absorption or response frequency is caused by a slight difference in the electronic (or chemical) character of the hydrogen and is called the *chemical shift*. The physical basis for the chemical shift is as follows.

Suppose that we take a molecule with several different types of hydrogen, such as ethanol, CH_3CH_2OH. This molecule has three types of hydrogen: the hydrogens on the terminal CH_3, those on the CH_2, and the one on the OH. Consider the nuclei of the different types of hydrogen. Each one is surrounded by orbiting electrons. That is, the orbitals may vary in shape or in the length of time the electrons spend near the hydrogen nucleus. Let us suppose that we place this molecule in a strong magnetic field B_0. The electrons associated with the nuclei will be rotated by the applied magnetic field B_0. This rotation, or *drift*, generates a small magnetic field σB_0, which opposes the applied magnetic field B_0. The nuclei are shielded slightly from the magnetic field by the orbiting electrons. The extent of the shielding depends on the drift or movement of the electrons caused by the magnetic field (not by the simple orbiting of the electrons). If the extent of this shielding is σB_0, then the nucleus is not exposed to the full field B_0, but to an *effective* magnetic field at the nucleus, B_{eff}:

$$B_{eff} = B_0 - \sigma B_0 \tag{4.12}$$

where

B_{eff} = effective magnetic field at the nucleus
B_0 = applied field
σB_0 = shielding afforded by the drift of local electrons

In the case of ethanol, σB_0 is different for each type of hydrogen; therefore the effective field B_{eff} is different for each type of hydrogen. In order to get absorption at frequency ν, we must compensate for this variation by varying B_0. In other words, resonance of the different types of hydrogens take place in slightly different *applied* magnetic fields. Another way of expressing this relationship is that if the applied field is kept constant, the nuclei absorb at slightly different frequencies ν. This change in frequency of absorption because of shielding is the chemical shift.

Usually, in order to avoid measuring the shift in absolute numbers, one measures it relative to a standard. A popular standard is *tetramethylsilane*

(TMS), which has the chemical formula $Si(CH_3)_4$. In this compound all 12 hydrogens are equivalent; that is, they are all exposed to the same shielding and give a single absorption peak. The relative chemical shift for other hydrogen nuclei is represented as follows:

$$\text{relative chemical shift} = \frac{B_s - B_r}{B_r} \tag{4.13}$$

where

B_s = magnetic field at which the sample absorbs
B_r = magnetic field at which the reference (in this case TMS) absorbs

The chemical shift is the difference in the observed shift between the sample and the reference divided by the resonance magnetic field of the reference. It is a simple number and therefore dimensionless. Typical values range for B_s between 10^{-6} and 10^{-5}. In order to make these numbers easier to handle, they are usually multiplied by 10^6 and then expressed in parts per million (ppm). (This unit should not be confused with the concentration expression parts per million used in trace analysis.) The symbols for the chemical shift is δ. It is expressed as

$$\delta = \frac{B_s - B_r}{B_r} \times 10^6 \text{ ppm} \tag{4.14}$$

Shielding by the drifting electrons is modified by other nuclei in their vicinity. These in turn are affected by the chemistry, geometry, and electron density of the system. Consequently, some deshielding takes place, and we are able to distinguish between different functional groups in the neighborhood of the nucleus.

The effect of functional groups on chemical shifts is most important since it allows us to identify which functional groups are present in a molecule or a mixture. For example, it is easy to distinguish between aromatic and aliphatic hydrogens, ethylenes, acetylenes, or hydrogens adjacent to ketones, esters, etc. Some secondary effects on chemical shifts are brought about by the solvents present or by the geometry of the molecules, particularly when hydrogens may be affected by being physically located above ring currents.

Chemical shift is important because it enables us to ascertain what types of hydrogen are present in a molecule by measuring the shift involved and comparing it with known standards run under the same conditions. Some typical examples of chemical shift are given in Table 4.3.

Figure 4.5 shows the low-resolution spectrum of ethyl alcohol (i.e., ethanol). This absorption spectrum, which is historically significant as one of the first NMR spectra to be recorded, discloses that there are three types of hydrogens present in the ethanol molecule. Because absorption spectra provide valuable information about a molecule's structure, they are very helpful in

Table 4.3 The Chemical Shift of Typical Hydrogen Types

Type of hydrogen	Chemical shift (ppm)			
	a	b	c	d
Tetramethylsilane (TMS) $CH_3-Si(CH_3)_2-CH_3$		0		
$CH_3CH_2CH_2CH_3$ (a b)	1.0	1.2		
$CH_3CH_2-C(CH_3)(CH_3)-CH_2CH_3$ (a b, c = CH_3 on central C, b a)	1.0	1.2	2.0	
Substituted paraffins				
CH_3CH_2Br (a b)	1.2	3.43		
CH_3CH_2I (a b)	1.8	3.2		
CH_3CH_2OH (a b c)	1.32	3.7	5.3	
$CH_3C(=O)OH$ (a, OH b)	2.1	11.4		
$CH_3CH_2CH_2NH_2$ (a b c d)	0.92	1.5	2.7	1.1
$CH_3CH_2CH=CH_2$ (a b c d)	.9	2.0	5.8	4.9
$CH_3CH_2CH=CHCH_2CH_3$ (a b c c b a)	1.0	1.3	5.4	
$CH_3CH_2CH(NO_2)-CH_3$ (a b c d)	1.0	1.9	4.2	1.5
$CH_3C(=O)-CH_2CH_3$ (a b c)	2.1	2.5	1.1	
$CH_3CH_2S-CH_3$ (a b c)	1.3	2.53	2.1	
$CH_3CH_2-C\equiv CH$ (a b c)	1.0	2.2	2.4	
$CH_3CH_2CH_2C(=O)H(d)$ (a b c)	1.0	1.7	2.4	9.7

Table 4.3 Continued

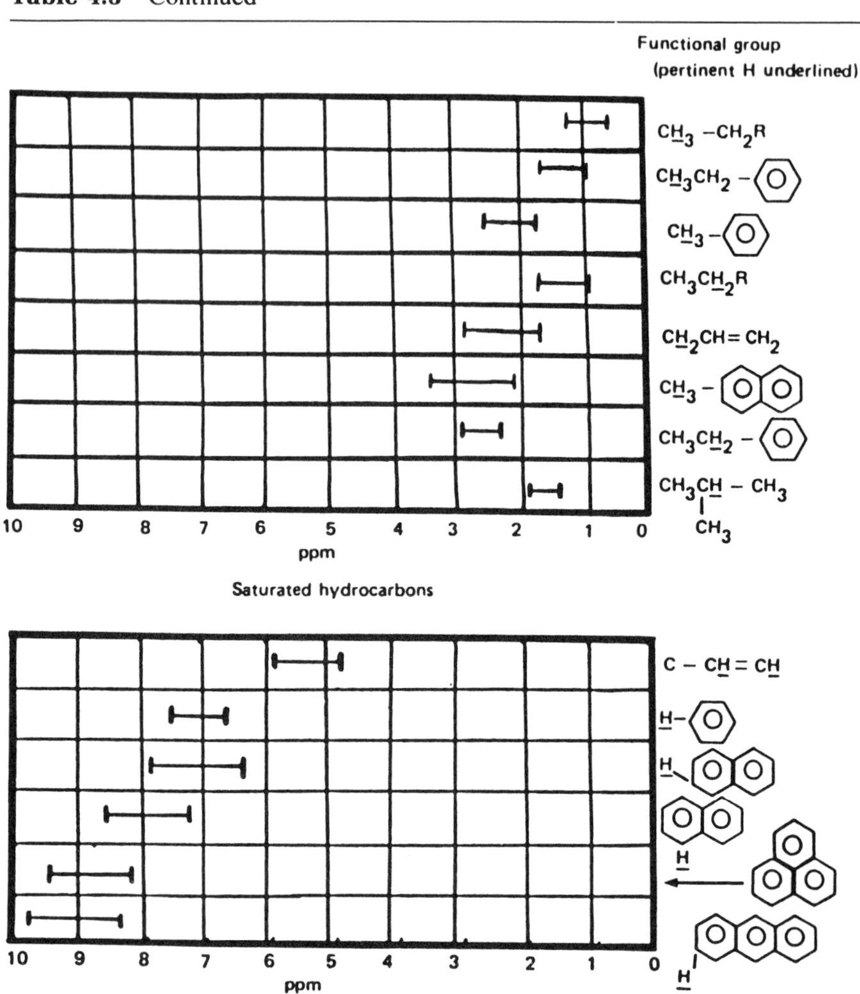

characterizing unknown compounds or compounds for which we know the empirical formula but not the structure.

With the improvement in instrumentation since it was first recorded, the absorption spectrum of ethyl alcohol has been found to be more complex than Fig. 4.5 indicates. When examined under high resolution, each peak in the spec-

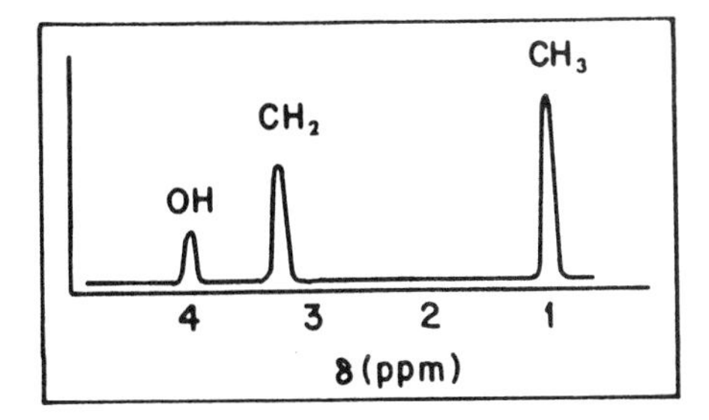

Figure 4.5 Absorption spectrum of ethanol.

trum can be seen to be composed of several peaks. This fine structure is brought about by spin–spin splitting.

E. SPIN–SPIN SPLITTING

As we have already seen, the hydrogen nuclei of an organic molecule are spinning, and the axis of rotation may be with or against the applied magnetic field. Since the nucleus is magnetic, it exerts a slight magnetic field, which may be either with or against the applied magnetic field.

Suppose that we have a molecule such as an aldehyde (I):

```
    H  H ←————————— type 1 hydrogen
    |  /
 R—C—C
    |  \\
    H   O
    ↖
     ————————————— type 2 hydrogen
   (I)
```

let us ignore the nuclei in the R group. The chemical shift causes differences in the frequency of absorption of the two types of hydrogen. The absorption is shown in Fig. 4.6. However, the hydrogen of the $\mathrm{-C}\Big\langle^{\mathrm{H}}_{\mathrm{O}}$ (CHO) group may be spinning either *with* or *against* the applied magnetic field. The spinning of the nucleus creates a small magnetic field, either with or against the applied magnetic field. This changes the effective field of the hydrogens of the adjacent CH_2 group. The CH_2 hydrogens next to a CHO hydrogen that is spinning *with* the field absorb at a slightly different frequency from that of the CH_2 hydrogens adjacent to a hydrogen spinning *against* the field. In practice, a sample will contain as many hydrogens spinning with the field as against it; so both groups will be equally represented. We can see that the oppositely spinning hydrogen of the CHO group splits the adjacent hydrogens in the CH_2 group into two groups.

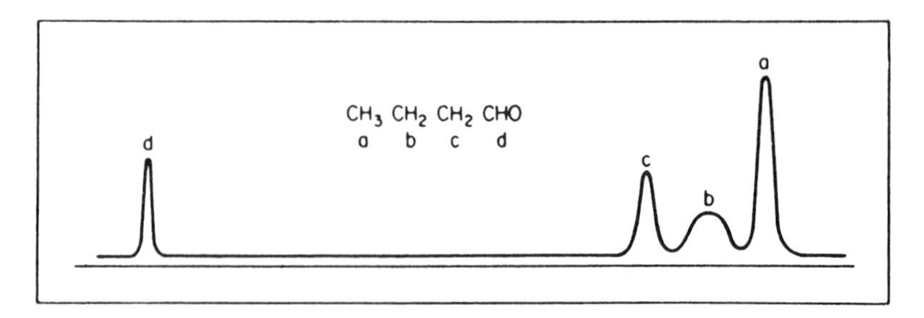

Figure 4.6 Spectrum of the aldehyde RCH_2CHO (no spin–spin splitting).

At the same time, the hydrogens of the CH_2 group are also spinning, and they affect the frequency at which the neighboring CHO hydrogen nucleus absorbs. In this case, each of the hydrogens of the CH_2 group can spin with or against the applied field (Fig. 4.7). Several spin combinations are possible. These combinations may be depicted as in Fig. 4.8, where the arrows indicate the directions of the magnetic fields created by the spinning nuclei. In a normal sample many billions of molecules are present, and each spin combination is represented equally by number. The three combinations (a), (b), and (c) in Fig. 4.8 modify or shield the applied field to three different degrees, and responses occur at three frequencies. It can be seen that there are two ways in which the combination of Fig. 4.8b can exist, but only one way for the combination of Fig. 4.8a or c. There will therefore be twice as many nuclei with a magnetic field equal to the combination of Fig. 4.8b. As a result, the CH_2 group will cause the neighboring H of the CHO group to respond at three frequencies with relative intensities in the ratio 1:2:1 (see Fig. 4.7). The NMR spectrum of $CH_3CH_2CH_2CHO$ will therefore appear as shown in Fig. 4.9.

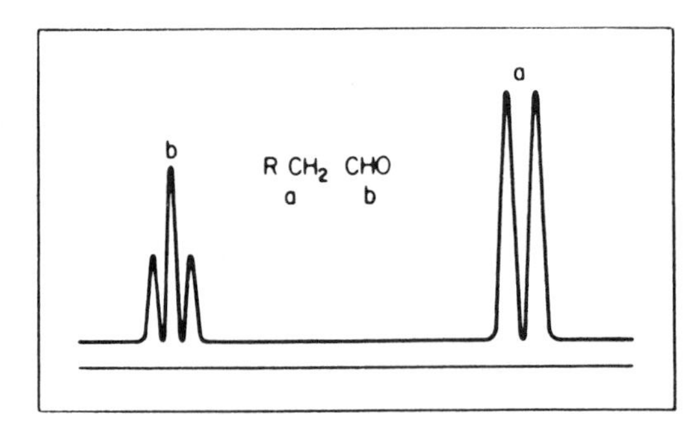

Figure 4.7 Spectrum of RCH_2CHO showing spin–spin splitting of CH_2 by the CHO group.

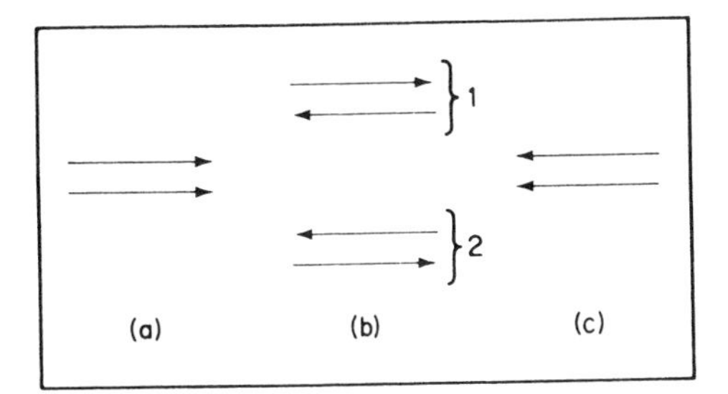

Figure 4.8 Direction of the magnetic field of pairs of hydrogen nuclei: (a) both with the field (B_0 appears to be increased), (b) one with and one against the field (there are two possible arrangements, 1 and 2, both with equal effects; the net effect on B_0 is zero), and (c) both against the field (B_0 appears to be reduced).

Spin–spin splitting is quite strong between hydrogens on *adjacent* carbons, but is generally negligible between hydrogens farther removed than this. It is important to remember that spin–spin splitting causes a change in the fine structure of the adjacent hydrogens on the molecule. For example, the CH_2 splits the CH, and the CH splits the CH_2, but the CH_2 *does not* split itself because interaction is forbidden by quantum theory.

Of course, it is always possible to deduce how many combinations of spins there are of the hydrogens on a given carbon. In practice, it is easier to remember that it is equal to $2nI + 1$, where n is the number of *equivalent hydrogens* on the carbon and I is the spin quantum number (in the case of hydrogen $I = 1/2$).

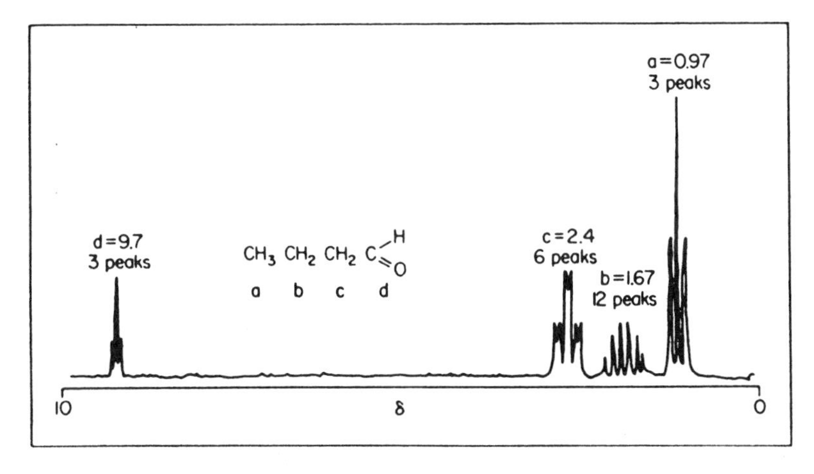

Figure 4.9 Spectrum of $CH_3CH_2CH_2CHO$ including spin–spin splitting.

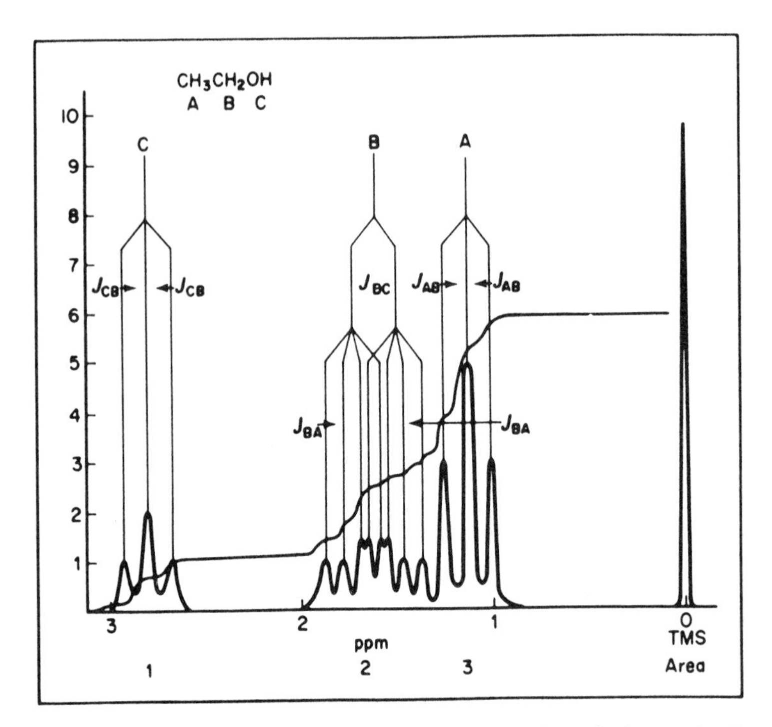

Figure 4.10 Spectrum of dry ethanol showing spin–spin interaction between hydrogens on carbons A, B, and C. Note the methylene hydrogens (B) are split by both A and C. The coupling between A and C is very small (not zero) and not shown under these conditions.

Hence we have the relationship for hydrogen. The number of bands in fine structure due to spin–spin splitting is

$$2nI + 1 \tag{4.15}$$

For hydrogen $I = 1/2$ and Eq. (4.15) simplifies to $n + 1$.

The number of bands is sometimes termed the *multiplicity*; n is the number of *equivalent* hydrogens on the adjacent group. In a methyl group (CH_3), $n = 3$ and the number of combinations is 4. This can also be stated as the multiplicity of the band is 4. If two different adjacent groups cause splitting, the multiplicity is given by $(2nI + 1)(2n'I + 1)$, where n and n' are the numbers of equivalent hydrogens in each group.

The NMR spectrum of ethanol would be expected to be as in Fig. 4.10. The CH_3 hydrogens are split by the CH_2 group into $(2 + 1) = 3$ peaks. The CH_2 is split by the OH into $1 + 1 = 2$ peaks, each of which is split by the adjacent CH_3 into 4 peaks, resulting in a total of 8 peaks; finally, the OH is split into 3 peaks by the adjacent CH_2 group. Equivalent hydrogens do not interact with each other, since such transitions are forbidden. Hence the three hydrogens in CH_3 do not split with each other.

Remember that n is the number of equivalent hydrogens taking part in spin–spin splitting. For example, on a methyl group, —CH_3, there are three equivalent hydrogens; on the —CH_2— group there are two equivalent hydrogens. Also, the number of peaks equals $n + 1$ [Eq. (4.15)] only in nuclei where $I = 1/2$, or H in this case.

The degree of splitting of a peak by adjacent hydrogens is a measure of the magnetic effect of that hydrogen. This magnetic effect is the coupling, or magnetic interaction, between the two hydrogens. It is termed the *J coupling constant*. It can be measured directly from the NMR spectrum. In Fig. 4.10 the coupling constant between the CH_2 and the OH is J_{BC} (as is the coupling between the OH and the CH_2). Similarly, the J coupling between the hydrogen on the CH_2 and the CH_3 is J_{AB}. The coupling constant is a measure of energy and is usually expressed in cycles per second. In practice, it is measured directly from the spectrum. The J coupling constant provides valuable information to physical chemists and to organic chemists interested in molecular interaction.

The magnitude of the J coupling constant between hydrogens on adjacent carbons is between 6 and 8 cps. This magnitude decreases rapidly as the hydrogens move farther apart. There is no coupling between hydrogens on the same carbon. For example, in a compound such as

$$\underset{A}{CH_2}{=}\underset{B}{CH}{-}\underset{C}{CH_2}{-}\underset{D}{CH_3}$$

the coupling between the hydrogens on carbon A is zero; the coupling between the hydrogen on carbon A and those on carbon B is about 6–8 cps (this is written $J_{AB} = 6–8$). The coupling between the hydrogens on carbon A and those on carbon C is never more than 1 cps, that is, $J_{AC} = 1$ cps, and J_{AD} is very small (not noticeable).

1. The Effect of H Exchange

If a solution of methanol (CH_3OH) is put in water, the hydrogen of the OH exchanges with hydrogen in the water (H_2O). If this exchange rate is rapid, the nearby nuclei see only the average position of the nucleus, and spin–spin splitting is lost The exchange rate is affected by temperature, increasing with increased temperature. The spectra of alcohols are therefore affected by any traces of water in the sample and by temperature. For example, the spectra of CH_3OH at 20°C and at −40°C are shown in Fig. 4.11.

Interpretation of NMR spectra can be very difficult if we consider all the multiplicities that are possible from the interactions of all the nuclei. This can be

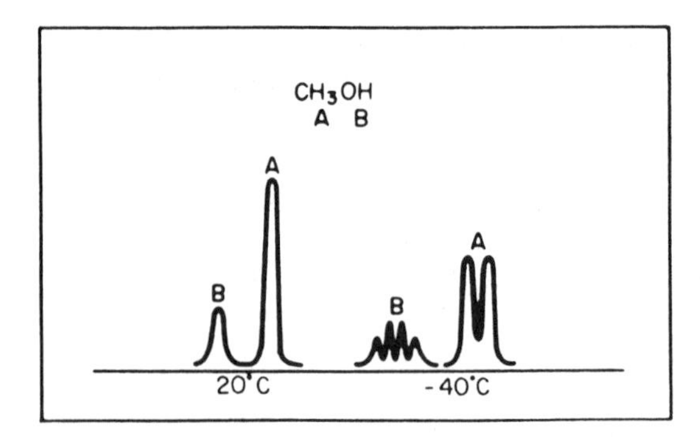

Figure 4.11 Spectra of CH_3OH with two different types of hydrogen, A and B. Note the loss of fine structure at 20°C compared with −40°C.

resolved using a computerized system, but even then interpretation is often difficult to undertake. If we confine the interpretation only to those nuclei that give reasonably strong coupling constants, the resolution is simplified. This is called utilizing *first-order spectra.* The more complicated systems involving other, weaker coupling constants are second-order spectra and will not be dealt with in this chapter.

The capital letters of the alphabet are used to define spin systems that have strong or weak coupling constants. For example, a system A_2B_3 indicates a system of two types of nuclei interacting strongly together of which there are two of type A and three of type B. A break in the alphabetical lettering system indicates weak or no coupling. For example, AB_2M indicates that there are three nuclei, one of type A and two of type B that interact strongly with each other but not with nucleus M. Similarly, a system $A_2B_2MNXY_2$ indicates two nuclei of type A and two of type B that interact strongly with each other, one of type M and one of type N that interact strongly with each other but not with nuclei of A or B, and one of type X and two of type Y that interact strongly with each other but not with types M, N, A, or B.

The rules for interpreting first-order spectra can be summarized as follows:

1. A proton spin coupled to any equivalent protons will produce $n + 1$ lines separated by J cps, where J is the coupling constant. The relative intensities of the lines are given by the binomial expansion $(r + 1)^n$, where n is the number of equivalent hydrogens. Splitting by one proton yields two lines, and by two protons yields three lines with a ratio of intensities 1:2:1. Three protons give four lines with a ratio of intensities 1:3:3:1. Four protons yield five lines with a ratio of intensities 1:4:6:4:1. And so on.
2. If a proton interacts with two different sets of equivalent protons, then the multiplicity will be the product of the two sets. For example, hydrogen split

by a methyl (CH_3) and a methylene (CH_2) group will be split into four lines by the methyl group. Each line will be split into three other lines by the methylene group. This will generate a total of 12 lines, some of which will probably overlap each other.

3. Equivalent hydrogens do not split each other, the transition being forbidden. In practice, however, interactions take place and can be seen in second-order spectra.

Some typical spectra will be shown in Figs. 4.13–4.18. It is not possible in this text to complete instructions on the interpretation of NMR spectra, but it is hoped that by understanding the rules involved, it can be understood that the NMR spectra can be interpreted to give important information. For example, the chemical shift indicates the functional groups that are present, such as aromatics, halides, ketones, amines, alcohols, and so on. Further spin–spin splitting indicates which groups are coupled to each other and therefore close to each other in the molecular structure. The multiplicity indicates how many equivalent hydrogens are in the adjacent functional groups. Information is also obtained on the geometry of the molecules. Bearing in mind the ring current or aromatic compounds, it is possible to obtain information on the three-dimensional character of molecules.

By using second-order spectra, it is possible to unravel the molecular formulas of large molecules. Nuclear magnetic resonance spectroscopy has become invaluable to organic chemists and biochemists and is continuing to make important contributions to both fields.

2. Double-Resonance Spectroscopy

It was shown in Section B.1 that a system of protons can become excited and saturated by applying a strong radio frequency (RF) field to the sample. When this occurs, the protons do not give an NMR signal; neither do they interact with nearby protons to generate spin–spin splitting. Advantage has been taken of this fact to simplify complicated NMR spectra. When a sample is first analyzed, it is not usually known which groups split which other groups. By a process of elimination, the spin coupling assignments are made and the structure of the sample elucidated. For example, from the spectrum of ethanol shown in Fig. 4.10, we may suspect that the hydrogen on the OH (C) is split by the hydrogen on the CH_2 (B). We can confirm this by running the NMR spectrum, but this time using two RF frequencies in the RF frequency generator. One frequency is used to sweep the field in the normal way and is kept at a low intensity; the second frequency is operated at high intensity and is at such a high frequency that when the first RF coil is in resonance with the OH group, the second RF coil is saturating the CH_2 group. The CH_2 group will therefore collapse and the spin–

spin fine structure of the OH caused by the CH_2 group will also collapse. This saturation process can be performed by using two frequencies that trail each other at a controlled frequency difference, in this case the frequency difference between the responses of the OH and CH_2. As an alternative method, the strong saturating frequency may be kept steady, in this case at the frequency of the CH_2, and the first RF scanned in the normal fashion.

By using double-resonance systems, one can greatly simplify the spectrum, and coupling between different types of protons is confirmed by both the disappearance of the spectrum of the saturated protons and the collapse of the fine structure of the coupled proton. In more advanced systems it is not uncommon to use double and triple resonance in order to further simplify the system sufficiently for interpretation.

3. The Nuclear Overhauser Effect

With the nuclear Overhauser effect (NOE), one nucleus is saturated and an increase in the signal from nearby nuclei observed. It is not necessary that the nuclei involved be coupled. For example, in a $^{13}C—^{1}H$ bond, the ^{1}H may be saturated and the intensity of the ^{13}C resonance observed. Saturating the ^{1}H increases the ground state population of the ^{13}C. This in turn increases the ^{13}C signal significantly.

The increased signal due to NOE is useful qualitatively, but the enhancement is quantitatively unpredictable. The NOE is proportional to r^{-6} when r is the internuclear distance. Consequently, adjacent $^{1}H—^{1}H$ do not display NOE unless they are closer than 3–5 Å.

F. EQUIPMENT

The most important parts of an NMR instrument are the permanent magnet, the RF generator, the RF detector, the sample holder, and the magnetic coils. These are arranged as shown in Fig. 4.12 and described as follows:

1. Sample Holder

The sample is held in a holder, which must be transparent to RF radiation, durable, and chemically inert. Glass or Pyrex tubes are commonly used. These are sturdy, practical, and cheap. They are usually about 6–7 in. long and approximately 1/8 in. in diameter.

2. Permanent Magnet

The magnet must be capable of producing a very strong magnetic field B_0. Its strength must be at least 10,000 gauss (G). It is common to express the operating

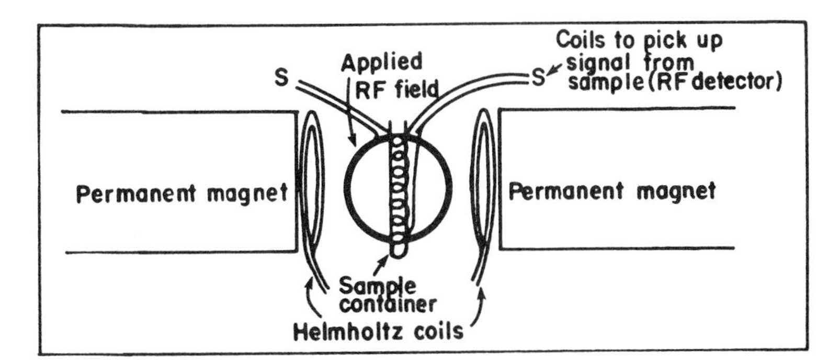

Figure 4.12 Schematic diagram of an NMR instrument. If B_0 is 14,000 G, the ν is 60×10^6 S. To measure the frequency to 1 cps, the magnet must be homogeneous to about 1 part in 10^8. This is a 60-MHz NMR instrument.

conditions in terms of the frequency ν rather than the magnetic field. There are related by Eq. (4.5). If B_0 is 14,092 G, the ν is 60×10^6 cps. To measure the frequency to 1 cps, the magnet must be homogeneous to about 1 part in 10^8. This is a 60-MHz NMR instrument. In practice, routine instruments operate at 60–100 MHz, and high-quality equipment for ^{13}C at 200–600 MHz using superconducting magnets.

It is also important that the magnetic field be constant over long periods of time. For this reason adequate temperature control is necessary to maintain the constant physical dimensions of the magnet. To do this, the magnet system is usually thermostatically controlled; also, it is advantageous to control the temperature of the room in which the instrument is housed. An entire room dedicated to the instrument is a distinct advantage.

In addition, the magnetic field should be *homogeneous* to a high order (1 in 10^8) over a space large enough to contain the sample. Homogeneous in this case means that the field does not vary in strength or direction from point to point over the space occupied by the sample.

Instruments are now being built with stronger magnets giving response at 600 MHz.

3. Magnetic Coils

We have already seen that there is a relation between the resonance frequency of the nucleus and the strength of the magnetic field in which the sample is placed. This relationship is expressed by Eq. (4.8):

$$\nu = \text{constant} \times B_0$$

For the nucleus to resonate, the precession frequency of the nucleus must equal the frequency of the applied RF radiation. If B_0 is kept constant, the precession frequency is fixed. To effect resonance, the frequency of the RF field must be changed to equal the resonance frequency. On the other hand, if the RF frequency is kept constant, the resonance frequency of the nucleus must be changed by varying B_0 [see Eq. (4.8)]. In practice, it is easier to vary B_0 than the RF frequency in order to effect resonance.

It is not easy or convenient to vary the magnetic field of a large, stable magnet; however, the problem can be overcome by superimposing a small variable magnetic field on the main field. This is done by using a pair of *Helmholtz coils* in the pole faces of the permanent magnet. These coils induce a magnetic field that can be varied by varying the current flowing through them. The small magnetic field produced is in the same direction as the main field and is added to it. The sample is exposed to both fields, which appear as one field to the molecules.

In commercial equipment the normal practice is to keep the RF radiation constant in attaining resonance. This is akin to scanning the spectrum by varying the wavelength in IR or UV spectroscopy. It is called *sweeping the field.*

4. Radiofrequency Generator

The radiofrequency radiation is created by using an RF oscillator or transmission coil (see Fig. 4.12). To get maximum interaction with the sample, the RF coil is wound around the sample container. The oscillator irradiates the sample with RF radiation. In the process of irradiation, the applied RF field must not change the effective magnetic field B_0. The oscillator coil is therefore wound perpendicular to the applied magnetic field.

5. Detector

The simplest form of detector is a second RF coil similar to the RF generator coil wound around the sample. This coil detects radiation that is absorbed by the sample and reemitted (or dispersed). This type of detector must not be affected by either the RF generator or the main magnetic field. It is therefore wound around the sample in such a way that it is perpendicular to both the RF source and the main magnetic field. The physical setup is depicted in Fig. 4.12.

To summarize, NMR equipment must include the following:

1. A high-power magnet. Most are permanent magnets, but high-resolution equipment such as that used for ^{13}C research uses superconducting electromagnets at liquid helium temperatures. The magnetic field must be very homogeneous, ideally constant to 1 part in 10^9.

2. A sweep generator used to vary the magnetic field over a small range.
3. A transmitter coil to emit at a fixed frequency and a receiver coil to receive at that frequency.

This system is a *field sweep system* because the magnetic field is varied and the coils are emitting and receiving at a fixed frequency. The Larmor equation [Eq. (4.8)] indicates an alternative approach whenever the magnetic field B_0 is kept constant and the radiofrequencies of the transmitter and receiver are varied. This is a *frequency sweep system*.

6. Fourier Transform Spectra

The time required to record an NMR spectrum is Δ/R (sec), where Δ is the spectral range scanned and R the resolution required. For ^{1}H-NMR the time required is only a few minutes because the spectral range is small, but for ^{13}C-NMR the chemical shifts are much greater; consequently the spectral range scanned is much greater and the time necessary to scan is very long. For example, if the range is 5 kHz and a resolution of 1 Hz is required, the time necessary would be (5000 sec)/1 = 83 min, an unacceptably long time.

This problem was overcome by using Fourier transform (FT) NMR. In this system the spectral range is not scanned continuously and slowly from one end to the other; rather, the entire range is split up into a suitable number N of small increments. Each of the N increments is exposed to a field for a very short time (e.g., 10 μsec). All increments are exposed to the field at the same time. If any nuclei are in resonance within any increment, the nuclei become excited and then relax, generating an NMR signal from each increment simultaneously. The combined signals are then converted using Fourier transforms to generate a normal NMR spectrum.

The advantage of the system is that the entire spectrum is taken in a single pulse. Naturally the intensity of the spectrum is very low and is usually scarcely discernible above the noise level obtained at the same time. The process may be repeated many times very rapidly, for example, 8192 times (2^{13} scans) in 0.8 sec. The signal increases linearly, but the noise increases only as the square root of the number of readings. The net effect is a significant improvement in the signal-to-noise ratio. This directly improves the sensitivity of the method, allowing better measurement of both strong and weak lines.

There is a practical limit to this procedure. To get a signal twice as strong as that obtained with 2^{13} scans, we would need twice as many scans, that is, 2^{14}, or 16,384. To obtain a further signal doubling would require 32,768 scans. Clearly, in practice there is a point of diminishing return, especially when it is recalled that noise simultaneously increases by $\sqrt{s}$ with each signal doubling. However, signals can be obtained that are orders of magnitude more sensitive

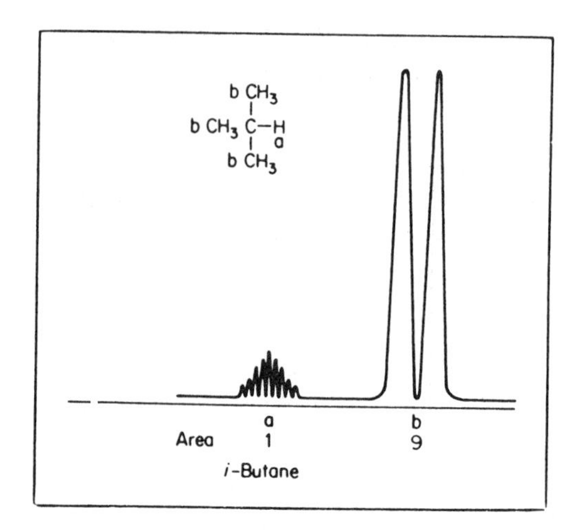

Figure 4.13 Typical NMR spectrum of the paraffin *i*-butane.

than those normally encountered with conventional NMR. This system is becoming increasingly useful in ^{13}C-NMR and in extending the sensitivity range of proton NMR.

G. TYPICAL SPECTRA: APPLICATIONS TO ANALYTICAL CHEMISTRY

We saw earlier (Sec. C.3) that solid samples can only be characterized directly by NMR by using special equipment involving the use of the "magic angle." Some solids can be examined by solvation prior to analysis, but not all solids can be treated thus. Furthermore, this technique is not sufficiently sensitive to be used to examine gas supplies. Therefore for analytical work we are usually directly, but other samples must be pretreated. Pretreatment includes such procedures as dissolving the sample in a suitable solvent or liquefying it by reducing its temperature. For the examination of liquid samples, the sensitivity is sufficient to determine concentrations down about 0.1%. More dilute samples must be preconcentrated. However, the spectra can be used for both quantitative and qualitative analysis.

Some typical NMR spectra of common compounds are shown in Figs. 4.13 and 4.14. It can be seen that subtle differences are caused by chemical shift and spin–spin splitting; they indicate the importance of this method in identifying organic molecules.

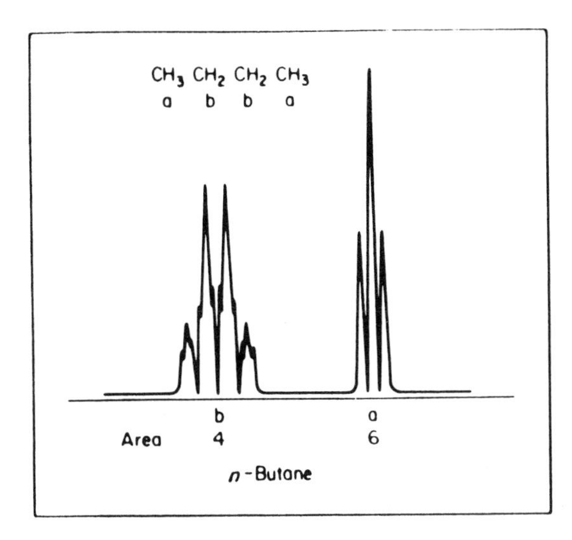

Figure 4.14 Typical NMR spectrum of the paraffin *n*-butane.

1. Relationship Between the Area of the Peaks and Molecular Formula

The area of the spectral peaks is directly proportional to the number of protons involved and not the multiplicity. In the spectra in Figs. 4.13–4.18, the relative area of each system of peaks is measured. For example, in a sample of ethanol (Fig. 4.10), the methyl group CH_3 may be split by the methylene group to give three peaks. The area of the methyl group is the area enclosed by all three peaks and the baseline. The two hydrogens of the methylene group are split by the methyl group into four peaks. Furthermore, each of the four peaks is split by the hydrogen of the OH into two peaks to give a total of eight peaks. The area contributed by the methylene hydrogens is that enclosed by all eight peaks and the baseline of the spectrum.

Notice that in Fig. 4.15 the protons on carbons a and b occur at the normal place for paraffinic hydrogens, but that those on carbon c are displaced downfield by the adjacent double bond. The splitting of the hydrogen on carbon d is quite complicated: First, it is coupled to the hydrogens on carbon c and therefore split into a triplet; second, it is coupled to the hydrogens on carbon e. However, these hydrogens are not equivalent, because one hydrogen is *cis* and the other is *trans* to the hydrogen on carbon d. The coupling between these hydrogens is therefore different. As a result, the hydrogens in the population are split by a *cis* hydrogen, and separately by a *trans* hydrogen. It will be remembered that this hydrogen was split into a triplet by the two hydrogens of carbon c. The triplet is

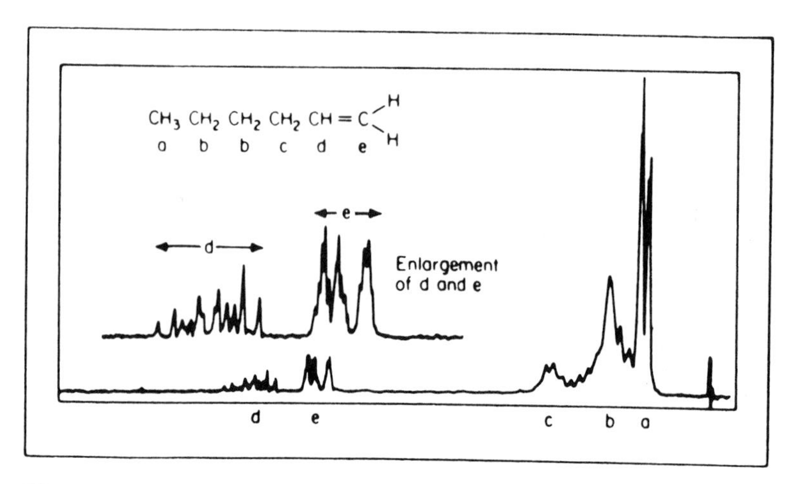

Figure 4.15 Typical NMR spectrum of an olefin.

split into 6 peaks by the *trans* hydrogen and 6 other peaks by the *cis* hydrogen of carbon e, resulting in a total of 12 peaks, which overlap considerably. In a similar manner, the hydrogens of carbon e are split by the single hydrogen on carbon d, but the coupling constants between this hydrogen and the *cis* and the *trans* hydrogens are not equal. Four peaks are formed, which overlap somewhat, as shown in Fig. 4.15.

In practice, the peak areas are measured by integrating the signal area and recording automatically as the NMR spectrum is recorded. Relative areas are shown Fig. 4.14.

a. Substituted Hydrocarbons

In the samples shown in Fig. 4.16, the hydrogens on the methyl groups are equivalent to each other. Recall that the multiplicity in the case of hydrogen spin–spin splitting is $n + 1$. In this example the number of equivalent hydrogens is 6. Hence the multiplicity of the hydrogen on carbon b is $6 + 1 = 7$. Also, since the two methyl groups are equivalent, they are split into one doublet by the hydrogen on carbon b. It should be noted that when an intervening methylene group is added to the molecule, as in methylethyl ketone, the methyl groups are no longer equivalent and act independently.

b. Aromatic Hydrocarbons

In the case of toluene (Fig. 4.17b), the coupling between the methyl group (CH_3) and the hydrogens on the aromatic ring is so weak that there is no apparent spin–spin splitting. The spectrum therefore appears to be two single peaks.

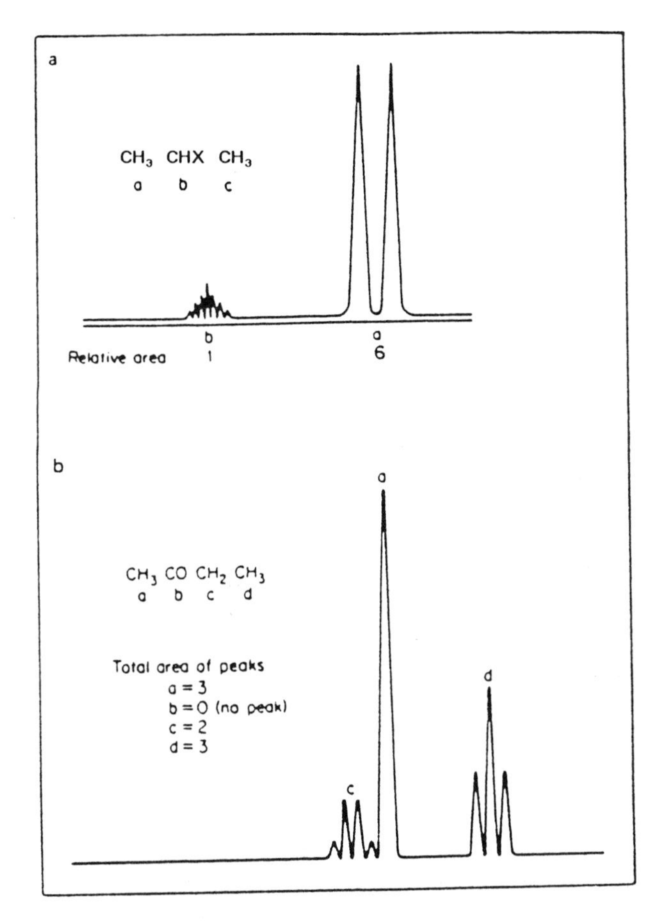

Figure 4.16 Typical NMR spectra of (a) a substituted hydrocarbon and (b) a ketone.

2. Analytical Applications

The term *analytical application* includes qualitative and quantitative analysis. It also includes the steps required to obtain data and evaluate them to understand certain aspects of the sample, such as its geometry or rate of chemical reaction. Some of the more useful applications of NMR are described in this section.

a. Qualitative Analysis

Nuclear magnetic resonance is particularly useful in the identification of organic compounds. The chemical shift indicates what types of hydrogens are present—e.g., methylene, methyl groups, olefins, aromatic compounds, and esters. The spin–spin splitting or multiplicity indicates what groups are next to each other in

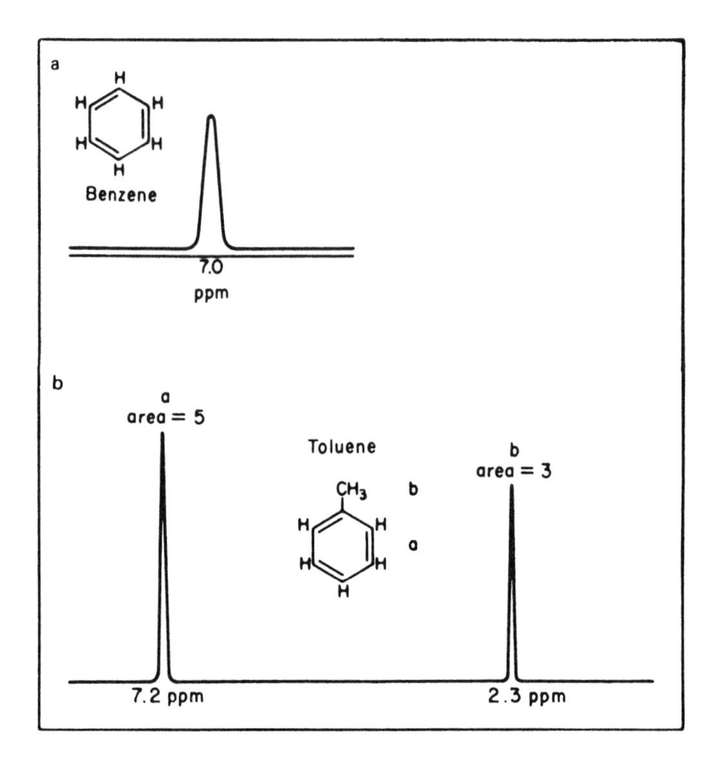

Figure 4.17 Typical NMR spectra of (a) benzene and (b) toluene.

the molecule. Another piece of information is obtained from the relative size or area of the absorption peaks in the spectrum, which tells us directly how many nuclei (or hydrogen atoms) are in each group. For example, in propane (CH_3—CH_2—CH_3) the ratio of the areas of methyl (CH_3) and methylene (CH_2) peaks would be 6:2; in butane (CH_3—CH_2—CH_2—CH_3) it would be 6:4.

Proper interpretation of the spectra should tell us much about the structure of an organic molecule. It does not tell us directly how much of that compound is present.

b. Quantitative Analysis

The number of protons (hydrogen nuclei) involved in absorption directly controls the area of the absorption peak (or peaks, if spin–spin splitting is involved). The total area of a methyl group (CH_3) will be 3/2 times as big as the total area of a CH_2 group. Mixtures of organic liquids can be analyzed quantitatively by using this fact.

For example, suppose that we have a mixture of *n*-octane and 1-octene. The combined spectrum is shown in Fig. 4.18. The formula for *n*-octane, which is

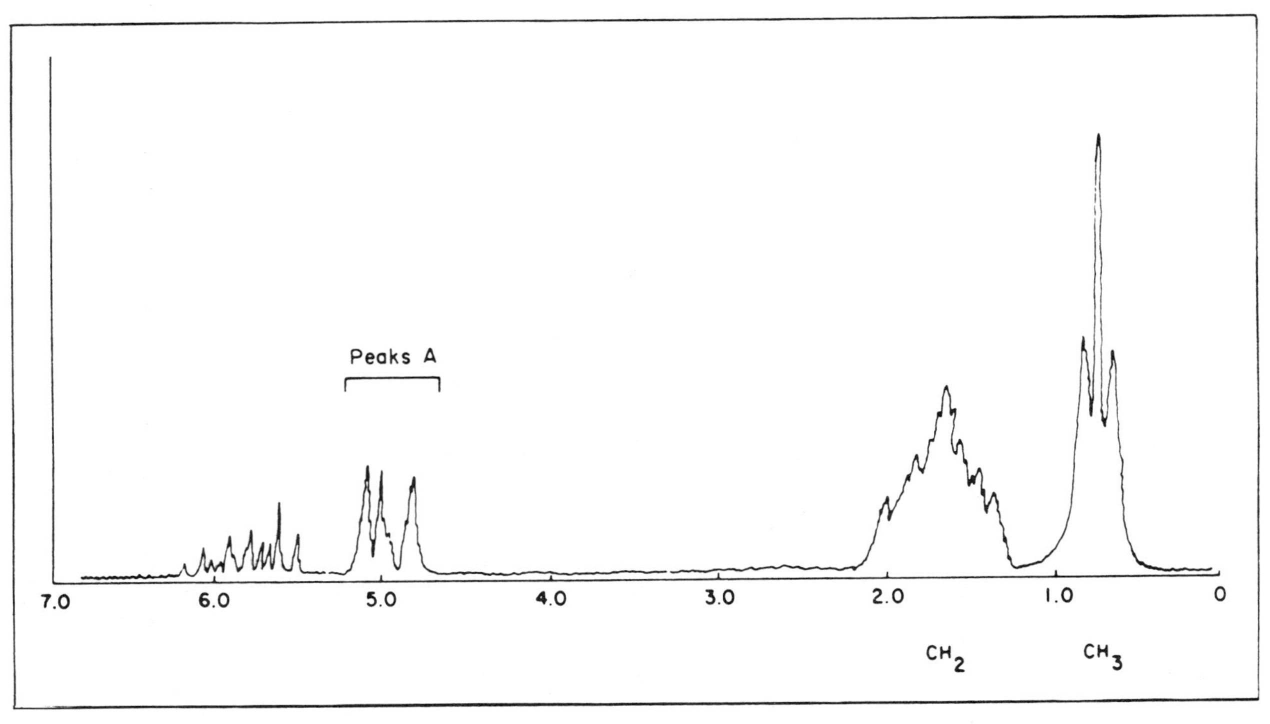

Figure 4.18 Spectrum of a mixture of *n*-octane and 1-octene. Olefin hydrogens on terminal carbons appear as peaks A.

composed of methyl and methylene groups and contains 18 protons, is

```
   H   H   H   H   H   H   H   H
   |   |   |   |   |   |   |   |
H--C - C - C - C - C - C - C - C--H
   |   |   |   |   |   |   |   |
   H   H   H   H   H   H   H   H
```

The formula for 1-octene is

```
   H   H   H   H   H   H        H
   |   |   |   |   |   |       /
H--C - C - C - C - C - C - C = C
   |   |   |   |   |   |   |    \
   H   H   H   H   H   H   H     H
```

It can be seen that there are two protons on the terminal carbon, which are olefinic in nature. These olefins will absorb at about 5.0 ppm because of the chemical shift. Quantitatively, they constitute 2 of 16 protons in the octene. If we measure the peaks at 5.0 ppm (peaks A), then the total area in the whole spectrum due to the presence of octene is equal to the area of peaks A times 8:

$$\text{area due to octene} = \text{area A} \times 8$$

$$\text{area due to octane} = \text{total area} - \text{octene area}$$

However, one molecule of octene contains 16 protons and one molecule of octane contains 18 protons. Hence, if these compounds were present in equimolar proportions, the ratio of the relative areas of the NMR absorption curves would be 16:18. A correction must be made for this difference in the final calculation. The mole ratio of octane to octene in the mixture would therefore be obtained as

$$\frac{\text{area of octane absorption curve} \div 18}{\text{area of octene absorption curve} \div 16}$$

that is,

$$\frac{\text{total area} - (\text{area of peak A} \times 8) \div 18}{\text{area of peak A} \times 8 \div 16}$$

From this result, the mole percent ratio of octane can be calculated by simply multiplying the ratio by 100.

Example. In an actual experiment involving a mixture of octene and octane, it was found that

$$\text{total area of all peaks} = 52 \text{ units}$$

$$\text{area of peak A} = 2 \text{ units}$$

From the data we find that

$$\text{area of octene protons} = 2 \times 8 = 16 \text{ units}$$

$$\text{area of octane protons} = 52 - 16 = 36 \text{ units}$$

$$\text{ratio of octane to octene} = \frac{36 \div 18}{16 \div 16} = \frac{2}{1}$$

Therefore the mole ratio of octane to octene is 2:1.

Rate of Reaction. In a normal chemical reaction, such as

$$C_8H_{18} \underset{\text{catalyst}}{\overset{\text{Pd}}{\rightarrow}} C_8H_{16} + H_2$$

one type of hydrogen (paraffinic) is consumed and another type (olefinic) is formed. Samples of reacting mixture can be taken at frequent intervals during the reaction, and the rate of disappearance or formation of the different types of hydrogens can be measured. The results can be used to measure the rate of the chemical reaction and the kinetics involved.

Solvents Used in NMR. Frequently the sample cannot be obtained in the pure state, or it may be in the solid or gas phase. It may be necessary to dissolve it or extract it from some other medium. In this case a solvent must be used. Several requirements must be met by a good solvent, including that it (1) be chemically inert toward the sample, (2) be easily recoverable if the original sample is required after examination, and (3) have a very simple NMR spectrum (or no absorption spectrum, whichever is preferable). This last requirement can be met by using solvents containing no hydrogen in the molecule.

Solvents that have been successful in the past include CCl_4, $CHCl_3$ (or better, $CDCl_3$, deuterated chloroform), C_6H_6, and D_2O (deuterated water). It should be noted that the deuterium nucleus does not respond in the same region as the hydrogen nucleus. Substitution of D for H in organic compounds frequently provides a means of eliminating the unwanted H signal.

c. Typical Analytical Applications

Perhaps the most important analytical application of NMR is in the determination of the structure of organic compounds synthesized or separated by organic chemists or of biological compounds isolated by biologists or biochemists. It is possible to determine quantitatively one type of organic compound in the presence of a different type, such as the percentage of alcohols in paraffins, amines in alcohols, aromatics in paraffin compounds, olefins in hydrocarbons or alcohols, organic halides in other organic compounds, and metallo-organic compounds in other organic compounds, or the number of side chains in a hydrocarbon. However, it is not possible to examine gaseous compounds directly. Only in a few cases can solids be analyzed directly. (An example of the latter is a polymer that has been soaked in a solvent to allow the molecules to rotate freely.) However, solids can be analyzed using magic angle NMR. It is also not possible to determine the molecular weight of a compound directly.

Nuclear magnetic resonance has found widespread use in the characterization of compounds containing hydrogen and some analytical use for compounds that contain fluorine or phosphorus. However, its analytical application to compounds not containing any of these elements is rare. Current research is aimed toward the use of ^{13}C for the elucidation of structure from the different types of carbons present. The work is handicapped by the low concentration of ^{13}C in the normal sample (1.1%).

Determination of Gasoline Octane Number by NMR. The octane number of gasoline is usually determined using a standard engine (similar to an auto engine.) The method is cumbersome, very time-consuming, and subject to operator error.

A paper has been published illustrating the determination of octane number directly by NMR. In this procedure, various types of H are measured quantitatively. A typical spectrum is shown in Fig. 4.19. The volume percentages are measured and the octane number can be calculated using an empirical formula. Such a procedure would be much faster, cheaper, and probably more reliable than the method currently in use.

3. Other Isotopes that Give NMR Signals

Table 4.1, relating mass A, atomic number Z, and the charge number of a nucleus, illustrates that all nuclei with an odd mass number have a fractional (1/2, 3/2, etc.) spin number. Also, all nuclei with an even mass number and an odd atomic number have a unit (1,2,3, etc.) spin number. In fact, only those nuclei with an even mass and an even atomic weight have a zero spin number and therefore give no NMR signal. Unfortunately, this includes ^{12}C and ^{16}O—two important nuclei in organic chemistry.

However, it can be readily seen that essentially all other elements have at least one isotope that can be examined by NMR. Of these, several have been studied extensively. The most important of these are listed in Table 4.4.

a. ^{13}C

There has been great incentive to develop ^{13}C-NMR because carbon is the central element in organic chemistry and biochemistry. However, useful applications were not forthcoming until the 1970s because of the difficulties in developing instrumentation.

Two major problems were involved. First, the ^{13}C signal was very weak because the abundance of ^{13}C was low (1.1% of the total carbon). Also, the gy-

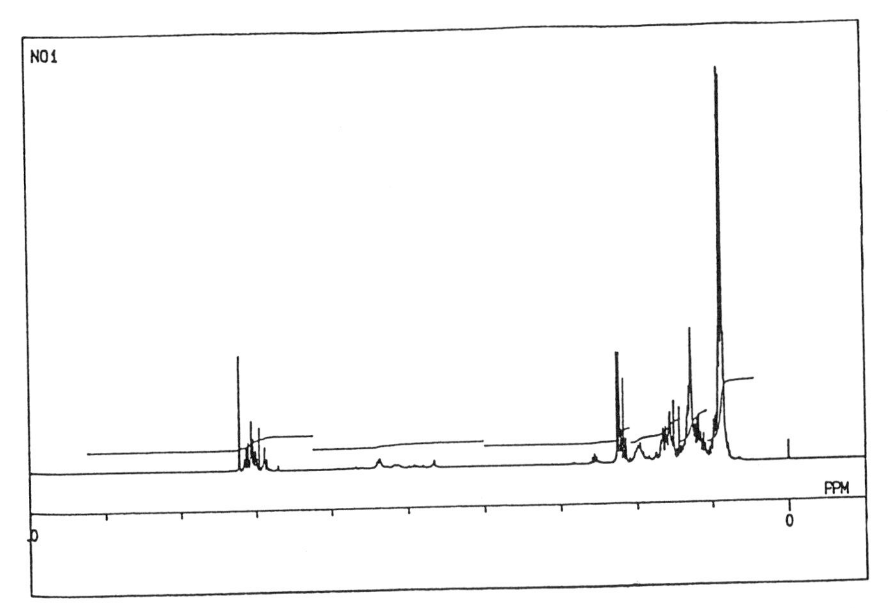

Figure 4.19 Spectrum of gasoline sample. (Proton NMR analysis of octane number for Motor Gasoline, by M. Ichikawa, N. Nonuka, H. Amano, I. Takada, S. Ishimori, H. Andoh, and K. Kumamoto. Applied Spectroscopy, 1992, 46, No. 2, p. 498).

Table 4.4

Nucleus	Natural abundance (%)	Spin	Sensitivity relative to ^{1}H
^{1}H	99.98	1/2	1.0
^{7}Li	92.6	3/2	0.3
^{13}C	1.1	1/2	0.0002
^{14}N	99.6	1	0.001
^{17}O	0.037	5/2	0.03
^{19}F	100	1/2	0.83
^{23}Na	15.9	3/2	0.09
^{25}Mg	10.1	5/2	0.003
^{27}Al	15.6	5/2	0.206
^{29}Si	4.7	1/2	0.008
^{31}P	24.3	1/2	0.07
^{33}S	0.8	3/2	0.002

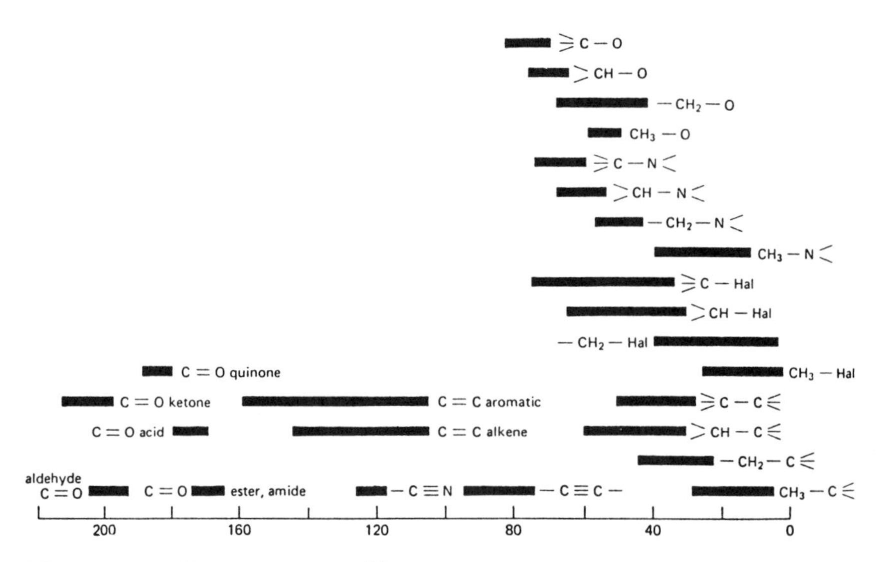

Figure 4.20 Chemical shifts for ^{13}C.

romagnetic ratio γ was low (0.25) compared to that for ^{1}H (1.0). The net result was a signal only 0.002 as intense as a comparative ^{1}H signal. Second, the chemical shift range was up to 200 ppm using TMS as a standard. This increased range precluded a simple "add on" to the ^{1}H NMR instruments already commercially available. However, the incentive to develop such instruments was great, and the information derived from them was very fruitful.

There are several advantages to ^{13}C NMR. First, the wide range over which chemical shift occurs greatly diminishes overlap between carbons in different chemical environments. Second, adjacent ^{12}C atoms do not induce spin–spin splitting, and the probability of two ^{13}C atoms being adjacent to each other is very low. Coupling between the ^{13}C and adjacent ^{1}H nuclei occurs, but there are techniques available to decouple these nuclei. Consequently, the ^{13}C NMR spectra are very simple. This enables interpretation of the spectra to be made easily. And, as we will see later, comparison of ^{13}C and ^{1}H spectra lead to data that can be interpreted with a high degree of confidence, thereby elucidating the structure of even very complicated molecules. Chemical shifts for ^{13}C are shown in Fig. 4.20.

Proton Decoupling. In proton decoupling, the sample is irradiated with a beam strong enough to saturate the ^{1}H nuclei [Eq. (4.9)]. The wavelength range is wide enough to saturate all the ^{1}H in the sample [Eq. (4.8)]. Under these conditions there is no coupling between the ^{13}C and any ^{1}H in the sample. This technique, known as proton decoupling, leads to a simplified, single-line ^{13}C spectrum.

Liquid samples are handled in much the same way as are samples in ^1H analysis. However, solid samples are much more difficult to handle. Line broadening arises because of the many orientations the different carbons have in a solid sample relative to the applied magnetic field. This can be reduced by using the magic angle technique. Proton decoupling further reduces the linewidth and increases its height. However, the relaxation time for ^{13}C is very long (several minutes), and this produces broad lines. A complicated technique called *cross polarization*, which will not be described here, can be used to reduce this effect.

In the final analysis, ^{13}C NMR is a very valuable characterization and will no doubt become increasingly important with time.

b. ^{19}F

Fluorine has a single isotope (19) with a spin of 1/2. In a magnetic field of 14.092 G at which ^1H resonate at 60 MHz, ^{19}F resonates at 56.4 MHz. Therefore, ^{19}F can be detected on instruments designed for ^1H NMR. However, the chemical shifts occur over a range of 300 ppm compared to 15 for ^1H. Therefore, to analyze ^{19}F compounds, some modification of a ^1H NMR instrument is necessary.

Interpretation of ^{19}F NMR is made more difficult by the fact that the solvent affects the resonance frequency and must be corrected for interpreting the spectra. It has been used for the analysis of fluorinated organic compounds and freon-type materials.

c. ^{31}P

Phosphorus is an important element in a number of biological fields. It is present in nerve tissue, enzymes, and many body tissues components. It is present in many insecticides, pesticides, fungicides, etc. It has been used in the studies of this phosphates, hydroxy methylamines and condensed phosphates.

4. Two-Dimensional NMR

Two-dimensional plots relating ^{13}C chemical shift on one axis and J spin–spin coupling constants on the other axis have been used to further simplify the interpretation of complex molecules. This technique involves a complicated pulsing sequence such as an RF pulse to excite the sample, a variable pause to permit some decay, a second pulse, followed by detection. This complicated sequence has generated two-dimensional data plots, which can be interpreted with greater certainty.

H. SOLID-STATE NMR TECHNIQUES

1. The Analysis of Polymers

Polymers today are used for a great variety of materials such as appliance, electrical and microelectrical telecommunications, aircraft, aerospace, automobiles, food packaging, and health care, to name a few. Adequate knowledge of these structures and their relationship to the performance of these polymers is crucial to their successful use. ^{13}C is the element most commonly examined, followed by ^{29}Si, ^{19}F, ^{15}N, and ^{1}H.

a. Magnetic Resonance Imaging (MRI)

Two areas of NMR have recently seen rapid progress. One is the development of techniques for the noninvasive imaging of the interiors of objects (magnetic resonance imaging, or MRI), which has been successfully utilized in the study of extruded polymers and foams, the bonding of adhesives and elastomers, and the study of spatial distributions of porosity in porous materials. It is likely that MRI will eventually become applicable to polymer processing. Another area of technical advancement, which at this point is fully developed, is the ability to obtain high-resolution NMR spectra for rigid materials.

b. Instrumentation

The NMR spectra of solids can be recorded on a modern Fourier transform (FT)-NMR spectrometer designed for solution samples. Only a few modifications are needed to enhance the resolution and the signal-to-noise ratio (S/N). These modifications include high-power decoupling, magic angle sample spinning (MAS), and cross polarization. Proton decoupling is a procedure in which the magnetic coupling between local protons and ^{13}C atoms is broken down by application of a strong field at the proton frequency. This greatly reduces peak width, particularly in ^{13}C NMR. Cross polarization is a pulsing technique that compensates for the slow relaxation times of ^{13}C. High-power proton decoupling is necessary for solids to eliminate ^{1}H-X (where X = ^{13}C, ^{19}F, or ^{29}Si) dipolar interactions. MAS is a device used to spin the sample at higher rotation frequencies (~5–15 kHz) around an axis forming an angle of 54.76° (magic angle) with the applied magnetic field. This technique increases the resolution observed in the spectrum for a solid by averaging the chemical shift anistropies to their isotropic values. Using a combination of these two techniques, considerable spectrum resolution can be achieved by reducing the resonance linewidths. Accessories for converting a standard NMR to one useful for solid-state studies are commercially available.

c. Morphology of Solid Polymers

Polymer chains may be ordered (crystalline) or random (amorphous) structures. In semicrystalline polymer systems both crystalline and amorphous regions are present. Their relative ratio depends on preparation conditions, such as temperature, time pressure, concentration, and stress. Proper quantitative characterization is essential because it affects the physical-mechanical properties.

The crystalline and noncrystalline phases of semicrystalline polymers can be distinguished by chemical-shift differences associated with them. It is possible to identify solid-state polymer chain conformations in each phase, that is, *trans*, *gauche*, helical, or random conformations. The results obtained through NMR methods are *complementary* to x-ray diffraction data.

However, one advantage of solid-state NMR over the diffraction method is the ability to provide solid-state conformational analysis for both amorphous and crystalline regions. For example, in polyethylene (PE), the chemical shift difference between the crystalline and amorphous resonance peaks is 2.36 ppm. Typical results show that PE chains in the crystalline regions are in the all-*trans* conformation. But in the amorphous region NMR results show 33% in gauche and 67% is *trans*.

d. Polymers on Surfaces and at Interfaces

Polymer composites or fiber-reinforced plastics (FRPs) have a wide range of applications in the aerospace and automotive industries. In these composite systems, silane-coupling agents are used to treat silica surfaces to improve bonding between silica and the matrix. High-resolution cross polarization magic angle (CP-MAS) NMR was used to observe structure, orientation, and silane-interface interactions of organo-silane coupling agents bound to silica surfaces. Recent applications of NMR show the changes in the molecular motion of polymer adsorbed on a silica surface compared to the bulk polymer resulting from hydrolysis of the coupling agent.

In a related study CP-MAS was able to detect polymer at a submonolayer level. Solid-state NMR is proving to be a powerful technique for the study of reactions at surfaces. For example, NMR has been used in catalysis studies for determining the structure of chemisorbed molecules and for monitoring changes occurring in those structures as a function of temperature.

Understanding the role of interfacial regions in semicrystalline polymers, polymer blends, polymer composites, and high-performance polymers enables suitable materials to be selected for industrial applications. Recently, ^{29}Si CP-MAS techniques were used to probe the interfacial regions of polyvinyl alcohol (PVOH) and silica. Solid-state NMR parameters were used to estimate the proximity of the backbone to the silica surface. This is probably the first example of probing the interfacial region of the polymer composite materials using such an elegant solid-state NMR technique.

e. Insoluble Polymers

Some new advanced materials may provide high performance under extreme mechanical, electrical, and environmental conditions: some of these polymers are solids, so characterization by high-resolution solution NMR is not possible. Such samples may be films, rods, powder, or fibers. Because of their amor phous nature, very little information can be obtained about their solid-state structures using other techniques. Solid-state NMR can reveal a correlation between the strength of polymers and the molecular chemistry of their three-dimensional network.

Elastomers are usually cross-linked in industrial applications. Solid-state NMR has been used to study the peroxide curing of both natural rubber and *cis*-polybutadiene to high cross-linking levels. New resonance peaks in these NMR methods have also been used to investigate sulfur vulcanization of natural rubber.

f. Phase Structure Analysis

The chemical shift of resonance peaks was used to identify crystalline and noncrystalline regions in polymer chain conformations. It is also useful to know polymer phase structure. For such studies, relaxation times have been studied. In multiphase systems this technique is particularly useful if there is no observable chemical shift difference among the phases. Each component of a multiphase system can be characterized using a relaxation parameter. The most common NMR parameters used in phase structure analysis are T_1 (spin–lattice relaxation time), T_2 (spin–spin relaxation time), T_{1p} (spin–lattice relaxation time in the rotating frame), T_{1p} (spin–lattice relaxation time in the rotating frame), and T_{DD} (dipolar dephasing time decay constant). These parameters are used to identify interfacial regions and defects in the crystalline regions.

g. Polymorphs

Thermal effects can be studied. For example, chemical shift difference between *a*-crystalline, *a*-quenched, and *b*-crystalline forms were observed in isotactic polypropylene. The chemical shift of ^{13}C is sensitive to changes in the form of the polymer.

h. Exposure to Radiation

It has been proposed that radiation breaks down amorphous regions more readily than other regions. This has been confirmed by solid-state NMR. Studies on mixtures of polymers have permitted quantitative measure of the composition of such mixtures. Other studies have been made on polyfluorinated compounds using ^{19}F and silyenes with ^{29}Si.

Another interesting application is the use of imaging to locate heterogeneous areas. This is done by soaking the sample in water and measuring the relaxation time, as in magnetic imaging used in medical diagnosis. The technique is nondestructive.

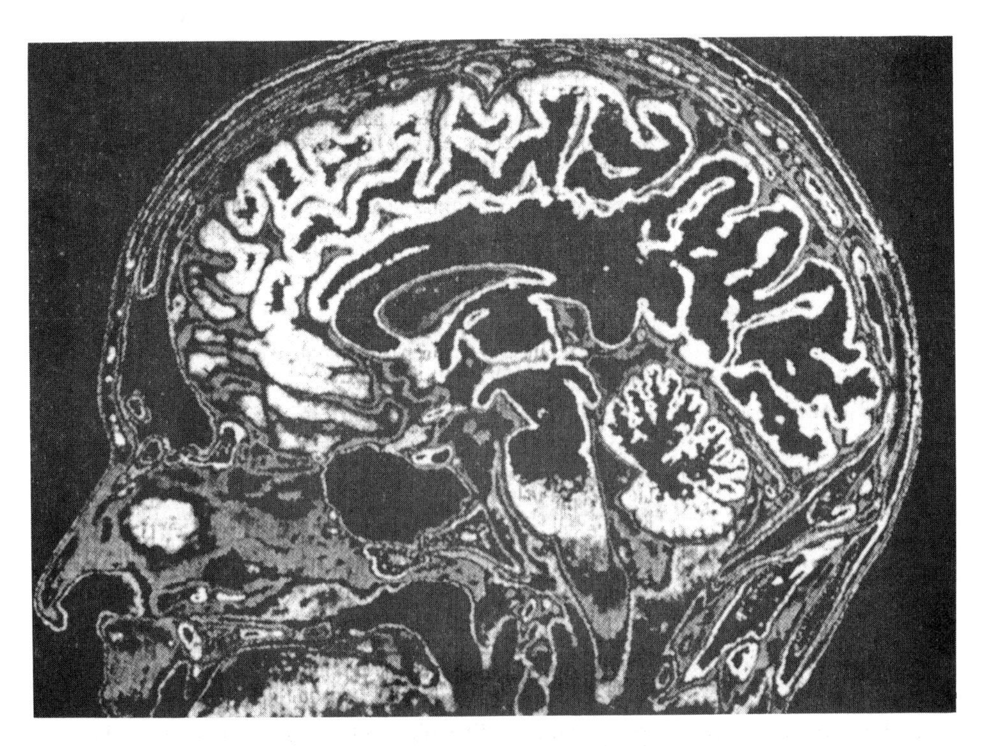

Figure 4.21 Magnetic resonance imaging of human head.

2. Medical Applications

MRI has found valuable applications in medicine. Assuming that cancerous cells are different from normal cells, the NMR properties of cancerous cells would be different from those of normal cells. This assumption has proven to be true and cancerous cells or tumors can be located by NMR. The method is noninvasive—no operation or cutting open is necessary. And since the method is nondestructive, it can be used on humans (Fig. 4.21).

MRI has been particularly useful for verifying and locating brain tumors. The technique not only locates, but gives the size and shape of the tumor. This is of immeasurable help in diagnosis and operating procedures.

X-rays have been used extensively for such studies in the past, but the contrast between different forms of tissue is low and difficult to read. Also, x-rays can have serious negative side effects.

In contrast, NMR—also a noninvasive procedure—has essentially no side effects. In this technique, a highly uniform magnetic field is used, but a magnetic gradient versus location is superimposed. ^1H nuclei therefore respond at different frequencies [see Eq. (4.8)] at different physical locations, thus locating the physical position of the nuclei.

By observing the NMR signal from numerous different angles simultaneously, the physical outline of the various body tissue components can be revealed. Any distortion or growth can then be located and measured in three dimensions. This has proven to be a very valuable noninvasive medical tool.

I. INTERPRETATION OF SPECTRA

NMR spectra depend on the chemical shift and spin–spin splitting.

Recall that the chemical shift is caused by the drifting, not orbiting or spinning, of nearby electrons under the influence of the applied magnetic field. It is therefore a constant depending on the applied field; i.e., if the field is constant, the chemical shift is constant. It is, therefore, an indication of the functional group involved, such as methyl, methylene, aldehydic H, alcohol, aromatics, etc. (Table 4.3).

As noted earlier, spin–spin splitting is caused by adjacent nuclei and is transmitted through the bonds. It is independent of the applied field. The multiplicity is therefore a function of the number of equivalent H nuclei in the adjacent functional groups. Numerically, it is equal to $(2nI + 1)$, where n is the number of equivalent H and I is the spin number (in this case 1/2). For two adjacent groups the number is $(2n\ 1 + 1)\ (2n\ ``I'' + 1)$ where n and n are the numbers of H nuclei in each separate group and I is the spin number of each.

Some typical spectra are shown in Figs. 4.22–4.35. All shifts are relative to tetramethylsilane (TMS), which is arbitrarily equal to zero.

J. ANALYTICAL LIMITATIONS OF NMR

The most serious problem with NMR is its lack of sensitivity. The minimum samples size that can be analyzed is about 0.1 ml, and usually about 0.5 ml is used. The minimum concentration that can be determined is about 1%. A second problem is that sometimes two different types of hydrogens absorb at similar resonance frequencies. This causes an overlap of spectra and makes interpretation difficult. Furthermore, in the characterization of organic molecules, no indication of molecular weight is given—only the relative number of different protons present. Finally, the method is most useful for liquids, but with special equipment some solids can be analyzed. In spite of these handicaps, NMR is one of the most useful analytical tools yet developed for the elucidation of many organic structural problems.

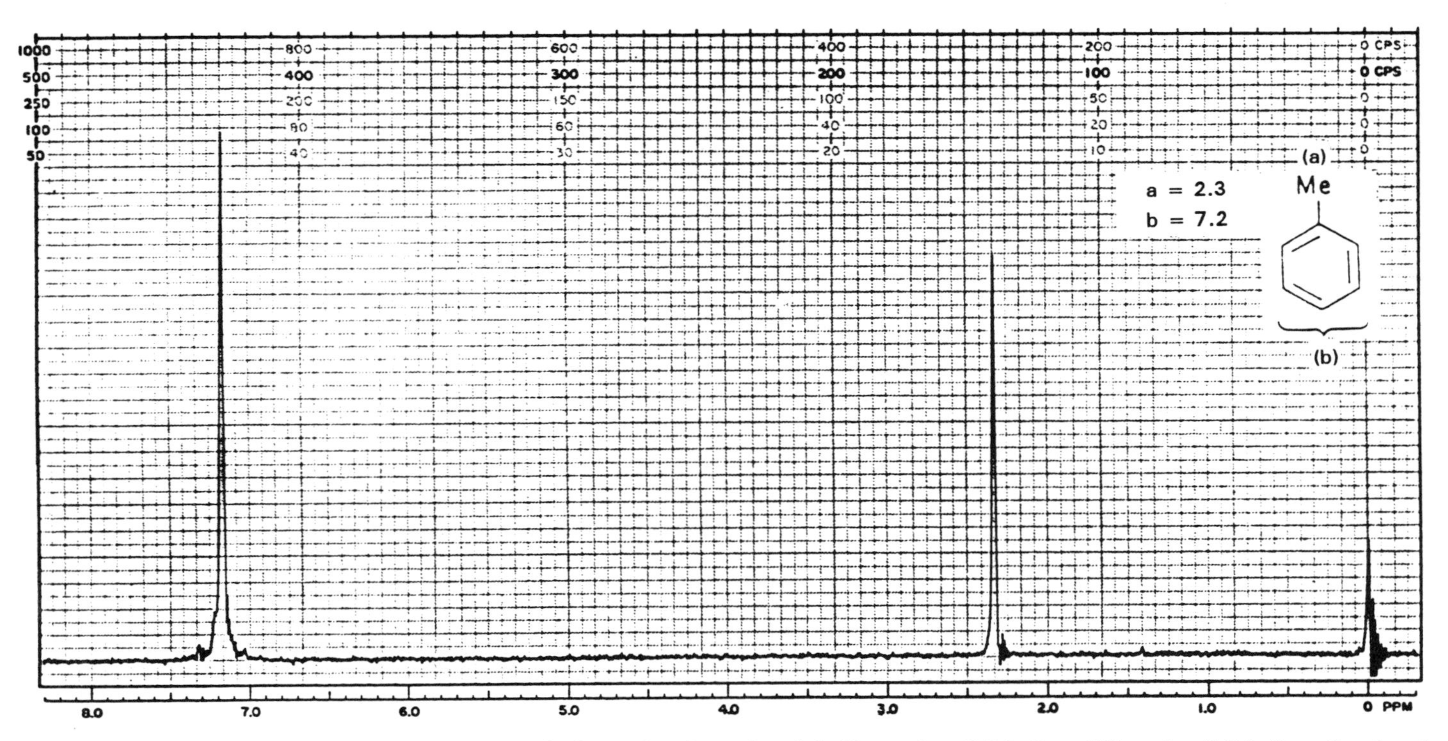

Figure 4.22 The spectrum of toluene shows two singlet peaks of area 3 and 5. The peak at 2.3 is from CH_3 and at 7.3 is from the phenyl hydrogens. Note no spin–spin splitting because the carbon to which the methyl is bonded has no hydrogens on it and spin–spin splitting is vanishingly small (not zero) when the groups are not immediately adjacent to each other.

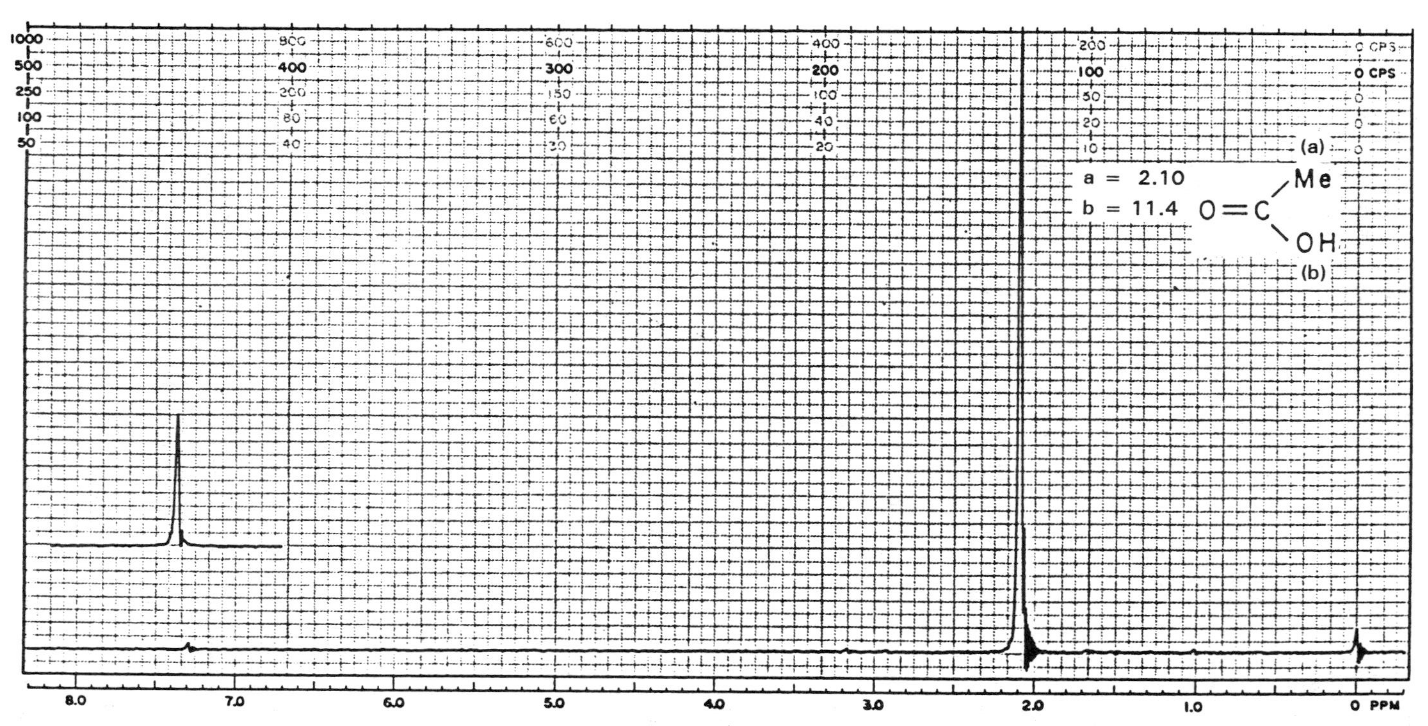

Figure 4.23 This spectrum of acetic acid shows a singlet at 11.4 (acid H) and the methyl at 2.1. Note no spin–spin splitting—the hydrogens are too far separated.

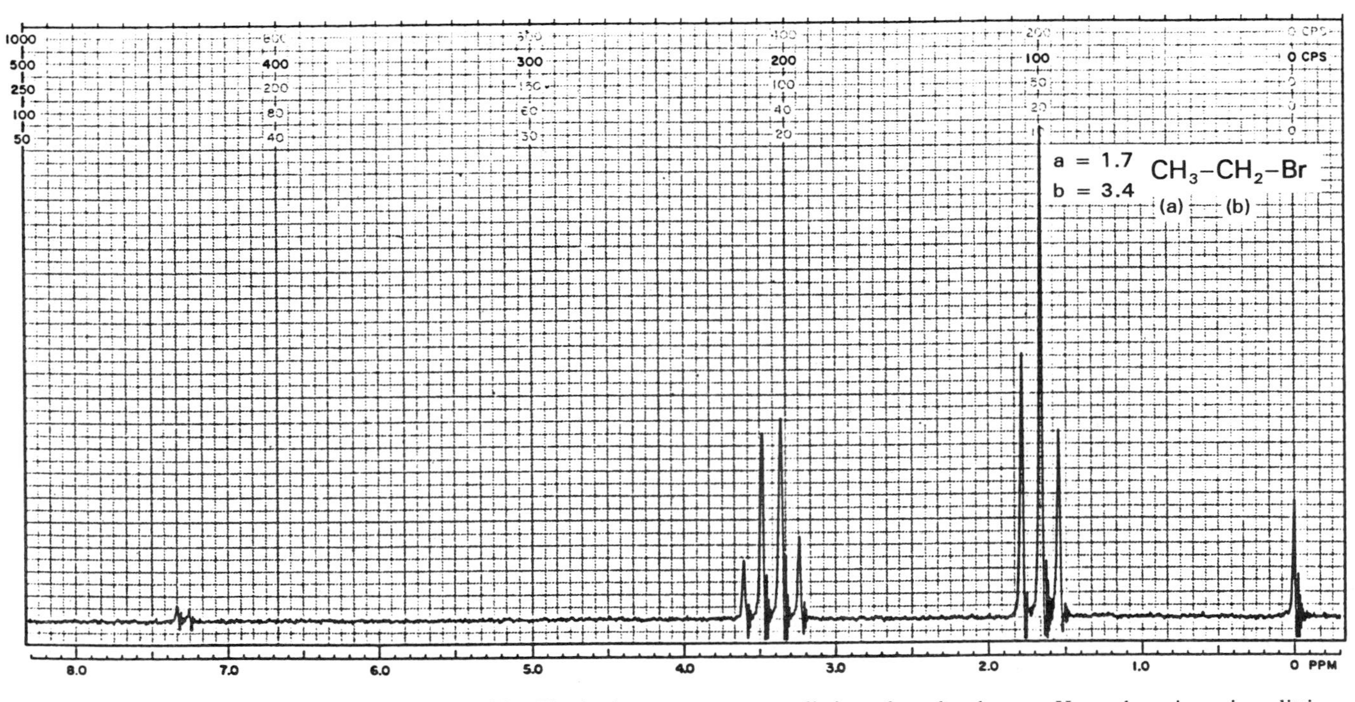

Figure 4.24 NMR spectrum for ethyl bromide. The hydrogens on a are split into three by the two Hs on b, spin–spin splitting = $2nI + 1 = 2 \times 2 \times 1/2 + 1 = 3$. The two Hs on b are shifted downfield by the Br and split into four by the three Hs on a, spin-spin splitting = $2nI + 1 = 2 \times 3 \times 1/2 + 1 = 4$.

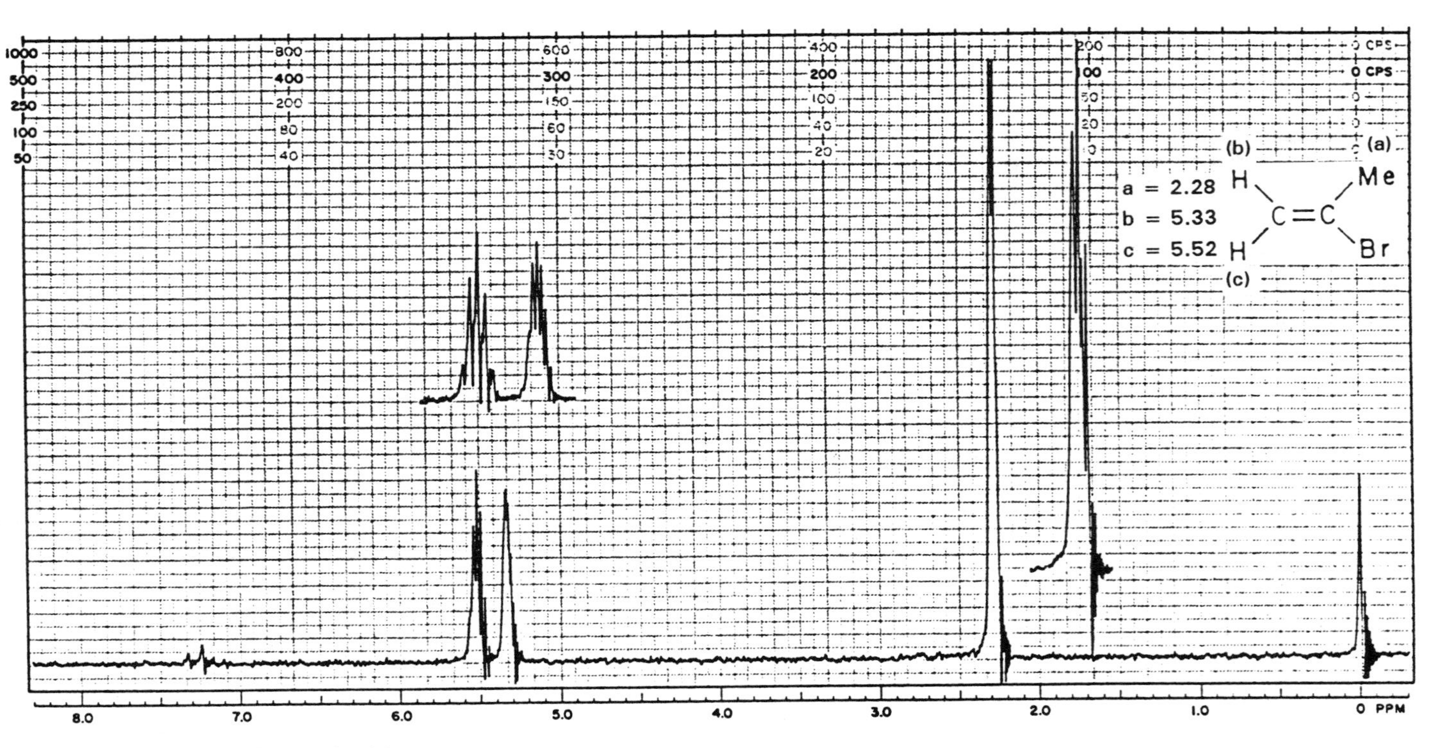

Figure 4.25 NMR spectrum for 2-bromopropene.

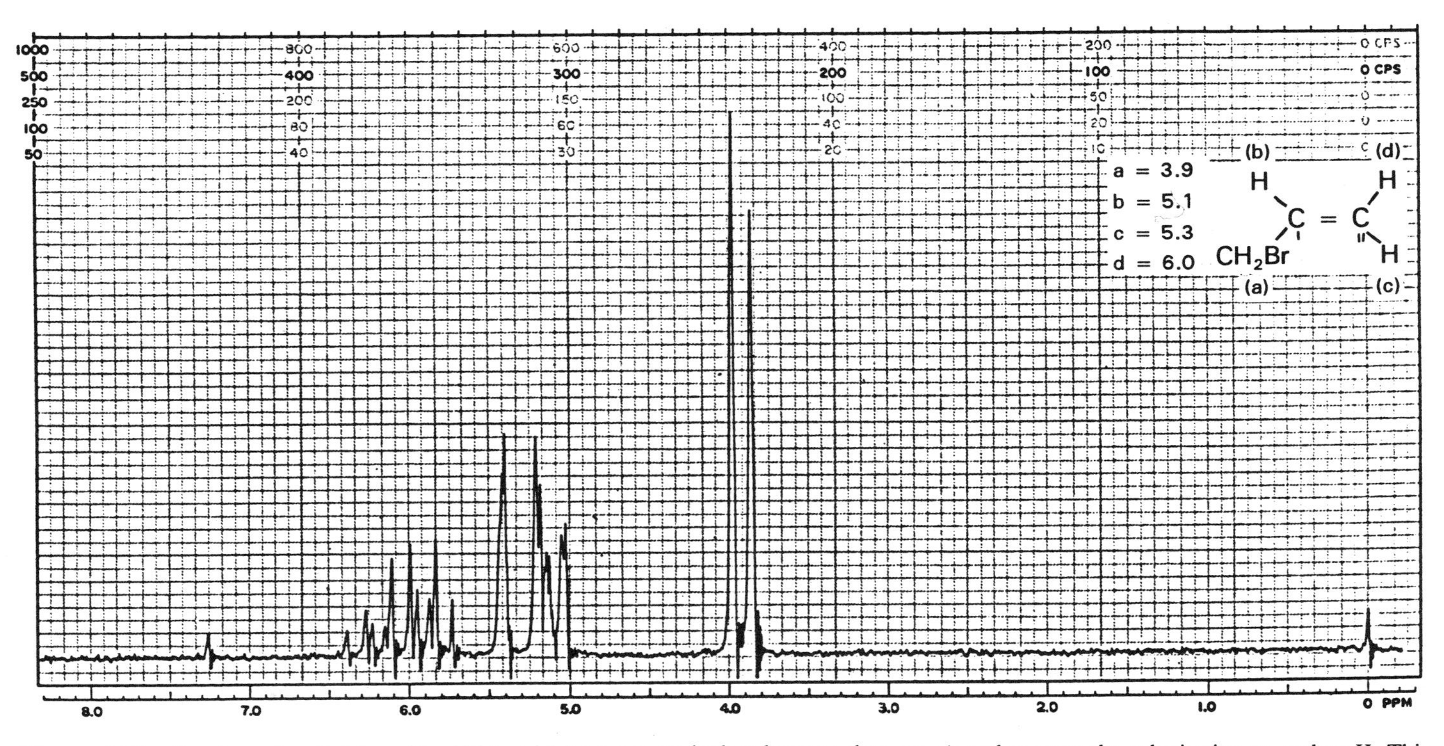

Figure 4.26 Spectrum for allyl bromide: H(b) and (c) are not equivalent because they are *cis* and *trans* to the substitution on carbon II. This leads to a different chemical shift and to separate spin–spin splitting. Again, the hydrogen on Carbon I are *cis* and *trans* and are not equivalent. Each is split into a doublet by H(d).

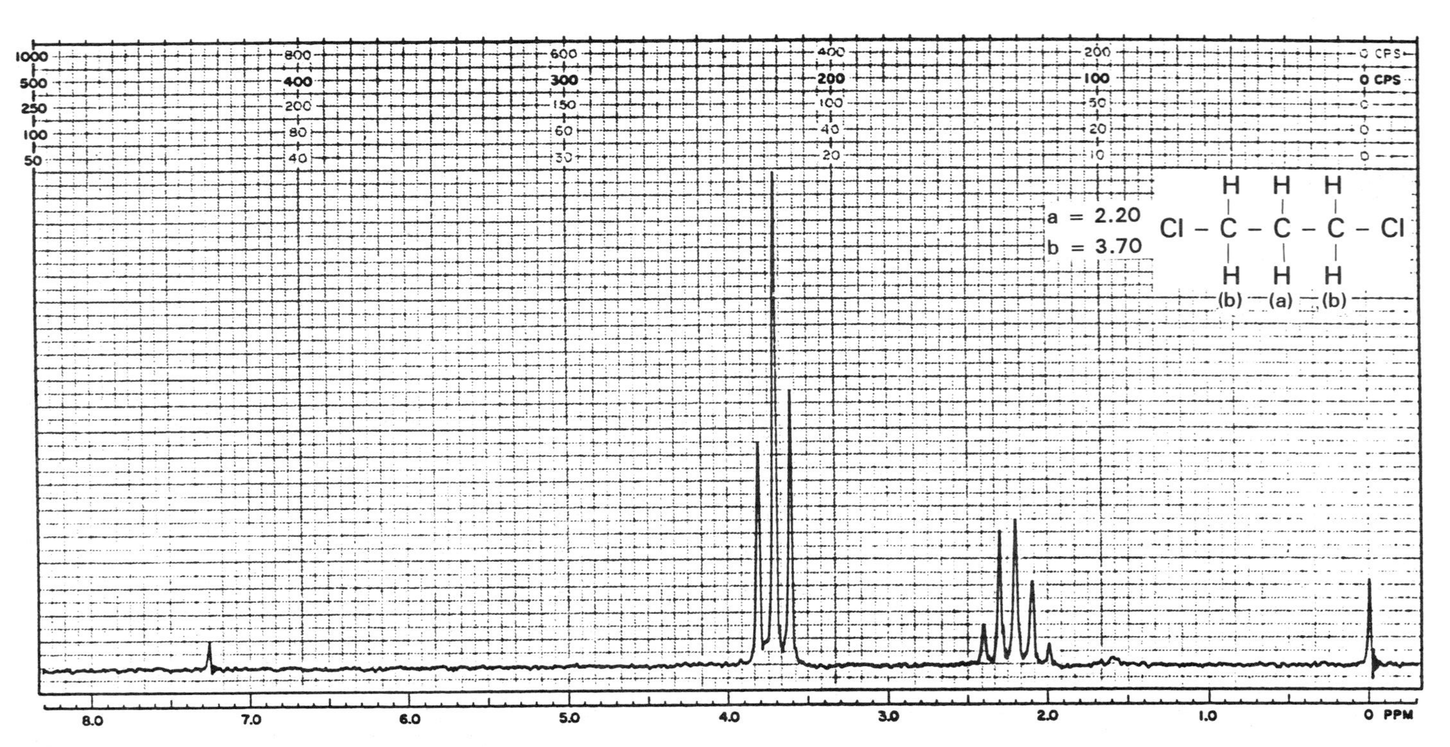

Figure 4.27 Spectrum for 1,3-dichlorpropane: H. The H on b are equivalent to each other, hence a is split into $2nI + 1 = 2x\ 4 \times 1/2 + 1 = 5$ peaks, whereas both b are split into $2nI + 1 = 2 \times 2 \times 1/2 + 1 = 3$ peaks.

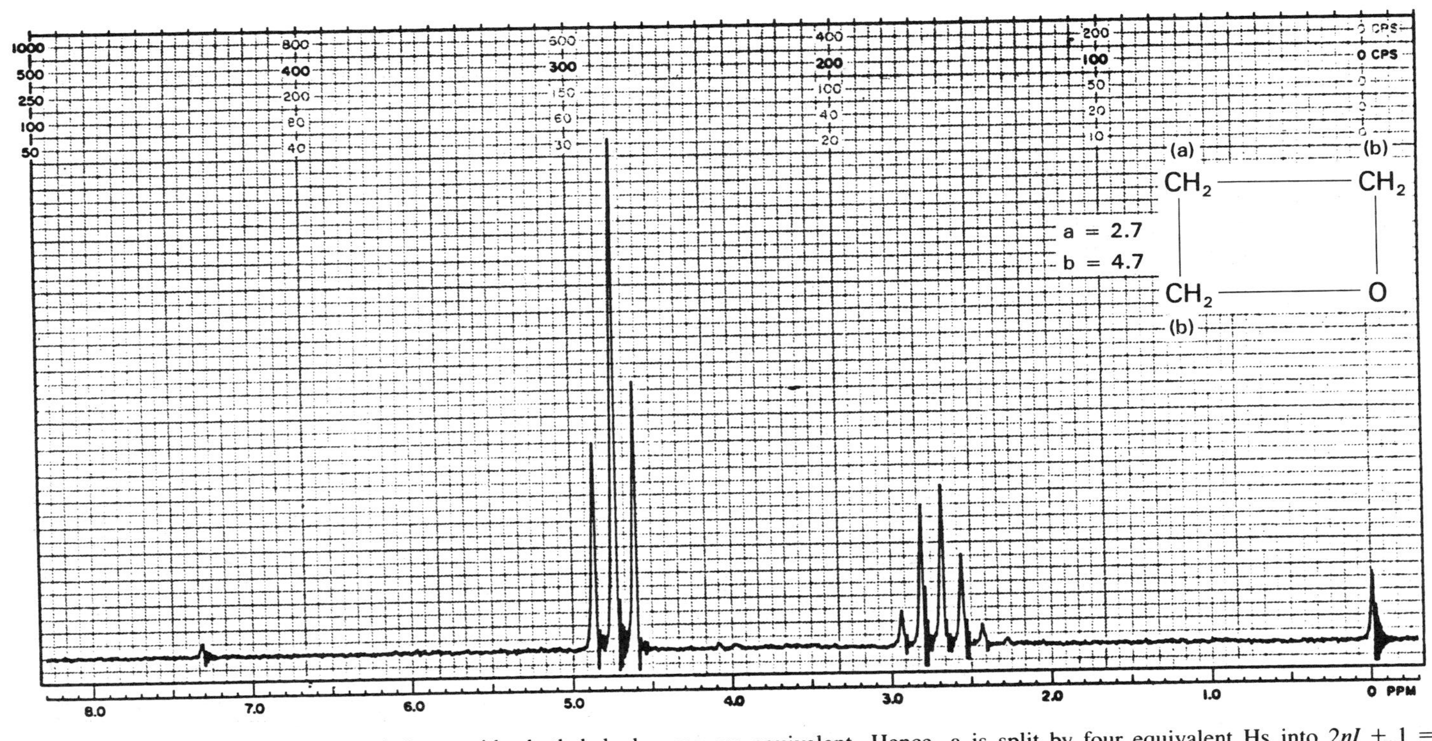

Figure 4.28 Spectrum for trimethylene oxide: both b hydrogens are equivalent. Hence, a is split by four equivalent Hs into $2nI + 1 = 2 \times 4 \times 1/2 + 1 = 5$, and both bs are split by a into $2nI + 1 = 2 \times 2 \times 1/2 + 1 = 3$ peaks.

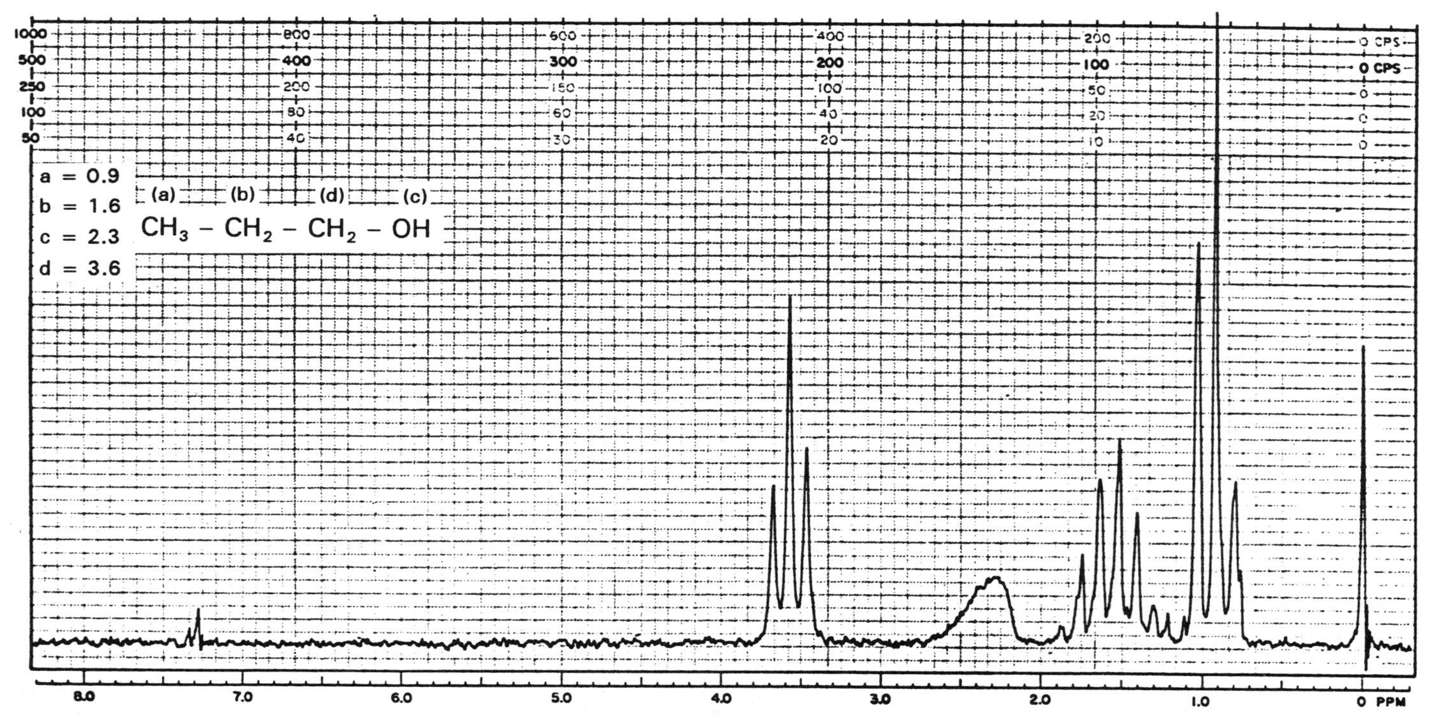

Figure 4.29 Spectrum for *n*-propyl alcohol: a is a triplet (b is 3H); b is 12 peaks (overlapping) (4×3 from a and d); d is here a triplet because the H on c (the OH) is exchanging so fast it does not split the adjacent H; c shows no spin–spin splitting because it is exchanging (probably with dissolved traces of water).

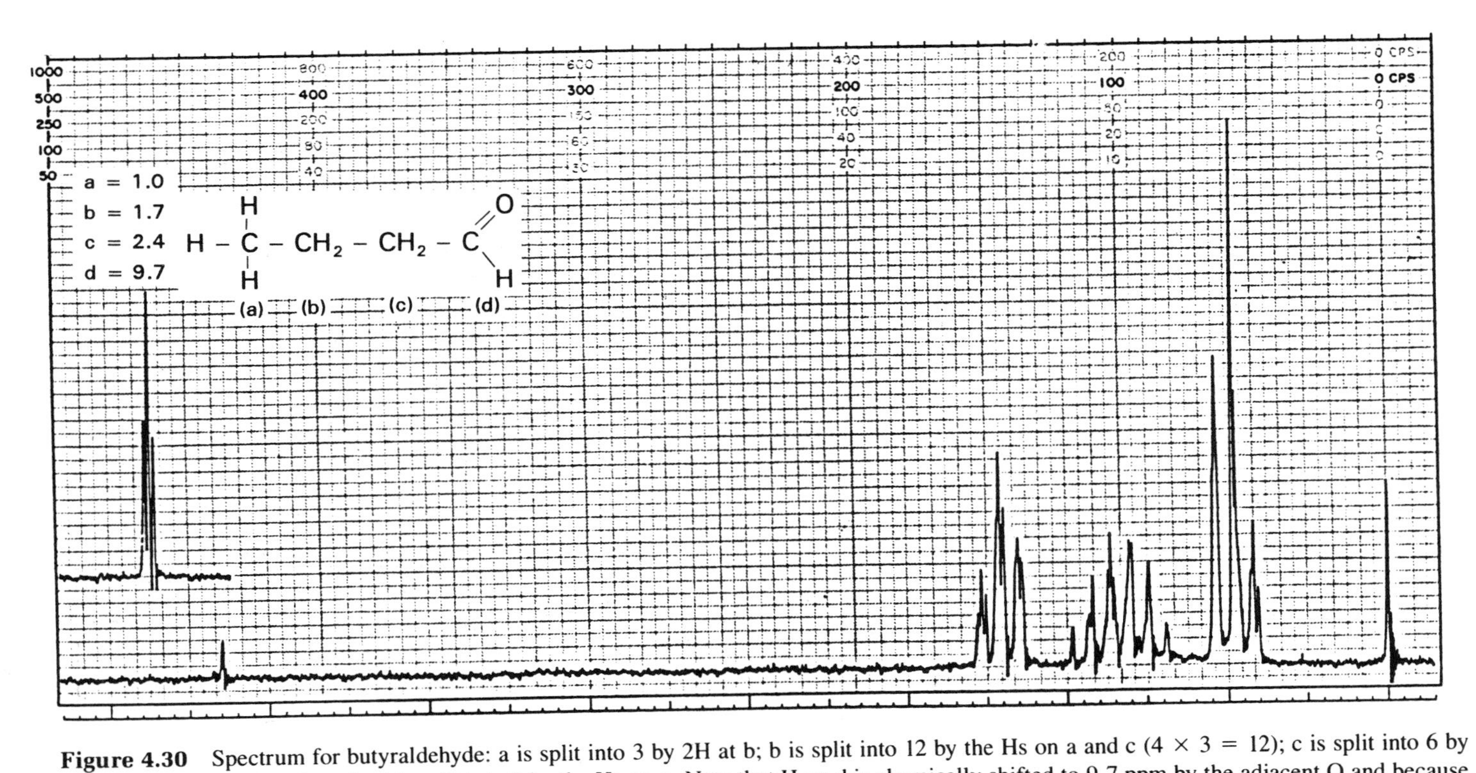

Figure 4.30 Spectrum for butyraldehyde: a is split into 3 by 2H at b; b is split into 12 by the Hs on a and c ($4 \times 3 = 12$); c is split into 6 by the Hs on b and d ($3 \times 2 = 6$); d is split into 3 by the Hs on c. Note that H on d is chemically shifted to 9.7 ppm by the adjacent O and because it is a labile H.

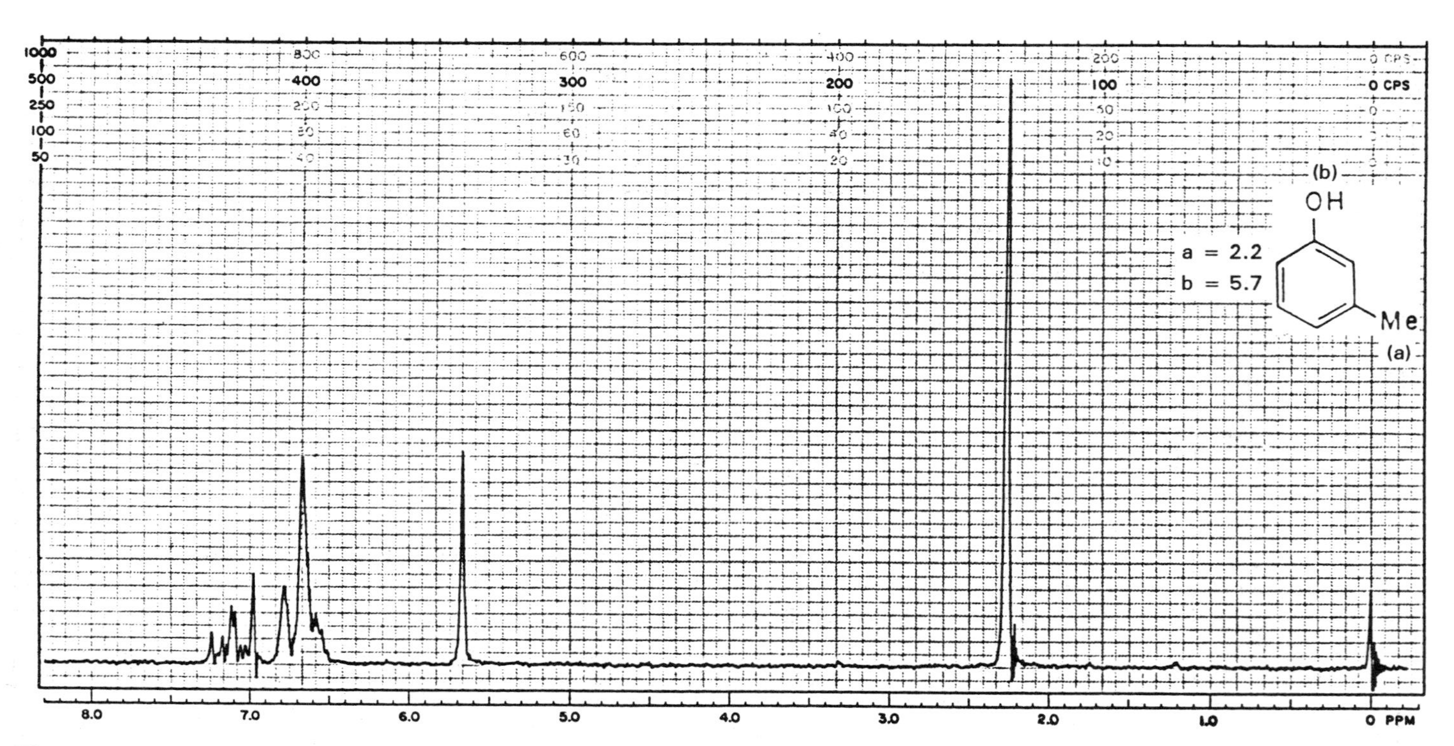

Figure 4.31 Spectrum for *m*-cresol: the spin–spin splitting on a and b is very small. However, the Hs on the ring are no longer equivalent and give rise to complex spectra between 6.7 and 7.2.

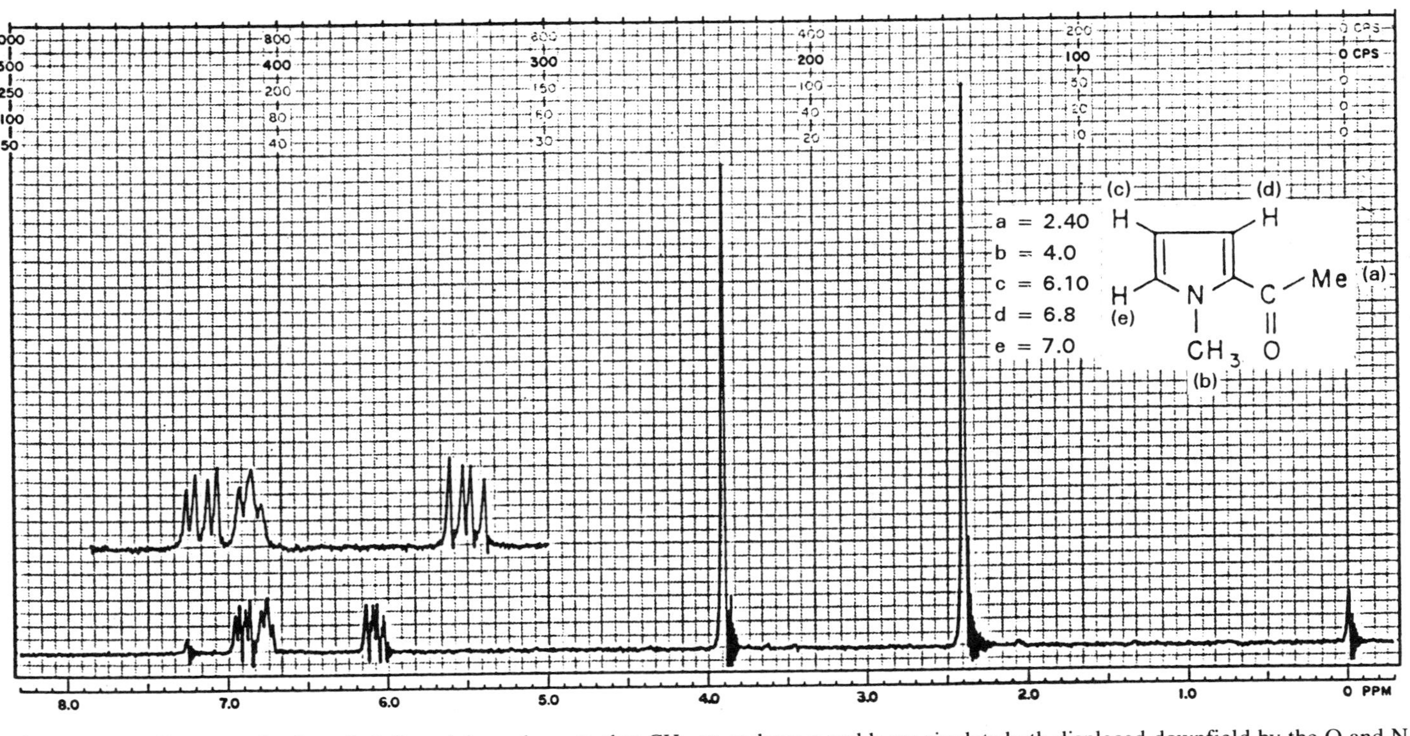

Figure 4.32 Spectrum for 1-methyl, 2-acetyl pyrole: note that CH_3 on carbons a and b are singlets both displaced downfield by the O and N, respectively. The hydrogens on carbons c, d, and e are not equivalent.

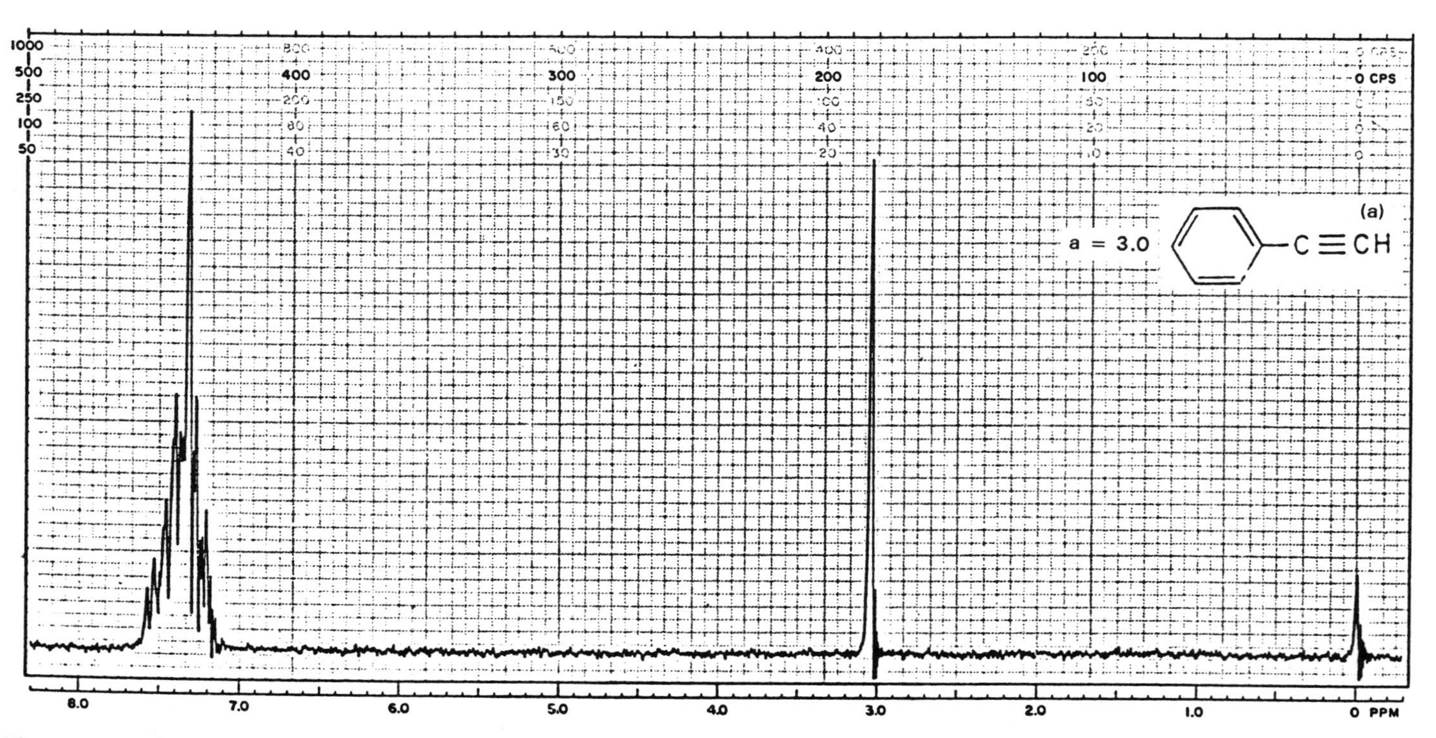

Figure 4.33 Spectrum for phenyl acetylene: note the hydrogens on the benzene ring are no longer equivalent, but are either *o*, *m*, or *p* to the substitution. This leads to complex multiplicity.

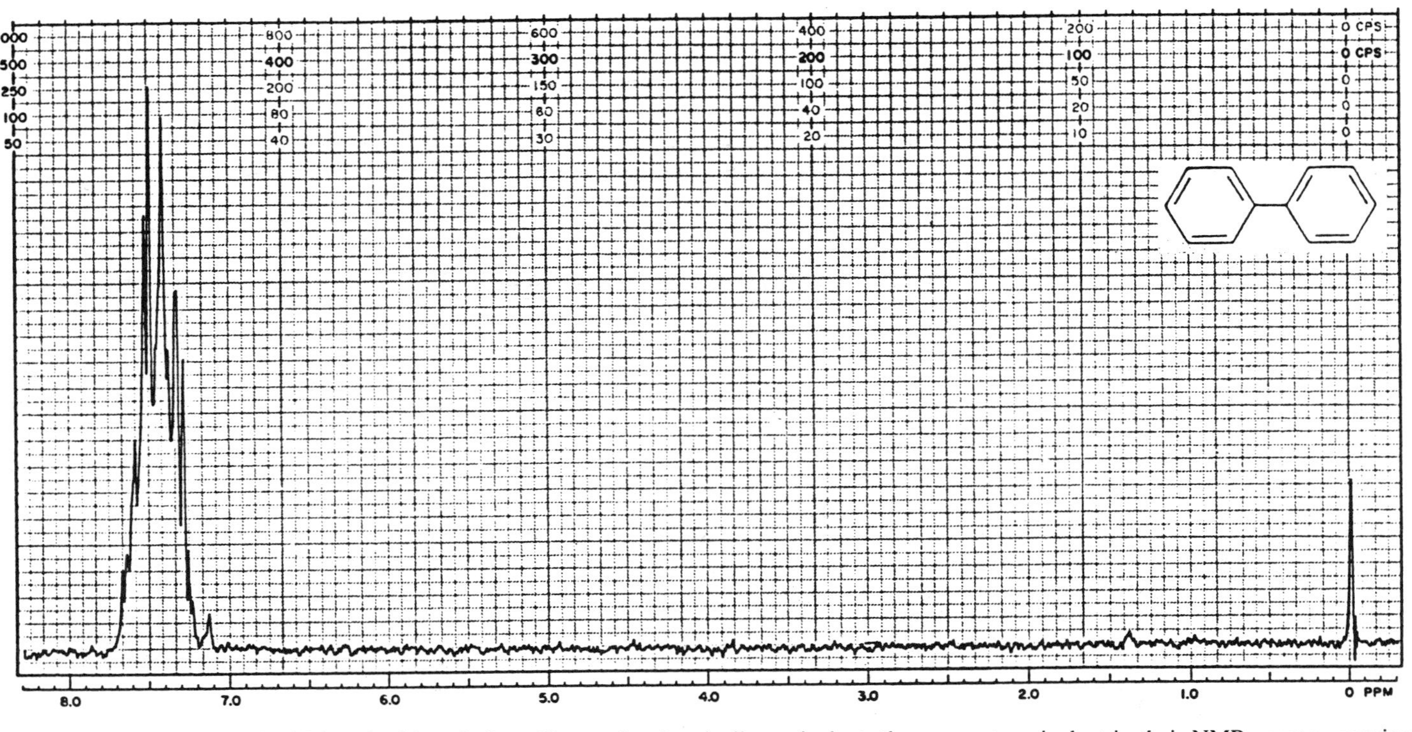

Figure 4.34 Spectrum for biphenyl: although these Hs may be chemically equivalent, they are not equivalent in their NMR spectra, causing complex multiplicity.

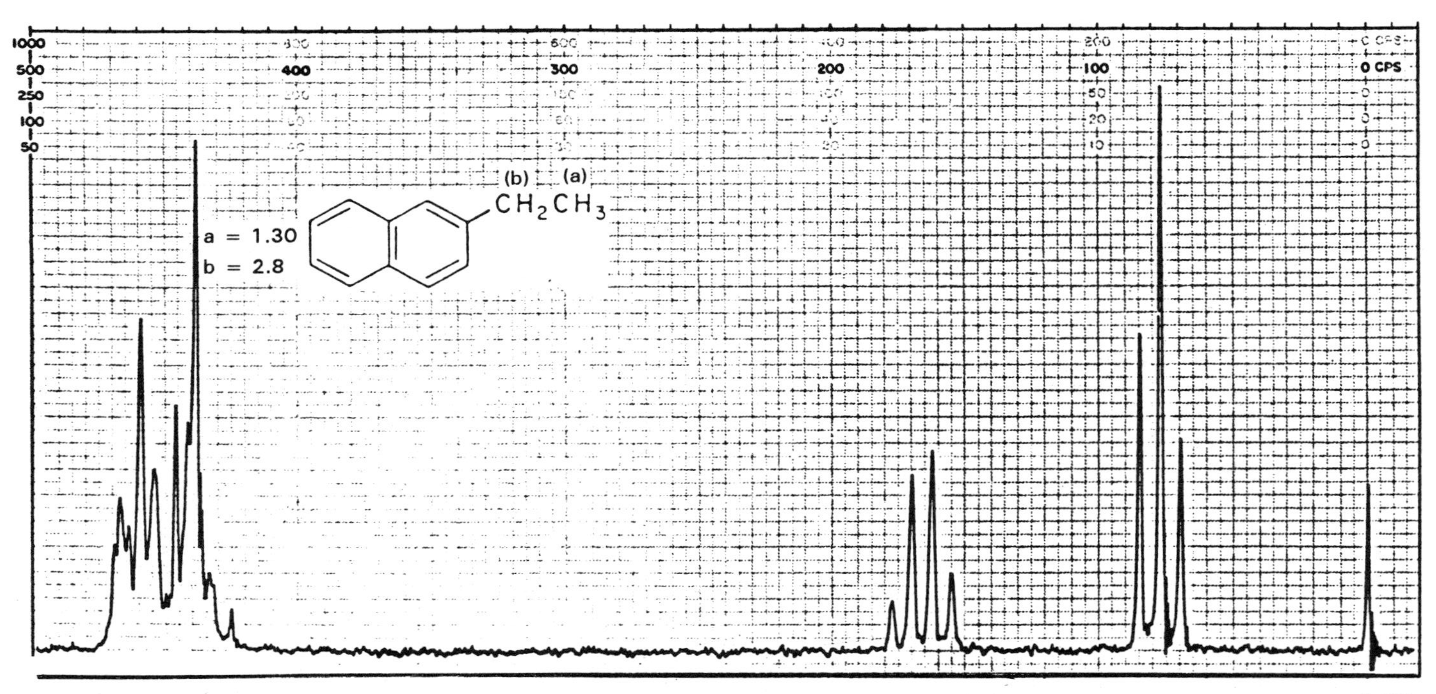

Figure 4.35 Spectrum for β-ethyl naphthalene: the multiplicity of a is $2nI + 1 = 2 \times 2 \times 1/2 + 1 = 3$ and of b is $2nI + 1 = 2 \times 3 \times 1/2 + 1 = 4$. The single substitution on the aromatic ring structure makes the rings not equivalent and the H on the rings not equivalent, leading to a complex spectra between 7 and 8 ppm (aromatic region).

BIBLIOGRAPHY

1. *Anal. Chem.*, Analytical Reviews, June 1993.
2. *Anal Chem.*, Application Reviews, April 1993.
3. Ando, I., Yamanobe, T., and Asakura, T., *Prog. NMR Spectrosc.*, (1990), *22*:349.
4. Choli, A. L., *Solid State NMR Spectroscopy*, 1992.
5. Creswell, C. J., and Runquist, O., *Spectral Analysis of Organic Compounds*, Burgess, Minneapolis, 1970.
6. J. N., Shoolery, NMR spectroscopy in the beginning, *An. Chem.*, (1993), *65*:731A.
7. Janusa, M. A., Wu, X., Cartledge, F. K., and Butler, L. G., Solid state deuterium NMR spectroscopy of d_5 phenol in white Portland cement. A new method for assessing solidification/stabilization, *Eng. Sci. Tech.*, (1993), *27*:1426.
8. Laupretre, F., *Prog. Polym. Sci.*, (1990), *15*:425.
9. NMR in industry, *Applied Spec. Rev.*, (1993), *28*(3):231.
10. Skloss, T. W., Kim, A. J., and Haw, J. F., High-resolution NMR process analyzer for oxygenates in gasoline, *An. Chem.*, (1994), *66*:536.
11. Silverstein, R. M., Bassler, G. C., and Morrill, T. C., *Spectrometric Identification of Organic Compounds*, 4th ed., Wiley, New York, 1981.
12. Shoolery, J. N., Spectroscopy in the beginning. *Anal. Chem.*, (1993), *65*(17):731A.
13. Non-medical application of NMR imaging, *An. Chem.*, (1994), *65*:1068A.
14. Wendesch, D. A. W., *Applied Spec. Rev.*, (1993), *28*(3):165.
15. What's new in instrumentation, *Spectroscopy*, (1993), *8*(4).
16. Yu, T., and Guo, M., *Prog. Polym. Sci.*, (1991), *15*:825.

SUGGESTED EXPERIMENTS

4.1 Demonstrate in the laboratory the principal components of an NMR instrument. Indicate the steps taken in tuning the instrument (e.g., choice of standard, ringing, sweep speed, and integration).

4.2 Obtain the NMR spectrum of a straight-chain paraffin, such as *n*-octane. Identify the methyl and methylene peaks. Note the chemical shift and the spin–spin splitting. Measure the total area of the methyl and methylene peaks and correlate this with the number of methyl and methylene protons in the molecule.

4.3 (a) Obtain the NMR spectrum of ethyl alcohol (1) wet and (2) very dry.
(b) Identify the methyl, methylene, and alcohol hydrogens. Note the chemical shift and spin–spin splitting.
(c) Measure *J*, the coupling constant, between the methyl and the methylene hydrogens. Note the effect of water on the alcohol peak. Explain this phenomenon.

4.4 Integrate the peak areas obtained in Experiment 4.3. Measure the ratios of the areas and the numbers of hydrogen nuclei involved in the molecules. What is the relationship between the area and the number of hydrogen nuclei?

4.5 Obtain the NMR spectrum of a mixture of ethanol and hexane. Integrate the peak areas of the different parts of the spectrum. From the ratio of the alcoholic hydrogen to the total hydrogen, calculate the percentage of ethanol in the mixture.

4.6 Obtain the NMR spectra of organic halides, olefins, aromatics, and substituted aromatics. Correlate the peaks with the different types of hydrogens present.

4.7 Identify an unknown compound from its NMR spectrum. The difficulty of the problem should match the level of competence of the student. A compound used in Experiment 4.2, 4.3, or 4.4 would be suitable for beginners, and a more complex compound for more advanced students.

4.8 Obtain the NMR spectra of (a) hexane and (b) heptane. Integrate the areas under the peaks in each spectrum. Would it be possible to distinguish between these compounds based on their NMR spectra? Compare Experiments 5.1 and 5.2.

4.9 Record the NMR spectrum of a sample of cooking oil. Measure the ratio of hydrogen in unsaturated carbon and that in saturated carbons. Compare the degrees of unsaturation of various commercial cooking oils.

4.10 Repeat Experiment 4.9 using margarine as the sample. First dissolve a known amount of margarine in CCl_4 and obtain the NMR spectrum. Compare the degrees of unsaturation of different brands of margarine.

PROBLEMS

4.1 What is chemical shift? How is it measured relative to tetramethylsilane?

4.2 (a) What is spin–spin splitting? (b) A methylene group (CH_2) is adjacent to a CH group. Into how many peaks is the CH_2 peak split by the adjacent single hydrogen?

4.3 How is the applied magnetic field B_0 varied in practice?

4.4 Indicate what spectrum you would expect from each of the following compounds.

(a) *n*-Butane (b) Tetramethylmethane

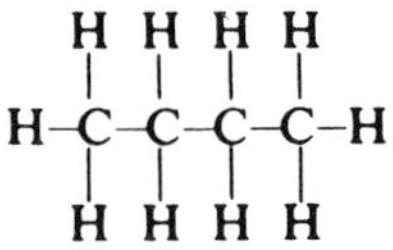

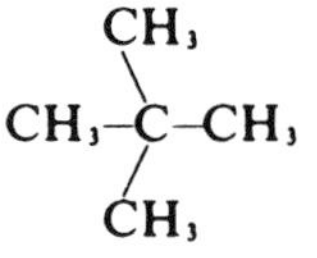

```
  H H H H            CH3
  | | | |             \
H-C-C-C-C-H      CH3-C-CH3
  | | | |             /
  H H H H            CH3
```

4.5 What spectrum would you expect for each of the following compounds?

(a) Benzene (b) Benzaldehyde

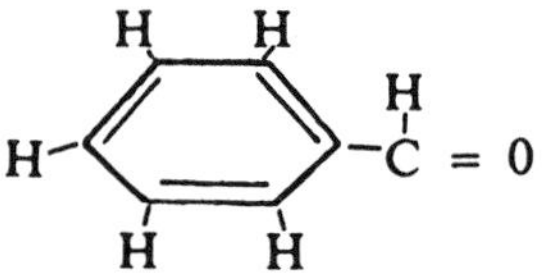

4.6 The total area of the peaks in

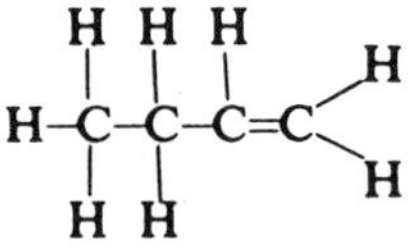

```
  H H H
  | | |    H
H-C-C-C=C
  | |      H
  H H
  a b c d
```

is 80 units. What will be the area of the peaks caused by (a) the CH_2 group on carbon b, (b) the CH group on carbon c?

4.7 Draw a schematic diagram of an NMR instrument.

4.8 Indicate the spectrum you would expect for ethanol (CH_3CH_2OH), a paraffin, types of olefins (terminal, internal substituted) in mixtures of olefins, and the HDO present in a sample of D_2O. In general, NMR can be used for the characterization of mixtures of organic compounds and for the elucidation of the structure of previously unidentified compounds.

4.9 What is the spin angular momentum of a nucleus?

4.10 What is the nuclear magnetic momentum of a nucleus? In the Stern–Gerlach experiment, what is the energy difference between the quantized energy levels?

4.11 What is the frequency of absorption of a nucleus in an applied magnetic field (a) in radians, (b) in Hertz?

4.12 What are the rules for determining the spin number of an element?

4.13 Which of the following have a spin number = 0?
^{12}C, ^{13}C, ^{16}O, ^{17}O, ^{1}H, ^{2}H, ^{19}F, ^{28}Si, ^{35}Cl, ^{108}Ag, ^{96}Mo, ^{66}Zn ^{65}Zn

4.14 What is the spin number for (a) ^{1}H, (b) ^{2}D? What is the number of orientation for (a) ^{1}H, (b) ^{2}O in a magnetic field (spin number = $2I + 1$)?

4.15 What causes the chemical shift in a magnetic field?

4.16 How is the chemical shift measured?

4.17 How is the chemical shift defined?

4.18 What are the requirements for a standard reference material?

4.19 What causes spin–spin splitting?

4.20 How many peaks are the fine structure caused by spin–spin splitting of (a) adjacent nuclei, (b) two sets of adjacent nuclei?

4.21 What would be the spin–spin splitting caused by adjacent CH_3, CH_2, CF_3, CF_2, CH, and CFH.

4.22 What is (a) transverse relaxation, (b) longitudinal relaxation?

4.23 What else causes (a) relaxation, (b) O_2, (c) quadropoles?

4.24 How does the relaxation time affect linewidth?

4.25 What is the gyro magnetic ratio?

4.26 What is the chemical exchange rate at the temperature at which multiplicity is lost?

4.27 What is the effect of viscosity on T_1.

4.28 Draw a schematic drawing of an NMR.

4.29 What is saturation?

4.30 What is the saturation factor Z_0?

4.31 If $I = 1/2$, what is Z_{min}?

4.32 What are the requirements for solvents used for NMR studies of organic compounds?

4.33–4.45

Provide the structures for the following empirical formulas or molecular weight (pp. 152–164):

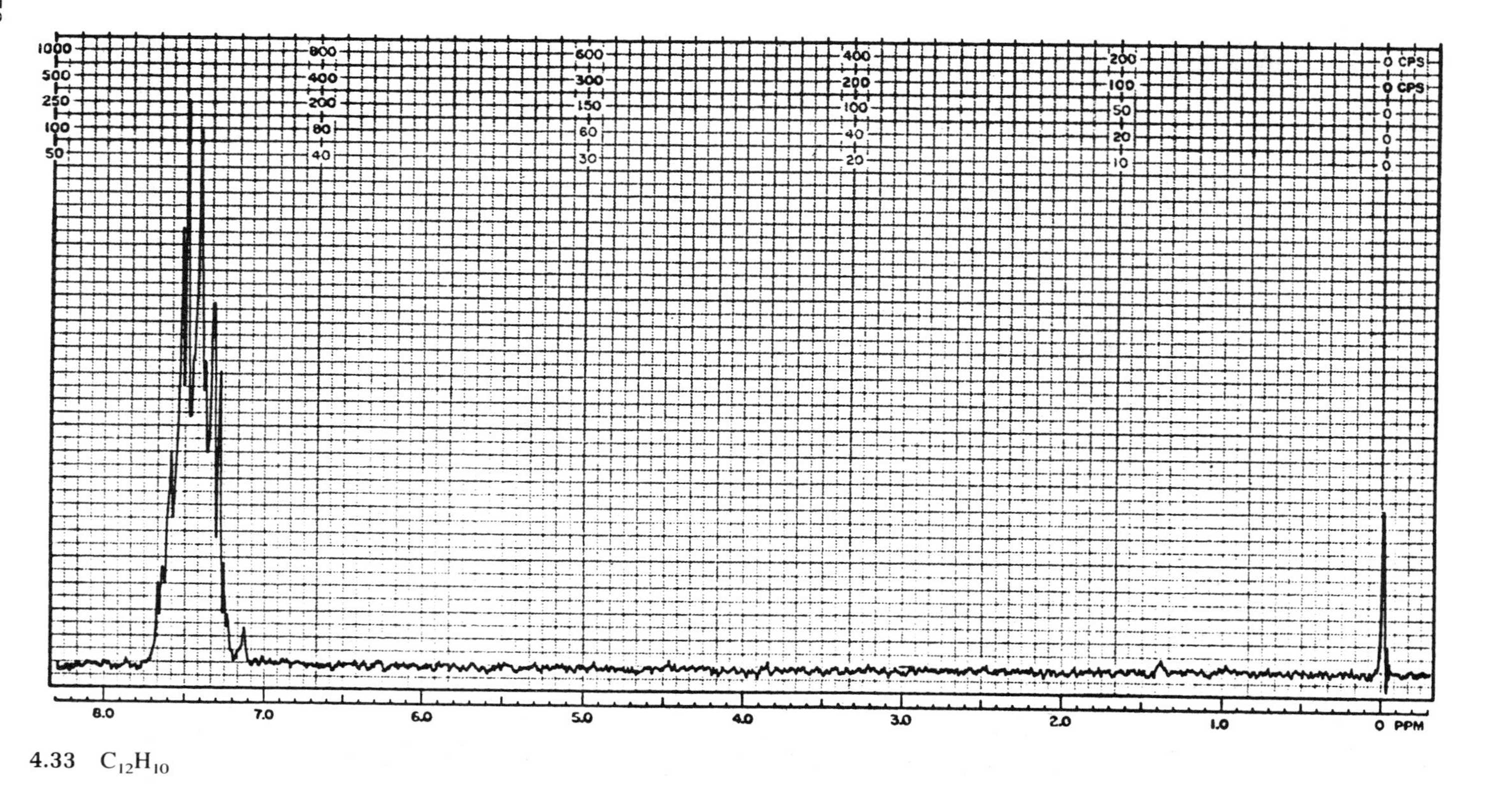

1000
500
250
100
50
800
400
200
80
40
600
300
150
60
30
400
200
100
40
20
200
100
50
20
10
0 CPS
0 CPS
0
0
0
8.0
7.0
6.0
5.0
4.0
3.0
2.0
1.0
0 PPM

4.33 $C_{12}H_{10}$

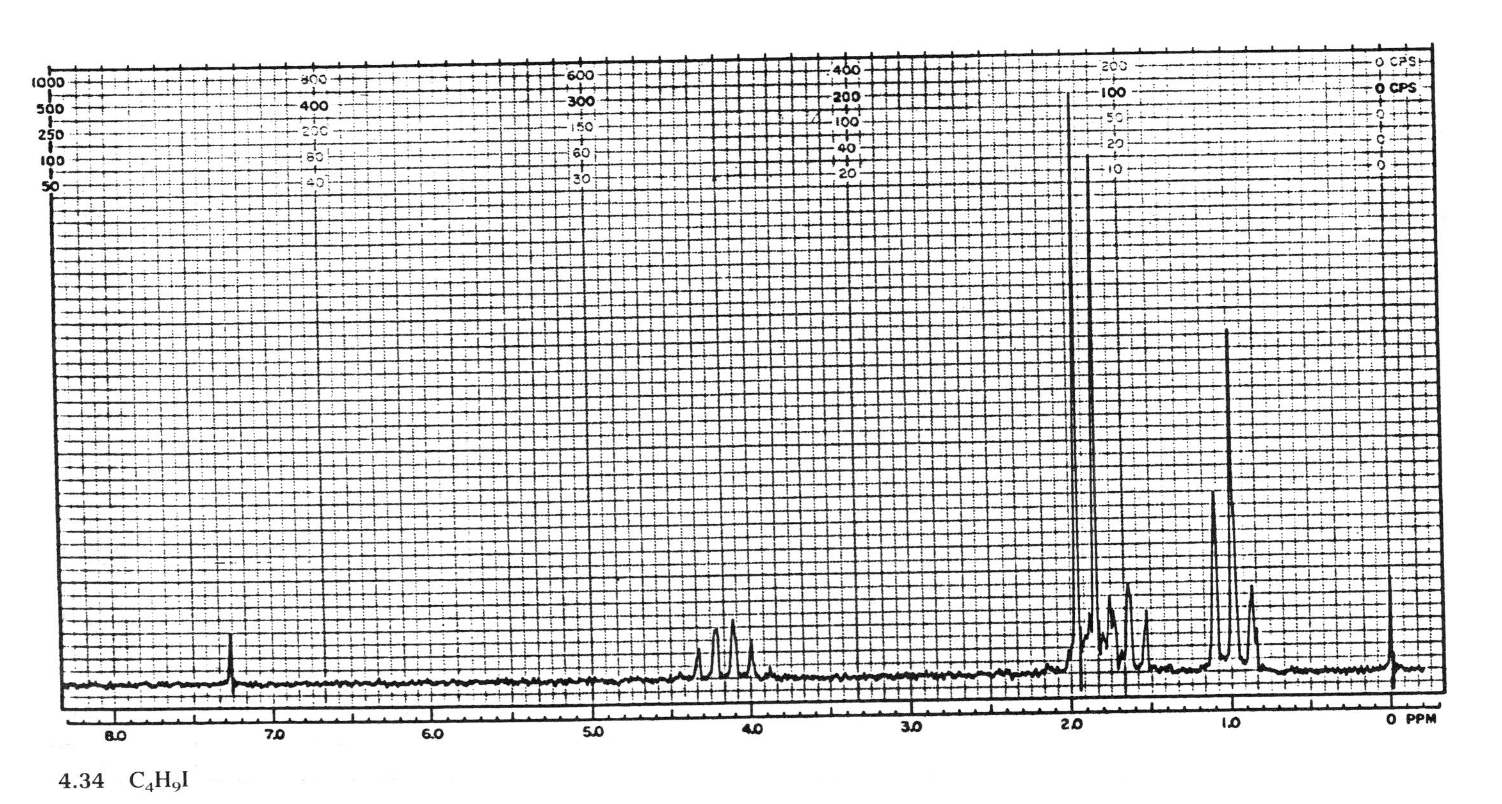

4.34 C_4H_9I

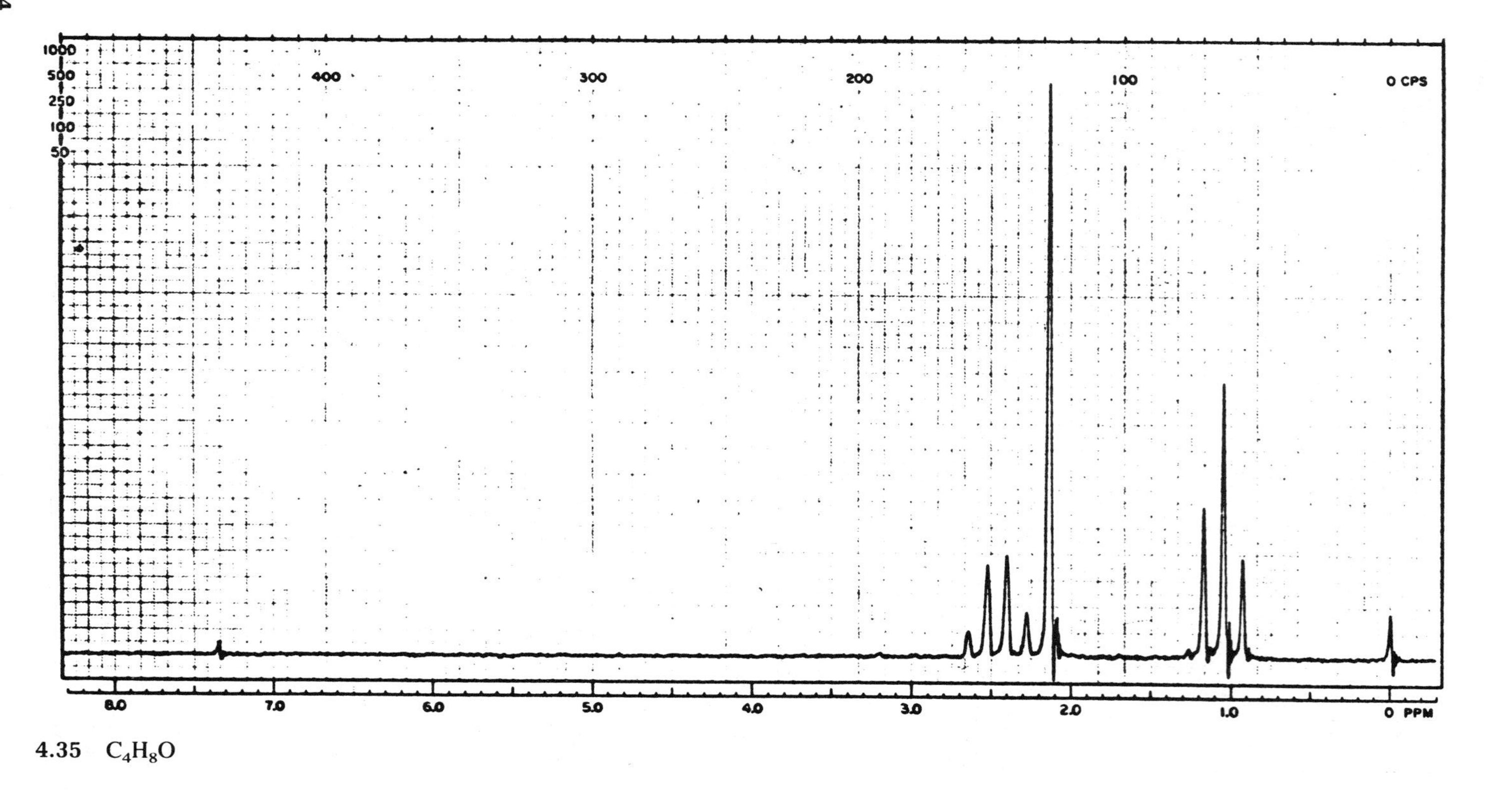
1000
500
250
100
50
400
300
200
100
0 CPS
8.0
7.0
6.0
5.0
4.0
3.0
2.0
1.0
0 PPM

4.35 C_4H_8O

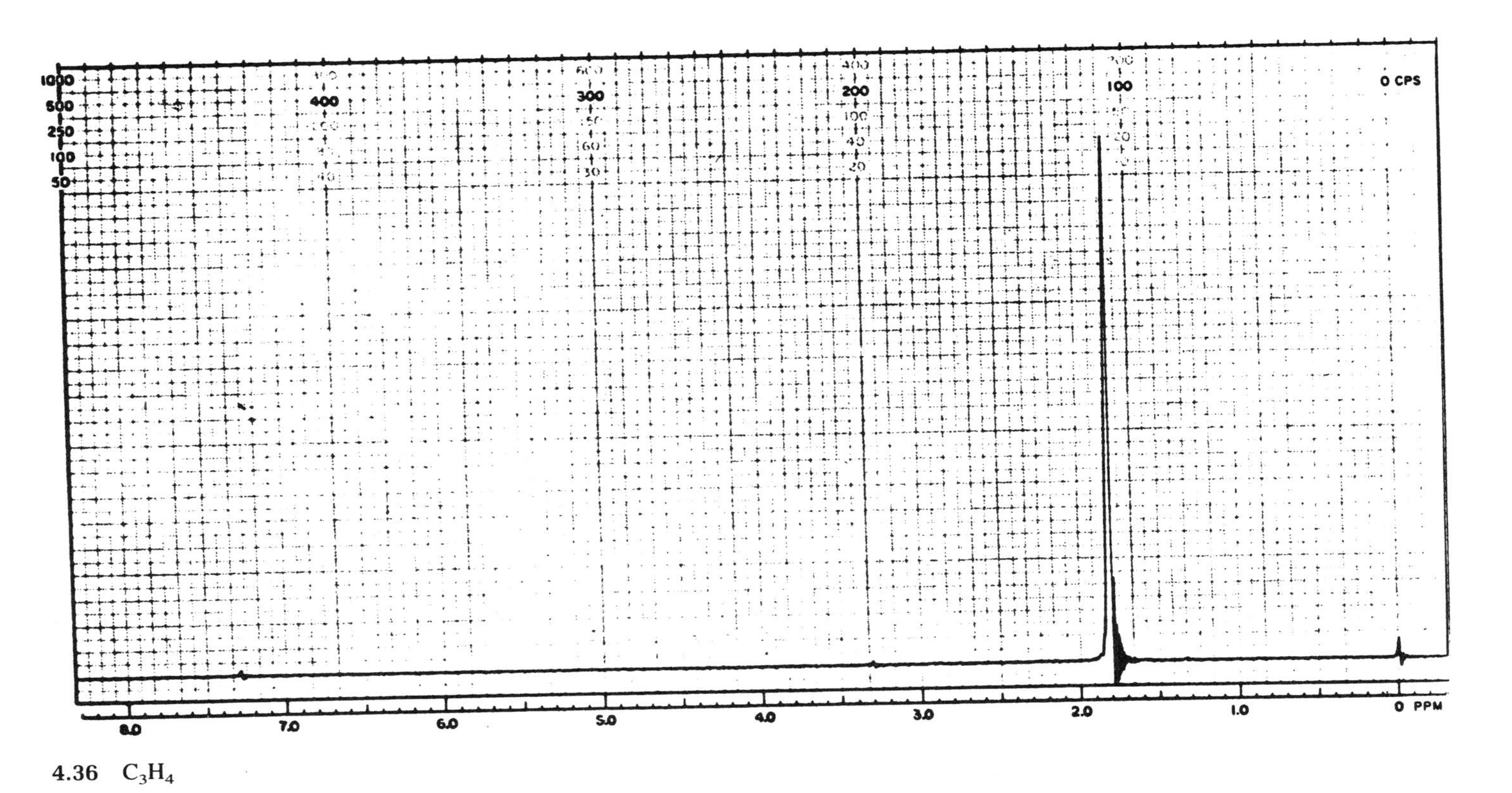
1000
500
250
100
50
400
300
200
100
0 CPS
8.0
7.0
6.0
5.0
4.0
3.0
2.0
1.0
0 PPM

4.36 C_3H_4

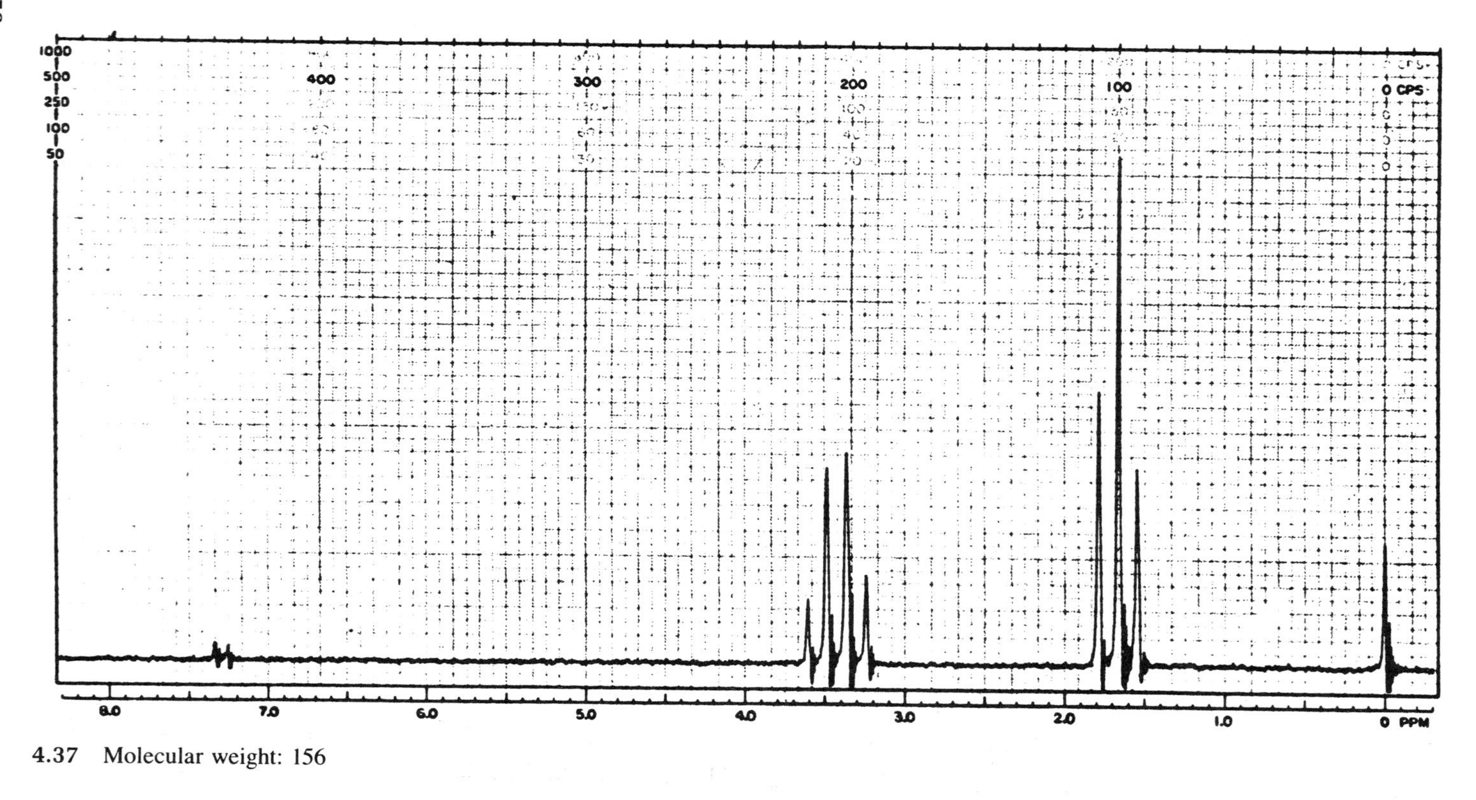

4.37 Molecular weight: 156

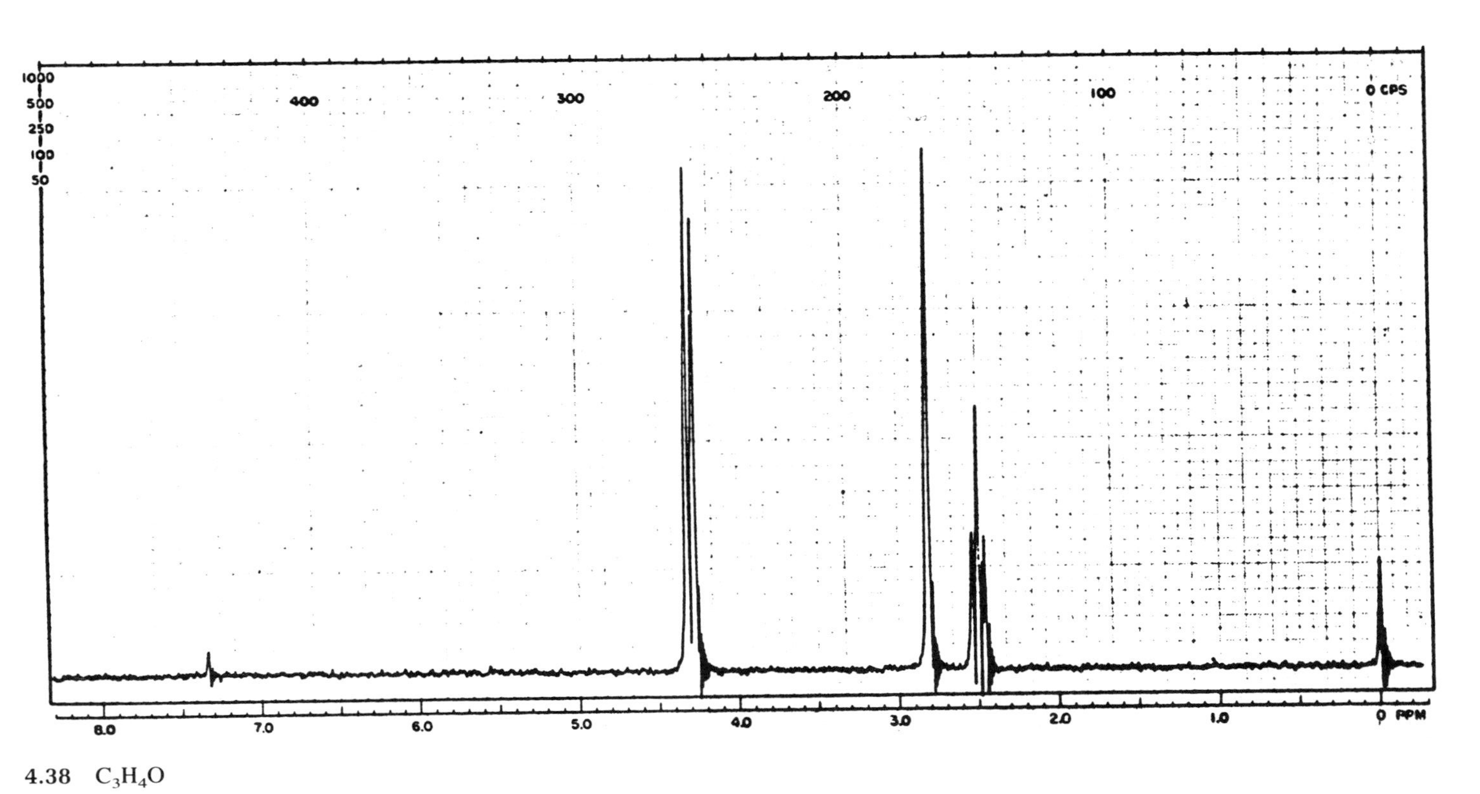
1000
500
250
100
50
400
300
200
100
0 CPS
8.0
7.0
6.0
5.0
4.0
3.0
2.0
1.0
0 PPM

4.38 C_3H_4O

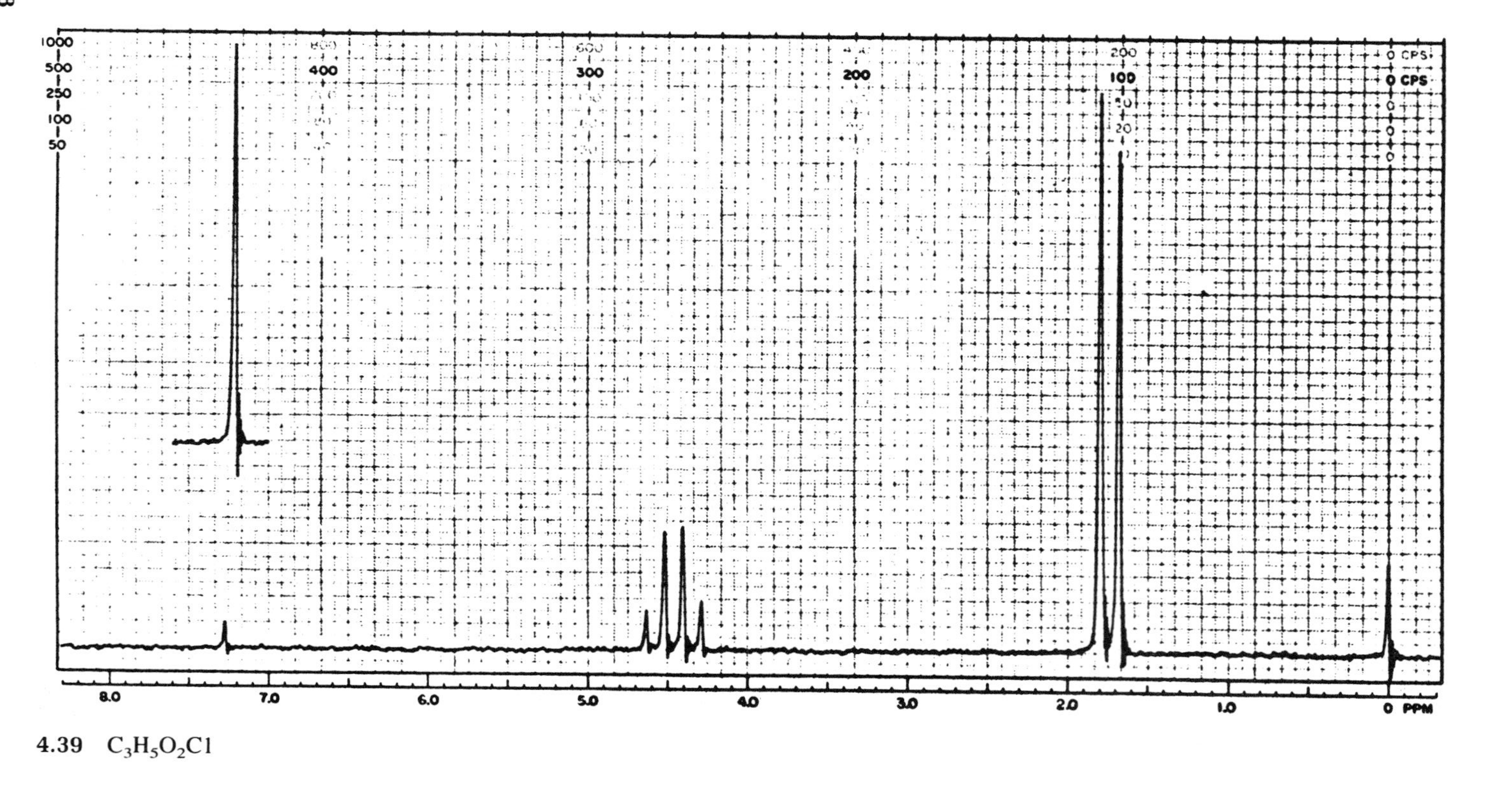

4.39 $C_3H_5O_2Cl$

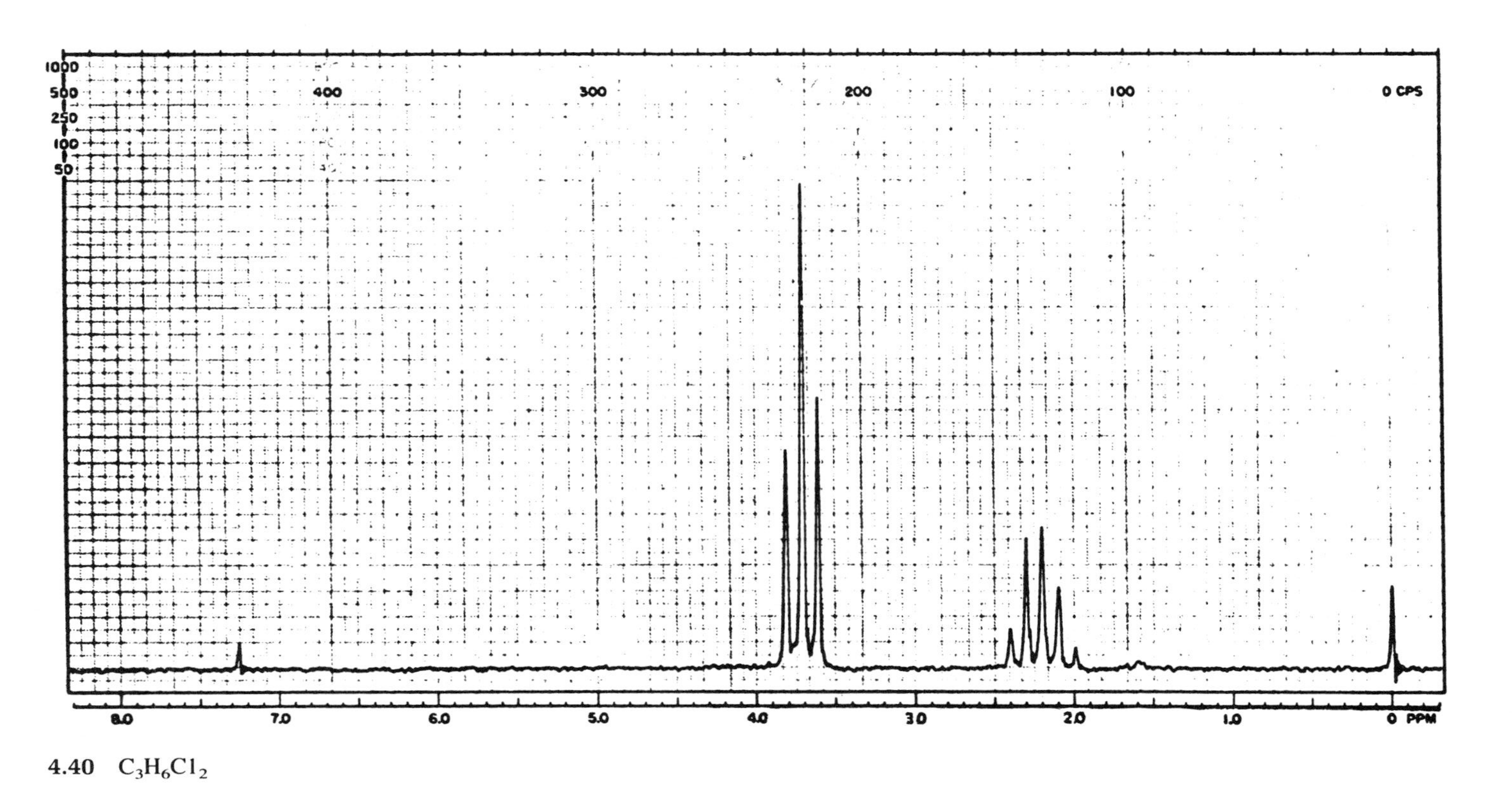
1000
500
250
100
50
400
300
200
100
0 CPS
8.0
7.0
6.0
5.0
4.0
3.0
2.0
1.0
0 PPM

4.40 $C_3H_6Cl_2$

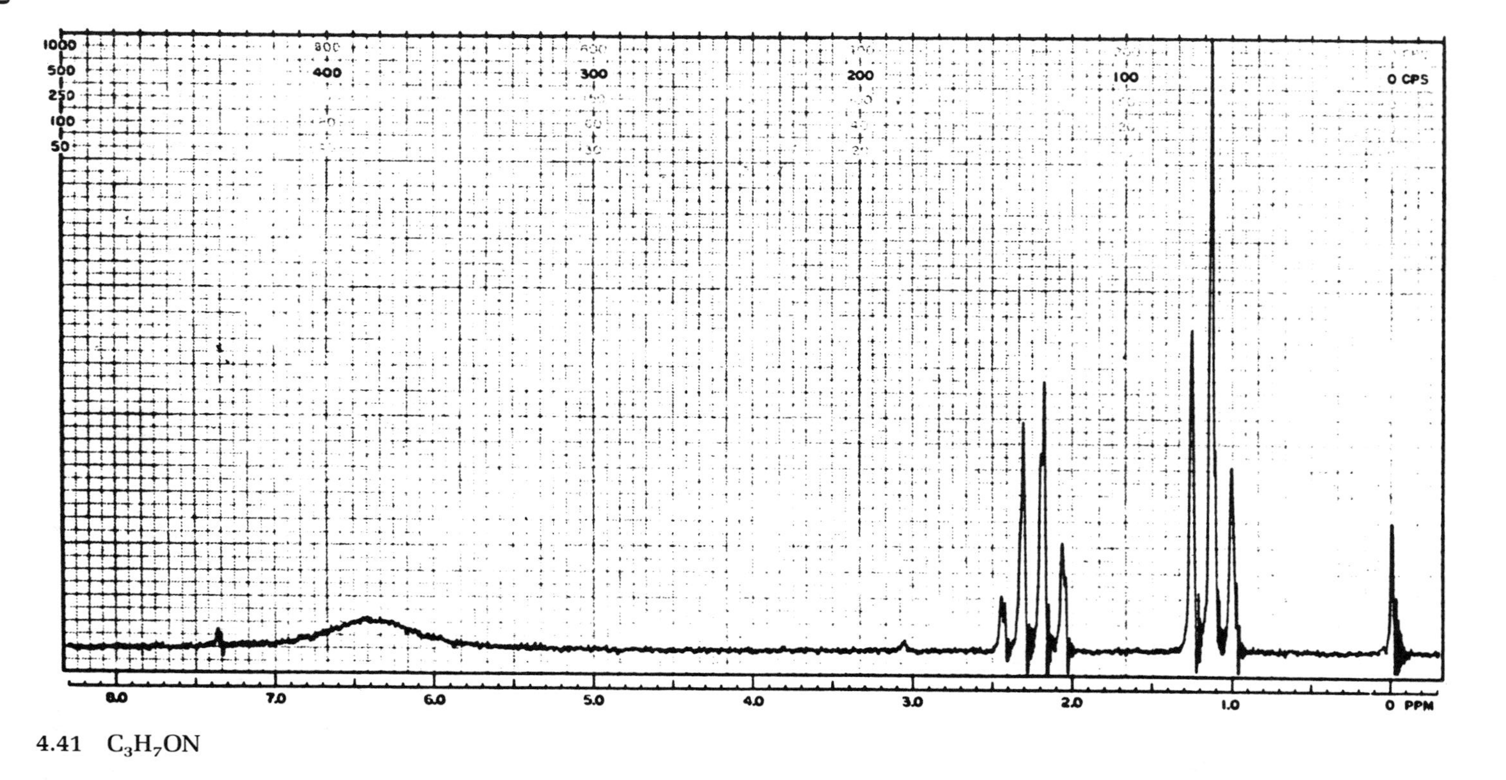
1000
500
250
100
50
400
300
200
100
0 CPS
8.0
7.0
6.0
5.0
4.0
3.0
2.0
1.0
0 PPM

4.41 C_3H_7ON

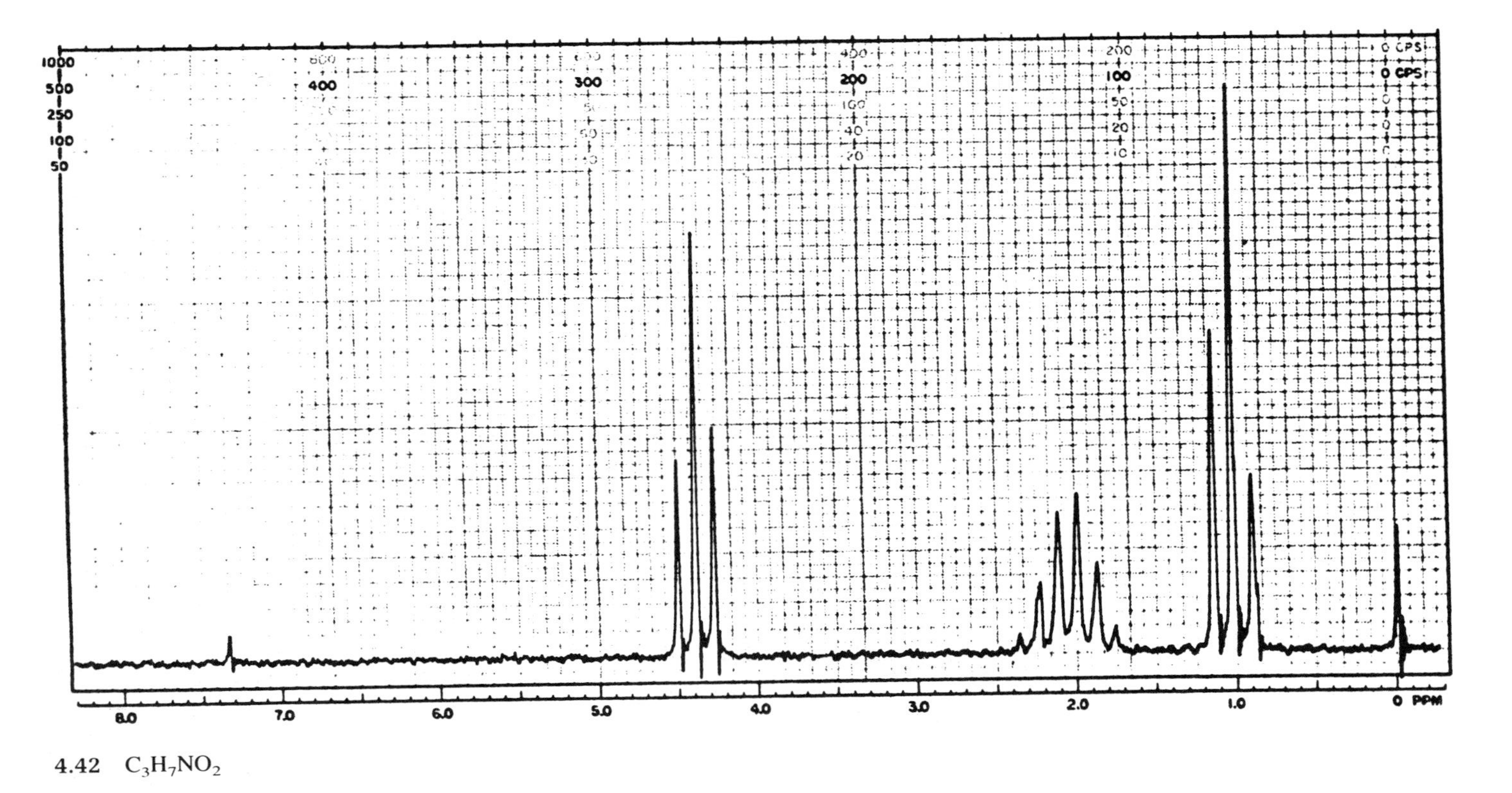
1000
500
250
100
50
400
300
200
100
0 CPS
8.0
7.0
6.0
5.0
4.0
3.0
2.0
1.0
0 PPM

4.42 $C_3H_7NO_2$

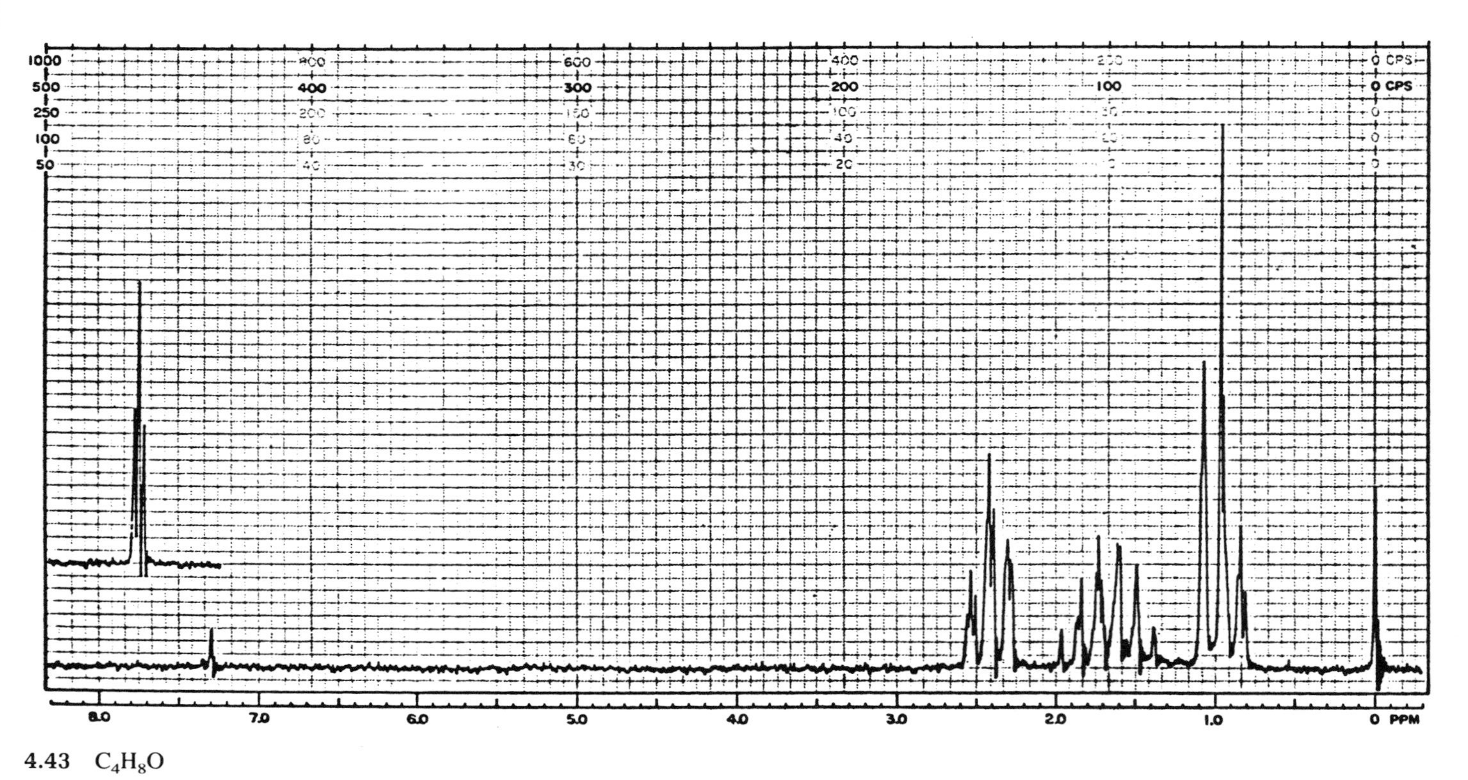

1000
500
250
100
50
800
400
600
300
400
200
100
0 CPS
0 CPS
8.0
7.0
6.0
5.0
4.0
3.0
2.0
1.0
0 PPM

4.43 C_4H_8O

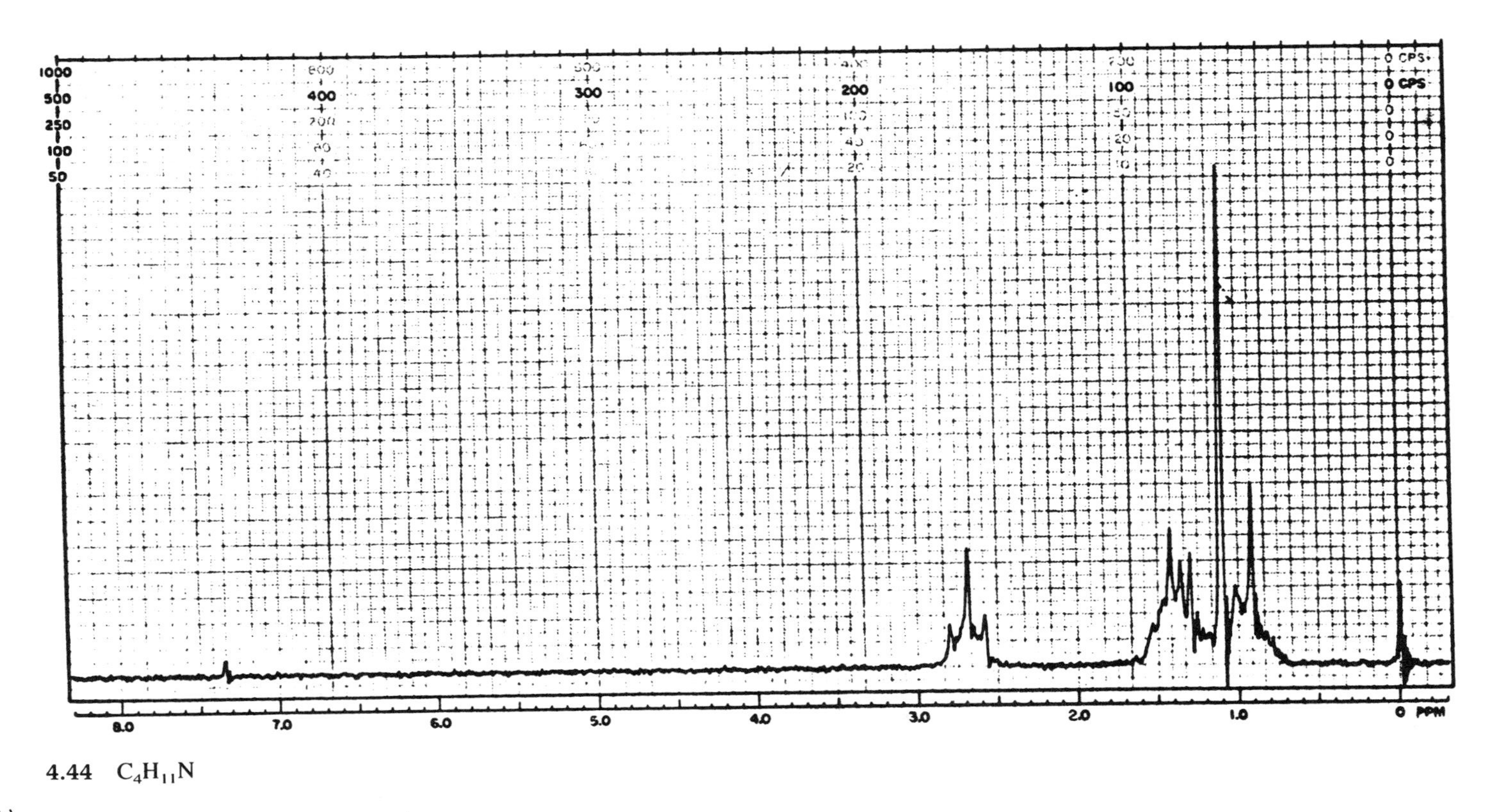

1000
500
250
100
50
400
300
200
100
0 CPS
8.0
7.0
6.0
5.0
4.0
3.0
2.0
1.0
0 PPM

4.44 $C_4H_{11}N$

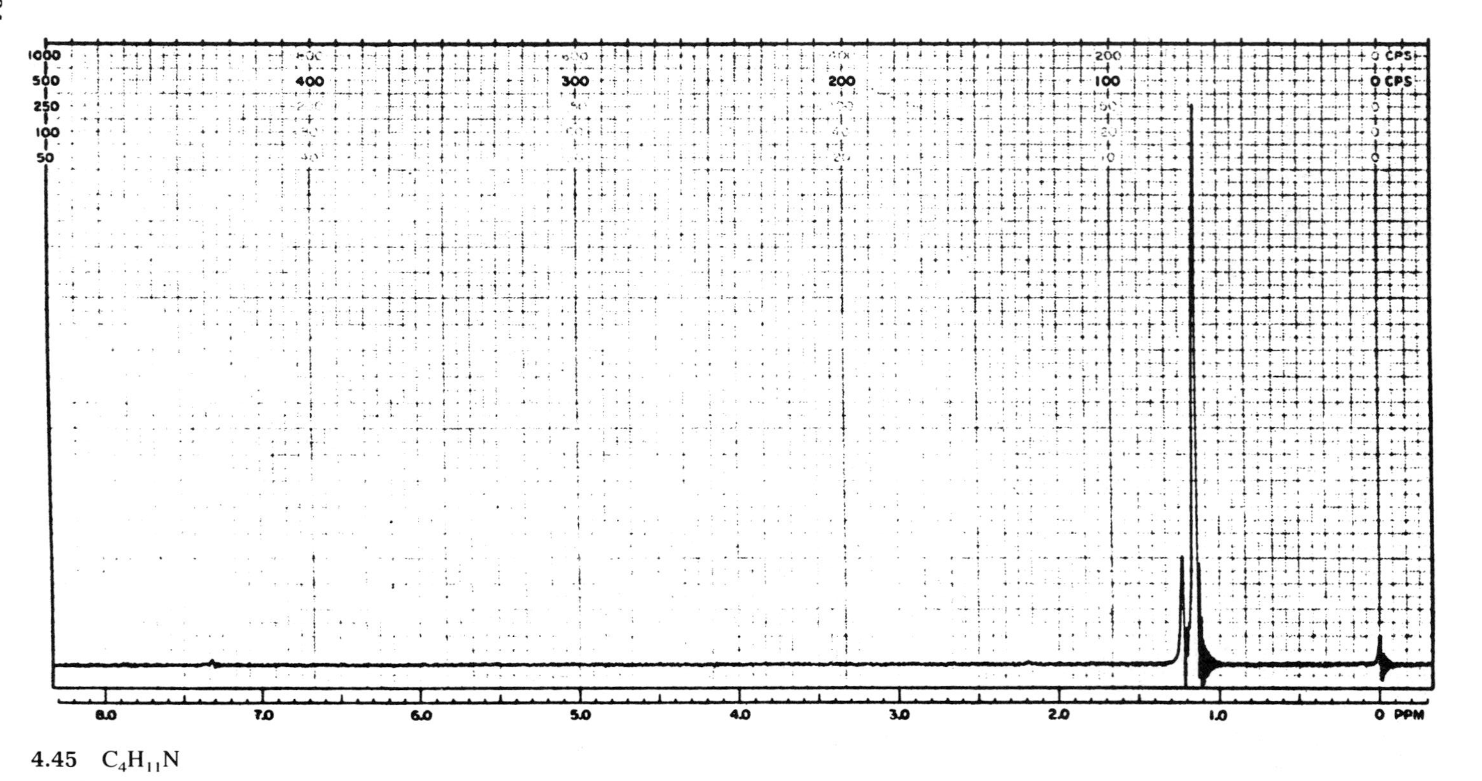

1000
500
250
100
50
400
300
200
200
100
0 CPS
0 CPS
8.0
7.0
6.0
5.0
4.0
3.0
2.0
1.0
0 PPM

4.45 $C_4H_{11}N$

5
Infrared Absorption

Infrared (IR) radiation was first discovered in 1800 by Sir William Herschel, who used a glass prism with blackened thermometers as detectors to measure the heating effect of sunlight within and beyond the boundaries of the visible spectrum. In 1882, further research was carried out by Aubrey and Festing on the absorption of organic compounds, and 10 years later by Julius, who investigated the spectra of 20 organic compounds using a rock salt (NaCl) prism and bolometer detector. At this time it was not clear whether the absorption was due to individual atoms, intermolecular effects, or intramolecular motions.

Coblentz laid the groundwork for IR spectroscopy with a systematic study of organic and inorganic absorption spectra. Experimental difficulties were immense; since each point in the spectrum had to be measured separately, it could take 4 hours to record the full spectrum. But from this work came the realization that each compound had its own unique IR absorption pattern and that certain functional groups absorbed at about the same wavelength even in different molecules.

However, the technique was rarely used by chemists until the commercial production of IR spectrometers during World War II. Then the potential of the technique was realized, and IR spectroscopy became widely used. Double-beam instruments were developed, better detectors were designed, and better dispersion elements, including gratings, were incorporated.

In the recent past the practice of IR has changed dramatically. Conventional double-beam, monochromator pen recorder systems are fast becoming obsolete, while Fourier transform IR (FTIR) insturmentation in becoming standard. Such systems are faster and more accurate than their predecessors.

Simultaneously, considerable attention has been paid to the application of near-IR. A third area of interest is the application of IR emission from flames as a method of elemental detection of chromatographic effluents.

The wavelength of IR radiation falls in the range 0.78–40 μm. For most people, the wavelength of visible light falls between 0.45 (violet light) and 0.75 μm (red light). Infrared radiation therefore is of lower energy than visible radiation, but of higher energy than radio waves. Visible radiation marks the upper energy end of the IR region; the other end is defined somewhat arbitrarily at about 40 μm. At longer wavelengths, the *far infrared* can be considered to extend to 200 μm, but the maximum wavelength end is quite arbitrary. Near IR covers the range of 0.75–2 μm. The IR wavelength range tells us the IR frequency range from the equation (introduced in Chapter 2)

$$\nu = \frac{c}{\lambda} \tag{5.1}$$

where ν the frequency, c is the speed of light, and λ is the wavelength. Furthermore, we have seen [Eq. (2.2)] that $E = h\nu$. Hence, when the frequency is high (i.e., λ is short), the energy of the radiation is high. The frequency range of IR radiation is 1.4×10^{14} to 2.5×10^{16} cps.

It is common to use the term *wave number*, $\bar{\nu}$, in describing radiation used in IR. The wave number is the number of waves of the radiation per centimeter, whereas frequency is waves per second. The normal range is from 13,000 to 250 cm^{-1}. This corresponds to a wavelength range of 0.78–40 μm. Most chart paper is calibrated in wave numbers and wavelengths. The two units are related by the fact that

$$\text{wavelength (in cm)} \times \text{wave number} = 1 \text{ cm}$$

or

$$\text{wavelength (in } \mu\text{m)} \times \text{wave number} = 10{,}000$$

A. REQUIREMENTS FOR INFRARED ABSORPTION

1. Wavelength of Absorbed Radiation

After other requirements are met, molecules absorb radiation when some part of the molecule (i.e., the component atoms or groups of atoms) vibrates at the same frequency as the incident radiant energy. After absorbing radiation, the molecules vibrate at an increased amplitude.

The atoms comprising a molecule can vibrate in several ways (see Sec. B). Consider, for example, the carbon and hydrogen atoms in formaldehyde (Fig.

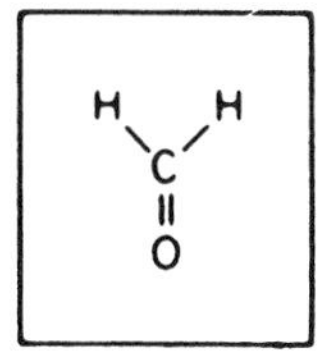

Figure 5.1 The structure of formaldehyde.

5.1), which can vibrate toward and away from each other. The rate at which these component atoms vibrate is quantized and can take place only at well-defined frequencies that are characteristic of the atoms concerned. That is, formaldehyde absorbs radiation that causes its carbon, oxygen, or hydrogen atoms to vibrate in this manner. Since only radiation with certain well-defined frequencies can produce this effect, its wavelengths are characteristic of formaldehyde absorption. If we measure the molecular absorption, we record the frequencies at which the formaldehyde molecule absorbs. This record is the basis of the IR spectrum of formaldehyde; the complete IR spectrum, however, is also modified by the molecule's rotational energy (see Sec. B.2).

Mathematically, the energy of the photon absorbed must equal the energy differences between the vibrational energy states. This can be summarized as

$$\Delta E = E_{\text{vib},1} - E_{\text{vib},2} = h\nu$$

In terms of the wavelength of the light,

$$\Delta E = \frac{hc}{\lambda}$$

2. Molecular Dipoles

For a molecule to absorb IR radiation, it must have a *changeable dipole*. A molecule has a dipole where there is a slight positive and a slight negative electrical charge on its component atoms. These slight charges are not equal to the charge of a whole electron or proton, but represent a slight excess or depletion of electrons in some areas. This excess or depletion has the effect of producing a fractional charge. Two adjacent (but opposite) fractional charges create a dipole. This dipole must change in strength as a result of the vibrational excitation, or IR absorption will not take place.

Figure 5.2 shows the fractional charges between the carbon and hydrogen atoms of formaldehyde. (In addition, the oxygen in formaldehyde is slightly negative compared to the carbon.) We can see that the molecule has several dipoles and that absorption of IR radiation can take place between the carbon and the other atoms in the molecules.

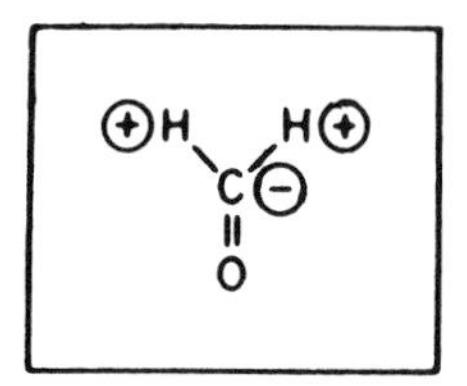

Figure 5.2 Fractional charges of the carbon and hydrogen atoms of formaldehyde.

If the slightly charged parts of a vibrating molecule move—for example, vibrate toward each other—the dipole changes. It has been observed that when the *rate of change* of the dipole during vibration is great, the absorption of radiation is intense. (The strength of the dipole, and therefore the rate of change of the dipole, is likely to be fast between atoms that are close to each other in a molecule, particularly if the charge is high). Conversely, if the rate of change of the dipole is low, the absorption is weak. This situation occurs when the dipole is weak or the vibrating atoms in the same molecule are far apart. For example, the dipole between a hydrogen atom and a carbon atom that are not combined would be weak. Mathematically, the *absorption* v is *proportional to the square of the rate of change of the dipole*. In summary, if the charge is large and the vibration rapid, the rate of change of the dipole is rapid and the absorption of radiation intense.

B. ENERGY LEVELS IN VIBRATING AND ROTATING MOLECULES

Molecules at room temperature vibrate and rotate continuously. If a molecule rotates slowly or is light, it has less rotational energy than if it rotates quickly or is heavy. Rotational energy levels are very small compared to vibrational energy levels. If a molecule had no rotational energy before and after it became vibrationally excited, the radiant energy would be the difference between the vibrational energy levels of the molecule. Because molecules do rotate, however, they usually rotate either faster or slower after vibrational excitation. The energy required to rotate the molecule must be obtained from the radiation. Therefore molecules absorb at wavelengths that cause them to vibrate more quickly and to rotate at a different rate. The net result is that the rotational energy is usually simply added to or subtracted from the vibrational energy during absorption. The total radiant energy absorbed by the molecule is thus equal to the algebraic sum of the molecule's vibrational energy and its rotational energy change (Fig. 5.3).

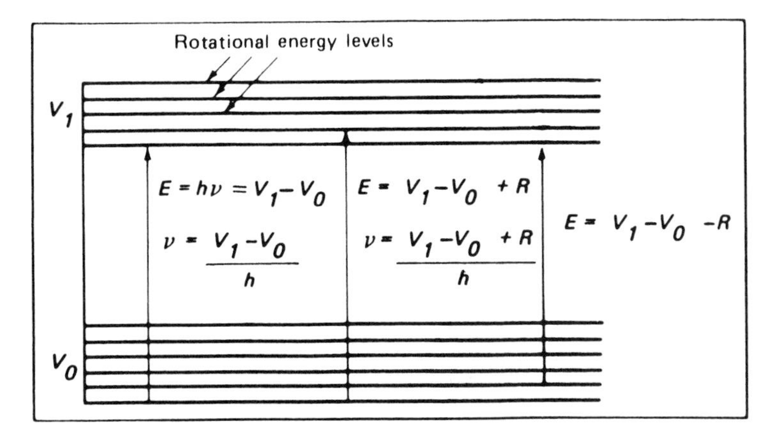

Figure 5.3 Diagram of vibrational and rotational energy levels. The energy difference between the vibrational energy levels is $V_1 - V_0$. The energy of transition can be modified by rotational energy changes in the molecule. But if the rate of change of dipole is small, the degree of absorption is small.

The excitation from the ground state V_0 to the first excited state V_1 is called the *fundamental transition*. It is the most likely transition to occur. Fundamental absorption bands are strong compared to overtones (transitions from the ground state V_0 to upper excited states V_2, V_3, etc.) or combination bands (see next section). Transitions from the ground state V_0 to the second excited state V_2 are *overtone transitions*.

The requirements for the absorption of IR radiation can summarized as follows.

1. The natural frequency of vibration of the molecule must equal the frequency of the incident radiation.
2. The frequency of the radiation must satisfy Eq. (2.2), $E = h\nu$, where E is the energy difference between the vibrational states involved.
3. The change in vibration must stimulate changes in the dipole moment of the molecule.
4. The degree of absorption is proportional to the square of the rate of change of the dipole during excitation.
5. The energy difference between the vibrational energy levels is modified by rotational energy levels, which add to or subtract from the vibrational energy levels.

The modification of vibrational energy levels by rotational energy levels leads to broad band spectra. Rotational lines around the vibrational bands are often observed as band spectra, especially if the molecular complex has many rotational lines.

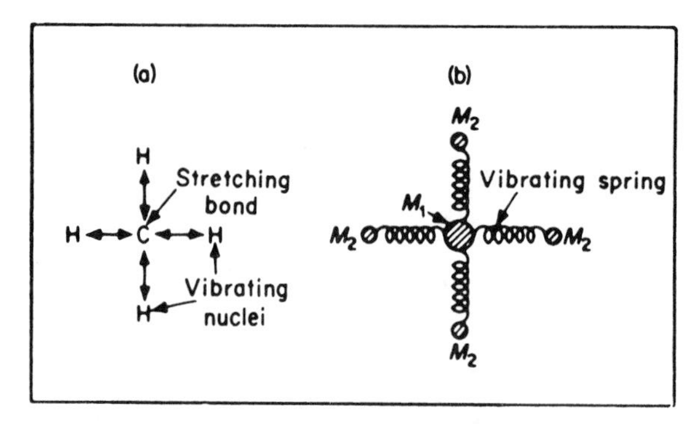

Figure 5.4 (a) Atoms and chemical bonds presented as (b) system of weights and springs.

1. Vibrational Motion

A molecule is made up of a number of atoms joined by chemical bonds. Such atoms vibrate about each other in the same way as weights held together by springs (Fig. 5.4). Hooke's law states that two masses joined by a spring will vibrate such that.

$$\omega = \sqrt{\frac{f}{\mu}} \tag{5.2}$$

where ω is the frequency in rad/sec, but since $\omega = 2\pi\nu$, we have

$$\nu = \frac{1}{2\pi}\sqrt{\frac{f}{\mu}} \tag{5.3}$$

where ν is the frequency of vibration, f is a measure of the binding strength of the spring, and μ is the reduced mass, or

$$\mu = \frac{M_1 M_2}{M_1 + M_2} \tag{5.4}$$

where M_1 is the mass of one vibrating body and M_2 the mass of the other. But ν is in cycles per second (cps). During this time light travels a distance measured in cm/sec (i.e., the speed of light). Therefore, if one divides ν by c, the result is the number of cycles per cm. This is $\bar{\nu}$, the wavenumber, and

$$\bar{\nu} = \frac{\nu}{c}$$

It can be deduced that

$$\bar{\nu} = \frac{1}{2\pi c}\sqrt{\frac{f}{\mu}} \tag{5.5}$$

Two atoms joined by a chemical bond vibrate according to the same equation, except that in this case f is a measure of the binding force of the chemical bond and μ is the reduced mass of the vibrating atoms. Here f is the oscillator strength of the absorption band and is proportional to the bond order (single bond, double bond, etc.), the electronegativity of the vibrating atoms, and the mean distance between them. These are all physical constants and properties of the molecule. Since f and μ are constant for any given sets of atoms and chemical bonds, the frequency of vibration ν is also constant. *The radiation absorbed by the system has the same frequency and is constant for a given set of atoms and chemical bonds*, that is, for a given molecule. The absorption spectrum is therefore a physical property of the molecule.

There are several ways in which two atoms of a large molecule can vibrate in relation to a third atom. A typical example is the vibration of hydrogen about carbon in an organic compound. The principal modes of vibration between carbon and hydrogen in paraffinic compounds are shown in Fig. 5.5, where the arrow indicates the direction of motion of the hydrogen nucleus in the plane of the paper. The signs ⊖ and ⊕ indicate motion perpendicular to the plane of the paper, ⊕ meaning approaching and ⊖ meaning receding. In wagging motion (Fig. 5.5e) both hydrogen atoms move together; in twisting (Fig. 5.5f) they

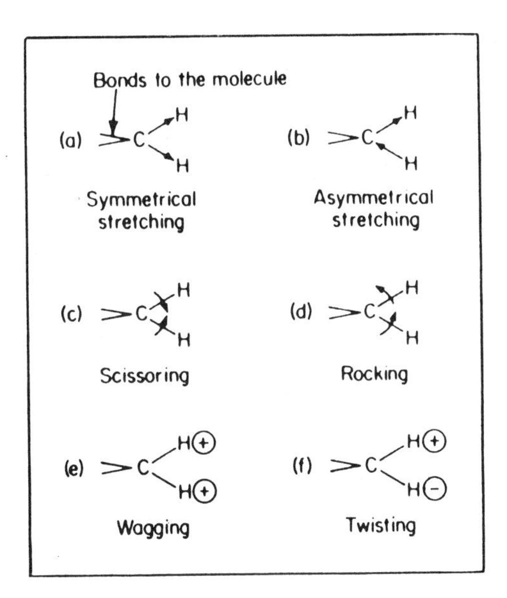

Figure 5.5 Principal modes of vibration between carbon and hydrogen in a paraffinic compound: (a) symmetrical and (b) asymmetrical stretching and the bending vibrations, (c) scissoring, (d) rocking, (e) wagging, and (f) twisting.

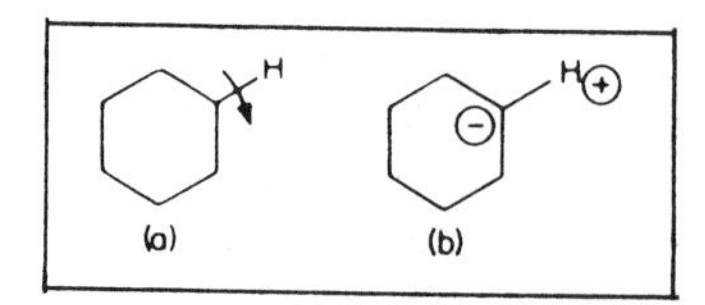

Figure 5.6 Two of the modes of vibration of aromatic compounds.

move in opposite directions. Each of the modes of vibration depicted absorbs radiation at a different wavelength.

The modes of vibration of aromatic compounds are shown in Fig. 5.6. Similar motions are made by other chemical atoms or groups of atoms. The absorption wavelengths of some common functional groups are shown in Table 5.1.

2. Overtones and Combination Bands

In Fig. 5.3, it was illustrated how a molecule can be excited from V_0, the ground state, to V_1, the first excited state. However, there are even higher vibrationally excited states—V_2, V_3, etc.—that can be populated. It is therefore possible to be excited from V_0 to V_2 or V_0 to V_3, as shown in Fig. 5.7. Such a transition is called an *overtone*. More energy is required (almost two or three times that required for V_0 to V_1), and the absorption wavelength is shorter. However, since the probability is less for these transitions, only strong fundamental bands have overtones. Many occur in the near IR.

A combination band occurs when a single photon can excite two vibrations. The energy of the photon is the sum of the energy required to excite both transitions. These bands also occur in the near IR.

3. Rotational Movement

At the same time as the components of a molecule vibrate toward each other, the molecule as a whole rotates in the same way as a ball spins. This rotation may

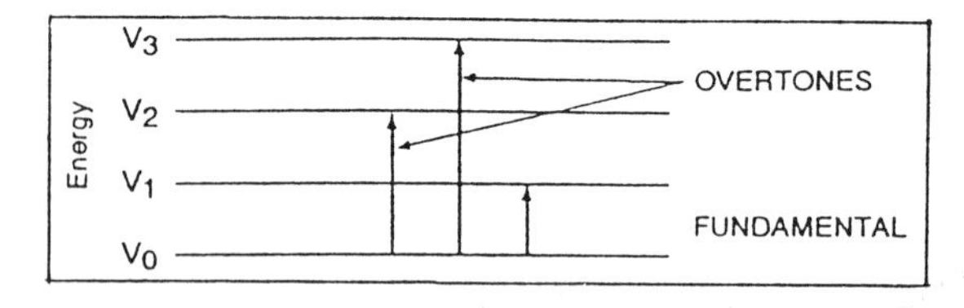

Figure 5.7 Note transitions from the ground state to higher excited states are overtones.

Table 5.1 Molecular Vibrations and Absorption Wavelengths (μm)

Molecular group	Structure	Stretching	Bending		
CH bonds		C → H	C ⟂ H		
Alkane methylene	R—CH₂—R	3.42	6.83	7.8	8.8
Methyl	R—CH₃	3.38	6.9	7.3	
Alkene	R—CH=CH₂	3.33	5.5	6.1	
Benzene	C₆H₆	3.2	6.7	14.8	
Aldehyde	R—CHO	3.68			
Alkyne	R—C≡CH	3.1	16.0		

Table 5.1 Continued

Molecular group	Structure		Stretching		Bending
Acid	R—C(=O)—OH	3.4–4.4.0 (broad)	5.8	7.1	7.6
Ether	R—CH₂—O—CH₂—R		C → 0 9.0		
Nitrogen bonds					
Primary amine	R—N(—H)—H	C → N 7.6	N → H 2.8–3.0		N ⊥ H 6.1–6.3
Secondary amine	R_1—NH—R_2	7.5	2.9		6.0
Tertiary amine	R_1, R_2, R_3—N	7.0			

Oxygen bonds		$O \rightarrow H$	$C \Rightarrow O$	$C \rightarrow O$		O ╪ H
Water	H—O—H	2.8				6.8
Primary alcohol	$R-CH_2-OH$	2.9		9.8		7.2
Secondary alcohol	R_2CH-OH	2.9		9.0		7.2
Oxygen bonds						
Tertiary alcohol	$R_1R_2R_3C-OH$	2.9		8.3	7.2	
Phenol	C_6H_5-OH	2.9	8.4			
Ketone	$R_1R_2C=O$		5.8			
Aldehyde	$RHC=O$		5.8			

Table 5.1 Continued

Molecular group	Structure	Stretching		Bending
Carbon bonds				
Alkanes	$R-CH_2-CH_2-R$	$C \rightarrow C$ 11.3	$C \Rightarrow C$	$C \nleftrightarrow CH_3$ 7.3
Alkene	$R-C-C(H)=CH_2$	11.0	7.1	
Toluene	$C_6H_5-CH_3$	9.6	6.2	6.9
Chloroalkanes	$R-C(Cl)(H)-R$	$C \rightarrow Cl$ 8.2		$C \nleftrightarrow Cl$ 13.3

be fast or slow and may take place about different axes of the molecule. As we indicated earlier, the energy involved in spinning a molecule is very small compared to that required to make it vibrate.

Molecules rotate at many rates of rotation and have many rotational energy levels. Hence, instead of a single vibrational absorption line, we observe many absorption lines separated by rotational energies. We call these close-packed absorption lines an *absorption band*. The absorption band for any particular pair of vibrating functional groups may extend over a wide wavelength range. For example, the C → H stretching vibration pictured in Fig. 5.5b extends over the range 3.0–3.32 μm. Under extremely high resolution such a band can be resolved into numerous single lines. In practice, however, the single "continuous" band is not resolved because the extra information obtained by high resolution is not particularly useful to the analytical chemist and would require an expensive high-resolution instrument. These unresolved bands are normally used in IR absorption spectroscopy, because they are almost as valuable analytically as high-resolution IR spectra. Typical absorption spectra are shown in Figs. 5.27–5.33.

C. EQUIPMENT

The optical diagram in Fig. 5.8 is that of a typical single-beam IR absorption spectrometer. Radiation from the source passes through the sample and then through the entrance slit to the dispersion element. The desired wavelength is chosen and passed through the exit slit to the detector, which measures the intensity of the remaining radiation. Knowing the original intensity of radiation, we can measure how much radiation has been absorbed by the sample. By measuring the degree of absorption at different wavelengths, the absorption spectrum of the sample can be obtained.

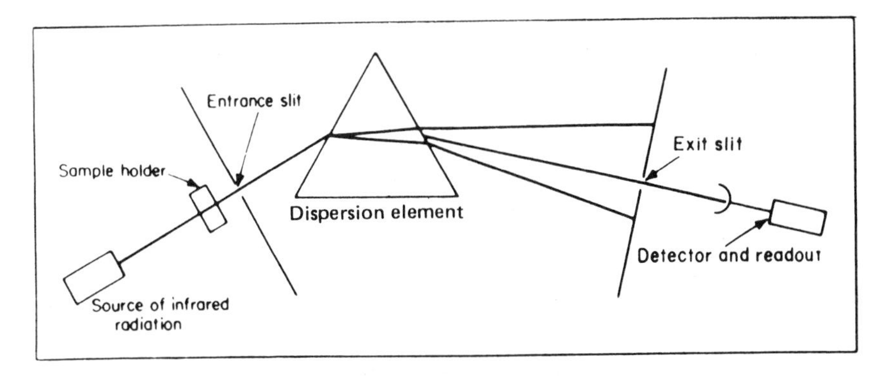

Figure 5.8 Schematic diagram of a single-beam IR absorption instrument.

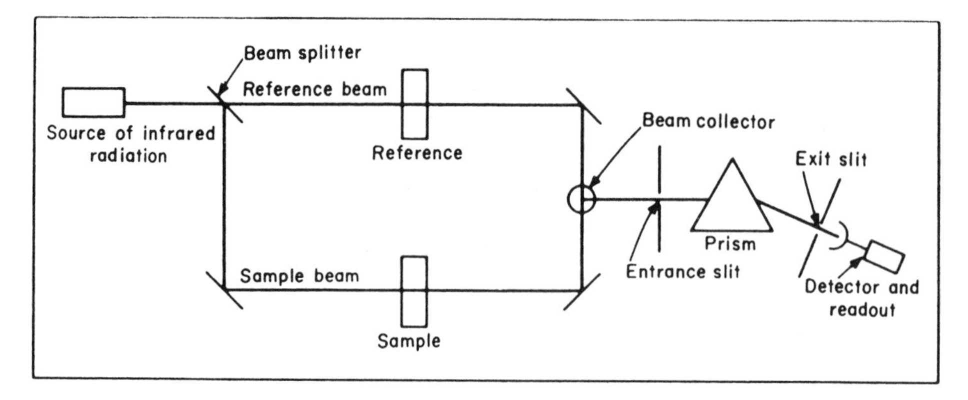

Figure 5.9 Schematic diagram of a double-beam IR absorption instrument.

For quantitative analysis, the intensity of radiation I_0 is first measured with no sample in the light beam. The reading from the detector and readout system is then taken as I_0. The sample is then put into the light beam and absorbs radiation. The intensity falling on the detector is the measure of I_1 [see Eq. (2.14)]. The ratio I_1/I_0 is the transmittance T of the Beer–Lambert equation [Eq. (2.19)]. With this ratio quantitative analysis of the sample can be made.

In practice, a major problem frequently arises. The intensity of the radiation source varies slowly over long periods of time, or even rapidly over short periods of time. In these circumstances the value of I_0 varies constantly and it is difficult to measure the ratio I_1/I_0 accurately. This change in I_0 is called the *drift*. It may also arise from a change in detector response or movement of the dispersion element. It is a source of error that is uncorrected in single-beam optics. Also, CO_2 and H_2O, which absorb in air, are recorded in single-beam optics, causing an apparent change in I_0 with no sample in the system. This change arises with atmospheric conditions.

In practice, single-beam optics of this type are rarely, if ever, used. But it should be pointed out that FTIR, a very common instrument, is a single beam system (Sec. C.4.i). Quantitative results are therefore subject to error.

To overcome these problems we use a *double-beam system* (Fig. 5.9). This instrument separates the source beam into two half-beams, radiating light down each beam alternately and in rapid succession, e.g., 50 cps. The sample is placed in the sample beam, and a reference material, such as the solvent used in the sample, is placed in the reference beam. The two half-beams are recombined to form a continuous team mode up of two half-beams, which now passes along the optical path to the detector.

Since there is no sample in the reference beam, radiation reaches the detector unabsorbed and is a measure of I_0. If there is no sample in the sample cell, the half-beam traveling along the sample beam is not absorbed either and is therefore equal to the reference beam. When these two equal half-beam recombine, a steady signal falls on the detector. When the sample half-beam is absorbed, however, it decreases in intensity. After the two half-beams are recombined, they then produce an oscillating signal. The greater the absorption of the sample beam, the greater the amplitude of oscillation. The detector system is designed to measure the amplitude of oscillation, which then becomes a direct measure of the degree of absorption by the sample and the ratio I_1/I_0. From this ratio the concentration of the sample can be determined.

The degree of oscillation is a measure of the ratio I_1/I_0. Any variation in the intensity of the source I_0 simultaneously decreases I_1 but does not change the ratio I_1/I_0. Large changes in I_0 may be noticeable, but if the variation is small, the error is greatly reduced. Analytical errors caused by variations in the source intensity are therefore greatly reduced.

Single-beam and double-beam systems are also discussed in Chapter 3. The signals falling on the detector are shown in Fig. 5.10. The components used for IR absorption studies are described below.

1. Radiation Source

The two most popular sources of IR radiation are *Nernst glowers* and *Globars*. The Nernst glower is a bar composed of zirconium oxide, cerium oxide, and thorium oxide that is heated electrically to a temperature between 1000 and 1500°C. The Globar is a bar of sintered silicon carbide, which is heated electrically to a similar temperature. Electrically heated Nichrome wire has also been used successfully as a light source. At these elevated temperatures, each source strongly emits IR radiation. Each also fulfills two important requirements

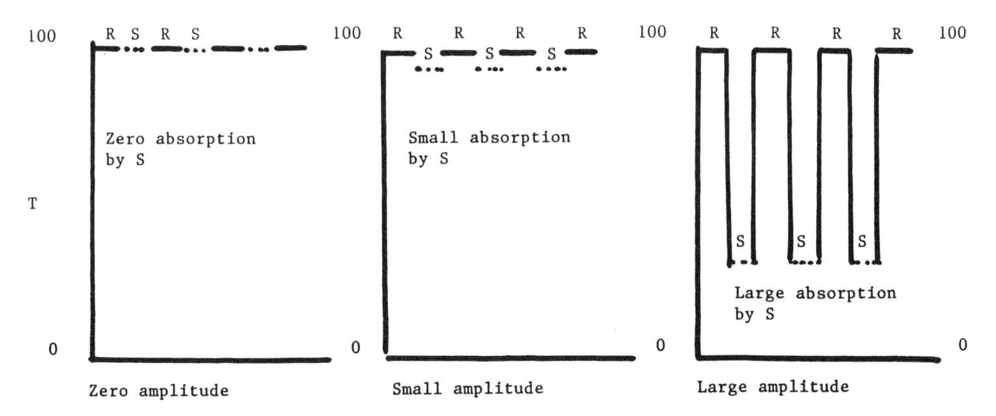

Figure 5.10 Signals falling on detection using double-beam optics.

of a radiation source, namely, that the intensity of radiation be (1) continuous over the wavelength range used and (2) constant over long periods of time. In addition, the radiation from these sources extends over a wide wavelength range. Unfortunately, however, the intensity of the radiation is not exactly the same at all wavelengths (see Fig. 5.11), but the rate of change is gradual. For most practical purposes, this variation is no handicap, because compensations can be made if necessary and are made automatically in double-beam optics. Also, it should be remembered that the absolute value of the radiation is seldom measured. Usually it is the fraction of the radiation absorbed that is important. This is determined by measuring the ratio I_1/I_0 rather than the absolute values of I_0 and I_1.

2. Monochromators

The radiation emitted by the source covers a wide frequency range. However, the sample absorbs only at certain characteristic frequencies. In practice, it is important to know what these frequencies are. To obtain this information we must be able to select radiation of any desired frequency from our source and eliminate that at other frequencies. This is done by means of a monochromator, which consists of a dispersion element and a slit system. The two main types of dispersion elements are prisms and gratings. Monochromators are discussed in Chapter 3, Section A.1.b.

a. Prisms

Prisms split up radiation beams in exactly the same way as glass prisms or rain droplets split up visible white light into the colors of a rainbow. The glass prism separates the visible light according to its wavelength. The same system works

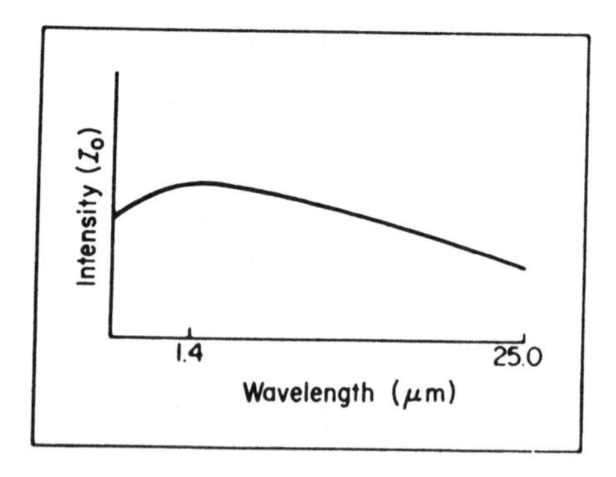

Figure 5.11 Intensity of radiation from a Globar at different wavelengths.

with IR radiation (see Fig. 3.2). Prisms are used for various forms of radiation, including visible, IR, and ultraviolet.

The material used to make the prism must be carefully selected. First, it must be transparent to IR radiation. This requirement eliminates common materials, such as glass and quartz, which are not transparent to IR at wavelengths longer than 3.5 μm. Second, the material used must be strong enough to be shaped and polished.

The most common materials used are metal salts, such as potassium bromide, calcium fluoride, sodium chloride (rock salt), and thallous bromide. The final choice among the compounds is determined by the wavelength range to be examined; for example, sodium chloride is transparent to radiation between 2.5 and 15 μm. This wavelength range is sometimes termed the *rock salt region*. Potassium bromide or cesium bromide can be used over the range 2.1–26 μm, and calcium fluoride in the range of 2.4–7.7 μm. The limiting wavelength ranges of materials commonly used for prisms and sample containers are shown in Table 5.2.

To make a prism, a single large crystal of the metal salt is cut and polished to the required shape. The larger the crystal, the better the resolution (see Chap. 3, Sec. A.1.b). The surfaces must be smooth to prevent random scattering of the radiation by the prism surfaces. It is also important that the crystal be of high quality. Any faults or other imperfections will lead to scattering of the radiation when it passes through the prism, and hence a loss of signal and resolution.

It should be pointed out that the metal salts used to make prisms are water soluble. If the surface becomes wet, it dissolves. On drying it is left etched. For this reason, it is necessary to keep the prisms dry. Installing the instrument in an air-conditioned room is very advantageous. If this is not possible, the prism should be kept in a desiccated atmosphere. Commercial equipment alleviates the problem by using small heaters to keep the prism above room temperature at all times, thus preventing condensation of water vapor from the atmosphere on its surface. Should water condense on the prism, its surface would become rough etched. Radiation would then be scattered by the prism and the data would no longer be reliable. For this reason many IR instruments are left on "stand-by," that is, not switched off or unplugged when not in use. It should be pointed out that prisms are used less and less in this field, now dominated by gratings.

b. Gratings

In recent years, gratings have become increasingly popular in IR spectroscopy. Their principle of operation is discussed in Chapter 3. One of the important advantages of gratings is that they can be made with materials, such as aluminum, that are stable in the atmosphere and not attacked by moisture. This is in contrast to metal salt prisms, which, as we just noted, are subject to etching from at-

Table 5.2 Materials Used in IR Optics for Studying Organic Compounds

Material	Transmission range (μm)	Solubility (g/100 g water)	Refractive index	Comments
Sodium chloride (NaCl)	0.25–16	36	1.49	Most widely used; reasonable range and low cost
Potassium chloride (KCl)	0.3–20	35	1.46	Wider range than NaCl; used as a laser window
Potassium bromide (KBr)	0.25–26	65	1.52	Extensively used; wide spectral range.
Barium fluoride (BaF_2)	0.2–11	0.1	1.39	Extremely brittle
Cesium iodide (CsI)	0.3–60	160	1.74	Transmits to 60 μm
Cesium bromide (CsBr)	0.3–45	125	1.66	
Thallium bromoidide	0.6–40	<0.05	2.4	For internal reflection in the far infrared when moisture is a problem
Stontium fluoride (SrF_2)	0.13–11	1.7×10^{-3}		A substitute for BaF_2; resistant to thermal shock.
Silver chloride	0.4–25	1.5×10^{-4}	2.0	Darkens under ultraviolet light
Silver bromide	0.5–35	1.2×10^{-5}	2.2	Darkens UV light
Germanium	2–11	Insoluble	4.00	
Fused silica	.2–4.5	Insoluble	1.5 at 1 μm	Most useful for the near-IR and UV
Irtran 1	0.5–9	Insoluble	1.34 (at 5 μm)	Pressed magnesium fluoride
Irtran 2	0.4–14.5	Insoluble	2.2	Pressed zinc sulfide
Irtran 3	0.4–11.5	Insoluble	1.3	Pressed calcium fluoride
Irtran 4	0.5–22	Insoluble	2.4	Pressed zinc selenide
Irtran 5	0.4–9.5	Insoluble	1.6 (at 5 μm)	Pressed magensium oxide
Irtran 6	0.9–31	Insoluble	2.7	Pressed cadmium telluride

mospheric moisture. Another advantage of gratings is that they can be used over a wide wavelength range. This is in contrast to salt prisms, which have smaller wavelength ranges in which they can be used; outside this range they absorb so much radiation that they cannot be used. A further important advantage of gratings is that their resolution is constant over the operable range. This provides improved resolution at long wavelengths and greatly facilitates calibrating the wavelength of absorption bands. A problem with gratings is that different orders (see Chap. 3) of a wavelength travel the same light path upon leaving the face of the grating. Since several wavelengths are superimposed on each other in this way, several displaced spectra are superimposed on each other. The problem can be solved by using a small prism in conjunction with the grating. This prism acts as an *order sorter*. Only low resolution is required to separate orders, hence prisms used for this purpose are smaller and less expensive than those used as principal dispersion elements.

Several filters, each transparent over limited wavelength ranges, can be used as order sorters in conjunction with gratings. Such filters are built into the grating system so that the latter is rotated to change wavelengths, the pertinent filter is mechanically moved into place to eliminate different orders from the light path. Such mechanical movement takes time; for this reason, the pen carriage controlling the monochromator system must not be forced to move too fast, or the filter system will be broken.

3. Slit Systems

The radiation from the source is dispersed by the dispersion element into a fan-shaped beam, only a small part of which is used. This small section is separated from the rest of the beam by the exit slit. The slit system is discussed in detail in Chapter 3, Section A.1.c.

There are special problems involved in IR spectroscopy. At longer wavelengths, the energy of each proton is decreased ($E = hc/\lambda$). Furthermore, at long wavelengths, the power from the light source drops off. Also, if prisms are used they lose transparency at longer wavelengths. Therefore, the detector operates at reduced light levels. To make matters worse, the response of the detector decreases at these longer wavelengths. The sum total of these problems is to reduce the signal available at long wavelengths significantly. Although these effects are true to some extent with all types of spectroscopy, they are particularly severe with IR. The problem is not eliminated but is somewhat reduced by the use of a *programmed slit system*. In this system the slit opens up automatically to permit a wider wavelength range to reach the detector and therefore to increase the intensity of the signal. The extent to which the slit opens as a function of wavelength change is programmed by the manufacturer, who arrives at a suitable program by taking into account all the problems mentioned above.

The advantage of the programmed slit system is that more radiation falls on the detector and the instrument responds more reliably. The disadvantage is that the use of wider slits results in an automatic loss in resolution. This may lead to interference from unresolved bands in quantitative analysis and loss of fine structure in qualitative analysis.

If necessary, resolution can be regained by operating in the *manual* mode at fixed slit widths, but lack of response from the detector may be encountered for the reasons stated. The choice is made by the operator and is dictated by the sample to be analyzed and the information needed.

4. Detectors

For many years the most common types of detectors used in IR spectroscopy were *bolometers*, *thermocouples*, and *thermistors*, but new and faster detectors are now becoming more popular. The more important detectors are described below.

a. Bolometer

A bolometer is a very sensitive electrical resistance thermometer that is used to detect and measure feeble thermal radiation. Consequently, it is especially well suited as an IR detector. The bolometer usually consists of a thin metal conductor, such as platinum wire. Incident radiation, such as IR radiation, heats up this conductor, which causes its electrical resistance to change. The degree of change of the conductor's resistance is a measure of the amount of radiation that has fallen on the detector. In the case of platinum, the resistance change is 0.4% per °C. The change in temperature depends on the intensity of incident radiation and on the thermal capacity of the detector. It is important to use a small detector and to focus the radiation on it. The *rate* at which the detector heats up or cools down determines how fast the detector responds to a change in radiation intensity as experienced when an absorption band is recorded. Ths constitutes the *response time* of the detector. For bolometers, the response time is long, that is, of the order of seconds. Consequently a complete spectrum scan may take 20 minutes. This precludes the use of a bolometer when fast responses are necessary such as when recording the real-time IR spectra of gas chromatographic effluents.

b. Thermocouples

A thermocouple (Fig. 5.12) is made by welding together at each end two wires made from different metals. If one welded joint (called the *hot junction*) becomes hotter than the other joint (the *cold junction*), a small electrical potential develops between the joints.

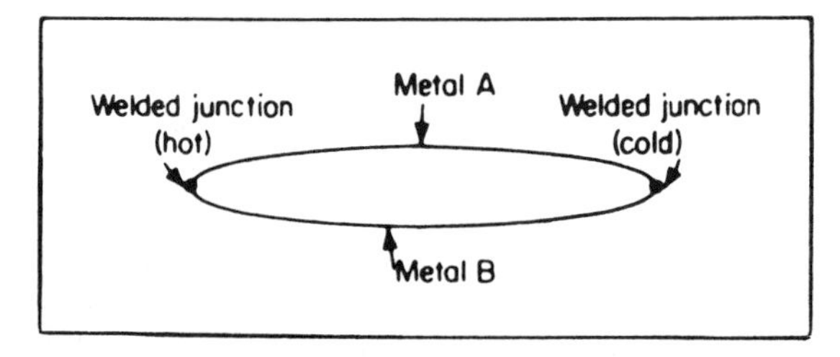

Figure 5.12 A thermocouple.

In IR spectroscopy, the cold junction is carefully screened in a protective box and kept at a constant temperature. The hot junction is exposed to the IR radiation, which increases the temperature of the junction. The potential difference generated in the wires is a function of the temperature difference between the junctions and, therefore, of the intensity of IR radiation falling on the hot junction. As in the bolometer, the response temperature of the thermocouple is slow.

c. Thermistors

A thermistor is made of a fused mixture of metal oxides. As the temperature of the mixture increases, its electrical resistance decreases (as opposed to the bolometer). This relationship between temperature and electrical resistance allows thermistors to be used as IR detectors in the same way as bolometers. The thermistor typically changes resistance by about 5% per °C. Its response time is also slow.

d. Golay Detectors

The Golay detector is a small hollow cell filled with a nonabsorbing gas such as xenon. In the center of the cell is a blackened film. Radiation falls on the blackened film and is absorbed, causing the film to heat up. In turn, the film heats the enclosed gas in the confined space. This causes the internal pressure of the cell to increase. One wall of the cell is a thin convex mirror that is part of an optical system. As the pressure inside the cell increases, the mirror bulges. This in turn modifies the quantity of light reaching the detector of the optical system. In short, a change of radiation intensity falling on the Golay detector causes a change in the readout from the optical detector. The instrument operates best when the radiation falling on the Golay detector is chopped at a frequency between 10 and 15 cps.

An important advantage of this detector is that its useful wavelength range is very wide. The response is linear over the entire range from the ultraviolet through the visible and infrared into the microwave range to wavelengths about as long as 7.0 mm. The Golay detector's response time is about 10^{-2} sec, much faster than that of the bolometer, thermistor, or thermocouple.

e. Semiconductor Detectors

Semiconductors are materials that are insulators when no radiation falls on them but become conductors when radiation falls on them. Exposure to radiation causes a very rapid change in their electrical resistance and therefore a very rapid response to the IR signal. The basic concept behind this system is that an IR photon displaces an electron in the detector, changing its conductivity greatly. In order to do this, the photon must have sufficient energy to displace the electron. For this reason, early semiconductors were limited in the wavelength range over which they were applicable. For example, lead sulfide detectors were sensitive to radiation with wavelengths as long as 3 μm but were insensitive to longer wavelengths. Such detectors, of course, had a very limited application, except in the near IR. More recently, other materials, such as lead telluride, indium antimonide, and germanium doped with copper or mercury, have been used as semiconductor detectors. Doped germanium detectors cooled in liquid helium are sensitive to radiation with wavelengths as long as 120 μm and for this reason are very popular. This range covers more than the entire IR range. There are other semiconductors that can be used at room temperature which are sensitive to a lower range of wavelengths but are nevertheless sensitive over the entire IR range.

In general, a semiconductor detector is fabricated with the semiconductor material deposited on glass in a sealed, evacuated envelope. Exposure to radiation causes a rapid change in the material's conductivity. The response time of this detector is the time required to change the semiconductor from an insulator to a conductor, which if frequently as short as 1 nsec. This fast response time is a valuable attribute and has enabled fast-scan infrared to become reality.

Semiconductor detectors are very sensitive, very fast, and are finding wide acceptance in the field of IR spectroscopy. The spectral responses of some semiconductor detectors are shown in Figs. 5.13 and 5.14.

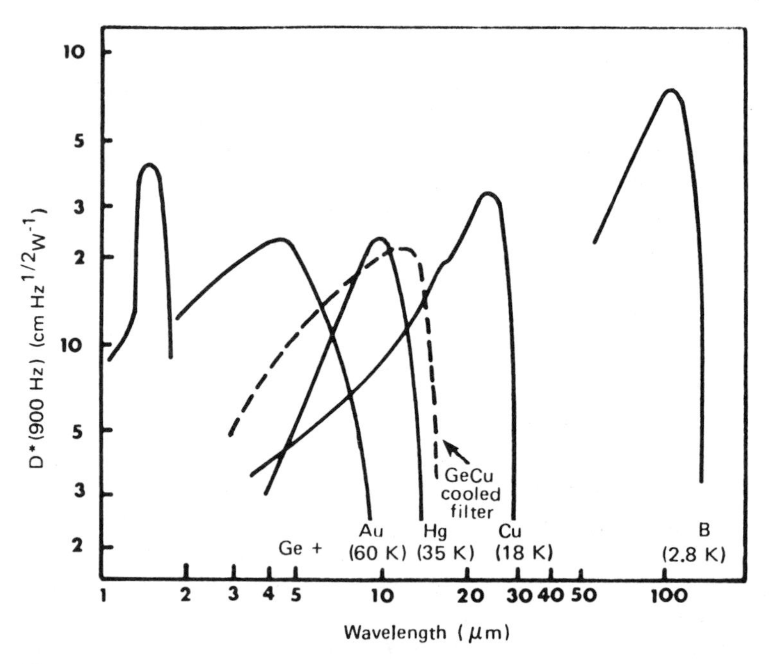

Figure 5.13 Spectral responses of typical doped germanium detectors.

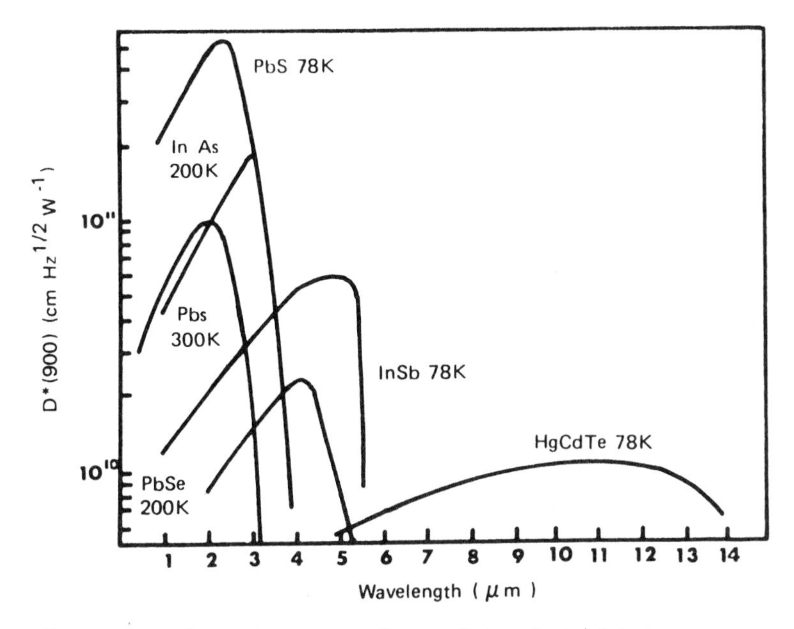

Figure 5.14 Spectral responses of some simiconductor detectors.

f. Pyroelectric Detectors

A dielectric placed in an electrostatic field becomes polarized, depending on the dielectric constant. If the field is removed, the polarization usually disappears, except with ferroelectric compounds, which retain a strong residual polarization. Sometimes their residual polarization is temperature sensitive. Such materials are *pyroelectric*.

A pyroelectric detector consists of a thin dielectric flake on the face of which an electrostatic charge appears when the temperature of the flake changes. This happens upon exposure to infrared radiation. Electrodes attached to the flake collect the charge, creating a voltage.

A pyroelectric flake is cut from a single crystal and is very small. It varies in size from about 0.25 to 12.0 mm^2. The radiation from the IR instrument must be focused on this small flake for operation, which is often demanding and never convenient.

The most common pyroelectric is triglycerine sulfate (TGS); however, its response rapidly deteriorates above 45°C and is lost above the Curie point (49°C). For this reason it is usually cooled in liquid N_2. Recently, deuterated TGS detectors have become available that can be used at room temperature. They are the detector of choice with FTIR, superseding Golay detectors. Their response is fast, so they can be used for multiplex scanning, a process often employed in FTIR.

g. Photon Detectors

Semiconductors have become very useful as detectors for IR. These detectors operate in the same manner as those discussed under semiconductors (see Chapter 6, Sec. C. p. 290), but they must respond to IR radiation, which has less energy per photon than UV light. A p-type InSb semiconductor forms a thin layer over an n-type semiconductor. IR falling on the detector generates a voltage, which is a function of the light intensity. This system works at wavelengths down to 5.5μm and must be operated in liquid N_2.

Lead telluride detectors are sensitive down to 18 μm and consequently are useful for the mid-IR range. Mercury cadmium telluride (MCT) detectors are more sensitive than lead telluride detectors, and response times as low as 20 nsec have been achieved. This permits very rapid scanning and improved sensitivity. These detectors are now widely used in FTIR.

h. The Effect of Response Time on Recorded Signal

The length of time that a detector takes to reach a steady signal when radiation falls on it is called the *response time*. This varies greatly with the type of detector used and has a significant influence on the design of the IR instrument. Thermal detectors such as thermocouples, thermistors, and bolometers have very slow response times, on the order of seconds. Consequently, when a spectrum is being scanned, it takes several seconds for the detector to reach an equilibrium point and thus give a true reading of the radiation intensity falling on it. If the detector is not exposed to the light long enough, it will not reach equilibrium and an incorrect absorption trace will be obtained. Consequently, it is normal for routine IR instruments to take on the order of 15 min to complete an IR scan. Attempts to decrease this time result in errors in the intensity of the absorption bands and recording of distorted shapes of the bands.

The slow response time is due to the fact that the temperatures of the systems must change in order to generate the signal to be measured. When there is a change in radiation intensity, the temperature at first changes fairly rapidly, but as the system approaches equilibrium, the change in temperature becomes slower and slower and would take an infinitely long time to reach the true equilibrium temperature. It should also be remembered that when an absorption band is reached, the intensity falling on the detector decreases and the response depends on how fast the detector cools. Naturally, cooling in liquid nitrogen speeds up cooling, exaggerates the temperature change on heating, and improves response time, and this approach is often used.

Semiconductors operate on a different principle. When radiation falls on them, they change from a nonconductor to a conductor. No temperature change is involved in the process; only the change in electrical resistance is important. This takes place over an extremely short period of time. Response times of the order of nanoseconds are common. This enables instruments to be designed with very short scanning times. It is possible to complete the scan in a few seconds using such detectors. These kinds of instruments are very valuable when put onto the end of a gas chromatograph and used to obtain the IR spectra of the effluents. Such scans must be made in a few seconds and be completely recorded before the next component emerges from the gas chromatographic column.

i. Fourier Transform Systems

If we regard light as a wave, then we accept the fact that it has a wavelength. If two beams of light of the same wavelength are brought together in phase, the beams reinforce each other and continue down the light path. However, if the two beams are out of phase, destructive interference takes place. This interference is at a maximum when the two beams of light are 180° out of phase. Advantage is taken of this fact in the Fourier transform spectrophotometer. This system is based on the Michelson–Morley experiment used to measure the influence of the earth's rotation on the speed of light. A schematic diagram of the equipment is shown in Figure 5.15. The system consists of four optical arms, usually at right angles to each other, with a beam splitter at their point of intersection. Radiation passes down the first arm and is separated by a beam splitter into two perpendicular half-beams of equal intensity that pass down into other

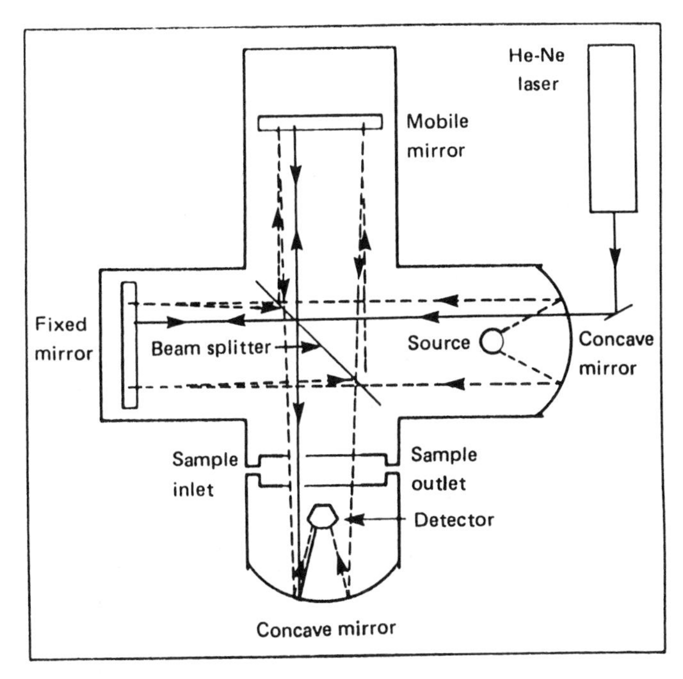

Figure 5.15 Schematic of a Fourier transform detector system.

arms of the spectrometer. At the ends of these arms the two half-beams are reflected by mirrors back to the beam splitter, where they recombine and are reflected together onto the detector.

If the initial radiation is at one wavelength and in phase with itself (which it always is) and if the side arm paths are equal in length, then, when the two half-beams are recombined, they will still be in phase, reinforcing each other, and the maximum signal will be obtained on the detector. If the mirror in one arm is moved up by one-quarter of a wavelength, then the half-beams will be one-half of a wavelength out of phase with each other; that is, they will interfere with each other. In practice, the mirror in one arm is kept stationary and that in the second arm is moved slowly in the direction of the beam splitter. As the moving mirror moves, the net signal falling on the detector is a cosine wave with the usual maxima and minima when plotted against the travel of the mirror. The frequency of the cosine signal is equal to

$$f = \frac{2\nu}{\lambda}$$

where

f = frequency
ν = velocity of the moving mirror
λ = wavelength of radiation

The frequency of modulation is therefore proportional to both the wavelength of the incident radiation and the velocity of the mirror.

In practice, however, repeated bursts of radiation of all wavelengths to be considered are obtained form the source and travel down the arms of the interferometer. They proceed through the sample holder before reaching the detector, where absorption by the sample can take place.

In practice, it is mechanically difficult to move the reflecting mirror at a controlled, known, steady velocity. The original position of the moving mirror must be highly reproducible each run. The position and the velocity are controlled by using a laser beam that is shone down the light path producing an interference pattern with itself. The cosine curve of the interference pattern of the laser is used to regulate the velocity of the moving mirror.

There are a number of advantages and disadvantages of the Fourier transform spectrophotometer. Among the advantages is the fact that the sample is exposed to the entire radiation beam at the same time and absorption effects at all wavelengths modify the final signal. This is known as the *multiplex* advantage, and it greatly increases the sensitivity and accuracy in measuring the wavelengths of absorption maxima. The Fourier transform spectrophotometer is considerably more accurate than the monochromator system normally employed. Another aspect of the multiplex advantage is that the intensity of the radiation falling on the detector is much greater. Resolution is dependent on the linear movement of the mirror. As the total distance traveled increases, the resolution improves. It is not difficult to obtain resolutions of 0.5 cm^{-1}. A list of the advantages of FTIR over dispersive IR is given in Table 5.3.

Background Correction. The two main sources of background absorption (i.e., absorption from material other than the sample) are the solvent and the air in the optical light path. In a conventional double-beam system, both are automatically eliminated by comparing the sample beam to the reference beam and recording the differences. However, FTIR is a single-beam system and both contribute to the signal, so corrective measures must be taken.

Table 5.3 Comparison of Dispersive and Fourier Transform IR

Dispersive IR	Fourier Transform IR
1. Many moving parts result in mechanical slippage	1. Only mirror moves during an experiment.
2. Calibration against reference spectra required to measure frequency	2. Use of laser provides high frequency accuracy (to 0.01 cm^{-1})
3. Stray light within instrument causes spurious readings	3. Stray light does not effect detector, since all signals are modulated
4. In order to improve resolution, only small amount of IR beam is allowed to pass through the slits	4. Much larger beam may be used at all times; data collection is easier
5. Only narrow-frequency radiation falls on the detector at any one time	5. All frequencies of radiation fall on detector simultaneously; improved S/N ratio obtained quickly (*Felget advantage*)
6. Slow scan speeds make dispersive instruments too slow for monitoring systems undergoing rapid change (e.g., gas chromatographic effluents)	6. Rapid scan speeds permit monitoring samples undergoing rapid change
7. Sample subject to thermal effects from the focused beam	7. Beam is not focused, hence sample is not subject to thermal effects
8. Any emission of IR radiation by sample will fall on detector	8. Any emission of IR radiation by sample will not be detected
9. No multiplex advantage	9. Simultaneous interference by all wavelengths on all other wavelengths gives more complete spectra (multiplex advantage)

Solvent Absorption. The solvent absorption spectrum may be taken directly by putting pure solvent in the sample container and recording it on a computer (spectrum A). The absorption spectrum of a sample of that solvent is then taken and recorded (spectrum B). Spectrum A is then subtracted from spectrum B, leaving the net spectrum of the sample.

Air Absorption. CO_2 and H_2O in air are both strong IR absorbers. Both occur in air in the optical light path and contribute to any IR signal obtained in the instrument. This background signal may be corrected for in one of two ways. First, the air spectrum may be recorded by running a spectrum with no sample present. This constitutes the "blank" spectrum and is recorded and stored on tape. Samples may be run and their total spectrum (air + sample) recorded. The air spectrum may then be subtracted, leaving the net sample signal. Any suspected changes in humidity or CO_2 content can be corrected by updating the blank spectrum at regular intervals. The second method, *optical purging*, is more difficult. The optical system can be purged with dry air or N_2 by removing CO_2 and H_2O in the process. This eliminates the necessity of correcting for the blank signal derived from impurities in the air. The purge gas may be N_2 from a dewar flask or purified air.

Recent FTIR models have tightly enclosed optics. The frequency of purging may be reduced to as little as once every several months, with a warming light alarm denoting the necessity to purge. This system greatly reduces the cost of running the instrument.

Typical background spectra for purged and unpurged cells are shown in Fig. 5.16. It can be seen that H_2 and CO_2 absorb extensively and cover up the regions 2800–2100 cm^{-1} and 1800–1500 cm^{-1}, making interpretation of spectra in these regions difficult. Note that there is a small amount of absorption because the space containing the sample compartment contains a little CO_2 and H_2O. If

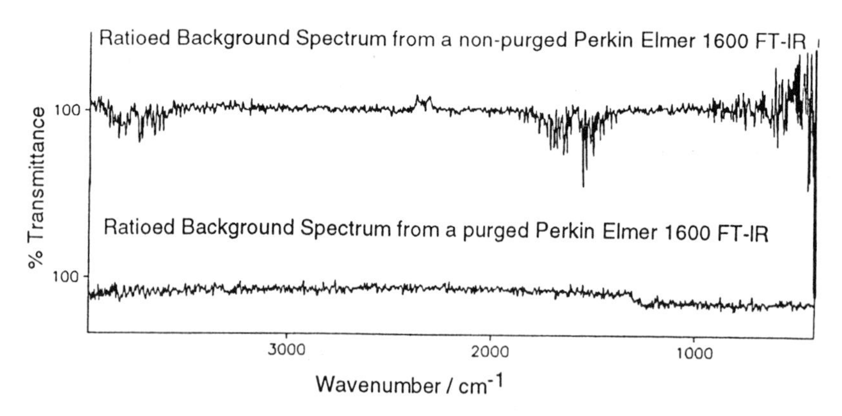

Figure 5.16 Typical background spectra for purged and unpurged cells.

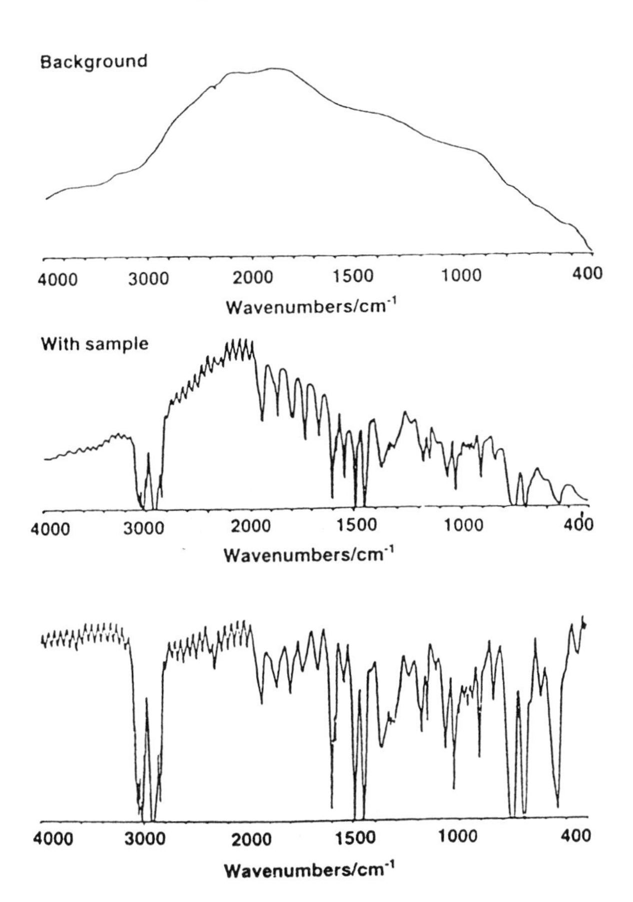

Figure 5.17 Raw FTIR spectra, including (a) instrumental background, (b) background plus sample, and (c) sample spectrum achieved by subtracting (a) from (b).

necessary, this can be corrected for by taking a blank spectrum of CO_2 and H_2O and subtracting from the sample spectrum using the computer program. A typical spectrum is shown in Fig. 5.17.

A schematic diagram of a commercial instrument is shown in Fig. 5.18. The sample area is close to a focal point in the optical system. It can be seen that, although there are two beams that are related to each other, this is a single-beam system. Any absorption by the air in the light path or the solvent contributes to the final signal.

Repetitive Scans. FTIR has the advantages that a spectrum can be scanned and rescanned in a few seconds and the spectra superimposed on a tape. In this manner, the IR signal accumulated is additive, but the noise level (N) in the signal is random. The S/N ratio increases with the square root of the number of scans. (i.e., if 64 scans are accumulated, the S/N ratio increases 8 times).

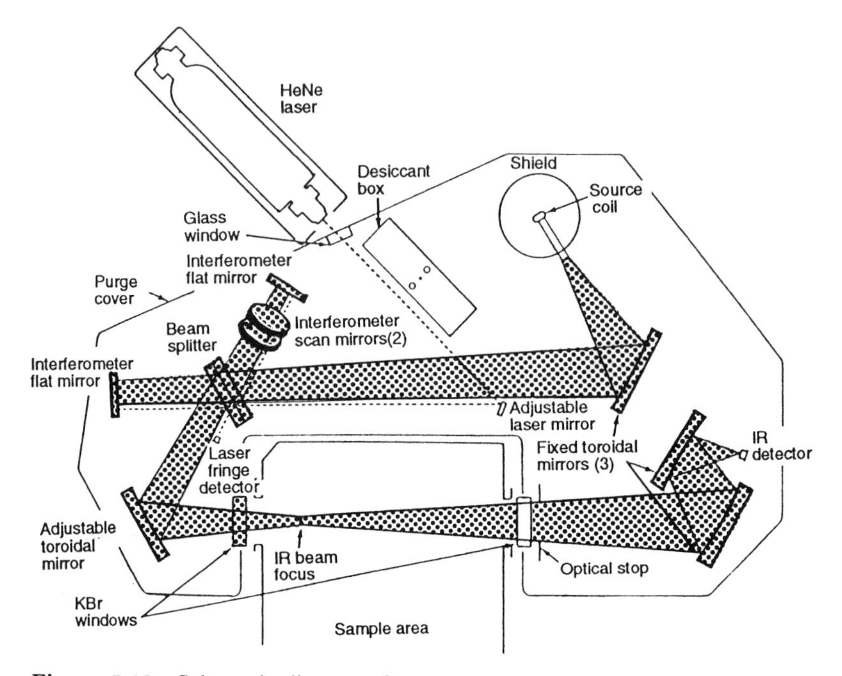

Figure 5.18 Schematic diagram of a commercial instrument.

FTIR has the potential to be many orders of magnitude more sensitive than conventional IR. However, in practice the improvement is limited by the sheer number of scans necessary to continue to improve the sensitivity. For example, 64 scans improve sensitivity 8 times, but 256 scans are required to improve 16, 1024 to improve 32, 4096 times to improve 64, and 16,380 to improve 128 times. It would take a considerable time to run 65,380 scans, prohibitive number. A practical limit of 1 to 2 orders of magnitude sensitivity increase is therefore normal unless extenuating circumstances merit the time and expense. However, the potential for high sensitivity is available using FTIR.

FTIR Microscopy. The use of FTIR has been extended to the examination of small samples. This has been achieved with great success by combining FTIR with an optical microscopy. Best results are achieved by cutting a very thin slice of the sample through which radiation can penetrate. Such absorption spectra are generally better than spectra obtained by examining the light reflected from the sample.

This system has several distinct advantages over previous models:

1. The design allows optical viewing of the sample, permitting precise location of the area to be examined.

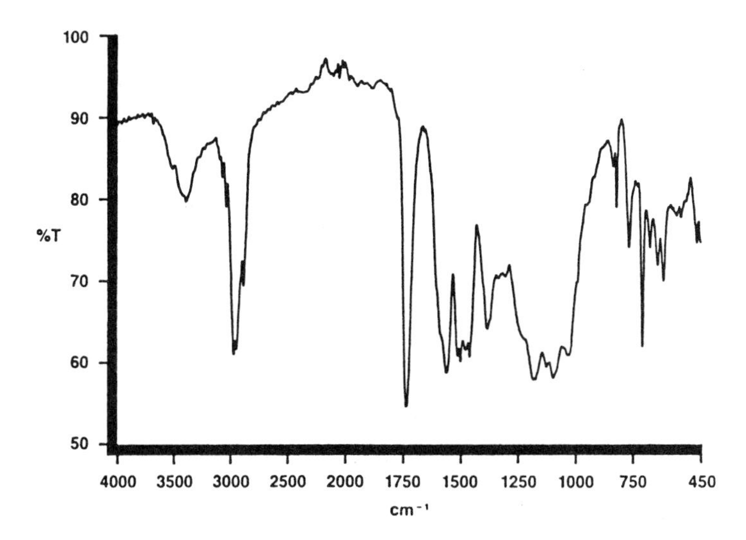

Figure 5.19 Transmission spectrum of a blue paint chip from an American car measured using a miniature diamond anvil cell and the Perkin-Elmer 5X beam condenser at 4-cm^{-1} resolution and 64 scans.

2. The use of a mercury cadmium telluride detector increases the sensitivity of the technique.
3. The use of FTIR allows more reliable assignment of the absorption wavelength. Also, repetitive scans permit further sensitivity improvement.

Uses of FTIR Microscope. A prime example of the use of FTIR microscopy is in the examination of polymers. Polymers have become very important materials in many commercial areas. Their physical properties are very dependent on their chemistry and their physics. Different polymers such as polyethylene and nylon are used for different things. The presence of impurities, gels, crystalline patches, etc., greatly affect their function.

Contamination spots on paper and other surfaces have been studied nondestructively by FTIR. Food-packaging materials may be made up of several layers of different polymers rolled into each other to provide a single plastic sheet with the desired properties. Typical layers are 200-μm, 15-μm, and 10-μm thick. And thin as they are, it is possible to obtain acceptable spectra and identify the polymers involved.

In forensic sciences, FTIR has been used to examine paint chips from automobile accidents. A typical spectrum is shown in Fig. 5.19. Hit-and-run drivers frequently leave traces of paint on cars with which they collide. This information can lead to the driver's arrest. Other uses include the examination of different fibers, drugs, and traces of explosives.

Additional applications of IR to forensic science microscopy include characterization of *pharmaceuticals*, catalysts, minerals, *semi-precious stones*, *adhesives*, composites, processed metal surfaces, and *semi-conductor* material. Little or no sample preparation is necessary. The sensitivity obtainable is subnanogram quantities.

In biological samples, FTIR has been used to examine plant leaves and stems, animal tissue, and other biomaterials. Frequently, such information cannot be obtained by any other means.

Failure of electronic equipment can be brought about by contaminants. The identity and thereby the source of such contaminants has been delineated by FTIR microscopy. This permits identification of very small quantities of such materials.

Another unusual application is the identification of fossilized organic material in coal. Because it is possible to view the sample being characterized, it is possible to characterize exactly the portion of the sample of interest to the geologist or archaeologist.

In the medical field, an important application is the study of human tumor cells. It has been found that absorption bands at 1240 cm^{-1} and 1080 cm^{-1} differ in normal and cancerous cells. In cancerous cells the absorption at these wavelengths is about twice as strong as normal cells. The chemical interpretations of these spectral changes is not clear, but they appear to involve the symmetrical vibrations of PO_2 associated with DNA and perhaps the Amide I band. Studies such as these help us understand cancer formation from a scientific viewpoint based on chemical reactions rather than from treatment based on empirical data (does this drug work?). Scientific understanding is a necessary step in controlling eratic cell growth.

5. Sample Cells

Infrared spectrometry deals with the measurement of the IR spectrum. It is one of the few branches of analytical chemistry that can be used for the characterization of solid, liquid, and gas samples. Naturally, samples of different phases must be treated differently. The material used to contain the sample must always be transparent to IR radiation. This limits the selection to certain salts, such as NaCl or KBr. A final choice of the material used depends on the wavelength range to be examined. A list of commonly used materials is shown in Table 5.2.

a. Solid Samples

Three techniques are available for preparing solid samples for IR spectrometry. First, the sample may be ground to a powder. The powder can be made into a thick slurry, or mull, by grinding it with a greasy, viscous liquid, such as Nujol (a paraffin oil) or chlorofluorocarbon greases. This method is good for qualitative studies, but not for quantitative analysis.

The second technique is the KBr pellet method, which involves mixing a finely ground solid sample with powdered potassium bromide. The mixture is compressed under very high pressure (at least 25,000 psig) to form a small disk about 1 cm in diameter and 1–2 m thick. Inexpensive presses are commercially available for this. The disk is transparent to IR radiation and may be analyzed directly.

In the third method the sample is deposited on the surface of the KBr or NaCl cell by evaporating a solution of the solid to dryness. Infrared radiation is then passed through the thin layer deposited. It is difficult to carry out quantitative analyses with this method, but it is useful for rapid qualitative and semiquantitative analyses.

In general, analyses of solid samples do not yield spectra as fine or as quantitative as those of liquid samples. Solid samples are therefore avoided if possible.

b. Cells for Liquid Samples

The easiest samples to handle are liquid samples. These may be poured, frequently with no preparation, into circular cells made of NaCl, KBr, and so on (see Table 5.2), and their IR spectrum determined directly (Fig. 5.20). For double-beam work, "matched" cells are used. One cell is used to contain the sample, and the other cell to contain a reference material (e.g., the solvent used in the sample). Matched cells are similar in total thickness and in cell wall thickness. All cells must be protected from water, because they are water soluble. Organic liquid samples should be dried before being poured into the cells; otherwise, the cell surfaces become opaque and cause erroneous results to be obtained.

Such etching is frequently a very serious problem, particularly when quantitative analyses are to be performed. After continued use, even for a relatively short period of time, the surfaces become etched and the distance between the surfaces will not be constant throughout the cell. This can be determined as follows.

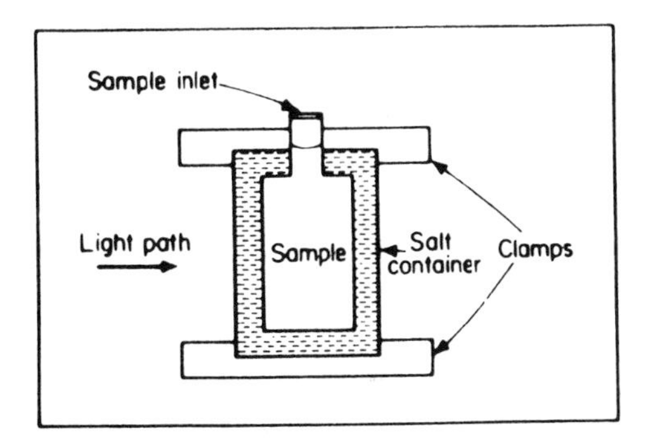

Figure 5.20 Liquid cell for IR absorption studies.

It will be remembered that Beer's law indicates that the absorbance = abc, where b is the path length through the sample, or in this case the diameter of the empty cell. In order for quantitative data to be reliable, b must be a constant, or at least measurable and correctable. A measurement of b may be performed by using a procedure based on interference fringes. An empty and dry cell is put into the light path, and the instrument is scanned over a suitable wavelength range. Partial reflection of the light takes place at the inner surfaces, forming two beams. The first beam passes directly through the sample cell, and the second beam is reflected twice by the inner surfaces before proceeding to the detector. The second beam therefore travels an extra distance $2b$ compared to the first beam. If this distance is a whole number of wavelengths ($n\lambda$), then the two emerging beams remain in phase and no interference is experienced. However, if $2b = (n + 1/2)\lambda$, interference is experienced. and the signal reaching the detector diminished.

A typical signal is shown in Fig. 5.21. The signal is a sine wave, and each wave indicates an interference fringe. The path length of the sample holder can be measured by using the formula.

$$b(\mu\text{m}) = \frac{n}{2\eta}\left(\frac{\lambda_1\lambda_2}{\lambda_2 - \lambda_1}\right)$$

where

n = number of fringes
η = refractive index of the sample (or air, if empty)
λ_1, λ_2 = wavelengths between which the number of fringes is measured

If λ is measured in micrometers, b is also micrometers. In Fig. 5.21 it can be seen that $n = 14$, $\lambda_1 = 2$, and $\lambda_2 = 20$. Therefore

$$b = \frac{14}{2}\frac{2 \times 20}{20 - 2} = 15.5\ \mu\text{m} \quad (\text{assuming that } \eta = 1)$$

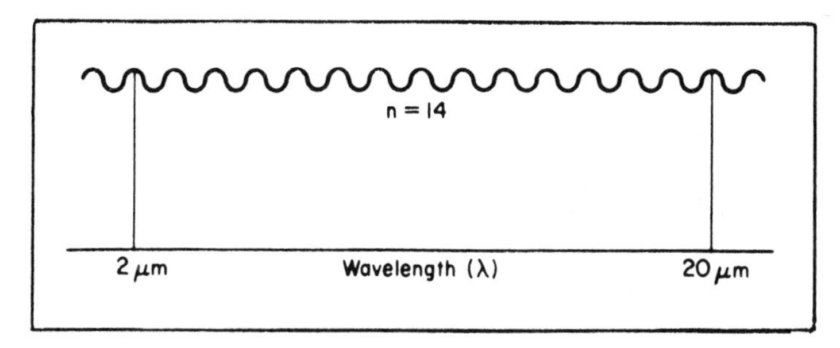

Figure 5.21 Typical interference pattern used to measure cell thickness.

For quantitative analysis it is often necessary to measure the path length in order to use calibration curves obtained with the same sample but at different times. If the cell becomes too etched, the interference pattern becomes noisy and the cell will have to be removed and repolished with a soft chamois cloth to regain a flat, shiny surface.

c. Gas Sample Cells

The gas sample cell is similar to the cell for liquid samples inasmuch as the surfaces in the light path are made of KBr, NaCl, and so on. To compensate for the small number of molecules of a sample contained in a gas, however, the cells are larger; usually they are about 10 cm long, but they may be up to 1 m long. Multiple reflections can be used to make the effective path length as long as 40 m, so that constituents of the gas can be determined. Such a cell is shown in Fig. 5.22.

The gas must not react with the cell windows or the reflecting surfaces. Gas analyses are performed with IR, but the method is not commonly used because of its lack of sensitivity.

d. Attenuated Total Reflectance

It is sometimes necessary to measure the absorption spectra of materials that are not transparent, such as varnishes, or materials that are hard, flat surfaces and cannot be put into an infrared cell, such as the paint on a door or wall or a piece of art. These types of samples can be analyzed by using attenuated total reflectance (ATR). A schematic diagram of equipment is shown in Fig. 5.23.

This system is an attachment that can be added to a conventional IR system and is used only for sampling. It consists of a thimble of material that is transparent to infrared, such as sodium chloride or potassium bromide. Radiation is reflected from the light source onto the thimble, where it penetrates to the far wall. The angle at which the light reaches the far wall must be less than the

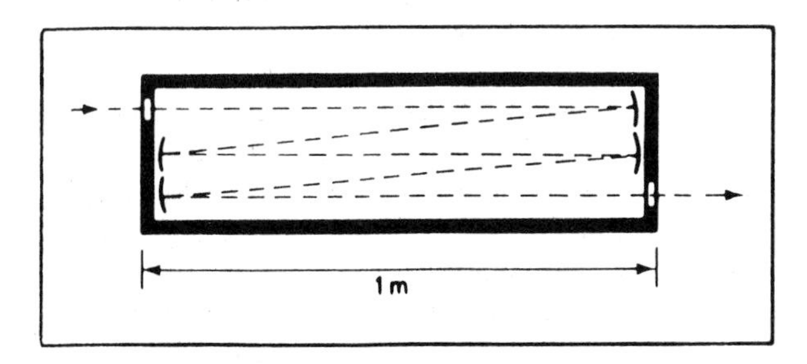

Figure 5.22 Gas absorption cell. Reflection of the light beam makes the effective path length several times longer than the cell length.

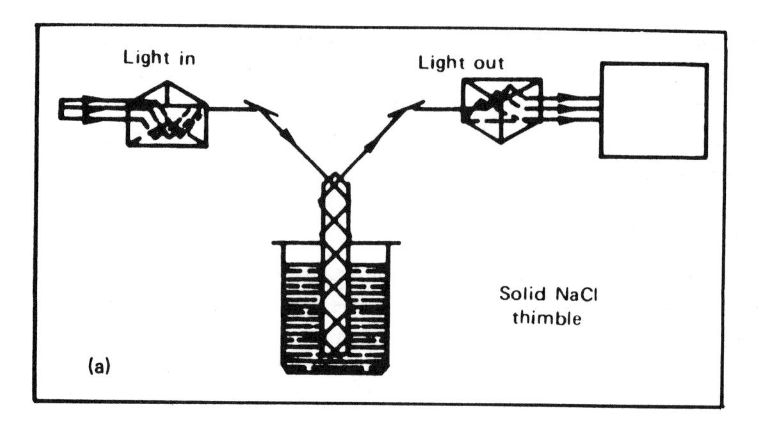

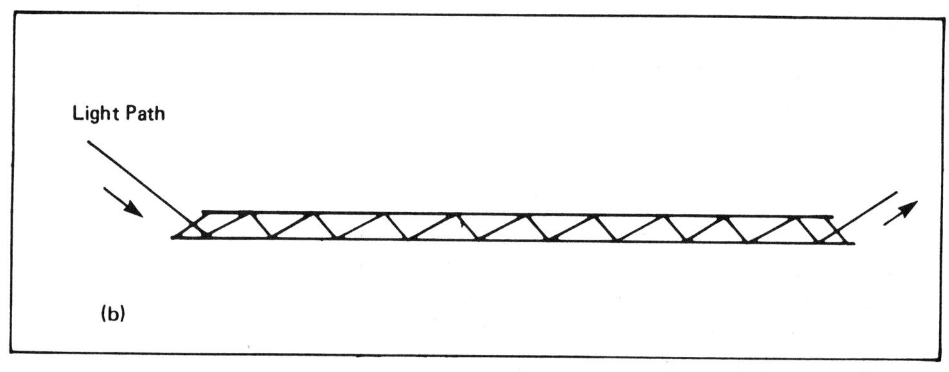

Figure 5.23 (a) Attenuated total reflectance cell. The cell is made of solid NaCl and radiation is reflected at the interface between the cell and the sample in which it is immersed. (b) Attenuated total reflectance cell with an aperture at each end.

critical angle so that total reflection takes place. A series of reflections therefore takes place down the thimble and back out again into the infrared light path. The thimble is then immersed in the sample and scanned again. If the sample absorbs at a particular wavelength, say, 3.3 μm (C → H), when radiation of that wavelength hits the interface between the thimble and the sample, coupling occurs and there is a loss in radiation caused by an interaction similar in many ways to absorption. By completing the scan, an IR absorption spectrum of the sample can be obtained. A second system uses only an aperture, as shown in Fig. 5.23b. It is claimed that the system simplifies the optics and doubles sensitivity.

The most essential practical feature of this system is that solid samples must be in actual molecular contact with the thimble. To achieve this, the thimble may be pressed against the surface of a sample such as a painting and the IR spectrum of that painting obtained without destroying the material. This is essential in examining artwork. The procedure can be used to obtain the IR spectrum of any solid surface and is often used in forensic science.

e. Diamond Cells

For longer wavelengths, sodium chloride cells are not transparent. For the study of metal surfaces on metal oxides, the pertinent wavelengths are often longer than 20 μm. Diamond is transparent at th wavelengths and offers the advantage that it is hard, stable, and not affected by water.* It has been used to study TiO_2, Ag_2O, and other metal oxides.

ATR can also be used to monitor organic reactions. For example, if the thimble is put into a mixture of reacting organic compounds, one particular wavelength can be monitored to indicate the disappearance of one of the reactants or the appearance of a product as the reaction proceeds. This eliminates the need to remove samples from the reaction vessel in order to obtain an IR spectrum and permits continuous monitoring of the reaction without disturbing the system.

A major use of ATR is in qualitative and quantitative IR analyses of opaque material. Samples such as inks, glues, stains, paints, and varnishes, which cannot normally be handled by IR methods, can be analyzed.

An interesting application is in the study of fossils. IR spectra can be obtained on the surface of fossilized plants or animals from areas as small as 2.5 X 1.0 mm. The method is nondestructive, and the samples need not be removed from the fossil surface. The method is of particular interest to paleontologists and archeologists. Fossilized leaves, amber, bone, fish, trilobile teeth, and bone from mesohippos and many other sample types have been examined.

f. Infrared Emission

Some samples are not amenable to absorption analysis or ATR. Such samples can be characterized by IR emission. The Boltzmann distribution [Eq. (2.7)] indicates that even at room temperature some molecules will possess vibrational energy. Consequently, some of the molecules will be in a vibrationally excited state. They may therefore emit IR molecules and return to the ground state. If the molecules are heated, many more will occupy the excited state and emit radiation, returning to the ground state. The radiation emitted is characteristic of the IR spectrum and can be used to identify the emitting sample. A schematic diagram of commercial equipment is shown in Fig. 5.24.

Some typical applications include analysis of gases, remote flames, discharges, photochemistry, photofragmentation, change in the composition of lu-

*C. E. Gigola and G. L. Haller, Diamond interval reflection cell for IR measurements on metal–metal oxide film. *Appl. Spec.*, *44.1*: 159 (1990).

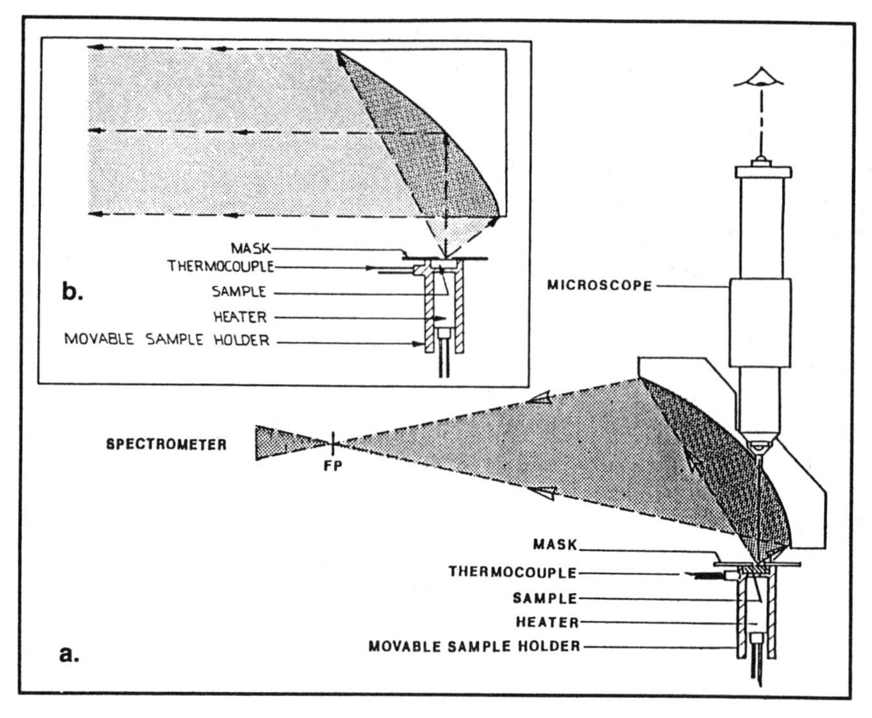

Figure 5.24 Sample configuration used in the Harrick Scientific ES-1 emission attachment. Ray diagram of IR emission collecting accessories: (a) ellipsoidal collector and (b) paraboloid collector. (Reprinted with permission from *Optical Spectroscopy: Sampling Techniques Manual*, Harrick Scientific Corporation, Ossining, New York, 1987.)

bricants during their use, molten salts, glasses, polymer films, silicon wafers, water deposits on various surfaces, and polymer coatings.

The technique has been coupled with FTIR to examine microscopic samples. The FTIR gives increased sensitivity.

In general, IR emission is useful for qualititive identification by quantitative analysis, is subject to many variables, and in many cases only semiquantitative data can be obtained.

g. Nondispersive Infrared

In industry it is sometimes necessary to monitor quality on a continuous basis. Infrared may be the method of choice, but it is usually not feasible to use conventional IR instruments under industrial-production conditions because they are too delicate. One hard hat containing a sandwich is usually enough to sicken a conventional IR spectrophotometer. Nondispersive processes have therefore been developed that are much sturdier and can be left running continuously.

A schematic diagram of a positive-filter nondispersive IR instrument is shown in Fig. 5.25. The system consists essentially of the radiation source and two mirrors that reflect two beams of light, which pass through the sample and reference, respectively, to two detectors. These detectors are hollow boxes similar in design to the Golay detector; each contains the gas phase of the compound being determined. The detector is selective for the compound.

If there is no sample in the sample call, detectors A and B will absorb the IR radiation and consequently their temperatures will increase. The temperature difference between the two detectors is measured and is at a minimum when there is no sample in the light path. When sample material is introduced into the sample cell, it absorbs radiation. The light falling on detector A is therefore decreased in intensity and the temperature of the detector decreases. The temperature difference T_2-T_1 between the two detectors increases. As the concentration of sample increases, the temperature of detector A decreases and T_2-T_1 increases. The relationship between T_2-T_1 and sample concentration is positive in slope—hence the term positive filter. The reference call is usually a sealed cell containing N_2 gas, which does not absorb in the IR.

A negative-filter system is shown in Fig. 5.26. In this case detector A contains a nonabsorbing gas such as N_2 and detector B the compound being determined. With no sample in the sample cell, the temperature of detector B is a maximum and detector A is at room temperature; T_2-T_1 is a maximum. As the concentration of sample is increased, light is absorbed in the sample cell at the absorption bands of the sample. Less light falls on detector B, which only absorbs at absorption bands of the sample, and its temperature decreases; that is, T_2-T_1 decreases. Hence the relationship between T_2-T_1 and sample concentration is negative in slope. The response of the system is usually better if the light beams are chopped, and it is common to chop the light as it leaves the

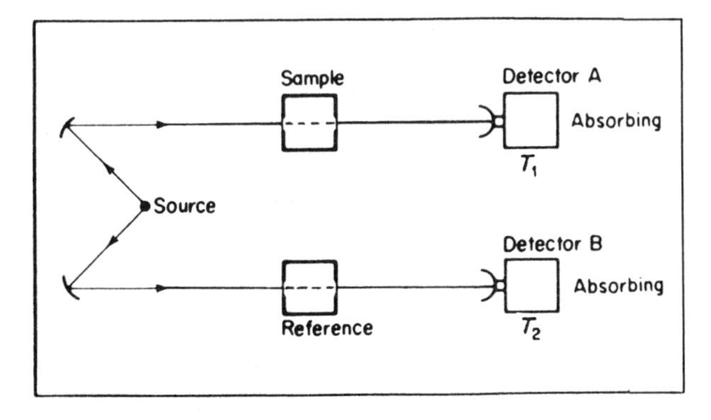

Figure 5.25 Positive-filter nondispersive IR.

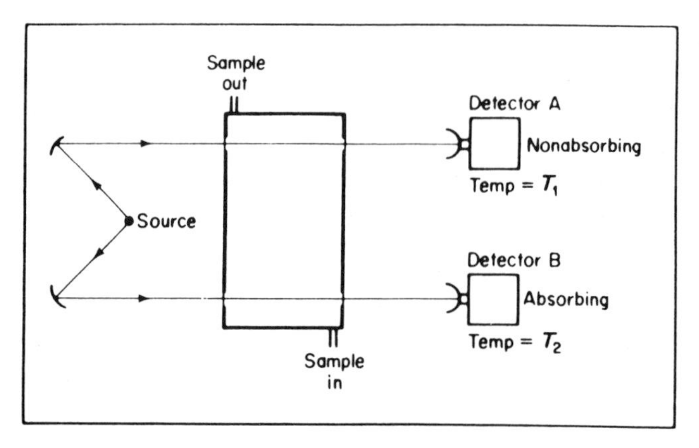

Figure 5.26 Negative-filter nondispersive IR.

source so that the sample and reference beams are chopped equally. This helps correct for changes in room temperature during operation.

There are several modifications to these two systems, particularly in determining the temperature difference as a measure of sample concentration. For example, the pressure difference in the detector cells generated by the temperature change can be measured, and a quicker response is obtained. An alternative method is to use the two parallel walls of each cell as a condenser in an electrical circuit. The condenser is charged to a given voltage. As the temperature of one cell increases, its pressure increases and the wall bulges outward, decreasing the capacity of the condenser. This results in a change in the voltage, which can be measured continuously and quite accurately.

A problem may be encountered if an interfering material is present in the sample that has absorption bands overlapping those of the sample. This will result in a direct interference in the measurement. The problem can be overcome by placing a cell containing a pure sample of the interfering material in the sample arm. In this fashion all the absorbable radiation at a common wavelength is absorbed and this eliminates any variation in absorption due to the impurity in the sample because all the light at this wavelength has been removed.

Nondispersive IR systems are good for measuring concentrations of specific compounds under industrial and other similar circumstances. As can be readily understood, they are not generally used as research instruments and do not have scanning capabilities. However, they are robust and enduring and can be used of the continuous monitoring of specific compounds.

h. Photothermal Beam Deflection Spectroscopy (PBDS)

An ingenious method for surface analysis has been developed by A. C. Boccara (1980) and refined by M. J. D. Low (1984) for the nondestructive analysis of surfaces. In this process infrared radiation is shone onto the sample continuously and a laser beam probe is directed across (but not onto) the sample surface to a detector. The beam is as close to the surface as possible. If the sample absorbs radiation, it becomes hot, heating the nearby gases and consequently changing their refractive index. The change in refractive index alters the direction of the laser beam. This deflection is monitored against the wavelength of the irradiating infrared, and from this the IR spectrum of the surface can be obtained. Using FTIR it is possible to improve sensitivity by scanning 400–500 times.

The advantages of this method are that it is rapid and nondestructive. This fact is very important if the samples are valuable—such as works of art—or rare—such as fossils.

D. ANALYTICAL APPLICATIONS

The introduction and widespread use of FTIR has resulted in considerable extension of the uses of IR in analytical chemistry. With regard to wavelength assignment, speed of analysis, and sensitivity, FTIR has opened new fields of endeavor. Some of these uses are described below.

The two most important analytical applications of IR spectroscopy are in the qualitative and quantitative analysis of organic compounds and mixtures. We pointed out at the beginning of this chapter that the frequency of the radiation absorbed by a given molecule is characteristic of the molecule. Moreover, functional groups, such as $—CH_3$, C═O, and C—OH, act as separate groups and have characteristic absorption spectra. This fact, which enables us to identify many of the functional groups that are important in organic chemistry, provides the *qualitative* basis of IR absorption spectroscopy. Since different molecules have different IR spectra, depending on the structure and mass of the component atoms, it is possible, by matching the absorption spectra of unknown samples with the characteristic absorption frequencies or wavelengths of different functional groups (given in Table 5.1), to identify the samples. In addition to identifying a molecule, or its functional groups, we can acquire information about its geometry by analyzing its IR spectrum.

Earlier (Sec. A.2) we said that the degree of absorption is controlled by the rate of change of the dipole during vibration. This rate of change is a physical property of the molecule and does not vary under normal conditions. We can measure the extent of absorption by a sample by using a solution of the sample with a known concentration. If now we were to measure the extent of absorption by a solution of the sample of unknown concentration, the result could then be compared and we could determine that sample's concentration. Thus, as a *quantitative tool*, IR spectroscopy enables us to measure the concentration of various samples.

1. Qualitative Analysis

Qualitative analysis is a major part of the work of an analytical chemist. Since it is better to give no answer than an incorrect answer, most analytical chemists perform qualitative analyses using an array of techniques that overlap and confirm each other, providing in the sum more information than could be obtained with the separate individual methods. The most commonly used methods are IR to tell which groups are present; NMR to indicate their relative positions in the molecule; mass spectrometry to define the molecular weight; chromatography to determine the number of compounds present, separate and purify the compound of interest, and indicate impurities; and UV to study unsaturated or substituted compounds. Each technique provides an abundance of valuable information on molecular structure, but a combination of methods is now used often for more reliable confirmation of results.

It must also be said that the value to qualitative analysis of prior knowledge about the sample cannot be overemphasized. Before trying to interpret an IR spectrum, it is important to find out as much as possible about the sample, For example, to identify the products of an organic reaction, it is very valuable to have information about the materials that were present before the reaction started, the compounds the reaction was expected to produce, the possible degradation products that may come about after the reaction, and so on. Armed with as much of this information as possible, we may be able to identify the molecules in the sample.

The general technique for qualitative analysis is based on the characteristics of molecular structure and behavior mentioned at the beginning of the chapter. That is, the frequency of vibration of different parts of a molecule depends on the weight of the vibrating atoms (or groups) and the bond strength. These physical properties of the molecules do not vary significantly under normal circumstances. Since the absorption frequency is the same as the vibration frequency, the frequency of radiation absorbed by different groups is characteristic of that group. Table 5.1 is a list of functional groups and the relevant absorption frequencies.

Qualitative analysis is carried out by matching the wavelengths of the absorption bands in the spectrum with the relevant functional groups. This comparison is best done by combining prior knowledge of the sample with information from tables such as Table 5.1. Before a positive identification can be made, all the absorption bands typical of a functional group must be observed.

Three typical IR absorption spectra are shown in Figs. 5.27–5.29. Note that all the absorption bands in Fig. 5.27 are caused by the stretching or bending of C—H bonds. These absorption bands are common in all organic compounds containing paraffinic hydrocarbons. The spectrum in Fig. 5.28 shows an extra absorption band at 3.0 μm caused by the stretching of the O—H bond of water. The C—H absorption bands are similar to those evident in Fig. 5.27.

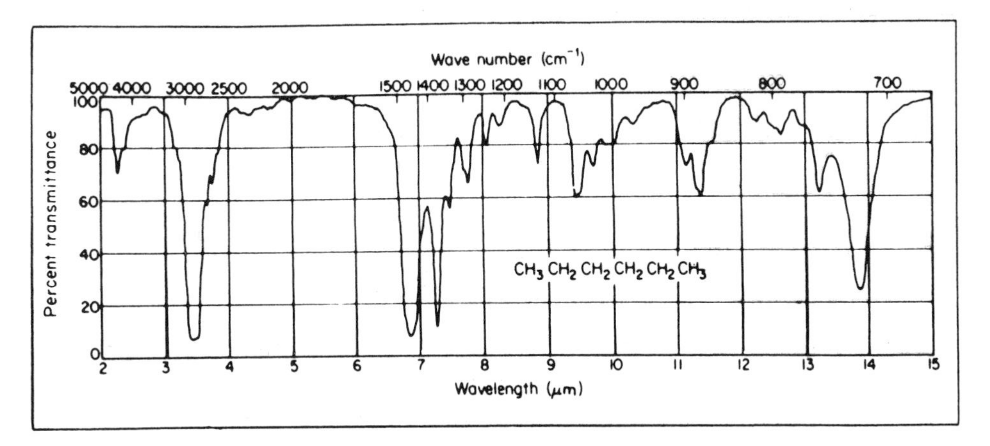

Figure 5.27 Absorption spectrum of hexane (a saturated paraffin).

A group found in alcohols is the OH band also found in water. The C—H absorption bands are similar to those evident in Fig. 5.27. The group typifying alcohols is the C→OH group. The absorption spectrum in Fig. 5.29 shows the bands due to O→H at 3.0 μm and those caused by C → OH vibration at 9.4 μm. Alcohols must be identified by the presence of both these bands in the absorption spectrum. The presence of a 3.0-μm band can also be attributed to the OH bands in water, as shown in Fig. 5.28 and does not confirm the presence of alcohols.

Figure 5.30 is a spectrum that was obtained from an unknown compound. Using the information available in Table 5.1, we can interpret this spectrum. If

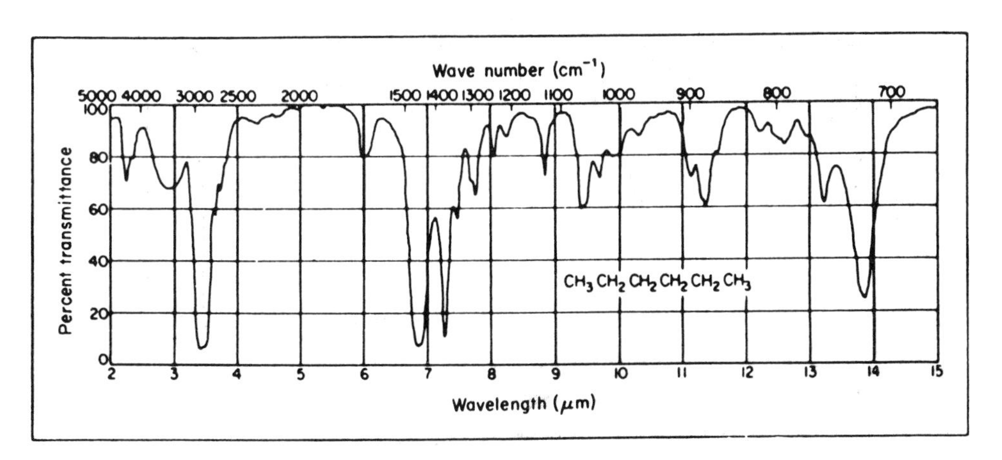

Figure 5.28 Absorption spectrum of wet Nujol.

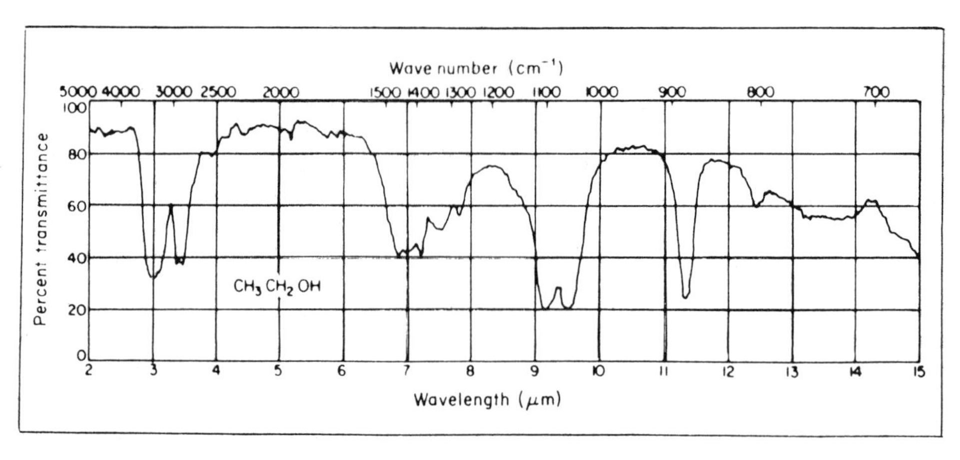

Figure 5.29 Absorption spectrum of ethanol.

our interpretation is correct, we will have identified all the major peaks as arising from CH_3, CH_2, C—O, C═O, and OH bonds. Figure 5.30 could therefore be the spectrum of

$$CH_3(CH_2)_nC\begin{matrix}{/\!\!/}O \\ \backslash OH\end{matrix}$$

that is, a long-chain fatty acid. Note that previous studies have shown that the acidic character of the carboxylic acid group distorts the OH absorption band

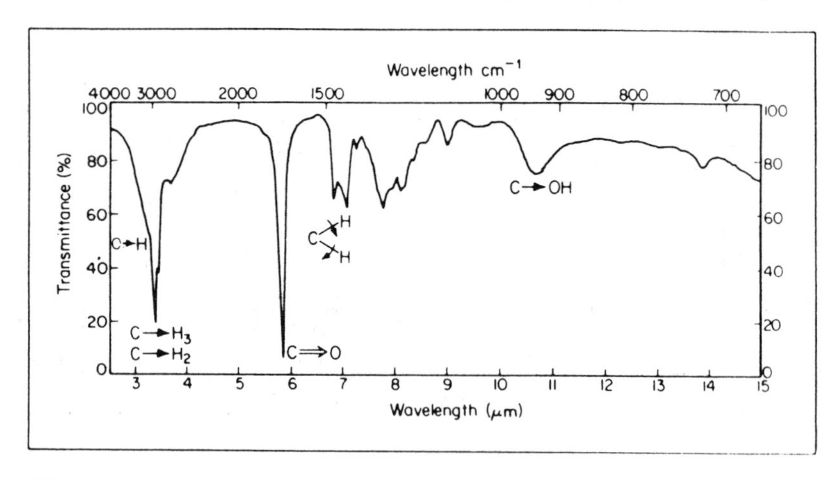

Figure 5.30 Absorption spectrum of an unknown compound.

at 3.0 μm compared to the OH absorption bands of alcohols, and is strong confirmation of the proposed structure of the unknown compound. These subtleties in the shape of absorption spectra are very revealing to the skilled spectroscopist.

It should be noted that the length of the paraffinic side chain cannot be identified with certainty, but it is safe to say that it contains at least four carbon atoms together with the necessary hydrogen atoms (the presence of an absorption band at 13.7 μm confirms this). To ascertain the molecular weight and to make final positive identification of the molucule, we would have to use other means.

Other spectra are shown in Figs. 5.31–5.33.

2. Quantitative Analysis

The quantitative determination of various compounds by IR absorbtion is based on the determination of the concentration of one of the functional groups of the compound being estimated. For example, if we have a mixture of hexane and hexanol, the hexanol may be determined by measuring the amount of absorption that takes place at 3.0 μm by the OH band—or, better, by the absorption by the C → OH group at 9.5 μm. From this the concentration of alcohol can be calculated. Whenever possible, an absorption band unique to the sample molecule should be used for measuring purposes. This reduces interference and the problem of overlapping bands. The final calculation is based on Beer's law.

It will be remembered [Eqs. (2.8), (2.16), and (2.19)] that

$$T = I_1/I_0$$

$$A = -\log (I_1/I_0)$$

$$= abc$$

where

T = transmittance
I_0 = intensity of radiation before entering the sample
I_1 = intensity of radiation after leaving the sample
A = absorbance
a = absorptivity of the sample
b = internal path length of the sample cell
c = concentration of the solution

If the same sample cells are used throughout, b is a constant. Also, the absorptivity a is a property of the molecular species being determined and can be taken as a constant. Therefore A is proportional to c.

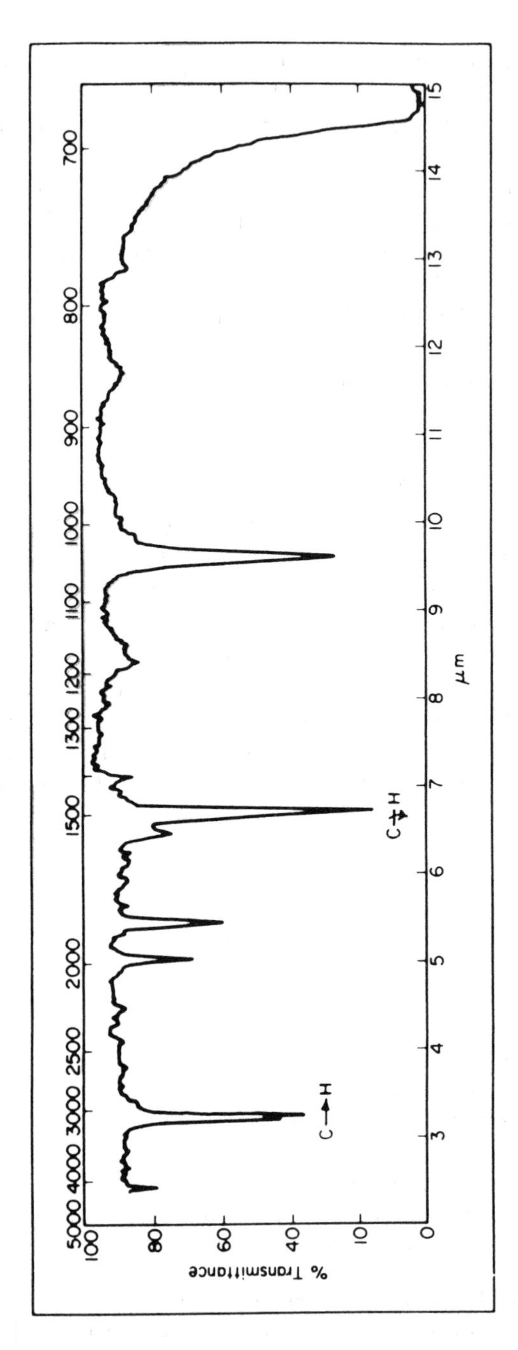

Figure 5.31 Absorption spectrum of benzene.

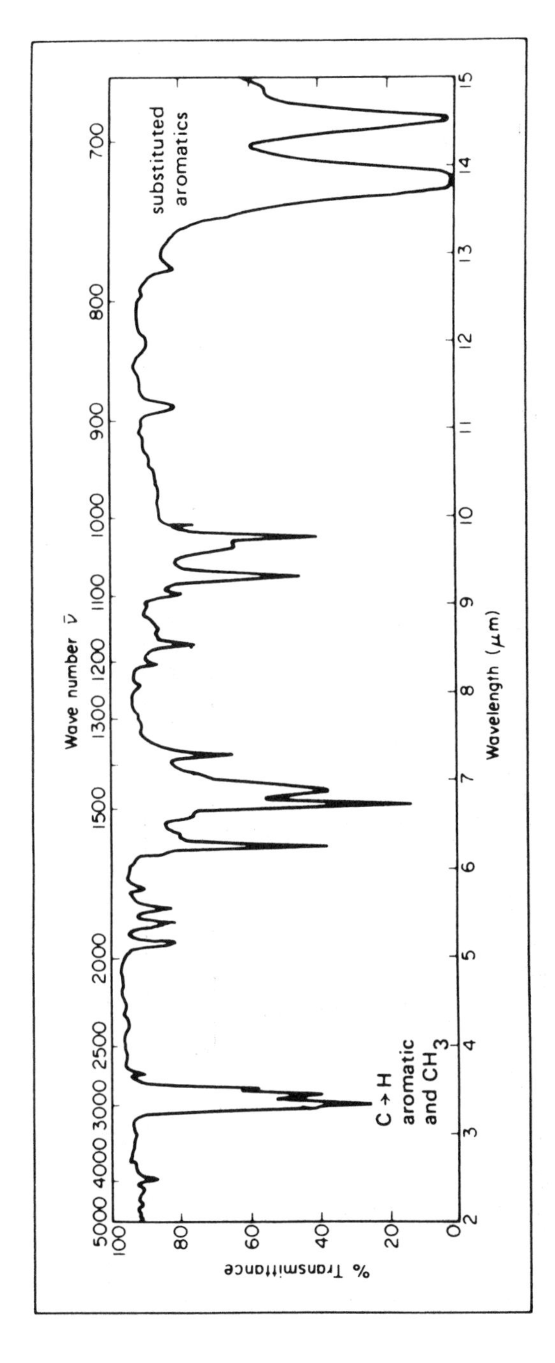

Figure 5.32 Absorption spectrum of toluene.

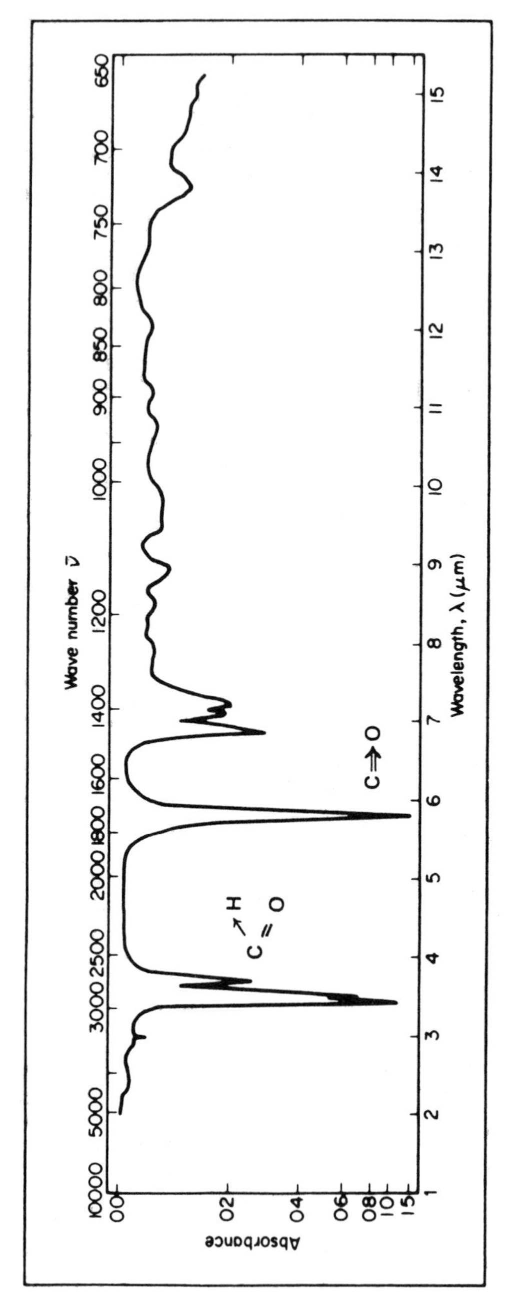

Figure 5.33 Absorption spectrum of $C_6H_{13}CHO$. Note the aldehydic H at 3.7 μm and the carbonyl band at 5.8 μm.

Most modern recorders are built to use logarithmic paper and record A directly. To find the proportionality constant between A and c, calibration curves are usually at different known concentrations. From these data the relationship between absorption and concentration can be obtained. To carry out quantitative analysis, the absorption A of a sample of unknown concentration is measured. By comparison with the calibration curve, the sample composition can be determined. The procedure is illustrated in Chapter 2, Section F.

Quantitative analysis using IR absorption is not as easy to obtain or as accurate as that using UV absorption. The sample cells in IR absorption must be made with material transparent to IR radiation, such as NaCl or KBr. These materials are very soft and easily distorted; as a result, the sample thickness may vary from one sample to the next. Also, the salt surface is subject to chemical attack from the solvent, particularly if traces of water or similar substances are present. Such materials etch the surface of the sample cell, which must then be repolished. This interaction makes it difficult to maintain a cell of constant dimensions and sample thickness. It will be remembered that in quantitative analysis $A = abc$, where b is the sample thickness; hence any error in the measurement causes an error in the calculation of c, the concentration, from the measurement of A. In practice, these problems make it difficult to maintain a reliable relationship between the calibration curve and the sample concentration.

3. Analytical Shortcomings

It is not possible (except under certain special circumstances) to determine molecular weight by IR spectroscopy. Also, it is not generally possible to get information on the relative positions of different functional groups of a molecule. Another difficulty is that a simple IR spectrum of an unknown substance will not differentiate between mixtures of compounds and pure compounds. For example, a mixture of paraffin and alcohols will give the same spectrum as higher molecular weight alcohols. It should be pointed out, however, that with prior knowledge of the components of a mixture and their spectra, many analyses of mixtures are performed daily by IR spectroscopy, which despite these shortcomings, has proved to be one of the most valuable methods for characterizing, both qualitatively and quantitatively, the multitude of organic compounds and mixtures of compounds encountered in research and industry.

4. Analyses Carried Out by Infrared Spectroscopy

The applications of IR spectroscopy to analytical chemistry are vast and are still growing in number. Most are concerned with the identification and determination of organic materials. The widespread use of IR spectroscopy arises from the fact that functional groups, such as carboxylic acids and ketones, can be identified by this method. From this information, the composition and structure of new and unknown compounds can be deduced.

Typical analyses include the detection or determination of paraffins, aromatics, olefins, acetylenes, aldehydes, ketones, carboxylic acids, phenols, esters, ethers, amines, sulfur compounds, halides, and so on.

From the IR spectrum it is possible to identify the odor and taste components of food, distinguish one polymer from another, or determine the composition of mixed polymers or the solvents in paints. Atmospheric pollutants can also be identified while still in the atmosphere. Another interesting application is the examination of old paintings and artifacts. It is possible to identify the varnish used on the painting and the textile comprising the canvas, as well as the pigments in the paint. From this information fake "masterpieces" can be detected. Modern paints use materials that were not available when many masterpieces were painted. The presence of new paints confirms that the painting must have been done recently.

Frequently, paints and varnishes are measured by *reflectance analysis*, a process wherein the sample is irradiated with IR light and the reflected light is introduced into an IR instrument. The paint, or other reflecting surface, absorbs radiation in the same manner as a traversed solution. Its IR spectrum can be obtained from the reflection. This technique can be used to identify the paint on appliances or automobiles without destroying the surface. Scraps of paint from automobiles involved in wrecks can be examined. From the data obtained, the make and year of the car can be determined.

In industry, IR spectroscopy has three important uses. First it can be used to determine impurities in raw material. This is necessary to ensure good products. Second, it can be used for quality control, that is, for continuously checking the composition and the percent present of the required product. This is the cheapest way to produce satisfactory products. The third use is for the identification of materials made in industrial research laboratories or of materials made by competitors. For example, a manufacturer was interested in identifying the transparent wrapping used by a competitor. Infrared absorption indicated that it was polyethylene.

Beeswax, carnuba wax, and various waxes used for coating floors and furniture can be identified by IR spectroscopy. Furthermore, it is possible for a manufacturer to ascertain whether a sample of beeswax is pure or whether it has been diluted with cheaper petroleum wax.

Another application of IR is identification by comparison. It is understood that the major application of IR includes qualitative analysis based on functional group analysis and quantitative analysis. But often the question arises, Is this material the same as that? An algorithm has been developed.

5. Near-IR

The near-IR (NIR) covers the range 0.78 μm (1300 cm^{-1}) to 2.5 μm (4000 cm^{-1}). This is the range from the long-wavelength end of visible light (red) to 2.5 μm. It is outside the normal or midrange of IR. Only overtones and combination bands occur here, and consequently the absorption coefficients are weaker and the degree of absorption or sensitivity is reduced.

However, there are some inherent advantages to working in the range. High-intensity sources such as tungsten-halogen lamps give strong, steady radiation over the entire range. Also, very sensitive detectors, such as lead sulfide, silicon, or mercury cadmium telluride, can be used because they are responsive to the relatively high-energy photons. These detectors need not be operated in liquid N_2. The third important advantage is that quartz or glass can be used both in optical systems or as sample containers.

Optical fibers are available which are transparent to the NIR. These can be used for remote sampling or even continuous monitoring of bulk flowing strains of commercial products. This results in increased instrumental sensitivity and reduced operating costs. On the negative side, the molecular absorption coefficients are reduced, resulting in reduced sensitivity. This, however, is sometimes an advantage because a greater sample thickness can be used giving more representative results. Also, in this region there are no fundamental frequencies to interfere or overlap with the spectra, but there are numerous overtones and combination bands that overlap each other. However, using multivariant techniques such overlapping can be resolved.

NIR has been used to characterize polymers and other products using two beams of light, polarized perpendicular to each other (see Chap. 10). The two beams are absorbed to different degrees, depending on the orientation of the sample or crystals in the sample.

The field has also found wide acceptance in the area of NIR reflectance analysis. In this method, the sample is irradiated with NIR, the reflected light is gathered and scanned, and the spectra are used for qualitative and quantitative analysis. Absorption takes place at characteristic wavelengths according to the equation:

$$\mathrm{Ar} = -\log\left(\frac{R_1}{R_0}\right)$$

where

Ar = pseudoabsorption associated with reflectance
R_0 = Intensity of incident beam
R_1 = Intensity of reflected beam

The successful use of this empirical relationship, has led to widespread use of NIR reflectance analysis. The equation mimics Beer's law, where by definition:

$$A = -\log T = -\log\left(\frac{I_1}{I_0}\right)$$

a. Spectra Interpretation.

The most important bands are C → H (2.2 μm), ^{1}O → H (1.4 μm), and N → H (1.5 μm). These bands enable the characterization of paraffin, olefins, amines, polymers, body issues such as fatty acids, proteins, etc., and have lead to analysis of many types of materials, which will be discussed below.

b. Polymers

Polymers can be made either from a pure monomer, such as polyethylene, or from mixtures of olefins, such as ethylene propylene and styrenes. An important parameter of such copolymers is the relative amount of each present. This can be determined by NIR.

Nitrogen-containing polymers, such as nylon, have been characterized by using polarized light, and other N-containing polymers, such as polyurethanes, polyamic acids, and urea formaldehyde resins, by using NH bands at 1.4 μm (7150 cu^{-1}), 1.95 μm (5128 cu^{-1}), and 2.3 μm (4350 cu^{-1}), respectively.

Block polymers—which are typically made of a "soft block" of polyethers of polyesters and a "hard block" containing aromatics, urethanes, and urea groups—have been analyzed. These have utilized C ⇒ O (1.98 μm) bands, aromatic C → H (1.14 μm, 1.63 μm, and 2.15 μm) bands, and methyl (1.72 μm) and methylene (1.23 μm) bands.

Diene polymers such as polybutadiene polyisoprene and styrene-butadiene copolymers have have been examined using the H_2C (1.64 μm) band as well as aromatic (2.18 μm) and aliphatic (2.35 μm) bands to determine the styrene content.

a. Fibers and Textiles

The composition of fibers has been performed using reflectance NIR. The identity of fiber has been achieved. In addition, the uptake of dyes, the presence of processing oil in polyester yarns, and the distinction between nylon 6 and nylon 66 has been performed. Frequently starch or wax is added to fiber in "sizing." The presence of these add-ons can be determined.

b. Moisture Determinators

The presence of absence of water is important in polymer chemistry. Water loss may cause polymer dehydration or be a chain stopper during polymer growth. Further, it can cause H bonding with nitrogen polymers, modifying the polymer properties. Its determination is therefore very important. This has been achieved by examining the O → H at 1.93 μm, 1.41 μm, and 1.46 μm.

An important application is the detection of components separated by thin layer chromatography. The chromatograms are mounted directly on a microscope stage and examined by micro-FTIR. Lipids and other natural products can be identified using NIR. The sensitivity obtained is 1.0 μg of material.

c. Conclusion

These NIR techniques are rapid, require little sample preparation, and are nondestructive. In addition, the equipment is stable, sensitive, and can be used for many other types of analysis. No doubt the application of NIR will expand in the future.

E. RAMAN SPECTROSCOPY

Raman spectroscopy is a field that in many ways complements IR absorption spectroscopy. It is similar to IR spectroscopy in that vibrational energy changes in the molecule are involved. If there is a change in dipole with a change of vibration, the mode is IR active. If there is no change in dipole, the mode is Raman active and undergoes a change in polarization, brought-about by a distortion of the electron cloud.

It is not uncommon for molecules to be both IR and Raman active. For example, CO_2 can be represented as $O{=}C \Rrightarrow O$. The vibration of one oxygen about the C=O is IR active. But vibration of the carbon about both oxygens $O \Lleftarrow C \Rrightarrow O$ is not IR active, but Raman active. Many molecules have numerous modes of vibration, some of which are Raman active and some IR active.

All modes are either IR active or Raman active. Many molecules have modes that are IR active and other modes that are Raman active. Raman spectroscopy thus permits us to examine the vibrational spectra of compounds that do not lend themselves to IR absorption spectroscopy.

1. Principles

When UV radiation is passed through a sample, some of the radiation is scattered, or dispersed, by the molecules present. For simplicity, it is best to use radiation of only one wavelength; minimal absorption of the radiation must take place and the beam is merely dispersed in space. The wavelength of the light dispersed is mostly the same as the UV radiation falling on the molecule. This wavelength is called the *Rayleigh wavelength*, after Lord Rayleigh, who spent many years studying light scattering. However, if the scattered radiation is studied closely, it can be observed that slight interaction of the incident beam took place. After interaction with the radiation, the molecules are excited to a *virtual* (nonquantized) *energy level*. It may cause the molecule to vibrate faster in the

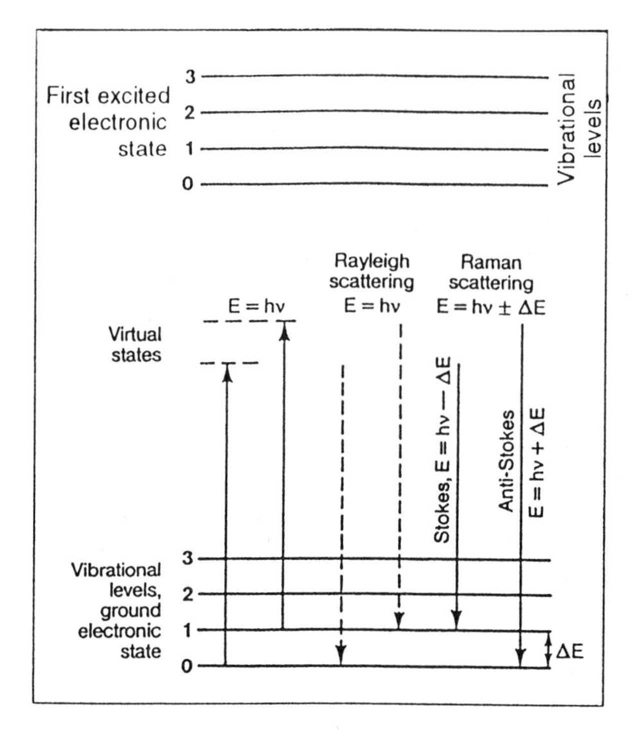

Figure 5.34 Origin of Raman and Rayleigh scattering.

process. On returning to the ground state, the emitted radiation may be at the same wavelength as the source. This is Rayleigh scattering. Moreover, radiation at slightly longer wavelengths than the Rayleigh wavelength can be observed. These spectral lines are called *Raman lines*, after Sir C. V. Raman, who first observed them. The slight shift in wavelength is caused by the molecules of the sample. After radiation, they vibrate faster. The energy required to cause faster vibration is absorbed from the incident radiation, which results in a slight decrease in the energy of the latter. This is illustrated in Fig. 5.34.

The energy of the Rayleigh wavelength is given by the familiar expression

$$E = h\nu$$

where E is the energy of the incident (Rayleigh) radiation, ν is the frequency of the incident radiation, and h is Planck's constant.

The energies of the Raman lines are $E - \Delta E$, where ΔE represents the various vibrational energy changes of the molecule. Also from this relationship,

$$E - \Delta E = h(\nu - \nu_1)$$

where v is the wavelength of the incident radiation and ν_1 is the frequency shift due to an energy change ΔE. Several vibrational levels are involved, resulting in several lines of energy $h(\nu - \nu_1)$, $h(\nu - \nu_2$, $h(\nu - \nu_3)$, and so on. This results in several simultaneous shifts in frequency from the Rayleigh frequency.

When the molecule *increases* in vibrational energy after the radiation passes through the samples, the molecule has absorbed energy. The remaining radiation experiences a shift from the Rayleigh frequency to a lower frequency. These lines are called *Stokes–Raman lines*, after Sir George Gabriel Stokes, who observed a similar phenomenon in fluorescence.

Less commonly, the molecule *decreases* in vibrational energy after the radiation passes through the sample. In this case, it has given energy to the radiation. The latter experiences a shift to frequencies higher than the incident radiation. These lines, which are called *anti-Stokes–Raman lines*, have important theoretical implications, but to the analytical chemist they are less important than the Stokes–Raman lines because of their much lesser intensity, although the use of lasers has improved sensitivity.

It is convenient to record the Raman shifts in wave numbers v, because these can be related directly to IR spectra (in $\bar{\nu}$) and avoid the complication of including the wavelength of the exciting UV radiation. Typical shifts are seen in Fig. 5.35.

2. Analytical Procedure

In practice, the beam of radiation that is passed through the sample is at one wavelength only, the Rayleigh wavelength, which is normally in the UV region of the spectrum. The spectrum of the radiation that penetrates the sample is then recorded. A schematic diagram of the equipment is shown in Fig. 5.36. The spectrum appears as in Fig. 5.37.

The wavelengths and intensities of the Stokes–Raman and anti-Stokes–Raman lines are measured. The wavelength shift can be correlated directly with

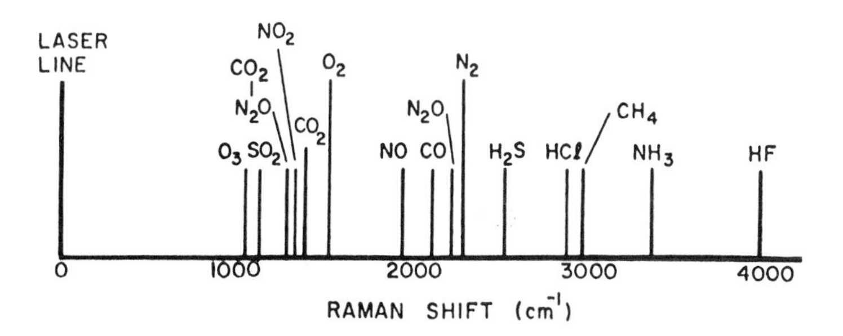

Figure 5.35 Typical Raman shifts using a laser source.

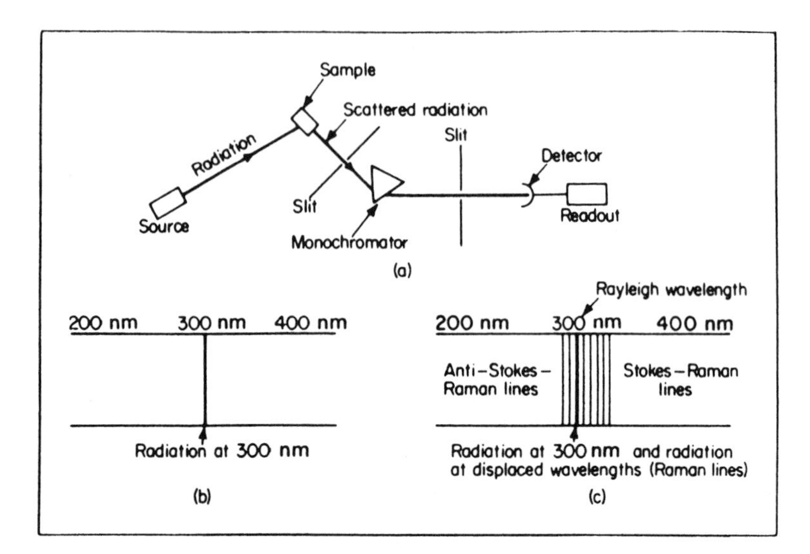

Figure 5.36 (a) Schematic diagram of the equipment used in Raman spectroscopy and an example of a Raman spectrum (b) before scattering and (c) after scattering. A common light source is the radiation from a continuous-wave laser.

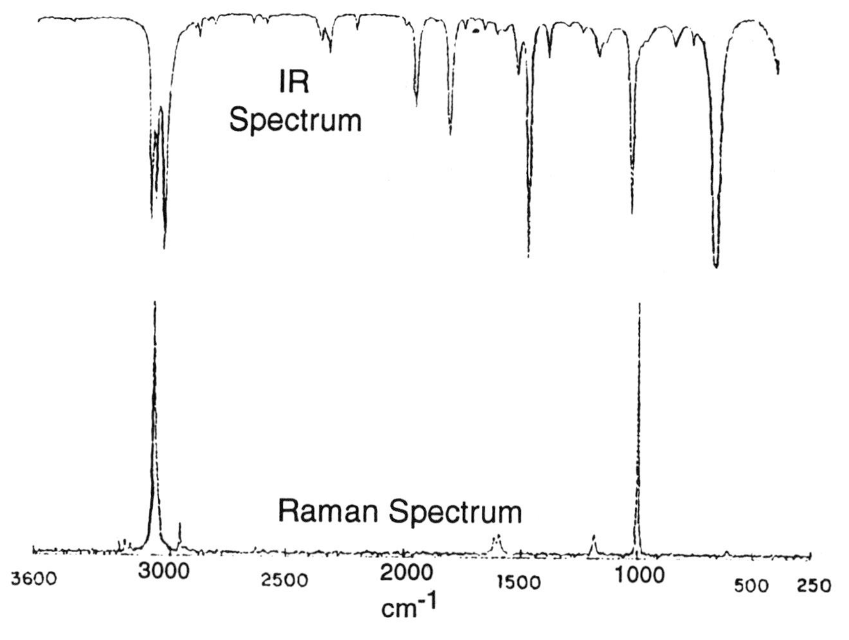

Figure 5.37 IR and Raman spectra of benzene.

the structure of the molecule in the same way that IR spectroscopy is used. The intensities of the Raman lines provide a quantitative measure of the concentration of the scattering material present in the sample. In practice, it is best to avoid the use of radiation at a wavelength that can be absorbed by the molecule to give rise to electronic transitions. When simple UV absorption takes place, as in this instance, the Raman spectra become much more difficult to interpret.

The light sources used originally were simple UV light sources; however, these were relatively weak and only weak Raman lines were observed. More recently laser beams have been used for this purpose. Their use has greatly expanded the applications of Raman spectroscopy, because of the dramatically increased intensity of light and a simultaneous improvement of the signal-to-noise ratio.

The intensity of Raman scattering is proportional to fourth power of the excitation wave number cm^{-1}. It is also directly proportional to the power of the light source. It has therefore become popular to use tunable lasers as light sources. These are intense and of variable wavelengths.

3. Applications: The Use of Laser Beams as Sources

The analytical applications of Raman spectroscopy are not as widespread as those of IR spectroscopy. It can be used, however, to examine compounds with no changeable dipole, such as O_2 and N_2. Raman spectroscopy has become increasingly important for the characterization of polymers: It reveals physical properties such as how the molecules are arranged with respect to each other in the polymer and the polymer's crystallinity, tacticity, and amorphous character, which in turn affect structure and properties such as strength and durability. Normal IR absorption spectroscopy is difficult to use without dissolving the sample. After a sample is dissolved, its original character is destroyed and necessary information lost. If the sample is not dissolved, it is frequently so dense that very little radiation passes through it unabsorbed. Hence Raman spectroscopy on the original polymer is most valuable.

By using a laser beam we can greatly increase the intensity of radiation penetrating the sample. As a result, it is easier to detect and measure the wavelength shifts and intensities of the Stokes and anti-Stokes lines of the Raman spectrum. From this the vibrational energy and therefore the structure of the polymer can be deduced.

An added advantage of Raman spectroscopy is that the optical system is built to operate in the UV part of the spectrum. In this region of the spectrum the equipment includes a detector (a photomultiplier), a monochromator (prism or grating), a sample holder (quartz), and a source (an intense laser). Each of these items is generally easier to handle than its counterpart in IR spectroscopy.

Consequently, Raman spectroscopy has become increasingly important in recent years, particularly for polymers and other solid samples. It has also been used for direct air pollution analysis using O_2 and N_2 in the air as a wavelength and concentration standard.

4. Use of Near-IR and Diode Array in Raman Spectroscopy

Raman spectroscopy has been expanded by utilizing the *near-IR region*. Several factors have contributed including the use of quartz optics which are more stable than other IR-transparent materials. In addition, *diode arrays* are now available that can give fast response to present the entire Raman spectra for unstable compounds, intermediate compounds, or compounds in low concentration. Studies have been performed on aromatics and other compounds.

Although the advantage of fast analysis is important for unstable compounds and perhaps as a sophisticated L.C. detector, there is a loss of resolution and increase in noise level when the diode array is used. This is illustrated in Fig. 5.38, showing much increased background emission.

F. PHOTOACOUSTIC SPECTROMETRY

The photoacoustic effect was discovered in 1880 by Alexander Graham Bell; however, it lay forgotten until 1973, when it was revived in photoacoustic spectrometry by researchers at Bell Telephone Laboratories and Johns Hopkins University. Photoacoustic spectroscopy is a technique for obtaining IR or UV absorption spectra similar to those obtained by conventional transmittance or reflectance spectrophotometry. It is primarily used for surface analysis, particularly of powders. A salient feature of the photoacoustic technique is its capability to analyze samples that are otherwise difficult to process, such as opaque samples and samples that scatter light, including powders and turbid liquids.

The photoacoustic effect is the generation of an acoustic signal by a sample exposed to modulated light. It is an indirect but efficient process. The sample is placed in an enclosed chamber with a coupling gas, typically air, and exposed to chopped monochromatic light. The sample is heated to the extent that it absorbs the incident light and converts it through nonradiative processes to heat. Because the incident light is chopped, the temperature rise is periodic at the chopping frequency. This periodic temperature rise at the surface of the sample in turn causes a similar periodic change in the pressure of the gas in the enclosed chamber. This pressure modulation is an acoustic signal at the chopping frequency, and it is detected by means of a sensitive microphone suitably coupled to the chamber. This setup is shown in Fig. 5.39.

The magnitude of the acoustic signal is related to the UV or IR absorption coefficient of the sample. Therefore, an absorption spectrum can be obtained

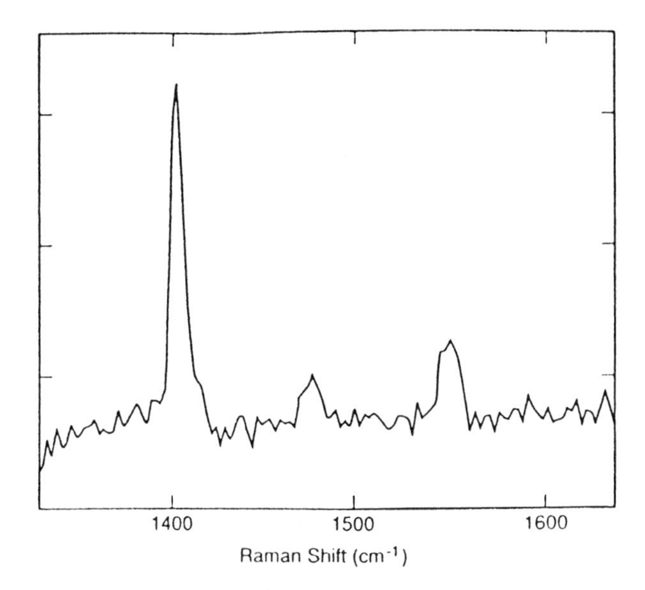

(a)

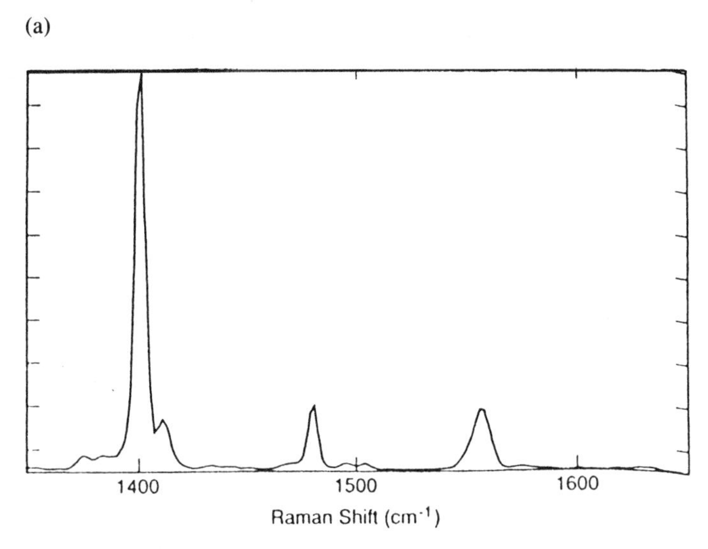

(b)

Figure 5.38 (a) Raman spectrum of anthracene using a diode array. Total measurement time is 0.28 sec. (b) FT-Raman spectrum of anthracene using FT Raman. Total measurement time is 120 sec. Note noisy signal in (a).

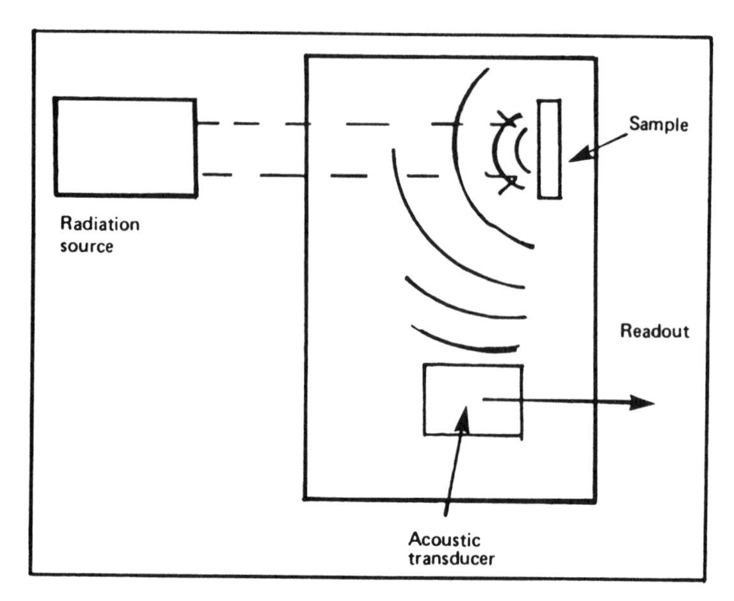

Figure 5.39 Schematic diagram of a photoacoustic spectrometer.

by scanning the wavelength of the chopped monochromatic light falling on the sample.

The depth of the sample surface that contributes to the signal is limited. At too great a depth, even if the sample is heated, it does not transfer its energy to the coupling gas because it is insulated by the outer layer. This effective depth is called the *thermal diffusion depth*. It is equal to $2(k/\omega)^{1/2}$, where k is the thermal diffusivity of the sample (surface only) and ω is the modulation frequency of the light; that is, $\mu = (2k/\omega)^{1/2}$.

1. Equipment

A schematic diagram of a Princeton Applied Research Model 6001 is shown in Fig. 5.40. The *light source* is a 1-kW xenon arc lamp of the type used in motion picture projectors. This lamp produces an image of the arc at the entrance slit of the monochromator. The lamp is electronically modulated (rather than chopped by a rotating aperture disk) to minimize acoustic noise and extend the range of available modulation frequencies. This dispersion element is a concave holographic grating covering the spectral range from 200 nm to 3.0 μm. At the exit slit of the monochromator is an order-sorting filter. The signal is detected using a pyroelectric detector.

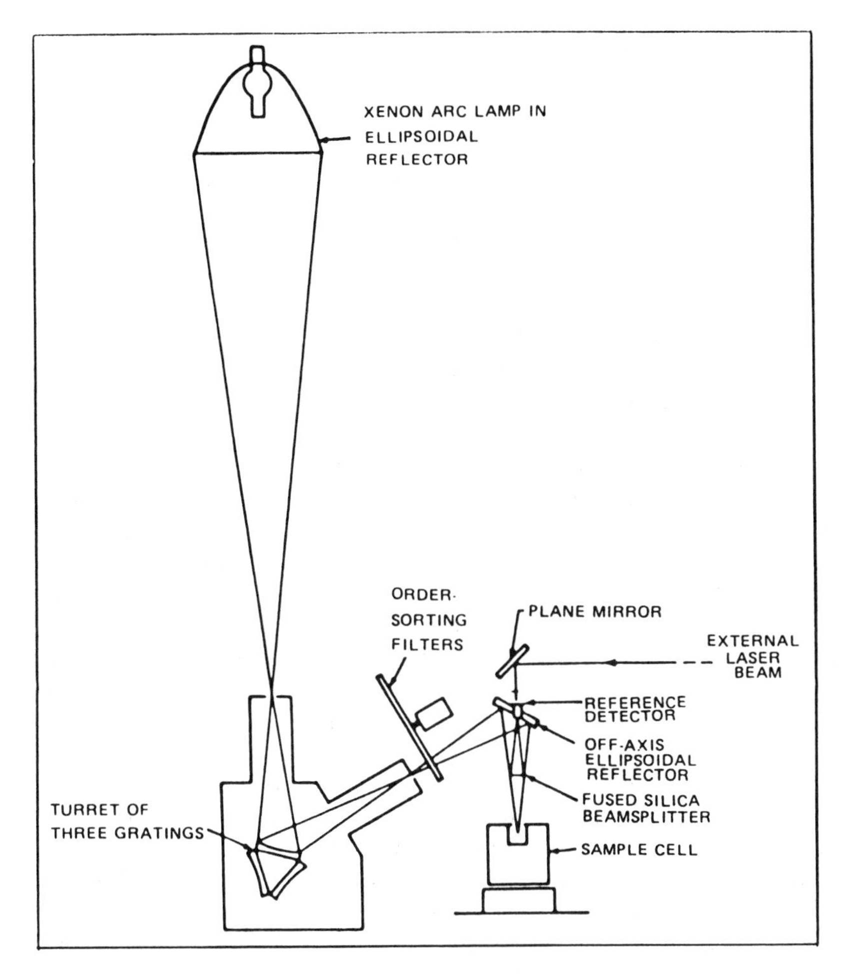

Figure 5.40 Schematic diagram of a Princeton Applied Research Model 6001 photoacoustic spectrometer.

2. Analytical Applications

The most useful application of photoacoustic spectroscopy is in analyzing samples that scatter or reflect light and that are unsuited to analysis by transmittance spectrophotometry. Examples are powders and turbid liquids. The technique is insensitive to the effects of scattering or reflection because light that is simply reflected or scattered does not heat the sample. Only that portion of the incident light that is actually absorbed contributes to heating the sample, and hence to the generation of the photoacoustic signal. Although reflectance spectrophotometry can also be used for such samples, it is generally found that the photoacoustic technique is faster, requires less sample preparation, and yields more reproducible results. A second broad class of samples that can be analyzed by this technique consists of those too opaque for measurement in transmittance. Because the photoacoustic spectroscopic measurement process does not involve the measurement of either transmitted or reflected light, it offers a means of analyzing these types of samples.

The photoacoustic signal depends on the total amount of light absorbed in the sample within the thermal diffusion depth or the actual sample thickness, whichever is less. The photoacoustic signal is not very different whether the light is totally absorbed in the first monolayer at the sample surface or within the thermal diffusion length.

There are limits to the range of absorption coefficients that can be measured by photoacoustic spectroscopy; the upper range of these limits is defined by "signal saturation." *Signal saturation* arises because at a given chopping frequency there is a characteristic thermal depth to which the photoacoustic process responds. For example, at a modulation frequency of 40 Hz, the thermal diffusion length in an aqueous sample is about 30 μm. Then, even though the actual sample thickness may be considerably greater than 30 μm, the photoacoustic detection process can only yield a measure of the extent to which light is absorbed within this 30-μm thermal diffusion depth.

There are several approaches to reduce the effects of signal saturation. First, and usually most satisfactory, is to make the sample "optically thin," that is, thinner than the diffusion depth. Solids can be ground to a fine powder or deposited as a thin film onto a reflecting or transparent substrate such as silica or alumina. The second approach is to make the sample "thermally thin" by increasing the modulation frequency, thereby decreasing the thermal diffusion depth. For example, with the aqueous sample described above, if the chopping frequency is increased to 2 kHz, the thermal diffusion length (which varies as the square root of the chopping frequency) decreases to about 5 μm. The disadvantages to this approach are that (1) the photoacoustic signal varies inversely with the modulation frequency, so signal-to-noise is sacrificed at the higher frequency, and (2) relatively large changes of modulation frequency are required to effect modest reductions of the thermal diffusion length.

In other applications, photoacoustic spectroscopy can be used as a depth probe. By systematically changing the modulation frequency, the depths to which certain thin layers of components respond can be measured. In addition, thermal diffusivity can be measured by means of a special sample cell arranged so that heat generated by the irradiation of one surface of the sample must diffuse through a known sample thickness before reaching a second surface exposed to the coupling gas.

G. INTERPRETATION OF IR SPECTRA

Infrared absorption spectra depend on the stretching and bending modes of vibrations of the components of a molecule relative to each other, as previously explained. As such, they provide us with information on the functional groups (alcohols, ketones, etc.) present in the molecule.

Molecules can be identified from their spectra using one of two techniques. The first uses comparison with known spectra. Memory banks of known spectra are available. These memory banks may contain spectra of 5000 or more compounds pertinent to various fields of endeavor, such as pollution, pharmaceuticals, food additives, clinical, plant, or animal tissue, forensic science, etc. The unknown spectrum or some predetermined number of the strongest absorption bonds, may be fed into the computer, which compares it with stored spectra. It then retrieves all compounds that may match the unknown spectrum. The operator then identifies the spectrum based on matching two spectra and common sense in rejecting impossible matches.

Frequently, however, a memory bank is not available or the spectra don't match exactly, and the operator must identify the compound from first principles. This process is described below.

1. Modes of Vibration

Typical modes of vibration were shown in Fig. 5.6, and typical absorption wavelengths of different groups are listed in Table 5.1. These will now be examined in greater detail.

a. Paraffins

Functional group	CH vibrations	Wavelength (μm)
Methyl	C → H	3.3–3.5
Methylene	C → H	3.4–3.43
Methyl	C ↕ H	6.9–7.3
		(sym) (assym)
Methylene	C ↕ H	6.8 7.4 13.9
Terminal methyl	C → CH_3	7.3
Gem dimethyl	C ↗ CH_3, C ↘ CH_3	(sym) (assym) 7.1 7.4

Figure 5.41 shows the spectrum of *n*-hexane with the major peaks identified, and Fig. 5.42 shows the spectrum of cyclohexane. Note the difference in the spectra. The C → H is narrower (no CH_3), there is no peak at 7.3 μm, and there are two peaks at 11.1 and 11.6 μm denoting boat-and-chain interchanging.

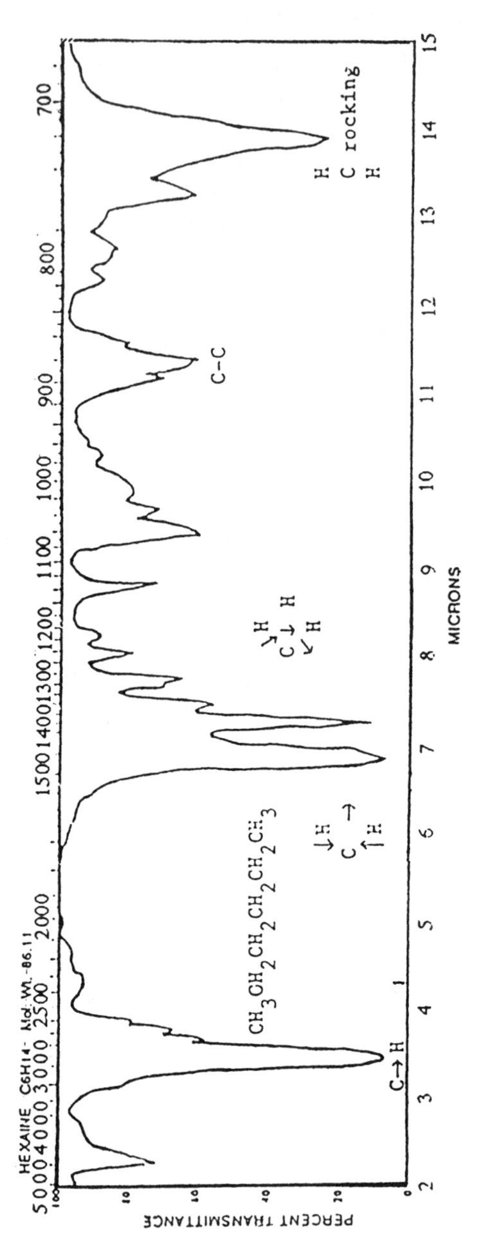

Figure 5.41 Spectrum of *n*-hexane, with peaks identified.

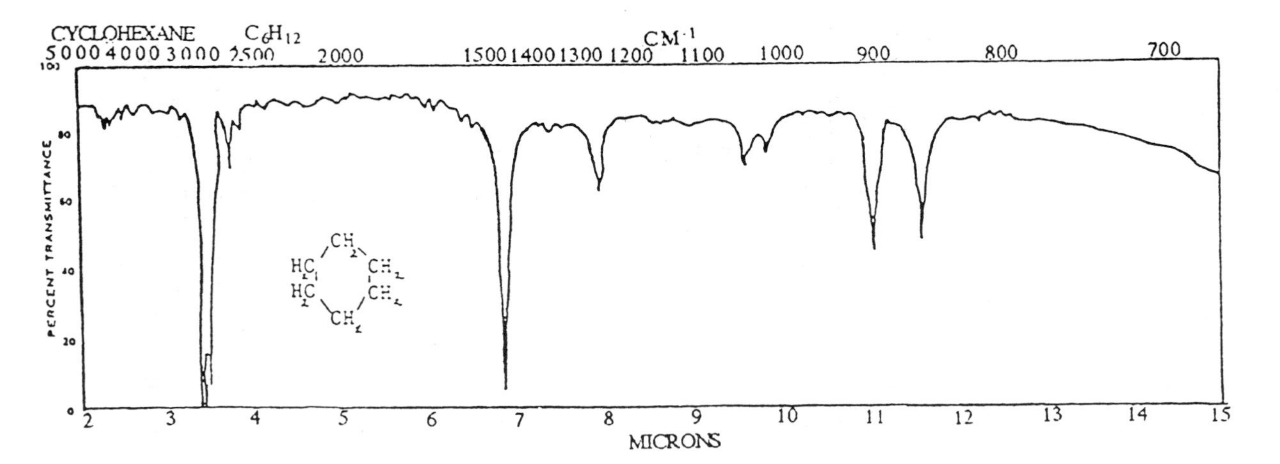

Figure 5.42 Spectrum of cyclohexane. Note the difference between this spectrum and the one in Figure 5.41. The C → H is narrower (no CH_3), there is no peak at 7.3 μm, and there are two peaks at 11.1 and 11.6 μm, denoting boat-and-chain interchanging.

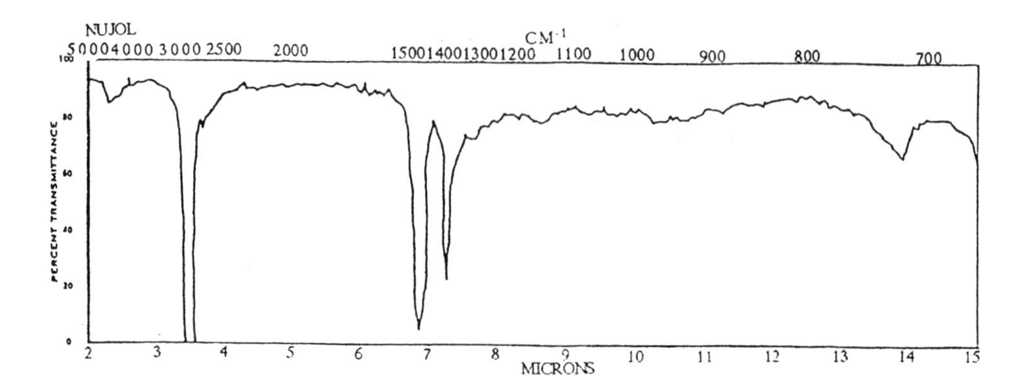

Figure 5.43 The spectrum for dry Nujol shows only C → H and C ↓ H and is much clearer than the spectrum for wet Nujol. This material, a saturated hydrocarbon with a range of molecular weight, is frequently used as a solvent. Only the C→H and C ↓ H are covered by the solvent spectrum, and it can usually be assumed that these are present. Other functional groups can be observed without interference.

The spectrum of dry Nujol (Fig. 5.43) shows only C → H and C ↓ H and is much clearer than the spectrum for wet Nujol. This material, a saturated hydrocarbon with a range of molecular weight, is frequently used as a solvent. Only the C → H and C ↓ H are covered by the solvent spectrum, and it can usually be assumed that these are present. Other functional groups can be observed without interference.

Unsaturation mode	Typical wavelengths μm
C = C → H	3.3
C ≡ C → H	3.06
C ≡ C ↓ H	>15.0 overtone at 8.0
C ⇛ C	6.1
C ⭆ C	4.74

The spectrum shown in Fig. 5.44 is for an olefin, $C_8H_{17}C = CH_2$. Note the C → H bond is made more complex by the addition of the C → H adjacent to the olefin. Also, C ⇛ C at 6.1 μm is a sharp but weak peak.

Figure 5.45 shows the spectrum for $C_4H_9{-}C{\equiv}CH$. Note C≡C → H at 3.0 μm and C≡C at 4.7 μm. Very little else occurs in this part of the spectrum except C≡N.

b. Aromatics

The IR spectrum of aromatics is quite complex. The C → H bands are at about 3.3 μm but are sharp and weak. The C ↓ H bands are at wavelengths varying

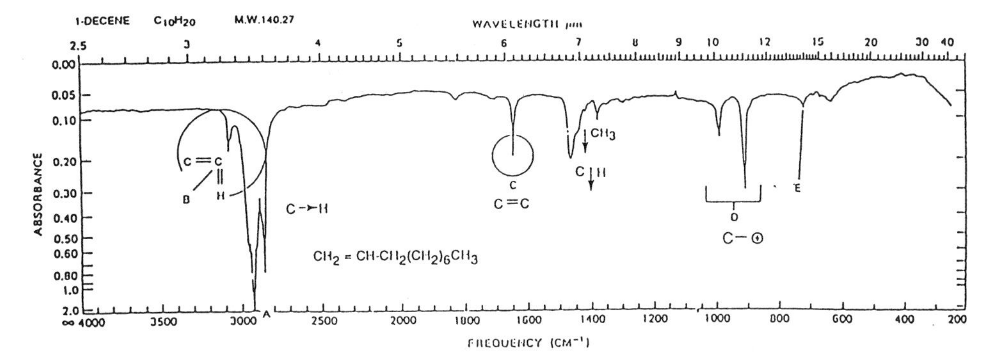

Figure 5.44 The spectrum for an olefin, $C_8H_{17}C = CH_2$. Note the C → H bond is made more complex by the addition of the C → H adjacent to the olefin. Also, C ⇒ C at 6.1 μm is a sharp but weak peak. (A) C = H stretch; (B) olefinic C — H stretch, 3049 cm^{-1} (3.28 m); (C) C = C stretch, 1645 cm^{-1} (6.08 m); (D) out-of-plane C — H bend, 986 cm^{-1} (10.14 m), (olefinic) 907 cm^{-1} (11.03 m); (E) methylene rock, 720 cm^{-1} (13.88 m).

from 11.0 to 14.0 μm depending on substitution of the rings. These are shown in Table 5.4. However, there is a very important region between 5 and 6 μm where a number of weak overtone peaks occur. These are indicative of the aromatic substitution of the compound. They are illustrated in Fig. 5.46, showing

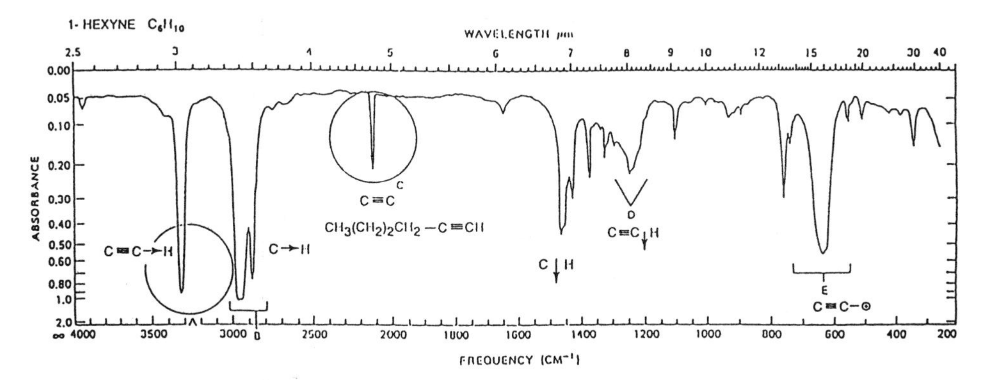

Figure 5.45 The spectrum for C_4H_9—C≡CH. Note C≡C→H at 3.0 μm and C≡C at 4.7 μm. Very little else occurs in this part of the spectrum except C≡N. (A) C≡C stretch, 3268 cm^{-1} (3.06 m); (B) Normal C—H stretch, 2857–2941 cm^{-1} (3.4–3.5 m); (C) C = C stretch, 2210 cm^{-1} (4.74 m); (D) = C — H bend overtone, 1247 cm^{-1} (8.02 m); (E) = C — H bend fundamental, 630 cm^{-1} (15.87 m).

Table 5.4 C—H Out of Plane Bonding and Benzene Ring Substitution

Substitution positions	Band positions (cm^{-1}, μm)	Band strength
Benzene	(14.9)	*s*
Monosubstitution	(12.99–13.70)	*s*
	(14.08–14.49)	*s*
1,2 Disubstitution	(12.99–13.61)	*s*
1,3 Disubstitution	(11.11–11.63)	*m*
	(12.35–13.33)	*s*
	(13.74–14.71)	*m*
1,4 and 1,2,3,4 Substitution	(11.63–12.50)	*s*
1,2,3 Trisubstitution	(12.50–12.99)	*s*
	(13.89–14.60)	*m*
1,2,4 Trisubstitution	(11.63–12.50)	*s*
	(11.11–11.63)	*m*
1,3,5 Trisubstitution	(11.11–11.63)	*m*
	(11.56–12.35)	*s*
	(13.70–14.81)	*s*
1,2,3,5–1,2,4,5 and 1,2,3,4,5 Substitution	(11.11–11.63)	*m*

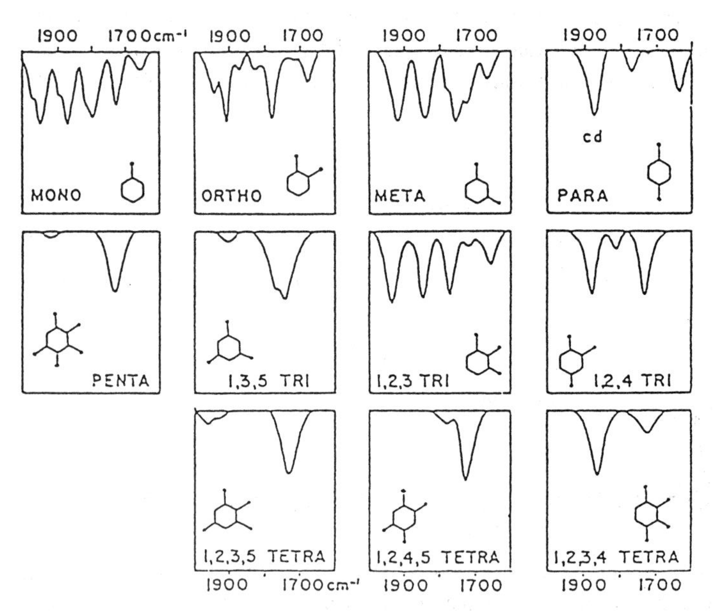

Figure 5.46 Characteristic absorption bands for various substituted benzene rings in the 5- to 6-μm region.

absorption spectra of substituted aromatics in the 11- to 13- μm region. Note that these bands are usually strong.

Spectra are shown for *o*-xylene (Fig. 5.47) and polystyrene (Fig. 5.48). Note the complex and only moderately region at 3.3 μm, the overtone region at 5.1–6.0 μm, and the strong band at 13 μm.

Figure 5.48 is particularly important since polystyrene is a stable transparent polymer with well-defined sharp peaks at reproducible wavelengths. It is used as a standard to check wavelength calibration in IR equipment. Note the well-defined peaks at 5–6 μm showing monosubstitution confirmed by the two peaks at 13.2 and 14.25 μm. Also, note the sharp C → H peaks in the 3.4-μm region.

2. C-O Bonds

a. Alcohols

The absorption frequencies for this group are as follows:

Functional group	Mode	Wavelength (μm)
Alcohol	CO → H(free)	2.8 Vapor 3.0 Solution
	CO ↓ H	7.2
	C—CO → H	9.5 Primary
	C—C → OH C	9.0 Secondary
	C C—C → OH | C	8.5 Tertiary
Acid	O // —CO → H	3.3 Broad

The spectrum for methanol is shown in Fig. 5.49. In it can be seen the strong broad CO → H at 3.0 μm, the weak CO ↓ H at 7.5 μm. and the strong C→ OH at 9.6 μm. This spectrum is simplified by the absence of CH_2 bands.

Figure 5.50 shows the absorption spectrum of phenol. Note that the CO → H is very broad and the C → H is a short sharp peak typical of aromatics. The C → OH is at 8.4 μm, displaced from the 9.5 μm of a primary alcohol but similar to a tertiary paraffinic alcohol. Characteristic bands at 5–6 μm and the bands at 12.4 and 13.4 μm typify monosubstituted benzene.

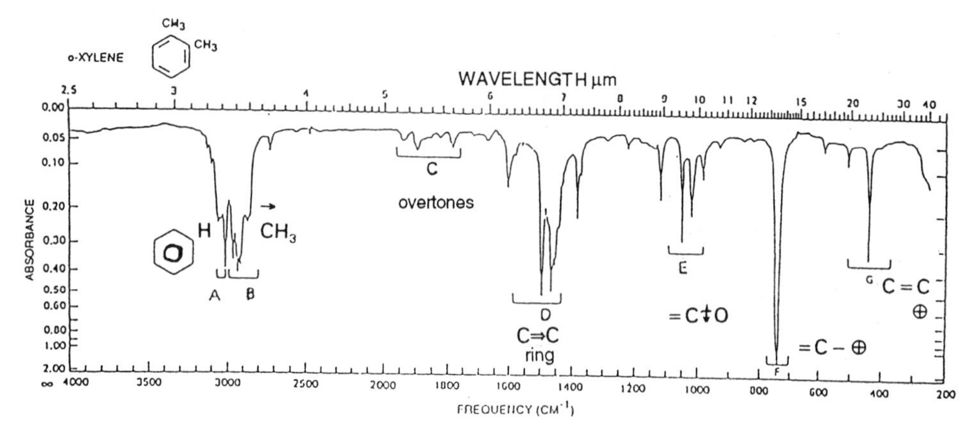

Figure 5.47 Absorption spectrum of *o*-xylene. Note the complex and only moderately strong region at 3.3 μm, the overtone region at 5.1-6.0 μm, and the strong band at 13 μm.

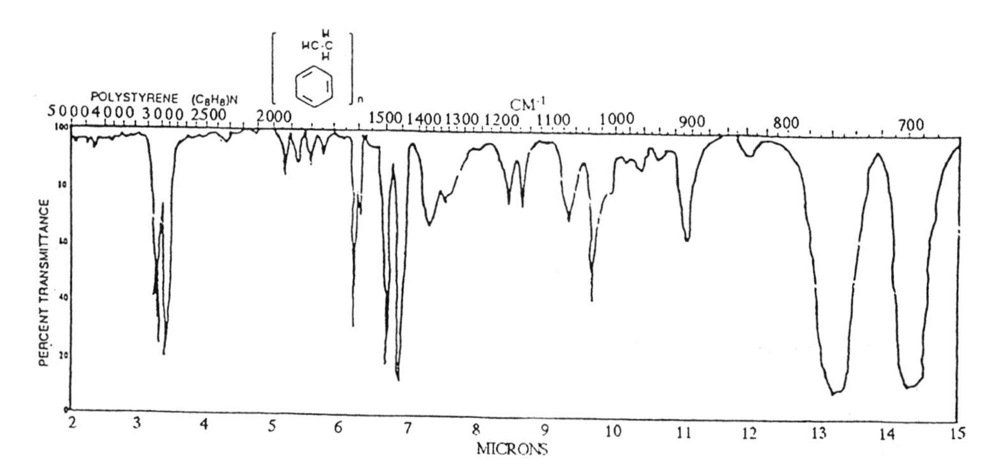

Figure 5.48 Absorption spectrum of polysterene. This figure is particularly important since polystyrene is a stable transparent polymer with well-defined sharp peaks at reproducible wavelengths. It is used as a standard to check wavelength calibration in IR equipment. Note the well-defined peaks at 5-6 μm showing monosubstitution confirmed by the two peaks at 13.2 and 14.25 μm. Also, not the sharp C → H peaks in the 3.4-μm region.

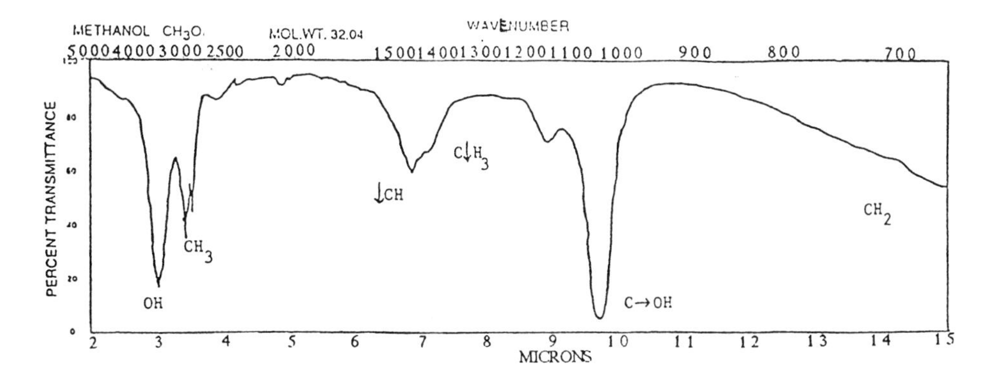

Figure 5.49 Absorption spectrum of methanol. Note the strong broad CO → H at 3.0 μm, the weak CO ↓ H at 7.5 μm, and the strong C → OH at 9.6 μm. This spectrum is simplified by the absence of CH_2 bands.

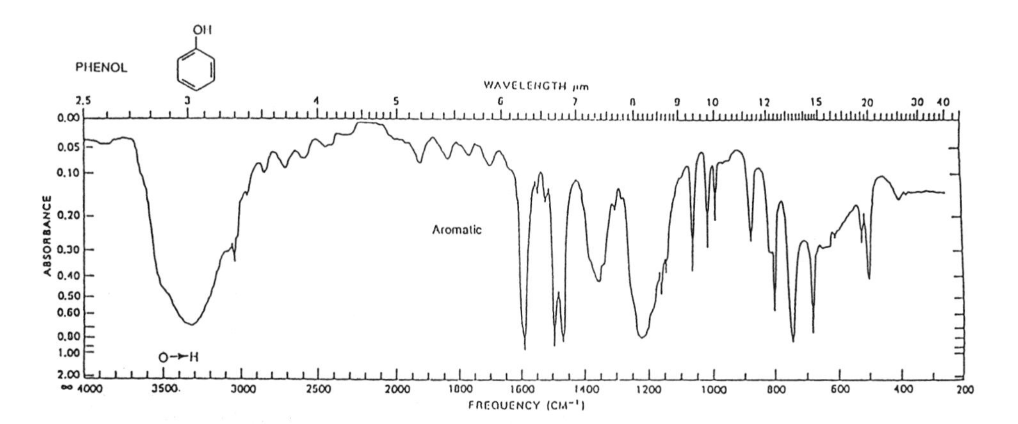

Figure 5.50 Absorption spectrum of phenol. Note that the CO → H is very broad and the C → H is a short sharp peak typical of aromatics. The C → OH is at 8.4 μm, displaced from the 9.5 μm of a primary alcohol but similar to a tertiary paraffinic alcohol. Characteristic bands at 5-6 μm and the bands at 12.4 and 13.4 μm typify monosubstituted benzene. This spectrum is typical of IR spectroscopy, including the characteristic absorption bands of the functional groups, in this case, aromatic CH and C — OH. The spectra are additive.

This spectrum is typical of IR spectroscopy including the characteristic absorption bands of the functional groups, in this case, aromatic CH and C—OH. The spectra are additive.

b. Ketones and Aldehydes

These are typified by the C ⇛ O bands, which occur as follows:

Functional group	Mode	Wavelength (μm)
Aliphatic	C ⇛ O	5.8
Aromatic		5.9
Aliphatic 6-membered ring		5.9
Smaller ring		5.0–5.7
Aldehydes	C ⇛ O	5.8
	C → H	3.7
Acid	C ⇛ O	5.8

In esters, the carbonyl occurs at shorter wavelengths

Functional group	Mode	Wavelength (μm)
Esters O↗C↘O	Stretch	7.0 (assymetrical) weak
		7.6–8.3 (symmetrical) strong
β-Diketone	C ⇛ O	6.1–6.5
Enol OH	O → H	3.8–4.0

The spectrum for acetone is shown in Fig. 5.51.

The spectrum for phenylpropionaldehyde (Fig. 5.52) shows the simple paraffinic spectrum plus the strong carbonyl C ⇛ O at 5.8. This complex spectrum can be broken down to see the carbonyl C ⇛ O at 5.8 μm, the aromatic and aliphatic, C → H peaks at 3.3–3.6 μm with the aldehyde C → H at 3.6 μm in addition to the CH bands at 13.3 and 14.3 μm typical of monosubstituted aromatic. In short, the spectrum shows aromatic plus paraffinic plus carbonyl plus aldehydic H.

The spectrum for heptanoic acid ($C_6H_{13}C(=O)$—OH) (Fig. 5.53) shows the carbonyl C ⇛ O at 5.8 μm, C → OH at 7.8 μm, and a very broad HCO → H from 2.8 to 3.8 μm. This broad CO → H is caused by dimerization and by traces of water in the sample. It is very typical of carboxylic acids.

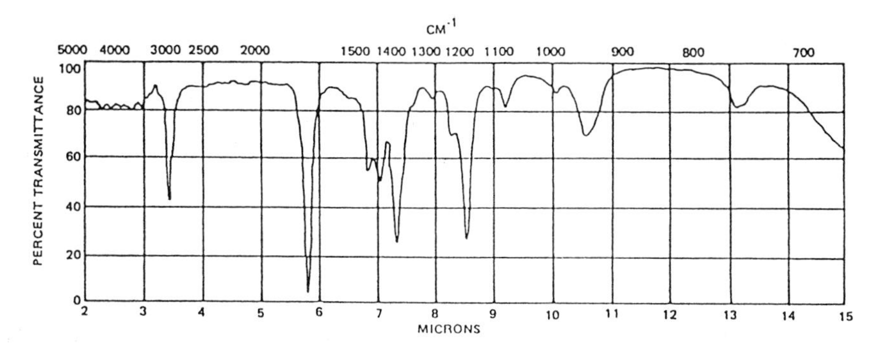

Figure 5.51 Absorption spectrum of acetone. Note the simple paraffinic spectrum plus the strong carbonyl C ⇒ O at 5.8 μm.

Phenyl propionaldehyde $CH_3CH(C_6H_5)CHO$

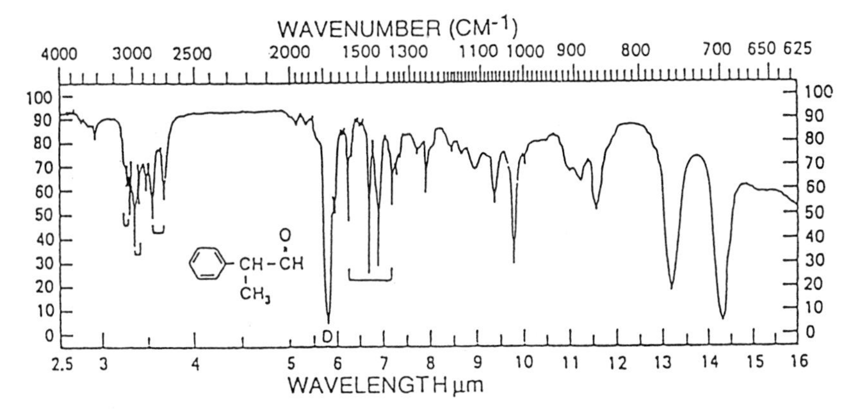

Figure 5.52 Absorption spectrum of phenyl propionaldehyde. This complex spectrum can be broken down to see the carbonyl C ⇒ O at 5.8 μm, the aromatic and aliphatic, C → H peaks at 3.3-3.6 μm with the aldehyde C → H at 3.6 μm in addition to the CH bands at 13.3 and 14.3 μm typical of monosubstituted aromatic. In short, the spectrum shows aromatic, plus paraffinic, plus carbonyl, plus aldehydic H.

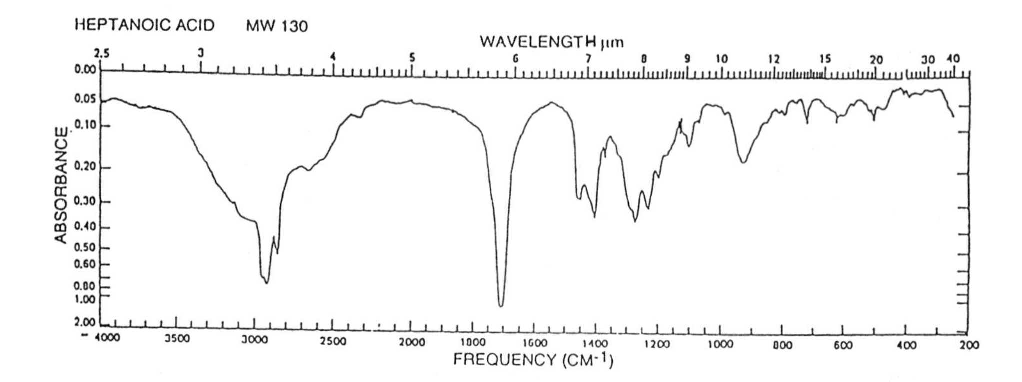

Figure 5.53 Absorption spectrum of heptanoic acid ($C_6H_{13}C(=O)$—OH) shows the carbonyl C ⇒ O at 5.8 μm, C → OH at 7.8 μm, and a very broad HCO → H from 2.8 to 3.8 μm. This broad CO → H is caused by dimirsation and by traces of water in the sample. It is very typical of carboxylic acids.

3. Nitrogen Compounds

Functional group	Mode	Wavelength (μm)
Aliphatic *Amines*	*N → H*	
	Primary R-NH_2	2.8 3.0 Doublet
	Secondary R_2>NH	3.0 Singlet
	Tertiary R_3>N	No NH
	NH_3^+	3.0 3.1 Doublet
	NH_2^+	3.73 3.74 Doublet
	NH^+	4.5 Singlet
Aromatic Amines	Primary N ↓ H	6.1–6.3
	Sec	6.1–6.3
	NH⊕	11.5–15.4
	C → N	8–10
	C ⇒ N	5.9–6.1
	C≡N nitrile	4.4–5.0
	N≡C isonitrile	4.6–4.7

Figure 5.54 shows the spectrum of butyl amine, $C_4H_9NH_2$. Two N → H peaks can be seen at 2.8 and 3.0 μm. They are weak, but distinctive. The C—N band is weak and not very useful.

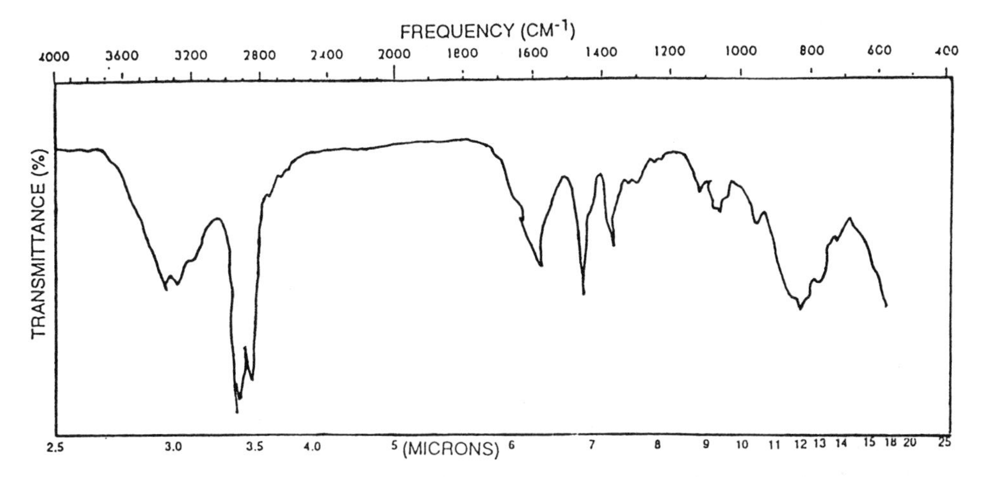

Figure 5.54 Absorption spectrum of butyl amine, $C_4H_9NH_2$. Two N → H peaks can be seen at 2.8 and 3.0 μm. They are weak, but distinctive. The C—N band is weak and not very useful.

Figure 5.55 is the spectrum for dibutyl amine $(C_4H_9)_2$ NH. The single NH peak is weak but distinctive.

The spectrum for butyronitrile (Fig. 5.56) shows paraffic C → H and a strong C≡N band at 4.3 μm.

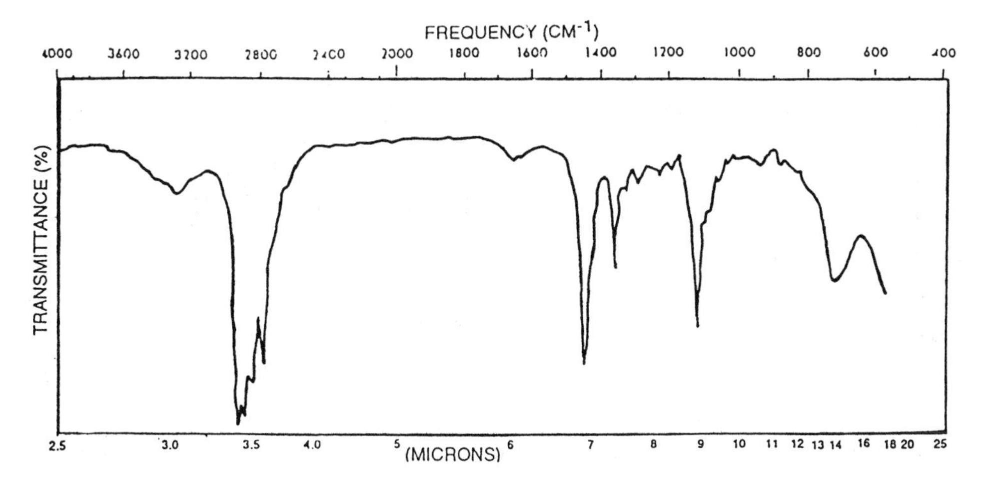

Figure 5.55 Absorption spectrum of dibutyl amine $(C_4H_9)_2$ NH. The single NH peak is weak but distinctive.

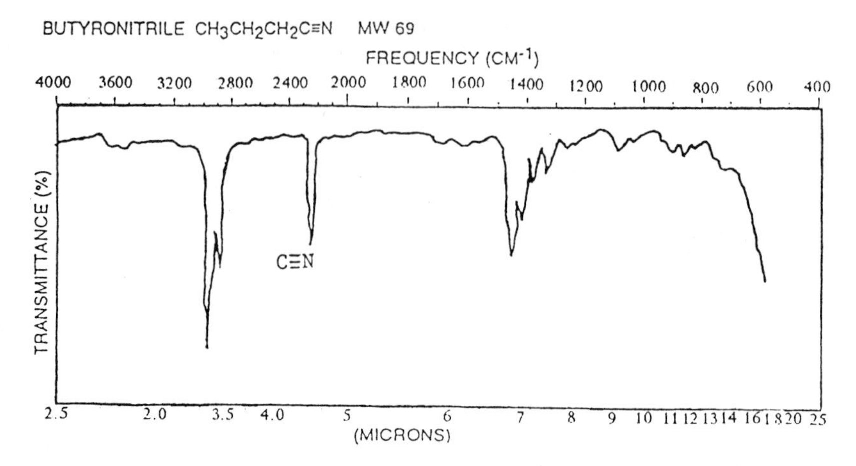

Figure 5.56 Absorption spectrum of butyronitrile shows paraffic C → H and a strong C ≡ N band at 4.3 μm.

4. Nitrogen–Oxygen Compounds

These nitro compounds, nitrates, and nitramines all contain NO_2. They give N⇉O/⇉O symmetrical and asymetrical stretch absorption bands as follows:

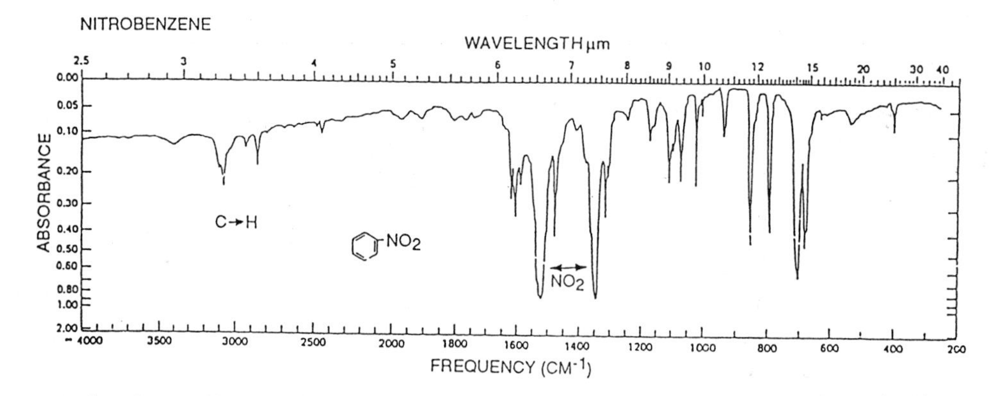

Figure 5.57 Absorption spectrum of nitrobenzene shows strong symmetrical and asymmetrical N⇉C/⇉C stretch at 6.4 and 7.4 μm. The typical aromatic bands are also excellent.

Functional group	Mode	Wavelength (μm)
Primary NO_2 $N\begin{smallmatrix}\nearrow C\\ \searrow C\end{smallmatrix}$	Asymmetrical	6.4–7.3
Secondary	Symmetrical	
Teritiary	Slightly larger wavelength	
	NO	6.0–7.8
	N ⇒ O	6.1
	O—N ⇒ O	14.5–17.9

The spectrum for dibutyl amine shows the strong symmetrical and assymetrical N → O stretch at 6.4 and 7.4 μm. The typical aromatic spectrum can be seen.

5. Amino Acids

These are the building blocks of protein and as such are very important in biochemistry and natural products chemistry. Important absorption bands are as follows.

Mode	Wavelength (μm)
N → H	3.2
N ↓ H	6.1
	7.7
C ⇒ O	6.2 and 7.1

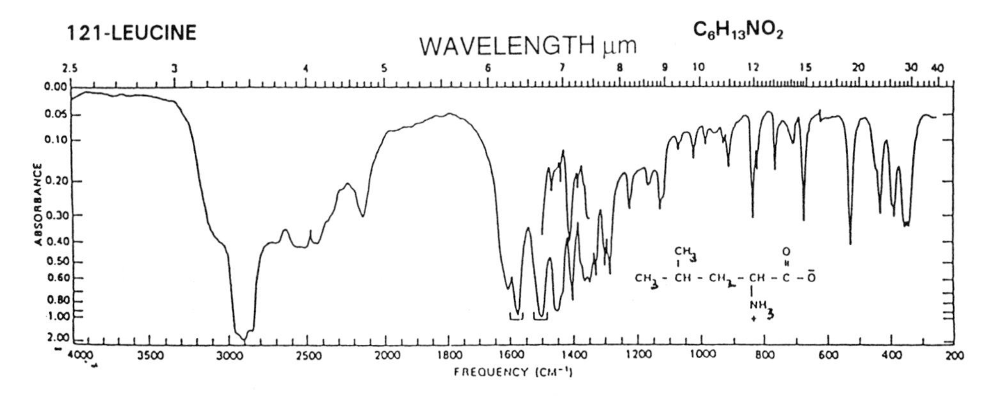

Figure 5.58 Absorption spectrum of leucine shows the broad irregular band between 3.2 and 5 μm caused by C → H, CO → H, N → H, and the effect of H bonding between different molecules in the system. This always causes broadening, as observed with carboxylic acids.

The spectrum for leucine (Fig. 5.58) shows the broad irregular band between 3.2 and 5 μm caused by C → H, CO → H, N → H, and the effect of H bonding between different molecules in the system. This always causes broadening, as observed with carboxylic acids.

6. Hetero Groups

a. Halides

The C → halide stretch absorption bands are in the fingerprint region, but the bending modes are at longer wavelengths and are not quoted here. Several halides on the same C cause an increase in absorption and a shift to longer wavelengths.

Functional group	Mode	Wavelength (μm)
Aliphatic halide	C → F	7.1–13.7
	C → Cl	11.8–18.2
	C → Br	14.5–19.2
	C → I	16.7–20.0
Aromatic		
(benzene ring) → F	C → F	8.0–9.1
(benzene ring) → Cl	C → Cl	9.1–9.2

The spectrum shown in Fig. 5.59 is of chloroform, $CHCl_3$. C → H at 3.3 μm, C ↓ H at 8.2 μm and the very strong C → Cl at 13.2 μm can be seen.

b. Miscellaneous Absorption Bands

The absorption bands of the hetero atoms S, P, and Si are listed below.

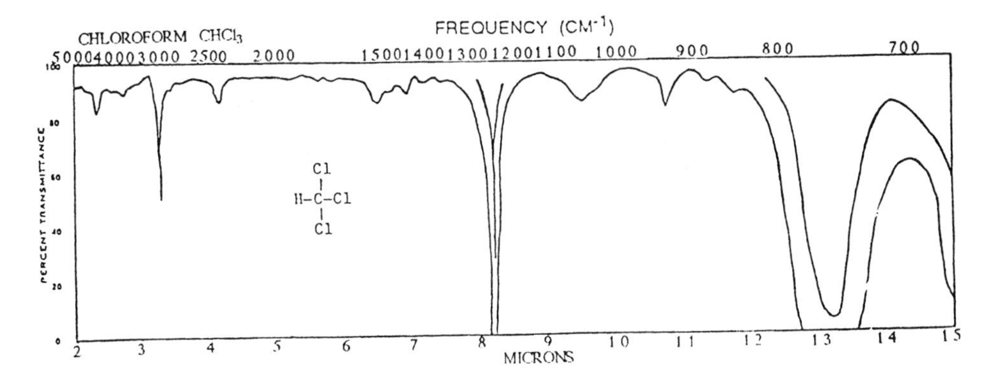

Figure 5.59 Absorption spectrum of chloroform, $CHCl_3$. C → H at 3.3 μm, C ↓ H at 8.2 μm and the very strong C → Cl at 13.2 μm can be seen.

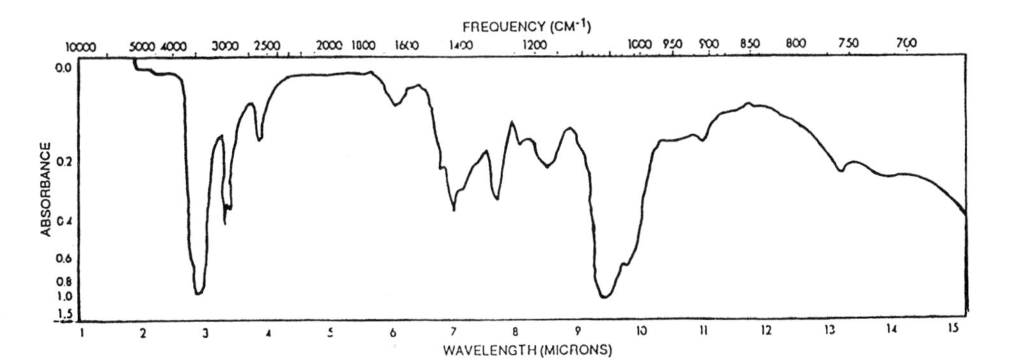

Figure 5.60 Absorption spectrum of ethylthioalcohol, $OHCH_2CH_2SH$, shows clearly the different absorption bands for CO → H and CS → H at 3.0 and 3.9 μm, respectively. The C → SH bands are weak and not useful.

Sulfur

Mode	Wavelength (μm)	Bonding
S → H	3.85–3.92	H bonding weak
C → S	14.3–16.5	Weak
S → S	25.0 Not in rock salt region	
C ⇒ S	6.6	
	15.4	
	5.97	
C═C—C═S	8.65	
Sulfonide		
S═0	9.5	
Sulfone		
O═S═0	8.6–8.9	Intense
	7.4–7.7	
Sulfonate	7.0–7.5	
	8.3–8.7	
Sulfhate	6.9–7.4	
	8.1–8.7	

Phosphorus

Mode	Wavelength (μm)
C → P	14–15 μm
P — Benzene	7–10 μm
P — O — C	10 μm
P ⇒ O	8 μm

Silicon

Mode	Wavelength (μm)
Si →F	10 μm
Si → H	4.8 μm
Si → C	12–14 μm
Si → OR	9–10 μm
Si → O→ Si	9–10 μm
Si → O → Arom	10.3–10.8 μm
Si → OH	11–12 μm

Mode	Sulfur	Wavelength (μm)
S → H	Mercaptan	3.85–3.92 μm
C → S	(Sulfides)	14.3–16.7 μm
S → S	(Disulfide)	20–25 μm
C ⇒ S	Thio carbonyl	8.0–9.8 μm
Me O		
S	Sulfoxide	9.3–9.7 μm
Me O		

Fig. 5.60 shows the spectrum for ethylthioalcohol, $OHCH_2CH_2SH$ and shows clearly the different absorption bands for CO → H and CS → H at 3.0 and 3.9 μm, respectively. The C → SH bands are weak and not useful.

H. INTERPRETATION OF SPECTRA PROCESSES

A fruitful approach is as follows:

1. Determine whether the compound is aromatic or aliphatic (see C → H and and band(s) at 11–14 μ).
2. Identify functional groups present such as C═O, C → OH, NH_2, C═C, C═N, etc. Note their formula weight (FW).
3. Subtract FW from molecular weight of compound given.
4. Accommodate the difference as paraffin or aromatic components. For example, phenol (MW 94) IR shows OH (FW 17) and an aromatic (94 − 17 = 77).
   Hence,

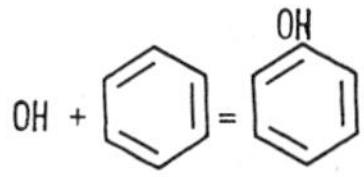

   OH + [benzene ring] = [phenol, OH on ring]

   Acetic acid (MW 60) IR shows C ⇒ O and OH as acid COOH:

   FW of COOH = 45

   MW − FW = 60 − 45 = 15

   15 is probably CH_3. Hence, CH_3 and COOH is CH_3COOH.

BIBLIOGRAPHY

Anal. Chem., Application Reviews, June 1993.
Anal. Chem., Applications Reviews, April 1984.

Benedetti E., Teodori, L., Trinca, M. L., Vergamini, P., Salvati, F., Mauro, F., and Spremolla, G., *Appl. Spec.*, (1990), *44* (8):1276.

Boccara, A. C., Fourier, D. and Badoz, J. App. Phys. Letters, (1980), 36, 130.

Bundig-Lee, K. A., Hood, A. L., Clobes, A. L., Schroeder, J. A., Ananth, G. P., and Hawkins, L. H., IR spectroscopy for indoor air monitoring, *Spectroscopy*, (1993), *8*(19):24 .

Burns, D. R., and Curezak, E. W., eds. *Handbook of Near Infrared Analysis*, Marcel Dekker, New York, 1993.

Busch, M. A., and Busch, K. W., Signal/Noise comparison of flame/furnace I/R emission spec. with room-temperature non-dispersive IR absorption spectroscopy, *Appl. Spec.*, (1993), *47*(7):912.

Calthup, N. B., Daly, L. H., and Wiberley, S. E., *Introduction to Infrared and Raman Spectroscopy*, Academic Press, New York, 1975.

Carney, J. M., Landrum, W., Mayes, L., Zou, Y., and Lodder, R. A., NIR spectrometric monitoring of stroke-related changes in the protein and lipid composites of whole gerbil brain, *Anal. Chem.*, (1993), *65*:1305.

Cassis, L. A., and Lodder, R. A., NIR imaging of atheromas in living arterial tissue, *Anal. Chem.*, (1993), *65*:1247.

Chase, B., and Talmi, Y., *Appl. Spec.*, (1991), *45*(6):929.

Coffey, M. T. and Martin, W. G., F. T. spectroscopy for stratospheric research, *Spectroscopy*, (1993), *7*:22.

Creswell, C. J., and Runquist, O., *Spectral Analysis of Organic Compounds*, Burgess, Minneapolis, Minn., 1970.

De Blase F. J., and Compton S. IR Emission Spectroscopy: A theoretical and experimental Review, *Appl. Spec.*, (1991), *45*(4):611.

FTIR microscopy of fossilized organic material, *Spectroscopy*, (1991), (Sept):24.

Graham, P. B., and Miller, R. S., *Spectroscopy*, (1993), *8*:31.

Lai, E. P. C., Chan, B. L., and Hadjmohammadi, *Appl. Spec. Rev.*, (1985), *21*(3):179.

Lee, K. A. B. *Appl. Spec. Rev.*, (1993), *28.3*:231.

LeVine, S. M., and Wetzel, D. L. B., *Applied Spec. Reviews*, (1993), *28*:385.

Low, M. J. D., and Morterra, C., *Appl. Spectrosc.* (1984), *38*:807.

McClelland, J. F., Photoacoustical spectroscopy, *Anal. Chem.*, (1983), *55* (1):89A.

Marbach, R., Kosehinsky, Th., Gries, F. A., and Heise, H. M., Noninvasive blood glucose assay by NIR diffuse reflectance spectroscopy of the human inner lip, *Appl. Spec.*, (1983), *47*(7):875.

Metzel, D. L., and LeVine, S. M., In situ FTIR miscroscopy and mapping of normal brain tissue, *Spectroscopy*, (1993), *8*(4):40.

Mirabella, F. M., *Internal Reflection Spectroscopy*, Marcel Dekker, New York, 1992.

Morris, M. D., *Microscopic and Spectroscopic Imaging of the Chemical State*, Marcel Dekker, New York, 1993.

Near IR spectroscopy of synthetic polymers, *Appl. Spec. Rev.* (1991), *26*(4):277.

Noda, I., Dowrey, A. E., and Marcott, C., Recent developments in two-dimensional infrared (2D IR) correlation spectroscopy, *Appl. Spec.*, (1993), *47*:1317.

Redd, D. C. B., Feng, Z. C., Yue, K. T. and Gansler, T. S., Raman spectroscopic characterization of human breast tissues: implications of breast cancer diagnosis, *Appl. Spec.*, (1993), *47*(6):787.

The Sadtler Standard Spectra, Sadtler Research Labs., Inc., Philadelphia. Silverstein,

R. M., Bassler, G. C., and Morrill, T. C., *Spectrometric Identification of Organic Compounds*, 4th ed., Wiley, New York, 1981.

Smith, A. L., *Infra Red Spectroscopy, Treatise on Analytical Chemistry*, Part 1, Vol. 7, Wiley, 1981.

Tilotta, D. C., Lam, C. K. Y., Busch, K. W., and Busch, M. A., Evaluation of thermospray and cross-flow pneumatic nebulization as means of interfacing a flame infrared emission (FIRE) radiometer to a high-performance liquid chromatograph, *Appl. Spec.*, (1993), *47*(2):192.

Ultra low volume gold light pipe cell for IR analysis of dilute organic solutions. *Appl. Spec.*, (1990), *44*:1092.

What's new in instrumentation, *Spectroscopy*, (1993), *8*(4):.

Workman, J., and Andren, H. S., NIR spectroscopy for at-line meat product measurements, *Am. Lab.*, (1992), (December):20R.

SUGGESTED EXPERIMENTS

5.1 Record the IR absorption spectrum of hexane. Identify the absorption bands caused by the C $\rightarrow$ H stretching frequency, the C $\dotplus$ H bending frequency, and the C $\rightarrow$ C stretching frequency.

5.2 Record the IR spectrum of heptane. Note the similarity with the spectrum obtained in Experiment 5.1. Would it be possible to distinguish between these compounds based on their IR spectra?

5.3 Record the IR spectrum of *n*-butanol. Note the O $\rightarrow$ H stretching peak and the C $\rightarrow$ OH stretching peak. Repeat with *i*-butanol and *t*-butanol. Note the change of C $\rightarrow$ OH stretching peak but that there is little change in the O $\rightarrow$ H stretching peak. Could this change be used to distinguish among primary, secondary, and tertiary alcohols?

5.4 Record the IR spectrum of *n*-butylamine. Note the N $\rightarrow$ H and C $\rightarrow$ N stretching peaks. Repeat with *sec*-butylamine and *tert*-butylamine.

5.5 Record the IR spectra of acetone, chloroform, terminal olefins, internal olefins, and cis- and trans-substituted olefins. Identify the vibrational modes of the molecule responsible for the relevant absorption bands.

5.6 Make up several different solutions of known concentration of ethanol in carbon tetrachloride over the concentration range 1–30%. Measure the absorption of the C—OH absorption band (9.10 μm). Plot the relationship between absorbance at this wavelength and the concentration of ethanol.

5.7 Repeat Experiment 5.6, using acetone as the sample. Note the sharp C $\Rightarrow$ O stretching band and use the absorbance of this band for quantitative studies as suggested in Experiment 5.6.

5.8 Take a sample of unsaturated cooking oil and record the IR absorption spectrum. Note the C $\Rightarrow$ C stretching frequency. Repeat with several brands of cooking oil. Based on the IR absorption spectrum, which brand was the most unsaturated?

PROBLEMS

5.1 What types of vibrations are encountered in organic molecules?

5.2 What materials are used for making prisms for use in IR spectroscopy? Why must special precautions be taken to keep these materials dry?

5.3 Why do organic functional groups resonate at characteristic frequencies?

5.4 How can primary, secondary, and tertiary alcohols be distinguished by their IR absorption spectra?

5.5 Indicate which C and H stretching and bending vibrations can be detected in an IR absorption trace of range of 2.5–16 μm. What would be the frequency of each vibration?

5.6 A solution is known to contain acetone and ethyl alcohol. Indicate the expected IR absorption curve for each compound separately. Which absorption band could be used to identify the presence of acetone in the mixture?

5.7 What requirements must be met before a molecule will absorb IR radiation?

5.8 In preparing a calibration curve for the determination of methyl ethyl ketone,

$$CH_3\underset{\underset{O}{\|}}{C}—CH_2—CH_3$$

solutions with different concentrations were prepared. The measured absorbance A for each solution is given below.

Concentration (%)	Absorbance
2	0.18
4	0.36
6	0.52
8	0.64
10	0.74

(a) Does the relationship between A and sample concentration deviate from Beer's law?

(b) The absorbance of several mixtures of methyl ethyl ketone was measured at the same wavelength as that used for the calibration curve. The results were as follows:

Sample	Absorbance
A	0.27
B	0.44
C	0.58
D	0.69

What were the concentrations of the solutions?

5.9 What are the advantages of double-beam optics over single-beam optics in IR absorption analysis?

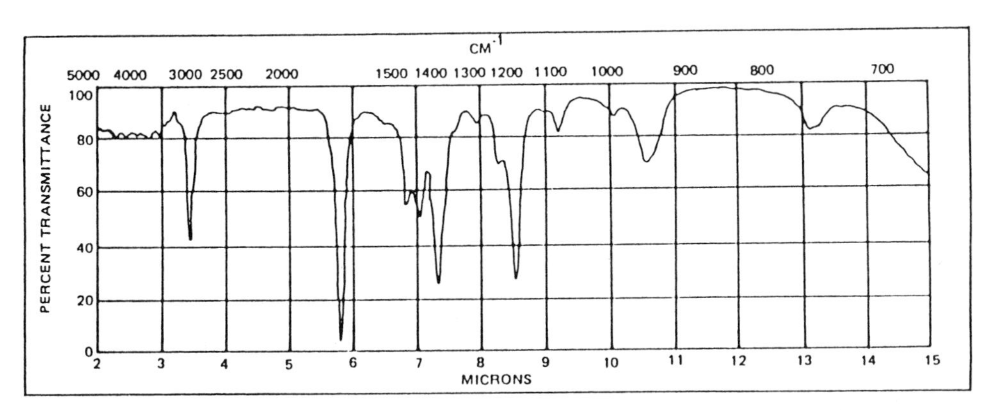

5.10

5.10 Identify the compound that gives the IR spectrum shown above.
5.11 What are (a) fundamental and (b) overtone absorption bands?
5.12 Is FTIR single beam or double beam? How is background correction achieved?
5.13 How do Ge-doped detectors work? What is their advantage over thermisters and thermocouples?
5.14 Describe FTIR. Which detectors are used?
5.15 List the advantages of FTIR over standard IR.
5.16 Describe ATR.
5.17 Describe infrared flame emission.
5.18 What wavelength range is covered by near-IR? What bands occur here?
5.19 What is the advantage of using near-IR?
5.20 Describe Raman spectroscopy.
5.21 What are the requirements for a molecule to be Raman active?
5.22 Describe photoacoustic spectroscopy.

For Problems 5.23–5.37, recall the approach discussed in Section H, Interpretation of Spectra Processes, p. 245. Based on the information given, identify the following compounds.

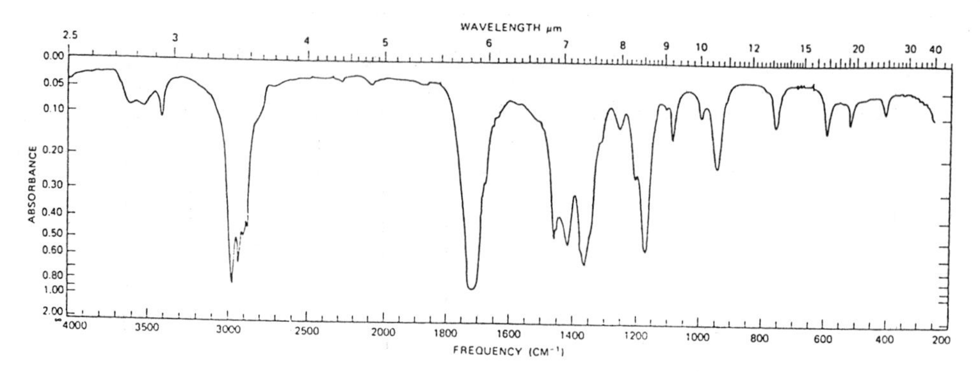

5.23 MW 72

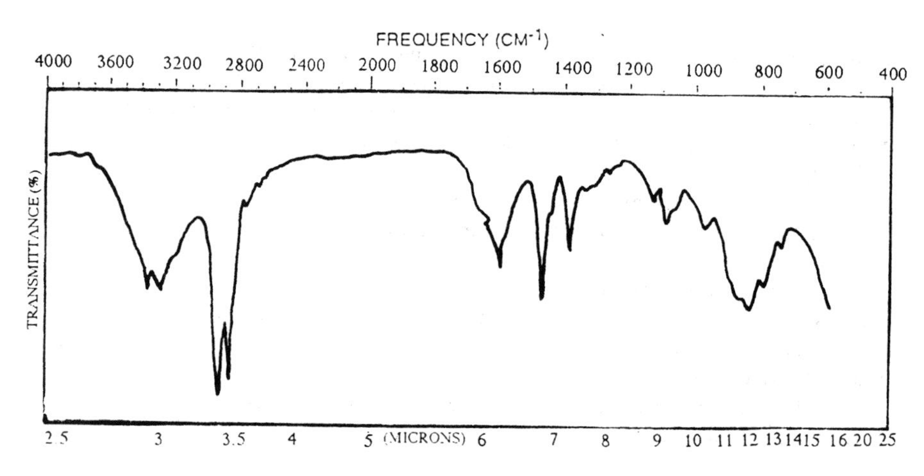

5.24 MW 73

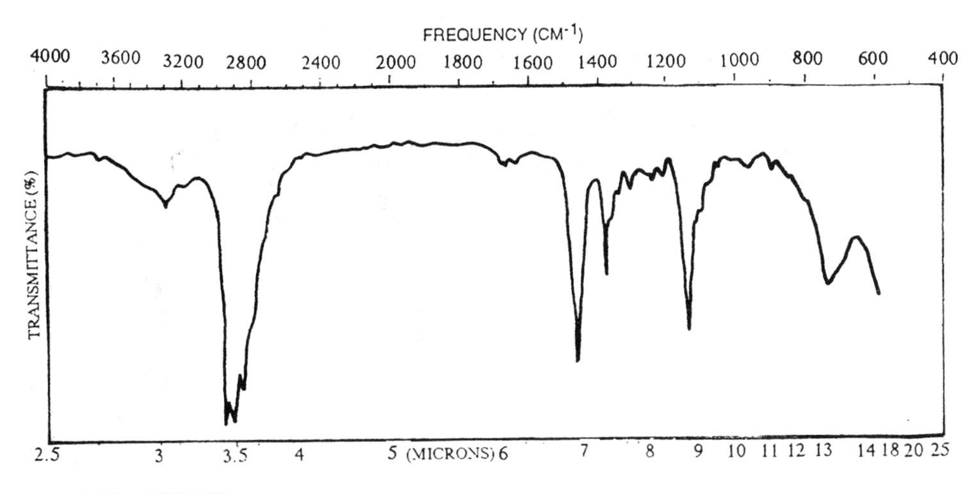

5.25 MW 129

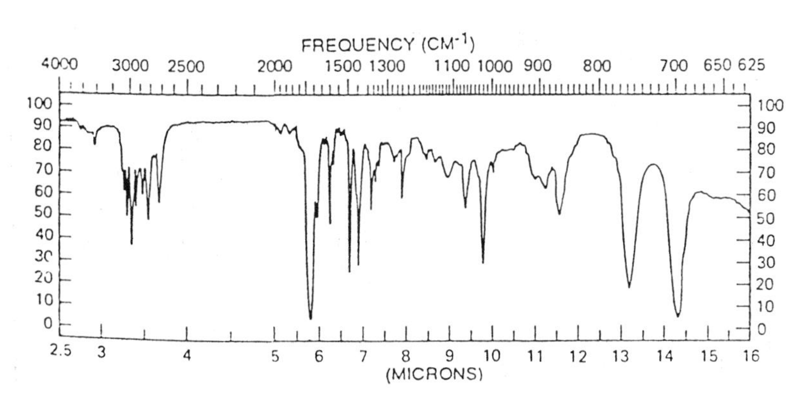

5.26 MW 134

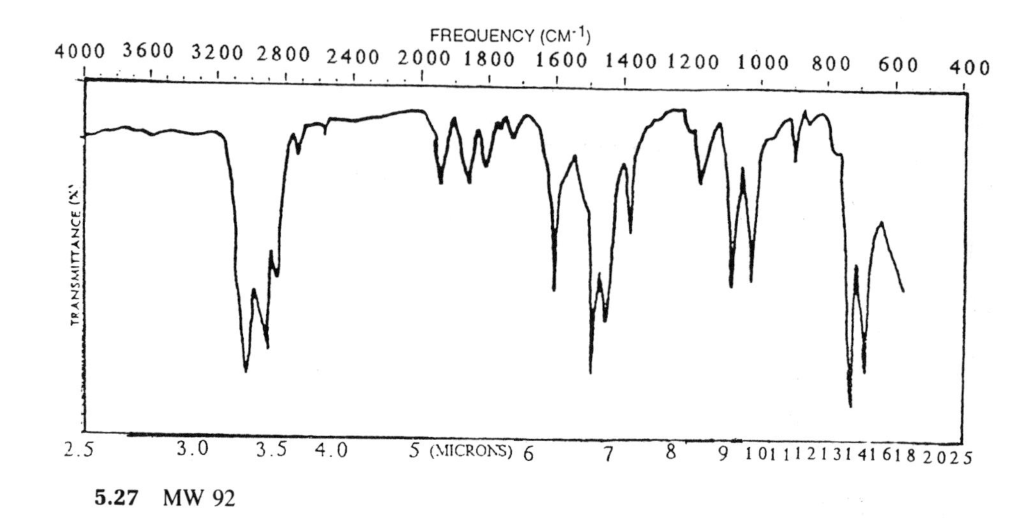

5.27 MW 92

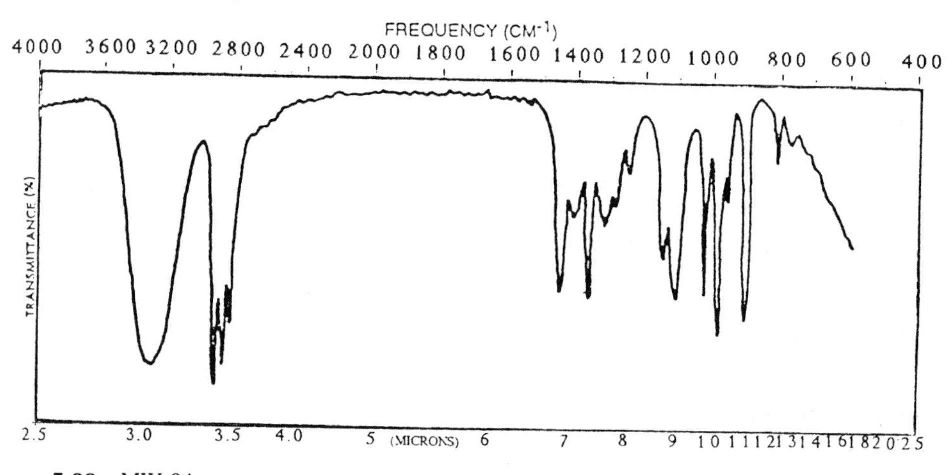

5.28 MW 94

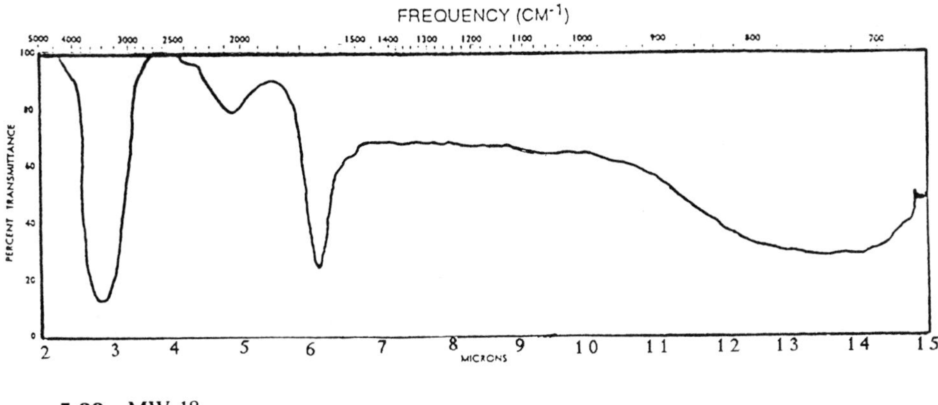

5.29 MW 18

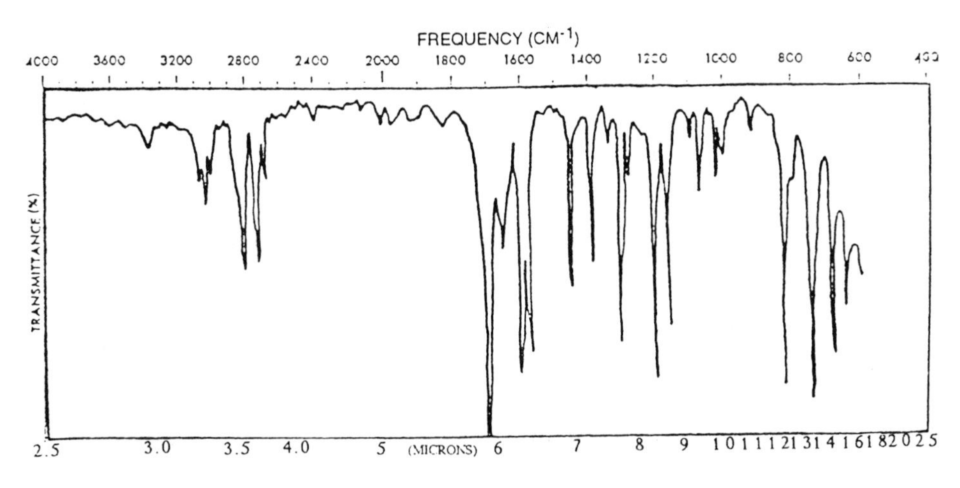

5.30 MW 106

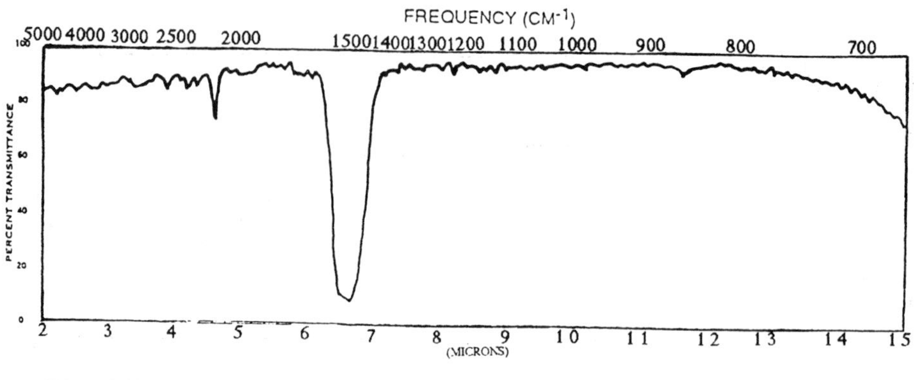

5.31 MW 76

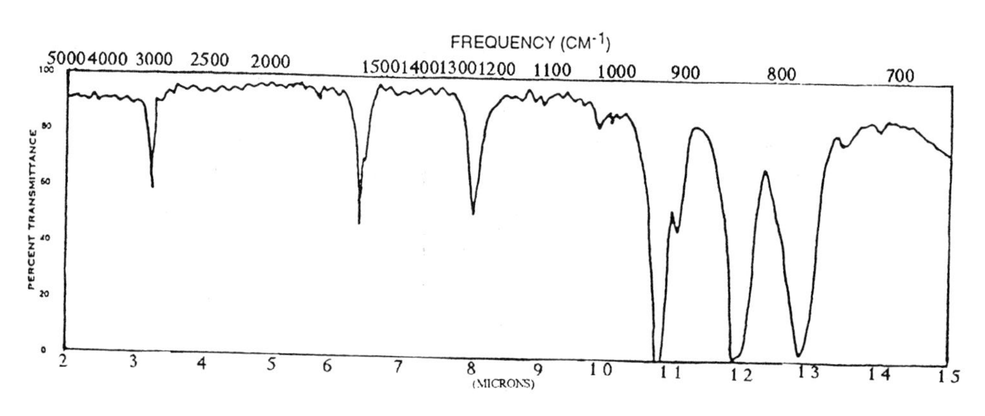

5.32 MW 130

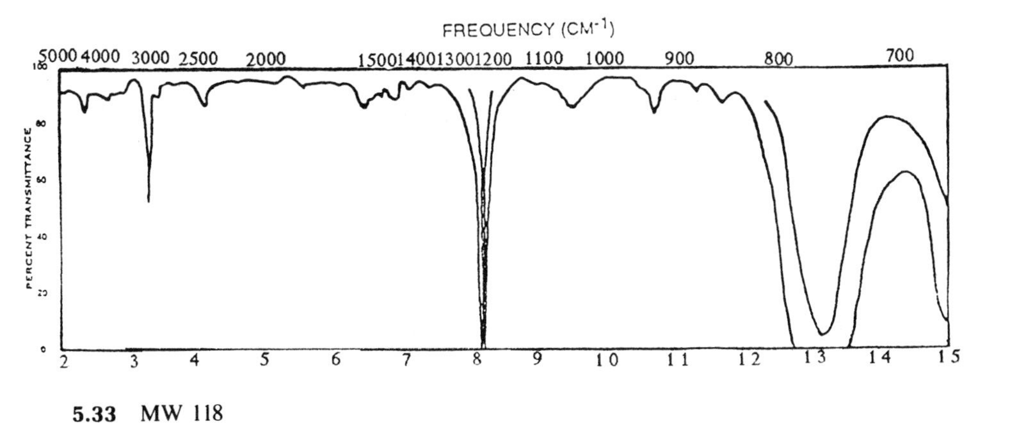

5.33 MW 118

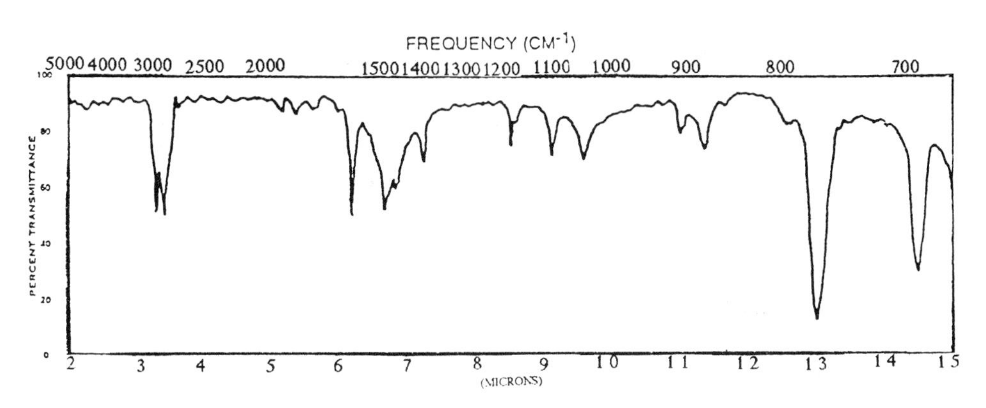

5.34 MW 108

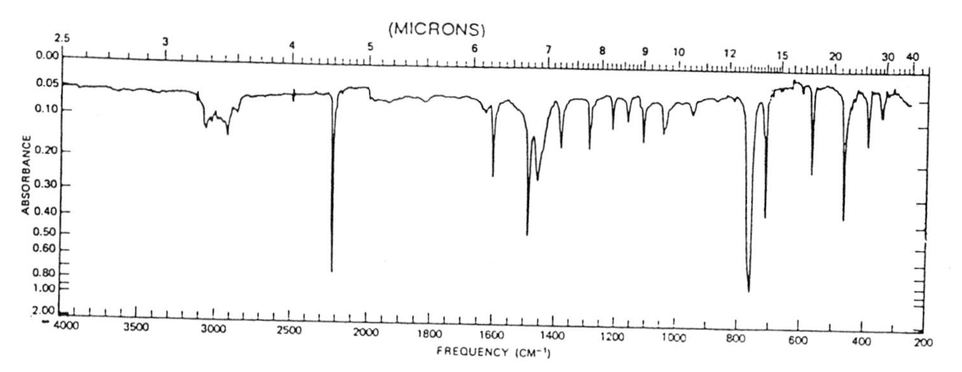

5.35 MW 117

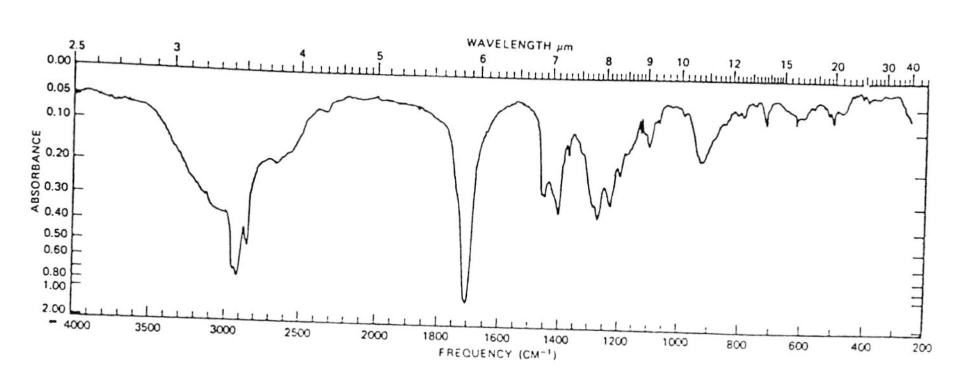

5.36 MW 130

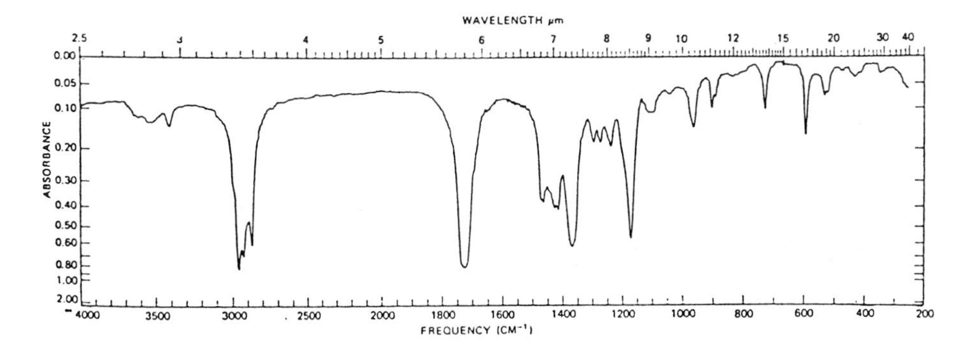
WAVELENGTH μm
2.5
3
4
5
6
7
8
9
10
12
15
20
30
40
ABSORBANCE
0.00
0.05
0.10
0.20
0.30
0.40
0.50
0.60
0.80
1.00
2.00
4000
3500
3000
2500
2000
1800
1600
1400
1200
1000
800
600
400
200
FREQUENCY (CM⁻¹)

5.37 MW 86

6
Visible and Ultraviolet Molecular Absorption Spectroscopy

A. INTRODUCTION

The wavelength range of ultraviolet (UV) radiation starts at the blue end of visible light (about 400 nm) and ends at 200 nm. The radiation has sufficient energy to excite valence electrons in many atoms or molecules; consequently, UV is involved with electronic excitation. Spectroscopically, visible light acts in the same way as UV light; hence it is generally considered part of the electronic excitation region. For this reason we find that commercial UV equipment often operates with wavelengths between 800 and 200 nm. The seven major fields of analytical chemistry operating within this wavelength range are listed in Table 6.1

1. Electronic Excitation

Atoms and molecules are composed of nuclei and electrons. As we saw in Chapter 2, electrons move in orbitals at definite energy levels around the nucleus as defined by quantum theory. When the energy of the electrons is at a minimum, the molecules are in the lowest energy state, or ground state. The molecules can absorb radiation and move to a higher energy state, or *excited state*. When the molecule becomes excited, an outer shell electron moves to an orbital of higher energy. The process of moving electrons to higher energy states is called *electronic excitation*. For radiation to cause electronic excitation, it must be in the visible or UV region of the electromagnetic spectrum.

Table 6.1 Spectroscopy

Function	Analytical field	Analytical application
	Atomic UV Spectroscopy	
Absorption of UV radiation	Atomic absorption	Quantitative elemental analysis
Emission of radiation	Flame photometry	Quantitative analysis of alkali metals, alkaline-earth metals, and other metals
Emission of radiation	Emission spectrography Plasma emission	Qualitative and quantitative multielemental analysis
	Molecular UV Spectroscopy	
Absorption of UV radiation	UV absorption	Qualitative and quantitative determination of aromatics and unsaturated compounds, including natural products
Emission of UV radiation	Molecular fluorescence	Detection of small quantities (10 μg) of certain aromatics and natural products
Emission of UV radiation	Molecular phosphorescence	Analysis of gels and glasses

The frequency and the energy of radiation are related by Eq. (2.2), $E = h\nu$. The actual amount of energy required depends on the difference in energy between the ground state E_0 and the excited state E_1 of the electrons. The relationship is described by

$$E_1 - E_0 = h\nu$$

where E_1 is the excited state and E_0 is the ground state.

Three distinct types of electrons are involved in organic molecules. First are the electrons involved in saturated bonds, such as those between carbon and hydrogen in paraffins. These bonds are called *sigma* (σ) *bonds*. The amount of energy required to excite electrons in σ bonds is usually more than UV photons possess. For this reason, paraffinic compounds do not absorb UV radiation and are therefore frequently very useful as transparent solvents.

The second type is the electrons involved in unsaturated bonds. These bonds usually involve a *pi* (π) *bond*. Typical examples of compounds with π bonds are conjugated olefins and aromatic compounds (Fig. 6.1). Electrons in π bonds are excited relatively easily and absorb in the UV.

Electrons that are not involved in bonding between atoms are the third type of electrons involved in molecules. These are called *n* electrons. In saturated

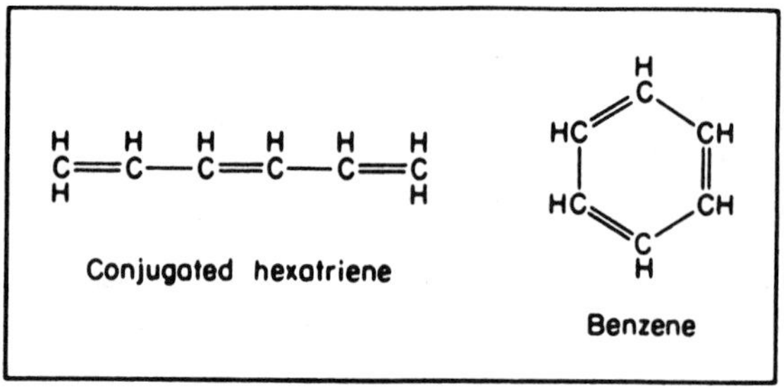

Conjugated hexatriene

Benzene

Figure 6.1 Examples of molecules with π bonds.

hydrocarbons the outer shell electrons of carbon and hydrogen are all involved in bonding; hence these compounds do not have any *n* electrons. Organic compounds containing nitrogen, oxygen, sulfur, or halogens, however, frequently contain electrons that are nonbonding (Fig. 6.2). Because *n* electrons are usually excited by UV radiation, most compounds that contain *n* electrons absorb UV radiation.

2. Absorption by Molecules

When molecules are electronically excited, the electrons go from a bonding to an antibonding orbital. Electrons in a σ bond are excited to antibonding σ orbitals. Electrons in π bonds are excited to antibonding π orbitals, and *n* electrons are excited to either σ antibonding orbitals or π antibonding orbitals.

Inorganic molecules exhibit the same kind of phenomenon. The presence of *n* electrons explains why some compounds, such as potassium permanganate, are highly colored. The change of electron density and therefore the energy necessary to cause excitation have provided the basis for colorimetric analysis and pH indicators. Molecular groups that absorb visible or UV light are called *chromophores*.

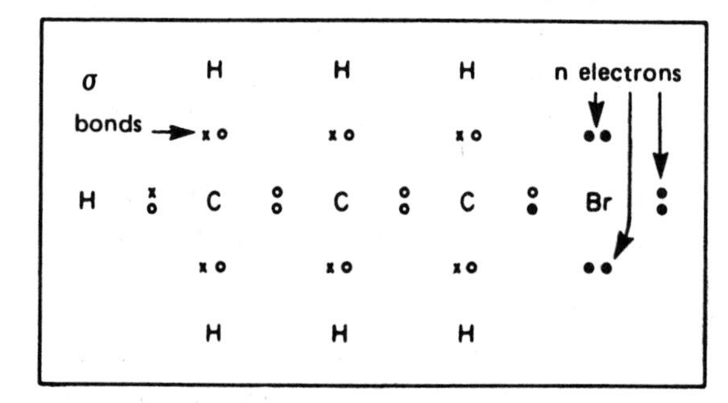

Figure 6.2 Nonbonding *n* electrons.

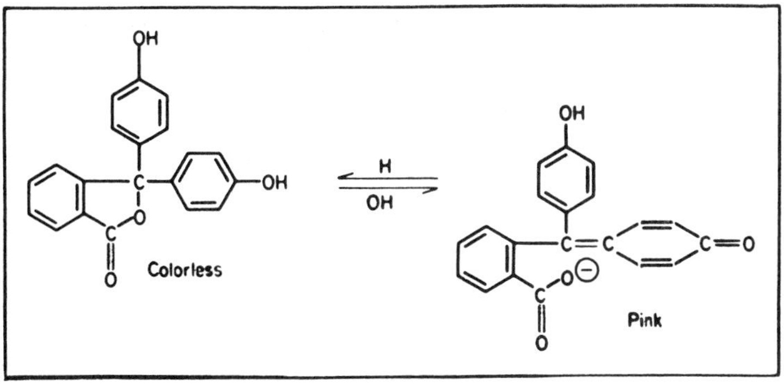

Figure 6.3 Color change in phenolphthalein brought about by a change in H^+ concentration, that is, pH. This is the basis for its use as a pH indicator.

Figure 6.3 shows the structure of phenolphthalein in an acid solution and in a basic solution. As can be seen, there is a change in the structure of the molecule and a significant change in the energy levels of the chromophores. This results in the molecule absorbing visible radiation when it is in an alkaline solution, but not in an acid solution. Such structure changes and energy level changes are the basis of many acid–base indicators.

If a hydrogen ion is replaced by a metal ion, the same kind of rearrangement of the bonds and energy levels may take place. Reagents have been developed that are specific for certain metals and cause color changes in their presence. Spectrophotometry and colorimetry are generally based on a specific reaction between the metal and an organic compound that is essentially unaffected by other metals present. A change in color or the UV absorption spectrum is observed denoting the presence of the metal of interest. If the metal is absent, no reaction occurs and no change in spectrum is observed.

We can summarize by saying that UV radiation may be absorbed by organic compounds that contain nitrogen, oxygen, sulfur, halogen atoms, or unsaturated hydrocarbons (such as aromatics and conjugated olefins). Functional groups that contain these systems and absorb radiation in the UV region are called *chromophores*. A list of compounds and the wavelengths of their absorption maxima is given in Table 6.2. Compounds such as paraffins contain only σ bonds, which do not absorb radiation in the visible or UV region. They can be detected by special equipment designed to operate in the *vacuum UV region*, which stretches from 200 nm to the x-ray region (about 100 Å). Between these wavelengths, nitrogen and oxygen of the atmosphere absorb radiation. Consequently, any studies carried out in this region must operate with the light beams in a vacuum. This requirement generated the name *vacuum UV* for this part of the electromagnetic spectrum. Analytically, the region is of minor importance because of the difficulties inherent in equipment requiring a

Table 6.2 Absorption Wavelengths of Typical Organic Functional Groups

Chromophore	System	Wavelength of absorption maximum (nm)
Amine	$—NH_2$	195
Bromide	—Br	208
Iodide	—I	260
Thioketone	> C=S	205
Ester	R—C(=O)—OR	205
Aldehyde	R—C(=O)—H	210
Carboxylic acid	R—C(=O)—OH	200–210
Nitro	$—NO_2$	210
Nitrite	—ONO	220–230
Azo	—N=N—	285–400
Conjugated olefins	$(—C{=}C)_2$	210–230
	$(—C{=}C)_3$	260
	$(—C{=}C)_5$	330
Benzene		(1) 198
		(2) 255
Napthalene		(1) 220
		(2) 275
		(3) 314

vacuum and the minor potential of the region compared to UV. In addition, radiation sources and detectors that work well in this region are not readily available commercially.

3. Molar Absorptivity

The absorptivity of a molecule defines how much radiation will be absorbed at a given concentration of that compound. If the concentration is molar, the absorptivity is said to be the *molar absorptivity*. This is defined by the probability of a photon at a given wavelength being absorbed when passing through the solution, and is therefore the probability of an electronic transition taking place when the radiation is passed through a solution of this concentration. This can be defined as

$$\frac{\delta I_1}{I_0} = \frac{1}{3} gcNA_x \left(\frac{\delta b}{1000}\right)$$

where

I_1 = radiant power after passing through the sample
I_0 = radiant power before passing through the sample
c = concentration
N = Avogadro's number
A_x = cross section of the molecule
g = electronic transition probability
⅓ = statistical correction for random orientation
b = path length

But

$$A = abc$$

or

$$\log \frac{I_0}{I_1} = abc$$

or

$$a = \log \frac{I_0}{I \times bc}$$

If b is 1 cm and c is molar, then a is the molar absorptivity and is given the symbol ϵ. Commonly $\epsilon = 10^5$. It can be seen that these numbers are all physical properties of the molecules and, therefore, the absorption coefficient is a constant and a physical property of that compound.

4. The Shape of UV Absorption Curves

The time taken for an electron to move from a ground state to an excited state is about 10^{-15} sec. During this time the molecule is, of course, vibrating and rotating. An average time for a molecule to vibrate is about 10^{-12} sec, and to rotate about 10^{-10} sec. Therefore, during electronic excitation a molecule goes through about 1/1000 of a vibration and 1/100,000 of a rotation. Essentially, the molecule is stationary during electronic excitation. This means that the molecule does not vibrate or rotate during excitation, although a change in vibrational or rotational energy may occur in the process.

The molecules are already in various states of vibration. For example, if we have a million molecules, some have a vibrational energy of zero (i.e., they are not vibrating), some will be vibrating at the first permitted vibrational energy level, some will be at vibrational energy level 2, some at energy level 3, and so on, up to high vibrational levels. The radiant energy required to cause electronic excitation of each group varies and is modified by the vibrational energy levels and further modified by rotational energy changes. For this reason, when UV radiation falls on the million molecules, it is absorbed at numerous wavelengths. The total range of the absorption wavelengths may stretch over 10 mm. Ultraviolet radiation is therefore absorbed in absorption bands rather than at discrete wavelengths. An absorption band is made up of numerous absorption lines that can be separated only under very high resolution. The effect of the rotational energy of the molecule is to add even more absorption lines to the single band. The increased number of absorption lines makes the lines even closer together, but it does not appreciably increase the total range of the band. This is because the energy involved in rotation is very small compared to vibrational energy and extremely small compared to electronic excitation energy.

The transitions can be expressed mathematically as follows. If no vibrational energy change takes place during electronic excitation, the frequency absorbed is given by Eq. (2.2) as $E_2 - E_0 = h\nu_1$, where E_1 and E_0 are the upper and lower electronic energy states, respectively. If the vibrational energy decreases, then the frequency is given by $E_1 - (E_0 - V) = h\nu_2$, where V is the vibrational energy change. Similarly, if the rotational energy R decreases, the absorption frequency is given by $h\nu_3 = E_1 - (E_0 - V - R)$. However, vibrational energy and rotational energy may increase during electronic excitation and, accordingly, $h\nu = E_1 - (E_0 \pm V \pm R)$.

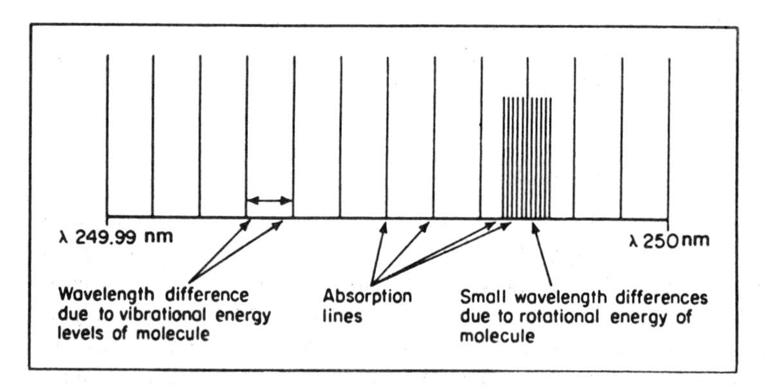

Figure 6.4 Illustration of a UV absorption band greatly expanded.

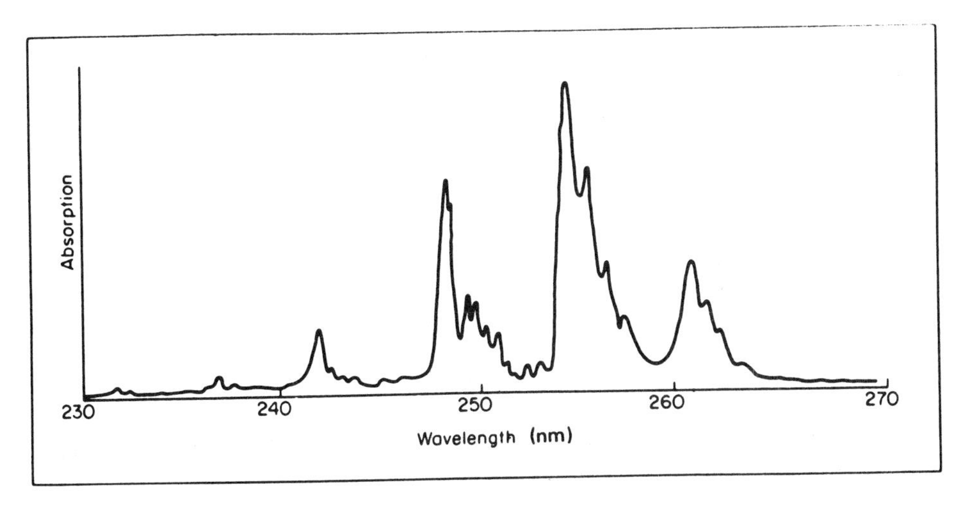

Figure 6.5 Absorption spectrum of toluene.

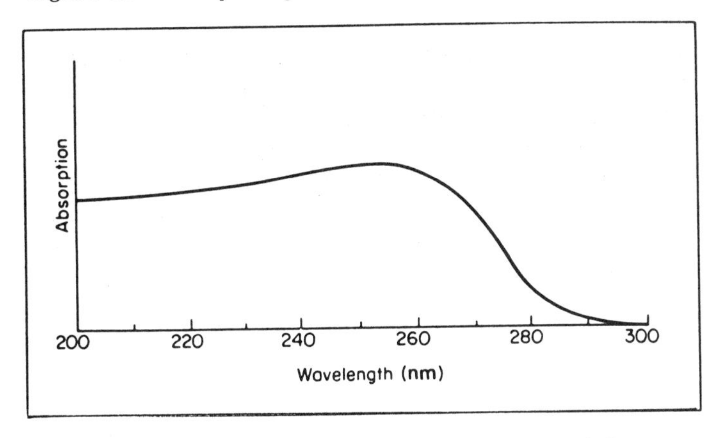

Figure 6.6 Absorption spectrum of a conjugated polyolefin.

The electronic excitation line is split into many sublevels by vibrational energy. Each sublevel is split by rotational energy, as seen in Fig. 6.4. The gross effect is to produce an absorption band rather than an absorption line. Typical absorption spectra are shown in Fig. 6.5 and 6.6.

5. Interpretation of UV Absorption Spectra of Molecules

An energy diagram of two *s* electrons combining to form a σ bond is shown in Fig. 6.7. The antibonding orbital is noted by σ^*; the energy difference between

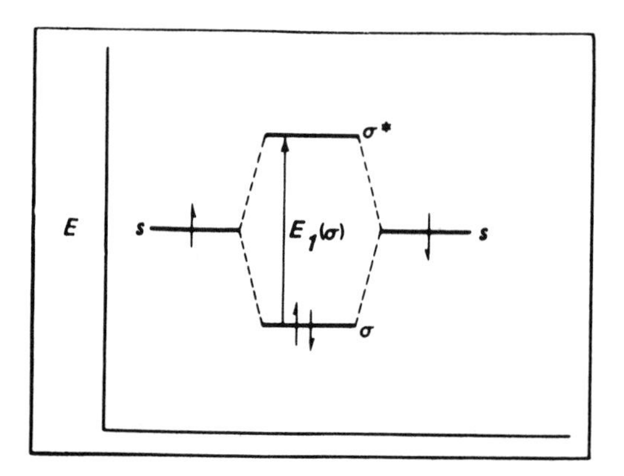

Figure 6.7 Energy diagram of two s electrons forming a σ bond. σ* is the energy of the antibonding σ orbital (excited σ state).

σ and σ* is equal to $E_1(\sigma)$. The energy diagram for the formation of a π orbital is shown in Fig. 6.8, and the energy difference between the π orbital and the antibonding π* orbital is $E_1(\pi)$. In a similar fashion, Fig. 6.9a illustrates an energy diagram where two atoms A and B combine, forming a π bonding orbital, but leaving one electron in the p orbital. The energy difference between the π bonding and the antibonding orbital is equal to $E_1(\pi)$. If four electrons are involved, two may form a π bonding orbital, leaving a pair of unbonded electrons. These may constitute a lone pair of n electrons. The energy diagram is shown in Fig. 6.9b.

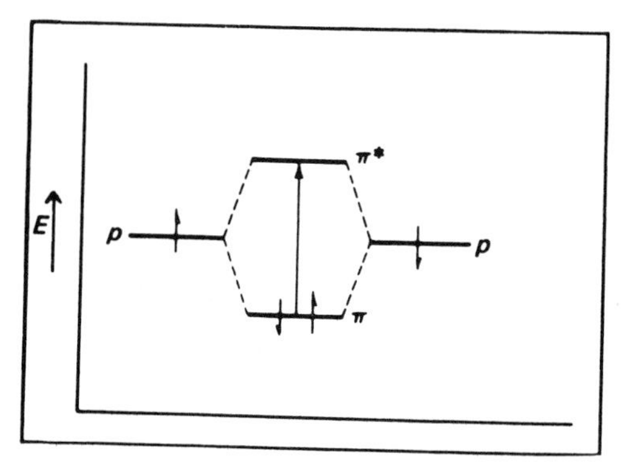

Figure 6.8 Energy diagram of two p electrons forming a π bond. π* is the energy of the antibonding π orbital (excited π state).

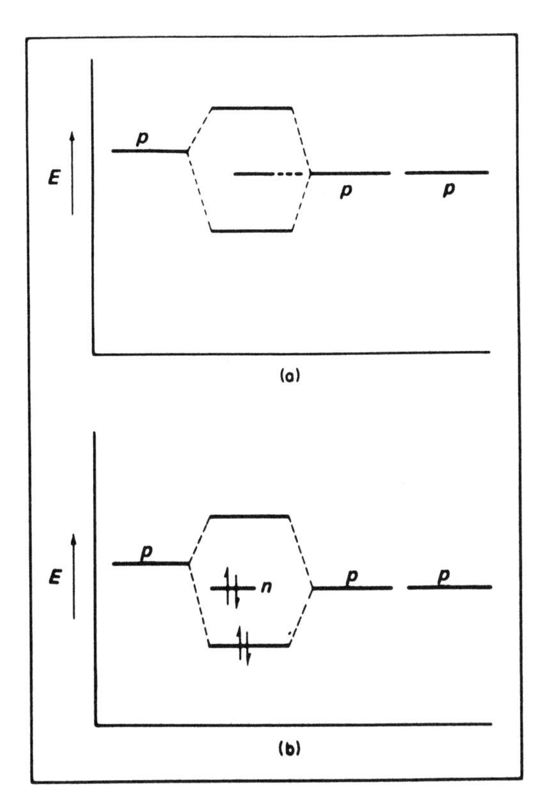

Figure 6.9 (a) Energy diagram of three *p* electrons forming a π bond and leaving an unpaired electron. (b) Energy diagram of four electrons combining to form a π orbital and a lone pair (*n* electrons).

From these energy diagrams a general energy diagram of σ, π, and *n* electrons can be deduced and is shown in Fig. 6.10. It can be seen that the energy required to excite an electron from a σ to a σ* orbital is considerably greater than that required to excite an electron from a π to a π* orbital or an *n* electron to either a σ* or a π* orbital.

The energy E_1 required to bring about excitation is directly related to the frequency of the absorbed energy by the equation $E = h\nu$. As a consequence, the energy frequency necessary to excite σ electrons to σ* orbitals is greater than that available in the ultraviolet, but usually UV radiation is sufficient to excite electrons in π orbitals to π* antibonding orbitals or *n* electrons to π* or σ* antibonding orbitals.

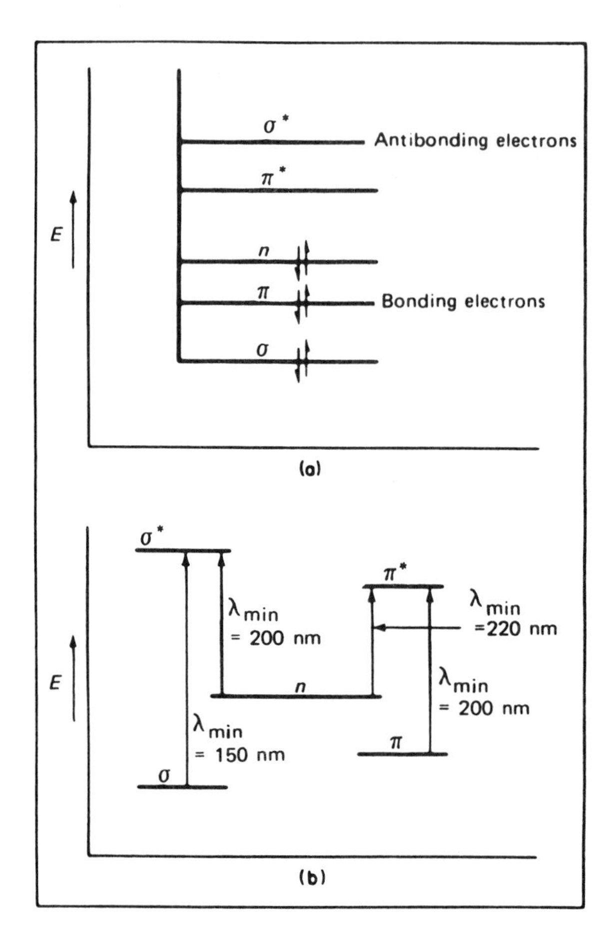

Figure 6.10 (a) Relative energy levels of bonding, antibonding, and *n* electrons. (b) Transition of ground-state σ, *n*, and π electrons to σ* and π* orbitals.

B. EFFECTS OF SOLVENTS ON ABSORPTION WAVELENGTHS

1. Red Shift

A molecule will dissolve in a solvent if solution leads to a lower energy system, that is, if energy changes favor solution. This can be expressed on an energy diagram as in Fig. 6.11.

Electrons in a π* antibonding orbital are more polar than electrons in a π bonding orbital. Therefore, if a molecule is dissolved in a polar solvent, the energy level of the π* antibonding orbital will decrease more than the energy level of the π bonding orbital. This is also illustrated in Fig. 6.11. The energy dif-

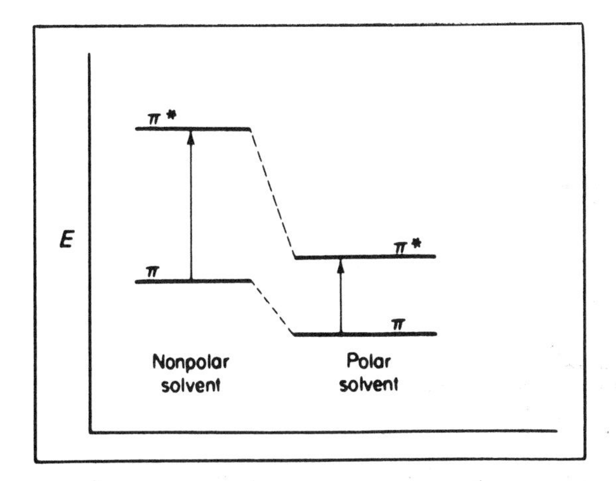

Figure 6.11 The energy difference between π and π* levels is decreased in the polar slovent; the absorption wavelength increases. This is the red shift.

ference in the polar solvent is less than the energy difference when the molecule is in a nonpolar solvent. As a consequence, the absorption maximum is changed to a longer wavelength in a polar solvent.

This leads to the general observation that molecules that absorb UV due to a $\pi \rightarrow \pi^*$ transition exhibit a shift in the absorption maximum to a longer wavelength when the molecule is dissolved in a polar solvent compared to a nonpolar solvent. The shift to a longer wavelength is called the *red shift*. This information can be used to confirm the presence of $\pi \rightarrow \pi^*$ transitions in a molecule. The confirmation is carried out by dissolving a sample in a nonpolar solvent such as hexane and then in a polar solvent such as alcohol. If the absorption maximum in the alcohol exhibits a red shift, a $\pi \rightarrow \pi^*$ transition is present in the molecule.

2. Blue Shift

The *n* electrons are known to be susceptible to hydrogen bonding. They are more susceptible to hydrogen bonding than electrons in a π orbital. The energy levels of *n* electrons decrease more than the energy levels of electrons in the π* orbital if hydrogens are available in the solvent. A solvent that may provide hydrogens for hydrogen bonding is ethanol, which contains a labile hydrogen on an OH. A typical sample containing *n* electrons on the lone pair is pyridine.

An energy diagram of such a system is shown in Fig. 6.12. The energy involved in the transition $n \rightarrow \pi^*$ when the solvent is nonpolar is less than the

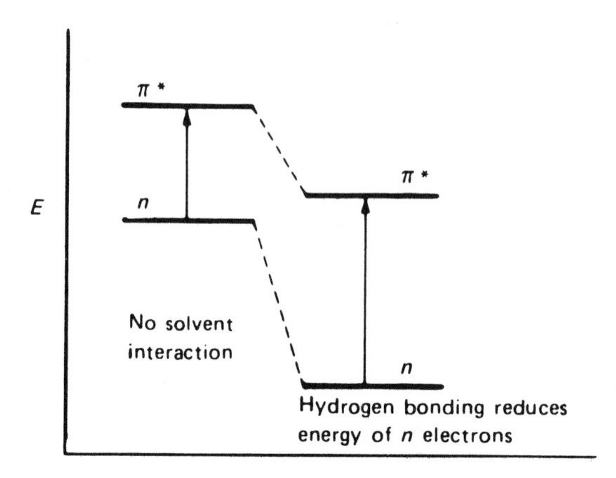

Figure 6.12 Energy differences between n and π^* levels increase in solvents that can provide H to the lone pair. The absorption wavelength decreases (blue shift).

energy involved in the same transition when the molecule is in a hydrogen-donating solvent. As a consequence, the absorption maximum in ethanol will move to a shorter wavelength. This is called the *blue shift*. (Again, blue shift does not mean the absorption becomes blue, merely that the absorption wavelength becomes shorter.) This information can be used to confirm the presence of n electrons in a molecule. The confirmation is carried out by dissolving a sample in a non–hydrogen-donating solvent such as hexane and also dissolving a similar sample in a hydrogen-donating solvent such as alcohol. If, upon comparing the absorption spectrum of the second solution with the first, the absorption spectrum exhibits a blue shift, n electrons are present in the sample molecule.

A molecule that contains both π orbitals and n electrons may exhibit both a red shift and a blue shift.

In general, $\pi \rightarrow \pi^*$ transitions absorb approximately 10 times more strongly than $n \rightarrow \pi^*$ transitions. Molecules that contain both π orbitals and n electrons would be expected to have an absorption spectrum in a nonpolar non–hydrogen-donating solvent such as hexane, as shown in Fig. 6.13. If this were the absorption spectrum of an unknown compound, we would suspect that the absorption at 250 nm was due to a $\pi \rightarrow \pi^*$ transition and the absorption at 350 nm was due to an $n \rightarrow \pi^*$ transition. This is because the absorption coefficient of the $\pi \rightarrow \pi^*$ transition is considerably greater than that of the $n \rightarrow \pi^*$ transition, thus generating a higher degree of absorption. Also, in general, the $n \rightarrow \pi^*$ transitions occur at longer wavelengths because the energy difference for this transition is lower.

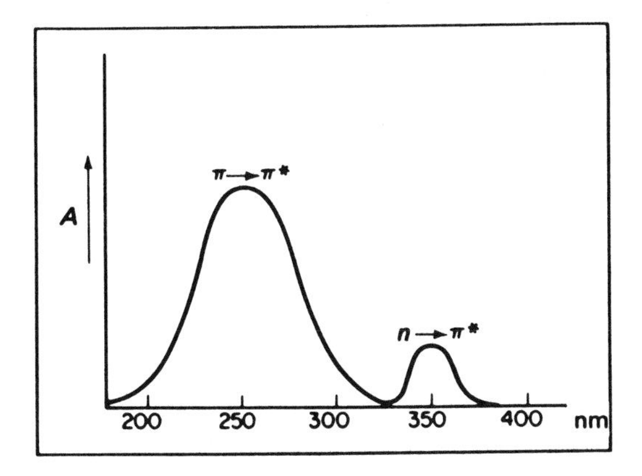

Figure 6.13 Expected absorption spectrum of a molecule undergoing $\pi \rightarrow \pi^*$ and $n \rightarrow \pi^*$ transitions. Note the different degrees of absorption in a nonpolar solvent.

If we now needed to confirm these assignments, we would put the sample in a solvent such as ethanol, which is both polar and capable of donating hydrogens. The polar nature would induce a red shift in the $\pi \rightarrow \pi^*$ transition and hydrogen donation would induce a blue shift in the $n \rightarrow \pi^*$ transition. If our assignments were correct, then the expected absorption spectrum in ethanol would be as shown in Fig. 6.14.

The combined evidence of the relative degree of absorption and the blue and red shifts occurring in ethanol strongly supports the idea that the molecule contains both π bonds and n electrons.

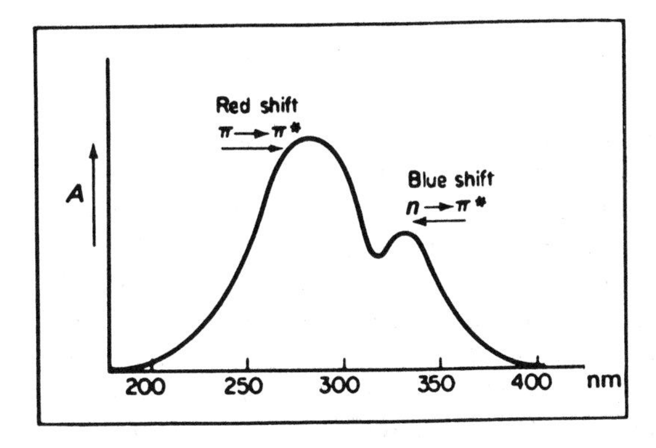

Figure 6.14 Absorption spectrum in ethanol of the same molecule as in Fig. 6.13. There is a red shift in the $\pi \rightarrow \pi^*$ transition and a blue shift in the $n \rightarrow \pi^*$ transition.

Table 6.3 Empirical Rules for Calculating the Absorption Maxima (in nm) of Conjugated Dienes

Absorption of parent diene system C═C—C═C	217
Shift to longer λ	
Double bond extension	30
Diene system within a ring	36
Exocyclic nature of any double bond	5
Each alkyl substituent or ring residue	5
Auxochrome is	
O-acyl	0
O-alkyl	6
S-alkyl	30
N-alkyl$_2$	60
Cl, Br	5

With permission from A. I. Scott, *Interpretation of the Ultraviolet Spectra of Natural Products*, Pergamon, Oxford, 1964.

3. Interpretation of UV Spectra

Empirical rules based on laboratory observations have been developed over the years relating the wavelengths of the UV absorption maxima to the structures of molecules. There are essentially four molecular systems. The principal parent systems are conjugated dienes, monosubstituted benzene rings, disubstituted benzene rings, and conjugated carbonyl systems. The method of calculation is to take a parent system and assign an absorption maximum. The parent system is then modified by the presence of other systems within the molecule. From these modifications, the absorption maximum can be calculated. These systems are discussed below.

a. Conjugated Diene Systems

The parent diene of conjugated diene systems is C═C—C═C. This system in a hexane solvent absorbs at 217 nm. If the conjugated system is increased, the wavelength of the absorption maximum is increased by 30 nm for every double bond extension. Similarly, an alkyl group attached to the conjugated system increases the absorption maximum by 5 nm. Other substitutions, such as O-alkyl, cause an increase in the absorption maximum, as does the inclusion within a ring or the exocyclic character of a double bond. These shifts are listed in Table 6.3. The conjugated diene system C═C—C═C absorbs at 217 nm in heptane and is considered to be the parent diene. Other modifications increase the wavelength of the absorption maximum as shown in Table 6.3. It should be emphasized that these results are empirical and not theoretical. They result from experimental observations.

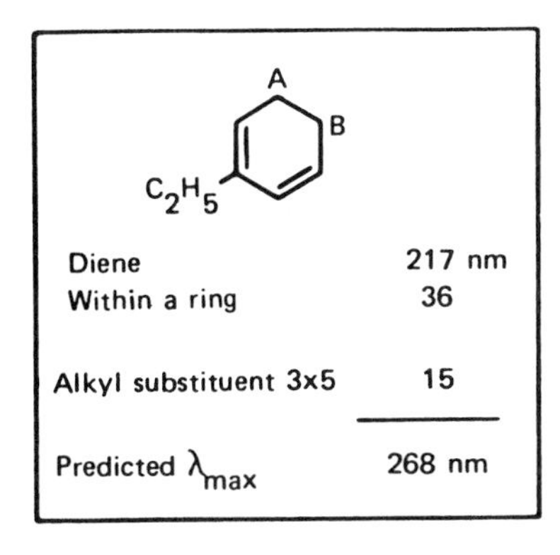

A
B
C_2H_5
Diene 217 nm
Within a ring 36
Alkyl substituent 3x5 15
Predicted λ_{max} 268 nm

Example 6.1

Some applications of these laws are shown below. The first is shown in Example 6.1. The predicted absorption maximum is calculated to be 268 nm.

In Example 6.1, the value assigned to the parent diene is 217 nm. Furthermore, this diene is within a ring; therefore 36 nm is added to the absorption maximum. It is not so clear that they are three alkyl substituents on this compound. One can be seen as the ethyl group, but the groups in positions A and B are also alkyl substituents. In each case carbons at A and B are attached to the diene system and therefore contribute to the electron density and to the shift to a longer wavelength. The fact that carbons A and B are attached to each other does not change their effect on the shift of the absorption maximum.

The calculation of the predicted wavelength of the absorption maximum of Example 6.2 is as follows. The parent diene is at 217 nm. A conjugated system is within a ring, causing the addition of 36 nm. There is one double bond extension to the system because the conjugated system consists of three double bonds rather than two double bonds. This adds a further 30 nm to the absorption maximum. There is one exocyclic double bond between carbons C and D. This double bond is touching the ring system and is within the conjugated system. A double bond between carbons E and F is within a ring, but neither carbon *touches* the other ring system, so this double bond is *not* exocyclic to a ring. There are five alkyl substitutes as follows: two on carbon A, one on carbon B, one on carbon D, and one on carbon F. The alkyl substituents on carbons D and F are designated by an asterisk.

Example 6.3 is a molecule with the same empirical formula as that in Example 6.4, but the double bonds are in a different position. The wavelength of the absorption maximum is 309 nm, which is 54 nm longer than in Example 6.4. The calculation of the absorption maximum in Example 6.4 is similar to that in Example 6.3, but the double bond between carbons 6 and 7 is not part of the conjugated system. The two compounds in Examples 6.3 and 6.4 would be extremely difficult to differentiate by any other analytical technique, such as

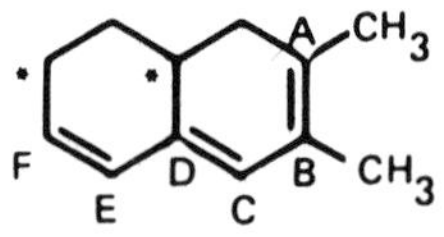

Diene	217 nm
Within a ring	36
Double bond extension	30
Exocyclic double bond	5
Alkyl substituents (5)	25
λ_{max}	313 nm

Example 6.2

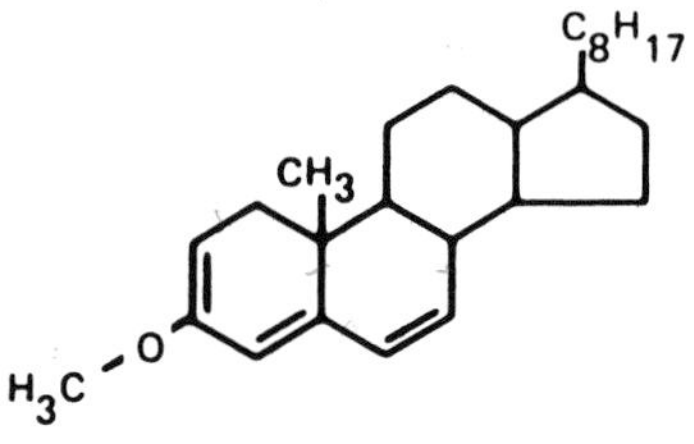

Parent diene	217 nm
Within a ring (1)	36
Double bond extension (1)	30
Exocyclic double bond (1)	5
Alkyl substituents 3x5	15
$O_{\lambda}CH_3$	6
λ_{max}	309 nm

Example 6.3

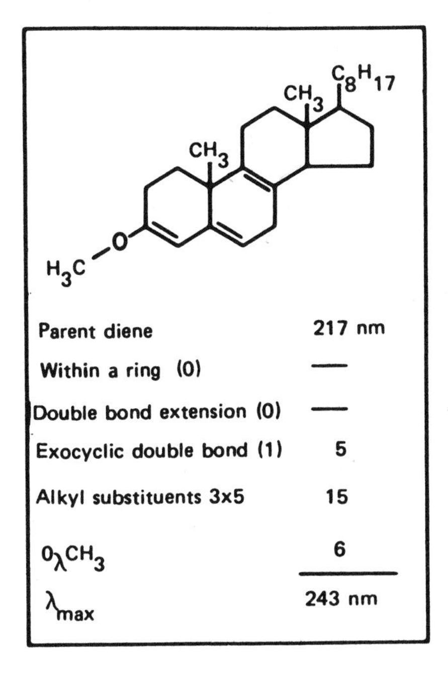

Example 6.4

NMR, IR, or mass spectrometry. However, with UV absorption they can readily be distinguished.

With these rules the absorption maximum in Example 6.5 would be calculated as follows:

Parent diene	217 nm
One exocyclic double bond (A)	5
Three alkyl substituents (C, D, E)	15
Calculated λ_{max}	237 nm

The experimentally observed value is 235 nm. It should be noted that there is no homoannular diene system because the complete diene system is not contained

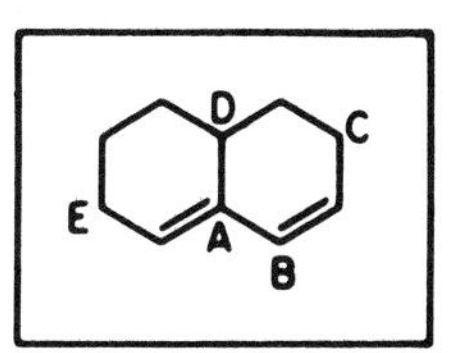

Example 6.5

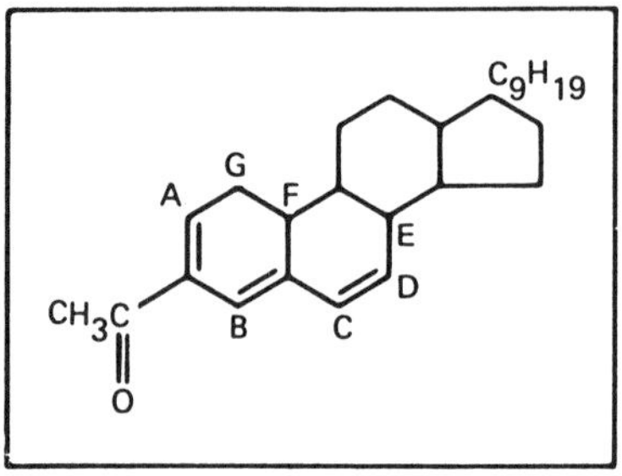

Example 6.6

in a single ring. Also, double bond B is not exocyclic to the ring, because it is not attached to a carbon that is part of another ring.

A more difficult example is Example 6.6. The absorption maximum is calculated as follows:

Parent diene (A, B)	217 nm
One exocyclic double bond (B)	5
Homoannular system (A, B)	36
Extension of conjugated system (C, D)	30
Three alkyl substituents (E, F, G)	15
Calculated λ_{max}	303 nm

The experimentally observed maximum is 304 nm. Note that the substitution between A and B does not affect the absorption maximum, and the ring systems are not touched by the conjugated system.

Similar rules for calculating absorption maxima for α, β-unsaturated ketone systems and for benzene systems have been developed and are shown in Tables 6.4 and 6.6. It can be seen that the absorption maxima can be calculated with reasonable accuracy. A complete study of this problem is beyond the scope of this book. A satisfactory treatment can be found in C. J. Creswell and O. Runquist, *Spectral Analysis of Organic Compounds*, Burgess, Minneapolis, 1970.

b. Conjugated Ketone Systems

The parent system is

C═C—C═C—C═O
δ γ β α

The absorption coefficient assigned to this parent system is 215 nm. In a manner similar to that for conjugated dienes, the wavelengths of the absorption maxima for conjugated ketones are modified by extension of the double bond substitution and position relative to rings and relative to the carbonyl group. The carbons are labeled α, β, γ, and δ, and substitutions in these positions change the shift of the absorption maximum. The empirical values used for calculating the absorption maxima of different compounds are shown in Table 6.4.

An example of the calculation of the wavelength of the absorption maximum using Table 6.4 is shown in Example 6.7. In this example the calculations

Table 6.4 Rules for α,β-Unsaturated Ketone and Aldehyde Absorptions

δ γ β α
C═C—C═C—C═O

Value assigned to parent α,β-unsaturated six-ring or acyclic ketone: 215
Value assigned to parent α,β-unsaturated five-ring ketone: 202
Value assigned to parent α,β-unsaturated aldehyde: 207

Increments added for	Shift to longer λ
Each exocyclic double bond	5
Diene within a ring	39
Double bond extending the conjugation	30
Each alkyl substituent	
α	10
β	12
γ and higher	18
Each	
OH	
α	35
β	30
δ	50
O-Acyl	
α,β,δ	6
O-Me	
α	35
β	30
γ	17
δ	31
S-alkyl	
β	85
Cl	
α	15
β	12
Br	
α	25
β	30
NR_2	
β	95

With permission from A. I. Scott, *Interpretation of the Ultraviolet Spectra of Natural Products*, Pergamon, Oxford, 1964.

Parent ketone	215 nm
Alkyl substituent γ(1)	18
δ(2)	36
Within a ring (1)	39
Double bond extension	30
Calculated λ_{max}	338 nm

Example 6.7

are similar to those used in the conjugated diene system. It should be noted that the β carbon has two alkyl substituents and therefore increases the absorption maximum by 36 nm. It should also be noted that the double bond between the γ and δ carbons is not an exocyclic double bond: it is within two different rings, but is not exocyclic to either of them. It should also be noted that the OH group is not attached to the conjugated system and therefore does not contribute to its spectrum.

An example of an isomer of the molecule used in Example 6.7 is shown in Example 6.8. The calculated absorption maximum is 286 nm. This spectrum is

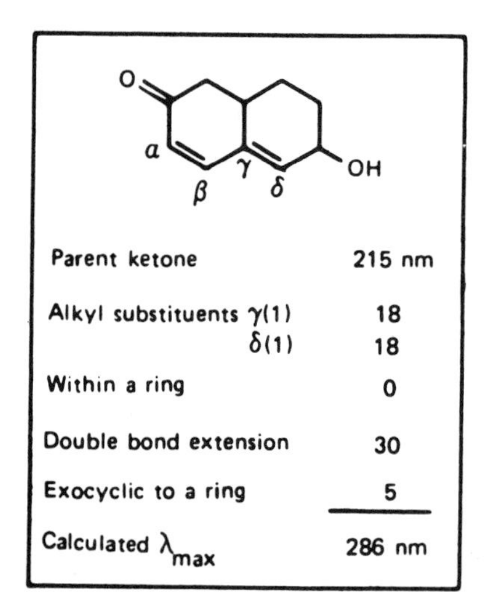

Parent ketone	215 nm
Alkyl substituents γ(1)	18
δ(1)	18
Within a ring	0
Double bond extension	30
Exocyclic to a ring	5
Calculated λ_{max}	286 nm

Example 6.8

different in several ways from Example 6.7. For example, the carbon at the δ position has only one alkyl substitution. Also, the double bond between the γ and δ positions is exocyclic to a ring in this case and therefore increases the wavelength of the absorption maximum. Furthermore, the conjugated system is not within a ring, and therefore contributions from ring currents are not observed.

A more complicated molecular system is shown in Example 6.9. The calculation of the absorption wavelength maximum is shown to be 286 nm. The calculation follows the steps used in previous examples. It should again be noted that the parts of the molecules in the upper two-ring system A and B do not contribute to the UV absorption spectrum. As before, this makes prediction of the spectrum easier to carry out, but by the same token gives no information as to the structure of that part of the molecule.

c. Substitution of Benzene Rings

Benzene is a strong absorber of UV radiation and particularly in the gas phase shows considerable fine structure in its spectrum. Substitution on the benzene ring causes a shift in the absorption wavelengths. It is not uncommon for at least two bands to be observed, and frequently several more. The observed wavelengths of the absorption maxima of some substituted benzene rings are shown in Table 6.5. These are experimental data and may be insufficient to completely

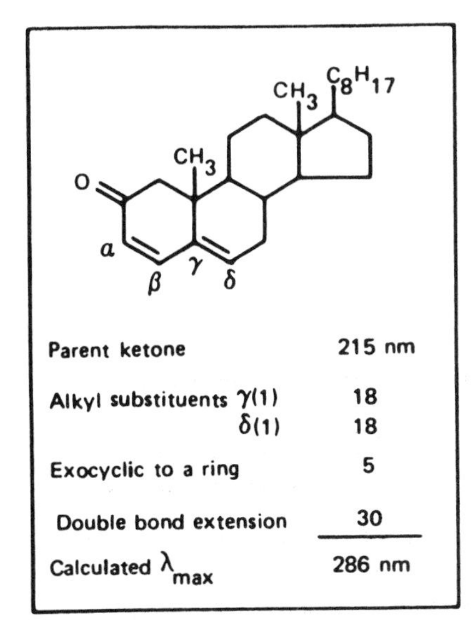

Example 6.9

Table 6.5 Absorption Maxima of Monosubstituted Benzene Rings Ph—R

R	λ_{max}(nm) (solvent H_2O or MeOH)	
	Band 1	Band 2
—H	203	254
$—NH_3^+$	203	254
—Me	206	261
—I	207	257
—Cl	209	263
—Br	210	261
—OH	210	270
—OMe	217	269
$—SO_2NH_2$	217	264
$—CH_3$	224	271
$—CO_2$	224	268
$—CO_2H$	230	273
$—NH_2$	230	280
$—O^-$	235	287
—NHAc	238	
—COMe	245	
$—CH{=}CH_2$	248	
—CHO	249	
—Ph	251	
—OPh	255	
$—NO_2$	268	
$—CH'{=}CHCO_2H$	273	
$—CH'{=}CHPh$	295	

From H. H. Jaffe and M. Orchin, *Theory and Applications of Ultraviolet Spectroscopy*, Wiley, New York, 1962.

identify unknown compounds. If the benzene ring is disubstituted, then calculations are necessary to predict the absorption maximum, because a list containing all the possible combinations would be very long and unwieldy and would need further experimental supporting evidence. There are some rules that help us understand disubstitution of benzene rings. These are as follows.

1. An electron-accepting group, such as NO_2, and an electron-donating group, such as OH, situated *ortho* or *para* to each other tend to cancel each other out and provide a spectrum not very different from the monosubstituted

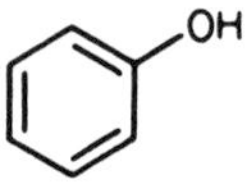

OH

Parent system
λ_{max} 230nm

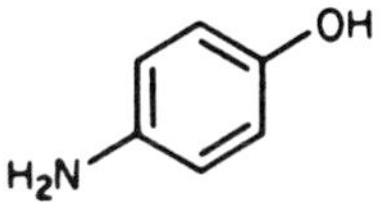

OH

H_2N

Parent system	230 nm
Para sub NH_2	58
Calculated λ_{max}	288 nm

Example 6.10

Table 6.6 Absorption Maxima of Disubstituted Benzene Derivatives

Substituent	Orientation	Shift for each substituent, γ, in EtOH (nm)
R		
X = alkyl or ring residue		246
X = H		250
X = OH or O-alkyl		230
R = alkyl or ring residue	*o*−, *m*−	3
	p−	10
R = OH, O-Me, O-alkyl	*o*−, *m*−	7
	p−	25
R = O	*o*−	11
	m−	20
	p−	78
R = Cl	*o*−, *m*−	0
	p−	10
R = Br	*o*−, *m*−	2
	p−	15
R = NH_2	*o*−, *m*−	13
	p−	58
R = NH-Ac	*o*−, *m*−	20
	p−	45
R = NH-Me	*p*	73
R = NMe_2	*o*−, *m*−	20
	p−	85

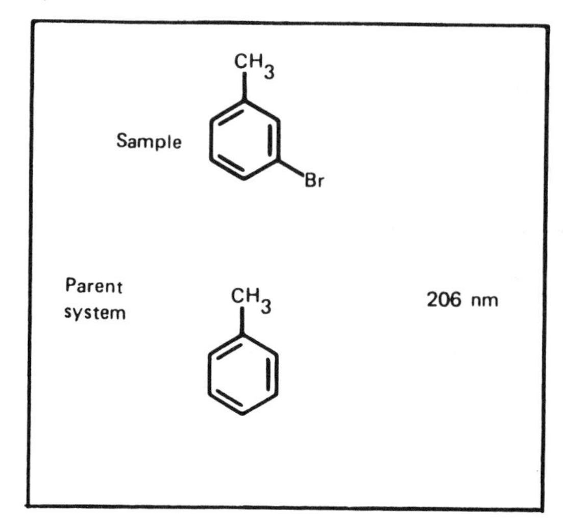

Example 6.11

benzene ring spectrum. An example of a disubstituted benzene ring can be seen in Example 6.10.

2. Two electron-accepting groups or two electron-donating groups *para* to each other produce a spectrum little different from the spectrum of the monosubstituted compound.
3. An electron-accepting group and an electron-donating group *para* to each other cause a shift to longer wavelengths.

The predicted absorption maximum and the effect of substitution is given in Table 6.6.

Example 6.11 is another illustration of the system. Note that the substitution of bromine makes little difference to the absorption maximum of the original toluene.

It can be seen from these examples that the science of interpretation of UV spectra is based on theoretical and practical considerations. More extensive information is available in the literature, and the reader can become an expert in this field by continuous practice and using expanded tables.* Other systems have been developed for aromatic and acid systems, but these will not be discussed again in this presentation.

---

*C. J. Creswell and O. Runquist, *Spectral Analysis of Organic Compounds*, Burgess, Minneapolis, 1970.

C. EQUIPMENT

1. General Optical System

Both the single-beam and the double-beam systems (schematic diagrams of which appear in Chaps. 3 and 5) are used in UV molecular absorption spectroscopy.

a. Single-Beam System

In the single-beam system, UV radiation is given off by the source. A convex lens gathers the beam of radiation and focuses it on the inlet slit. The inlet slit permits light from the source to pass, but blocks out stray radiation. The light then reaches the monochromator, which splits it up according to wavelength. The exit slit is positioned to allow light of the required wavelength to pass through. Radiation at all other wavelengths is blocked out. The selected radiation passes through the sample cell to the detector, which measures the intensity of the radiation reaching it. By comparing the intensity of radiation before and after it passes through the sample, it is possible to measure how much radiation is absorbed by the sample at the particular wavelength used. The output of the detector is usually recorded on graph paper, tape, or screen.

One problem with the single-beam system is that it measures the total amount of light reaching the detector, rather than the percentage absorbed. Light may be lost at reflecting surfaces or may be absorbed by the solvent used to dissolve the sample. Furthermore, the source intensity may vary with changes in line voltage. For example, when the line voltage decreases, the intensity of the light coming from the source may decrease unless special precautions are taken. Consequently, the intensity of radiation may be constantly changing.

Another problem is that the response of the detector varies significantly with the wavelength of the radiation falling on it. Even if the light intensity is constant at all wavelengths, if the wavelength is steadily increased from 200 to 750 nm, the signal from the detector starts at a low value, increases to a value that is steady over a wide range, and then decreases once more. This relationship between the signal from the detector and the wavelength of radiation is called the *response curve* (Fig. 6.15). It indicates the wide variation in signal that can be expected from the detector even though the light intensity falling on it is constant. Different detectors respond differently at different wavelengths. For example, the 1P28 is not useful at 800 nm, but the R136 and gallium arsenide detectors respond in this range. The detector selected must operate over the desired range of the experiment. The problem of instrument variation can be largely overcome by using the double-beam system.

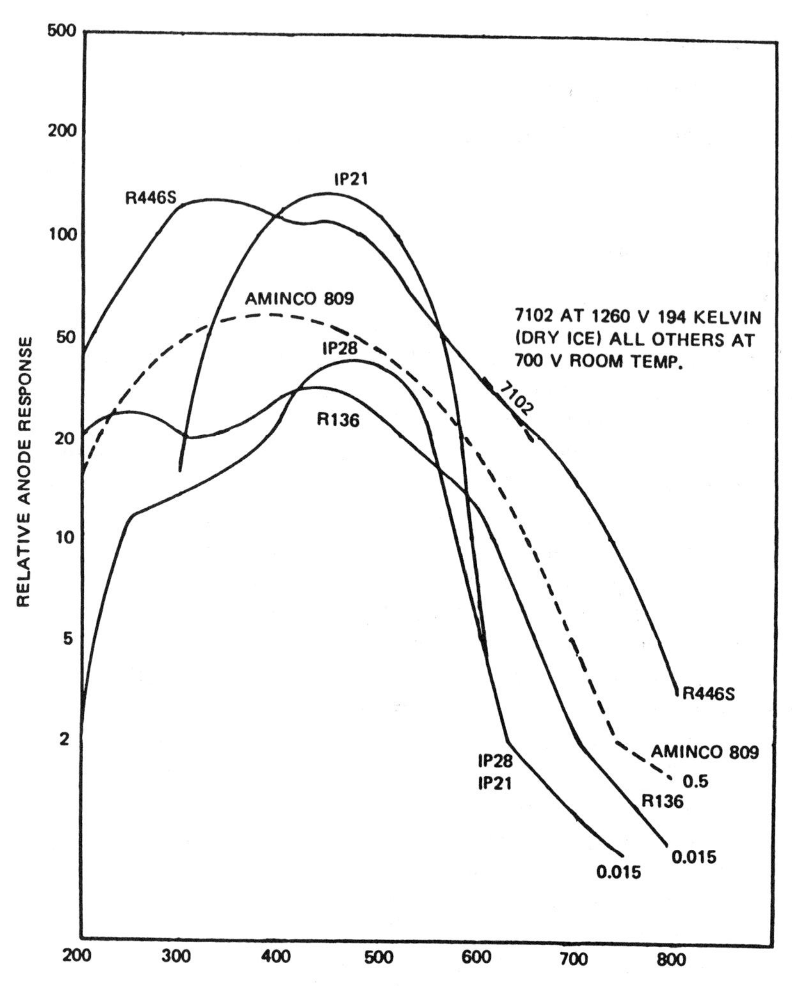

Figure 6.15 Response curves for various types of photomultipliers. Note the variable range between models and the sharp dropoff response outside the usable range.

b. Double-Beam System

In the double-beam system, the source radiation is split into two beams of equal intensity. The two beams traverse two light paths identical in length; a *reference* cell is put in one path and the *sample* cell in the other. The intensities of the two beams after passing through the cells are then compared. The radiation lost to the optical system (e.g., cell surfaces) should be equal for both beams, allowing a correction for this source of error. Furthermore, absorption by the solvent should be about the same for each beam and cancel out. In this manner a correction can be made for absorption by the solvent. The difference in intensity of the two beams is a direct measure of the absorption by the sample. For a more complete explanation see Chapter 3.

2. Components of the Equipment

a. Radiation Source

The two most common radiation sources are tungsten lamps and hydrogen discharge lamps. The *tungsten lamp* is similar in functioning to an electric light bulb. It is a tungsten filament heated electrically to white heat. It has two shortcomings: the intensity of radiation at short wavelengths (<350 nm) is small; furthermore, to maintain a constant intensity, the electrical current to the lamp must be carefully controlled. However, the lamps are generally stable, robust, and easy to use. Typically, the emission intensity varies with wavelength as shown in Fig. 6.16a. It is most useful over the visible range and is therefore used in spectrophotometry.

The *hydrogen discharge lamp* consists of hydrogen gas under relatively high pressure through which there is an electrical discharge. The hydrogen molecules are excited electrically and emit UV radiation. The high pressure brings about many collisions between the hydrogen molecules, resulting in pressure broadening. This causes the hydrogen to emit a continuum (broad band) rather than a simple hydrogen line spectrum. The lamps are stable, robust, and widely used. If deuterium (D_2) is used instead of hydrogen, the emission intensity is increased by as much as a factor of 3 at the short-wavelength end of the UV range. *Deuterium lamps* are more expensive than hydrogen lamps but are used when higher intensity is required.

Another type of lamp used is the *mercury discharge lamp*. As in the hydrogen lamp, the mercury vapor in the discharge lamp is under high pressure. It does not exhibit an even continuum and is not used for general UV spectroscopy. *Xenon lamps* work in the same manner as hydrogen lamps. They provide very high radiation intensity and are widely used in the visible and long-wavelength end of the UV range.

The radiation intensity of the lamps varies greatly at different wavelengths. This is illustrated in Fig. 6.16b. If the lamps were ideal black body radiators, their radiant energy would vary as T^4 and their intensity would be as shown in Fig. 6.16a.

b. Monochromators

The purpose of the monochromator (as described in Chap. 3) is to disperse the radiation according to wavelength. Prisms and grating are used extensively for this purpose in UV spectroscopy. The most popular materials for making prisms

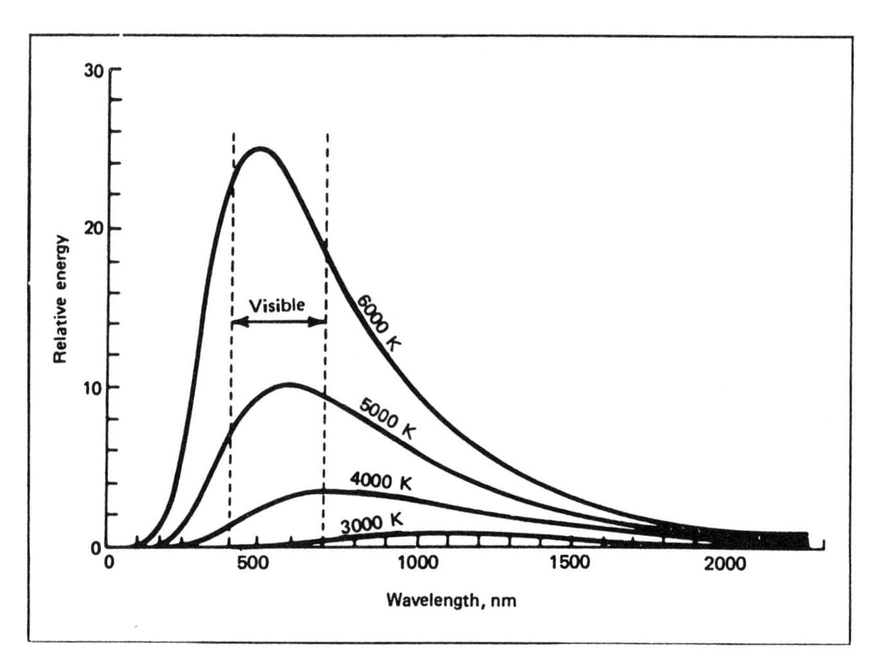

Figure 6.16a Theoretical emission intensity of black body radiation at various temperatures. In practice, the highest available is about 3500K. (6000K is equivalent to a xenon lamp, 3000K to a tungsten lamp.)

are glass, quartz, and fused silica. Of these, glass has the highest resolving power; that is, it disperses the light strongly, particularly over the visible region of the spectrum. But it is not transparent to radiation with wavelengths between 350 and 200 nm, because glass absorbs strongly and therefore cannot be used over this wavelength range. It is found, in practice, that the refractive index of transparent materials increases if the wavelength of the radiation is close to an absorption band of the transmitting material. For this reason, glass disperses visible light more than quartz. At the same time, the presence of the absorption band limits the use of glass prisms to the longer-wavelength end of the UV range.

Quartz and fused silica prisms, which are transparent over the entire UV range, are used in UV spectrophotometers. Fused silica prisms are somewhat more transparent in the short-wavelength region than quartz, but because they are more expensive than quartz prisms, they are used only when highly intense radiation is necessary. However, their resolution is not good and varies with wavelength. Also, they absorb some radiation at all wavelengths. Gratings are now most commonly used in research and routine analysis. The resolution is

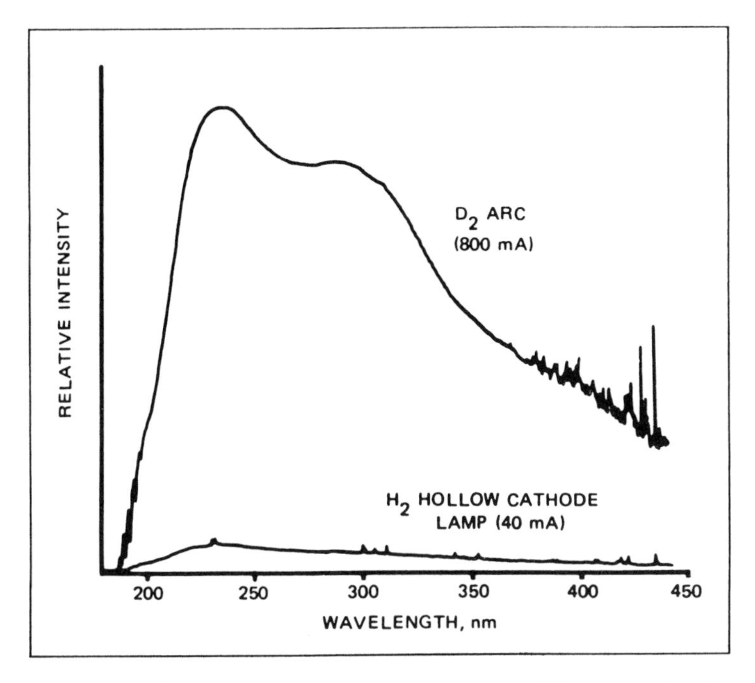

Figure 6.16b Intensity of a hydrogen lamp at different wavelengths. Note that at wavelengths greater than 350 nm, the source becomes discontinuous, and loss of H_2 emission occurs at higher wavelengths.

constant over the whole wavelength range. Also, since radiation is diffracted from the surface, no absorption problems are encountered.

c. Detectors

There are four common forms of detectors: the barrier layer cell, the photocell, the photomultiplier, and semiconductor detectors.

Barrier Layer Cell. In the barrier layer cell, silver is coated onto a semiconductor, such as selenium (see Fig. 6.17), that is joined to a strong metal base, such as iron. To manufacture these cells, the selenium is placed in a container and the air pressure reduced to a vacuum. Silver is heated electrically, and its surface becomes so hot that it melts and vaporizes. The silver vapor coats the selenium surface, forming a very thin but evenly distributed layer of silver atoms. Any radiation falling on the surface generates electrons at the selenium–silver interface. A barrier seems to exist between the selenium and the iron that prevents electrons from flowing into the iron. The electrons are therefore collected by the silver. The accumulation of electrons creates an electrical voltage difference between the silver surface and the base of the cell. The voltage cre-

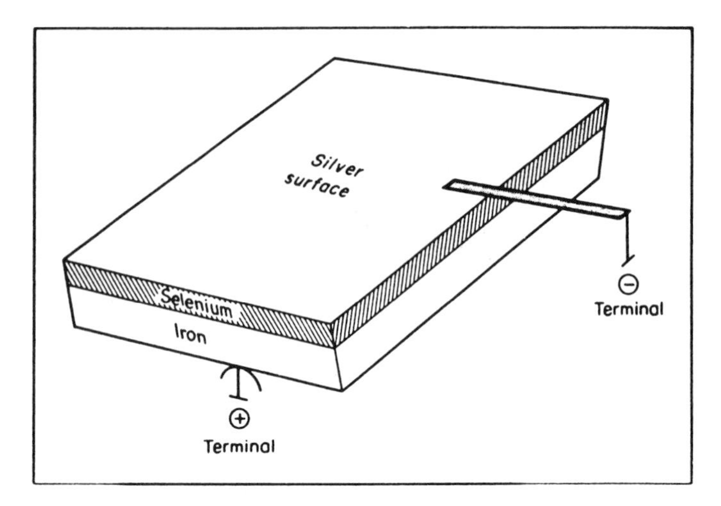

Figure 6.17 Barrier layer cell.

ated is a measure of the intensity of the radiation falling on the detector if the intensity is less than 5 foot-candles (fc). Barrier layer cells are used as light meters in cameras but are used only on inexpensive routine analytical equipment because of their limited response range.

Photocell. The photocell (see Fig. 6.18) consists of a metal surface (cathode) that is sensitive to light. When light falls on it, the surface gives off electrons. The metal is kept in a vacuum and the electrons given off are attracted and

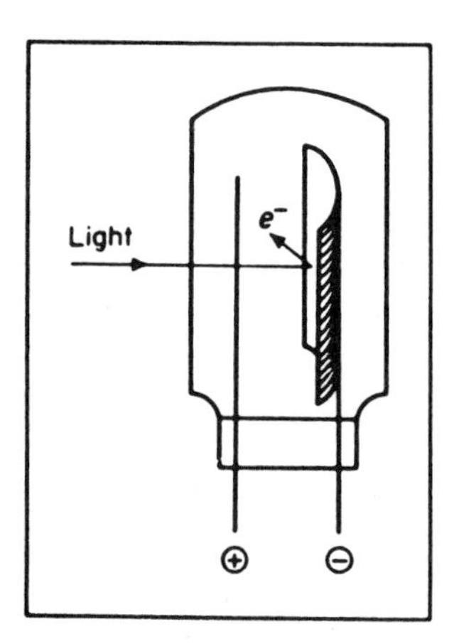

Figure 6.18 Photocell.

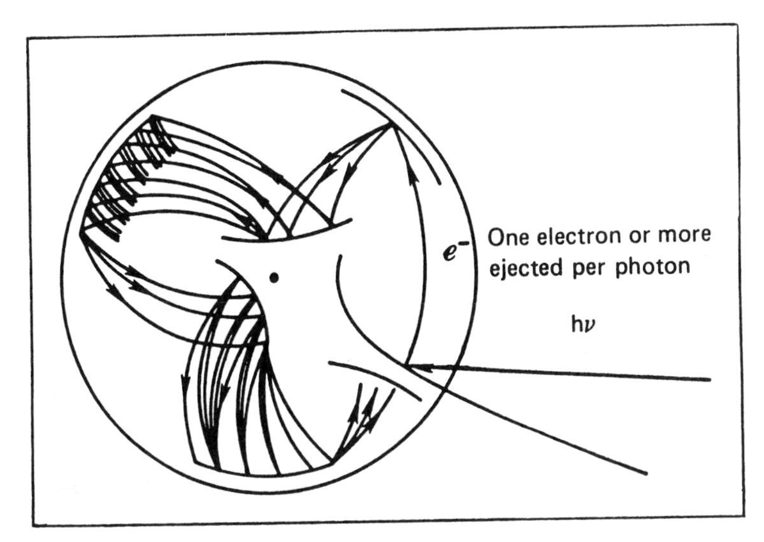

Figure 6.19 Photomultiplier. Impinging photons liberate electrons from a light-sensitive metal. The liberated electrons are accelerated to a second electrode, each liberating several electrons on impact.

collected by a positively charged anode. The current that is created between the anode and the cathode is a measure of the intensity of radiation falling on the detector. Photocells have been superseded by photomultipliers.

Photomultiplier. The most common detector is the photomultiplier (Fig. 6.19), which operates as follows. Radiation falls on a metal surface, which emits electrons in a fashion similar to the photocell. The electrons are attracted to a dynode that is maintained at a positive voltage. Upon arrival at the dynode, each electron strikes the dynode's surface and causes several more electrons to be emitted from the surface. These emitted electrons are in turn attracted to a second dynode, where similar electron emission occurs. The process is repeated several times until a shower of electrons arrives at the last dynode, which is the collector. The number of electrons falling on the collector is a measure of the intensity of light falling on the detector. In the process, a single photon may generate many electrons and give a high signal. Increasing the voltage between dynodes increases the signal, but if the voltage is made too high, stray radiation causes a signal and the signal from the detector becomes erratic or *noisy*. The dynodes are therefore operated at an optimum voltage that gives a steady signal. It is common for there to be 10–15 stages in a commercial photomultiplier. The gain may be as high as 10^9 electrons per photon, but this is ultimately limited by the noise level of the detector system and, in practice, lower gains and lower noise levels may be preferable.

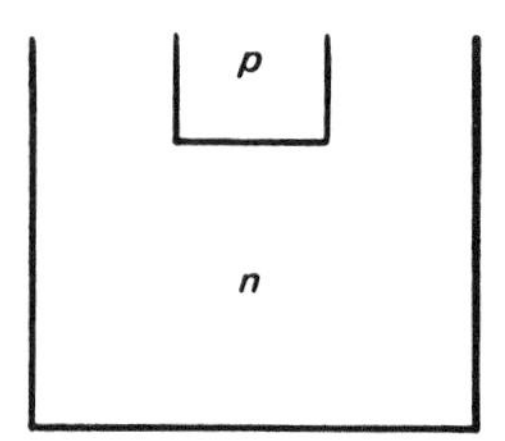

Figure 6.20 *p*-Type semiconductor embedded in an *n*-type semiconductor forming a *pn* junction.

Semiconductor Detectors—Diodes and Diode Array Systems. Silicon and germanium both belong to group IV of the periodic table. Their electrons are all in the valence band and are therefore nonconducting. However, a few electrons can be thermally shaken free from the valence band, and as free electrons they become conductors of electricity. When an electron leaves the valence band, it leaves behind a positive hole that is also mobile, thus producing an electron–hole pair.

The effect can be increased by doping either one of these elements with a group V element, such as arsenic or antimony, or a group III element, such as indium or gallium. A group V element has an extra electron and is likely to generate a negative charge. This is an *n*-type semiconductor. Similarly, adding a group III element leads to a depletion of electrons and the generation of positive holes. This is a *p*-type semiconductor. In an *n*-type semiconductor the electron is mobile, and in the *p*-type the positive hole is mobile, but less so than the electron.

Diodes. If we put together a *p*-type semiconductor and an *n*-type semiconductor, the junction provides a *pn* junction, as shown in Fig. 6.20. It can be formed by diffusing at a particular location an excess of a group III element into an *n*-type semiconductor.

This junction can be presented as in Fig. 6.21. If we apply a positive bias to the *p*-type and a negative potential to the *n*-type, as shown in Fig. 6.22, positive charges flow from *p* to *n* and negative charges flow from *n* to *p*. This is called a *forward bias*, and under these conditions current flows easily. These conditions are shown in Fig. 6.22.

However, if applied voltage is in the reverse direction, the current flow would be in the opposite direction, as shown in Fig. 6.23, and current does not flow easily. These are the conditions of *reverse bias*, and, in short, the *pn* junction acts as a rectifier.

If we plot the current flowing in these circuits versus voltage, the relationship is shown in Fig. 6.24. Under the conditions of reverse bias the current flowing is minimal until a sudden break occurs and current flow increases rapidly. This is called the *Zener breakdown*. At this point the potential that permits elec-

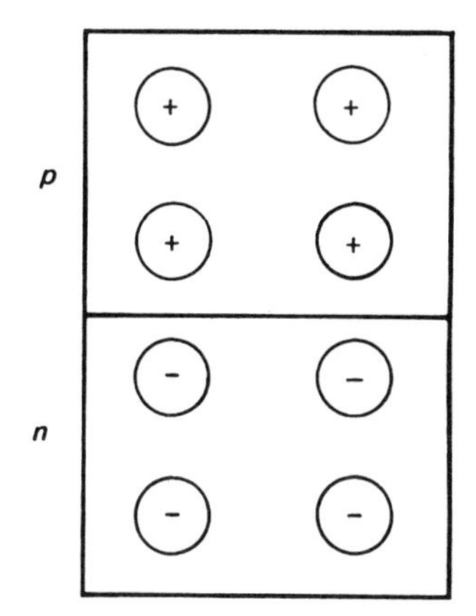

Figure 6.21 *pn* junction with electrons (−) in the *n*-type semiconductors and holes (+) in the *p*-type semiconductors.

trons to be torn out of the covalent bond and become conducting has been reached. The voltage required is from 2 to 200 volts, depending on composition. Too high a reverse voltage will break down the diode.

Transistors (amplifiers). There are two types of transistors in common use: the *bipolar transistor* and the *field-effect transistor*.

1. The bipolar transistor is symbolized as *pnp*, or *npn*, indicating three joined semiconductors in the order symbolized. They constitute two back-to-back diodes. A *pnp* transistor may be made by inserting a very thin layer of

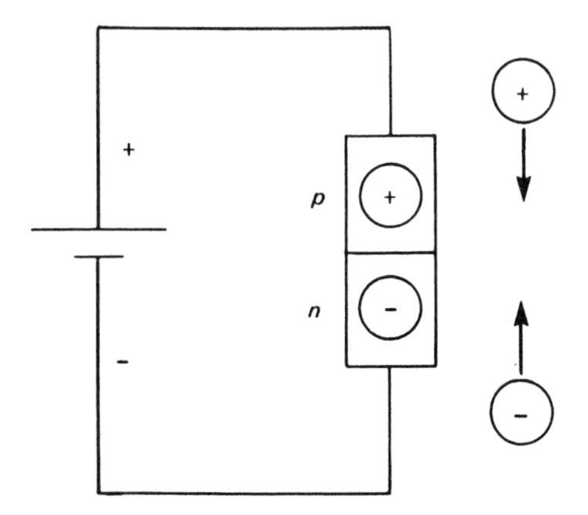

Figure 6.22 Direction of flow of holes and electrons on applying a positive potential to the *p*-type semiconductor (forward bias).

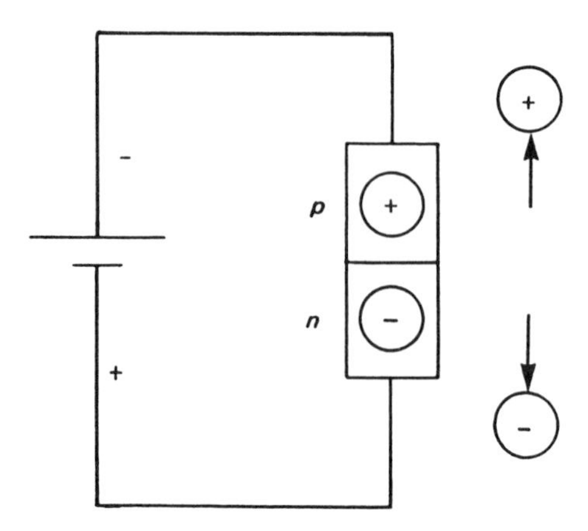

Figure 6.23 Direction of flow of holes and electrons on applying a positive potential to the *n*-type semiconductor (reverse bias).

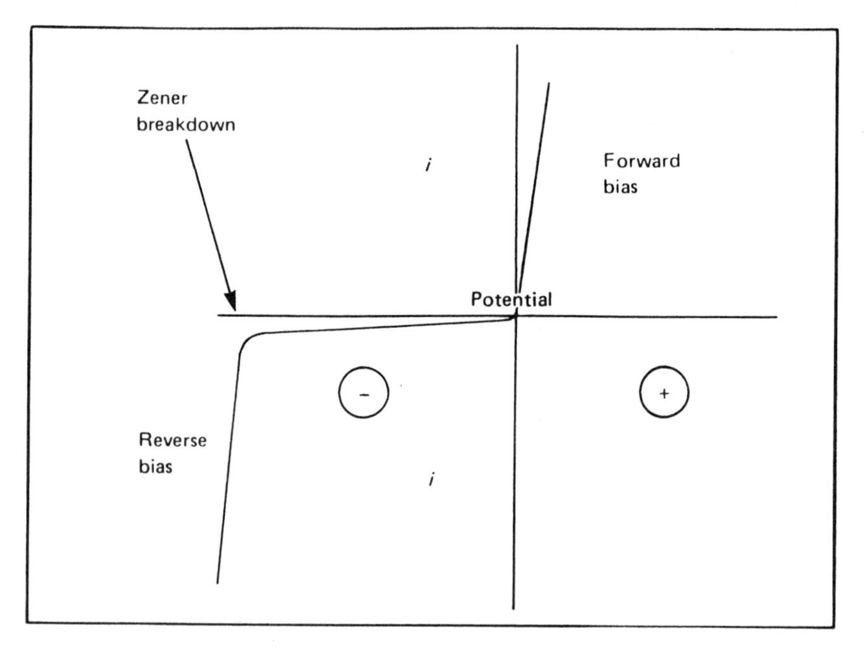

Figure 6.24 Relationship between potential and current using forward bias or reverse bias. Note rapid current increase at point of Zener breakdown.

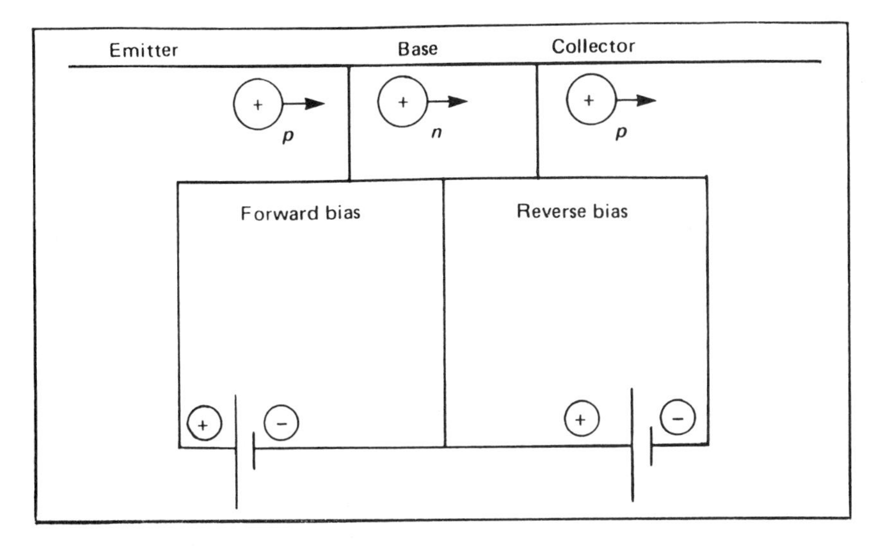

Figure 6.25 Bipolar transistors with forward bias and reverse bias. Note the direction of flow of the positive holes. Flow is sensitive to changes in the potential of the bias.

n-type semiconductor such as silicon into a *p*-type semiconductor, as shown in Fig. 6.25. Both the *pnp* and the *npn* transistors work in the same manner, but it is considered that in the *pnp* transistor the positive holes are the main current carrier and in the *npn* transistor electrons are the main current carrier.

A forward bias is placed across one *pn* junction, as shown in Fig. 6.25. The *p*-type semiconductor is in this case the *emitter*. The *n*-type is the *base*. The positive holes move toward the junction (see Fig. 6.22). A reverse bias is placed across the other *np* junction, and this *p*-type semiconductor becomes the *collector*. The positive holes move away from the *np* junction (see Fig. 6.23).

In practice, current flows across the *pn* junction carried by the positive holes under the influence of the forward bias. The base, however, is very thin and poor in total number of conduction electrons. Consequently, the holes proceed through the base to the collector.

The current flowing through the base is very sensitive to potential and any potential changes of the base. This provides the basis for amplification.

The ratio I_c/I_b is the amplification factor β, which can be expressed as

$$\beta = \frac{I_c}{I_b}$$

where

β = amplification factor
I_c = current in the collector
I_b = current in the base

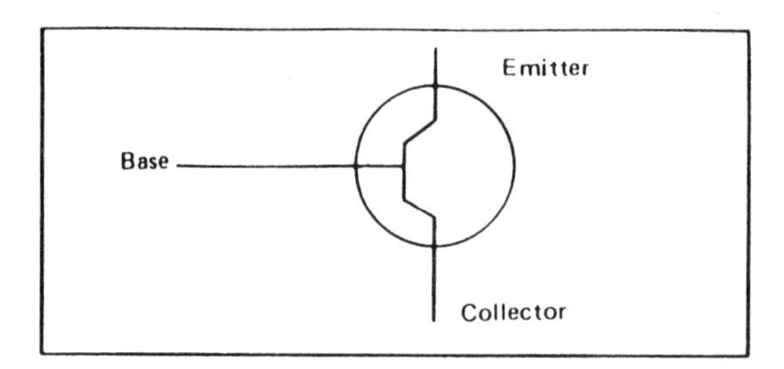

Figure 6.26 *pnp* diode. Arrow indicates direction of flow of positive holes.

Amplification of 50- to 200-fold is common. A *pnp* bipolar transistor is shown in Fig. 6.25. The transistor is often presented in simplified form as indicated in Fig. 6.26.

2. The field-effect transistor (FET) works on a totally different principle, as shown in Fig. 6.27. A field-effect transistor is effectively a bar of an *n*-type semiconductor equipped with terminals designated as the "source" and the "drain." The bar of *n*-type semiconductor is sandwiched between two *p*-type semiconductors, called collectively a "gate," which are connected and maintained at the same electrical potential. The transistor may be arranged as a cylinder of a *p*-type semiconductor around a tube of an *n*-type semiconductor. Under normal circumstances a current reaching the source passes through the channel to the drain to complete the circuit. If there is no voltage on the gate, current flows with fairly low resistance. However, if a reverse bias is applied to

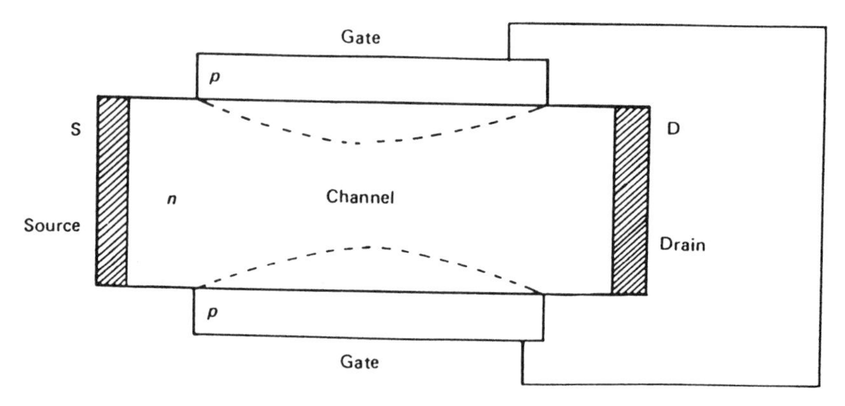

Figure 6.27 FET transistors. By applying a reverse bias to the gate, the channel is reduced and the electrical resistance between the source and the drain increases. The latter resistance is sensitive to small changes in the reverse bias.

the gate, electrons are depleted near the *pn* interface, as indicated by the dotted line. Under these conditions the material becomes nonconducting. The resistance of the channel is greatly increased since the effective conducting cross section is greatly decreased. The change in gate potential produces a much larger change in output voltage. This is the source of the amplification.

If a metal oxide semiconductor is deposited as a thin film between the gate and the channel elements, it forms a rectifying junction. Therefore, the gate can be given either polarity without drawing any current. This transistor has the highest impedence of all transistors and is known as a MOSFET (metal oxide semiconductor FET).

Diode arrays. In the conventional UV-visible part of the spectrum, a complete absorption spectrum can be obtained by scanning though the entire wavelength range and recording the spectrum. This takes considerable time using the conventional scanning monochromator system, and the absorption at one wavelength is measured at a different time from that at another wavelength. In emission spectroscopy an alternative procedure is to place a series of photomultipliers at predetermined positions on a Rowland circle or on a base if Echelle optics are used. Finite wavelengths only are monitored, which is usually satisfactory for elemental analysis where the pertinent wavelength is known with some degree of certainty.

However, there are two conditions under which these optical systems do not work very well. The first is when there is a rapid chemical reaction taking place and conventional scanning is too slow to follow the reaction. The second is when the sample is available only for a limited time and complete scanning is not possible. An example of the latter is the eluant from a liquid chromatogram. Modern-day UV detectors in HPLC only permit detection at particular wavelengths but do not present the entire spectrum.

The linear photodiode array (LPDA) is a technique developed to enable simultaneous measurement of light intensity and hence absorption at many wavelengths. The optical system is shown schematically in Fig. 6.28. In this system radiation from the large source, which may be a deuterium light or other UV light source, passes through the sample through a grating, where the radiation is separated by wavelengths and directed to a diode array system. The grating in this case acts as a polychromator. No exit slit is used, and no particular wavelength is chosen for measurement.

The diode array consists of a number of semiconductors embedded in a single crystal. A common procedure is to use a single crystal of doped silicon which is an *n*-type semiconductor. A small excess of a group III element, such as arsenic, is embedded onto the surface at regular intervals. This creates a local *p*-type semiconductor. The entire system then ideally has a cross section such as that shown in Fig. 6.29. The *p*-type semiconductor forms a series or an array of diodes. The individual diodes are called *elements*, *channels*, or *pixels*.

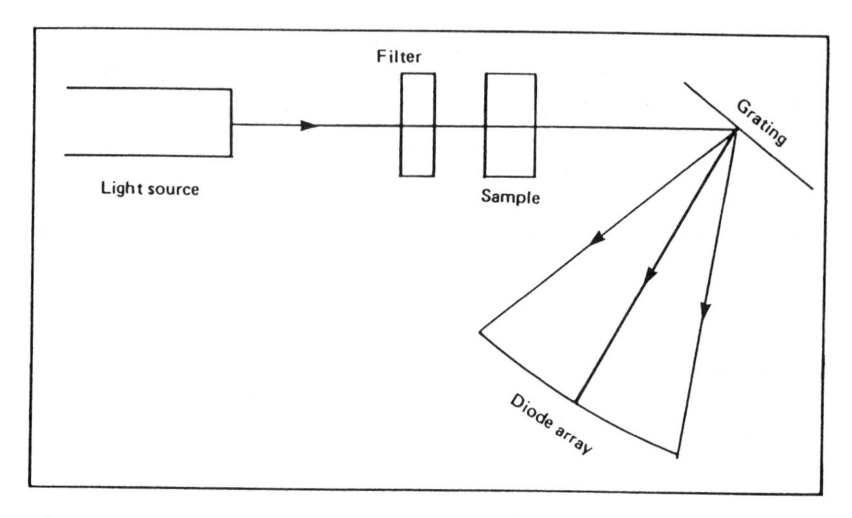

Figure 6.28 Schematic diagram of a diode array system.

If radiation falls on the system, the photon will displace an electron in the *n*-type silicon. The electrons will then go to the nearest *p*-type semiconductor under the influence of the potential of a reversed-biased *pn* junction. The reversed-biased junction is nonconducting, and a potential builds up on the particular element. The charge depends on the light intensity falling on the system in the immediate vicinity of that particular element.

At the end of the predetermined time, the charge is measured by reversing the voltage and discharging it. By measuring the charges on each individual

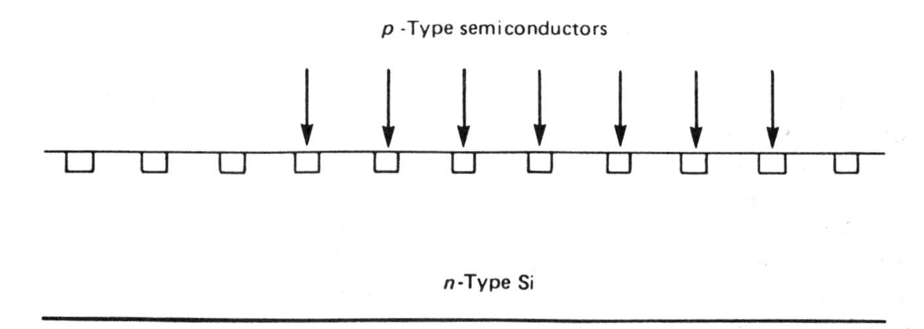

Figure 6.29 Diode array made by inserting *p*-type semiconductors (pixels) in *n*-type silicon.

element, it is possible to get a measurement of light intensity versus wavelength of the whole range. This is tantamount to a digital UV absorption spectrum.

The system is speeded up by using a computer that stores the signal on each separate pixel and reads out on demands. Using multiple and rapid measurements, it is possible to measure the absorption spectrum in a very short time and to determine changes in the absorption spectrum over time.

Alternative procedures have been developed that use interferometers and permit the interference pattern to fall on the diode array. The interference pattern can then be treated using a Fourier transform and the original absorption spectrum obtained.

By using multiple measurements the signal can be accumulated and sensitivity considerably increased.

The useful wavelength range is between 190 and 1100 nm. Simultaneous use of the entire wavelength range provides the multiplex advantage and improves the resolution of the system. The resolution of the system is limited by the number of pixels involve. Typical diode spacing is 0.025 mm. Each one can be thought of as covering a finite spectral range. Systems have been developed with as many as 4096 pixels in the array. The measuring and recharging of each pixel can be accomplished in less than a microsecond. With as many as 4000 data points per microsecond, computer capacity can be rapidly taxed, and this can provide a limiting factor to the system.

Applications. The most important applications for diode array systems are in molecular spectroscopy, since in general they do not have the resolution necessary for atomic spectroscopy or emission measurements. In molecular spectroscopy the most useful areas of application are for (1) scanning fast reactions, (2) applications involving low light levels, and (3) detectors for HPLC.

In rapid scanning for fast reaction, many complex biological reactions take place quickly. This is particularly true of redox chemistry of both organic and inorganic compounds. These reactions can be studied by using a combination of electrochemistry and spectroscopy. Diode array systems are usefully employed when transparent thin electrodes are used to study these reaction mechanisms.

By taking the absorption spectra in rapid succession and accumulating the data, it is possible to detect and measure the spectra of intermediates formed in such complex reactions. This is much more reliable than using absorption at a single wavelength to measure the reactions, since the choice of the single wavelength is often made with the assumption that the intermediates and end products are well known and suitable absorption wavelengths are therefore easily chosen. This is often not the case. Using the diode array system, the complete UV absorption spectra can be obtained, and much more information on the identity and concentration is therefore available.

There are also important applications to HPLC. HPLC is becoming increasingly valuable in the separation and study of complicated organic mixtures, particularly those involving high-molecular-weight biological samples.

One of the difficulties with conventional HPLC is that the detectors using UV detection normally operate at a single wavelength, such as the 254-nm atomic mercury line. The detectability of many compounds is low at this wavelength and zero if the compound does not absorb at 254 nm. In addition to having variable detectability, conventional HPLC detectors provide no indication of the rest of the UV absorption spectrum.

This problem is overcome by using the diode array system. The absorption spectrum can be obtained at predetermined, regular intervals, sorted on a computer, and displayed as necessary. Not only is the sample detected, but its entire UV absorption spectrum is obtained; this provides much more qualitative information than using a single wavelength.

As an alternative to obtaining the entire UV spectrum, the total array signal can be integrated and used as a single detector. Under these circumstances the total signal from all elements is measured and a increase in sensitivity of up to 500-fold is experienced.

Finally, we discuss applications involving low-light-intensity detectors. The emission from low levels of radiation can be accumulated on the photodiode array by rapid scanning and accumulation on the system. After some predetermined time, the accumulated signal can then be displayed, giving the spectrum of the sample. Of course, the average spectrum over the period of integration is obtained.

In conclusion, it can be seen that photodiode array systems hold numerous advantages over the single-wavelength detector systems. Their applications are expected to increase with time, particularly as better techniques for creating arrays and accumulating and processing data become available.

d. Sample Cells

The cells used in UV absorption must be transparent to UV radiation. The most common materials used are quartz or fused silica. They are readily available even in matched pairs, where the reference cell is almost identical to the sample cell. Quartz and fused silica are also chemically inert, which makes them sturdy and dependable in use.

For spectrophotometric analysis in the visible region of the spectrum, glass cells may be used. These are less expensive but cannot be used at shorter wavelengths. Typical transparencies are shown in Fig. 6.30.

It is important that cells be treated correctly in order to achieve best results and to prolong their lifetime. To that end, the operator should (1) always choose the correct cell, (2) keep the cell clean, check for stains, etch marks, or scratches that change the transparency of the cell, (3) hold cells on the nontrans-

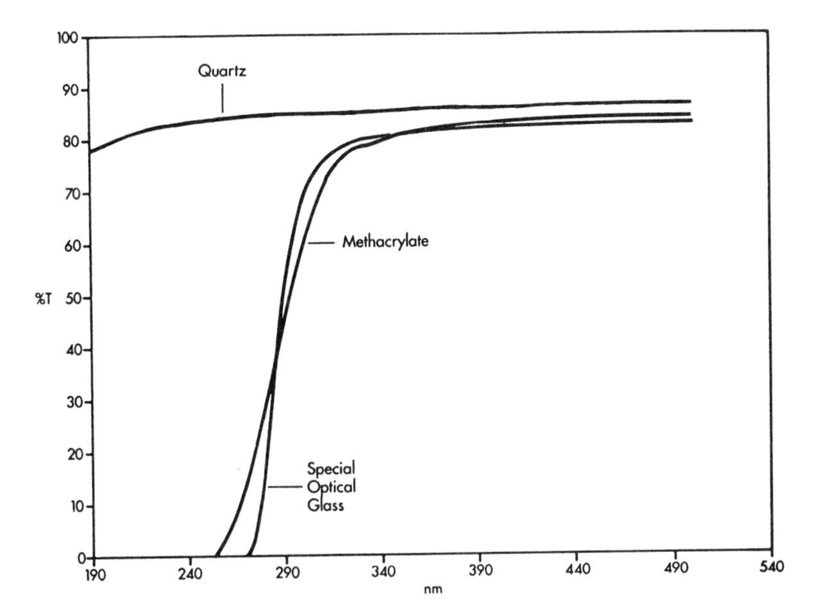

Figure 6.30 Transparencies of materials used in UV analysis.

parent surface, (d) clean cells thoroughly, wash out with a small portion of sample to remove previous wash solution, and (e) do not put strong basic solutions (e.g., NaOH) into the cells.

e. Matched Cells

When double-beam instrumentation is used, two cells are needed: one for the reference and one for the sample. It is normal for the absorption by these cells to differ slightly. This causes a small error in the measurement of the sample absorption and can lead to analytical error. For most accurate work, *matched cells* are used. These are cells in which the absorption of each one is equal to or very nearly equal to the absorption of the other. A large number of these cells are manufactured at one time and their respective absorptivities measured. Those with very similar absorptivities are put together and designated as matched cells. Naturally, the cost of a pair of matched cells is greater than the cost of two unmatched cells. It should also be noted that if one matched cell is broken, it cannot be used with another matched cell from another pair, because it is unlikely that their absorptivities will be equal to each other.

At all times when not in use, cells should be kept clean and dry. Any sample left in a cell will tend to dry out and cause a stain on the cell walls, and this will lead to analytical error and eventual destruction of the cell.

D. ANALYTICAL APPLICATIONS

1. Qualitative Analysis

The types of compounds that absorb UV radiation are those with nonbonded electrons (*n* electrons) and conjugated double bond systems (π electrons) such as aromatic compounds and conjugated olefins. Unfortunately, such compounds absorb over similar wavelength ranges, and the absorption spectra overlap considerably. Furthermore, the absorption curve is influenced by the whole molecule as well as by the particular group that contains the absorbing electrons. As a first step in qualitative analysis, it is necessary to purify the sample to eliminate absorption bands from impurities. Even when pure, however, the spectra are often broad banded and frequently without fine structure. For these reasons, UV absorption is less useful for the qualitative identification of functional groups, or particular molecules, than some other analytical methods, such as mass spectrometry, infrared, and NMR. It is very useful, however, in detecting aromatic compounds and unsaturated conjugated systems. Also, if the compounds are reasonably pure, they can be qualitatively identified. Ultraviolet absorption is most used in the fields of natural products and biochemistry.

Identification is carried out by comparing a compound's absorption spectrum with the spectra of known compounds. A record of UV absorption traces can be found in the Sadtler indices or American Petroleum Institute (API) indices.

The absorption spectrum can be obtained by measuring how much radiation is absorbed at various wavelengths. In practice, the slit and the rest of the optics are kept stationary while the dispersion element is rotated and the absorption measured. The plot of the absorption by the molecule against the wavelength of radiation is the absorption spectrum (Fig. 6.31).

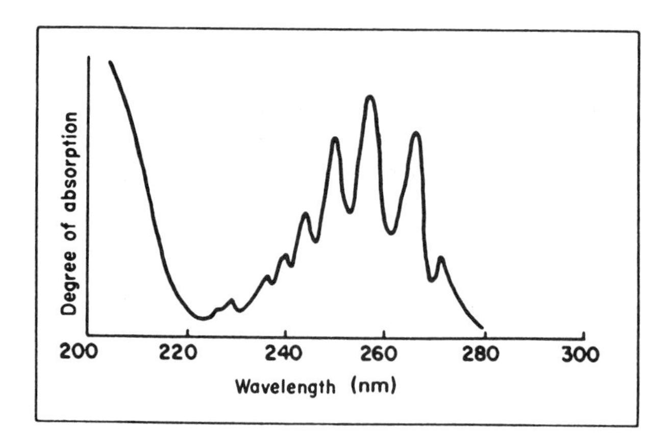

Figure 6.31 Ultraviolet absorption spectrum of benzene.

2. Quantitative Analysis

Although UV absorption spectroscopy is not a widely used tool for qualitative analysis, it is a powerful tool for quantitative analysis. It is used in chemical research, biochemistry, chemical analysis, and industrial processing. Quantitative analysis is based on the relationship between the degree of absorption and the concentration of the absorbing material. Mathematically, it is described by Beer's law, as indicated in Chapter 2. Recall that [Eq. (2.8)] $T = I_1/I_0$, where I_0 is the intensity of the incident radiation, T = transmittance, and I_1 is the intensity of the radiation after passage through the sample; moreover, the absorbance [Eq. (2.11)] $A = -\log T = \log(I_0/I_1)$ and [Eq. (2.18)] $A = abc$, where a is the absorptivity, b is the path length, and c is the concentration. Beer's law shows that there is a direct relationship between c and A.

Most commercial equipment is built to read out absorbance directly, although some read transmittance directly. Therefore, if a and b are known, the concentration c can be calculated by measuring A. However, the relationship is valid only as long as Beer's law is obeyed. In more concentrated solutions, this is usually not the case.

It is frequently found experimentally that the relationship between A and c is not simple. The relationship should be determined experimentally using the equipment that will be used for subsequent analytical determinations. This is done using a calibration curve.

a. Calibration Curves

Calibration curves are obtained by preparing several aliquots of the test material, each at different known concentrations. At a predetermined wavelength, the UV absorption of each aliquot is measured, and a curve relating absorption and concentration is traced over the desired wavelength range. The preparation of calibration curves is described in Chapter 2, Section F.

b. Sensitivity

Ultraviolet absorption analysis is frequently quite sensitive. It is not unusual to detect compounds in concentrations as low as 1 ppm. For example, 1 ppm = 1 mg/ml. If the molecular weight is 100, the concentration is 10^{-5} M. From Beer's law, if $A = 10^5$, the sensitivity is 1 ppm. The sensitivity determines the lowest concentration levels that can be determined quantitatively by this method. Concentrations lower than the sensitivity limits cannot be measured. Often reliable results are obtained only at concentrations 10 times as high as the sensitivity limits. The highest concentration level that can be measured is limited by deviations from Beer's law. It is very important that the detectors receive no stray light, because this displaces the curve. All solutions of high concentration deviate from Beer's law. Also, when the absorption is high, the absorbance changes only slightly, even if the sample concentration varies greatly. Small errors in the measurement of the absorption lead to large errors in the determined concentration. This problem can frequently be overcome by diluting the sample to a more desirable concentration range.

In summary, molecular UV absorption spectroscopy is useful for the determination of molecules containing π bonds or n electrons. Included among these compounds are conjugated olefins, aromatic compounds, and many natural products from both animals and plants. The method is sensitive and reliable, but limited in its application to qualitative analysis.

3. Typical Applications

UV/visible spectrophotometry is the most widely used spectroscopic technique. It has found use everywhere in the world from the south pole to the northern extremities.

Some typical applications of UV absorption spectroscopy include the determination of (1) in environmental testing, polynuclear aromatic compounds that may cause cancer; the concentration of phenols, sulfates and nitrites in drinking water; (2) natural products, such as steroids or chlorophyll; (3) dyestuff materials; and (4) vitamins, such as vitamin A. This technique is also used to distinguish between chemical forms, such as substituted aromatics, conjugated olefins, vitamins, or aldehydes. Ultraviolet analysis has been used to identify the compounds found in various parts of plants, such as leaves, pollen, and roots, as well as decomposition products, such as humic acids and folic acids derived from leaf decay. It has also been used in studies of the chemical components, such as protein, of animal tissue and fluids. Other compounds, such as vitamins and hormones, have been identified and their chemical structure elucidated with the help of UV absorption spectroscopy. Interaction with UV light is an essential feature of photosynthesis and hence of living tissue. Many naturally occurring molecules are light sensitive, so the use of UV spectroscopy has become a cornerstone in the study of living systems. It has been used to determine the purity of DNA samples and to study the kinetics of many biochemical reactions catalyzed by enzymes.

In addition, the UV method has been used for the determination of impurities in organic samples, such as industrial plant streams. For example, it can be used to determine traces of conjugated olefins in simple mono-olefins or aromatic impurities in pure hexane or similar paraffins. It has also been used in the detection of possible carcinogenic materials in foods, drinks, cigarette smoke, and other air pollutants. It has been used to characterize the reflectance of energy-efficient glass windows as well as the reflectance from surfaces coated with paint designed to resist light, heat, and salt water.

In the field of agriculture, UV spectroscopy can be used for the determination of pesticides on plants, in polluted rivers, and in fish and animals that eat or drink polluted food, as well as of vitamins and minerals (or amino acids) in foods or tablets.

In the medical field, it can be used for the analysis of enzymes, vitamins, hormones, steroids, alkaloids, and barbiturates. These measurements are used in the diagnosis of diabetes, kidney damage, and myocardial infarction, among other ailments. In pharmacy, it can be used to measure the purity of drugs during manufacture and of the final product.

In common with other spectroscopic techniques, UV spectroscopy can be used to measure the kinetics of chemical reactions. For example, suppose that two compounds A and B react to form a third compound C. If the third compound absorbs UV radiation, its concentration can be measured continuously. The original concentrations of A and B can be measured at the start of the experiment. By measuring the concentration of C at different time intervals, the kinetics of the reaction $A + B \rightarrow C$ can be calculated.

Ultraviolet absorption spectroscopy can also be used in conjunction with liquid chromotography. The eluant is passed through the UV absorption cell. This allows monitoring of fractions as they elute from the column. The passing of a UV-absorbing component is detected by absorption and measured by a spectrophotometer. Diode arrays can also be used to give a complete absorption spectrums of each component of the eluant of a liquid chromatogram as they emerge from the column.

E. UV FLUORESCENCE

If a "black light" (UV light) is shone onto certain paints in the dark, they give off visible light. These paints are said to *fluoresce*. An energy diagram of this phenomenon is shown in Fig. 6.32. In the process, absorption of energy takes place and the molecules are electronically excited from E_0 to E_1. While in the excited state the molecules lose vibrational energy and descend to the lowest vibrational energy level. They then fluoresce, emitting radiation of a longer wavelength than that absorbed, such as visible, and descend to the lower electronic energy level once again. This process is very rapid, but if for some reason a molecule collides with other molecules while it is excited, it is possible for it to transfer its energy during this collision and not emit radiation. This is called *quenching*. Quenching in fluorescence is often a serious problem, but with care it can be eliminated. Quenching by collision with the solvent molecules can be reduced by decreasing the temperature, thus reducing the number of collisions per unit time. The same can be achieved by increasing the viscosity—for example, by adding glycerine. Dissolved oxygen is a strong quenching agent. It can be removed by bubbling nitrogen through the sample.

If, when the molecule is excited, the excited electron changes its direction of spin and enters the triplet state, the transition back to the ground state is forbidden and the lifetime of the excited molecule is extended. The electrons may

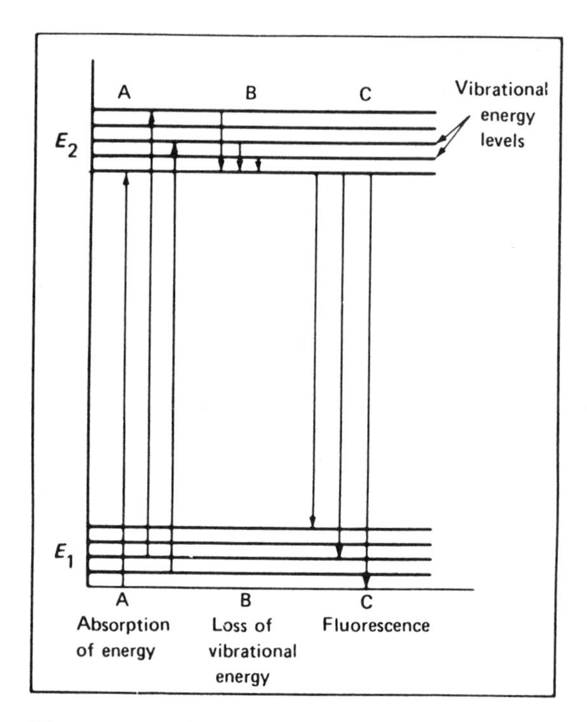

Figure 6.32 Energy transitions taking place during UV fluorescence.

eventually descend from the triplet state to the ground state, emitting radiation in the process. This is the phenomenon of *phosphorescence*. Since the molecule in a triplet state has an extended lifetime in the excited state, it is quite likely that it will collide with some other molecule and lose its energy of excitation and not emit radiation. Phosphorescence is very susceptible to quenching and not analytically useful unless extensive steps are taken, such as converting the sample into a glass to prevent collision.

1. Relationship Between Fluorescence Intensity and Concentration

The intensity of fluorescence F is proportional to the amount of light absorbed. We know that

$$T = \frac{I_1}{I_0}$$

and

$$A = -\ln\left(\frac{I_1}{I_0}\right)$$

Also, from Beer's law,

$$A = abc$$

Therefore

$$-\ln\left(\frac{I_1}{I_0}\right) = abc$$

and

$$\frac{I_1}{I_0} = e^{-abc}$$

so

$$1 - \frac{I_1}{I_0} = 1 - e^{-abc}$$

We multiply each side by I_0:

$$I_0 - I_1 = I_0(1 - e^{-abc})$$

But

$$I_0 - I_1 = \text{amount of light absorbed}$$

and

$$F = (I_0 - I_1)\Phi$$

where ϕ is the quantum efficiency and F is the fluorescence intensity. Therefore,

$$F = I_0(1 - e^{-abc})\Phi$$

When abc is small (low concentration),

$$F = I_0 abc\Phi$$

That is, F, total fluorescence, is proportional to I_0 and c. Hence $F = kI_0c$, where k is a proportionality constant. But only a portion of the total fluorescence is monitored or measured; therefore,

$$F' = Fk'$$

where

F' = measured fluorescence

and

$$F' = kI_0c$$

where k' = instrumental constant and includes k, a, b, and ϕ.

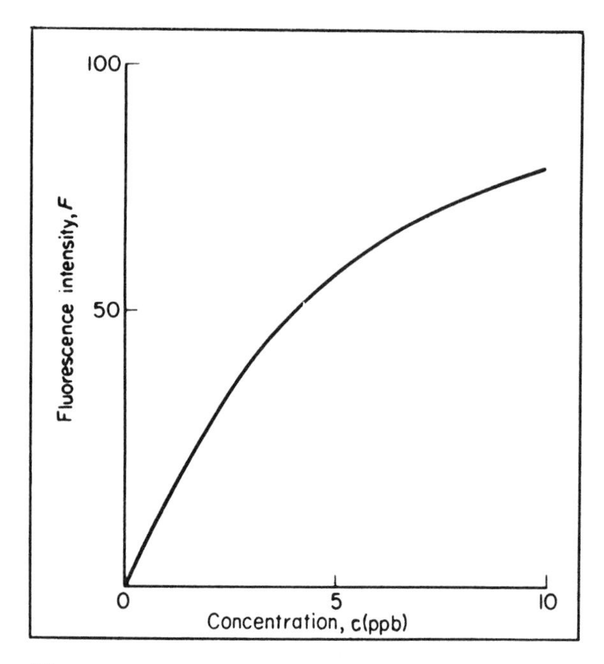

Figure 6.33 Relationship between fluorescence and concentration at low concentrations.

Therefore at low concentrations F is proportional to $kc - k''c^2$ where k'' is the proportionality factor of the self-absorption effect controlled by the concentration squared. If we plot this relationship, it will be found to be similar to that shown in Fig. 6.33.

At higher concentrations the term for self-absorption must be included. That is, F is proportional to $kc - k^1c^x$, and the shape of the curve relating F and c then becomes as shown in Fig. 6.34. It can be seen that at higher concentrations the fluorescence intensity actually decreases because the molecules on the outer part of the sample absorb the fluorescence generated by those in the inner part of the molecule. In practice, it is necessary to correct for this effect. It is not easy to tell directly if the fluorescence measured corresponds to concentration A or concentration B as shown in Fig. 6.34. Both concentrations would give the same fluorescence intensity. The dilemma can be solved by diluting the sample by exactly 50%. In this case, if the original concentration were A, then the fluorescence intensity would sharply decrease to approximately 50% of its original value. On the other hand, if the concentration were B, then the fluorescence would actually increase. In practice, when the concentration is unknown, suf-

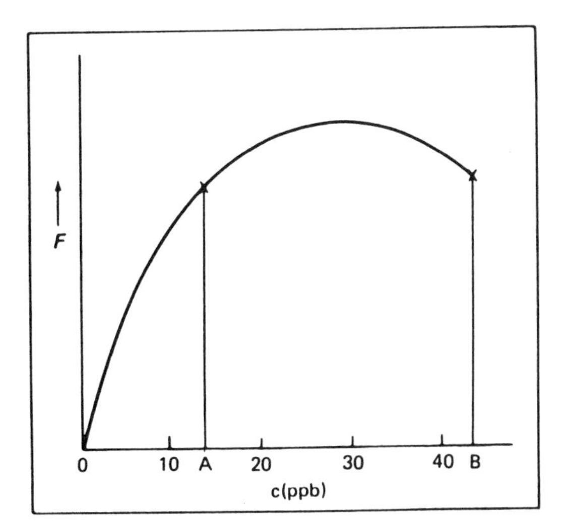

Figure 6.34 Fluorescence at high sample concentrations. Note the reversal of fluorescence: concentrations *A* and *B* give the same fluorescence intensity and could not be distinguished from a single measurement.

ficient successive dilutions are carried out until the fluorescence intensity drops off 50% for a 50% dilution of the sample.

2. Equipment

A schematic diagram of fluorescence equipment is shown in Fig. 6.35. The steps taken to measure the fluorescence are as follows:

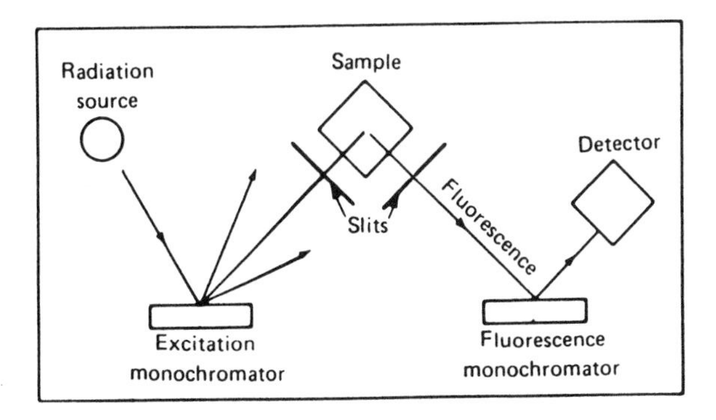

Figure 6.35 Schematic diagram of UV fluorescence equipment.

1. Set the excitation monochromator at some suitable short wavelength, such as 250 nm.
2. Record the fluorescence spectrum emitted by the sample by rotating the fluorescence dispersion element and recording the intensity of the emission at different wavelengths.
3. Set the fluorescence monochromator at the wavelength that gives the maximum fluorescence intensity.
4. Scan the excitation monochromator and record the excitation spectrum on the detector. This will be the fluorescence intensity as the excitation wavelengths varies.
5. Set the excitation monochromator at the excitation wavelength that gives the maximum fluorescence.
6. Record the fluorescence spectrum with the excitation monochromator set at its maximum and the fluorescence monochromator set at its maximum.

The process described provides two spectra: the excitation spectrum and the fluorescence spectrum. This is a distinct advantage over UV absorption analysis, which only provides one spectrum. Another advantage over UV absorption is that not all samples fluoresce; therefore, qualitatively many types of samples are eliminated from consideration. The disadvantage, of course, is that many samples cannot be examined by fluorescence.

3. Components

The components of the equipment are the same as those used in UV absorption. The sources used are hydrogen lamps, deuterium lamps, mercury arc lamps, or xenon lamps. Of course, it should be pointed out that the fluorescence intensity is a function of the amount of light absorbed, and therefore the more powerful the light source, the more intense will be the fluorescence. This is in contrast to UV absorption, where the percentage of light absorbed is independent of the intensity of radiation falling on it. For this reason tunable dye lasers have been used to advantage.

The monochromators used are prisms or gratings, although, for specific known compounds, filters can be used—one that transmits shorter wavelengths for excitation and one that transmits longer wavelengths for fluorescence. The cells used are made of quartz, and the detectors are photomultipliers or diode arrays.

4. Analytical Applications

The types of samples used are limited to liquid samples for quantitative work. The system has the advantage of using two spectra—the excitation and fluorescence spectra—and is therefore more reliable in qualitative analyses. The method is very sensitive and the concentration levels are low: 10^{-9} g/ml can easily be detected. The types of samples analyzed include vitamins, thymine, riboflavin, phenols, aromatic amines, alkaloids, estrogens, and flavins. Also, fluorescence indicators are commonly used in analysis.

a. Interferences in Analytical Applications

Other compounds that fluoresce must be removed from the system because of the method's high sensitivity and difficulty in discriminating between overlapping spectra. This can be done, for example, by column chromatography. Reversal of fluorescence intensity, or self-quenching, at high concentrations is a problem in quantitative analysis but can be eliminated by successive dilutions. Dissolved oxygen is a very powerful quenching agent. Any dissolved air or oxygen must be removed by bubbling nitrogen through the sample. Deactivation by collision with solvents reduces the fluorescence intensity. A decrease in temperature reduces this problem, as does an increase in viscosity. Changes in pH can frequently change structure and thereby change fluorescence and must therefore be controlled.

BIBLIOGRAPHY

Anal. Chem., Analytical Reviews, June 1993.
Anal. Chem., Analytical Reviews, June 1994.
Brode, W. R., *Chemical Spectroscopy*, Wiley, New York, 1943.
Creswell, C. J., and Runquist, O., *Spectral Analysis of Organic Compounds*, Burgess, Minneapolis, 1970.
Ewing, G. W., *Instrumental Methods of Chemical Analysis*, 5th ed., McGraw-Hill, New York, 1985.
Silverstein, R. M., Bassler, G. C., and Morrill, T. C., *Spectrophotometric Identification of Organic Compounds*, 4th ed., Wiley, New York, 1981.
What's New in Instrumentation, *Spectroscopy*, (1993) (May) *8*(4).

SUGGESTED EXPERIMENTS

6.1 Add a drop of benzene to a UV absorption cell. Record the absorption spectrum of benzene vapor using a comparatively n arrow slit. Repeat using a wider slit. (Adjust the recording conditions of the instrument if necessary.) Note the loss of resolution when wider slits are used. Explain.

6.2 Record the absorption spectrum of a solution of pure octane. Add benzene to the octane and record the absorption spectrum of dissolved benzene. Compare with Experiment 6.1.

6.3 Record the absorption spectrum between 400 and 200 nm of (a) 1-octene, (b) butadiene, and (c) a nonconjugated diolefin. Note the effect of a conjugated system on the absorption spectrum.

6.4 Record the absorption spectrum of a polynuclear aromatic compound such as anthracene and of a quinonoid such as benzoquinone. Note the effect of the quinonoid structure of the spectrum.
6.5 Record the absorption spectrum of neat benzene (liquid A). Dilute the benzene to a concentration of 50% benzene with octane. This is liquid B. Record the absorption spectrum of liquid B. Dilute liquid B to 50% with more octane (it is now 25% benzene). This is liquid C. By a series of 50% dilutions, prepare solutions D, E, F, G, and so on. Record their spectra.
6.6 Using the curves obtained from Experiment 6.5, plot the relationship between the absorbance A and concentration c of benzene at the wavelength of each of the absorption maxima. Indicate the useful analytical range for each curve.
6.7 Prepare a standard solution of quinine by dissolving a suitable quantity of quinine in water. Record the UV absorption spectrum of the solution between 500 and 200 nm. Record the absorption spectra of several commercial brands of quinine water (after allowing the bubbling to subside). Which brand contained the most quinine?

PROBLEMS

6.1 What part of a molecule is excited by UV radiation?
6.2 Indicate which of the following molecules absorb UV radiation and explain: (a) heptane, (b) benzene, (c) butadiene, (d) water, (e) heptene, (f) chlorohexane, (g) ethanol, (h) ammonia, and (i) *n*-butylamine.
6.3 Draw a schematic diagram of a double-beam spectrophotometer. Briefly explain the function of each major component.
6.4 What are the principal components of a spectrophotometer?
6.5 Radiation with a wavelength of 640nm is dispersed by a simple grating monochromator at an angle of 20°. What are the other wavelengths of radiation that are dispersed at the same angle by this grating (lowest wavelength 200 nm)?
6.6 What is the principle of the photomultiplier?
6.7 What are the limitations of UV absorption spectroscopy as a tool for qualitative analysis?
6.8 (a) Prepare a calibration curve for the determination of monochlorobenzene from the data listed below. (b) Three samples of chlorobenzene were brought in for analysis. The samples transmitted (1) 90%, (2) 85%, and (3) 80% of the light under the conditions of the calibration curve just prepared. What was the concentration of chlorobenzene in each sample?

Concentration (ppm)	Absorbance
1.2	0.24
2.5	0.50
3.7	0.71
5.1	0.97
7.2	1.38
9.8	1.82

6.9 Several samples of monochlorobenzene were brought to the laboratory of an analysis. The absorbance of the samples is listed below.

Sample	Absorbance
A	0.400
B	0.685
C	0.120
D	0.160
E	2.0

(a) What were the respective concentrations?
(b) How could the analysis of sample E be obtained?

6.10 Which of the following absorb in the UV?

a) N_2 f) C_2H_4
b) O_2 g) I
c) O_3 h) Cl_2
d) CO_2 i) Cyclohexane
e) CH_4 j) C_3H_6

6.11 Why does phenolphthalein change color when going from an acid to a basic solution?

6.12 Why are UV absorption spectra broad band?

6.13 What are the blue shift and the red shift?

6.14 Why do H_2 lamps and D_2 lamps emit a continuum and not H_2 or D_2 line spectra?

6.15 How does the diode work?

6.16 How does a bipolar transistor work?

6.17 How does a field effect transistor (FET) work?

6.18 Describe a diode array.

6.19 Describe the process of UV molecular fluorescence.

6.20 What is the relationship between fluorescence and excitation light intensity I_0?

6.21 Explain the reversal of fluorescence intensity with increase in analyte concentration. How is the this source of error corrected?

6.22 Describe the equipment used for measuring UV fluorescence intensity.

6.23 What interferences are encounter in UV fluorescence?

What is the wavelength of the absorption maximum of the following compounds?

6.24

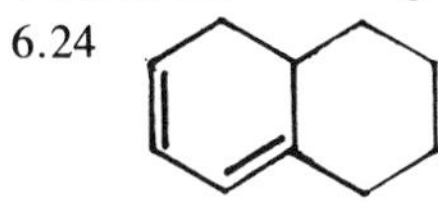

6.25

6.26

6.27

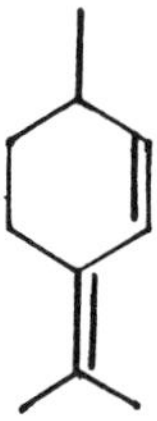

6.28

6.29

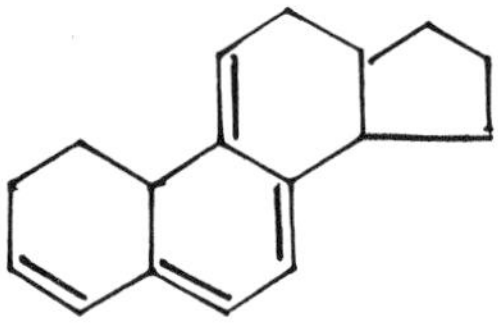

6.30

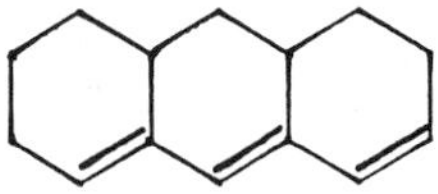

6.31

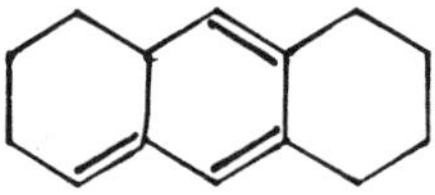

6.32

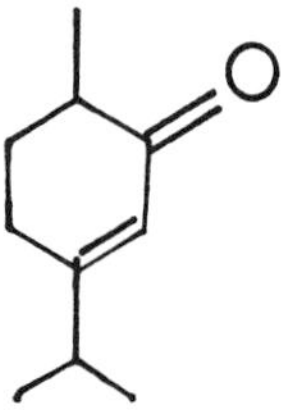

O

6.33

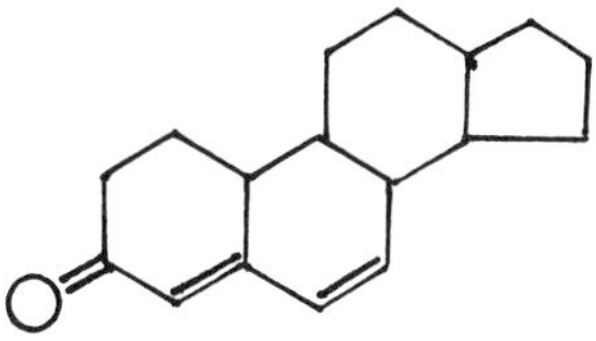

6.34

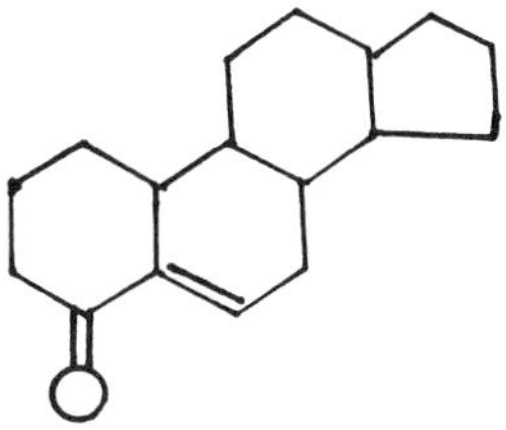

6.35

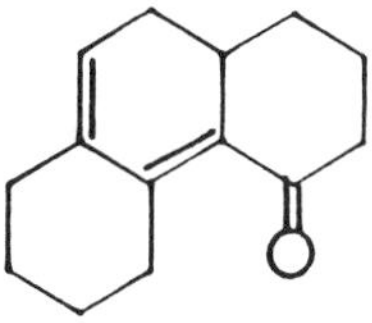

6.36

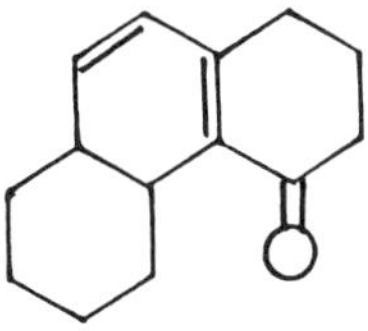

6.37  Cl

6.38 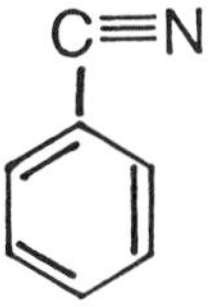 C≡N

6.39 OH

Cl

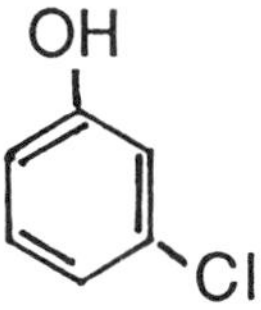

6.40 OH

Cl

6.41 OH

NH_2

6.42 CH_3

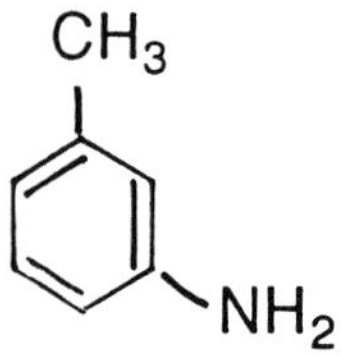

NH_2

6.43 CH_3

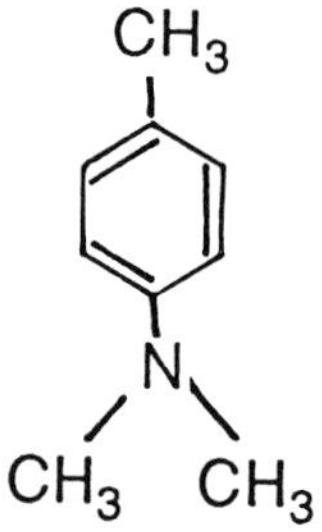

N

CH_3 CH_3

7
Atomic Absorption Spectroscopy

Atomic absorption spectroscopy involves the absorption of very narrow absorption lines by free atoms. It is a reliable method of elemental analysis, and is particularly useful for determining trace metals in liquids. Most importantly, it is practically independent of the chemical form of the metal in the sample. For example, we can determine the total cadmium concentration of a water sample—it does not matter whether the cadmium exists as a chloride, nitrate, sulfate, or other salt. The method is very sensitive and can detect different metals in concentrations as low as and frequently lower than 1 ppb. As an analytical technique, it has become increasingly important because of its high sensitivity and the comparative ease with which quantitative results can be obtained. One of its principal advantages is that determinations can be carried out in the presence of many other elements, which do not interfere. This freedom from interference makes it unnecessary to separate the test element from the other elements from the other elements present in the sample. Not having to separate out the test element saves a great deal of time and in the process eliminates numerous sources of error. A disadvantage of the method is that only one element can be determined at a time. A change in light sources and a change of analytical wavelength are necessary to determine a second element. As a consequence, the method has limited use for qualitative analysis.

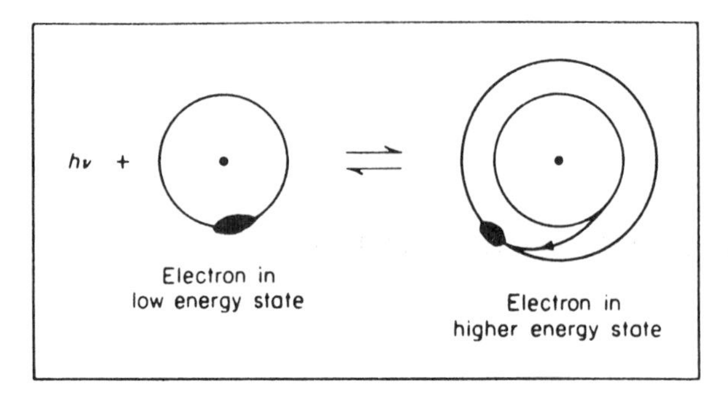

Figure 7.1 Absorption of radiation by atoms.

A. ABSORPTION OF RADIANT ENERGY BY ATOMS

Atomic absorption spectroscopy is the study of the absorption of radiant energy by atoms. In the process of absorption, an atom changes from a low energy state to a higher energy state. This transition is illustrated in Fig. 7.1. Atoms do not vibrate in the same sense that molecules do. Also, they have virtually no rotational energy. Hence no vibrational or rotational energy is involved in the electronic excitation of atoms. As a result, atomic absorption spectra consist of a few very narrow absorption lines.

The frequency of the absorption lines is derived from Eq. (2.2), $E = h\nu$, where E, the energy of excitation of the atom, equals $E_1 - E_0$ (E_0 is the energy of the atom's low energy level and E_1 is the energy of its higher energy state) and ν is the frequency. Hence E in Eq. (2.2) is the difference in energy between the low energy state and the high energy state involved.

Quantum theory defines the electronic orbitals in an atom. For example, sodium is an atom with complete inner shells but one electron in the outer shell. Its configuration is therefore $1s^2$, $2s^2$, $2p^6$, $3s^1$. This is frequently abbreviated to $3s$. In atomic absorption, UV radiation is absorbed that does not have sufficient energy to excite the inner shell electrons at the principal quantum numbers 1 and 2. Consequently, in sodium, only the $3s$ electron, sometimes called the *optical electron*, is involved. This is true of all elements: Only the optical electrons are involved in atomic absorption.

In the case of sodium and all other elements, the upper orbitals predicted by quantum theory are empty. Thus the $3p$, $4s$, $3d$, $4p$, $5s$, and so on, orbitals of an unexcited sodium atom are empty. These empty orbitals constitute the upper energy levels of the atom. If a valence electron is moved to one of these upper levels, the atom is in an excited state. The number of energy levels can be predicted from quantum theory. The actual energy differences of these levels have been deduced from studies of atomic spectra. These levels have been graphed in *Grotrian diagrams*. A partial Grotrian diagram for sodium is shown in Fig. 7.2.

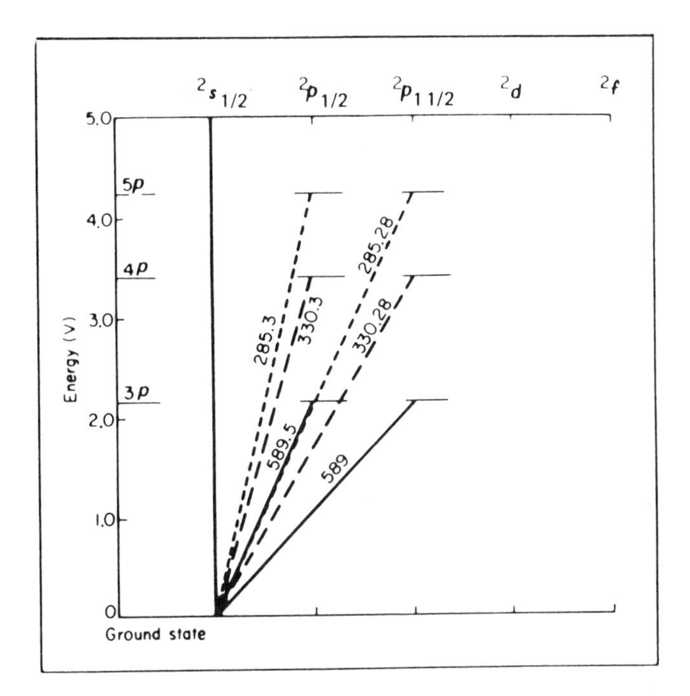

Figure 7.2 Partial Grotrian diagram for sodium.

The energy levels are split because the electron itself may spin one way or another, resulting in two similar energy levels and therefore two lines (a doublet) rather than a single line (a singlet). For the transition from the ground state to the first excited state of sodium, the electron moves from the 3*s* orbital to the empty 3*p* orbital. The latter is split into two levels $p^{1/2}$ and $p^{3/2}$ by the electron spin, so two transitions are possible. The differences in energy are $[E(3s) - E(3p^{1/2})] = h\nu_1$ and $[E(3s) - E(3p^{3/2})] = h\nu_2$. The wavelengths are associated with these transitions are 589.5 and 589.0 nm, the well-known sodium D lines.

Suppose that the energy states of a given atom are E_o, E_1, E_2, E_3, E_4, . . . , where E_0 is a ground state and E_1, E_2, E_3, E_4. . . . , are excited states. The absorption frequency may correspond to the transition from the ground state to an excited state. The frequency may correspond to the transition from the ground state to an excited state. The frequency ν is given by $E_1 - E_o = h\nu_1$ or $E_2 - E_o = h\nu_2$, or $E^3 - E_o = h\nu_3$, and so on. If the transition were between two excited states, ν would be given by $E_2 - E_2 - E_1 = h\nu_4$ or $E_1 = h\nu_5$, and so on.

Under normal conditions atoms exist in their lowest possible energy state, the ground state. Very few atoms are normally in the higher energy states. For example, it can be calculated from the Boltzmann distribution law [Eq.(2.6)] that if zinc vapor with a resonance line at 213.8 nm is heated to 3000 K, there will be only one atom in the first excited state for every 10^{10} atoms in the ground state. Zinc atoms need a considerable amount of energy to become excited. On the other hand, sodium atoms are excited more easily than the atoms of most other elements. Nevertheless, at 3000 K only 1 sodium atom is excited for every 1000 atoms in the ground state. At room temperature the ratio of the number of excited to unexcited atoms is greatly reduced. In a normal atom population, therefore, there are very few atoms in states E_1, E_2, E_3, ... The total amount of radiation absorbed depends, among other things, on how many atoms are available in the lower energy state to absorb radiation and become excited. Consequently, the total amount of radiation absorbed at frequencies ν_2, ν_3, and so on, is exceedingly small and not useful analytically.

For practical purposes, all absorption is by atoms in the ground-state level. This greatly restricts the number of absorption lines that can be used in atomic absorption. Quite frequently only three or four useful lines are available in the UV spectral region for each element, and in some cases fewer than that. The wavelengths of these absorption lines can be deduced from the Grotrian diagram of the element being determined.

1. Detection of Nonmetals

Grotrian diagrams correctly predict that the energy required to reach even the first excited state of *nonmetals* is so great that they cannot be excited by ultraviolet radiation. Their resonance lines are in the vacuum UV. Consequently, using flame atomizers, atomic absorption cannot be used for the direct determination of nonmetals, although with the use of a capillary discharge lamp, iodine has been measured directly. However, nonmetals have been determined by indirect methods. For example, chlorides can be precipitated as silver chloride; from the subsequent determination of silver, the chloride can be calculated.

2. Degree of Absorption

The total amount of light absorbed by atoms is given by

$$\text{Total amount of light absorbed at } \nu = \int K_\nu \, d_\nu = \frac{\pi e^2}{mc^2} Nf \qquad (7.1)$$

where

K_ν = absorption coefficient at frequency ν
e = electronic charge
m = mass of the electron
c = speed of light
N = total number of atoms that can absorb at frequency ν in the light path
f = oscillator strength or ability of each atom to absorb at frequency ν

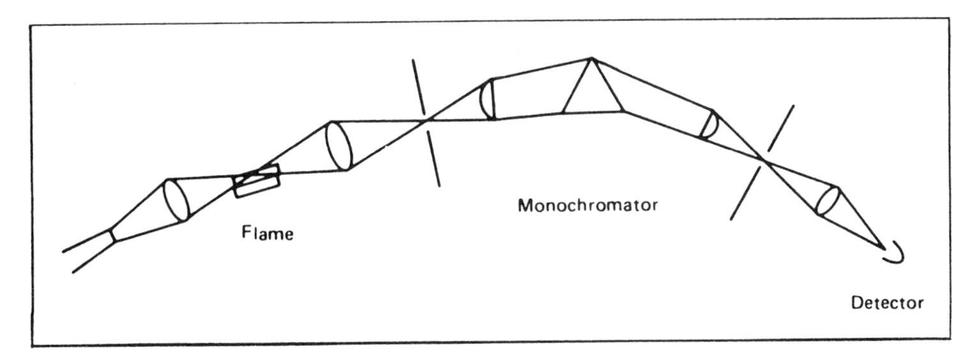

Figure 7.3 Schematic diagram of the equipment used for atomic absorption spectroscopy.

Note that π, e, m, and c are constants; therefore Eq. (7.1) can be simplified to

$$\text{total amount of radiation absorbed} = \text{constant} \times Nf \qquad (7.2)$$

Two important features of this expression should be noted: First, there is no term involving the wavelength (or frequency) of absorption other than that indicating the actual absorption wavelength; second, there is no term involving the temperature. The absence of these terms clearly indicates that absorption by atoms is independent of the wavelength of absorption and the temperature of the atoms. This makes the method relatively free of interference from temperature changes in the atomizer and of problems involved in utilizing different regions of the spectrum. These two features give atomic absorption distinct advantages over other atomic spectral methods such as emission spectroscopy and flame photometry.

It should be pointed out, however, that although the temperature does not affect the *process* of absorption by atoms, it does affect the efficiency with which free atoms are produced from a sample and therefore indirectly affects the signal quite significantly. Furthermore, some atoms, particularly those of the alkali metals, easily ionize at high temperatures. Ions do not absorb at atomic absorption wavelengths. Atoms that become ionized are effectively removed from the absorbing population, resulting in a loss of signal.

B. EQUIPMENT

A schematic design of the equipment used for atomic absorption spectroscopy is shown in Fig. 7.3. The principle of the equipment is similar to that which holds for other spectroscopic absorption methods (see Chap. 3). Light from a suitable

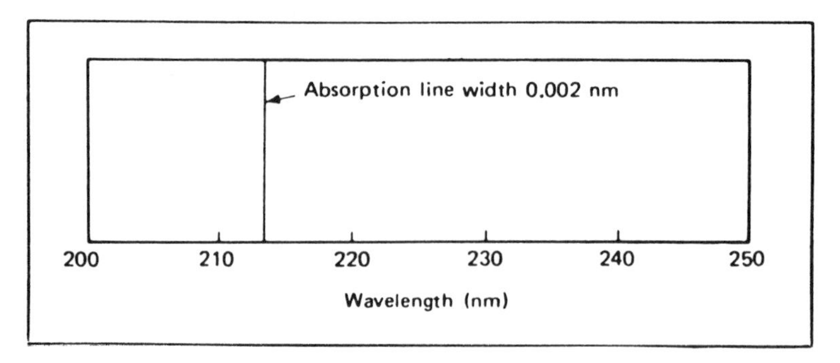

Figure 7.4 Absorption of atomic absorption lines (Zn, 213.9 nm) from a hydrogen lamp.

source is directed through the sample and a monochromator–slit system to a detector. The latter measures how much light is absorbed by the sample by measuring the intensity of radiation before and after it passes through the sample. The components of the equipment are as follows.

1. Radiation Sources

a. Hollow Cathode

As we have already mentioned, atomic absorption lines are very narrow (about 0.002 nm). They are so narrow that if we were to use a continuous source of radiation, such as a hydrogen lamp, it would be very difficult even to detect the lines. Absorption of a narrow band from a continuum is illustrated in Fig. 7.4, which shows the absorption of energy from a hydrogen lamp by zinc atoms absorbing at 213.9 nm. The width of the zinc absorption line is exaggerated for illustration purposes. The wavelength scale for the hydrogen lamp in Fig. 7.4 is 50 nm wide. If the absorption band were 0.002 nm wide, its width would be $0.002 \times 1/50 = 1/25{,}000$ of the scale shown. Such a narrow line would be detectable only under extremely high resolution, which is not encountered in conventional equipment.

With the use of slits and a good monochromator, the band falling on the detector can be reduced to about 0.2 nm. If a band 0.002 nm wide were absorbed from this, the signal would be reduced 1%. Since this is about the absorption linewidth of atoms, even with complete absorption of radiation by atoms, the total signal would change by only 1%. This would result in an insensitive analytical procedure of little practical use. The problem of using such narrow absorption lines was solved by adopting a *hollow cathode* (Fig. 7.5) as the radiation source.

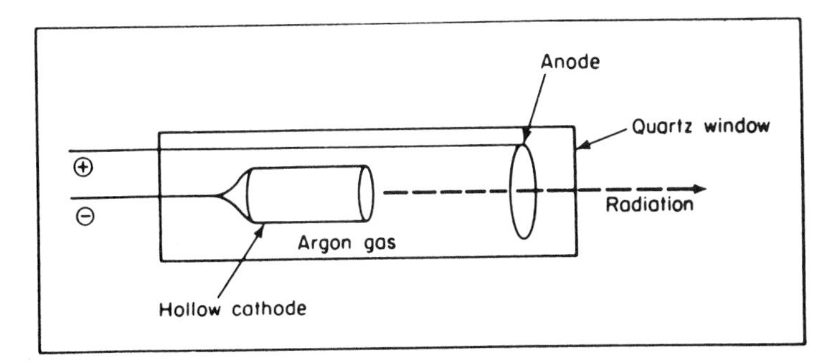

Figure 7.5 Schematic diagram of a hollow cathode.

The hollow cathode works as follows: A high voltage is put across the anode and cathode. Atoms of the filler gas (argon or helium) become ionized at the anode and are attracted and accelerated toward the cathode. The fast-moving ions strike the surface of the cathode and physically dislodge the surface metal atoms into the space above. The displaced atoms are excited and emit the characteristic spectrum of the metal used to make the cathode. Hollow cathodes emit spectra with very narrow lines. The lines are so narrow that they can be almost completely absorbed by the absorption of the atoms. Using this light source, atomic absorption is easily detected and measured.

Each hollow cathode emits the spectrum of metal used in the cathode. For example, copper cathodes emit the copper spectrum and zinc cathodes emit those lines originating from the zinc energy states as shown in the Grotrian diagram. Only copper atoms absorb copper resonance lines; only zinc atoms absorb zinc resonance lines. For this reason, a different hollow cathode must be used for each different element to be determined. This is an inconvenience in practice. The handicap is more than offset, however, by the advantage of the narrowness of the spectral lines and the specificity of the method. The emitted spectrum consists of all the emission lines of the metal, including many lines that are not absorption lines. In general, however, they do not interfere with the method of analysis.

Multielement hollow cathodes were first developed in 1958 (J. W. Robinson, unpublished work). The original cathode was made of three elements (Fe, Ni, V) for the simultaneous determination of these elements in oils. They are poisons to platinum catalysts used in gasoline production from cat crackers. In practice it was found that one element in the hollow cathode was always more volatile than the others and slowly completely coated the inside of the cathode during sputtering. This is a serious problem because the light intensity from these elements is continually changing, one element increasing in intensity the other two decreasing.

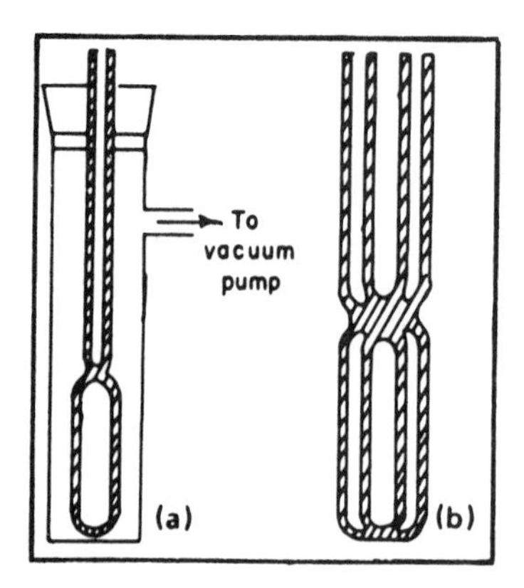

Figure 7.6 Electrodeless discharge lamps with vacuum jackets: (a) dismountable and (b) permanent. (From T. S. West and M. Cresser, *Appl. Spectrosc. Rev.*, *7*:79 (1973).)

However, simultaneous multielement analysis is very attractive because it saves samples and saves labor costs—very attractive features for routine labs. Later, multielement hollow cathodes were made by pressing powdered metals together and fusing them.

The most number of elements determined simultaneously was 12. However, this elaborate system involved a Rowland Circle, and the costs were prohibitive.

Recently the simultaneous analysis of four elements has been described, switching automatically to a second four elements. It involves the use of an eight-element hollow cathode available from Perkins-Elmer Corp. This multielement facet of atomic absorption is an attractive feature because it saves time and sample. It is especially useful because of the high degree of freedom from interference of atomic absorption. But it should be pointed out that both ICP and ICP-MS can more easily accomplish simultaneous multielement analysis, although with increased interference in the results obtained.

b. Electrodeless Discharge Lamps

It is difficult to make stable hollow cathodes from certain elements, particularly those that are volatile, such as arsenic, germanium, or selenium. An alternative light source has been developed in the electrodeless discharge lamp (EDL) (Fig. 7.6). It consists of an evacuated tube in which the metal of interest is placed. The tube is filled with argon at low pressure and sealed off. The sealed tube is then placed in a microwave discharge cavity. Under these conditions the argon becomes a plasma and causes excitation of the metal sealed inside the tube. The emission from the metal is that of its spectrum, including the resonance line. The intensity of these lamps is very high, and they have been made quite stable in recent years.

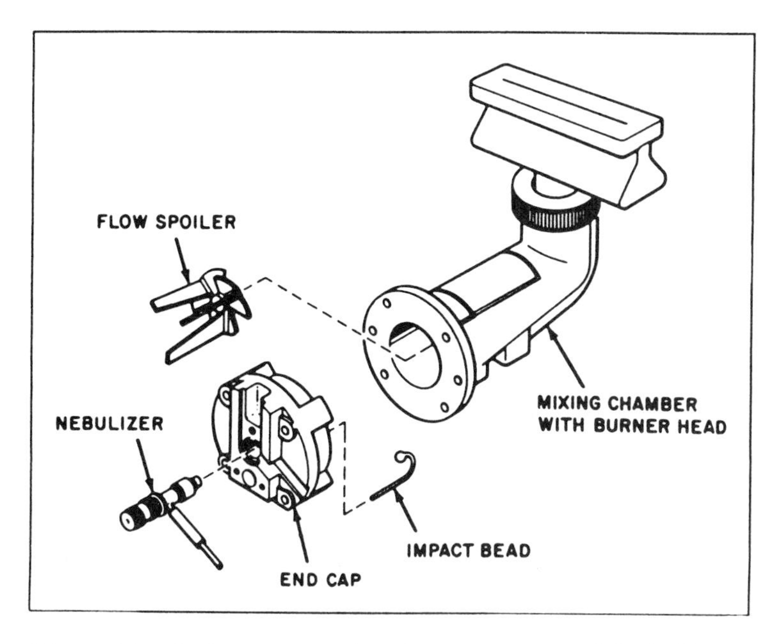

Figure 7.7 Exploded presentation of the components of the Perkin-Elmer (Norwalk, Conn.) atomizer based on Lundegardh's original design.

2. Atomizer

In order to achieve absorption by atoms, it is necessary to reduce the metal to be determined in the sample to the free atomic state. The most common atomizer is the flame. The burners used to produce the flames are basically the same as those used in flame photometry. They are elongated along the light path to increase the number of atoms exposed to the radiation. A typical flame atomizer is shown in Fig. 7.7.

In order to be injected into a flame, the sample must be a liquid or gas. In liquids the element to be determined will probably be in the state of either an ion in aqueous solution or an organic compound in organic solution. The aspirator nebulizes (break into cloudlike particles) the liquid sample into droplets, which are then introduced into the base of the flame. Here droplets are evaporated (or burned), and the sample element is left in the residue. The residue is then decomposed by the flame. In the process some of the sample element is reduced to atoms. This atomization is not very efficient, but flames are easy to use and produce results that are satisfactory for most purposes. Most burner designs are based on the original design by Lundegardh. The design incorporates two distinct consecutive steps, involving nebulization and atomization of the sample. The common oxidants used are oxygen, air, and nitrous oxide. The most com-

Table 7.1 Maximum Temperatures Obtained in Various Flames (°C)

	Oxidant		
Fuel	Air	O_2	N_2O
H_2	2100	2900	2900
Acetylene	2200	3100	3200
Propane	1900	2800	
Butane	1900	2800	
Cyanogen	2600	4500	

mon fuels are acetylene and hydrogen. The maximum temperatures achieved with these mixtures are shown in Table 7.1.

The oxidant and fuel are premixed in the barrel of the burner. The sample is aspirated into the same barrel, where nebulization and evaporation of the solvent take place. The evaporated sample and small droplets are swept together with a combustion mixture into the base of the flame. Any components of the sample that are not nebulized form large droplets that collect on the sides of the vaporization chamber and drain away.

The combustion mixture and vaporized sample enter the base of the flame. Here combustion and atomization take place, particularly in the hot reaction zone of the flame. After free atoms are formed, they are rapidly oxidized in the hostile chemical environment of the hot flame. As can be seen from Fig. 7.7, the flame is elongated, being approximately 10 cm long, and is in the shape of a slot. The light path is along the length of the flame.

If we measure the degree of absorption against flame height, we arrive at a relationship such as that shown in Fig. 7.8a. As can be seen, the signal starts off low at the base of the flame, increases to a maximum in a region close to the reaction zone of the flame, and then diminishes to zero. This curve is brought about by the complicated reactions that take place in a flame. A summary of these reactions is shown in Table 7.2. A more complete illustration of the absorption distribution is shown in Fig. 7.8b.

The sample introduced into the base of the flame goes through the process of evaporation, leaving behind the solid residue, which is then decomposed by the hot flame to liberate free atoms. The latter then oxidize, forming metal oxides in the upper parts of the flame.

The rate of formation of atoms depends on the flame temperature and on the chemical form of the sample. If the flame temperature is high, the rate of formation is increased. If the metal exists in the sample in a stable chemical form, it is difficult to decompose, and atomization efficiency is decreased. On the other hand, if the metal is in a chemical form that is easily decomposed, the rate of atomization is increased and the number of atoms formed is increased.

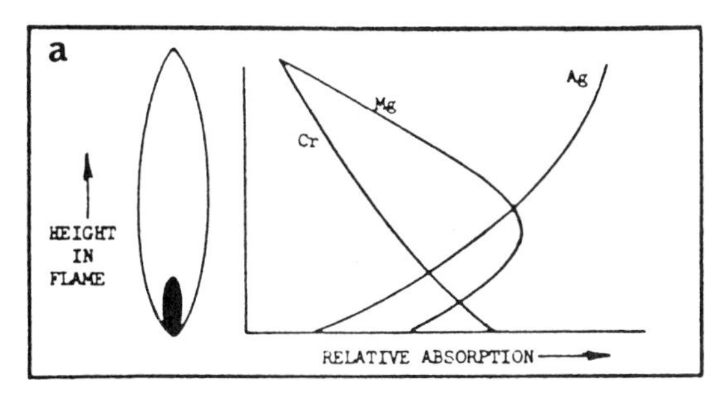

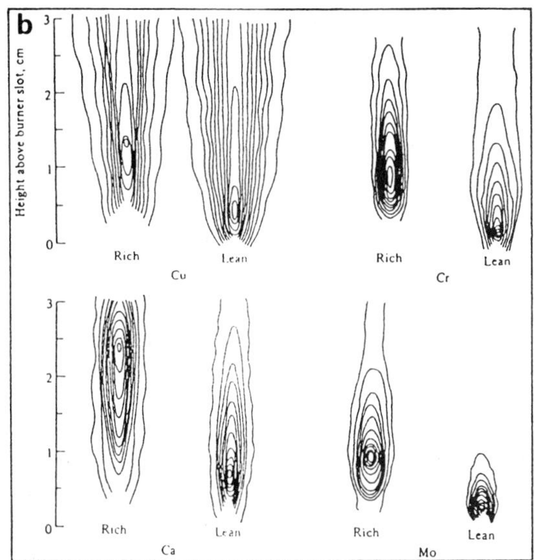

Figure 7.8 (a) Flame profile Cr, Mg, and Ag: relative absorption versus flame height. (b) Distribution of atoms in a 10-cm air/acetylene flame. Fuel-rich and fuel-lean results are shown. Maximum absorbance is at the center.

Table 7.2 Factors Affecting Flame Profiles

Part of flame	Physical form of sample in flame	Reaction	Factors controlling reaction
Outer mantle	Oxide	No reaction or reduction	Stability of metal oxide, flame composition
Reaction zone	Atoms	Accumulation or oxidation	Flame composition, stability of atoms
Inner core	Solid particles	Disintegration	Stability of compound, anions, flame temperature, ultraviolet light emitted from the flame
Base	Droplets	Evaporation	Droplet size, solvent, flame temperature feed rate, combustibility

The rate of loss of free atoms is also a function of the chemistry of the sample. If the oxide of the element being examined is very stable, the free atoms will rapidly form oxides and the population of free atoms will simultaneously decrease. This is the case with elements such as aluminum, molybdenum, tungsten, and vanadium. If, on the other hand, the metal is very stable and the oxide easily decomposed, the free atoms exist for a prolonged period. This is particularly so with the noble metals platinum, gold, and silver.

For most of the transition elements, a profile such as that shown in Fig. 7.8 is arrived at. The maximum absorbance signal depends on the number of free atoms in the light path. These free atoms are in dynamic equilibrium with the flame; they are produced continuously by the flame and lost continuously in the flame. The number produced depends on the original concentration of the sample and the atomization efficiency (Table 7.3). The number lost depends on the rate of formation of oxides. The variation of atomization efficiency with the chemical form of the sample is called the *chemical interference*. It is the most serious interference encountered in atomic absorption spectroscopy and must always be taken into account. There are methods of overcoming chemical interference, such as by complexing the sample and producing a complex such as Pb–EDTA that is independent of the original form of the lead compound.

a. Thermospray Atomizers

An alternate approach is to heat the injected sample solution before it enters the base of the flame. This is done by heating directly as it passed through an electrically heated stainless steel capillary tube (thermospray). The sample is converted into a vaporized jet, which is introduced into the flame. The vaporized jet normally contained a fine mist of sample residues, which constituted an aerosol. The aerosols produced were more easily atomized, which has desirable properties for sample introduction into a flame atomizer. The heating process that occurs within the thermospray capillary tubing has been modeled. Based on these projections, a burner was designed using a thermospray. It is shown in Fig. 7.9.

Construction of Interface. The sample passed through a 0.15 mm i.d. stainless steel capillary, silver soldered and concentric to a tubular electrical wire resistance heater. The capillary was heated by conduction. The current flowing, and therefore the temperature, was controlled with a Variac.

As shown in Fig. 7.9, the flashback safety diaphragm was removed from the burner chamber and the thermospray was inserted so that the jet of vapor produced was directed to the bottom of the burner head. In this way, contact between the vapor and the burner chamber wall was avoided. This minimized possible memory effects and sample loss. The heat from the thermospray was conducted to the burner chamber so that no extra heating was required to prevent vapor condensation. Optimum conditions were determined experimentally.

Table 7.3 Efficiencies of Atomization[a] of Metals in Flames

Metal	Flame	
	Air–C_2H_2	N_2O–C_2H_2
Ag	1.0	0.6
Al	<0.00005	0.2
Au	1.0	0.5
B	<0.0005	0.004
Ba	0.001	0.2
Be	0.00005	0.1
Bi	0.2	0.4
Ca	0.07	0.5
Cd	0.5	0.6
Co	0.3	0.3
Cr	0.07	0.6
Cs	0.7	—
Cu	1.0	0.7
Fe	0.4	0.8
Ga	0.2	0.7
In	0.6	0.9
K	0.4	0.1
Li	0.2	0.4
Mg	0.6	1.0
Mn	0.6	0.8
Na	1.0	1.0
Pb	0.7	0.8
Rb	1.0	—
Si	—	0.06
Sn	0.04	0.8
Sr	0.08	0.03
Ti	—	0.2
Tl	0.5	0.56
V	0.01	0.3
Zn	0.7	0.9

[a]The efficiency of atomization in the flames has been measured as the fraction (atoms + ions) produced ÷ Total element introduced into flame.

c. Effect of Energy Input to the Thermospray

The thermospray produced a jet of vapor containing a fine aerosol. If the energy supplied to the heater was too high, vaporization was completed inside the capillary and superheated vapor emerged. This vapor did not scatter laser light; thus, no track was visible. In this condition no absorption signal was observed. The

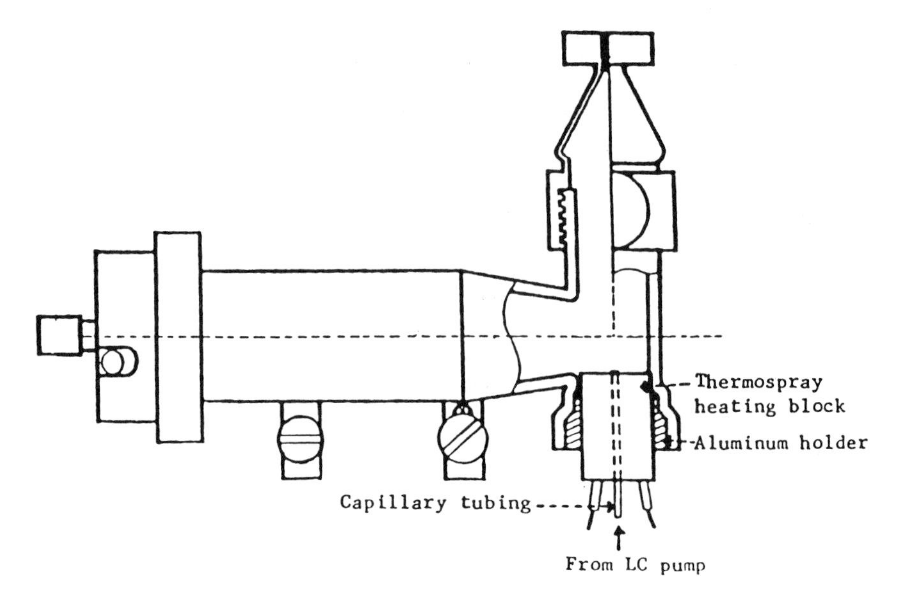

Figure 7.9 Schematic diagram of thermospray interface.

nonvolatile analyte deposited preferentially inside the walls of capillary tubing. Thus, the emerging vapor contained no analytes, explaining why no absorption signal was observed. If the energy supplied to the heater was too low, a portion of liquid was not vaporized and emerged as a wet aerosol, which scattered visible laser radiation strongly. Under optimum condition, the vaporization was completed at the tip of the capillary tubing and a dry aerosol emerged, which barely scattered light but did not condense on a solid surface upon contact. Thus the thermospray acted as a nebulizer and preliminary desolvation apparatus.

It was shown that the optimum performance of the thermospray was observed when the heat supplied resulted in very nearly complete vaporization. At this point, the mean aerosol droplet size was presumed to be the smallest, and the highest atomization signal was observed. At either side of this point, the response signal decreased.

d. Thermospray Flame Atomizer for AA Improved Sensitivity

The pattern of the signal from LC/FAA was somewhat different. As the energy (measured by voltage) to the thermospray (or thermospray tip temperature) increased, the absorption signal from flame atomic absorption (FAA) increased in an exponentially shaped curve. This is shown in Figs. 7.10 and 7.11, which depict the relationship of absorption signal to the voltage applied to the thermo-

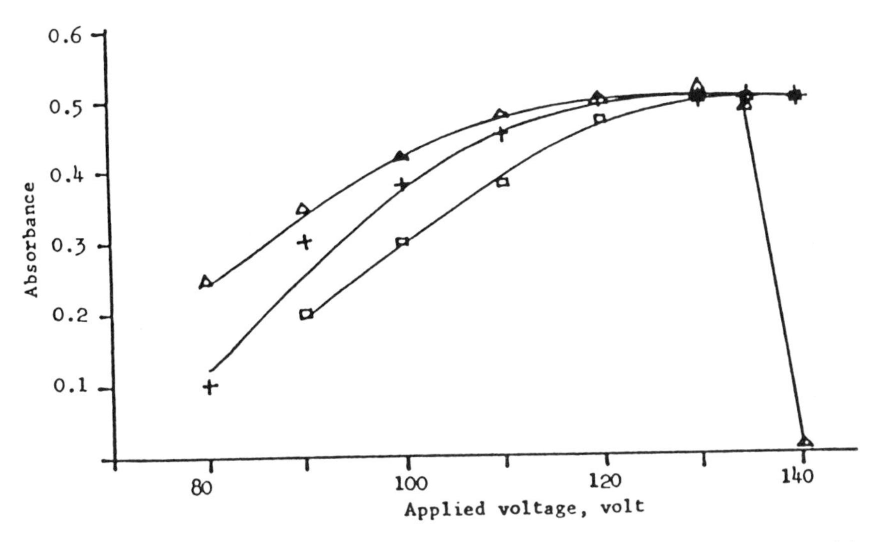

Figure 7.10 Effect of energy input to thermospray nebulizer on absorbance (sensitivity) using different solvents at 1.5 mL/min.□: water; +: 50% methanol in water;△:50% acetonitrile in water.

spray. It was noted that as the voltage increased beyond the maximum vaporization point, the absorption signal rapidly dropped to nearly zero.

e. Elimination of Solvent Effect on Sensitivity

Several different solvents were studied: water, methanol, and acetonitrile. Each solvent has different physical properties, such as boiling point and specific heat, and modifies the vaporization process. Normally, this affects the sensitivity of flame AAS. The results using the thermospray are shown in Figs. 7.10 and 7.11. The optimum applied voltage used. For example, a 50% acetonitrile in water mixture required less energy for steady vaporization than others. Improper choice of energy input to the thermospray therefore may lead to inaccurate results. However, there was a voltage range in which the sensitivity was independent of the solvent system.

The correct choice of voltage to the thermospray produced uniform sensitivity regardless of the solvent used. This eliminated the solvent effect, which is a major source of AA interference at least for these solvents. This enhances the potential of using the system as a device for interfacing HPLC to flame AAS as it would not restrict the solvent used, e.g., gradient elution could be possible without a continuous change in sensitivity. However, for optimum operation, it

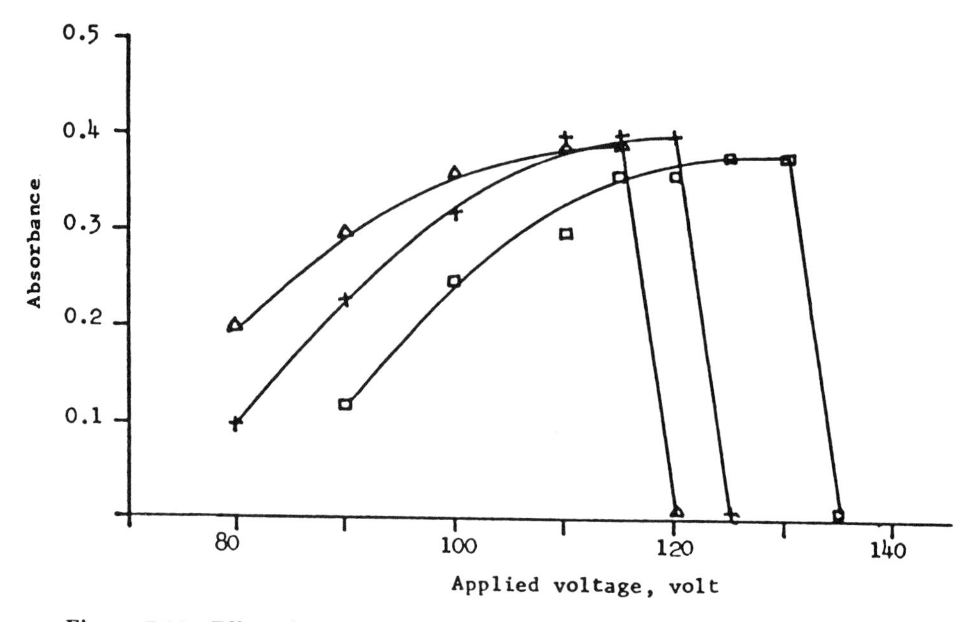

Figure 7.11 Effect of energy input to thermospray nebulizer on absorbance (sensitivity) using different solvents at 1.0 mL/min.□: water; +: 50% methanol in water;△:50% acetonitrile in water.

is necessary to monitor and keep constant the temperature of the tip of thermosprayer to prevent the variation of degree of vaporization with different solvents and flow rates.

f. Peak Broadening, Memory Effects, and Sample Volume

A perennial problem is the loss of chromatographic resolution caused by interfacing LC to FAA. This results in a loss of analytical sensitivity. To measure this effect, peaks were recorded using the same HPLC system but with (a) an AA detector and (b) a UV detector. Copper complex solution (copper dithiazone) was injected, and the absorption signals from UV-VIS detector at 525 nm and flame AAS were compared. The response signals from each detector were recorded at fast chart speed to exaggerate any differences. The attenuation of the detectors were adjusted to produce similar peak height. The results are shown in Fig. 7.12. The two set of peaks were similar, and the peak-broadening effect seemed to be minimal.

FAA sometimes suffers from memory effects due to the sample solution left in the mixing chamber. In this study, the thermosprayer was placed directly below the burner head, permitting direct introduction of the vapor into the flame. Contact of the vapor with the burner chamber was minimized and the memory

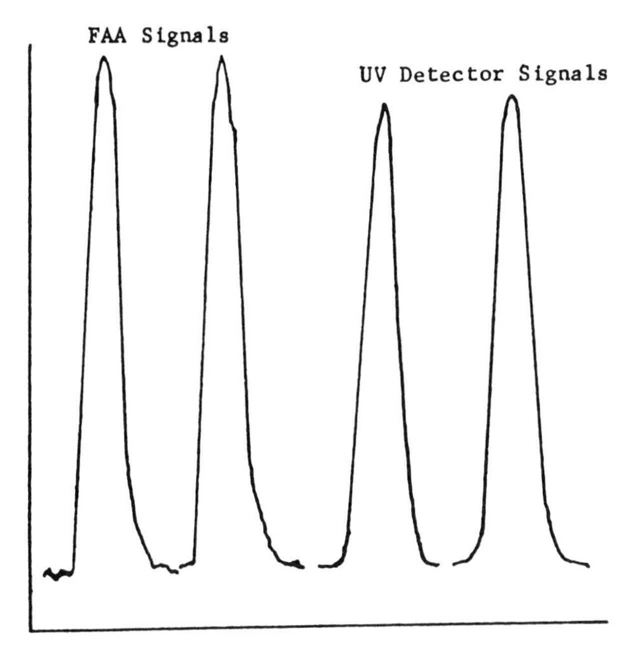

Figure 7.12 Comparison of response signals of thermospray interfaced with (a) FAA and (b) UV detectors.

effects observed were minimal. The results are shown in Fig. 7.13. The peaks were obtained by injecting 2 M HC solution after operating about one hour. The signal fell back to baseline very rapidly and no further significant peaks were observed. Presumably, the vapor is consistently washing off the analyte and did not significantly accumulate in practice.

Figure 7.13 shows calibration curves obtained by this technique. Precision obtained using consequent measurement is less than 3% RSD.

Figure 7.14 shows the relationship between peak height and the volume of sample injected per min. At low injection volumes a linear relationship was observed, but as the injection volumes increased, non-linearity with the peak height was observed. However, the peak area seemed to have a direct relationship with the volume injected. Therefore, if the peak height is used for analysis, a constant volume must be injected, but if peak areas are used, a wide range of volumes is permissible.

g. Response of Different Metals

The behavior of the thermospray varied depending on the metals studied. Generally, the absorption signal was increased as the liquid flow rate was increased. The results were compared with the data from ultrasonic nebulizer and concen-

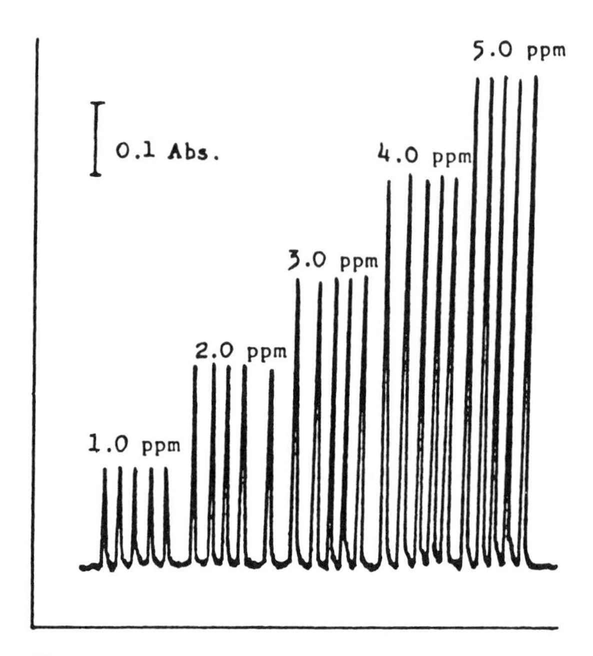

Figure 7.13 Calibration data of Cu at 1.5 ml/min.

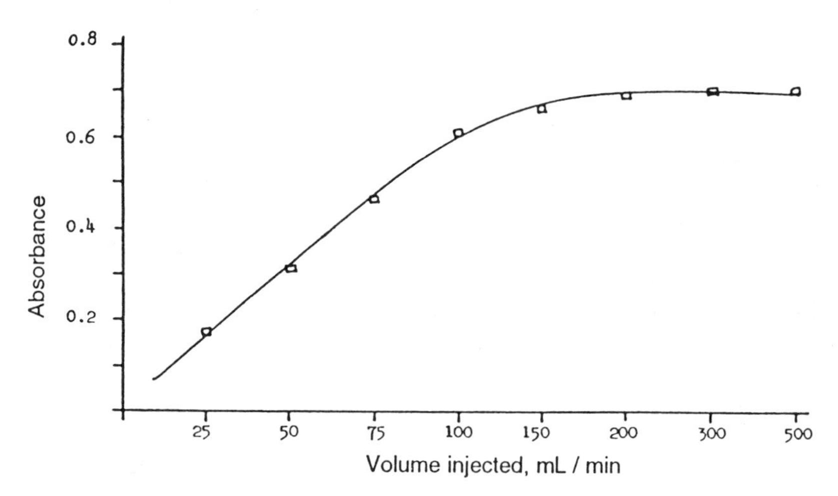

Figure 7.14 Effect of sample volume injected on absorbance of 5.0 ppm Cu at 1.5 ml/min.

Table 7.4 Comparison of Sensitivity[a]

Nebulizer type/Metal	Ag	Ca	Cd	Cu	Mg	Zn
Pneumatic	0.15	0.3	0.06	0.4	0.02	0.03
Ultrasonic	—	0.07	0.01	0.07	0.006	0.007
Thermosprayer	0.04	0.04	0.01	0.03	0.003	0.005

[a]1% Absorption at 1.5 mL/min flow rate (ppm).

tric pneumatic nebulizer and are summarized in Table 7.4. In most of the metals studied, the thermospray gave about 10-fold sensitivity improvement in sensitivity compared to the commercial burner.

It was concluded that a thermospray nebulizer improved the analytical sensitivity of FAA when used in a modified commercial flame atomizer. When used correctly, the sensitivity is independent of the solvent tested. The thermospray was successfully used to interface the flame atomizer with the effluent from an HPLC without loss of chromatographic resolution.

3. Dispersion Elements

The monochromator is required to separate the absorbing line from other spectral lines emitted from the hollow cathode. The most common dispersion elements used in atomic absorption spectroscopy are prisms and gratings. For many elements high dispersion is not necessary, but for a number of transition metals the emission spectrum from the hollow cathode is so complicated and rich that high dispersion is essential to distinguish the resonance absorption line from the other lines emitted by the metal of the hollow cathodes and by excited filler gas (Ar). Instruments that are intended for general use therefore utilize high-dispersion monochromators. Usually gratings are used.

4. Detectors

For UV spectroscopy the common detectors are film and photomultipliers. Film is quite unsuitable for atomic absorption spectroscopy. The signal from the source is intense, and after absorption it is slightly less intense. The resonance line emission from the hollow cathode appears as dark lines on the film and a comparison of the intensities of two dark lines that are generated before and after absorption is difficult and leads to poor accuracy and low precision. Therefore only photomultipliers are used for atomic absorption. They perform satisfactorily and are stable enough to enable us to compare intense lines with each other. Their construction has been described earlier (Chap. 6, C.2.c).

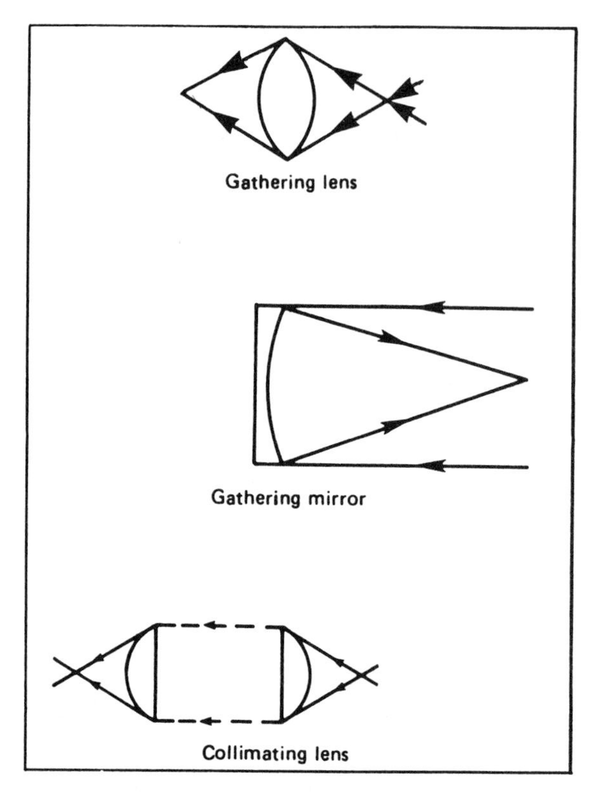

Figure 7.15 Light-gathering systems.

5. Slits and Lenses

Lenses are used to gather and focus the radiation at different parts of the optical system. This avoids losing too much signal as a result of the light beams being nonparallel and focuses the light beam along the flame so that all the light passes through the sample. Any light that does not pass through the sample cannot be absorbed and results in a loss in sensitivity. Quartz lenses have been found satisfactory. Front-faced concave mirrors, which reflect light from their faces and do not absorb much radiation in the process, can be used to some advantage. In a light-gathering system (Fig. 7.15), light from the first lens is focused on the sample and through a second lense to the entrance slit. This slit is used to prevent stray radiation from entering the light path in general and the monochromator in particular. Light passes from the entrance slit to the dispersion element. After dispersion by the dispersion element, the radiation is directed toward the exit slit. At this point the desired absorption line is permitted to pass, but the

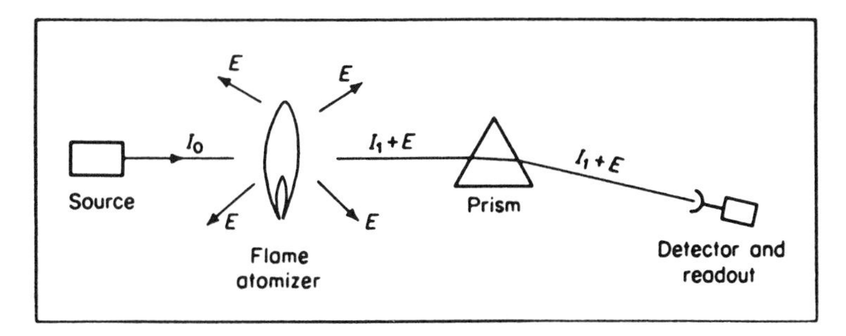

Figure 7.16 Emission by metals at their absorption frequency.

other lines emitted from the source are blocked from proceeding down the light path. They system of slits and dispersion elements enables one to choose the wavelength of radiation that reaches the flame atomizer.

6. Modulation

Many metals, when placed in a flame, emit strongly at the same frequency at which they absorb. The emission signal can cause a serious error when the true absorption is to be measured. This problem is illustrated in Fig. 7.16. When a flame atomizer is used, emission by the metal in the flame is at precisely the same wavelength as the absorption wavelength of the metal because the same electronic transition is involved. Better resolution cannot improve the situation. Furthermore, since we are trying to measure the degree of absorption $I_0 - I_1$, the interference by the emission from the flame will result in a direct error. Unless a correction is made, the signal recorded will be $I_0 - (I_1 + E)$ rather than $I_0 - I_1$, where E is the emission intensity.

The normal procedure in spectroscopy is to measure I_0, with no sample in the light path, and I_1, with the sample in place. For example, if $I_0 = 100$ and $I_1 = 60$, but the sample emits 10 units, the signals would be those shown in Fig. 7.17. The true absorption was 40%, but the simultaneous absorption and emission (of 10 units) produced the erroneous result of 30% absorption.

When modulated, the signal from the source was chopped, making it AC. Hence the signal for both I_0 and I_1, is AC but the sample emission E is DC. Measuring each signal consecutively and continuously produces the signal shown in Fig. 7.18. Its amplitude equals 40, and this is a measure of absorption.

If, as in our example, the sample simultaneously emits 10 units, the signal is DC and is superimposed on I_0 and I_1. The signal becomes that shown in Fig. 7.19. The amplitude is again 40 units, indicating 40% absorption. Whatever the

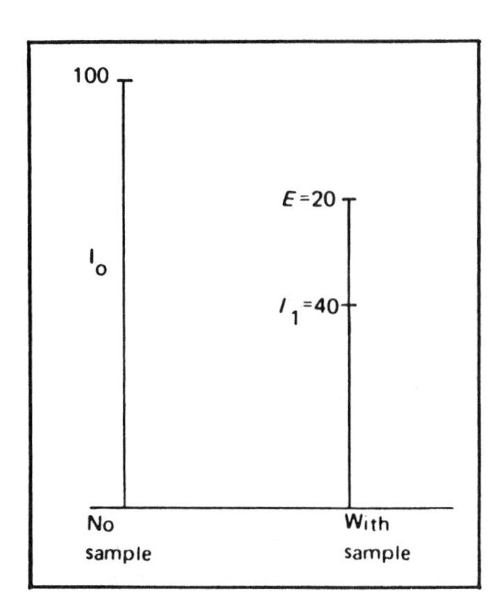

Figure 7.17 Measuring intensity falling on the detector with and without the sample in place. No modulation signals were obtained.

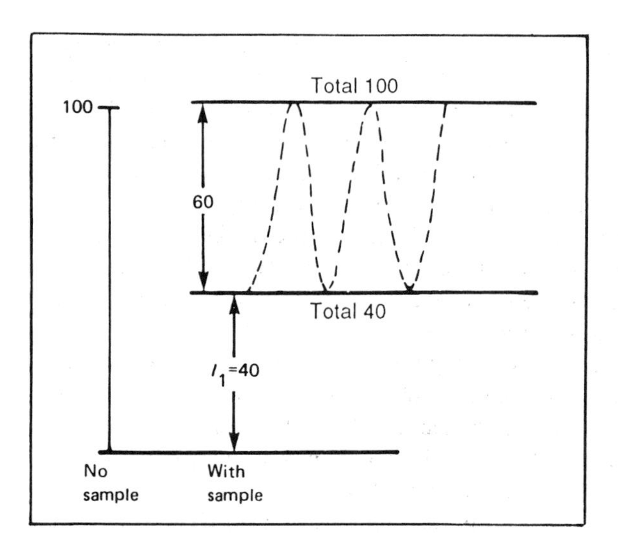

Figure 7.18 Signals obtained with modulation but no emission from the sample. The amplitude of the AC signal is 60, indicating 60% absorption.

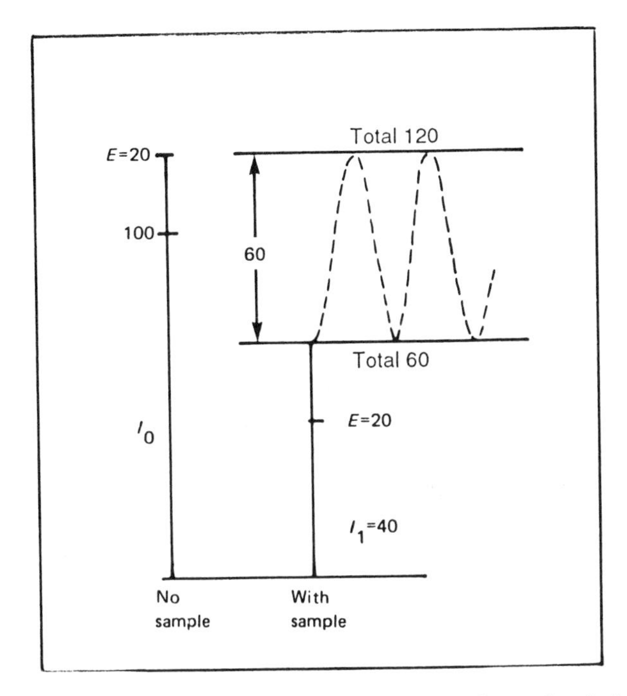

Figure 7.19 Signals obtained with modulation and emission from sample. The amplitude is 60, indicating 60% absorption. The amplitude is independent of emission intensity.

value of E, it is added to I_0 and I_1, and as long as it is constant it does not interfere with the measurement of the absorption signal.

The detector does not register the steady, nonintermittent emission signal from the flame. The net result is that the emission signal from the flame is not registered by the detector and is therefore eliminated as a source of error. Modulation is quite inexpensive and, for most metals, is a necessity if reliable results are to be achieved. It is now designed into virtually all commercial instruments.

C. ANALYTICAL APPLICATIONS

1. Qualitative Analysis

The radiation source used in atomic absorption is a hollow cathode, and a different hollow cathode is needed for each element to be tested. Only that element used in the hollow cathode can be detected with the system. Inasmuch as only one element (sometimes up to three) is usually measured at any one time, atomic absorption spectroscopy is not suitable for qualitative analysis unless specific elements are being tested for. For a sample of unknown composition, other techniques such as X-ray fluorescence, plasma-mass spectrometry, or optical emission spectrography are usually much more useful.

2. Quantitative Analysis

Quantitative measurement is one of the ultimate objectives of analytical chemistry. Atomic absorption is an excellent quantitative method. It is deceptively easy to use, particularly when flame atomizers are utilized. Calibration methods and the reliability of results is treated on page 30.

a. Quantitative Analytical Range

Like all other spectroscopic quantitative analytical procedures, there is a maximum and a minimum to the concentration range of application. The *minimum* of the range is a function of the detection limits of the element under optimum conditions. The ultimate limiting factor is the noise level of the instrument being used. The *maximum* of the analytical range is determined by the degree of absorption by the sample. At high concentrations the degree of absorption is very high. Small changes in concentration of the sample produce virtually no changes in absorption by the sample. Hence, it is difficult or impossible to measure absorption changes caused by concentration changes in the sample. The optimum is determined by the Ringbom plot (see 2.F.3.).

b. Calibration Curves

Calibration curves are prepared from solutions of known concentrations of the sample element. For example, if we wanted to make a calibration curve for copper determination in the range of 2–20 ppm, we would make up several solutions of copper. The concentrations of these solutions must vary regularly over the complete range of 2–20 ppm. The samples would then be atomized and the absorption measured.

Typical results from such a series are shown in Table 7.5. From these results, a calibration curve like that in Fig. 7.20 is prepared. It should be noted that the relationship between absorbance and concentration is linear over the range 2.0–10 ppm, but at higher concentrations the relationship is curved. Nevertheless, provided that the slope of the curve is fairly steep, it is still possible to distinguish between two solutions with similar concentrations, such as 18 and 20 ppm. (The preparation of calibration curves is described in Chap. 2, Sec. F.)

The calibration curve has two limits, the upper limit and the lower limit. The upper limit is reached when a significant increase in copper concentration causes only a slight increase in absorbance which is difficult to measure. The lower limit is controlled by the analytical sensitivity of the particular element being determined. These limits vary from one element to another.

Table 7.5 Absorption Data from Standard Solutions

Concentrations of standard solution of copper (ppm)	Absorbance $A = \log(I_0 / I)$
2	0.13
4	0.25
6	0.35
8	0.48
10	0.60
12	0.70
14	0.79
16	0.87
18	0.94
20	1.00

When a quantitative analysis is to be performed, the sample is atomized and the absorbance measured under exactly the same conditions as those used when the calibration curve was prepared. The concentration of the unknown copper solution is then determined from the calibration curve. For example, suppose that the absorbance reading of the sample solution was 0.60. Using the calibration curve shown in Fig. 7.20, it can be deduced that the concentration of the solution is 8.5 ppm.

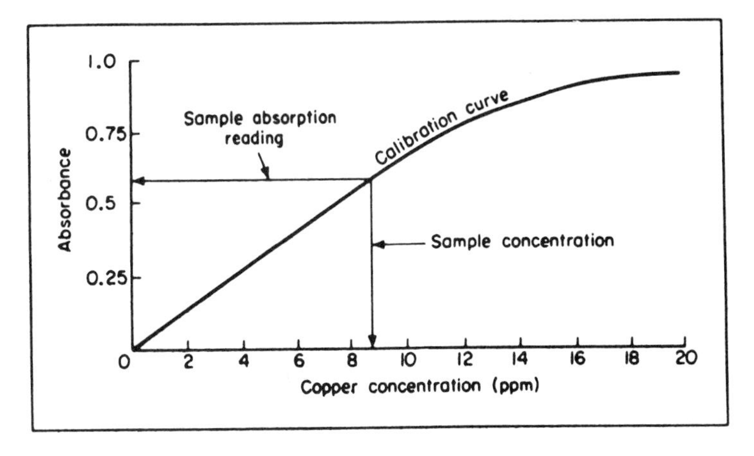

Figure 7.20 Calibration curve for the copper 342.8-nm line.

It is most important in preparing calibration curves that the samples and the standards be as similar as possible. To obtain reliable quantitative data the following must be the same for the sample and standard.

1. The same solvent (e.g., water, alcohol, acetone)
2. The same predominant anion (e.g., sulfate chloride) and the same concentration
3. The same type of flame (N_2O–acetylene, air–H_2)
4. The same pressure in the flame gases
5. Absorption measured at the same height in the flame
6. Background correction should be carried out on each sample using the same correction technique

c. Interferences

Chemical Interferences. The most serious source of interference is chemical interference. This is the effect of the predominant anions present with the metal ions in the sample. The anion affects the stability of the metal compounds formed during atomization, and this, in turn, affects the efficiency with which the atomizer produces metal atoms. For example, a solution of calcium chloride, when atomized, decomposes to calcium atoms more easily than a solution of calcium phosphate, because calcium phosphate is more difficult to break down than calcium chloride. Two solutions with equal concentrations of calcium will therefore absorb different amounts of radiation, depending on the predominant anion present. To compensate for this interference, it is important to prepare calibration curves from solutions that have the same anion present in the same concentrations as in the samples being analyzed.

Solvent Interferences. Another potential source of interference is the solvent. Metals in aqueous solutions invariably give lower absorbance readings than the same concentration of such metals in an organic solvent.

In order to produce free atoms from a solution containing molecular compounds, as opposed to ions, it is first necessary to drive off the solvent, leaving a residue containing the metals of interest. The residue is then decomposed, generating free metal atoms. The rate at which a solvent is driven off depends on the temperature of the atomizer and the volatility of the solvent. If a flame is used, the hotter the flame, the more rapidly the solvent is driven off. If the solvent is water, evaporation is relatively slow and tends to decrease the temperature of the flame. This in turn retards the entire atomization process. If the solvent is an organic solvent, such as acetone, alcohol, ether, or a hydrocarbon, the solvent not only evaporates rapidly, but may also burn, thus increasing the flame temperature. The atomization process is more efficient. More free atoms are produced from this system. An higher absorption signal is registered from organic

Table 7.6 Interferences in Solvent Combustion

	step 1	step 2	step 3	step 4
Combustion of an Aqueous Solvent				
Ions in water →	Evaporation leaving a hydrated residue →	Residue (dehydrated atoms formed) →	Excited atoms (emission) Neutral atoms (absorption) →	Oxide formation Loss of atoms forming oxide
Combustion of an Organic Solvent				
Metals in organic solvent →	Solvent burns →	Organic addend burns →	Excited atoms (emission) Neutral atoms (absorption) →	Oxide formation

solvents than from aqueous solutions, even though the metal concentration in the two solutions is equal. This process is shown in Table 7.6.

Flame Interference. Radiation interferences occur when absorption of the hollow cathode resonance line occurs by species other than the element being determined. For example, the Pb 217.0-nm line may be absorbed by the components of a flame even though no lead is present in the sample or in the flame.

The resonance absorption lines of the various elements are very narrow and at discrete wavelengths. Overlap between absorption lines of different elements is so rare that it can usually be ignored as a source of error. Absorption by the wings of the absorption lines of interfering elements present in high concentrations has been observed, but this is also a rare occurrence.

A much more common occurrence is absorption by molecules or fragments of molecules in the flame. This type of interference is more commonplace at short wavelengths (<250 nm), where many compounds absorb. Incompletely burned organic molecules in the atomizer can cause serious problems. If a flame atomizer is used, incomplete combustion of the solvent may take place, particularly if the flame is too lean. The extent of the interference depends on flame conditions (lean or oxidizing), the flame temperature used, the solvent used, and the sample feed rate. Interference is much more severe when carbon atomizers are used.

3. Background Absorption and Its Correction

We have seen that atoms absorb over a very narrow wavelength range and that overlapping absorption from other atoms of the resonance lines is extremely unlikely. However, it is not uncommon for molecular absorption to be observed in flame atomizers, and particularly the more recent carbon atomizers. The molecular absorption may come from hydroxyl ions generated from water in the flame, incompletely burned organic solvent, residues, metal oxides, and so on. If these broad absorption bands overlap the atomic absorption lines, they will absorb the resonance line from the hollow cathodes and cause direct interference with the data. There are several ways of measuring this background and therefore correcting for it.

a. Blank Specimen

The easiest method is to take a "blank" sample that is similar to the sample being analyzed in every way except that it does not contain the element being determined. Frequently the solvent used in the sample can be used as the blank. This blank is aspirated into the atomizer and the molecular absorption at the pertinent wavelength is measured directly. The extent of the interference depends on the flame temperature, the solvent used, and the sample feed rate. Consequently, the flame conditions, solvent, and solvent feed rate must be constant for both blank and measured sample. The value is then used to correct the absorption observed in the sample, which is the sum of the atomic absorption plus the background absorption. No special equipment is necessary for this correction, but the method suffers from the problem that it is sometimes difficult to get a blank specimen that exactly duplicates the sample with all other dissolved materials present.

b. Deuterium or Hydrogen Lamp Background Corrector

In this method a hollow cathode lamp is used to measure the total signal of atomic absorption plus background absorption. The hollow cathode is then removed and a hydrogen lamp is put in its place, but the monochromator and slit systems are left untouched. This allows radiation from the hydrogen lamp at the same wavelength as the resonance line to reach the detector. However, with the normal slit system the spectral slit width will be about 0.2 nm wide. The atomic absorption line has a total width of about 0.002 nm. Consequently, if the atoms absorb all of the radiation over that linewidth, they will absorb only 1% of the radiation from the hydrogen lamp falling on the detector. All light within the 0.20-nm bandwidth, but not within the 0.002-nm absorption line, will reach the detector and not be absorbed by the sample. Consequently, the effect of atomic absorption on the absorption from the hydrogen lamp is negligible. Any absorption of the radiation from the hydrogen lamp observed is broad banded, since it will be absorbed over the entire 0.2 nm. This is therefore a fairly accurate measure of background absorption. An advantage of this method is that the background is measured at exactly the same wavelength as the resonance line. The system is illustrated in Fig. 7.21.

A further advantage of this system is that it can be automatic. Commercial instruments now commonly use the hollow cathode lamp and the hydrogen lamp

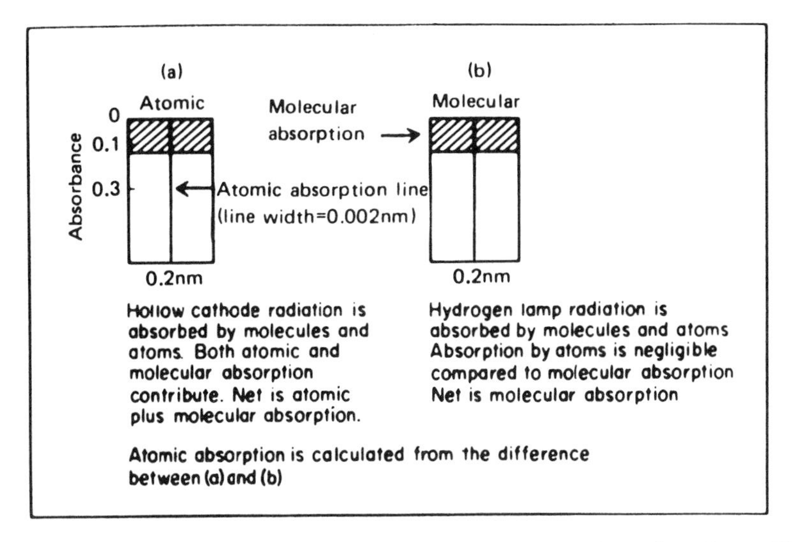

Figure 7.21 Correction for molecular absorption. (Reprinted from James W. Robinson, *Atomic Absorption Spectroscopy*, 2nd ed., Marcel Dekker, New York, 1975, p. 65.)

simultaneously in sequence, and the detector is able to measure the emission from each source and the absorption of each source, subtract the two readings from each other, and read out the net atomic absorption signal. Automatic background correction is a vital feature of carbon atomizers that will be discussed later.

A third method of measuring background correction that is often very convenient but less accurate uses the absorption of a nearby nonresonance line. The emission spectrum from a hollow cathode is quite rich and contains the resonance line of the element of interest plus many other lines. These other lines are not absorbed by the atoms being determined because they are not resonance lines; however, they are absorbed by molecular background. It is therefore convenient to measure the absorption of the nonresonance line closest to the resonance line, which may be slightly different from the resonance line and may introduce a small error; however, it remains a very convenient method since no change in light source, sample preparation, or complicated instrumentation is necessary.

c. Zeeman Background Correction

When an element such as mercury emits or absorbs radiation at a resonance line (e.g., 253.3 nm for Hg), the "absorption line" actually consists of several absorption lines, which are very close together and are characteristic of each iso-

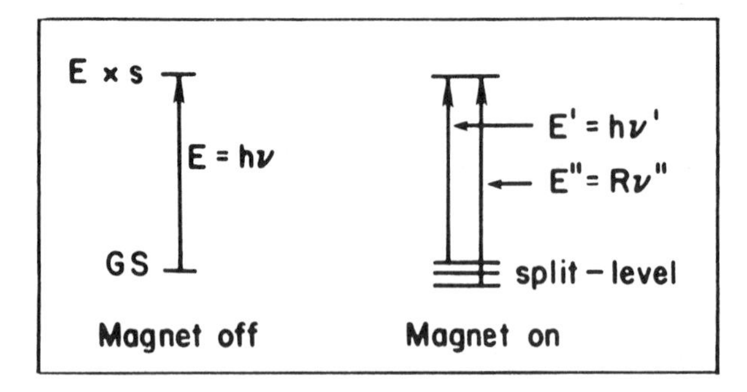

Figure 7.22 Zeeman background correction. Energy levels of valence electrons of atoms in a magnetic field and out of a magnetic field.

tope of mercury. If we direct our attention to one isotope, such as ^{204}Hg, it emits and absorbs at only one characteristic wavelength.

However, if we place the Hg lamp source in a magnetic field, the single line is split into two components by the Zeeman effect. The shift is about 10^{-5} nm as illustrated in Figs. 7.22 and 7.23.

Typically the strength of the magnetic fields is between 7 and 15 kgauss. The magnetic field has two important effects on the emission line. In the simplest case it causes splitting into two components: a π component, at the same wavelength as before, and a σ component, which undergoes a positive and negative shift in wavelength. The magnetic field also polarizes the light so the π component goes in one direction, the σ component in the opposite direction (see Fig. 7.23).

The unshifted radiation (no magnetic field) is at the natural wavelength of the metal resonance line and is absorbed by the Hg atoms of the sample and the molecular absorption. The Zeeman-shifted radiation (Fig. 7.23) (π component) is polarized and is not absorbed by the Hg atoms. However, this radiation is absorbed by the background molecular absorption. The difference between the two absorption signals is the net atomic absorption signal, which is thereby corrected for the molecular absorption signal.

Experimentally this has been achieved in two ways. First, the applied magnetic field may surround the hollow cathode and be alternating, causing splitting and nonsplitting at its frequency. By tuning the amplifier to this frequency it is possible to discriminate between the split and unsplit radiation. A major difficulty with the technique is that the magnetic field used to generate Zeeman splitting also interacts with the ions in the hollow cathode. This causes the emission from the hollow cathode to be erratic, which in turn introduces imprecision into the procedure. A better way is to put the magnetic field around the atomizer. The atoms absorb with no magnet on, but not with the magnet turned on.

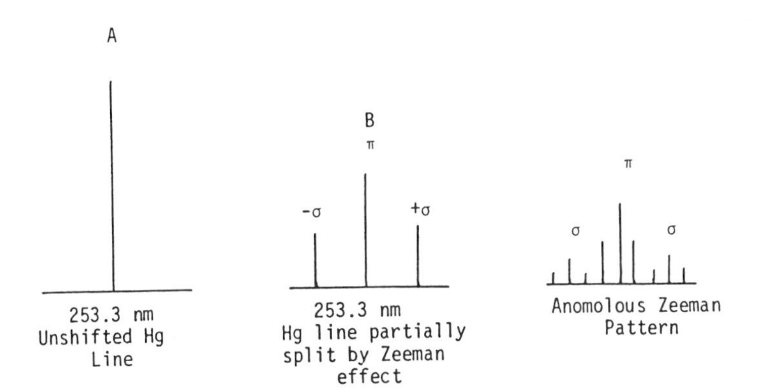

Figure 7.23 Zeeman effect causing shifting of emission lines. π component is at the original wavelength; the ±σ components are shifted upfield and downfield. In the anomolous pattern there is further splitting. Emission spectrum of Hg 253.3 nm (A) unshifted and (B) after putting the source in a magnetic field.

A second method is to take advantage of the fact that the split and unsplit light is polarized differently. The magnets may be run A/C or DC. If it is run DC, the π and σ components can be told apart using a polarizing filter. If the field is A/C, a fused silica filter is used. Fused quartz under pressure is birefringent, that is, light is diffracted differently when its axis of polarization is different. The quartz may be stressed at a controlled frequency to yield two axes of polarization and the two beams of light discriminated. By suitable amplification the background absorption—measured by Zeeman-shifted radiation—can be corrected for automatically.

The advantage of this technique is that only one source and only one detector are used. The background correction is made at a wavelength very close to the resonance line and is therefore accurate. The physical paths of the split and unsplit light are identical, as opposed to the use of the D_2 lamp, which may not pass along the identical path as the hollow cathode. Finally, the system has the advantage of being compact and relatively easy to operate.

The complexity of the emission or absorption spectra after Zeeman splitting depends on the magnetic moment of the nucleus and I_1, the spin quantum number. In its simplest form the original line is split into one π and two ν components. With anomolous atoms there may be several π and σ components. It is not uncommon for some isotopes of an element to absorb but not others. For this reason, the optimum wavelength for absorption measurements may be changed and the analytical sensitivity may be reduced. Measured values of Zeeman sensitivity rations (R_z are shown in Table 7.7.

Table 7.7 Measured Values of Zeeman Sensitivity Ratios

Element	Wavelength (nm)	Lamp current (mA)	R_z(%)	Element	Wavelength (nm)	Lamp current (mA)	R_z(%)
Ag	328.1	3	80	Nb[a]	334.9	20	80
Al[a]	309.3	5	75	Nd[a]	492.5	10	93
As	193.7	7	78	Ni	232.0	5	75
Au	242.8	4	72	Os[a]	290.9	20	87
B[a]	249.8	15	53	Pb	217.0	5	58
Ba	553.6	10	81		283.3	5	78
Be[a]	234.9	20	48	Pd	244.8	5	78
Bi	223.1	8	62	Pr[a]	495.1	8	85
Ca	422.7	3	84	Pt	265.9	10	70
Cd	228.8	3	77	Rb	780.0	15	86
Co	240.7	5	78	Re[a]	346.0	20	64
Cr	357.9	5	91	Rh	343.5	5	87
Cs	852.1	20	48	Ru	349.9	10	85
Cu	324.8	3	44	Sb	217.6	10	90
	327.4	3	70	Sc[a]	391.2	10	89
Dy[a]	421.2	15	89	Se	196.0	10	74
Er[a]	400.8	10	88	Si[a]	251.6	15	85
Eu[a]	459.4	10	88	Sm[a]	429.7	10	94
Fe	248.3	5	88	Sn[a]	235.5	5	67
Ga[a]	294.4	4	76		286.3	5	77
Gd[a]	368.4	25	88	Sr	460.7	10	80
Ge[a]	265.2	5	87	Ta[a]	271.5	20	75
Hf[a]	307.3	10	92	Tb[a]	432.7	15	88
Hg	253.7	3	67	Te	214.3	8	93
Ho[a]	410.4	15	85	Ti[a]	364.3	20	77
In	303.9	5	39	Tl	276.8	20	69
Ir	208.9	20	88	Tm[a]	371.8	15	80
K	766.5	5	91	U[a]	358.5	20	92
La[a]	550.1	10	82	V[a]	318.5	10	75
Li	670.8	5	36	W[a]	255.1	20	81
Lu[a]	336.0	10	88	Y[a]	410.2	10	88
Mg	285.2	3	58	Yb[a]	398.8	5	91
Mn	279.5	5	80	Zn	213.9	5	60
Mo	313.3	10	92	Zr[a]	360.1	20	86
Na	589.0	5	88				

[a]Determined in a nitrous oxide-acetylene flame. All other elements determined in air-acetylene.

The Zeeman sensitivity ratio is derived as follows: Beer's law states that

$$A_N = \log \frac{I_0}{I_1}$$

where

A_N = normal absorbance

but

$$A_z = \log \frac{I_H}{I_1}$$

where
A_z = Zeeman absorbance
I_H = measured intensity with magnet on
I_1 = measured intensity with magnet off

From this,

$$R_z = \frac{A_z}{A_N}$$

where
R = Zeeman sensitivity ratio

The use of a modulated field instead of a fixed field gave a better sensitivity for elements exhibiting anomolous splitting. Also as the magnetic field increased, the splitting of the π component increased causing an increase in R_z. This is in contrast to a fixed magnetic field where increased field strength increases splitting of the v component resulting in reduced R_z values. These effects are complicated and reduced if several isotopes are present or if hyperfine structure develops because of possible overlap of lines (Fig. 7.24).

The R_z values is also dependent on lamp current because this affects self-absorption in the lamp and therefore line shape. Narrower lines give better R_z values (Fig. 7.25).

It has also been established that if $R_z \leq 100\%$ and there is a component of unabsorbed light, then absorbance measurements will reach a maximum and reverse as concentrations increase (reflex) (Fig. 7.26). Reflexing is increased as the magnetic field increases. The problem of reflexing can be serious because two concentrations may give the same absorbance value. Steps must be taken to ensure that this error is eliminated.

The use of Zeeman effect also affects the noise level of the signal. Photon noise is increased, the effects of lamp flicker decreased, and flame noise decreased. This is because in a sense the Zeeman corrector system acts as a double-beam system in measuring sample and background signals.

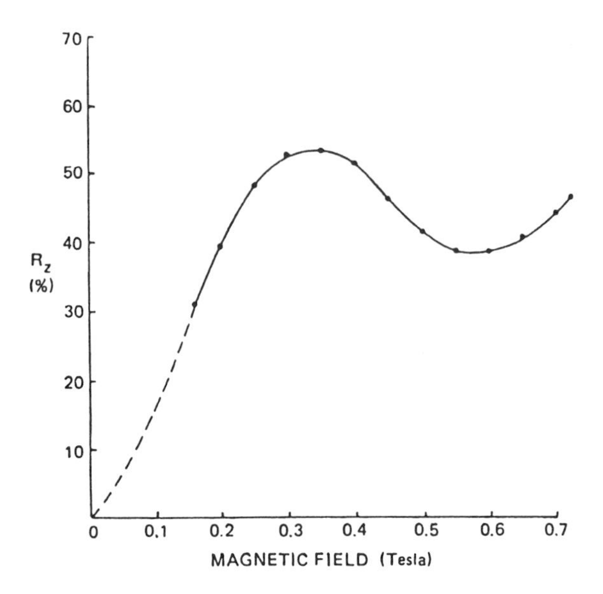

Figure 7.24 Relationship between R_z and magnetic field for copper. (Courtesy of Perkin-Elmer Corp.)

d. The Smith-Hiefje Background Corrector

It will be remembered that the hollow cathode functions by the creation of excited atoms which radiate at the desired resonance wavelengths. After radiating, the atoms form a cloud of neutral atoms which, unless removed, will absorb resonance radiation from other emitting atoms.

If the hollow cathode is run at a high voltage (overrun), an abundance of free atoms form. These free atoms absorb at precisely the resonance lines the hollow cathode is intended to emit. Such absorption is not easily detectable, because it is at the very center of the emitted resonance line and very difficult to resolve. It is, of course, radiation which is most easily absorbed by the atoms of the sample. In practice, if the hollow cathode is overrun, it affects the slope of the absorption-versus-concentration calibration curve (Fig. 7.27).

The Heifje-Smith background corrector has taken advantage of this phenomenon by alternately greatly overrunning the hollow cathode and then underrunning it. When the hollow cathode is underrun, a normal resonance line is emitted and the sample undergoes normal atomic absorption. When the hollow cathode is overrun, the center is self-absorbed, leaving only the wings of the emitted resonance line. Such a line is scarcely absorbed by the atoms of the sample. The wings of the line, however, are absorbed by the background. Conse-

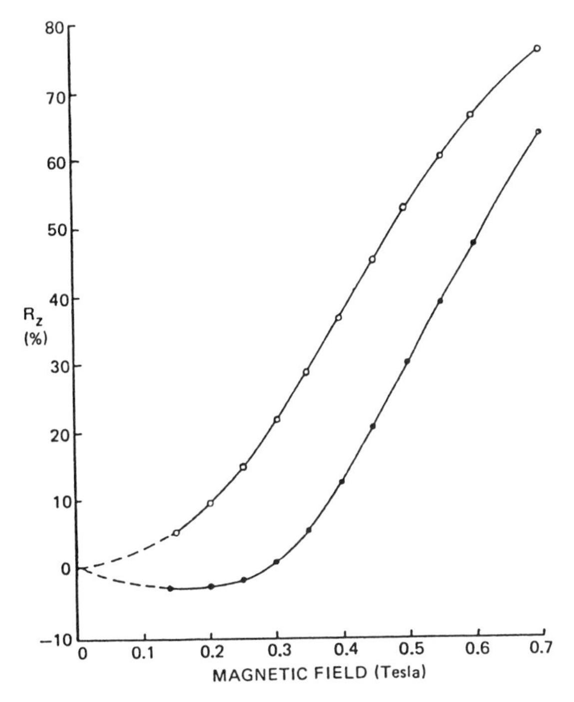

Figure 7.25 Relationship between R_Z and magnetic field for zinc at different lamp currents: (○) 3 mA; (●) 6mA. (Courtesy of Perkin-Elmer Corp.)

quently, the absorption of the wings is a direct measurement of the background absorption at the wavelength of the atomic resonance absorption.

By using a series of calibration curves the procedure can be used to automatically correct for background. In practice the low current is run for a fairly short period of time and measured. The high current is run for a very short, sharp burst, liberating intense emission and free atoms inside the hollow cathode. There is then a delay time to disperse the free atoms before the cycle is started again.

The advantage of the method is that it can be used in single-beam optics; it is not critical to align the beam measuring background absorption and atomic absorption since this is the same beam. This, of course, is in contrast to the use of the deuterium lamp, which uses different lamps for the two different measurements. In addition, the electronics is much simpler than that used in a Zeeman background correction system, where polarization is taken care of. An excellent comparison of the methods used for background corrector has been reported by Camrick and Slavin.

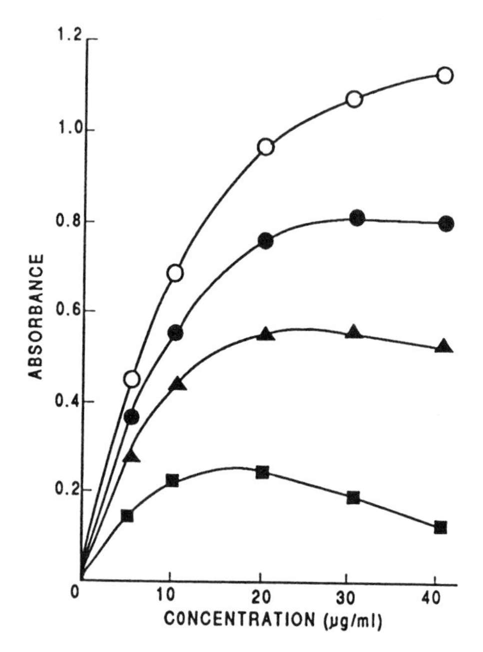

Figure 7.26 Analytical working curves for manganese at different magnetic fields: (○) A_N; (●) A_z at 0.7 T, (▲) A_z at 0.5 T, (■) A_z at 0.3 T. (Courtesy of Perkin-Elmer Corp.)

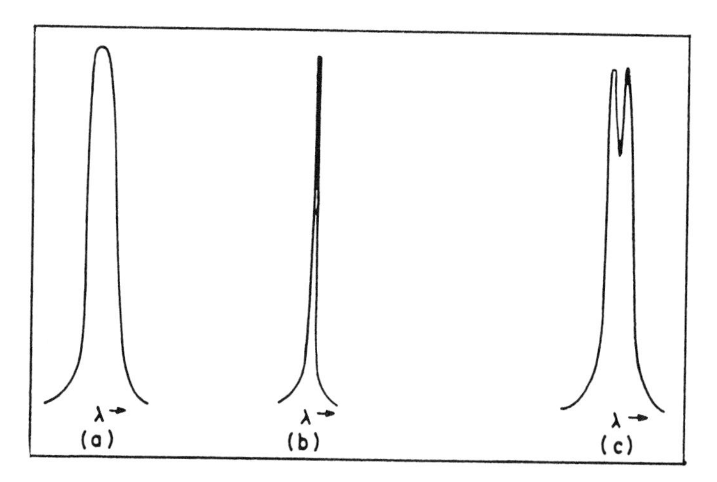

Figure 7.27 Distortion of spectral line shape in a hollow cathode. (a) Shape of spectral line emitted by a hollow cathode. (b) Shape of spectral energy band absorbed by cool atoms in a hollow cathode. (c) Shape of net signal emerging from a hollow cathode.

Table 7.8 Effect of Spectral Slit Width on Sensitivity

Mechanical slit width (mm)	Spectral slit width (Å) (halfwave)	Sensitivity (ppm, 1% absorption)
0.015	3.2 (measured)	5
0.05	3.2 (measured)	5
0.1	4.7 (measured)	6
0.4	7.5 (measured)	7
0.6	11.5 (measured)	10
1.0	19.5 (calculated)	50
1.5	29.5 (calculated)	100

e. Changes in Sensitivity with Unabsorbed Background Emission Radiation

Even if there is no emitted line from the hollow cathode close to the resonance line, there is always a small amount of background radiation that cannot be resolved from the resonance line. The effect of this unabsorbed radiation is exactly the same as that of an unabsorbed emission line. In general, the intensity is low and no serious error is involved. However, as a matter of principle, to avoid any unsuspected problem it is better to operate at as narrow a spectral slit width as is possible. An illustration of the effect of increasing slit width on sensitivity is shown in Table 7.8. There is rapid loss in sensitivity as the mechanical slit width, and hence the spectral slit width, is increased. Under there circumstances, sensitivity is worse and the calibration curve is much flatter than desirable. If it is not possible to resolve two lines that are very close together, the problem sometimes can be overcome if the unabsorbed line originates with the filler gas rather than the metal. In this case a different filler gas such as helium or neon can be used. But if the unabsorbed line is from the element being analyzed, then the only other alternative is to move to a different resonance line which, hopefully, will not be close to an unabsorbable line from the spectrum.

In commercial equipment fixed slits are often used. In this case the manufacturers have built in a fixed mechanical slit width that is considered acceptable for most purposes. For most cases, particularly research work, a manually adjustable slit is desirable where the slit width can be increased or decreased at will depending on the particular sample being analyzed or studied.

It is always good policy to operate with a maximum of dispersion and a minimum of spectral slit width. But there is no hard and fast rule concerning dispersion and slit width that applies to all elements. The spectrum from each element and filler gas must be examined separately. If there is an unabsorbable line in the immediate vicinity of the resonance line, then some other precaution must be taken—such as changing the filler gas or moving to another resonance line.

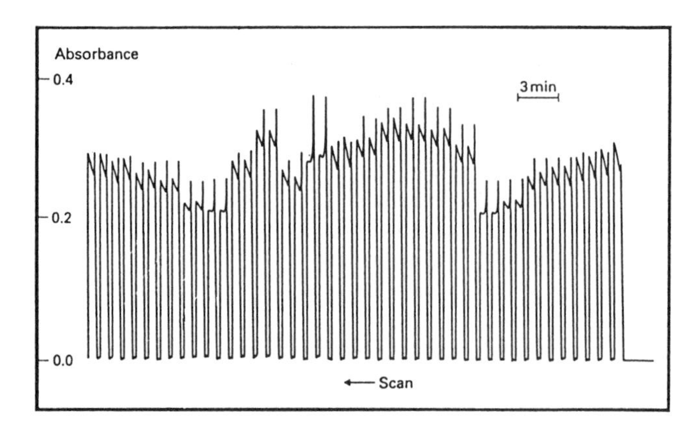

Figure 7.28 Stopped-flow determination of alcohol in blood. Series is six standards, 10 samples, and six standards, all in duplicate.

4. Typical Analyses

a. Flow Injection Systems

One of the cherished aims of analytical chemists is to provide a system that will automatically carry out many analyses consecutively without manned supervision. For example, the analyst may take test tube samples for 40 or 50 samples, load these into a rack, and the instrument will automatically take out a prescribed amount from each and complete the analysis, repeating the performance until all samples are analyzed. Results are then printed out automatically and without attention. An ultimate goal is to perform automatically continual analysis on samples such as plant streams.

An important step in achieving these objectives was reached with flow injection analysis. In this system the solvent is fed continuously into the atomizer. At predetermined intervals, a measured sample is injected into the solvent stream where it is swept into the atomizer. Absorption measurements are made and using a program that includes a calibration curve, the concentration is calculated and reported on a tape. This is a continuous flow system. If a chemical treatment step is necessary, such as dilution or the generation of a metal hydride, a *stop-flow* system is used. This distorts the signals somewhat, but the results are quite acceptable. Some typical results obtained using flow injection and stopped flow are shown in Figs. 7.28 and 7.29.

b. Analysis of Solid Samples

For many years, solid samples were analyzed by dissolving and analyzing the solution. This process is time consuming and entails the possibility of introduc-

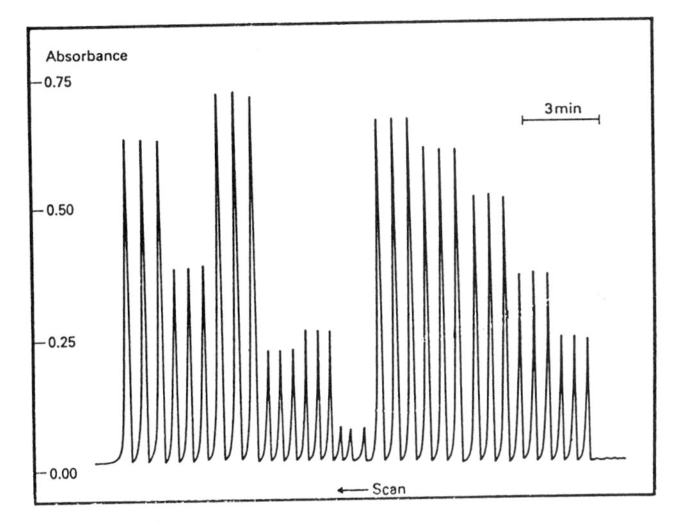

Figure 7.29 Continuous-flow determination of alcohol in beverages. Series is left-to-right, six samples followed by six standards (0–0.4%), all in triplicate.

ing impurities and causing error, particularly at low levels. Types of samples that have been analyzed after dissolution include metals, alloys, soils, animal tissue, plant material, fertilizer, ores, cements, and ash.

One approach to overcoming this problem was to form a slurry with a suitable solvent and introduce the slurry directly into the flame. A key step was to dispense the sample with an ultrasonic vibration during sample introduction. Commercial accessories to do this were available from Perkin-Elmer Corp.

A more recent approach is to use an Atomsource Atomizer from Analyte Corp. This is a vacuum chamber into which the sample is placed and bombarded with six streams of ionized argon.

This process sputters the sample, releasing free atoms into the light path. The method does not rely on a high temperature for atomization, and refractory compounds or elements such as boron, tungsten, zirconium, niobium, uranium, etc. can be easily determined as well as less fractory materials. This technique offers a real advance in solids analysis. A schematic diagram of the equipment is shown in Fig. 7.30.

c. Liquids

Frequently liquids can be analyzed directly. Typical samples that have been analyzed directly include blood, urine, electroplating solutions, petroleum products, wines, and pollutants in water. A calibration curve should always be prepared from a solution of the same solvent as the sample. If the samples are too concentrated, they may be diluted prior to analysis. If they are too dilute,

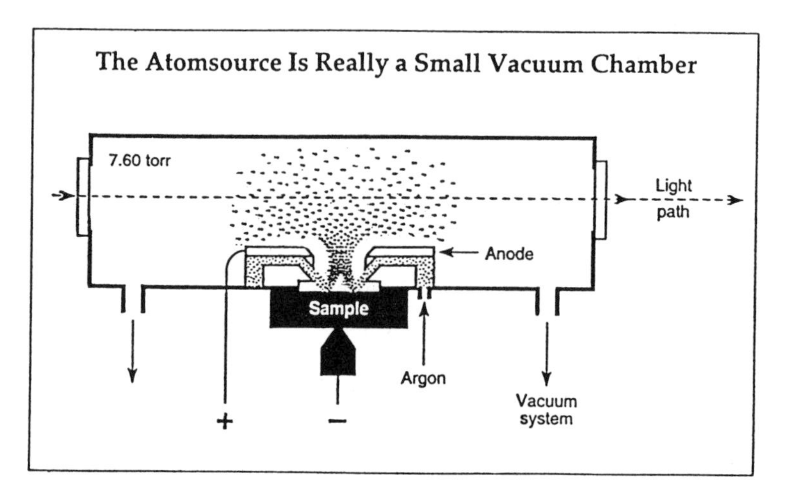

Figure 7.30 Six jets of ionized argon bombard the solid sample and sputter atoms into the sample beam where they absorb light. Detection limits for elements in the solid vary from 0.2 to 2 ppm. The system is remarkably free from background and matrix interferences. (Courtesy of Analyte Corp.)

they may be evaporated down or concentrated by solvent extraction. In all cases the calibration curve must be prepared the same way as the sample.

d. Gas Samples

In conventional samples the metal components must first be collected from gas samples by absorption or by trapping in a solution. The absorbant or solution may then be analyzed. Metals in air samples have been analyzed by atomic absorption after first trapping the metal component in a suitable solvent.

e. Hydride Analysis

Some metals are first converted to a gaseous hydride. These are introduced directly into the atomizer and analyzed the normal way using special equipment. Metals commonly analyzed this way include arsenic, selenium, and tellurium.

5. Sensitivity Limits

Sensitivity is defined as that concentration that gives 1% absorption of the signal. The elements that can be determined by atomic absorption are shown in Table 7.9. Shown are the lowest concentrations of these elements that have been detected experimentally. These sensitivities are reported by different workers in the field and may vary somewhat with equipment and experimental conditions. In general, quantitative analysis can be carried out on samples that contain be-

Table 7.9 Sensitivity Limits, 1% Absorption (Including Nitrous Oxide–Acetylene Flame)

Element	λ (nm)	Sensitivity (ppm)	Element	λ (nm)	Sensitivity (ppm)
Al	309.2	1.0	Mo	313.3	0.1
Sb	217.6	0.1	Nd	463.4	10.0
As	193.7	1.0	Ni	232.0	0.1
Ba	553.5	0.2	Nb	334.9	20.0
Be	234.9	0.1	Pd	247.6	0.5
Bi	223.1	0.1	Pt	265.9	1.0
B	249.7	30.0	K	766.5	0.1
Ca	442.7	0.05	Pr	495.1	10.0
Cs	852.1	0.1	Re	346.0	15.0
Cr	357.9	0.1	Rh	343.5	0.1
Co	240.7	0.1	Rb	780.0	0.1
Cu	324.7	0.1	Ru	349.9	0.8
Dy	421.2	1.0	Sm	429.7	10.0
Er	400.8	1.0	Sc	391.2	1.0
Eu	459.4	2.0	Se	196.1	1.0
Gd	368.4	20.0	Si	251.6	0.8
Ga	287.4	1.0	Ag	328.1	2.5
Ge	265.2	2.0	Na	589.0	2.5
Au	242.8	1.0	Sr	460.7	0.1
Hf	307.2	10.0	Ta	471.4	10.0
Ho	410.4	2.0	Te	214.3	0.5
In	304.0	0.1	Th	377.6	0.4
Fe	248.3	0.1	Sn	235.4	0.5
La	392.8	75.	Ti	364.3	1.0
Pb	217.0	0.05	W	400.9	1.0
Li	670.7	0.03	U	351.5	100.
Mg	285.2	0.001	V	318.4	1.0
Mn	279.5	0.05	Y	398.8	2.0
Hg	253.7	1.0	Zn	218.9	0.01
			Zr	360.1	50.0

tween 5 and 10 times the amount shown as the sensitivity level. For example, the sensitivity level for copper is 0.1 ppm. We should be able to get accurate determinations at levels greater than 1.0 ppm. Estimations can be made between 0.1 and 1.0 ppm, but the results are less reliable.

In summary, atomic absorption provides an accurate and sensitive method of elemental analysis of many elements in the periodic table, particularly met als. It does not generally indicate the molecular form of the element. The method is sensitive, and since it is subject to few interferences, it is usually more accurate than many analytical procedures used for similar determinations.

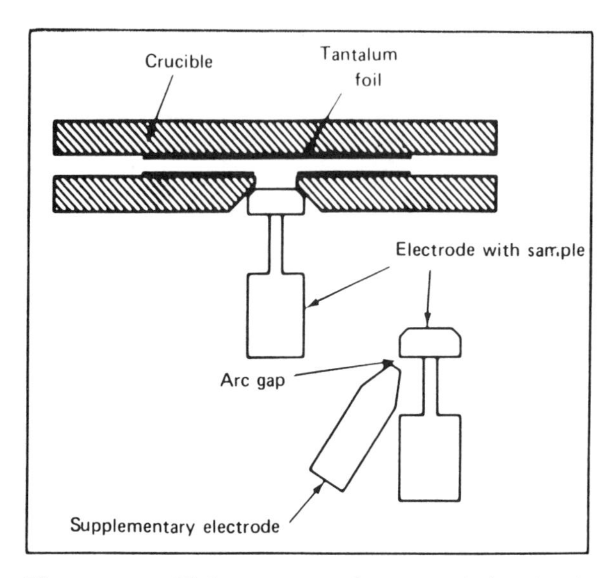

Figure 7.31 High-temperature furnace as designed and used by L'vov.

D. CARBON ATOMIZERS

In 1961 B. V. L'vov built an atomizer using a carbon rod heated with an electrical discharge. This is illustrated in Fig. 7.31. His system was orders of magnitude more sensitive than flame atomizers, but was difficult to control enough to provide quantitative data. The method was refined by other workers in the field, particularly West, Massman, and Robinson. Two typical commercial atomizers are illustrated in Figs. 7.32 and 7.33. In each atomizer the process used was similar. A small sample, usually of the order of 2–30 μl, was loaded onto the carbon atomizer. It is then warmed gently to remove the solvent. The temperature was then increased under controlled conditions to ash the sample and remove most of the organic material present. Finally the sample was heated rapidly to very high temperatures to cause atomization. The free atoms were vaporized from the carbon atomizer into the optical light path, where their absorption is measured.

The process of atomization is extremely fast and must be rigidly controlled. The temperature program is therefore very carefully controlled, both with respect to the times used for each section of the heating program and the temperature range involved in each section. It is vital to avoid loss of sample during the first two stages but it is also extremely important to eliminate as much organic and other volatile matrix material as possible. The final measurement is taken in a very short period of time (less than 1 sec), so recording of the absorption must be very fast in order to be useful. This has been done by improvement of equip-

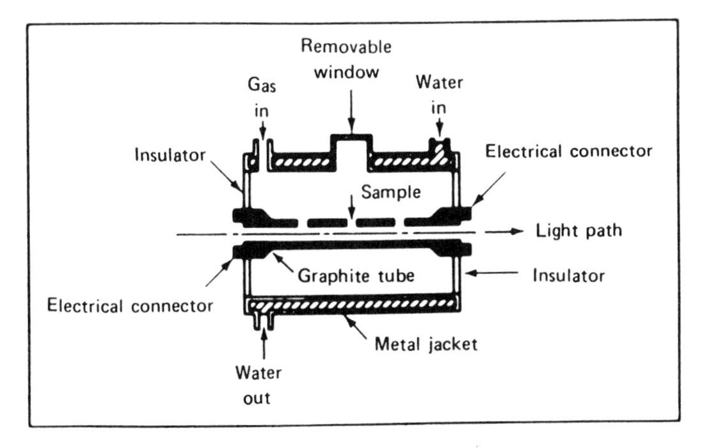

Figure 7.32 Heated graphite tube atomizer.

ment utilizing fast detector systems and a computerized readout system. The operator has complete control of the temperature program, which varies from one element to the next and from one sample type to the next. Here the program is carried out automatically by the computer-controlled instrument. In many cases, the instrument is preprogrammed by the manufacturer to get the best results. Although this relieves the operator of this formidable task, it leaves him unaware of breakdowns in the system caused by unanticipated interferences. Checking with a standard sample at regular intervals helps relieve this anxiety. A good deal of skill is necessary to make this system work. As one industrial manufac-

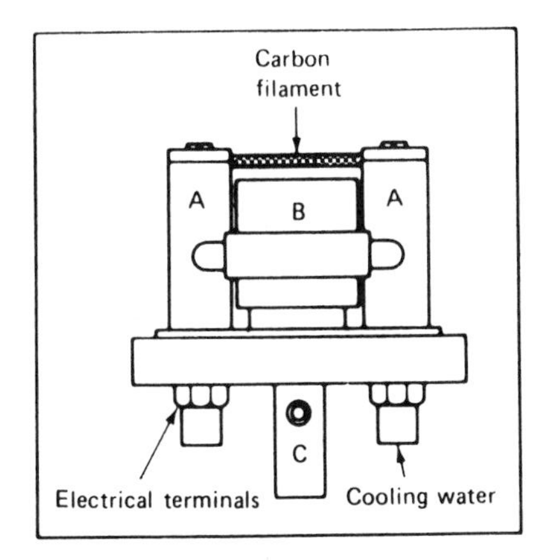

Figure 7.33 Carbon filament atom reservoir. (A) Water-cooled electrodes; (B) laminar flow box; (C) inlet for shield gas.

turer observed, "the difference between using carbon atomizers and flame atomizers is as different as playing tennis and watching tennis." *Answers* are easy to come by. *Accurate answers* are difficult to come by.

There is invariably a small amount of less volatile residue left when the sample is atomized. This generates a high and variable background. The variability of the background makes it vital to measure it directly rather than measuring the background of a blank, which may vary considerably from the sample. Automatic background correctors have been developed utilizing hollow cathode and hydrogen lamps simultaneously, which permit the simultaneous measurement and background correction for the particular sample being run. This is programmed onto a computer system that gives a net readout to the operator. In practice, this is very convenient, since it eliminates many of the problems encountered in manual operation and provides a net absorption signal. The major problem with the system is that the operator is frequently unaware of errors that may have crept into the system because he does not know how high the background is for any sample. Faulty operation of the instrument often results in errors in the atomization sequence with the generation of high background if the temperature is too low, or loss of sample if the temperature is too high. Nevertheless, the use of the carbon atomizer has greatly extended the range of analytical chemistry. It has proved particularly useful in the life sciences, because very small samples can be analyzed—of the order of microliters or micrograms. This is very valuable when body tissue, which is difficult to obtain from living patients, is being analyzed. Because of the high sensitivity of the method, metal determinations at concentration levels as low as 10^{-9}g/g are possible using ultrasmall samples. Typical sensitivities using these small samples are shown in Table 7.10. The method is subject to many kinds of error, particularly chemical interferences. With skill and attention, however, it can be used to analyze samples that cannot be analyzed by other techniques.

1. L'vov Platform

The precision of the carbon atomizer has been improved by the use of the L'vov platform, which is shown in Fig. 7.34. In this system a carbon platform is inserted into the standard atomizer. During the ashing and atomization step the metal atoms tend to condense on the platform, which is cooler than the electrically heated furnace.

After a short delay the platform becomes heated by radiation from the inside of the furnace and its temperature rises. At the increased temperatures the condensed metal atoms are revaporized and entered into the light path. At this time the background has been reduced somewhat, increasing the accuracy of the method. Also, the reproducibility of the procedure is improved by the L'vov platform.

Table 7.10 Sensitivities (1% Absorption) in the Graphite Tube Furnace (HGA-70)

Element	Absolute sensitivity (g × 10^{-12})	20 μl solution (μg/ml)
Al	150	0.007
As	160	0.008
Be	3.4	0.0002
Bi	280	0.014
Ca	3.1	0.05
Cd	0.8	0.00004
Co	120	0.006
Cr	18	0.01
Cs	71	0.004
Cu	45	0.02
Ga	1,200	0.06
Hg	15,000	1.5
Mn	7	0.01
Ni	330	0.10
Pb	23	0.001
Pd	250	0.013
Pt	740	0.02
Rb	41	0.002
Sb	510	0.15
Si	24	0.10
Sn	5,500	0.2
Sr	31	0.0015
Ti	280	0.5
Tl	90	0.1
V	320	0.2
Zn	2.1	0.0001

A comparison of the relative volatilization from the walls of the atomizer and from the L'vov platform is shown in Figure 7.35. Note that the signal from the L'vov platform is greater in size; also, the delay in time helps diminish the intense background signals encountered in carbon atomizers.

2. Background Absorption Measurements

With the carbon atomizer the background is usually very high, often 90% of the signal. With automatic background correctors such as the deuterium lamp or the Zeeman corrector, it is possible to get a reading of the atomic absorption in spite of this high background.

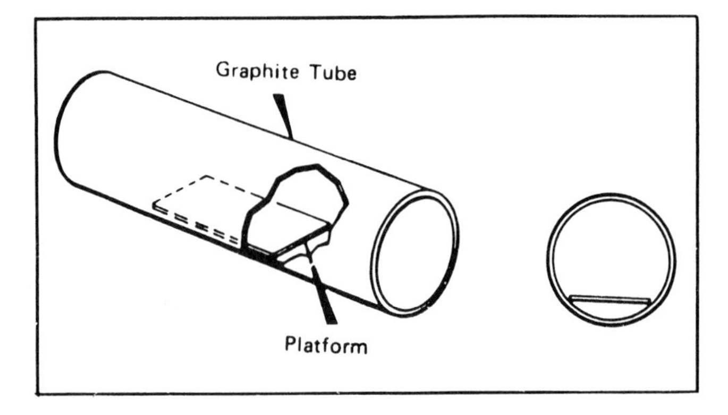

Figure 7.34 L'vov platform. The sample is loaded onto the platform and is atomized after a delayed time. A comparison of the time intervals and signals with and without the platform is shown.

One potential problem with these methods is that with an automatic background corrector the operator is usually unaware of the extent to which the background is present. It should be remembered that when $A = 2$, the transmittance $T = 1/100$ and the absorbed radiation is 99%. Therefore the total signal falling on the detector is only 1% of I_0. The quantitative analysis is carried out on this very small portion of the total signal from the sample and is subject to major error. A 1% error in measuring the background may be a 100% error in measuring the atomic absorption by the sample. Measurements of atomic absorption

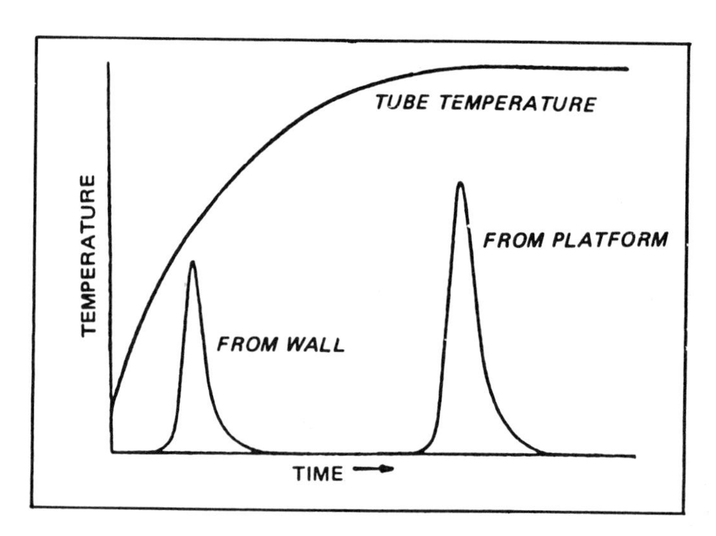

Figure 7.35 Volatilization of atoms from the wall and from the L'vov platform.

correcting backgrounds in excess of 99% are claimed by some manufacturers, but operators should be very wary of using such data because, as stated, even a small error in measuring or correcting for background becomes a major error in the net atomic absorption measurement.

3. Atomic Spectroscopy Detection Limits

The detection limits achievable for individual elements represent a significant criterion of the usefulness of an analytical technique for a given analytical prob-

Table 7.11 Comparison of Detection Limits of Atomic Spectroscopic Techniques by Perkin-Elmer[a]

Element	Flame atomic absorption	HGA graphite furnace	Inductively coupled plasma emission	Hydride
Aluminum	30.0[b,c]	0.01	20.0	
Antimony	30.0[d]	0.15		0.1
Arsenic	140.0[d]	0.2	50.0	0.02
Barium	8.0[a,b,e]	0.04	0.5	
Bismuth	20.0[d]	0.1		0.02
Boron	700.0[b,c]	15.0	4.0	
Cadmium	0.5[d]	0.003	4.0	
Calcium	1.0[d]	0.05		
Chromium	2.0[d]	0.01	5.0	
Cobalt	6.0[d]	0.02	6.0	
Copper	1.0[d]	0.02	3.0	
Gold	6.0[d]	0.1		
Iron	3.0[d]	0.02	3.0	
Lead	10.0[d]	0.05		
Lithium	0.5[d]	0.3		
Magnesium	0.1[d]	0.004		
Manganese	1.0[d]	0.01	1.0	
Mercury	170.0[d]	2.0		0.001[f]
Molybdenum	30.0[b,c]	0.02	8.0	
Nickel	4.0[d]	0.1	10.0	
Niobium	750.0[b,c,e,g]		20.0	
Phosphorus	53,000.0[b,c]	30.0	50.0	
Platinum	40.0[d]	0.2		
Selenium	70.0[d]	0.5	50.0	0.02
Silicon	60.0[b,c]	0.1		
Silver	0.9[d]	0.005		
Sulfur			50.0[h]	
Tantalum	500.0[b,c,g]		20.0	
Tellurium	19.0[d]	0.1		0.02
Thallium	9.0[d]	0.1		

Table 7.11 Continued

Element	Flame atomic absorption	HGA graphite furnace	Inductively coupled plasma emission	Hydride
Tin	110.0[b,c]	0.2	30.0	0.5
Titanium	50.0[b,c]	0.5	2.0	
Tungsten	1,200.0[b,c]		40.0	
Uranium	30,000.0[b,c]		50.0	
Vanadium	40.0[b,c]	0.2	5.0	
Zinc	0.8[d]	0.001	2.0	
Zirconium	300.0[b,c,g]		4.0	

[a]Detection limits depend on reproducibility. Sensitivity limits (Table 7.10) are for 1% absorption.
[b]Nitrous oxide flame used.
[c]Flow spoiler used.
[d]Impact bead used.
[e]With 1000 mg/liter potassium added to control ionization.
[f]Amalgamation accessory of MHS-20 used.
[g]With 0.2% aluminum and 1.0% hydrofluoric acid added to increase sensitivity and improve linearity.
[h]Purge required.
Source: Perkin-Elmer, 1981.

lem. Table 7.11, by Perkin-Elmer (Norwalk, Conn.), lists detection limits for four spectroscopic techniques: flame atomic absorption, furnace atomic absorption, inductively coupled plasma emission, and mercury-hydride generation atomic absorption. Perkin-Elmer defines its detection limits very conservatively, with a 95% confidence level. This means that if a concentration at the detection limit were measured very many times, it could be distinguished from zero in at least 95% of the determinations.

Generally, the best detection limits are attained using the graphite furnace. However, for elements that form refractory carbides in the furnace, such as uranium, zirconium, tungsten, and boron, inductively coupled plasma emission is the optimum technique. For those elements that form hydrides the generation technique offers unmatched detection limits, but the quantitative data are suspect.

E. ABSORPTION WAVELENGTH, PREFERRED FLAMES, AND SENSITIVITIES FOR FLAME ATOMIC ABSORPTION

No data from carbon atomizers have been included in this section. The author wishes to thank Allied Chemical Company, particularly Dr. Fred Brech, and the Perkin-Elmer Corporation for permission to use their publications on Recommended Procedures. A considerable part of the following information is based on these publications.

Abbreviations Used for Flame Composition (pp. 364–383)

A.A.	air acetylene
N.A.	nitrous oxide acetylene
O_2A.	oxygen acetylene
O_2.H.	oxygen hydrogen
O_2.air	oxygen air
ox.	oxidizing flame (excess oxidant air, oxygen or nitrous oxide)
red.	reducing flame (excess fuel, acetylene, or hydrogen)

text continues on page 383

ALUMINUM

Absorption wavelength (nm)	Preferred flame	Sensitivity 1% ab. (ppm)	Analytical range (ppm)
309.3	N.A. red.	1.0	5–50
396.1	N.A. red.	1.3	
308.2	N.A. red.	1.4	
394.4	N.A. red	2.0	
237.3	N.A. red.	3.3	
236.7	N.A. red.	4.0	
257.5	N.A. red.	8.8	

ANTIMONY

Absorption wavelength (nm)	Preferred flame	Sensitivity 1% ab. (ppm)	Analytical range (ppm)
217.6	A.A. (ox.)	0.5	3–40
206.8	A.A. (ox.)	0.7	
231.1	A.A. (ox.)	1.2	

ARSENIC

Absorption wavelength (nm)	Preferred flame	Sensitivity 1% ab. (ppm)	Analytical range (ppm)
189.0	A. H. (ox.)	1	5–50
193.7	A. H. (ox.)	2	50
197.2	A. H. (ox.)	3	

BARIUM

Absorption wavelength (nm)	Preferred flame	Sensitivity 1% ab. (ppm)	Analytical range (ppm)
553.5	N. A. (red.)	0.4	3–25
350.1	N. A. (red.)	5.0	

BERYLLIUM

Absorption wavelength (nm)	Preferred flame	Sensitivity 1% ab. (ppm)	Analytical range (ppm)
234.8	N. A. (red.)	0.02	0.2–4

BISMUTH

Absorption wavelength (nm)	Preferred flame	Sensitivity 1% ab. (ppm)	Analytical range (ppm)
223.1	A. A. (ox.)	0.4	2–50
222.8	A. A. (ox.)	1.5	
306.7	A. A. (ox.)	2.1	
206.2	A. A. (ox.)	5.5	

BORON

Absorption wavelength (nm)	Preferred flame	Sensitivity 1% ab. (ppm)	Analytical range (ppm)
249.7	N. A. (red.)	40	100–900
249.6	N. A. (red.)	100	

CADMIUM

Absorption wavelength (nm)	Preferred flame	Sensitivity 1% ab. (ppm)	Analytical range (ppm)
228.8	A. A. (ox.)	0.03	0.2–2.0
326.1	A. A. (ox.)	1.0	

CALCIUM

Absorption wavelength (nm)	Preferred flame	Sensitivity 1% ab. (ppm)	Analytical range (ppm)
422.7	N. A. (red.)	0.04	0.2–7
239.8	N. A. (red.)	20	

CERIUM

Absorption wavelength (nm)	Preferred flame	Sensitivity 1% ab. (ppm)	Analytical range (ppm)
520.0	N. A.	30	

CESIUM

Absorption wavelength (nm)	Preferred flame	Sensitivity 1% ab. (ppm)	Analytical range (ppm)
852.1	A. Coal Gas or A. A. (ox.)	0.2	2.0–15
894.3	A. Coal Gas or A. A. (ox.)	0.2	
455.6	A. Coal Gas or A. A. (ox.)	20.0	
459.3	A. Coal Gas or A. A. (ox.)		

CHROMIUM

Absorption wavelength (nm)	Preferred flame	Sensitivity 1% ab. (ppm)	Analytical range (ppm)
357.9	A. A. (red.)	0.1	0.5–5
359.3	A. A. (red.)	0.1	
425.4	A. A. (red.)	0.2	

COBALT

Absorption wavelength (nm)	Preferred flame	Sensitivity 1% ab. (ppm)	Analytical range (ppm)
240.7	A. A. (ox.)	0.15	0.6–5
242.5	A. A. (ox.)	.20	
252.1	A. A. (ox.)	0.3	
241.1	A. A. (ox.)	0.3	
352.7	A. A. (ox.)	2.2	
347.4	A. A. (ox.)	7.5	

COPPER

Absorption wavelength (nm)	Preferred flame	Sensitivity 1% ab. (ppm)	Analytical range (ppm)
324.7	A. A. (ox.)	0.1	0.5–10
327.4	A. A. (ox.)	0.2	
217.8	A. A. (ox.)	0.4	
216.5	A. A. (ox.)	0.7	
222.6	A. A. (ox.)	1.5	
249.2	A. A. (ox.)	7.0	
224.4	A. A. (ox.)	16.0	
244.2	A. A. (ox.)	30.0	

DYSPROSIUM

Absorption wavelength (nm)	Preferred flame	Sensitivity 1% ab. (ppm)	Analytical range (ppm)
421.2	N. A. (red.)	0.9	4.0–2.0
404.6	N. A. (red.)	0.8	
418.7	N. A. (red.)	0.9	
419.5	N. A. (red.)	1.0	

ERBIUM

Absorption wavelength (nm)	Preferred flame	Sensitivity 1% ab. (ppm)	Analytical range (ppm)
40.8	N. A. (red.)	1.0	5.0–40
415.1	N. A. (red.)	2.3	
386.3	N. A. (red.)	2.0	
389.3	N. A. (red.)	2.4	
408.8	N. A. (red.)	6.0	
393.7	N. A. (red.)	6.5	
381.0	N. A. (red.)	7.0	
390.5	N. A. (red.)	20.0	
394.4	N. A. (red.)	21.0	
460.7	N. A. (red.)	22.0	

EUROPIUM

Absorption wavelength (nm)	Preferred flame	Sensitivity 1% ab. (ppm)	Analytical range (ppm)
459.4	N. A. (red.)	0.6	3.0–50
462.7	N. A. (red.)	2.0	
466.2	N. A. (red.)	2.0	
322.1	N. A. (red.)	7.0	
321.3	N. A. (red.)	9.0	
311.1	N. A. (red.)	9.0	
333.4	N. A. (red.)	12.0	

GADOLINIUM

Absorption wavelength (nm)	Preferred flame	Sensitivity 1% ab. (ppm)	Analytical range (ppm)
407.9	N. A. (red.)	20	100–1000
368.4	N. A. (red.)	25	
378.3	N. A. (red.)	25	
405.8	N. A. (red.)	25	
405.4	N. A. (red.)	25	
371.4	N. A. (red.)	30	
419.4	N. A. (red.)	40	
367.4	N. A. (red.)	43	
404.5	N. A. (red.)	50	
394.6	N. A. (red.)	100	

GALLIUM

Absorption wavelength (nm)	Preferred flame	Sensitivity 1% ab. (ppm)	Analytical range (ppm)
287.4	A. A. (ox.)	2.5	15.0–200
294.4	A. A. (ox.)	2.5	
417.2	A. A. (ox.)	4.0	
250.0	A. A. (ox.)	20.0	
245.0	A. A. (ox.)	20.0	
272.0	A. A. (ox.)	50.0	

GERMANIUM

Absorption wavelength (nm)	Preferred flame	Sensitivity 1% ab. (ppm)	Analytical range (ppm)
265.1	N. A. (red.)	2.0	20–200
265.2	N. A. (red.)	2.0	
259.2	N. A. (red.)	5.0	
271.0	N. A. (red.)	6.0	
275.5	N. A. (red.)	6.5	
269.1	N. A. (red.)	9.0	

GOLD

Absorption wavelength (nm)	Preferred flame	Sensitivity 1% ab. (ppm)	Analytical range (ppm)
242.8	A. A. (red.)	0.3	2.5–20
267.6	A. A. (red.)	0.4	
274.8	A. A. (red.)	250.0	
312.8	A. A. (red.)	240.0	

HAFNIUM

Absorption wavelength (nm)	Preferred flame	Sensitivity 1% ab. (ppm)	Analytical range (ppm)
307.3	N. A. (red.)	15	100–500
286.6	N. A. (red.)	15	
289.8	O_2. A.	70	
296.5	O_2. A.	80	
368.2	O_2. A.	80	

HOLMIUM

Absorption wavelength (nm)	Preferred flame	Sensitivity 1% ab. (ppm)	Analytical range (ppm)
410.4	N. A.	2.0	
416.3	N. A.	3.0	
405.4	N. A.	3.0	

INDIUM

Absorption wavelength (nm)	Preferred flame	Sensitivity 1% ab. (ppm)	Analytical range (ppm)
304.9	A. A.	0.7	
325.6	A. A.	0.7	
410.5	A. A.	2.0	
275.4	A. A.	20.0	

IRIDIUM

Absorption wavelength (nm)	Preferred flame	Sensitivity 1% ab. (ppm)	Analytical range (ppm)
208.9	A. A. (red.)	10.0	50–1000
264.0	A. A. (red.)	10.0	
766.5	A. A. (red.)	8.0	
769.9	A. A. (red.)	19.0	
404.4	A. H.	4000.0	

IRON

Absorption wavelength (nm)	Preferred flame	Sensitivity 1% ab. (ppm)	Analytical range (ppm)
248.3	A. A.	0.1	1.5
248.8	A. A.	0.2	
252.3	A. A.	0.2	
271.9	A. A.	0.3	
302.1	A. A.	0.4	
250.1	A. A.	0.4	
216.7	A. A.	0.5	
372.1	A. A.	0.7	
296.7	A. A.	1.0	
386.0	A. A.	2.0	
344.1	A. A.	2.5	
368.0	A. A.	10.0	

LANTHANUM

Absorption wavelength (nm)	Preferred flame	Sensitivity 1% ab. (ppm)	Analytical range (ppm)
550.1	A. A. (ox.)	35	200–2500
418.7	A. A. (ox.)	50	
495.0	A. A. (ox.)	50	
357.4	A. A. (ox.)	110	
365.0	A. A. (ox.)	150	
392.8	A. A. (ox.)	150	

LEAD

Absorption wavelength (nm)	Preferred flame	Sensitivity 1% ab. (ppm)	Analytical range (ppm)
217.0	A. A. (ox.)	0.2	1–10
283.3	A. A. (ox.)	0.5	2–20
261.4	A. A. (ox.)	5.0	
368.4	A. A. (ox.)	12.0	

LITHIUM

Absorption wavelength (nm)	Preferred flame	Sensitivity 1% ab. (ppm)	Analytical range (ppm)
670.8	A. A. (ox.)	0.03	0.1–3.0
323.3	A. A. (ox.)	10	
610.4	A. A. (ox.)	100.0	

LUTETIUM

Absorption wavelength (nm)	Preferred flame	Sensitivity 1% ab. (ppm)	Analytical range (ppm)
336.0	N. A. (red.)	6.0	50–500
331.2	N. A. (red.)	11.0	
337.7	N. A. (red.)	12.0	
356.0	N. A. (red.)	13.0	
398.9	N. A. (red.)	55.0	
451.9	N. A. (red.)	66.0	

MAGNESIUM

Absorption wavelength (nm)	Preferred flame	Sensitivity 1% ab. (ppm)	Analytical range (ppm)
285.2	A. A. (ox.)	0.007	0.01–0.5
202.6	A. A. (ox.)	0.2	
279.6	A. A. (ox.)	5.0	

MANGANESE

Absorption wavelength (nm)	Preferred flame	Sensitivity 1% ab. (ppm)	Analytical range (ppm)
279.5	A. H. (ox.)	0.05	0.5–5.0
280.1	A. H. (ox.)	0.08	
403.1	A. H. (ox.)	0.5	
279.2	A. H. (ox.)	50.0	

MERCURY

Absorption wavelength (nm)	Preferred flame	Sensitivity 1% ab. (ppm)	Analytical range (ppm)
185.0	N. A. (red.)	0.5	10–300
253.7	N. A. (red.)	1.0	

MOLYBDENUM

Absorption wavelength (nm)	Preferred flame	Sensitivity 1% ab. (ppm)	Analytical range (ppm)
313.3	N. A. (red.)	0.5	5.0–60.0
317.0	N. A. (red.)	0.8	
379.8	N. A. (red.)	1.0	
319.4	N. A. (red.)	1.0	
386.4	N. A. (red.)	1.2	
390.3	N. A. (red.)	1.6	
315.8	N. A. (red.)	2.0	
320.9	N. A. (red.)	4.3	
311.2	N. A. (red.)	10.0	

NEODYMIUM

Absorption wavelength (nm)	Preferred flame	Sensitivity 1% ab. (ppm)	Analytical range (ppm)
334.9	N. A. (ox.)	10	10–700
463.4	N. A. (ox.)	10	
492.4	N. A. (ox.)	14	
471.9	N. A. (ox.)	20	

NICKEL

Absorption wavelength (nm)	Preferred flame	Sensitivity 1% ab. (ppm)	Analytical range (ppm)
232.0	N. A. (red.)	0.1	1.0–1000
231.1	N. A. (red.)	0.2	
352.5	N. A. (red.)	0.3	
341.5	N. A. (red.)	0.4	
304.1	N. A. (red.)	0.4	
341.2	N. A. (red.)	0.7	
351.5	N. A. (red.)	0.8	
303.8	N. A. (red.)	1.2	
337.0	N. A. (red.)	1.8	
323.0	N. A. (red.)	3.0	
294.4	N. A. (red.)	5.5	

NIOBIUM

Absorption wavelength (nm)	Preferred flame	Sensitivity 1% ab. (ppm)	Analytical range (ppm)
334.4	N. A. (red.)	20	100–1000
358.0	N. A. (red.)	22	
334.9	N. A. (red.)	24	
408.9	N. A. (red.)	28	
335.0	N. A. (red.)	30	
412.4	N. A. (red.)	38	
357.6	N. A. (red.)	50	
353.5	N. A. (red.)	60	
374.0	N. A. (red.)	64	
415.3	N. A. (red.)	100	

OSMIUM

Absorption wavelength (nm)	Preferred flame	Sensitivity 1% ab. (ppm)	Analytical range (ppm)
290.9	N. A. (red.)	1.0	10–200
305.9	N. A. (red.)	1.6	
263.7	N. A. (red.)	1.8	
301.8	N. A. (red.)	3.0	
330.2	N. A. (red.)	3.5	
271.5	N. A. (red.)	4.0	
280.7	N. A. (red.)	4.5	
264.4	N. A. (red.)	5.0	
442.0	N. A. (red.)	20.0	
426.1	N. A. (red.)	30.0	

PALLADIUM

Absorption wavelength (nm)	Preferred flame	Sensitivity 1% ab. (ppm)	Analytical range (ppm)
244.8	A. A. (red.)	0.3	5.0–50.0
247.6	A. A. (red.)	0.2	
276.3	A. A. (red.)	1.3	
340.5	A. A. (red.)	1.5	

PHOSPHORUS

Absorption wavelength (nm)	Preferred flame	Sensitivity 1% ab. (ppm)	Analytical range (ppm)
178.3	N. A. (red.)	5.0	2000–10,000
213.6	N. A. (red.)	250	
213.5	N. A. (red.)	250	
214.9	N. A. (red.)	500	

PLATINUM

Absorption wavelength (nm)	Preferred flame	Sensitivity 1% ab. (ppm)	Analytical range (ppm)
265.9	A. A. (ox.)	1.0	5–100
306.5	A. A. (ox.)	2.0	
283.0	A. A. (ox.)	3.5	
293.0	A. A. (ox.)	3.8	
273.4	A. A. (ox.)	4.0	
217.5	A. A. (ox.)	4.0	
248.7	A. A. (ox.)	5.0	
299.8	A. A. (ox.)	5.5	
271.9	A. A. (ox.)	9.0	

POTASSIUM

Absorption wavelength (nm)	Preferred flame	Sensitivity 1% ab. (ppm)	Analytical range (ppm)
766.5	A. A. (red.)	0.03	0.2–2.0
769.9	A. A. (red.)	0.05	
404.4	A. A. (red.)	10.0	
404.7	A. A. (red.)	10.0	

PRASEODYMIUM

Absorption wavelength (nm)	Preferred flame	Sensitivity 1% ab. (ppm)	Analytical range (ppm)
495.1	N. A.	15.0	

RHENIUM

Absorption wavelength (nm)	Preferred flame	Sensitivity 1% ab. (ppm)	Analytical range (ppm)
346.0	N. A. (red.)	15	100–1000
346.5	N. A. (red.)	25	
345.2	N. A. (red.)	35	

RHODIUM

Absorption wavelength (nm)	Preferred flame	Sensitivity 1% ab. (ppm)	Analytical range (ppm)
343.5	A. A. (ox.)	0.4	0.5–20
369.2	A. A. (ox.)	0.5	
339.7	A. A. (ox.)	0.7	
350.2	A. A. (ox.)	1.0	
365.8	A. A. (ox.)	1.8	
370.1	A. A. (ox.)	3.0	
350.7	A. A. (ox.)	10.0	

RUBIDIUM

Absorption wavelength (nm)	Preferred flame	Sensitivity 1% ab. (ppm)	Analytical range (ppm)
780.0	A. A. (ox.)	0.1	0.5–7.0
794.8	A. A. (ox.)	0.2	
420.2	A. A. (ox.)	12.0	
421.6	A. A. (ox.)	25.0	

RUTHENIUM

Absorption wavelength (nm)	Preferred flame	Sensitivity 1% ab. (ppm)	Analytical range (ppm)
349.9	A. A. (ox.)	0.5	3–50
372.8	A. A. (ox.)	0.8	
379.9	A. A. (ox.)	1.1	
392.6	A. A. (ox.)	6.0	

SAMARIUM

Absorption wavelength (nm)	Preferred flame	Sensitivity 1% ab. (ppm)	Analytical range (ppm)
429.7	N. A. (red.)	8.0	50.0–500.0
476.0	N. A. (red.)	20	
511.7	N. A. (red.)	20	
520.1	N. A. (red.)	25	
476.0	N. A. (red.)	40	
458.3	N. A. (red.)	50	

SCANDIUM

Absorption wavelength (nm)	Preferred flame	Sensitivity 1% ab. (ppm)	Analytical range (ppm)
391.2	N. A. (red.)	0.4	2.0–25.0
390.8	N. A. (red.)	0.4	
402.4	N. A. (red.)	1.0	
405.5	N. A. (red.)	1.1	
327.0	N. A. (red.)	2.0	
408.2	N. A. (red.)	3.0	
327.4	N. A. (red.)	5.0	

SELENIUM

Absorption wavelength (nm)	Preferred flame	Sensitivity 1% ab. (ppm)	Analytical range (ppm)
196.0	A. A. (ox.)	0.5	2.0–50.0
204.0	A. A. (ox.)	1.5	
206.3	A. A. (ox.)	6.0	
207.5	A. A. (ox.)	20.0	

SILICON

Absorption wavelength (nm)	Preferred flame	Sensitivity 1% ab. (ppm)	Analytical range (ppm)
251.6	N. A. (red.)	1.0	5.0–100
250.7	N. A. (red.)	3.0	
252.8	N. A. (red.)	3.5	
252.4	N. A. (red.)	4.0	
221.7	N. A. (red.)	4.5	
221.1	N. A. (red.)	8.0	

SILVER

Absorption wavelength (nm)	Preferred flame	Sensitivity 1% ab. (ppm)	Analytical range (ppm)
328.1	A. A. (ox.)	0.1	1.0–20
338.3	A. A. (ox.)	0.15	

SODIUM

Absorption wavelength (nm)	Preferred flame	Sensitivity 1% ab. (ppm)	Analytical range (ppm)
589.0	A. A. or A. H.	0.02	0.1–1.0
589.5	A. A. or A. H.	0.02	
330.2	A. A. or A. H.	0.30	
330.3	A. A. or A. H.	0.30	

STRONTIUM

Absorption wavelength (nm)	Preferred flame	Sensitivity 1% ab. (ppm)	Analytical range (ppm)
460.7	A. A. (red.) or A. H.	0.1	1.0–10.0
407.8	A. A. (red.) or A. H.	7.0	

SULFUR

Absorption wavelength (nm)	Preferred flame	Sensitivity 1% ab. (ppm)	Analytical range (ppm)
180.7	A. A.	9.0	No Data

TANTALUM

Absorption wavelength (nm)	Preferred flame	Sensitivity 1% ab. (ppm)	Analytical range (ppm)
271.5	N. A. (red.)	10.0	600–1200
260.8	N. A. (red.)	20.0	
260.9	N. A. (red.)	20.0	
264.7	N. A. (red.)	40	
293.5	N. A. (red.)	40	
255.9	N. A. (red.)	40	
265.3	N. A. (red.)	50	
269.8	N. A. (red.)	50	
275.8	N. A. (red.)	80	

TECHNETIUM

Absorption wavelength (nm)	Preferred flame	Sensitivity 1% ab. (ppm)	Analytical range (ppm)
261.4	A. A. (red.)	3.0	10–70
261.6	A. A. (red.)	3.0	
260.9	A. A. (red.)	12.0	
429.7	A. A. (red.)	20.0	
426.2	A. A. (red.)	25.0	
318.2	A. A. (red.)	30.0	
423.8	A. A. (red.)	33	
363.6	A. A. (red.)	33	
317.3	A. A. (red.)	300	
346.6	A. A. (red.)	300	
403.2	A. A. (red.)	300	

TELLURIUM

Absorption wavelength (nm)	Preferred flame	Sensitivity 1% ab. (ppm)	Analytical range (ppm)
214.3	A. A. or A. H.	0.5	2.0–23
225.9	A. A. or A. H.	4.0	
238.6	A. A. or A. H.	70.0	

TERBIUM

Absorption wavelength (nm)	Preferred flame	Sensitivity 1% ab. (ppm)	Analytical range (ppm)
432.6	N. A. (red.)	5.0	40–600
431.9	N. A. (red.)	6.0	
390.1	N. A. (red.)	8.0	
406.2	N. A. (red)	9.0	
433.8	N. A. (red.)	10.0	
410.5	N. A. (red.)	20.0	

THALLIUM

Absorption wavelength (nm)	Preferred flame	Sensitivity 1% ab. (ppm)	Analytical range (ppm)
276.8	A. A. (ox.)	0.4	2.0–20.0
377.6	A. A. (ox.)	2.0	
238.0	A. A. (ox.)	3.0	
258.0	A. A. (ox.)	10.0	

THORIUM

Absorption wavelength (nm)	Preferred flame	Sensitivity 1% ab. (ppm)	Analytical range (ppm)
324.6	N. A.	500	

THULIUM

Absorption wavelength (nm)	Preferred flame	Sensitivity 1% ab. (ppm)	Analytical range (ppm)
371.8	N. A. (red.)	0.4	3.0–60.0
410.6	N. A. (red.)	5.0	
374.4	N. A. (red.)	5.0	
409.4	N. A. (red.)	6.0	
418.8	N. A. (red.)	6.0	
420.4	N. A. (red.)	15.0	
375.2	N. A. (red.)	20.0	
436.0	N. A. (red.)	25.0	
341.0	N. A. (red.)	30.0	

TIN

Absorption wavelength (nm)	Preferred flame	Sensitivity 1% ab. (ppm)	Analytical range (ppm)
224.6	A. A. (red.)	0.2	5.0–100
286.3	A. A. (red.)	1.0	
235.5	A. A. (red.)	1.0	
270.6	A. A. (red.)	2.0	
303.4	A. A. (red.)	2.0	
253.5	A. A. (red.)	3.0	
219.9	A. A. (red.)	4.0	
300.9	A. A. (red.)	5.0	
235.5	A. A. (red.)	5.0	
266.1	A. A. (red.)	15.0	

TITANIUM

Absorption wavelength (nm)	Preferred flame	Sensitivity 1% ab. (ppm)	Analytical range (ppm)
365.3	N. A. (red.)	1.0	6.0–150
364.3	N. A. (red.)	1.1	
320.0	N. A. (red.)	1.2	
363.6	N. A. (red.)	1.2	
335.5	N. A. (red.)	1.5	
375.3	N. A. (red.)	1.7	
337.2	N. A. (red.)	1.7	
399.9	N. A. (red.)	1.7	
399.0	N. A. (red.)	2.0	

TUNGSTEN

Absorption wavelength (nm)	Preferred flame	Sensitivity 1% ab. (ppm)	Analytical range (ppm)
400.9	N. A. (red.)	5.0	10–500
255.1	N. A. (red.)	5.0	
294.4	N. A. (red.)	10.0	
268.1	N. A. (red.)	10.0	
272.4	N. A. (red.)	10.0	
294.7	N. A. (red.)	10.0	
283.1	N. A. (red.)	20.0	
289.6	N. A. (red.)	25.0	
287.9	N. A. (red.)	35.0	
430.2	N. A. (red.)	80.0	

URANIUM

Absorption wavelength (nm)	Preferred flame	Sensitivity 1% ab. (ppm)	Analytical range (ppm)
351.5	N. A. (red.)	50	200–2000
358.5	N. A. (red.)	50	
356.7	N. A. (red.)	80	

VANADIUM

Absorption wavelength (nm)	Preferred flame	Sensitivity 1% ab. (ppm)	Analytical range (ppm)
318.4	N. A. (red.)	1.0	5–150
306.6	N. A. (red.)	3.0	
306.0	N. A. (red.)	3.0	
305.6	N. A. (red.)	5.0	
320.2	N. A. (red.)	10.0	
390.2	N. A. (red.)	10.0	

YTTERBIUM

Absorption wavelength (nm)	Preferred flame	Sensitivity 1% ab. (ppm)	Analytical range (ppm)
398.8	N. A. (red.)	0.7–7.0	
346.4	N. A. (red.)		
246.4	N. A. (red.)		
267.2	N. A. (red.)		

YTTRIUM

Absorption wavelength (nm)	Preferred flame	Sensitivity 1% ab. (ppm)	Analytical range (ppm)
410.2	N. A. (red.)	2.0	20–200
407.7	N. A. (red.)	5.0	
412.8	N. A. (red.)	5.0	
414.3	N. A. (red.)	2.8	
362.1	N. A. (red.)	4.0	

ZINC

Absorption wavelength (nm)	Preferred flame	Sensitivity 1% ab. (ppm)	Analytical range (ppm)
213.9	N. A. (red.)	0.01	0.1–1.2
307.6	N. A. (red.)	50.0	

ZIRCONIUM

Absorption wavelength (nm)	Preferred flame	Sensitivity 1% ab. (ppm)	Analytical range (ppm)
360.1	N. A. (red.)	10.0	50–600
354.8	N. A. (red.)	15.0	
303.0	N. A. (red.)	15	
301.2	N. A. (red.)	17	
248.2	N. A. (red.)	18	
362.4	N. A. (red.)	20	

F. CONCLUSION

Atomic absorption is an excellent method for the rapid determination of the metallic elements. Quantitatively it is sensitive, accurate, and precise. It is subject to two major sources of error. These arise from the principal anion present and the solvent used. Small samples (μg) can be analyzed for metals in low concentration (ppm) using carbon atomizers, but extra attention is necessary.

BIBLIOGRAPHY

Anal. Chem., Application Reviews, June 1994.
Blakley, C. R., and Vestal, M. L., Anal. Chem, (1983), *55*:750.
Browner, R. F., Boorn, A. W., and Smith, D. D., *Anal. Chem.*, (1982), *54*:141.
Camrick, G. R., and Slavin, W. *Appl. Spec.*, (1983), *37*(1):1.
Chandler, C., *Atomic Spectra and the Vector Model*, 2nd ed., Van Nostrand, New York, 1964.
Contenium, G. F., *Appl.Spec.*, (1993), *47*(10):1557.
Farrah, K. S., and Sneddon, J., The use of simultaneous flame A.A. for the detection of inorganics included in toxicity characteristics leaching procedure, *Am. Env. Lab.*, *4*:1 (1993).
Gaydon, A. G., and Wolfhand, H. W., *Flames*, Chapman and Hall, Ltd., London, 1960.
Kahn, H., *Research and Development*, (1992), *Sept.*:6.
Katz, E. D., and Scott, K. P., *Analyst*, (1985), *110*:231.
Kirkbright, G. F., and Sargent, M., *Atomic Absorption and Fluorescence Spectroscopy*, Academic Press, New York, 1974.
Lowe, M. D., Sutton, M. M., and Clinton, O. E., *Appl. Spectrosc.*, (1982), *36*:22.
McIntosh, S., Fernandez, F., Erlen, W., and Gus, T., *Instrumental Solutions*, (1991), Mar. 22.
Robinson, J. W., *Atomic Spectroscopy*, Marcel Dekker, New York, 1990.
Robinson, J. W., and Wu, J. C., *Spectrosc. Lett.*, (1985), *18*:399.
Ruzieka, J., and Hansen E. H., *Anal. Chim, Acta.*, (1975), *78*:145.
Slavin, W., and Schmidt, G. J., *J. Chrom. Sci.*, (1979), *17*:610.
Smith, D. D., and Browner, R. F., *Anal. Chem.*, (1982), *54*:533.
Smith, D. D., and Browner, R. F., *Anal. Chem.*, (1984), *56*:2702.
Walsh, A., *Spectrochim. Acta*, (1955), 7:108.

What's new in instrumentation, *Spectroscopy*, (1993), *8*(4).
Wichems, D. N., Calloway, Jr., C. P., Fernando, R., Jones, B. T., and Morykwas, M. J., Determination of silicone in breast tissue by graphite furnace continuum source atomic absorption spectrometry, *Appl. Spec.*, (1993), *47*:1577.

SUGGESTED EXPERIMENTS

7.1 Prepare solutions of zinc chloride in water containing known concentrations of zinc over the range 0.5–50.0 ppm. Aspirate each solution in turn into the burner. Measure the absorption of the zinc line at 213.9 nm. Plot the relationship between the absorbance and the zinc concentrations. Note the deviation from Beer's law. Indicate the useful analytical range of the calibration curve. Compare the sensitivity with Experiment 9.4.

7.2 Repeat Experiment 7.1 for the elements Cd, Cu, Na, Pb, and Ca.

7.3 Prepare a series of standard solutions containing sodium at concentrations of 1, 3, 5, 7, 9, and 10 ppm. Measure the absorption of each curve and plot the absorption against the sodium concentration. Take water samples from various sources, such as tap water, drinking water, distilled water, river water, distilled water stored in a polyethylene bottle, and distilled water stored in a glass bottle. Determine the sodium concentration in each sample.

7.4 Prepare a solution containing 20 ppm of Pb as $Pb(NO_3)_2$. Prepare similar solutions containing 50 ppm of Pb as (a) $PbCl_2$, (b) lead oxalate, and (c) lead acetate. Measure the absorbance by the Pb of the Pb resonance line at 283.3 nm. Note the change in absorbance as the compound changes. This is chemical interference. Add excess ethylenediaminetetraacetate (EDTA) to each solution and again measure the absorbance. Note the elimination of chemical interferences.

7.5 Prepare an aqueous solution containing 5 ppm of NaCl; also prepare five separate solutions containing 5 ppm of Na plus 550 ppm of (a) Ca, (b) Mg, (c) Fe, (d) Mn, and (e) K. Measure the Na absorbance at the Na resonance lines at 589.0 nm by solutions (a), (b), (c), (d), and (e) to which no Na has been added. The absorption was caused by Na impurity in the solutions. Based on Experiment 7.3, how concentrated was the impurity?

PROBLEMS

7.1 Why are atomic absorption lines very narrow?

7.2 Calculate the number of atoms N required to give 1% absorption when the oscillator strength of the line is 0.1. How many moles is this?

7.3 Why must hollow cathodes be used as the radiation source? Illustrate a hollow cathode.

7.4 Why is "modulation" necessary for accurate results? How is modulation achieved?

7.5 Why is atomic absorption not used for qualitative analysis?

7.6 What causes chemical interferences? Give three examples.

7.7 How are solid samples analyzed by atomic absorption spectroscopy?

7.8 Several standard solutions of copper were prepared. These were aspirated into a flame and the absorption measured with the following results. Prepare a calibration curve from the data.

Sample concentration (ppm)	Absorbance
0.5	0.045
1.0	0.090
1.5	0.135
2.0	0.180
2.5	0.225
3.0	0.270

7.9 Samples of copper solutions of unknown composition were brought to the lab. The absorption was measured. Using the calibration curve from Problem 7.8, complete the following table.

Sample	Absorbance	Concentration (ppm)
A	0.080	
B	0.105	
C	0.220	
D	0.250	

7.10 What is a Grotrian diagram?
7.11 Why can nonmetals not be determined by AA?
7.12 What is the relationship between the amount of light absorbed and the oscillator strength of the transition involved?
7.13 How can the population distribution of atoms in various energy levels be calculated?
7.14 What is the basis for concluding that at temperatures up to 3000 K the great majority of an atom population is in the ground state?
7.15 Describe an electrodeless discharge lamp (EDL).
7.16 Why are EDLs used?
7.17 What is a flame profile?
7.18 Describe the atomization process that takes place in a flame.
7.19 How does the stability of the oxide of the analyte affect the flame profile?
7.20 Describe the effects of solvents on the absorption signal.
7.21 Describe the atomization process during combustion of the principal solvent of the sample.
7.22 What is the source of background absorption?
7.23 How is the background corrected?
7.24 Describe the Zeeman automatic background corrector.
7.25 Describe the D_2 lamp background corrector.
7.26 Describe carbon atomizers.
7.27 Describe the L'vov platform.
7.28 What is the advantage of the L'vov platform?
7.29 What are the advantages of carbon atomizers?
7.30 What are the disadvantages of a C atomizer?

8

Spectrophotometry, Colorimetry, and Polarimetry

A. BACKGROUND OF SPECTROPHOTOMETRY: RELATED FIELDS

1. History of Spectrophotometry

Probably the first physical method used in analytical chemistry was based on the quality of the color in colored solutions. The first things we observe regarding colored solutions are their *hue*, or color, and the color's *depth*, or *intensity*. For a long time experimental work made use of the human eye to measure the hue and intensity of colors. This analytical system, generally called *colorimetry*, is simple, fast, and often very effective. However, color-blindness (with respect to certain or even all colors) can lead to erroneous results. Moreover, even with perfect eyesight, the best operator can have difficulty comparing the intensity of two colors having slightly different hues. As a result, instruments have been developed to perform these measurements more accurately and reliably than the human eye. These instruments are dependable, easy to operate, and quite inexpensive.

The intensity of a solution's color depends on how much radiation is absorbed by the sample. (This situation is implicit in our discussion of radiant energy at the beginning of Chap. 2.) When white light passes through a solution and emerges as red light, we say that the solution is red. What has actually happened is that the solution has allowed the red component of the white light to pass, whereas it has absorbed the complementary colors, yellow and blue. The

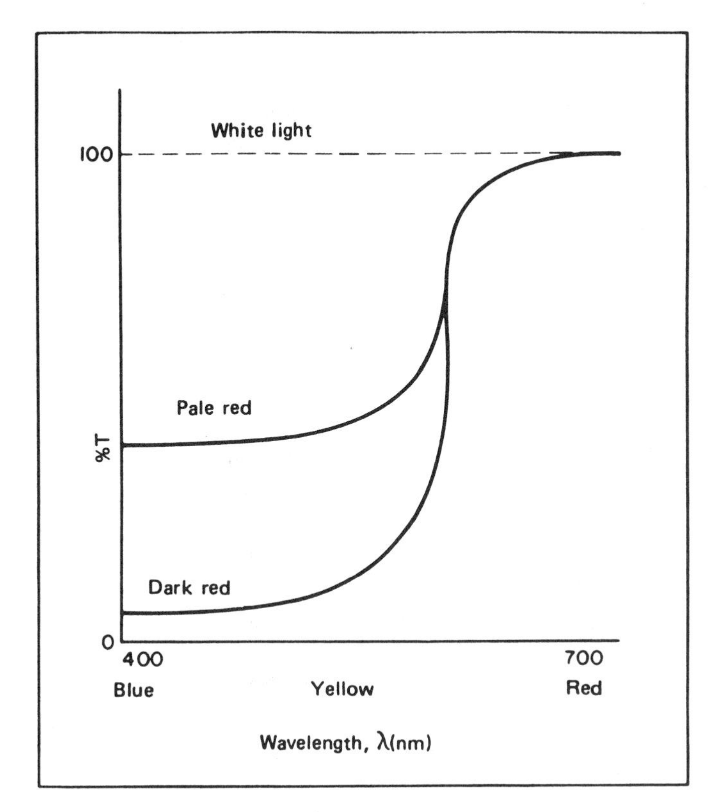

Figure 8.1 Absorption spectrum of a clear liquid, a pale red liquid, and a dark red liquid. Note that red light is not absorbed by any of them.

more concentrated the sample solution, the more yellow and blue light is absorbed and the more intensely red it appears to the eye.

The absorption spectrum of a red solution such as red ink is shown in Fig. 8.1. It can be seen that with white light, all radiation, including the wavelengths visually seen as blue, yellow, and red, is transmitted through the sample. If the light emerging from the sample is pale red, then some of the blue and some of the yellow have been absorbed, but all of the red has been transmitted. If the solution is dark red, then most of the blue and most of the yellow have been absorbed, but again, all of the red has been transmitted. The eye sees these different solutions as white, pale red, and dark red. The amount of red light falling on the eye is the same in each case, but the "depth of color" is modified because of the decreasing amount of blue and yellow light in the dark red solution. When all three colors pass through the sample equally, the eye mixes these three colors and a signal of "white light" is transmitted to the brain. When a small amount of yellow and blue is removed but all the red is transmitted, the signal transmitted to the brain is mostly white light with a little blue and yellow removed, leaving a pale red tint. In summary, the eye perceives and the brain interprets the transmitted light. This complex process that goes on in the eye and the brain may be disturbed if the operator has any degree of color blindness. Also, there is a distinct problem if a second solution is present that absorbs only yellow light. This changes the hue of the system, and the light transmits a different signal to the brain. This change in hue makes it very difficult to compare the color intensities of a standard and a sample.

In *spectrophotometry* the red in this example is ignored because the quantity of red light transmitted through the sample is constant in each case.The degree of absorption at some other wavelength, such as that of the blue light, is measured. The concentration of the solution is then calculated based on how much blue light is absorbed. Under these circumstances, if a compound that absorbs yellow is present in the sample, it will not interfere with the absorption of blue light and will therefore not interfere in the analysis. In practice, these problems are encountered frequently. The method of setting up a spectrophotometric experiment includes knowledge of possible interferences that may absorb at the same wavelength as the sample. The wavelength at which the absorption by the sample is measured is then chosen to be some wavelength remote from the absorption by interfering compounds.

The method is much less susceptible than colorimetry to operator error and to interferences from compounds encountered in the sample itself. Spectrophotometry is currently used extensively in all routine analytical labs and undoubtedly will be for many years to come. Elemental analysis has been used by many industrial operators, particularly when very rapid answers are necessary. For example, in steel manufacture, the steel sample may be in the molten state ready for pouring, but the carbon or phosphorus concentration must be within a certain specified range. It is possible for a skilled operator to take a quick sample, chill it, dissolve it in acid, and, by looking at the color of the solution, decide whether or not it is time to pour. This decision must be made accurately within minutes for satisfactory steel production. The operators are frequently skillful to the point of being artists. No doubt such methods of quality control will be used for decades to come.

Initially, colorimetry and spectrophotometry were used to ascertain the concentrations of simple colored solutions, such as potassium permanganate or dichromate. Over the years, however, it became recognized that many metals react with organic reagents to form intensely colored compounds. The reagents used have been studied and modified for many years. Very sensitive and selective analytical procedures have been developed for the determination of trace metal components in solutions with concentrations as low as 1 ppm and less. Obtain-

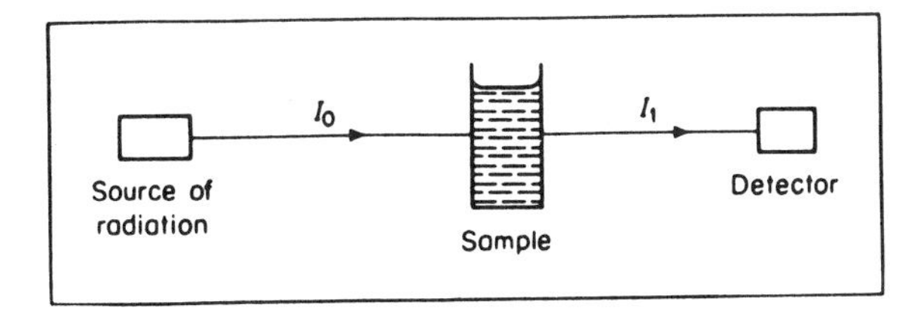

Figure 8.2 A simple spectrophotometric optical system.

ing reproducible results by means of spectrophotometric analysis often requires skill, loving attention, and a considerable knowledge of chemistry. On the other hand, the equipment is generally inexpensive, and most laboratories are able to handle this type of analytical work. Spectrophotometry provides the basis for many routine methods of analysis. It is often the quickest way to analyze an unusual sample that cannot be conveniently determined by the more rapid instrumental methods of analysis because of lack of calibration curves or knowledge of the interferences likely to be encountered in instrumental techniques. A schematic diagram is shown in Fig. 8.2.

2. Nephelometry and Turbidimetry

Much of the theory and equipment used in spectrophotometry apply with little modification to *nephelometry* and *turbidimetry*. These fields involve the scattering of light by nontransparent particles suspended in a liquid (examples are precipitates and colloidal suspensions). In *nephelometry* we measure the amount of radiation *scattered* by the particles, whereas in *turbidimetry* we measure the amount of light *not scattered* by the particles. These processes are illustrated in Figs. 8.3 and 8.4. The applications of nephelometry include the determination

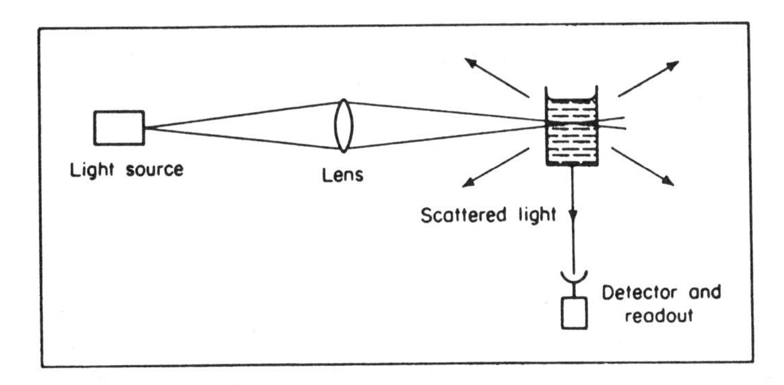

Figure 8.3 Optical system for nephelometry.

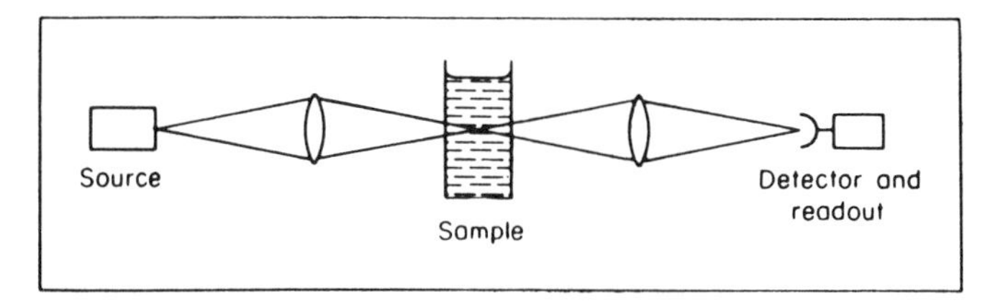

Figure 8.4 Optical system for turbidimetry.

of calcium or barium by precipitation as the phosphate or sulfate. The quantity of calcium or barium present is measured by the amount of radiation scattered by the precipitated metals. From the intensity of scattered radiation, the original concentration of calcium or barium can be determined.

When using nephelometry or turbidimetry for simple quantitative analysis, it is customary to prepare calibration curves from samples of known metal concentrations. The precipitate or suspensions must be prepared under rigidly controlled conditions. This is essential because the scattering of light depends on the *size* and *number* of the *particles* involved, as well as the concentration. It is necessary for a given solution to produce the same number of particles of the same size for the degree of light scattering to be meaningful. Furthermore, the wavelength of the light scattered most efficiently also depends on the physical size of the scattering particles. By controlling conditions, one can reproduce the particle size of the suspension, and useful analytical procedures may be developed. The methods are frequently difficult to use, because light scattering depends on the number of particles in the light path. Two identical samples of equal concentration will scatter light equally only if they form the same number of particles when they are precipitated. This depends on many experimental conditions, including the sample temperature, the rate at which the precipitant and the sample are mixed, the degree of agitation, and the length of time the precipitates are allowed to stand. Because of the uncertainty of the procedures, nephelometry and turbidimetry are not popular in many analytical labs unless they offer some advantage such as speed if a turbed plant stream is being monitored on a continuous basis.

B. THE ABSORPTION LAWS OF SPECTROPHOTOMETRY

The most important laws applicable to colorimetry are summarized in the Beer–Lambert law (often shortened to Beer's law) described in Chapter 2. The data shown in Table 8.1 were obtained by measuring the degree of absorption by the sample. This measurement is carried out on a spectrophotometer (i.e., a device for measuring the absorption of light by a solution). Plotting the results in Table

Table 8.1 Relationship Between Concentration and I

	Sample concentration (%)					
Sample conc. (%)	0.01	0.02	0.03	0.04	0.05	0.06
I	88.7	79.5	70.7	63.2	56.4	50.2
I_0	100	100	100	100	100	100
$I_0/I = 1/T$	1.128	1.259	1.413	1.585	1.778	1.995
$-\log T =$ $\log(1/T) = A =$	0.05	0.10	0.15	0.20	0.25	0.30

8.1 produces a figure similar to Fig. 8.5. Samples of unknown concentration can be determined by placing them in the spectrophotometer and measuring I. From this measurement, the concentration c can be calculated. For example, if the value of I after passing through the sample is 79.5, then $A = 0.1$ and the concentration of the sample is 0.02%. The relationship between absorbance and concentration is shown in Fig. 8.5. In practice, the use of calibration curves is the normal procedure for the quantitative application of spectrophotometry.

1. Methods of Operation in Colorimetry

Some of the many methods that have been used successfully to measure color intensities are described in the following sections.

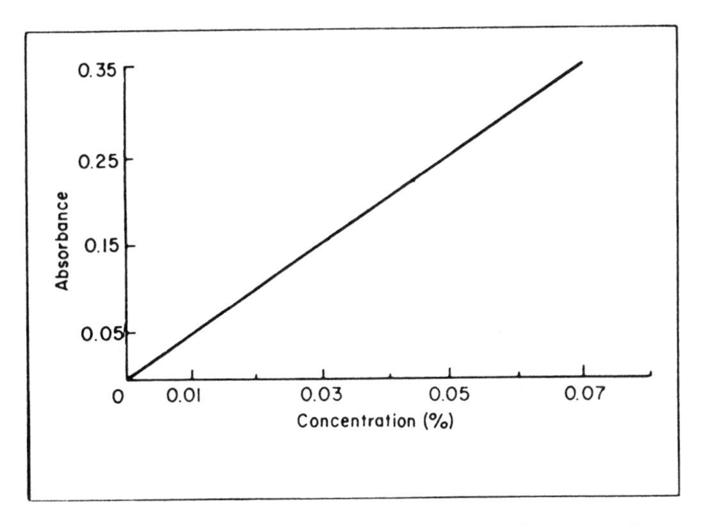

Figure 8.5 Relationship between absorbance and concentration.

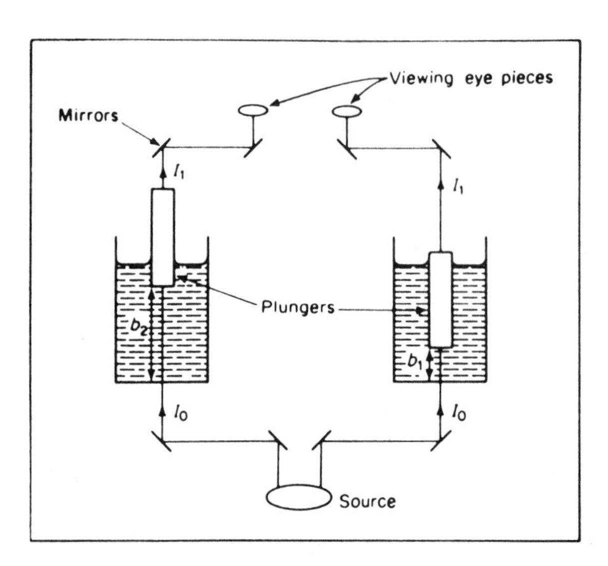

Figure 8.6 Optical balancing method.

a. Balancing Method

The balancing method is used for visual color balancing. A sample solution of unknown concentration is put into a cylinder with a flat, transparent base. A standard solution with known concentration is put into a second, similar cylinder. It is desirable for the two solutions to be similar in concentration. This can be determined by inspection, because the solutions should be similar in depth of color. A transparent plunger is placed in each cylinder. The plungers are moved up and down independently, varying the solution depth between the bottom of the plunger and the bottom of the sample cell. Increased depth increases the color intensity. The depths are varied until the colors seen from the top of each cylinder are identical. From the readings of the depth of the samples, the concentration of the unknown sample may be calculated. The process is illustrated in Fig. 8.6.

With this technique, results are calculated as follows. When the colors of the two solutions are balanced in intensity, the absorbance in each arm is equal. Intensity I_0 is common to both beams; therefore I_1 is equal in both beams. From the Beer–Lambert law [Eq. (2.19)], $b_1c_1 = b_2c_2$. Since b_1 and b_2 can be measured and c_1 is the concentration of the standard sample, c_2 can be calculated as b_1c_1/b_2. Hence the concentration of the colored component in the sample is given by the b_1c_1/b_2, where c_2 is the concentration of the standard solution and b_1/b_2 is the ratio of the depths of the solutions in the two arms of the instrument (see Fig. 8.6).

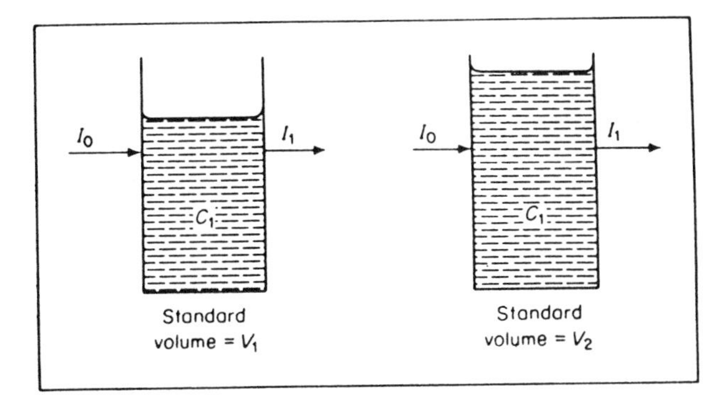

Figure 8.7 Dilution method.

b. Dilution Method

In the dilution method we take two cells of equal thickness and add to the first a standard solution and to the second a solution of unknown concentration. Light from the same source is passed through each cell, and the emergent beams are then compared. The more concentrated sample is progressively diluted and the volume changed until the light emerging from both cells is equal. At this point, the concentrations of the two solutions are equal (see Fig. 8.7).

When the two beams are equal in intensity, the absorbance A is equal in each case. From the Beer–Lambert Law [Eq. (2.19)],

$$A = -\log\left(\frac{I_1}{I_0}\right) = ab_1c_1 \qquad \text{(solution 1)}$$

$$= -\log\left(\frac{I_1}{I_0}\right) = ab_2c_2 \qquad \text{(solution 2)}$$

Therefore,

$$b_1c_1 = b_2c_2$$

But

$$b_1 = b_2$$

and thus

$$c_1 = c_2$$

Furthermore, because the total quantity of material in the sample remains equal to the original concentration times the original volume of sample,

$$c_1V_2 = c_2(\text{original}) \times V_1(\text{original})$$

But the final concentration of the sample c_1 equals the concentration of the standard; therefore

$$\text{concentration of original sample} = \frac{\text{conc. of standard} \times \text{volume of sample after dilution}}{\text{original volume of solution}}$$

Suppose that a calibration curve is reproduced using the data in Table 8.1. If the absorbances of three samples are 0.125, 0.175, and 0.25, then from the calibration curve the concentrations of the three samples can be shown to be 0.025, 0.035, and 0.05%, respectively.

The optical methods using calibration curves where the relationship between A or T and c is plotted are commonly used in colorimetry. Similar results can be obtained by measuring I_1 and preparing a calibration curve from the relationship between I_1 and c, the sample concentration. A typical curve is shown in Fig. 8.5. The calibration curve is used in exactly the same fashion as described in the preceding paragraph.

2. Methods Used in Spectrophotometry

a. Single-Beam Method

Perhaps the simplest spectroscopic instrument is the single-beam spectrophotometer. The optical system is schematically identical to that shown in Fig. 3.1. For many years the most widely used instrument in routine analytical labs was the Beckman DU spectrophotometer, later modified to the DBG grating spectrophotometer. The method of operation was quite simply based on the absorption spectrum of the sample: an absorption wavelength was selected such that the degree of absorption by the sample was at a maximum. This generally coincided with an absorption maximum in the spectrum of the sample. A series of standards of known concentrations were made up and the absorbance versus concentration relationship was measured and plotted, producing a curve similar to that observed in Fig. 8.5. A sample was then put into the sample holder and its absorbance measured. Based on the degree of absorption and the calibration curve, the concentration of the sample was calculated. Sometimes it was necessary to correct for absorption due to the solvent or other interfering bodies. This was done by running a blank standard in which all the ingredients of the sample *except* the sample element itself were added. The degree of absorption of this blank was subtracted from that of the sample, and the net absorption due to the sample was calculated and reported.

Single-beam optical systems are subject to drift, but in the hands of a trained operator, this does not generally present a problem because he or she checks I_0 repeatedly during operation.

The commonly used instruments cannot easily be used for measuring the absorption spectrum of a sample because the wavelength drive is manual, and only point-by-point spectra can be recorded. This is a laborious procedure that has long been outmoded by the automatic drive systems readily available commercially. Nevertheless, single-beam spectrophotometers have been widely used for several decades and will undoubtedly continue to be used until they are ultimately displaced by the more accurate but more expensive double-beam systems.

b. Double-Beam Method

More sophisticated equipment uses a double-beam system, in which the beam of light from the source is split into two beams of equal intensity. One beam is the sample beam, and its intensity after passing through the sample is I_1. The other is the reference beam; its intensity after passing through the reference cell is I_0. The two beams are recombined and the detector measures both beams in rapid succession on a continuous basis and reads out the ratio I_1/I_0 directly. By using a suitable graph paper, this ratio can be recorded or calibrated directly as $-\log(I_1/I_0$, $\log(I_0/I_1)$, or A. The double-beam system is described in Chapter 3.

The effect of interfering compounds can be reduced by placing a suitable quantity of the interferent in the reference cell. This causes an equal amount of absorption in the reference and sample beams, so the two "errors" cancel each other out. The method is very convenient for rapid analysis and is now used almost universally.

3. Absorption Spectra

The absorption spectrum of most solutions is seldom as simple as that illustrated in Fig. 8.1, but usually varies against wavelength. The sample may absorb strongly at some wavelengths and much less at others. This variation is responsible for the color of the solution. For example, a blue solution is one that absorbs red and yellow light but allows blue light to pass through. A graphical representation of the transmission spectrum of potassium permanganate is shown in Fig. 8.8.

When setting up an analytical procedure, the wavelength at which absorption measurements are made is usually the wavelength at which the sample is absorbed the most. In the case shown in Fig. 8.8, the best wavelength to choose would be 520 nm. A small error in the wavelength setting would cause minimal error, but a large error would cause significant error because of the variation in the absorption coefficient with wavelength.

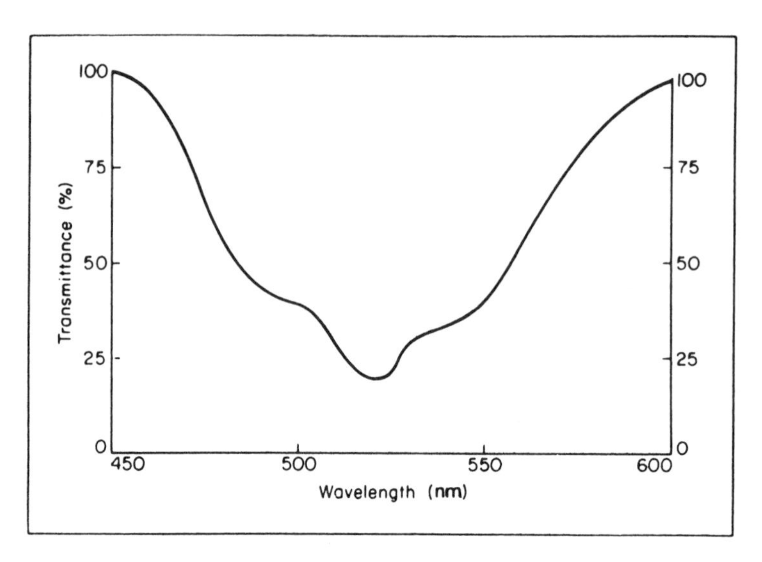

Figure 8.8 Transmission spectrum for potassium permanganate.

C. ERRORS AND RELATIVE ERRORS IN SPECTROPHOTOMETRY: THE RINGBOM PLOT

The actual readings made with a spectrophotometer are of the intensities of I_0 and I_1. The ratio of these two numbers gives the value of the transmittance. A typical transmittance–concentration curve is shown in Fig. 8.9. There are several sources of error in measuring I_1 (and therefore in measuring transmittance). One of the chief sources of error is the absorption of radiation by other compounds in the sample. Other sources of error include faulty readout from the detector, incorrect wavelength setting, variation in I_0 during measurement of I_1, and faulty sample preparation. In short, any measurement of transmittance or absorbance is subject to error.

Examination of Fig. 8.10 indicates the magnitude of the concentration error that results from a 1% error in measuring T. The smallest absolute error in transmittance occurs at the lowest concentration. The highest absolute error in concentration occurs where the curve is relatively flat. In this part of the curve, small errors in measuring T result in large errors in measuring concentration. However, the relationship between an error in measuring T and the *relative error* in measuring the concentration of the sample is different. In this instance,

$$\text{relative error} = \frac{\text{error in measuring the concentration}}{\text{true value of the concentration}}$$

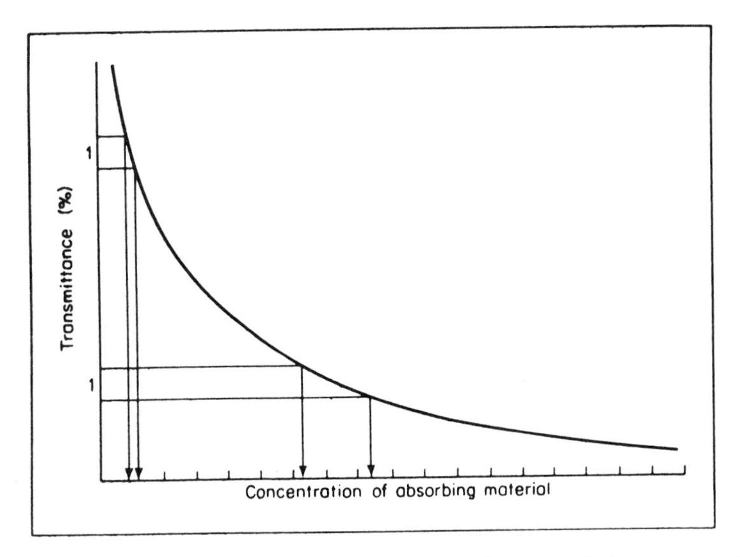

Figure 8.9 Relationship between the transmittance and the concentration of the absorbing material.

At the steep part of the curve in Fig. 8.10, 1% error in transmittance results in a low absolute error in concentration. However, the concentration is also very low. A small absolute error may therefore be large in relation to the concentration being measured.

The relationship first noted by Ringbom between the percent of relative error and the transmittance error can be calculated from Beer's law. The error in measuring the transmittance is termed the *photometric error*. Numerically, it can be shown that

$$\frac{\text{relative error (\%)}}{\text{photometric error (\%)}} = \frac{\Delta c/c3}{\Delta T}$$

where c is the concentration and T is transmittance; but from Beer's law

$$A = abc \quad \text{or} \quad ab = \frac{A}{c}$$

It can be shown that

$$\frac{\Delta c/c}{\Delta T} = \frac{1}{TA}$$

Using log base e,

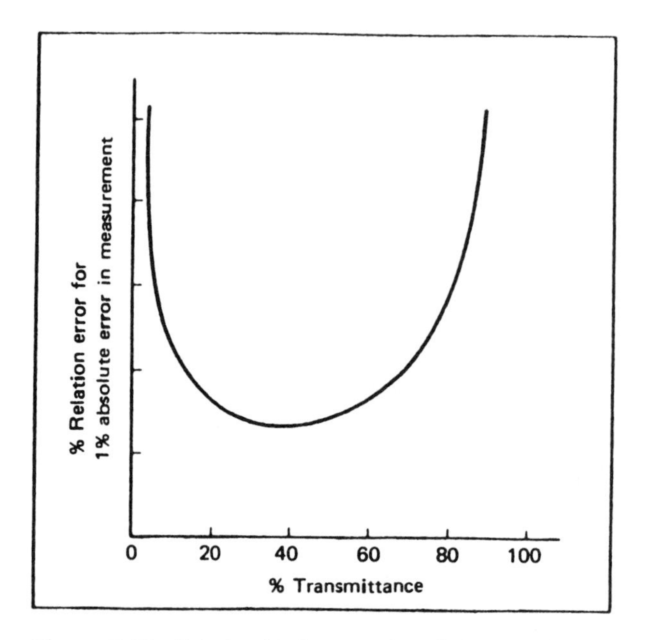

Figure 8.10 Relationship between the relative error in concentration (%) and the transmittance (Ringbom plot) with a constant error in T.

$$-\ln T = A = abc$$
$$\ln T = -abc$$
$$\frac{\Delta T}{T} = -ab\,\Delta c$$

or

$$\Delta c = -\frac{\Delta T}{T}\frac{1}{ab}$$

But $A = abc$; therefore,

$$\Delta c = \frac{\Delta T}{T}\frac{c}{A}$$
$$\frac{\Delta c}{c} = \frac{\Delta T}{TA}$$
$$\frac{\Delta c/c}{\Delta T} = \frac{1}{TA} = \frac{1}{T\ln T^{-1}}$$
$$= \frac{0.4343}{T\log T^{-1}} = \frac{0.4343}{T\log (I_0/I_1)}$$

The expression $\Delta c/c$ is the relative error in concentration resulting from an error ΔT in the measurement of transmittance. The relationship (shown in Fig. 8.9) discloses three important facts: (1) Relative error is very high at both very high and very low values of T (this occurs when the sample concentration is very low or very high, (2) the relative error is lowest over the 20–60% transmittance range, and (3) the relative error is a minimum at 36% transmittance (absorbance = 0.434).

For best results, it is advisable to work in the 20–60% transmittance range. If the sample is too concentrated (T too low), simple dilution may be effective. If the sample is too dilute (T too high), an extraction, evaporation, or other means of concentrating the sample may be required. Care should be taken not to introduce errors into the analytical method while performing such concentration steps. If none of these steps are possible, the results obtained should be reported with a note indicating the loss of reliability of the data under these conditions.

D. SPECTROPHOTOMETRIC EQUIPMENT

Commercial equipment is broken down into two broad categories, depending on the optics: single-beam and double-beam spectrophotometers. The individual components of a single-beam spectrophotometer, as shown in Figs. 8.11 and 8.12, are as follows.

1. Source

The wavelength range of visible light is between 400 and 750 nm. The most popular lamp for use over this range is the tungsten lamp, which is very similar to a household light bulb in that it consists of a piece of tungsten wire heated to white heat in a controlled atmosphere. In order to get best results it is important that the signal be constant over long periods of time. A steady power supply is therefore essential. The signal intensity from the lamp is not equal at different

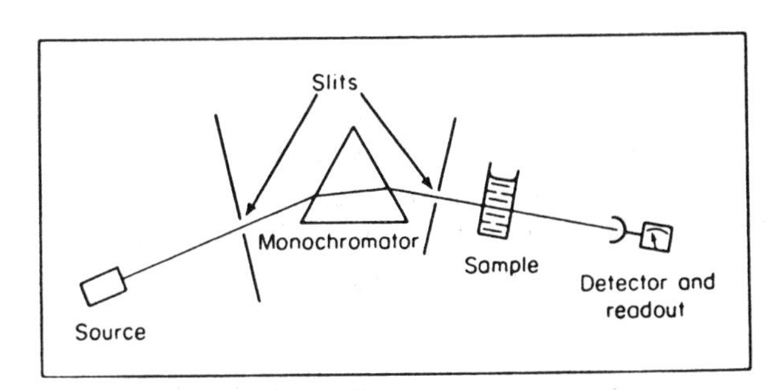

Figure 8.11 Schematic diagram of a single-beam colorimeter without lenses.

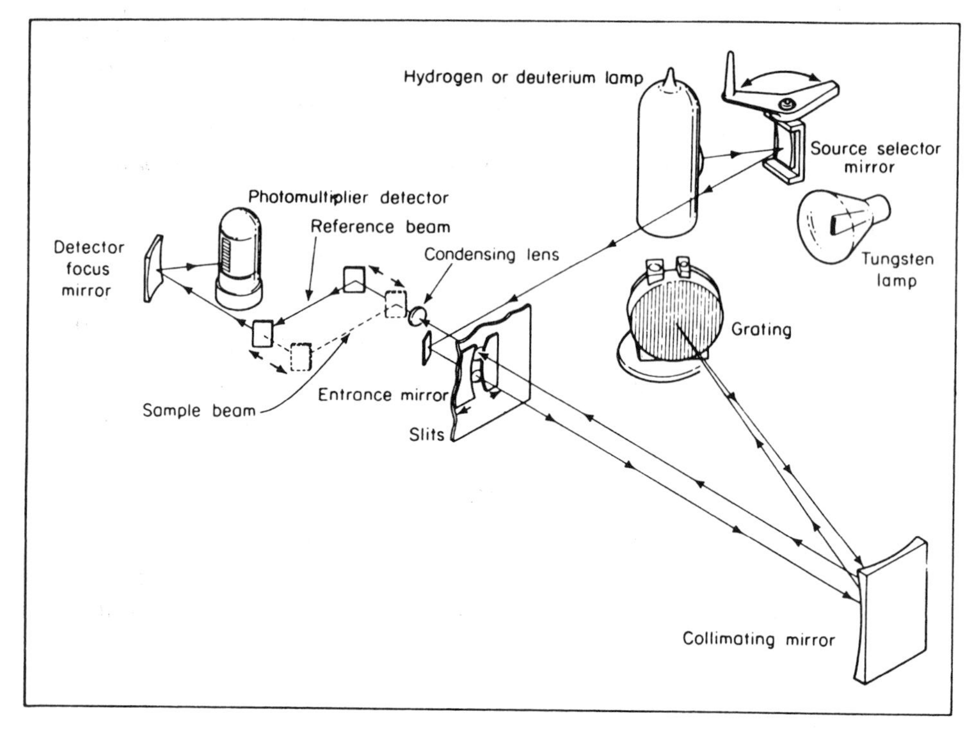

Figure 8.12 Optical diagram of a model DB-G grating spectrophotometer

wavelengths, but the change in intensity with wavelength is gradual and can be accommodated using a double-beam system or by using single-beam optics and measuring I_0 and I_1 at the wavelength used in the analysis. Hydrogen or deuterium lamps are widely used for absorption measurements at shorter wavelengths. They are discussed in Chapter 6.

2. Monochromator

A light filter allows light of the required wavelength to pass but absorbs light of other wavelengths and prevents it from complicating the analysis. On the simplest equipment, a particular filter may be used for a specific analysis. If the instrument is to be used for several different analyses, several filters may be used and interchanged. This method is very useful for a large number of repetitive determinations.

For analysis in the visible region, a *glass prism* may also be used as a dispersion element (see Chap. 3 Sec. A.1.b). Prisms are easier to use than filters because they can be tuned to any required wavelength range. The use of prisms greatly extends the versatility of the equipment when a wide variety of samples is to be analyzed and it is necessary to measure light absorption at a different wavelength for each type of sample. Gratings are also used extensively as monochromators for these instruments. For work in the UV region, quartz prisms are used. The dispersive properties of monochromators are discussed in Chapter 6.

3. Sample Cell

Suitable sample cells for use in the visible part of the spectrum are made from glass. Work in the infrared range must be done with cells made of materials such as sodium chloride or potassium bromide. Quartz cells may be used for studies involving the ultraviolet and visible regions of the spectrum as encountered in the broad field of spectrophotometry. For colorimetric analysis, glass cells are usually used.

4. Detectors

The most common detectors are photomultipliers or photocells. These convert the radiant energy to electrical energy, which may be measured and displayed on a dial or recorded on chart paper. Detectors are discussed in Chapter 6.

E. ANALYTICAL APPLICATIONS

Colorimetry and spectrophotometry are very widely used methods of quantitative analysis. Frequently special reagents can be used with different metals to form compounds with characteristic colors. Specific reagents have been developed for many metals in the periodic table, and in this respect colorimetry can be used to provide sensitive and accurate results. On the other hand, time-consuming chemical treatment is frequently necessary to prepare samples for analysis. The routine use of spectrophotometry depends on whether the accurate results obtainable justify the laboratory time necessary for sample preparation. It should also be pointed out that a great deal of loving care on the part of the operator is necessary. In routine industrial laboratories this is not always forthcoming. A list of metals and corresponding colorimetric reagents is shown in Table 8.2.

Colorimetric analysis can be used whenever the sample is colored. Many materials are colored without chemical reaction (e.g., inorganic ions such as dichromate, permanganate, cupric ion, and ferric ion) and need no further chemical reaction to form colored compounds. Colored organic compounds, such as dyestuffs, are also naturally colored. Solutions of such materials can be analyzed directly. The majority of metal ions, however, are colorless. The presence of these ions in a sample solution can be determined by first reacting with a

Table 8.2 Typical Reagents Used in Colorimetry

a. Aluminon

This compound forms a red lake with aluminum in a slightly acid solution (pH 4–5). It detects 0.1 μg of Al. Other elements, such as Be, Cr, Fe, Zr, and Ti, also becomes red with aluminon. These elements must be removed if a sample is being analyzed for Al. This reagent is an example of colorimetric analysis based on lake formation.

b. 4-Aminophenazone

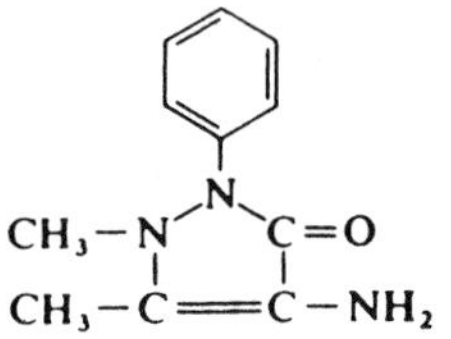

N
CH_3-N C=O
CH_3-C = $C-NH_2$

This compound reacts with phenols to give intensively colored compounds and will detect as little as 1.0 μg of phenol. This reaction, which must be carried out in alkaline solution, is an example of the determination of organic compounds by colorimetric analysis.

c. α-Benzoin oxime

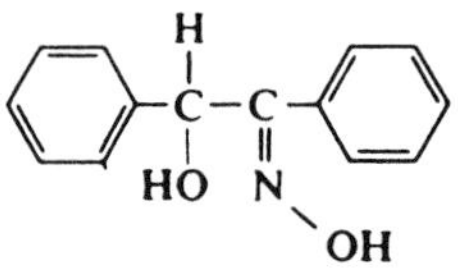

H
C–C
HO N
OH

This reagent forms precipitates with copper, molybdenum, or tungsten. The copper precipitate may be dissolved in chloroform and determined colorimetrically. The method is not highly sensitive, but will determine copper in more concentrated solutions starting at 0.05 μg/ml. This is an example of a low-sensitivity reagent.

d. Chloranilic acid

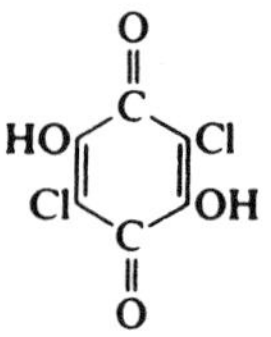

O
C
HO Cl
Cl OH
C
O

Chloranilic acid forms solutions that are intensely red. The addition of calcium to the solution precipitates the chloranilic acid and the intensity of the red diminishes. The change (loss) in color is a measure of the quantity of calcium added. Numerous other elements interfere with the procedure. This is an example of colorimetric analysis by loss of color after addition of the sample.

Table 8.2 *(continued)*

e. Quinalizarin

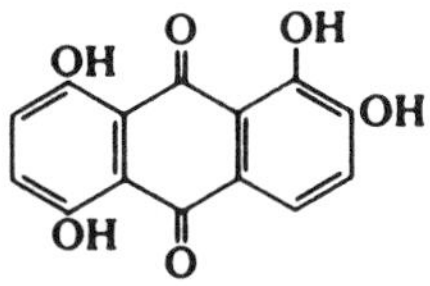

This reagent gives intensely colored solutions in aqueous solutions. In 93% w/wt H_2SO_4/H_2O, the color is red. The presence of borate causes the color to become blue. Numerous other ions, such as Mg^{2+}, Al^{3+}, and Be^{3+}, also react with quinalizarin. This is an example of a change of color of the reagent after reaction with the sample.

general organic reagent such as aluminon or quinalizarin. Ideally, the reagent should be selective; that is, it should react with only one ion under the conditions present. Second, the reagent should cause an abrupt color change when mixed with the metal ion. This imparts high sensitivity to the method. Third, this intensity of color should be related to the concentration of ions in the sample. Colorimetric reagents have been developed for almost all metal or acid ions that commonly occur in chemistry. Many of them are sensitive and selective and have provided the basis of many methods of elemental analysis.

F. POLARIMETRY

Polarimetry and spectrophotometry are similar only inasmuch as they utilize radiation in the same wavelength range. When we use the former technique, however, we measure the light's polarization rather than its hue or intensity. The components of the equipment used in polarimetry are the same as those of spectrophotometric equipment, with the addition of a polarizer.

1. Polarization

A beam of light consists of an oscillating electric field and therefore an oscillating magnetic field. The magnetic field can be resolved in any given plane (e.g., vertical, horizontal, or any intermediary plane). An unpolarized, or normal, light beam includes light waves with magnetic fields in all planes; these fields cancel each other out and produce no net field in any plane.

We can demonstrate the property of light called *polarization* and some of its effects by a simple experiment. First we remove the lens from a pair of polarized sunglasses. We then pass a beam of light through one of the lenses; all the light having a magnetic field vibrating in a particular plane (e.g., the vertical one) is

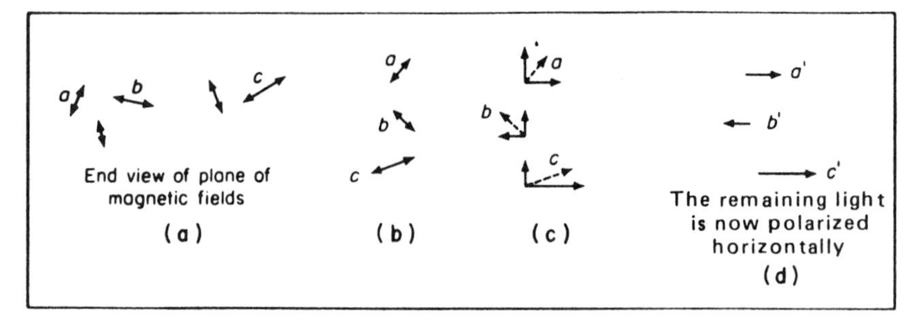

Figure 8.13 The selective removal of components from light: (a) ordinary light, showing various magnetic fields that comprise it; (b) selected radiation *a*, *b*, and *c*; (c) selected radiation resolved into vertical and horizontal components; and (d) light after the vertical component is removed (the remaining light is now polarized in the horizontal direction).

removed by the lens. In addition, the lens removes the *vertical component* of all other magnetic fields in the plane, as shown in Fig. 8.13. The light is now said to be polarized horizontally. If we now take the second lens, place it in the beam of light emerging from the first lens, and rotate it, we see that at one position of rotation no light emerges from the combination of the two lenses. The horizontal component of the light coming from the first lens has been removed by the rotated second lens; hence no light emerges from the lenses. One lens has removed the vertical component of the polarized light and the second lens has removed the horizontal component of the polarized light, with the result that no light from the original beam emerges from the second lens.

Polarized sunglasses show *optical activity*. This means that they possess the ability to remove one plane of polarization of a beam of polarized light. Some molecules, in either the pure state or in solution, also have this ability. Their presence and concentration may be detected by measuring the degree and direction in which they polarize light. The field devoted to the analysis of optically active compounds by the polarization of light is called *polarimetry*.

2. Optical Rotation

Let us consider a single beam of light polarized in a horizontal direction by a polarizing prism, such as the first polarized sunglass lens in our experiment above. Passage through an optically active solution will cause the axis of polarization to be rotated either to the right (dextrorotatory) or to the left (levorotatory). This is illustrated in Fig. 8.14. No absorption of light takes place, but the plane of polarization is changed. The extent of the change depends on several factors, such as the light wavelength, the temperature, and the solvent used,

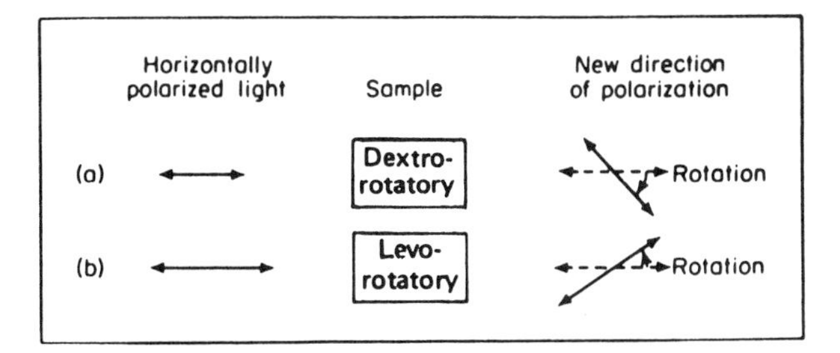

Figure 8.14 Diagram of the effect of an optically active solution on the direction of polarization of transmitted light.

all of which can be controlled. The sodium D doublet at 589.0 and 589.5 nm is often used as the source of radiation. It is polarized in one direction (e.g., vertical) by removing the complementary direction (horizontal). The specific rotation of a particular compound is designated as $[\alpha]_\lambda^{t_0}$, where t_0 is the temperature and λ is the wavelength used. It can be shown that the optical rotation of a solution is

$$[\alpha]_\lambda^{t_0} = \frac{100\alpha}{lC}$$

where α is the experimentally determined degree of rotation, l is the sample path length in decimeters, and C is the sample concentration in g/100 ml. When a pure undiluted sample is used, the relationship is modified to

$$[M]_\lambda^{t_0} = \frac{[\alpha]_\lambda^{t_0} M}{100}$$

where M is the molecular weight of the compound, $[M]_\lambda^{t_0}$ is the molar rotation, and

$$[\alpha]^{t_0} = \frac{\alpha}{ld}$$

where d is the density of the liquid.

3. Types of Molecules Analyzed

To be analyzed by polarimetry, a molecule must be optically active. If it is not, polarized light passes through it unaffected. The most common types of optically active molecules are those with an asymmetrical carbon. An asymmetrical carbon is one that is attached to four different groups. It can be shown that these groups can be put together in two arrangements that are mirror images of each

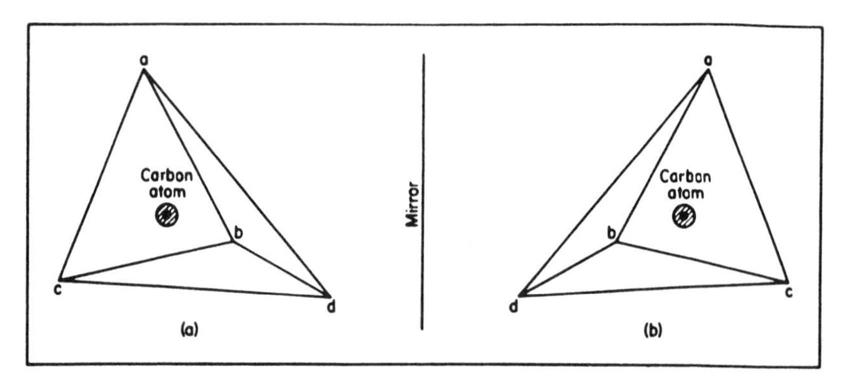

Figure 8.15 The possible arrangements of four groups of molecules that have an asymmetrical carbon.

other, as shown in Fig. 8.15. Such molecules resemble each other as do two shoes from a pair or two gloves from a pair. Chemically, they are virtually identical. However, molecules of the type shown in Fig. 8.15a form crystals that are mirror images of those in Fig. 8.15b. Pasteur laboriously proved this by crystallizing tartaric acid and separating the crystals by hand into the two forms, I and II, and illustrating their optical properties. Some examples of optically active components are shown in Fig. 8.16.

Polarimetry is particularly useful in sugar manufacture. Sugar is optically active, and the concentration of sugar in extracts can be rapidly determined by measuring the degree of rotation of light after it has passed through the sample.

Recently, it has been shown that pharmaceutical drugs are optically active and that the optical isomers have different physiological effects on the body. For example, the body can digest and interact with D active compounds but not L active compounds. Some results show that one optical isomer of a particular compound acts as a β-blocker whereas the other isomer does not. These isomers can be separated by chromatography using chiral active substrates. Such observations will have profound effects on the pharmaceutical industry, increasing the efficacy of drugs but also increasing the cost of manufacture.

4. Optical Rotary Dispersion

The molar rotary dispersion is a direct measure of the rotation of light by a molar solution under standard conditions. However, the molar dispersion varies with the wavelength of the light used. The relationship between M and the wavelength of light is the basis of optical rotary dispersion. A typical curve exhibits the shape shown in Fig. 8.17. This field is particularly useful in an increasing

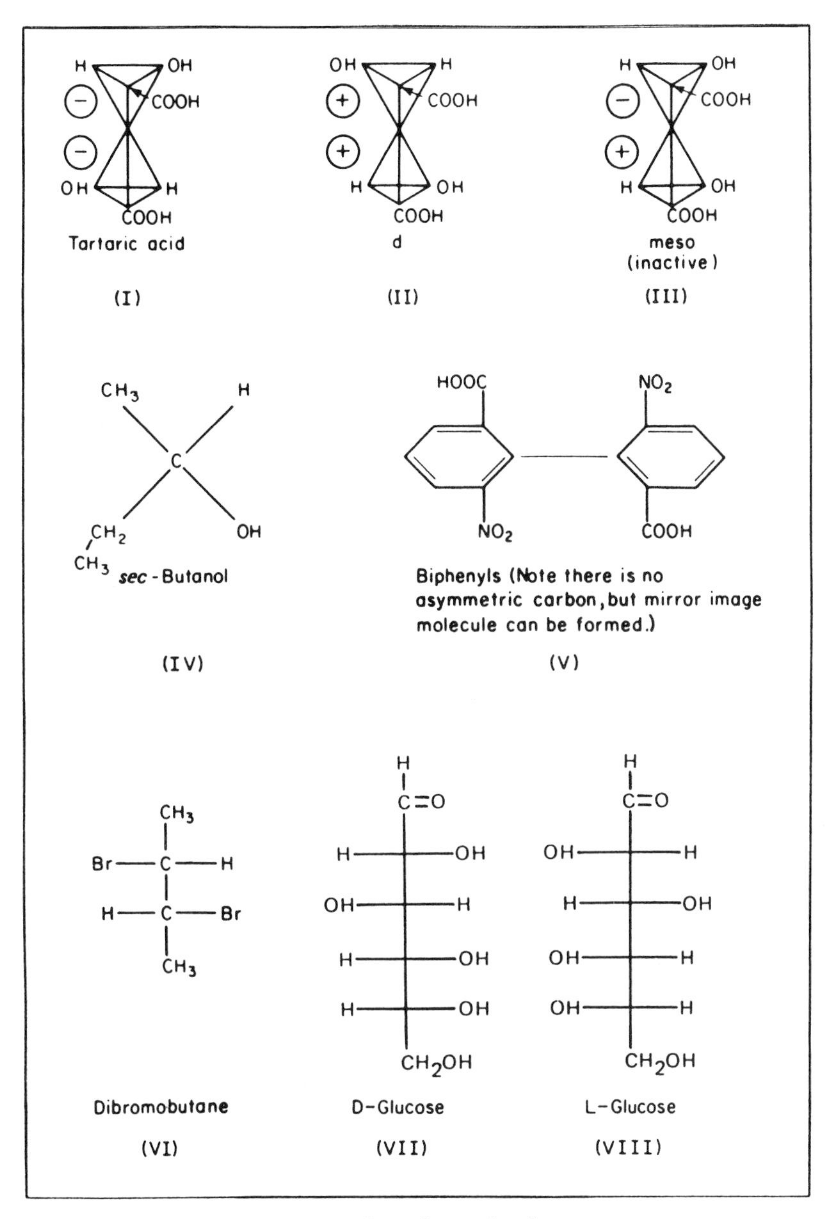

Figure 8.16 Examples of optically active molecules.

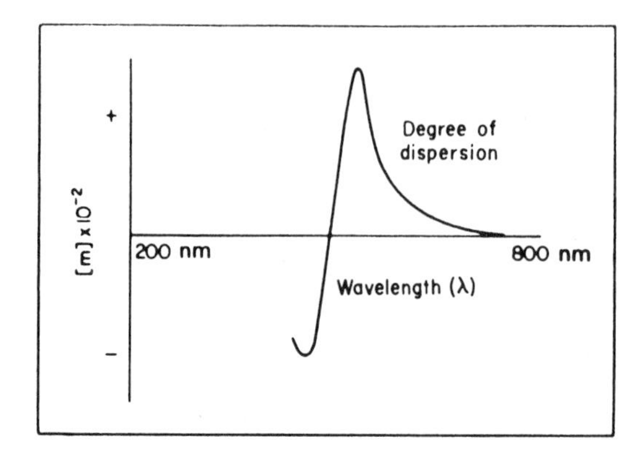

Figure 8.17 A typical curve showing the relationship between optical rotary dispersion and the wavelength of light.

number of branches of organic chemistry and biochemistry. Many natural products are optically active. Their presence in living tissues and fluids can be studied and confirmed by measuring the wavelength at which maximum rotation occurs—a physical characteristic of the compound and therefore the basis of qualitative identification. The degree of rotation is a quantitative measure of the concentration present. Prior separation from other compounds present is usually essential.

BIBLIOGRAPHY

Ayres, G. H., *Quantitative Chemical Analysis*, 2nd ed., Harper and Row, New York, 1968.

Anal. Chem., Application Reviews, April 1994.

Djerassi, C., *Optical Rotary Dispersion; Applications to Organic Chemistry*, McGraw-Hill, New York, 1960.

Meehan, E. J., Optical methods of analysis, in *Treatise on Analytical Chemistry* (P. J. Elving, E. Meehan, and I. M. Kolthoff, eds.), Vol. 7, Wiley, New York, 1981.

Sandell, E. B., *Colorimetric Determination of Traces of Metals*, 3rd ed., Wiley, New York, 1959.

Wong, K. P., *J. Chem. Ed.*, (1974), *51*: A573.

SUGGESTED EXPERIMENTS

8.1 Prepare a stock solution containing 1 g/liter of ferrous sulfate. By taking aliquots of the stock solution and diluting, prepare standard solutions containing Fe^{2+} in the concentration range 2–20 ppm. Reduce the iron by adding excess ascorbic acid. Add a slight excess of 1,10-phenanthroline and adjust to ph 7 with ammonium acetate buffer. Measure the absorbance at 512 nm. Correlate the absorbance with the concentration of iron in the sample and prepare a calibration curve.

8.2 Take a sample of mild steel and dissolve it in HNO_3. Add potassium persulfate to oxidize any carbon remaining, and sodium bisulfite to return all manganese to the bivalent state. Oxidize the manganese to permanganate by the addition of potassium periodate. Add phosphoric acid to complex the iron and remove the yellow color from solution. Measure the absorbance using the $KMnO_4$ absorption curves prepared in Experiment 1.1.

PROBLEMS

8.1 If a solution appears blue when a white light is passed through it, what colors has the solution absorbed?

8.2 What is Beer's law? Complete the following table:

Solution	Absorption (%)	*T*	A	Concentration (ppm)
1	1			1
2	13			6
3	30			15
4	55			34
5	80			69

8.3 Assuming the data obtained in Problem 8.2 were for a calibration curve and the same cell w as used for all measurements, complete the following table:

Solution	Absorption (%)	*T*	Concentration (ppm)
1	30		
2	3		
3	10		
4	50		
5	70		

8.4 What is the relationship between the absorption cell length b and the absorbance A? Complete the following table (the concentration c was equal in all cases):

Sample	Path length b (cm)	A
1	0.1	0.01
2	0.5	
3	1.0	
4	2.0	
5	5.0	

8.5 The dilution method was used to measure the concentrations of several solutions. The original volume of each solution was 1 ml. The volume of the standard was 1 ml and its concentration was 10 ppm. Complete the following table:

Sample	Final volume (ml)	Original concentration of sample
1	1.2	
2	2.0	
3	1.5	
4	1.7	
5	3.0	

8.6 What is the optimum transmittance range to be used in order to keep relative error to a minimum? If a sample shows 10% transmittance, how may it be treated to bring its absorbance inside the desired range?

8.7 Draw a schematic diagram of a single-beam spectrophotometer. Indicate the principal components.

8.8 Indicate three reagents used for colorimetric analysis.

8.9 In a Duboscq colorimeter, the colors of a standard and a sample matched at the solution depths shown below. Calculate the concentration of the sample solution.

Standard	Depth	Concentration (ppm)	Sample	Depth	Concentration (ppm)
A	25	5	1	20	
A	25	5	2	15	
B	32	10	3	28	
B	24	10	4	28	

8.10 What materials are used as prisms in spectrophotometry?

8.11 What detectors are used in spectrophotometry?

8.12 Describe optical rotation.

8.13 Describe optical rotary dispersion.

8.14 What is a typical relationship between optical rotary dispersion and the wavelength of light?

8.15 What kind of sample can be studied by optical rotary dispersion

9

Flame Photometry

Flame photometry is based on the measurement of the emission spectrum produced when a solution containing metals or some nonmetals such as halides, sulfur, or phosphorus compounds is introduced into a flame. In early experiments the visible color of the flame was used to confirm the presence of certain elements in the sample, particularly alkali metals and alkaline-earth metals. Later the whole ultraviolet and visible range was utilized using a spectrophotometer. The wavelength of the radiation indicates what element is present, and the radiation intensity indicates how much of the element is present. The method is useful for elemental analysis in general, but is particularly useful for the determination of the elements in the first two columns of the periodic table, including sodium, potassium, lithium, calcium, magnesium, strontium, and barium. The determination of sodium, potassium, calcium, and magnesium is often called for in medicine, agriculture, and animal science. Flame photometry can also be used for the determination of certain transition elements, such as copper, iron, and manganese. (Tables 9.3–9.5 (Sec. D.1) include a list of the elements that have been determined by this method.) Although flame photometry is a means of determining the total metal content of a sample, it tells very little about the molecular form of the metal in the original sample. It is not suitable for the direct detection and determination of the noble metals or inert gases.

A. ORIGIN OF SPECTRA

Before excitation can occur, an analyte element in a sample must literally be atomized, i.e., converted to the free atomic state. This is a complex process, which is subject to many contributing factors that influence the *atomization efficiency*, i.e., the efficiency of converting elements existing as molecules or ions into free atoms. (This complex process is described in Chap. 7, Sec. B.2 and Table 7.2.)

As we know, an atom is composed of a nucleus and orbiting electrons. The size and shape of the orbitals in which the electrons travel around the nucleus are not random. For each orbital, only certain permitted energies are possible. They are described mathematically by quantum theory, which allows their energies to be calculated. If the valence electrons are orbiting at the lowest permissible energy level, the atom is said to be in the ground state and is unexcited. However, the orbitals at higher energy levels, as described by quantum theory, are available but are unfilled. A valence electron may move to one of these orbitals. If the electron is in a higher orbit, the energy of the atom is increased and the atom is said to be in an excited state. Flame photometry deals with the emission of radiant energy that accompanies the relaxation of excited atoms.

When a valence electron moves from a higher excited state to a lower excited state (including the ground state), a photon of radiant energy is given off. When atoms are put into a flame, they become excited. The excited atom, which is unstable, quickly emits a photon of light and returns to a lower energy state, eventually reaching the ground state. The measurement of this emitted radiation is the basis of flame photometry. The permitted energy levels of all atoms can be presented in the form of a diagram like Fig. 9.1. Such diagrams are sometimes called Grotrian charts and are discussed in Chapter 7.

The permitted energy levels of the electrons of many elements are defined by quantum theory. A drop or transition of an electron from one level down to the next involves the loss of a well-defined energy $E_2 - E_1$, where $E_2 - E_1$ is the energy difference between the two levels. But $E = h\nu$, where h is Planck's constant and ν is the frequency of emitted light. Since $\nu = c/\lambda$, where c is the speed of light and λ is the wavelength of the radiation, from Eqs. (2.2) and (2.3) we have

$$E_2 - E_1 = \frac{hc}{\lambda}$$

or

$$\lambda = \frac{hc}{E_2 - E_1}$$

The energy levels are characteristic of the emitting element. Hence the wavelength of the radiation emitted is characteristic of the atoms of the particular/

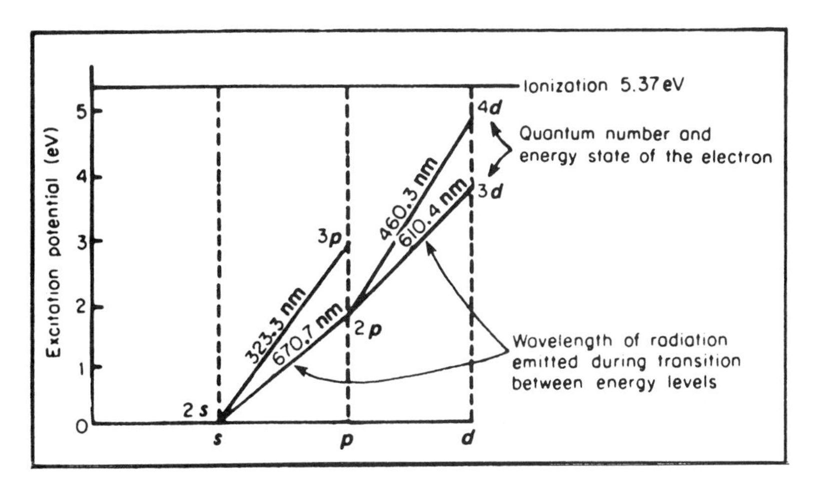

Figure 9.1 Partial energy level diagram of lithium.

element from which it was emitted. When flame photometry is used as an analytical tool, the wavelengths of the radiation coming from a flame allow us to know what elements are in that flame. Also, the intensity of the radiation tells us the concentration of these emitting elements present.

The energy available in a flame for exciting atoms is quite low compared to other excitation sources such as electrical discharges or plasma used in emission spectrography. For this reason atoms in a flame are usually excited only to low excitation levels. To raise an electron from the ground state to the upper excitation levels requires more energy than is usually available from the flame. In many cases, even the energy necessary for a transition between the ground state and the first excited state is more than is available in the flame. As a consequence, quite a number of elements in the periodic table cannot be usefully excited by flames, and flame photometry is not the method of choice for their analysis. However, elements in the first two columns of the periodic table require small amounts of energy in order to become excited. Included are sodium, potassium, lithium, calcium, beryllium, and so on, all of which give brightly colored flames. The Chinese were perhaps the first to observe these colored flames and used them to develop the colors in fireworks many centuries ago in the same way that we do today.

The more recent high-temperature flames such as the nitrous oxide–acetylene flame (see Chap. 7) have more energy available for excitation. More elements can be excited in these flames, and their emission spectra can be observed and used in analytical determinations. The use of the elongated burner has also increased the sensitivity achievable using flame photometry.

B. EQUIPMENT

A schematic diagram of the equipment used in flame photometry is shown in Fig. 9.2. The instrument has the same basic components as other spectroscopic apparatus, namely, a source, a monochromator, a slit system, and a detector system. It also includes a burner, which is used to burn the sample and thermally excite it in the process. The various components of the instrument are described in the following sections.

1. Burner

The central component of a flame photometer is the burner assembly, which has two functions: to vaporize the sample and then introduce it into the flame. Here free atoms are formed. The flame then excites the atoms and causes them to emit radiant energy. For analytical purposes, it is essential that the emission intensity be steady over reasonable periods of time (1–2 min). There are two types of burners in common use: the total consumption burner and the Lundegardh burner, the latter being more common.

In the *total consumption burner* the sample, which must be in the form of a liquid, is aspirated completely into the flame. A typical total consumption burner is illustrated in Fig. 9.3. In this burner the air or oxygen aspirates the sample into the base of the flame. Atomization and excitation of the sample then follow. (Note that *atomization* in flame photometry, as in atomic absorption spectroscopy, means *reduction to the free atomic state*.)

In the *Lundegardh burner* (Fig. 9.4), the sample, which again must be liquid, is first aspirated into a spray. Larger droplets coalesce on the side of the spray chamber and drain away. Smaller droplets and vapor are swept into the base of the flame in the form of a cloud. An important feature of the Lundegardh

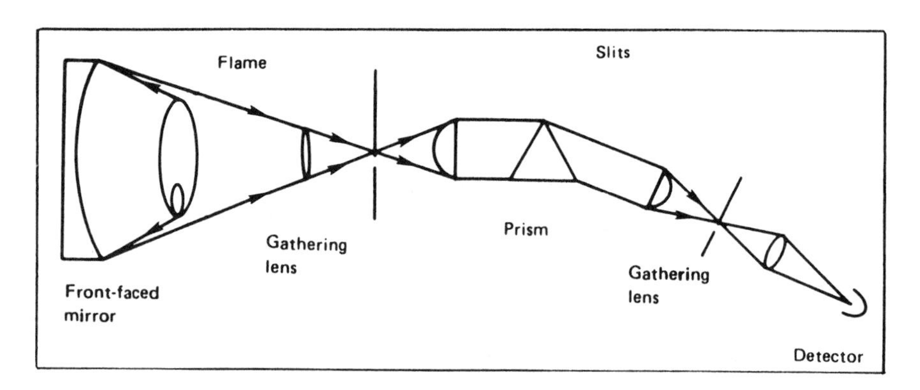

Figure 9.2 Schematic diagram of a flame photometer.

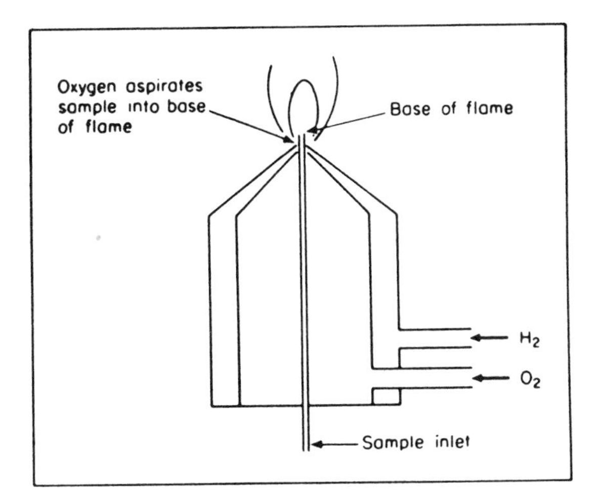

Figure 9.3 Beckman total consumption burner.

burner is that only a small portion (about 5%) of the sample reaches the flame. The droplets that reach the flame are, however, very small and easily decomposed. This easy decomposition results in an efficient atomization of the sample in the flame. The high atomization efficiency leads to increased emission intensity and increased analytical sensitivity. Furthermore, the Lundegardh burner is physically quiet to operate, which is a distinct advantage over the noisy total consumption burner.

Difficulties may arise if there is any selective evaporation of the sample in the spray chamber. For example, if the sample contains two solvents, the more volatile sample will preferentially evaporate in the spray chamber, leaving the

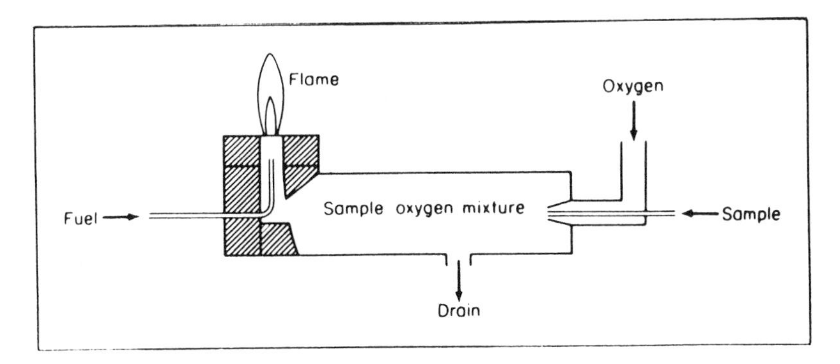

Figure 9.4 Lundegardh burner.

element of interest in the less volatile component. The latter may drain away, taking the sample with it. In this event, many of the sample atoms will never reach the flame. The emission intensity will be reduced and an incorrect analysis obtained.

2. Mirrors

The radiation from the flame is emitted in all directions in space. Much of the radiation is lost and a loss of signal results. In order to increase the amount of radiation falling on the detector, a mirror is located behind the burner to reflect radiation that would otherwise be lost. Both the reflected radiation and the direct radiation from the flame travel down the light path. The mirror is concave and is focused on the entrance of the monochromator. To get the best precision, the steadiest part of the flame is located at the center of the curvature of the mirror. This allows the steadiest possible emission signal to fall on the detector.

The reflecting surface of the mirror is on the front surface. If it were on the rear surface, as in normal household mirrors, the radiation would have to go through the support material (quartz or glass) twice, and considerable radiation would be lost, particularly at the shorter wavelengths. Front-surface mirrors are more efficient reflecting surfaces, but they are not protected by the glass. They are therefore easily scratched and are subject to chemical attack. Great care must be taken to protect them.

3. Monochromators

These consist of slits and dispersion elements. They are discussed in greater detail in Chapter 3, Section A.1.b.

4. Slits

With the best equipment, entrance and exit *slits* are used before and after the dispersion elements. The entrance slit cuts out most of the radiation from the surroundings and allows only the radiation from the flame and the mirrored reflection of the flame to enter the optical system. The exit slit is placed after the monochromator and allows only a selected wavelength range to pass through to the detector. For many purposes it is essential that this wavelength range be very narrow, that is, of the order of a few nanometers. This is necessary if emission lines from other components in the flame have a wavelength similar to those of the emission lines of the elements being determined. The slit must prevent such interfering lines from reaching the detector. The function of the slits is illustrated in Fig. 9.5.

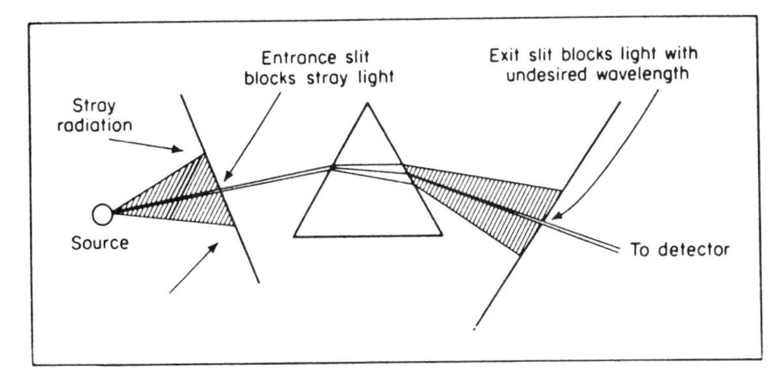

Figure 9.5 Monochromator system shown without collimator lenses for simplification.

5. Dispersion Elements

Grating dispersion elements are most commonly used, but a *prism* is usually used in less expensive models. Glass disperses visible light better than most transparent materials, but is not transparent in much of the UV region. Quartz is transparent over the entire UV region and is therefore the material most commonly used for making prisms, even though its dispersive power in the visible range is less than that of glass. A more complete description of monochromators is given in Chapter 3.

Some elements emit only few lines and therefore have a simple emission spectrum. In these cases wider wavelength ranges may be allowed to fall on the detector without causing serious error. In such circumstances the monochromator system may be replaced by an *optical filter*. Filters are built with materials that are transparent over a narrow spectral range. The transparent spectral range is designed to be the one in which emission from the sample occurs, making it useful for the determination of a specific element, such as sodium or potassium, in which case a sodium filter or potassium filter is used. For example, if a sample of blood contains both sodium and potassium and it is necessary to determine the sodium, a sodium filter allowing only sodium emissions to pass is used. To determine potassium in the same sample, a potassium filter is used, which is transparent to potassium emission but blocks out the sodium emission. When a filter is placed between the flame and the detector, radiation of the desired wavelength from the sample is allowed to reach the detector and be measured. Other radiation is absorbed by the filter and is not measured. Instruments that use filter monochromators are very convenient for simple repetitive analysis but are limited with regard to the number of elements for which they can be used unless a large number of filters are employed. These filters are specifically designed for the purpose.

Table 9.1 Process Generating Atomic Emission in Flames

Liquid sample → enters flame	Droplets of → liquid	Evaporation of droplets	
→Formation of → residue	Decomposition → of residue	Formation of excited atoms → and emission of radiation from atoms	Oxidation of atoms

6. Detectors

The function of the detector is to measure the intensity of radiation falling on it. As far as possible, it should be sensitive to radiation of all wavelengths that may be examined. Also, it should change the radiation signal into an energy form that can be readily measured and recorded. The most common types of detectors are photomultiplier detectors, which produce an electrical signal from the radiation falling on them. They are described in Chapter 6, Section C.2.c.

C. FLAMES

What is usually thought of as a flame is the exothermic chemical reaction between two gases. As a physical reaction, a great deal of heat is generated when a fuel burns. As a chemical reaction, an oxidation–reduction takes place, with the oxidant oxidizing the fuel. Common oxidants are air, oxygen, and nitrous oxide. Common fuels are hydrogen, acetylene, and propane. When a liquid sample is introduced into a flame, the process is very complicated. Table 9.1 is a concise description of the sequence of events that normally occurs.

In order to obtain a steady emission signal of constant intensity, each step in the process shown in Table 9.1 must be regulated. Modern equipment is able to do this with little attention from the operator. The skillful operator, however, is always aware of the chemical and physical processes involved and is thus able to recognize and correct any problems that may arise. Variations in emission intensity can be caused by several factors, including (1) blockage in the burner, preventing the flow of sample; (2) change of viscosity of the sample; (3) change of solvent in the sample; (4) change in the fuel or oxidant flow rate to the burner; and (5) change of the position of the burner in the instrument, causing the radiation to be displaced from the light path.

1. Emission of Atomic Spectra

Spectral emission lines are generated by the excited atoms formed during the process of combustion in a flame. The intensities and wavelengths of emission lines are measured in flame photometry. The *intensity* of the emission depends on several factors, including (1) the concentrations of the element in the sample and (2) the rate at which excited atoms are formed in the flame; in turn, the

Table 9.2 Maximum Flame Temperatures

Fuel	Flame temperature (°C)		
	Oxidant: Air	O_2	N_2O
H_2	2000	2800	
ArH_2	1600		
Acetylene	2200	3050	3000
Propane	1900	2800	
Coal gas	1800		

latter depends on (3) the rate at which the sample is introduced into the flame, (4) the temperature of the flame, and (5) the composition of the flame. The relationship between the emission intensity and the concentration of atoms in the sample provides the basis for quantitative analysis by flame photometry.

Flame temperature is probably the most important single variable in flame photometry. It is controlled by the type of fuel and oxidant used. Some typical flame temperatures are given in Table 9.2. In general, an increase in flame temperature causes an increase in emission intensity. This does not happen with elements that ionize easily, such as sodium, potassium, or lithium. If these elements are heated too strongly, they become ionized instead of excited, because the outer electrons move to higher and higher energy states until they leave the atom completely, forming an ion. If the atoms ionize, the valence electrons are lost and therefore cannot return to the ground state and emit atomic radiation in the process. A loss of atomic emission intensity results. These elements are therefore determined in low-temperature flames. However, for most metals the emission intensity increases with an increase in flame temperature. This follows as a consequence of the Boltzmann distribution, which is discussed in Chapter 2.

The relationship between intensity and temperatures is given by the equation

$$S = \frac{N_1 E}{\tau} = \frac{N_0 E}{\tau} \frac{g_1}{g_2} e^{-E/kT}$$

where

S = intensity of the emission line
N_1 = number of excited atoms
N_0 = number of unexcited atoms
τ = lifetime of the excited state
E = energy of excitation ($E = E_1 - E_0$)
g_1/g_2 = ratio of the statistical weights of the ground state and the excited state
T = absolute temperature
k = Boltzmann distribution coefficient

From this equation it can be seen that S is related directly to the number of atoms in the excited state. As this number increases, the intensity of radiation increases. However, the Boltzmann distribution tells us that for any system of atoms at temperature T there will be an equilibrium between the number of atoms in the ground state and the number of atoms excited. Therefore S is related directly to the number of atoms in the ground state times a constant $e^{-E/kT}$, where T is the absolute temperature. As the absolute temperature increases, the number of atoms in the excited state increases and the emission intensity increases (see Chap. 2).

At the same time as the energy required to cause excitation increases, it is more difficult to excite the atoms, and the number of atoms in the excited state decreases. But we have already seen that $E = h\nu = hc/\lambda$. As a consequence of this relationship, a decrease in the wavelength of the emission line indicates that more energy is required to excite the atom. The process becomes more difficult and fewer atoms are excited and the intensity of radiation decreases. Consequently, elements with emission lines in the short-wavelength part of the spectrum are very weak. This applies to elements such as zinc, cadmium, mercury, and arsenic. On the other hand, as the temperature of the system increases, more energy is available to cause excitation and more atoms are excited, resulting in a stronger emission signal.

High-temperature flames are generally favored, particularly for the transition elements and the alkaline-earth metals. However, such flames should not be used for the determination of elements that ionize easily, such as sodium, potassium, lithium, and cesium.

Another factor that influences emission intensity is the ratio of fuel to oxidant in the flame. The highest flame temperature is obtained when a stoichiometric mixture of the two is used. In a stoichiometric flame the number of moles of fuel and oxidant present exactly balance out and there is no excess of either after combustion. Any excess of oxidant or fuel results in a dilution of the gases and a decrease in the temperature. Furthermore, some atoms are unstable in different kinds of flames. For example, aluminum atoms oxidize very quickly to aluminum oxide. Aluminum oxide emits molecular radiation that is not at the same wavelength as the line emission associated with aluminum atoms. This results in a direct loss of atomic emission intensity. To prevent the formation of aluminum oxide, the flame is usually run in a "reducing state," that is, with an excess of fuel. The excess fuel mops up free oxidant and greatly slows down the oxidation of the aluminum in the flame. On the other hand, some elements emit more strongly in oxidizing flames. An example is the alkaline earths. In this case the metal oxides are unstable in the flame, and although they may form for a transient period of time, they are rapidly decomposed back to the free atoms. The excess oxygen helps decompose other materials present and reduces the molecular background. Consequently, an oxidizing flame is recommended for these elements. Manufacturers of flame photometric equipment have a list of recommended flame compositions for the elements that can be determined by this process.

A second factor is the solvent used in the sample. If the solvent is water, the process of atomization is endothermic and thus relatively slow. Atomization is therefore slowed up. On the other hand, if the solvent is organic, the reaction in the flame is exothermic and the reaction is faster. Atomization takes place more quickly when organic solvents are used. From a given sample, more free atoms are liberated and emission intensity is increased. This ''solvent effect'' greatly changes the emission intensity. It is therefore very important that calibration curves be prepared using the same solvent as that of the sample (see Chap. 7, Sec. C.2.b).

2. Emission of Molecular Spectra

In addition to the atomic spectra emitted by elements in a flame, we can also detect molecular spectra from excited molecules, such as metallic oxides. These are usually band spectra and are therefore not as intense as line spectra. They are, however, analytically useful under certain circumstances. For example, fragments such as BaOH, CaOH, SrOH, and MnOH, along with CaO and oxides of the rare earths, all emit molecular band spectra when introduced into a flame and can be used for the determination of Ba, Ca, Sr, Mn, and the rare earths, respectively.

3. Background Emission

The flame is a source of spectral energy arising from combustion of the fuel used to create the flame. Furthermore, the hot or burning solvent and other components of the sample also emit radiation over an extensive wavelength region. These two sources of radiation form the *background* radiation that is always present in the spectrum of the flame. If necessary, a correction must be applied to quantitative measurements of emission line intensities. A typical example of background emission from an oxygen–acetylene flame is illustrated in Fig. 9.6. The principal emission lines of nickel, sodium, and potassium are superimposed on this background for illustrative purposes. The background can be measured either from a blank at the emission line wavelength or by measuring the background of the actual sample very close to the emission line. The net emission from the sample is the total emission minus the background.

T. Rains at the U.S. National Bureau of Standards developed an ingenious method to overcome this problem. He designed a reflector plate that oscillated at a controlled frequency and through a controlled angle. At one end of the swing of the reflector plate the emission line from the metal falls onto the de-

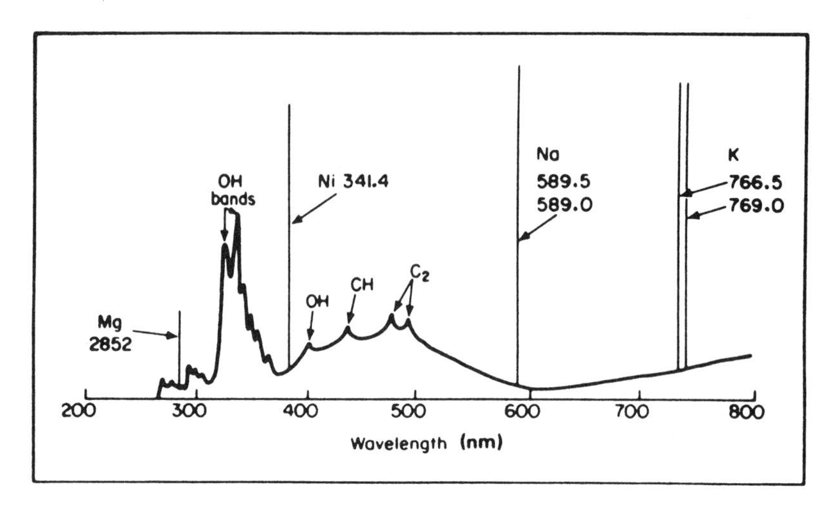

Figure 9.6 Total emission spectrum of a flame.

tector; at the other end of the swing the background emission falls on the detector. As a result, the detector is exposed alternately to the background and to the metal emission. It therefore generates an AC signal that is proportional to the difference between the two signals. Effectively, correction is made for the background emission and the intensity of the emission line is measured more accurately.

D. ANALYTICAL APPLICATIONS

1. Qualitative Analysis

The qualitative aspects of flame photometry are useful mostly for the detection of elements in groups I and II of the periodic table. These elements include sodium, potassium, lithium, magnesium, calcium, strontium, and barium. The presence of certain elements can be detected visually, as in the case of the yellow flame produced by sodium. It is generally much safer, however, to use a filter or a monochromator to separate radiation with the wavelengths characteristic of the different metals from other radiation present. The yellow radiation from sodium impurities in a sample is often intense enough to mask radiation from other elements present. If radiation of the characteristic wavelength is detected, this is taken to indicate the presence of the corresponding metal in the sample. The method is not as reliable as emission spectroscopy, where radiation at several wavelengths can be examined to confirm the presence of the pertinent element, but it is fast, simple, and, if carried out with care, quite reliable. No information about the molecular structure of the sample compound can be obtained. Furthermore, other nonradiating elements, such as carbon, hydrogen, and halides, cannot be detected except under special circumstances. For example, halides can be precipitated using an excess of a standard silver nitrate solution. The excess silver can be determined by flame photometry. From the results, the halide content can be calculated. The method is an indirect determination of the halide, although no radiation from the halide was detected at the wavelength used.

2. Quantitative Analysis

The most useful application of flame photometry is for the rapid quantitative determination of the elements in groups I and II of the periodic table. Using equipment with high optical resolution, other metallic elements may also be determined. Table 9.3 is a list of elements with their emission wavelengths and detection limits.

a. Experimental Procedures for Quantitative Analysis

To perform quantitative analysis, the sample is introduced into the flame and the intensity of radiation is measured at the pertinent wavelength. The concentration of the emitting metal in the sample is then calculated by one of two methods: (1) the use of calibration curves or (2) the standard addition method. The former method is more commonly used.

Calibration curves are prepared by making up standard solutions of known concentrations similar to those expected in the sample. For example, if we are to determine on a routine basis samples containing lithium, and their normal concentration is about 6 parts per million (ppm), a calibration curve would be prepared using standards ranging from 1 to 20 ppm or a similar suitable range. The actual standard solutions prepared may contain 0, 1, 2, 4, 6, 8, 10, 12, 14, 16, 18, and 20 ppm each. The sample is then analyzed, and the intensity of emission from each standard is measured and plotted on a curve. This curve relating emission intensity and the concentration of lithium is the calibration curve. When a sample is run, the emission intensity of the sample is measured and from the calibration curve the concentration in the sample can be calculated as shown in Fig. 9.7. Suppose, for example, that the emission intensity of the sample was 7.5. From the calibration curve it can be shown that this is equivalent to approximately 11 ppm.

It should be noted that an emission signal was detected even when no lithium was added to the standard. This is called the *background* or *blank emission signal*. Any variation in the blank signal must be corrected for in the final calculation. It can also be seen from the calibration curve that the relationship between emission intensity and the lithium concentration is linear at low concentration but deviates from linearity at higher concentrations. There is a relative decrease in intensity as concentration increases. This is quite common for flame photometric calibration curves.

Table 9.3 Flame Spectra of Some Common Elements

Element	Wavelength (nm)	Type of flame	Detection limits (ppm)
Aluminum	396.2	OA	0.1
	484.0	OA	0.5
Antimony	252.8	OA	1.0
Arsenic	235.0	OA	2.2
Barium	455.5	OH	3
	553.6	OH	1
Bismuth	223.1	OA	6.4
Boron	249.8	OA	7
	518.0	OA	3
Cadmium	326.1	AH	0.5
Calcium	422.7	OA	0.07
	554.0	OA	0.16
	662.0	OA	0.6
Cesium	455.5	OH	2.0
	852.0	OH	0.5
Chromium	425.4	OA	5.0
Cobalt	242.5	OA	1.7
	353.0	OA	4.0
Copper	324.7	OA	0.6
Gallium	417.2	OA	0.5
Gold	267.6	OA	2.0
Indium	451.1	OH	0.01
Iron	372.0	OA	2.5
	386.0	OA	2.7
	550.0	OA	0.5
Lanthanum	442.0	OA	0.1
	741.0	OA	4.5
Lead	405.8	OA	1.0
Lithium	670.8	OA	0.007
Magnesium	285.2	OA	0.8
	383.0	OA	1.6
Manganese	403.3	OA	0.01
Mercury	253.6	OA	2.5
Molybdenum	379.8	OA	0.5
Neodymium	555.0	OH	0.2
	702.0	OH	1.0
Nickel	352.4	OA	0.1
Niobium	405.9	OA	12
Palladium	363.5	OH	0.1
Phosphorus	253.0	OH	1.0
Platinum	265.9	OA	10

Table 9.3 *(Continued)*

Element	Wavelength (nm)	Type of flame	Detection limits (ppm)
Potassium	404.4	OH	1.0
	767.0	OH	0.01
Rhenium	346.1	OA	0.3
Rhodium	369.2	OH	0.1
Rubidium	780.0	OH	0.3
Ruthenium	372.8	OA	0.3
Scandium	604.0	OH	0.012
Silicon	251.6	OH	4.0
Silver	328.0	OH	0.1
	338.3	OH	0.6
Sodium	590.0	OH	0.001
Strontium	460.7	OA	0.01
Tellurium	238.6	OA	2.0
Thallium	377.6	OH	0.6
Tin	243.0	OA	0.5
Titanium	399.9	OA	1.0
Vanadium	437.9	OA	1.0
Yttrium	597.0	OA	0.3
Zinc	213.9	OA	77

AH, air-hydrogen flame; OA, oxygen-acetylene flame; OH, oxygen-hydrogen flame.

In the standard addition method, the sample is split into several aliquots. One aliquot is left untreated. To the other aliquots known amounts of the test element are added. For example, the test element may be calcium. The intensity is then plotted against the quantity of lithium added to each aliquot as shown in Fig. 9.8.

From Fig. 9.8 it can be seen that the intensity of lithium emission from the original untreated sample was 30 units. When 5 ppm of Li was added to the sample, the intensity increased to 33 units. From Fig. 9.8 we can see that the addition of 5 ppm of lithium produced an increase in the emission signal of 3 units. Since the emission intensity from the original sample was 30 units, it must have contained 30 × (5/3) ppm of lithium. This procedure is called the *standard addition method*. One of its most important advantages is that it compensates for any unexpected or unmeasurable interference to the method. Any such interference should affect the added lithium and the lithium originally present in the same manner, and the interference is thus compensated for. A correction should be made for the background emission from the flame. A background emission is produced that is not generated by the lithium present. The intensity of the background emission should be subtracted from the total emission from the sample

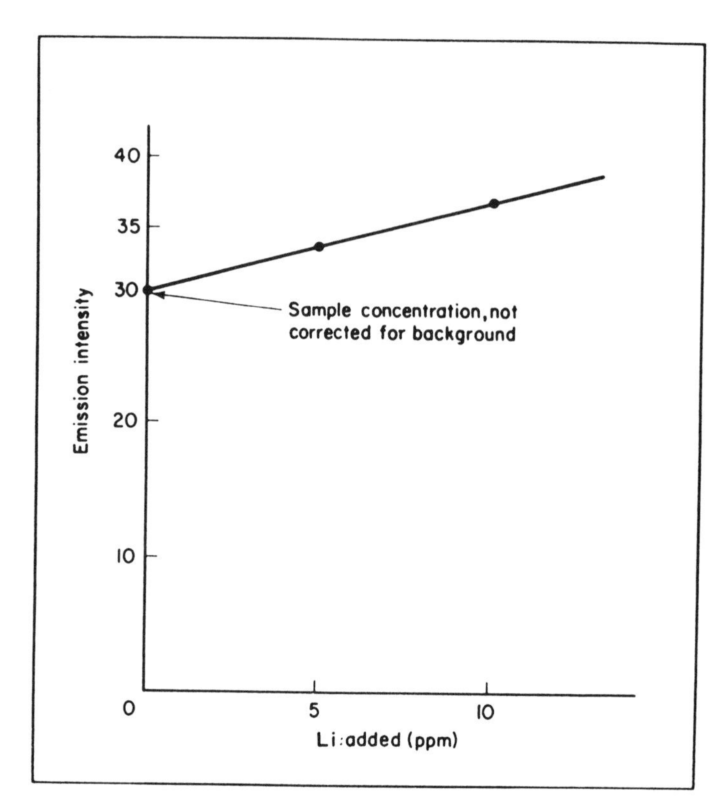

Figure 9.7 Plot of emission intensity against the quantity of lithium added to each aliquot, showing how the standard addition method is used in flame photometric quantitative analysis.

before the calculations of the lithium content are made. Another advantage of the standard addition method is that it can be used for samples that are rarely analyzed and that would not justify an analytical program to develop a new procedure.

An alternative method of treating the data is to extend the baseline to the left of the vertical axis and extend the line joining the data points until it intersects the baseline. At the point of intersection, the concentration of the sample can be read off. It must be corrected for background. This is illustrated in Fig. 9.9.

b. Interferences

The radiation intensity may not accurately represent the sample concentration because of the presence of other materials in the sample. These materials may cause interferences in the analytical procedure. Three principal sources of interferences are encountered in flame photometry.

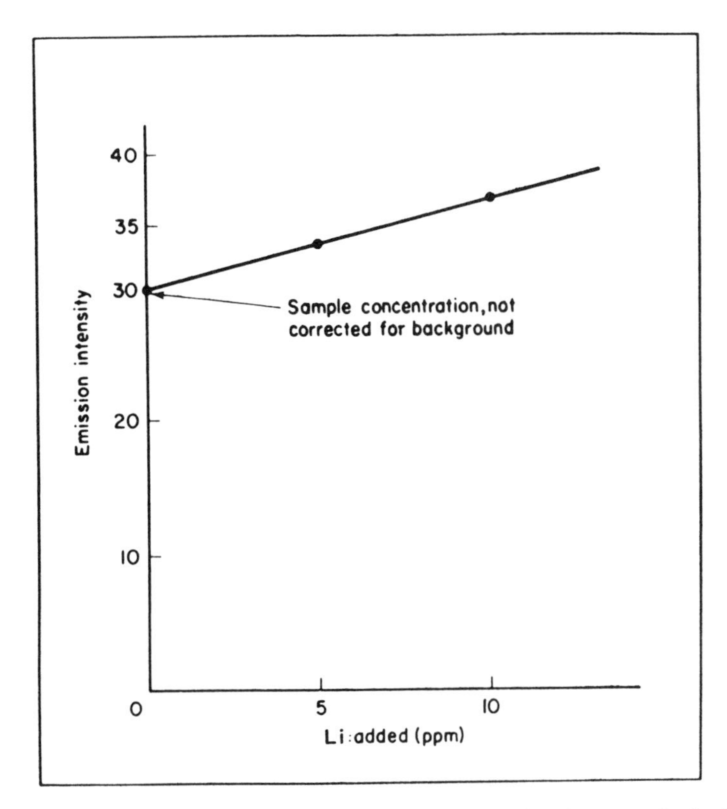

Figure 9.8 Determination of sample concentration from standard addition data.

Radiation Interferences. Two elements or compounds have different spectra, but their spectra may partly overlap and they may both emit at some particular wavelength. The detector cannot distinguish between the sources of radiation and reads out the total signal. If the emission wavelength of the interferent coincides with the wavelength used to measure the radiation intensity from the sample element, an incorrect answer will be obtained. This is a direct source of error. Sometimes it may be corrected by eliminating the effect of the interfering element (e.g., by extraction) or using calibration curves prepared from a solution containing similar quantities of the interfering material, which become part of the calibration signal. However, if the concentration of the interferent is not known, this process will not work.

Chemical Interferences. The emission intensity depends on how many excited atoms are produced in a flame. If the sample element is in the presence of anions with which it combines strongly, it will not decompose easily. But if the predominant anion combines weakly with the sample element, decomposition may be easier. For example, a given concentration of sulfate will give a

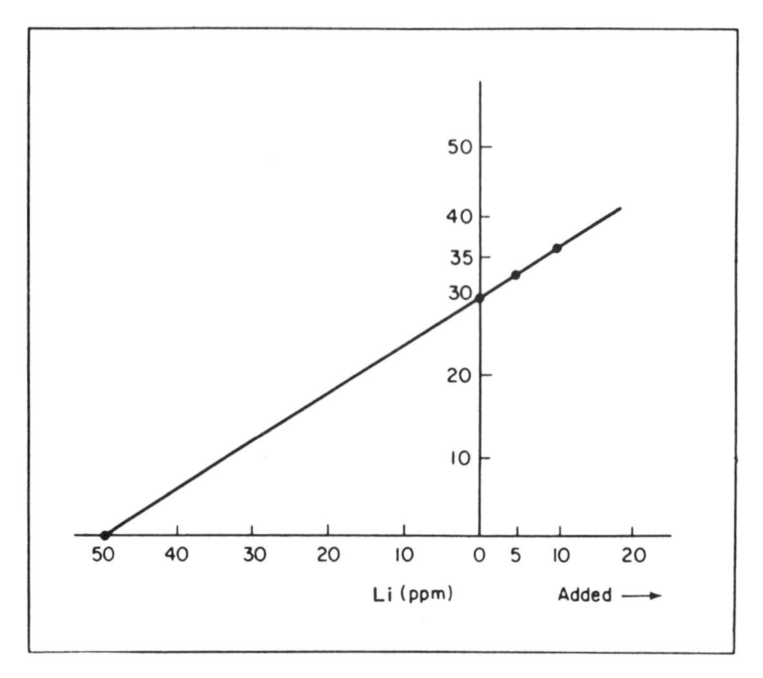

Figure 9.9 Alternative method of calculating the concentration using standard addition. The curve is extended backward to the point of intersection and the concentration read off the axis.

lower signal than the same concentration of barium chloride because barium chloride is broken down more easily than barium sulfate. This effect is called *chemical interference*. It is brought about by differences in the strengths of the chemical bonds between the metal and different prominent anions. Any change in an anion can result in chemical interference. The effect can be eliminated by extraction of the anion or by using calibration curves prepared with the same predominant anion at the same concentration as that found in the sample.

Excitation Interferences. When an atom is liberated in a flame, it reaches an excited state because of the high temperature of its surroundings. The population of the excited states is given by the Boltzmann distribution. In addition, a number of atoms may be ionized and generate ionic spectra. These neutral, excited, and ionized atoms are all in a state of dynamic equilibrium. The radiation emission intensity becomes steady and provides the basis for flame photometry. If atoms of another species are also present in the flame, they may affect the equilibrium. For example, they may absorb some of the thermal or radiant energy of the flame, thereby reducing the amount of energy available to the sample. This in turn will affect the intensity of both the atomic emission and the ionic radiation, creating an interference to quantitative analyses. To use another example, potassium atoms in the flame may become excited. If they collide with an unexcited sodium atom, their energy may be passed on, exciting the sodium in the process. This results in an increased number of excited sodium atoms and, therefore, increased radiation intensity, generating erroneously high results. This interference is called *excitation interference*. It has been reported to be particularly severe if the absorption and emission lines of the two elements overlap.

This type of interference is restricted to elements of the first group of the periodic table. It can be corrected by preparing calibration curves from standards that contain the interfering element in concentrations similar to that of the sample.

c. Operating Conditions

In order to obtain reproducible analytical results, it is important to operate under reproducible experimental conditions. To do this, it is necessary to determine the optimal flame conditions, including the ratio of fuel to oxidant, the wavelength, slit width, solvent used, and sample feed rate. The fuel-to-oxidant ratio of the flame greatly affects the intensity of emission from a sample with constant composition. For example, if there is excess fuel, it will not all be burned and the flame temperature will be decreased. This results in a drop in emission intensity. The wavelength at which the radiation intensity is measured must be kept constant. For example, sodium emits strongly at 589.5 nm; at 590.0 nm the intensity of radiation is low but not zero. The slit width controls the wavelength range of radiation falling on the detector. If the slit width is increased, interfering radiation from other elements reaches the detector. This causes high results to be recorded. Even if there are no other emission lines in the spectral vicinity of the sample line, the amount of background radiation reaching the detector increases with a corresponding increase in detector reading.

A change in solvent may drastically change the efficiency of producing excited atoms. Metals in organic solvents usually give an emission signal several times greater than that emitted by the same metal concentration in an aqueous sample. For this reason, the same solvent must be used in the calibration standards as in the unknown sample.

The effect of undesirable impurities may be eliminated either by extracting them from the sample or by extracting the sample element from the sample, thus leaving the impurities behind. The sample element is then put into a suitable pure solvent and analyzed. To obtain accurate and reproducible results, it is also essential that the samples be determined under exactly the same conditions as those under which the calibration curves were prepared.

d. Use of Nitrous Oxide–Acetylene Flames

During research in atomic absorption spectroscopy, Willis* and Amos† found that nitrous oxide–acetylene flames were superior to other flames for efficiently producing free atoms. This was particularly true for metals with very refractive oxides, such as aluminum and titanium. Later workers in the field of flame photometry‡ found that the same type of flame was very useful in flame photometry (see Table 9.4).

One problem encountered with this type of flame was the intense background emission, which makes measurement of the metal emission very difficult. However, the "wobbler" designed by Rains for background correction and the nitrous oxide–acetylene flame the Fisher Scientific Co. (Waltham, Mass.) has developed provide equipment capable of high sensitivity and accuracy. Results obtained by Fisher Scientific using this technique are shown in Fig. 9.10. Note the linear relationship between the emission intensity and the concentration of the metal. Note also the extended useful analytical range of the method. The calibration data are drawn on log-log paper for convenience.

e. Shielded Burners

T. S. West et al.** developed shielded burners in which the flame (particularly the reaction zone) was shielded from the ambient atmosphere by a stream of inert gas. This shielding leads to a quieter flame and to better analytical sensitivity. Table 9.5 shows results obtained with commercial equipment based on this technique and developed by the Beckman Instrument Co. (Fullerton, Calif.).

E. DETERMINATION OF NONMETALS

The determination of nonmetals by flame photometry has two major handicaps. First, the atomic state is usually not stable. For example, chlorine does not exist as Cl but as Cl_2. Second, the resonance lines are in the vacuum ultraviolet. High energy is necessary to cause electronic excitation—more energy than is available in conventional flames. Atomic emission of nonmetals has not exploited in flame photometry until recently.

---

*J. B. Willis, *Nature*, *207*:715 (1965).

†M. Amos, *Anal. Chim. Acta*, *32*:139 (1965).

‡V. A. Fassel, *Spectrochim. Acta*, *24B*:1494 (1969); E. Picket and S. R. Koirtyohann, *Spectrochim. Acta*, *24B*:325 (1969).

**T. S. West et al., International Symposium on Atomic Absorption Spectroscopy, Sheffield, England, 1969.

Table 9.4 Sensitivity Results Obtained with Nitrous Oxide–Acetylene Flames

	Sensitivity (ppm) obtained by		
Element	Fisher Scientific Co.[a]	Fassel[b]	Picket and Koirtyohann[c]
Al	0.1	0.03	0.005
Ba	0.08	0.002	
B	100		
Cd		2.0	
Ca	0.005	0.001	
Cr	0.02	0.02	
Co	0.5		
Cu	0.1	0.05	
Dy	0.5	0.07	
Eu	0.1	0.0006	
Gd	1.0	2.0	
Ga			0.2
Ge	4.0		0.5
In	0.1	0.03	0.002
Fe	1.0	0.7	
Pb	3.0	3.0	
Mg	0.05	0.2	
Mn	0.7	0.1	
Hg	100	40	
Mo	1.0	0.2	
Ni	5.0	0.6	
P	10.0		
K	0.02	0.003	
Na	0.001	0.0001	
Sr	0.02	0.004	
Th			0.02
Ti	0.3	0.4	
V		0.07	
Zr	0.5		

[a]Data courtesy of Fisher Scientific Company, Waltham, Mass.
[b]Data from V. A. Fassel, *Spectrochim. Acta, 24B*:1494 (1969).
[c]Data from E. Picket and S. R. Koirtyohann, *Spectrochim. Acta, 24B*:325 (1969).

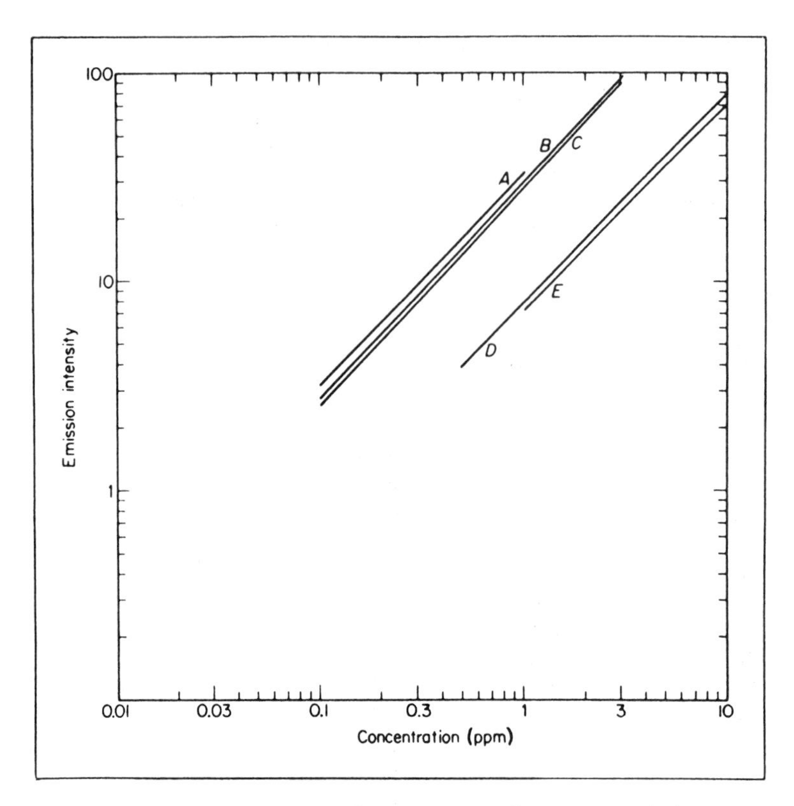

Figure 9.10 Calibration data (in nanometers) for trace metals in aqueous solution: A = Cr 475.4, $N_2O—C_2H_2$; B = V 440.8, $N_2O—C_2H_2$; C = Al 396.1 $N_2O—C_2H_2$; D = Mn 403.1, $N_2O—C_2H_2$; E = Ti 365.3 $N_2O—C_2H_2$. (Courtesy of Fisher Scientific Co., Waltham, Mass.)

Table 9.5 Results Obtained with Shielded Burners

Element	Sensitivity (ppm)
Ba	0.05
Bi	2
Ca	0.002
Cr	0.0007
Co	0.04
Pb	0.05
Mg	0.3
V	10

Source: Data courtesy of Beckman Instrument Co., Fullerton, Calif.

However, molecular emission has been observed. These spectra are band spectra rather than line spectra and are therefore less intense, analytically less sensitive, and more open to spectral interferences. Typical data are as follows.

1. *Br_2*: Orange-Red System. These orange-red bands occur in the outer cones of flames containing bromine. The system consists of a very large number of bands, which are degraded to red.

wavelength (nm)										
654,	647,	641,	636,	634,	631,	629,	626,	623,	622,	619,
617,	612,	607,	596,	594,	586,	583,	575,	572,	560,	559.

In the inner cone of organic flames containing Br_2 was a strong band near 290 nm.

2. *BrO (Ethyl Bromide Flame Bands)*. The bands are degraded to the red. The heads are slightly diffuse above the inner cone in flames containing ethyl bromide and in an oxygen/methyl bromide flame or an oxy-hydrogen flame containing Br_2.

wavelength (nm)		
396	414	439
400	418	453
403	422	467
407	427	481
411	434	485

3. *C_2: Swan System*. These bands have single heads and are degraded to the violet.

wavelength (nm)		
667	595	509
659	592	473
653	563	471
648	558	469
644	554	468
619	550	467
612	547	438
605	516	437
600	512	436

4. *CCl*. Observed in the inner cones of the flames of CH_3Cl, $CHCl_3$, and CCl_4. Each band has about four heads, degraded to shorter wavelengths.

wavelength (nm)		
286	279	277
285	278	272
284		271
284		

5. *CF*. Observed in flames containing fluorine compounds.

wavelength (nm)	
255	233
247	230
240	224

6. *CN: UV Bands*

wavelength (nm)	
421	387
419	386
418	385
416	385
415	358
388	358

7. *CN: IR Bands*

wavelength (nm)		
Infrared		Visible (C_2N_2 flame)
787	523	585
806	534	619
914	547	633
939	559	647
1093	573	663

8. *CS*. Degraded to the red.

wavelength (nm)	
250	260
252	262
253	266
257	267
258	269

9. *CaOH*. These bands, formerly attributed to CaO, are a persistent impurity in the spectra of flames, especially explosion flames. The green band is fairly narrow and shows some complex rotational structure. It is slightly degraded to the violet, but there are no sharp heads, strong lines, or maxima.

wavelength (nm) of green band						
555,	5547.4,	554,	554,	554.6,	554,	5539.

The orange-red band is quite diffuse and wider. It extends from 628 to 612, with a maximum intensity at 623 nm.

10. *Cl_2*. The inner cones of organic flames containing chlorine also show a wide diffuse band of Cl_2 at about 258 nm.

wavelength (nm)		
559	584	613
563	588	620
566	589	626
571	594	624
576	600	641
582	607	650

11. *ClO*

wavelength (nm)		
365	395	424
372	399	428
376	407	437
384	411	441
387	415	445

12. *CuCl*

wavelength (nm)	
425	475
428	478
433	484
435	488
441	494
443	498
449	515
451	526

13. *H_2O: Rotation-Vibration Spectrum*

wavelength (nm)				
568	598	645	716	927
571	616	646	729	933
580	618	649	809	944
586	620	651	891	948
588	622	657	897	955
590	625	662	912	961
592	632	691	918	966
594	637			

14. *OH*. The band has four heads.

wavelength (nm)	
260	294
261	
262	306
	306
267	307
268	308
269	
	312
281	
281	342
282	343
	345
287	347
289	

15. *I_2*. In the inner cones of organic flames containing iodine, there is a diffuse band of I_2 at about 342 nm.
16. *Sulfur*

wavelength (nm)		
336	358	404
341	364	415
346	373	419
350	383	431
355	393	443

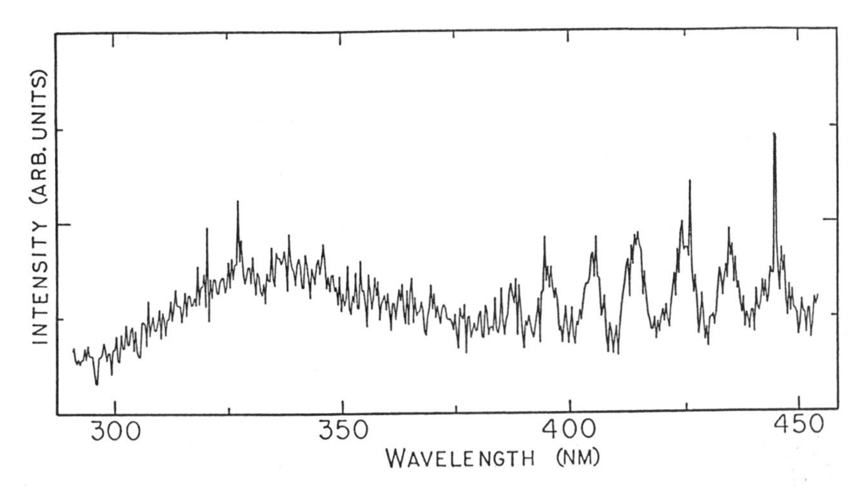

Figure 9.11 Emission spectrum of SO_2 (From H. R. Martin, R. J. Glinski, *App. Spec.*, *46*(6): 948 (1992).)

When organic sulfur compounds are introduced into a flame they are oxidized to SO_2. The SO_2 emits infrared radiation, which can be detected. The emission spectrum is shown in Fig. 9.11. Elemental sulfur can be detected at 257.6 and 256.0 nm. Sulfur can exist as S_2, the emission spectra of which is shown in Fig. 9.12. S_2 quickly polymerizes to S_4, S_8, etc., so like most flame emission species, the lifetime of the emitting species is transient. Flame conditions must be reproducible and steady to achieve reproducible results.

F. FLAME INFRARED EMISSION (FIRE)

Recently, the infrared radiation from flames, especially from the nonmetals, has attracted attention. These give molecular spectra in the near-IR. The emitting species are usually combustion products or intermediates rather than the original species and are therefore useful for elemental analysis. Typical spectra for methanol, HF, HCl, and CO_2 are shown in Fig. 9.13.

The use of this region of the electromagnetic spectra could make possible a rapid, simple, convenient method for nonmetal detectors and determination. For example, it can be used as a real-time detector for gas or liquid chromatography for these specific elements present in molecules.

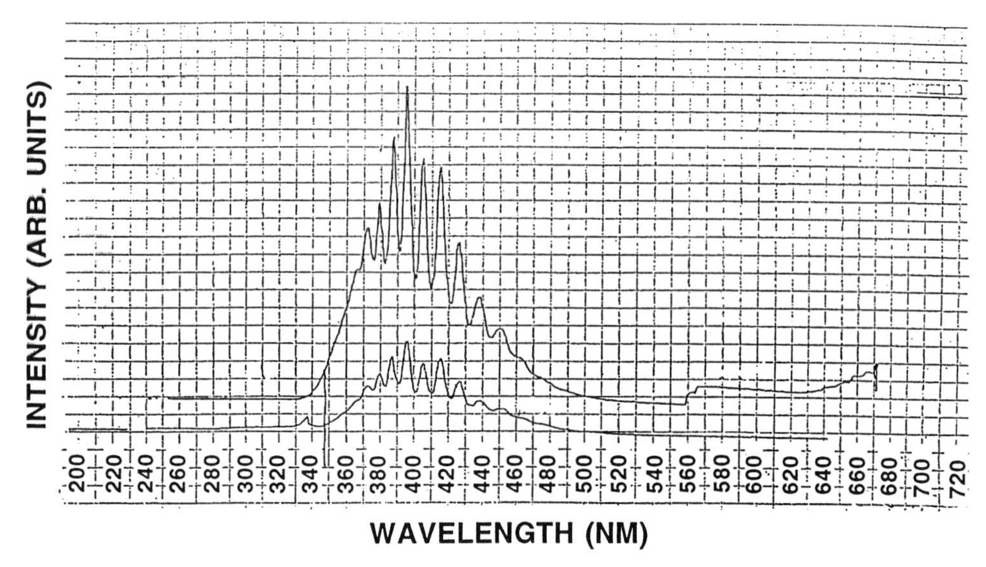

Figure 9.12 Emission spectrum of S compounds.

G. CONCLUSIONS

The principal analytical advantages of flame photometry include its simplicity, sensitivity, and speed. There is, however, a limitation to the number of elements that can be determined by this method. Also, only liquid samples may be used. Sometimes lengthy sample preparation steps are necessary.

Flame photometry gives no information on molecular analysis, but it is used widely for elemental analysis. In soil samples the elements sodium, potassium, calcium, and magnesium are frequently determined. These same elements are determined in plant analysis.

In the medical field, Na, K, Ca, and Mg are routinely determined in blood analysis by flame photometry. These four elements are also determined in urine and other excretions, as well as in body tissue.

In metallurgy, sodium is determined in aluminum and aluminum alloys, as well as in metallic calcium. Other examples include the determination of silver in blister copper; magnesium in slags; iron, copper, and cobalt in cobalt mattes; and calcium in cement. Boron has also been determined in various types of organic compounds.

In summary, flame photometry is a simple, rapid method for the routine determination of elements that are easily excited.

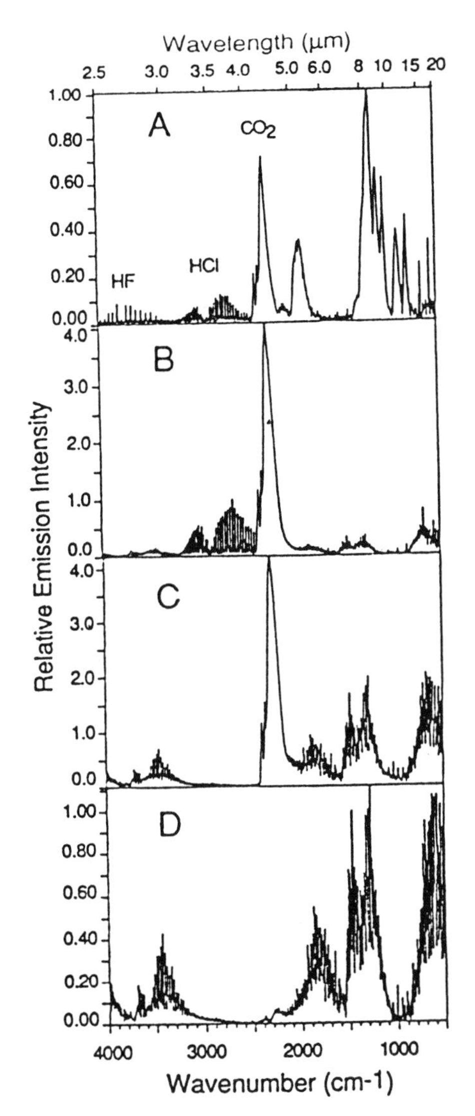

Figure 9.13 Flame infrared emission (FIRE) spectra from 4000 to 5000 cm^{-1} for; (A) Freon-113, (B) carbon tetracholoride, (C) methanol, and (D) hydrogen/air flame water background. Molecular emissions from terminal combustion products CO_2, HCl, and HF are indicated in (A).

BIBLIOGRAPHY

Anal. Chem., Application Reviews, April 1993.
Anal. Chem., Application Reviews, June 1994.
Dean, J. A., *Flame Photometry*, McGraw-Hill, New York, 1960.
Gaydon, A. G., and Wolfhand, H. G., *Flames*, 2nd ed., Chapman and Hall, London, 1960.
Herman, and Alkemade, *Chemical Analysis by Flame Photometry*, Interscience, 1963.
Robinson, J. W., *Atomic Spectroscopy*, Marcel Dekker, New York, 1990.
Robinson, J. W., ed., *Handbook of Spectroscopy*, CRC Press, 1974.
Syty, A., Flame photometry, in *Treatise on Analytical Chemistry* (P. J. Elving, E. Meehan, and I. Kolthoff, eds.), Vol. 7, Wiley, New York, 1981.

SUGGESTED EXPERIMENTS

9.1 Warm up the flame photometer. Using an oxygen hydrogen burner, set the flow rates of oxygen:hydrogen to (a) 1:1 vol vol (oxidizing flame), (b) 1:2 (stoichiometric flame); and (c) 1:3 (reducing flame). Note the change in flame shape and size. With each flame measure the emission at 589.0 and 589.5 nm. This is from sodium contamination. Aspirate into the flame (a) water from a plastic container and (b) pure water from a glass container. Note the sodium contamination of the water (emission at 589.0 nm).

9.2 Prepare aqueous solutions containing amounts of potassium varying from 1.0 to 100.0 ppm of K^+. Aspirate each sample into the flame. Measure the intensity of emission at 767.0 and 404.4 nm. Plot the relationship between the emission intensity of each line and the concentration of the solution aspirated into the burner.

9.3 Repeat Experiment 9.2, but add 50 ppm of Na^+ to each solution. Note the change in the emission intensity due to the potassium.

9.4 Prepare solutions containing 50 ppm of Mg, Ca, Zn, Na, Ni, and Cu. Measure the emission intensity of Mg at 285.2 nm, Ca at 422.7 nm, Zn at 213.9 nm, Na at 589.0, Ni at 341.4 nm, and Cu at 324.7 nm. Note the wide divergence in emission intensity, even though the metal concentration was constant. Compare the sensitivity of your results with that of the results of Experiment 7.1.

9.5 Repeat Experiment 9.4 with pure water aspirated into the flame. This is one source of radiation interference.

9.6 Take a sample containing an unknown quantity (about 3 ppm) of sodium. Split it into six 10-ml aliquots. To the first aliquot add 10 ml of standard sodium solution containing 2 ppm Na. To other aliquots add solutions containing 4, 6, 8, and 10 ppm of Na. Leave one aliquot undiluted. Measure the intensity of Na emission at 589.0 and 589.5 nm. Using the method of standard addition, calculate the Na concentration in the original sample.

9.7 Determine the sodium content of the drinking water in your department.

PROBLEMS

9.1 What are the two principal types of burners used in flame photometry? Draw schematic diagrams of them.

9.2 When a sample is introduced into a flame, what processes occur that lead to the emission of radiant energy?

9.3 What is the function of the slit system in flame photometry?

9.4 What is the standard addition method? Illustrate your answer.

9.5 In preparing a calibration curve for the determination of potassium, the following data were obtained:

Concentration of K (ppm)	Emission intensity
0.1	4.1
0.2	8.0
0.4	16.2
0.6	24.5
0.8	31.8
1.0	36.0
1.2	48.5

(a) Plot the results. (b) Which of the results appear to be in error? (c) What should the emission intensity have been? (d) Prepare a calibration curve using the corrected result.

9.6 The emission intensities from various samples were determined as 10.0, 22.0, 29.9, 40.0, and 36.0. Using the calibration curve prepared in Problem 9.5, calculate the potassium concentration of the related samples.

9.7 The following results were obtained using the standard addition method:

	Radiation intensity from calcium
Sample alone	5.1
Sample +0.1 ppm Ca	6.2
Sample +0.2 ppm Ca	7.3
Sample +0.5	10.6
Flame alone	0.7

What was the concentration of calcium in the original sample?

9.8 What preparation steps are necessary in the analysis of (a) gas samples and (b) solid samples by flame photometry?

9.9 In preparing a calibration curve for the determination of calcium, the following data were obtained:

Sample concentration	Radiation intensity
0.1	3.6
0.2	6.1
0.4	11.1
0.6	16.0
0.8	20.9
1.0	25.7

Prepare the curve. What was the radiation intensity of the flame with no calcium present? The radiation intensity from a sample containing calcium was 18.4. What was the calcium concentration in the sample? On the equipment used, the error involved in measuring the radiation intensity indicated that the value was between 18.3 and 18.5. What error in concentration measurement was introduced by this uncertainty?

9.10 List 10 elements and 4 types of samples that are commonly analyzed by flame photometry.

9.11 What is a Grotrian design?

9.12 What are the advantages of a Lundegardh burner?

9.13 What are the advantages of a total consumption burner?

9.14 What causes background emission in flame photometry?

9.15 Describe the major interferences encountered in flame photometry.

10

Emission Spectrography, Inductively Coupled Plasma Emission (ICP), and ICP–Mass Spectroscopy

In recent years emissions spectrography, which usually denotes emissions by samples excited by electrical discharge, has been increasingly superseded by inductively couple plasma (ICP) emission, which is much more reproducible and requires less skill to operate. Many ICP instruments are automated for minimum attention and manipulation by the operator. ICP–mass spectroscopy is rapidly becoming much more important as a method for elemental analysis, with very high sensitivity for the detectors of both metals and nonmetals.

All of these methods are discussed in this chapter.

A. EMISSION SPECTROGRAPHY

Emission spectrography is the study of the radiation emitted by a sample when it is introduced into an electrical discharge. Since each element emits a different spectrum, it is possible to determine what elements are present in a sample from the spectra that are emitted. This technique is a reliable method for elemental qualitative analysis. With it, all metallic elements can be detected in low concentration, as can the metalloids, such as arsenic, silicon, and selenium, but nonmetals are more difficult to handle. Liquids and solids can be analyzed quite easily, but gases are more difficult to analyze. If halides or inert gases are present in the pure gas form, they can be determined directly by the emission spectra generated by an electrical discharge through the gas. However, mixtures of gases present highly variable spectra that are difficult to interpret analytically and almost impossible to handle quantitatively.

With proper care, emission spectrography can be used for quantitative analysis of all the foregoing elements at concentration levels as low as 1 ppm. A high degree of skill is needed in order to obtain reproducible quantitative results with this technique. Both qualitative and quantitative analytical interpretation of emission spectra demand skill and considerable experience to obtain reliable results. In the hands of the right operator, however, extremely useful data can be obtained from the emission spectra of elements introduced into an electrical discharge.

B. ORIGIN OF SPECTRA

The outer electrons of an atom travel in orbitals with well-defined energy levels. From quantum theory we know that there are *s*, *p*, *d*, and *f* orbitals occupied by electrons. We also know that the sizes of these orbitals varies according to their principal quantum number, 1, 2, 3, ... Each of these orbitals represents a permitted energy level for an electron in an atom. As we move from one element to the next in the periodic table, the orbitals are filled progressively, with the orbitals of lowest energy being filled first. If an atom is unexcited, all of its electrons are in the lowest permitted energy levels. The energy levels depend on the particular atoms and are defined by the quantum numbers of the electrons involved. The upper orbitals are empty but represent vacant permitted first energy levels. If the atom becomes excited, its electrons, and in particular the valence electron, move from a low-energy orbital to an orbital with higher energy. The atom is said to be in an *excited state*. The excited atom rapidly emits a photon of energy and its electron returns to a lower-energy orbital, finally ending up in the original orbital with the lowest energy—the unexcited state or *ground state*. The emitted radiation from excited atoms is the basis of emission spectrography. Figure 9.1 is a partial representation of the energy states available to the valence electron of lithium. This energy representation is a partial Grotrian diagram. Grotrian diagrams are discussed in Chapter 7.

Numerous energy forms are available that can be used for exciting atoms, such as thermal energy, as in flame photometry, or spectral energy, as in atomic fluorescence. Emission spectrography involves the use of an electrical discharge as the means of excitation. This energy form produces more emission lines than flame photometry and therefore requires instruments with better resolution. Plasma can also be used to excite atoms, and plasma emission spectrography is rapidly becoming very important in analytical chemistry.

C. EQUIPMENT

In simple terms, an emission spectrography (Fig. 10.1) functions as follows. An electrical source produces a steady electrical discharge between two electrodes

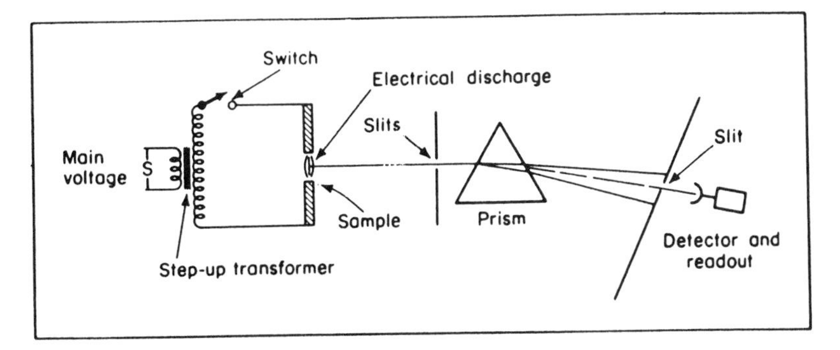

Figure 10.1 Schematic diagram of an emission spectrograph. The lenses are not included here.

designated as the *electrode* and the *counterelectrode*. The sample is put into the electrode and introduced into the discharge, where it becomes vaporized and excited. The excited atoms emits radiation, which is detected and measured by the detector readout system. The individual components of the equipment are described in the following sections.

1. Electrical Sources

There are three types of electrical sources: the DC arc, the AC arc, and the AC spark.

a. DC Arc

In this source a DC voltage is applied across the electrodes. A plasma forms into which the sample is vaporized by electrical and thermal energy. The atoms become excited and emit radiation. A common method of introducing the sample to the discharge is to present it as a powder. The electrical discharge is from the counterelectrode to the electrode, on the surface of which is the sample in the form of a powder. Generally, the plasma will link to a point or projection on the surface of the powder and will continuously erode the sample, vaporizing it in the process. A pit is formed in the sample by the erosion, and finally the pit becomes so large that the discharge can no longer be maintained with the electrode at that point. The discharge then wanders to another nearby point on the powder surface. In practice, the wandering of the discharge is quite fast and causes the signals to fluctuate and lose reproducibility. In practice, numerous such discharge points are created at the same time. Local hot spots are formed, which cause the sample in the immediate vicinity to evaporate rapidly and become excited easily; however, these hot spots are not continuous and soon die

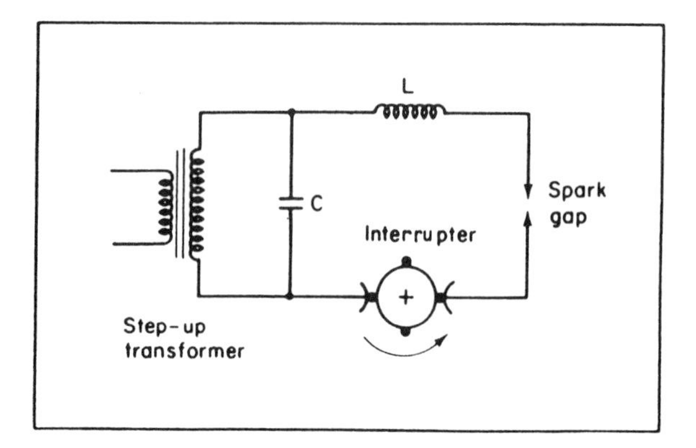

Figure 10.2 Schematic diagram of an AC spark system: $i = V\sqrt{C/L}$. L and C are variable.

away, and the local emission intensity is reduced. As a consequence of this behavior, the total emission intensity is somewhat erratic and the signals produced lack precision. On the other hand, the signals are very intense, and the analytical sensitivity is better than when other types of electrical discharges are used.

The DC arc therefore is used for qualitative analysis, particularly to detect the presence of trace materials that are not detectable using other types of discharge. However, the DC arc is avoided for quantitative use because of the poor reproducibility of the system. It is used, however, if quantitative analysis of very small quantities is required and other techniques will not produce a significantly strong signal for measurement. The choice of source depends on the operator, and skill and experience aid in the final choice.

b. AC Arc

The AC arc discharge is essentially an interrupted DC arc. The interruption of the discharge reduces wandering and local hot spots because the plasma is continually interrupted and reformed. In this process the discharge reaches a greater area than the local hot spots, which is the case using the DC arc.

c. AC Spark

The third source is the AC spark system. A schematic diagram of this is shown in Fig. 10.2. In this system the discharge is interrupted frequently under conditions of very high voltage and relatively low current. The discharge is considerably more stable than the DC or AC arc, but it is less sensitive. The improved stability is a desirable feature in quantitative analysis, so the AC spark is the source of choice for quantitative analyses, provided that high sensitivity is not required. Unfortunately signal intensity is lost with this technique, with a consequent loss of analytical sensitivity. The AC spark is therefore used when good precision is required but high sensitivity is not necessary.

In practice, the operator may vary the source from "arclike" to "sparklike" conditions. Selection is dictated primarily by the sample to be analyzed. If the elements to be determined are present in reasonably high concentrations (percent level), the operator may set the instrument in the AC spark mode. However, if the elements to be determined are very low in concentration (ppm level), then the AC spark system is not sufficiently sensitive and the operator may be forced to use DC arc conditions. The rule of thumb is to use DC arc conditions sufficiently sensitive to detect the element comfortably but to make them as sparklike as possible to improve the reproducibility of the data. The operator can do this by changing the inductance and the capacity of the condenser in the source system, thereby controlling the type of discharge all the way from DC arc to AC spark.

Discharges can be produced under many conditions between the extreme ones described above. For example, the discharge can be DC with a little AC characteristic superimposed. These AC arc conditions can be varied to be sparklike (pure AC) or arclike (pure DC), as desired. In between, the discharge is DC with an AC ripple. The spectra produced provide the highest reproducibility when sparklike and the greatest sensitivity when arclike. The choice depends on the sensitivity required and the precision to be achieved.

Two types of spectra are generated from these systems. First are the *atomic emission spectra*. These are designated by I. They arise from excited valence electrons descending from higher excitation levels to lower energy levels in the atom. However, under spark conditions, it is not uncommon for the atom to lose one electron completely during the excitation process, resulting in the formation of an ion. A second electron in the ion may then become excited and enter one of the higher energy states. From these states, the ion relaxes and emits a photon in the process. The energy levels of the ions are not the same as the energy levels of the atoms. They therefore emit different spectra, called *ionic spectra* and designated by II.

Ionic lines are less sensitive than atomic lines but are not subject to line reversal or self-absorption. They are used for quantitative analyses when a sufficient concentration of the sample is available. However, they are seldom, if ever, used for qualitative analysis because of their lack of sensitivity.

In summary, DC arc conditions are most useful for attaining high sensitivity and for qualitative analysis, but AC spark conditions are used for quantitative analysis because of better reproducibility.

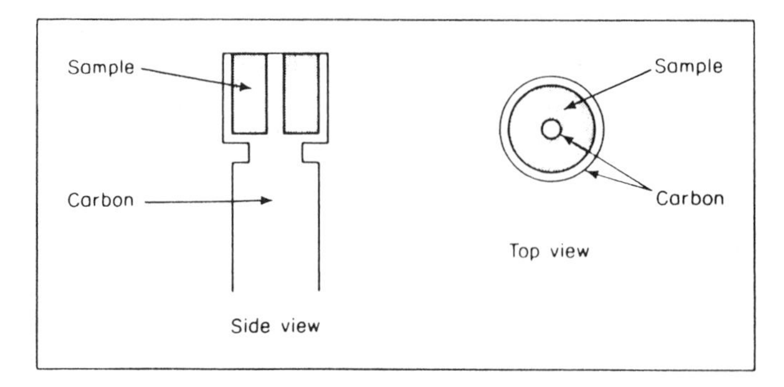

Figure 10.3 Sample holder for powders.

2. Sample Holder

The function of the sample holder is to introduce the sample into the electrical discharge. There are two types of sample holders, namely, those for solid samples and those for liquid samples.

a. Solid Sample Holders

Many kinds of samples may be determined by emission spectrography—different metals and alloys, wood, plastics, paint, polymers, soil, sea water, and so on. They all have different matrices and would give different emission intensities for the same metal concentrations. The problem is usually overcome by dissolving the sample in a known solvent such as a dilute acid, adding the solution to a common matrix, such as Al_2O_3, SiO_2, or C, mixing well, and then drying to from a powder. Consequently, many samples end up as solids after pretreatment. Many samples are therefore handled by being first reduced to a powder and then loaded into a carbon sample holder like that shown in Fig. 10.3. After it is loaded, the sample holder is placed in position as one of the electrodes used in the discharge. While electrical discharge takes place from the top surface of the electrode, the sample is vaporized into the plasma of the discharge and spectrographic emission takes place.

Organic compounds, such as animal tissue, body fluids, plant materials, and soil extracts, are frequently analyzed for metal components in this fashion. Usually the sample is first reduced to ash, either by burning in a furnace or by chemical decomposition. The ash may be mixed with a suitable powder, such as carbon powder, alumina, or silica. This powder is the major portion of the sample mixture that is actually fed into the discharge and therefore constitutes the matrix. Mixing with a matrix powder allows the sample to enter the electrical discharge at a steady rate. It avoids sudden "flares" of emission from the sample, which give rise to erratic results.

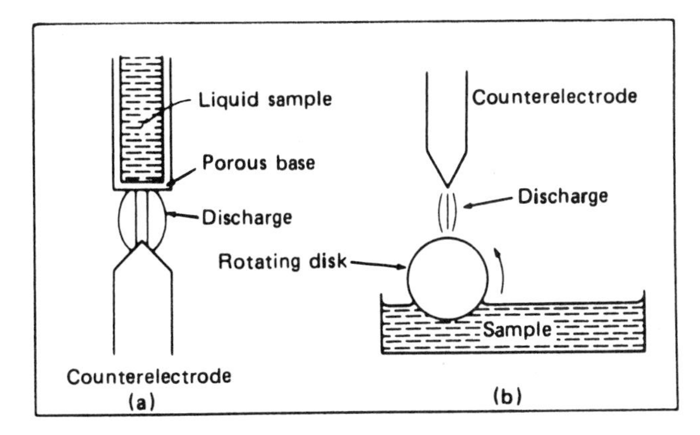

Figure 10.4 Holders for liquid samples: (a) porous cup container and (b) rotating disk electrode.

Metallic samples, such as sheets or rods of alloys or pure metals, may be used directly as one of the electrodes of the electrical discharge. In this case the only preparation required is to clean the sample and shape it to fit into the electrode holder.

b. Liquid Sample Holders

Liquid samples may be analyzed directly by several methods. Two commonly used sample holders for liquids are shown in Fig. 10.4. The porous base of the cup container shown in Fig. 10.4a permits the liquid sample to percolate through it at a steady rate into the electrical discharge. The sample slowly runs into the discharge, which can be sustained until the base of the holder is burned away. This may be several minutes, which is ample time to complete an analysis. One difficulty with this holder is that dissolved solids (e.g., NaCl) in the sample tend to form a crust at the base of the sample holder, causing erratic results. To obtain reproducible results, dissolved solids should be less than 1% of the sample. Samples with a high dissolved solids content should be dried and ashed and run as solid samples.

The rotating disk (Fig. 10.4b) rotates through the liquid sample, which wets the surface of the disk. The rotating wet surface carries the sample into the discharge at a steady rate. The disk rotates steadily for several minutes or as long as necessary, emitting the radiation from the excited sample in the discharge. In practice, it is found that initially the surface of the wheel is smooth and carries only a small quantity of sample into the discharge. After the first revolution the surface is roughened by exposure to the electrical discharge and carries an increased amount of liquid sample into the discharge. The emission intensity increases further with two revolutions; however, subsequent revolutions have little effect. The signal is then steady and can be measured at that time. Calibration data must be obtained under identical conditions and identical number of revolutions.

Each type of liquid sample holder is most useful for aqueous or inert solutions. Organic solutions tend to ignite in the discharge and cause erratic emission. This problem is particularly bad when the rotating disk electrode is used.

3. Monochromator

The function of the monochromator, as we know from our earlier discussions (see Chap. 3), is to separate the various lines of a sample's emission spectrum. Both prism and grating dispersion elements are used in emission spectrography. These are discussed below.

a. Prism Dispersion Elements

In recent years, the prism has been largely superseded by the grating. The latter has better resolution, is not subject to polarization, and is more stable. However, some prism instruments are still in use.

For a prism to be suitable for an emission spectrograph, it must be transparent to UV radiation. For this reason, prisms are usually made of quartz or fused silica. Unfortunately, quartz prisms split ordinary light into two beams of light that are polarized perpendicularly to each other.

In our discussion of polarimetry in Chapter 8, Section F, we saw that what appears to be a single beam of light may be several separate beams, each vibrating in a different direction. For example, a light beam contains photons, some of which are oscillating horizontally and some vertically (Fig. 10.5). The act of separating a light beam into two beams vibrating at right angles to each other is called *polarization*. Quartz polarizes ultraviolet and visible light. The refractive indices of the two polarized beams of light are slightly different.

The result of polarization by quartz prisms is that light of one wavelength, such as an emission line, emerges from the prism as two resolved lines. The lines are resolved not because they are of different wavelengths, but because they are polarized. This greatly complicates the interpretation of the spectrum and makes qualitative analysis based on the identification of emission lines very difficult. At the same time, beam splitting into two equal beams causes the loss of half the intensity of each beam, although the total intensity stays constant. This makes both qualitative and quantitative analysis difficult. These drawbacks can be overcome by using two half-prisms, one of which polarizes light in a right-hand fashion and the other in a left-hand fashion. The first half-prism splits the light into two beams; the second recombines them (Fig. 10.6).

An alternative arrangement is shown in Fig. 10.7. This is the Littrow prism, which is essentially a half-prism with a mirrored face. Light enters the sloping

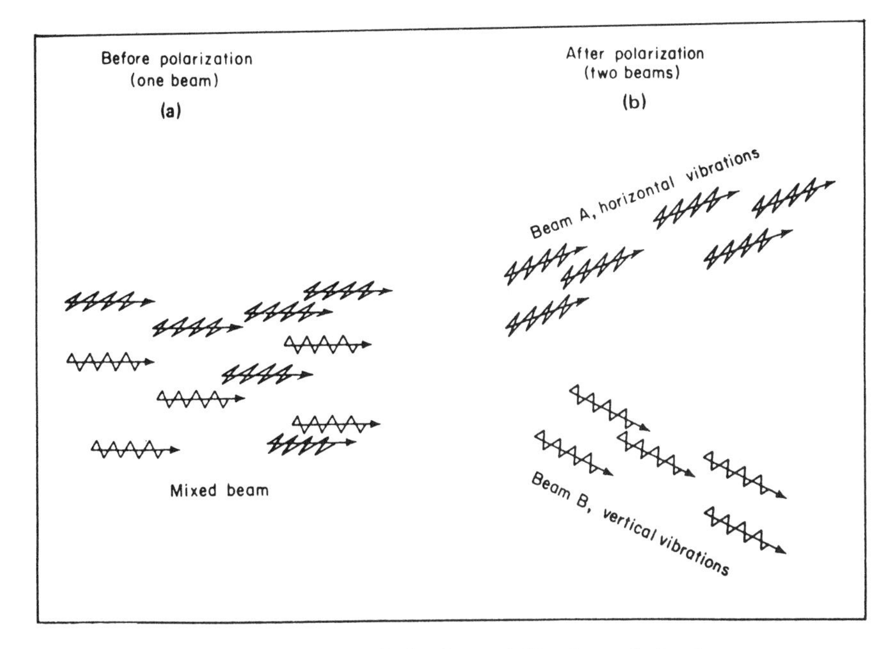

Figure 10.5 Light (a) prior to polarization and (b) after polarization.

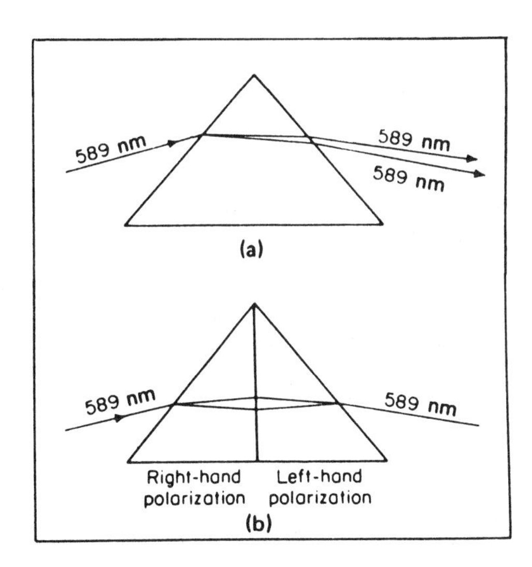

Figure 10.6 (a) An emission line split into two polarized beams by a quartz prism. (b) The splitting and recombination of an ordinary light beam affected by two half-prisms.

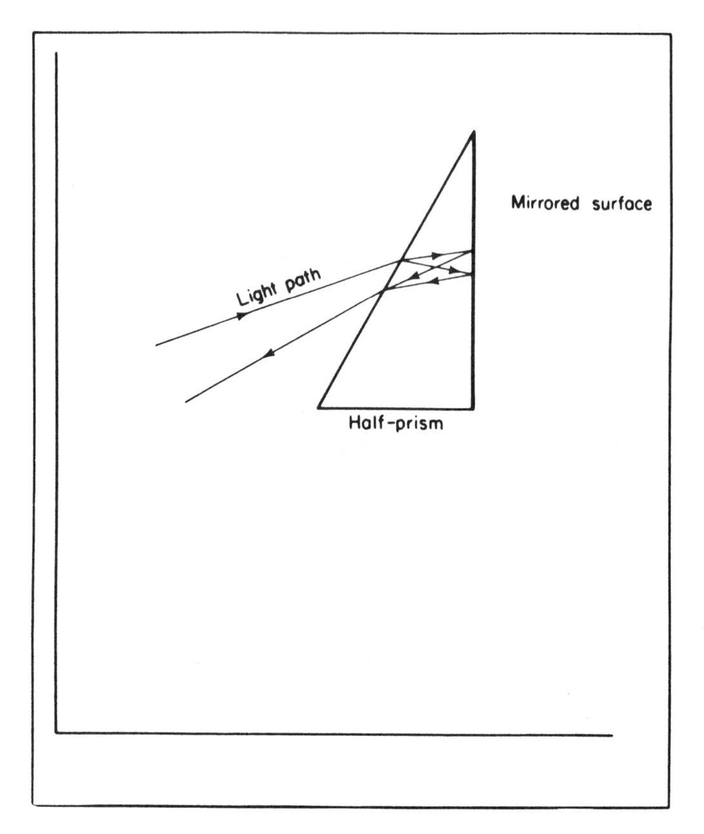

Figure 10.7 Recombination of polarized light using a Littrow prism.

face of the half-prism in the conventional manner and is dispersed according to wavelength. In this process the radiation is polarized, resulting in two lines separated very slightly. Each line then reaches the vertical mirrored surface and is reflected back into the half-prism. After reflection, the direction of polarization is reversed. Consequently, the two polarized beams of light are recombined but further dispersed from lines of other wavelengths. The net result is that the radiation emerging from the half-prism is resolved according to wavelength, but the splitting caused by polarization is corrected by reflection back through the half-prism.

b. Resolution

The resolving power of a prism is a measure of its ability to separate two emission lines. Mathematically, the resolution required to separate two lines of different wavelength is given by the relationship

$$\text{required resolution} = \frac{\text{average } \lambda}{\Delta\lambda} \tag{10.1}$$

where $\Delta\lambda$ is the difference in wavelength between the two lines and average λ is the average wavelength of the two lines. For example, suppose that a prism is able to separate two lines of wavelength 599.7 and 600.3 nm. Then

$$\text{required resolution} = \frac{\text{average } \lambda}{\Delta\lambda}$$

$$= \frac{(600.3 + 599.7 \text{ nm})/2}{600.3 - 599.7 \text{ mn}} = \frac{600.0}{0.6} = 1000$$

The resolution of the prism is given by

$$R = \frac{\delta\eta}{\delta\lambda} t$$

where

$\delta\eta$ = rate of change of refractive index
$\delta\lambda$ = rate of change of wavelength
t = thickness of the base of the prism

Light of different wavelengths is resolved only if the refractive index η is different; also, the larger the prism, the better. The resolving power of prisms is discussed more fully in Chapter 3.

Unfortunately, the resolution of a prism is not constant over a wide wavelength range. Emission lines of unknown wavelengths cannot be identified by simply measuring their dispersion. For this reason, standards with emission lines of known wavelengths must be used to identify the wavelengths of emission lines from the spectrum of an unknown sample. This is done by examining the emission spectrum of a known compound suspected to be in the sample. If they are the same, their wavelength must be the same; this is verified experimentally.

c. Grating Dispersion Elements

The increase in the popularity of grating monochromators has led to more sophisticated equipment with better resolution and to the development of direct-reading instruments. Chapter 3, Section A.1.b, describes grating monochromators and should be read before proceeding.

d. The Rowland Circle

Concave gratings can be used to conserve energy. This shape focuses a given wavelength to a point. For a different wavelength, the radiation falls at a

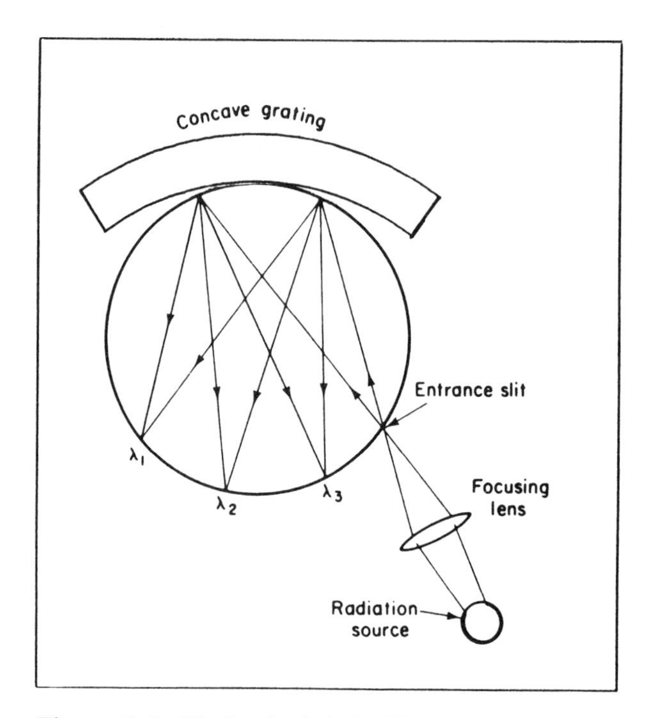

Figure 10.8 The Rowland circle. The entrance slit, curved grating, and focal points of the various wavelengths lie on a circle. The radius of curvature of the curved grating equals the diameter of the Rowland circle.

different point. The following important physical relationship has been found to exist between these points.

For a given grating, the entrance slit, center of the grating, and focal points of the various wavelengths fall on a circle called the *Rowland circle*. The *diameter* of the Rowland circle is equal to the radius of curvature of the surface of the concave grating. This relation greatly helps in locating the correct position for placing either a photographic film detector or a series of photomultipliers; that is, these must be positioned on the Rowland circle. The actual position can be found experimentally by moving the detector along the circle until the maximum emission signal is detected. The equipment is shown schematically in Fig. 10.8. In practice, photomultiplier detectors are put at the focal point of λ_1, λ_2, or λ_3 as desired. For qualitative work, a film or diode array may be located along the Rowland circle and all emitted lines recorded. However, the effective slit width of the diode array is too great-for general use because of loss of revolution.

e. Echelle Monochromators

In recent years echelle monochromators have become increasingly used in emission spectroscopy. Although this technique has been known for many years, it

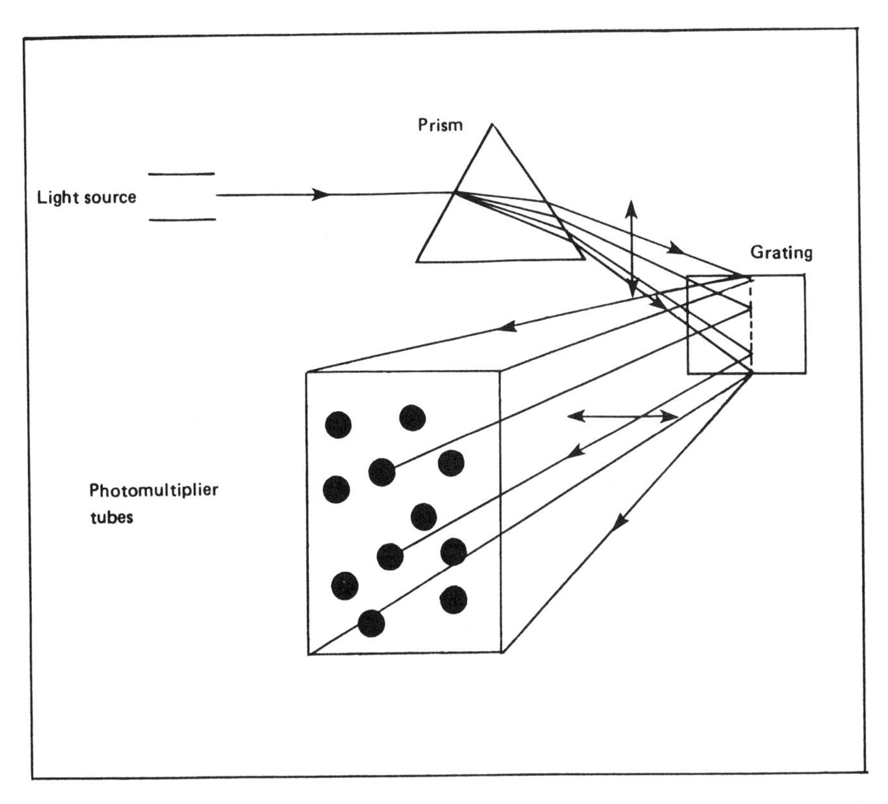

Figure 10.9 Schematic diagram of an echelle monochromator. The prism disperses in the vertical plane; the grating disperses in the horizontal plane.

was mostly of academic interest only. However, it has now been incorporated into some commercial plasma emission spectrometers.

The basic premise is to use two dispersive elements in tandem, a prism and a grating. These are placed so that the prism disperses the light in one plane. This light falls upon the grating, which disperses in a plane at right angles to the plane of dispersion of the prism. This is illustrated in Fig. 10.9. The net result is that the final dispersion into wavelengths takes place over a two-dimensional plane rather than along a single line as experienced in the Rowland circle optics. The radiation intensity is measured using numerous photomultipliers, which are located on a grid. It will be appreciated, of course, that the optics are much more complicated than illustrated in the schematic diagram of Fig. 10.9, because all focusing lenses and mirrors have been removed for simplicity of illustration.

In practice, the location of the photomultipliers is not as simple as in the Rowland circle, where it is known that the signal will fall on a fixed predictable line and the photomultipliers used for signal measurement can be moved at will to the correct position in the optical system. In the echelle system is it common to preselect the elements that will be monitored and to locate the points on the grid at which the radiation from these lines will fall. A hole is then drilled on the grid and the photomultiplier is permanently positioned.

The advantage of this technique is that the resolution is increased and the photomultipliers can be located at the most suitable blaze angle from the grating. This produces better resolution and higher intensity.

A disadvantage is that there is significant loss of freedom in locating and tuning the photomultiplier position on a day-to-day basis. In addition, monitoring new elements other than those predesigned into the instrument is a major undertaking not casually embarked upon. Frequently this means realigning the system, perhaps at the manufacturer's home laboratories.

Like many systems, the echelle monochromator has real advantages in providing data, but with increased expense and loss of versatility.

4. Slits

All spectrographs are fitted with entrance and exit slits. Entrance slits keep out stray light and permit only the light from the sample to enter the optical path. Exit slits placed after the monochromator stop all light but that in the desired wavelength range from reaching the detector. It is the image of the slit that reaches the detector. Emission lines from the various metals are the image of the slit illuminated by radiation from the sample at the relevant wavelength. Slits are usually created by the parting of two parallel metal strips. The width and height of the slit may be manually adjusted independently. A narrow slit is necessary to obtain resolution of two lines at similar wavelengths. If the slit is too wide, both lines pass through and reach the detector without being resolved. If the slit is too narrow, there is a loss in the intensity of radiation reaching the detector. A compromise based on experimental requirements is generally arrived at in practice. Slits are illustrated in Figs. 3.6 and 3.7. Typically the slits might be 1 cm high and 0.1 mm wide, although the width can be varied manually to meet the resolution required for the problem at hand.

5. Detectors

Two types of detectors are widely used: photomultipliers and photographic plates or films. Photographic plates or films are used only for qualitative analytical studies. The emission spectrum falls on the photographic film and is photographed. The emission spectrum of a standard, for example, iron, is photographed on the same plate immediately underneath the sample spectrum. By comparison one can measure the wavelengths of the radiation lines from the sample. This permits identification of the emitting elements and therefore provides a basis for qualitative analysis of the sample.

For quantitative analysis the intensity of one emission line per element is measured. The intensity of the emission line is correlated with the concentration of the pertinent element in the sample using standard calibration curves.

The line intensity can be measured by using a photomultiplier or a photographic film. When the latter is used, the intensity of the emission line is measured by the darkness of the line it produces on the photographic film. To make the measurement of the line's darkness meaningful, several problems must be taken into account. For example, the darkness of the line on the plate depends on the intensity of the emitted light and the wavelength of the light. Two lines of the same intensity may emit at 300.0 and 500.0 nm. The darkness of the line produced by these two lines on a film or plate may vary considerably, depending on the responses at the different wavelengths of the film used. No film has been developed with a response that is independent of wavelength. For example, ordinary black-and-white film responds slowly to the color red; in fact, dull red lights can be used in darkrooms because such light has so little effect on ordinary film. As a consequence, an emission line in the red part of the spectrum would not be detected by ordinary black-and-white film. Spectroscopic film responds faster and over a wider wavelength range than ordinary film, but is still limited in some spectral regions. Because of this variable response, calibration curves relating line darkness and sample concentration must be prepared at the wavelength range used for measuring the radiation intensity of the emission. Other emission lines from the same element at different wavelengths cannot be measured using the same calibration curve. Also, if two separate rolls of film are used for calibration purposes and for the subsequent analysis of different samples, they must respond equally to the emission from two samples with the same composition. To ensure equal response, these films must be stored, developed, and measured under identical conditions. The exacting requirements of emission spectrography call for skill and attention on the part of the operator. Self-developing film is now widely used. This film has been prepared so that its response and development are very reproducible. The film is built to fit automatically into commercial instrumentation, and its use has greatly simplified the problems of obtaining qualitative and, particularly, quantitative data from emission spectrography using film as the detector.

Photomultipliers (Chap. 6, Sec. C.2.c) are used only for quantitative work. Their response is less dependent on wavelength than that of film but is still somewhat dependant on wavelength (see Fig. 6.15). The relationship between photomultiplier response and wavelength is discussed in Chapter 6. Unlike the photographic plate, controlled development, storage, and production are not required. This saves much time and reduces the risk of error. Their immediate response and ease of interpretation have made photomultipliers the most commonly used detectors for quantitative determinations by emission spectrography.

D. ANALYTICAL APPLICATIONS OF EMISSION SPECTROGRAPHY

The unique spectra of different elements—particularly those of metals—make the emission spectrograph useful for qualitative and quantitative elemental analysis. But it provides very little direct information on the molecular form of the sample. In the process of analysis, the sample is destroyed by the electrical discharge. Hence it is a destructive method of analysis and cannot be used if the sample must remain intact.

1. Qualitative Analysis

Qualitative analysis is performed by recording the emission spectrum of the sample on a photographic film or plate. The sample elements are then identified by comparing their emission spectra with previously recorded spectra from known elements. Generally, for trace amounts, the *raies ultimes*, or *RU lines*, must be present and identified in the emission spectra. For higher concentrations, the RU lines are present together with many other emission lines.

a. Raies Ultimes

When a metal is excited in an electrical discharge, it emits a complex spectrum that consists of many lines, some strong, some weak. When the concentration of the metal is decreased, the weaker lines disappear. If the concentration of the metal is further reduced, some of the less strong lines disappear. Upon continuing dilution of the metal, more and more lines disappear until only a few are visible. The final lines that remain are called the raies ultimes or RU lines. These lines should be visible even at the lowest concentration of the metal concerned. Conversely, if the metal is present in a sample, whatever other spectral lines are emitted from the metal, the RU lines should always be detectable. Generally, for confirmation of the presence of a trace amount of a given metal, three raies ultimes should be detected. Confirmation based on a single RU line is unreliable because a line from another element present could be confused with such a line. If all the RU lines of a given element are not detected, that element cannot be present in detectable quantities. The RU lines are the lines arising from transitions between an excited state and the ground state. In atomic absorption they are the best absorption lines. They are the most intense emission lines but, as we shall see later, they are subject to reversal and not suitable for quantitative work at high concentrations.

b. Qualitative Analysis of Metals

The emission spectra encountered in emission spectrography are rich and intense for most elements in the periodic table, in contrast to those observed in flame photometry. Furthermore, the background interference encountered in flame photometry is greatly reduced when an electrical discharge is used as the excitation medium. As a result of these properties, emission spectroscopy is widely used in elemental qualitative analysis. Most metals can be detected at low concentrations (ppm) with a high degree of confidence. In practice, a photograph of the emission spectrum of the unknown sample is taken. Then a photograph of the emission spectrum of iron is taken on the same plate. A separate film containing the emission spectrum of iron and the RU lines of the metallic elements is then lined up with the sample spectrum using the iron as a standard. The presence of the RU lines of all the elements can then be checked by comparison between the sample and the standard. Iron contains a very rich emission spectrum. By comparison with the iron spectrum, the wavelengths of the lines from the sample can be determined, and, by comparison with the spectra of the other elements, the elements in the sample can be identified. Some of the elements and their RU lines and sensitivity limits are listed in Table 10.1.

2. Quantitative Analysis

Quantitative analysis is carried out by measuring the intensity of one emission line of the spectrum of each element to be determined. The choice of the line to be measured depends on the expected concentration of the element to be determined. For extremely low concentrations, the RU lines must be used. These are the most intense lines and must be used when high sensitivity is required. These lines are reversible, and their intensity does not increase linearly with concentration. Therefore, for greater concentrations a nonreversible, less intense line must be used. For high concentrations the intensity of a weak line would be measured in order to stay within the calibration range.

The measuring process may be effected by using a photographic plate as the detector. The plate must then be developed and the line intensity measured under highly controlled conditions. After the line's intensity has been measured, it must be compared to a calibration curve, and from the data the concentration is calculated.

We can also use a photomultiplier detector to measure the intensity of any emission line. Automatic analyzers commonly utilize as many as 20 photomultipliers simultaneously. Each is set at the wavelength of an emission line from a different element. Figure 10.10 shows the instrument's optics.

In practice, the sample is burned in the electrical discharge. Each photomultiplier measures the intensity of one line. After the sample is burned, the

Table 10.1 RU Lines of Common Metals

Element	Wavelength of RU lines (nm)	Element	Wavelength of RU lines (nm)	Element	Wavelength of RU lines (nm)
Ag	328.0	Co	345.3	Mo	379.8
	520.9		352.9		386.4
Al	396.1	Cr	425.4	Na	589.5
	394.4		520.6		568.8
As	183.1	Cs	852.1	Ni	341.4
	228.8		894.3		349.2
	237.0	Cu	324.7	P	253.5
Au	242.1		327.4		255.3
	280.2	Fe	358.1	Pb	405.7
B	249.7		379.1		363.9
	345.1	Ge	265.1	Pt	265.9
Ba	235.5		270.9		299.7
	553.5	Hg	185.0	Se	207.4
Be	234.8		253.6		473.1
	332.1		435.8		474.2
Bi	306.7	K	766.4	Si	288.1
	293.8		769.8		390.5
C	229.6	La	624.9	Sn	317.5
	247.8		593.0		452.4
Ca	442.6	Li	670.7	Sr	460.7
	443.4		610.3		483.2
Cd	228.8	Mg	285.2	Tl	498.1
	340.3		516.7		338.3
Ce	226.3	Mn	403.1	V	318.3
	404.0		403.4		437.9

photomultipliers "read out" in succession. Calibration curves must be prepared for each element separately. The concentrations of the metals in the sample can then be calculated from the photomultiplier readings and the corresponding calibration curves. Using a commercially available automatic readout system, the simultaneous determination of 20 elements in a sample can be carried out in a few minutes. With special equipment, as many as 60 elements have been determined at one time, but the mechanics of alignment are Herculean and the method is usually avoided.

The precision of the method depends on the concentrations of the metals being determined. At high concentrations (several percent), a precision of 1% relative error can be expected. At low concentrations (of the order of parts per

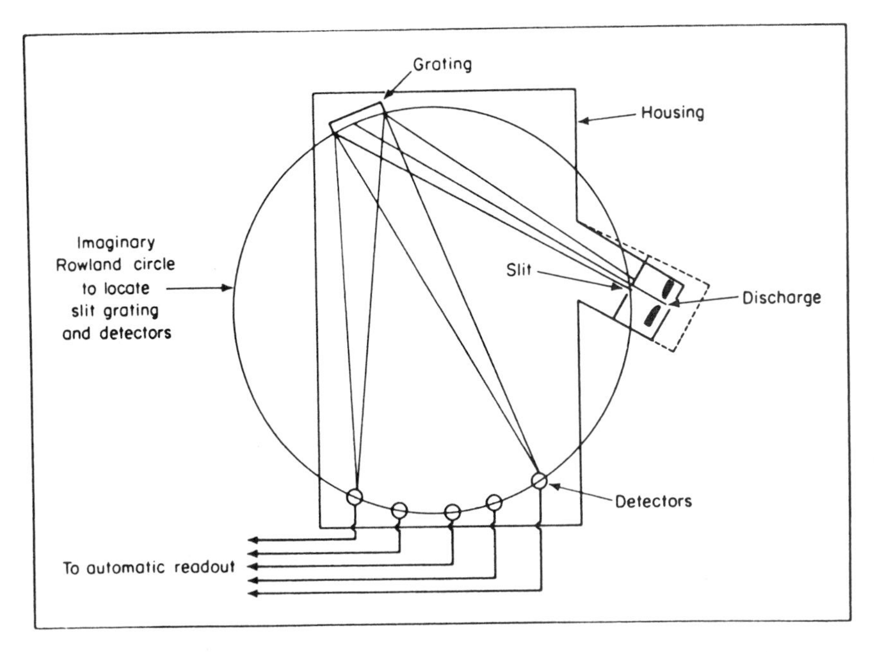

Figure 10.10 Optics of an emission spectrograph for simultaneous multielemental analysis.

million), a relative error of 5% is the best that can be expected on a routine basis. If the sample varies significantly in composition, the error may be as high as 20–30%.

We pointed out at the beginning of this section that it is essential in quantitative analysis by emission spectrography to select an emission line whose intensity can be measured accurately. Lines that are too intense are not suitable for measurement on film, because such lines become completely black and are not sensitive to small variations in concentration. Overly intense lines cannot be measured with the photomultiplier either, because the latter tends to saturate and become unstable when exposed to intense radiation. In practice, a particular line and exposure time are chosen with these factors in mind.

a. Line Reversal

The RU lines are the most intense lines and invariably involve an electronic transition from an excited state to the ground state. These lines can be detected even when small concentrations of the metal are present. When increased concentrations of the metal are present, there is an increase in the number of free unexcited atoms of the element in the discharge. These free atoms absorb radiation at the same wavelength as the emitted RU lines from other atoms of the same element since they involve the same transition (see Chapter 7, on atomic absorption). As a result, such emission lines are reabsorbed by neighboring unexcited atoms. There is a relative decrease in emission intensity; in fact, if the concentration is increased significantly, the absorption is so effective that the net intensity actually decreases. If the intensity is plotted against concentration, it goes through a maximum and then decreases. The intensity therefore ''reverses'' at higher concentrations. Such lines are said to be *reversible lines*. They cannot be used for quantitative analysis at higher concentrations, but they are useful for measuring very small concentrations, where reversal is not a problem.

b. Matrix Effects and Sample Preparation

The matrix, or main substance, of the sample greatly affects the emission intensity of the trace elements present. For example, the emission intensity of 50 ppm of iron in a sample of sand may be 10 times as intense as 50 ppm of iron in a sample of salt. This is because the matrix (i.e., the salt or the sand) affects the rate of vaporization of the iron in the electrical discharge. This, in turn, affects the number of iron atoms excited per unit time, and therefore the emission intensity from the iron. If the iron is more volatile than the matrix, it will vaporize rapidly and the spectral intensity will be intense. If the iron is less volatile than the matrix, it will vaporize slowly and the spectral intensity will be weak even if the concentration is the same.

To minimize this problem, it is common practice to pretreat samples chemically and reduce them all to a common matrix. For example, all samples requiring the determination of iron, irrespective of the origin or type of sample, may be dissolved and extracted to isolate the iron. The iron extract may then be mixed with an inert powder, such as alumina, and the mixture dried and analyzed in the normal fashion. Failure to reduce to a common matrix will result in errors, which may be gross. For this reason, any quantitative data on untreated samples should be used with caution. Errors may be involved that can produce results 10 times as great or 10 times as small as the true answer. The results are semiquantitative but may nevertheless be useful to give ballpark data in preliminary studies of samples of unknown composition.

c. The Use of Internal Standards

If the composition of a sample and matrix is unknown, it is not practical to determine an element's concentration from the measurement of a given line's intensity, even if the experimental conditions are rigidly controlled. This is because, apart from matrix affects, which are most important, such factors as physical packing and the size of particles cause a variation in the quantity of sample vaporized into the discharge from one sample to the next. These problems are overcome by using an *internal standard*.

A known amount of the element that is to serve as the internal standard is added to the sample. The following must be true for the internal standard used:

1. The element used as an internal standard should be similar in properties to the element to be determined.
2. The wavelength of the emission line from the internal standard should be similar to the emission line of the analyte.
3. The concentration of the internal standard in the sample should be zero before adding the standard sample.
4. The concentration of the internal standard should be similar to that of the analyte.

When the sample is exposed to the electrical discharge, the internal standard also emits radiation. The intensity of one of its lines is measured. Any variation in the quantity of the sample reaching the discharge causes a variation in the intensity of the measured line reaching the detector; this variation results in a variation in signal from the detector. The signal is a measure of the quantity of internal standard burned, and it can be used to predetermine how much sample is burned. By computer control, the same quantity of each sample may be burned and more precise data obtained.

We now have two measurements, that of the intensity of the sample emission line and that of the intensity of the internal standard emission line. We know the concentration of the internal standard and the ratio of the intensity of the two lines. From these data we can calculate the concentration of the element being determined. The ratio of the intensities of the two lines is not a simple one, in that if the concentrations of the two elements are equal, the two lines will not be equal, but will bear a simple relationship to each other. That is, one may be 85% as bright as the other at equal concentrations. This relationship can be determined experimentally by analyzing synthetic mixtures of known composition.

The use of an internal standard is widespread in quantitative work. The element used as the internal standard should resemble the elements being determined to the extent that their boiling points and chemical reactivity should be similar. The internal standard element should also have a measurable emission line in the same spectral vicinity as the sample emission line, and, as stated before, it must not be present in the original sample.

d. Preburn Time

When a sample is introduced into an electrical discharge, there is a lag in time before the emission signal becomes steady. In the first few seconds of discharge, the intensity of emission is erratic and difficult to control. In order to obtain reproducible results, it is advisable to ignore these erratic signals. This is done by not recording the emission until the signal is judged to be stable. By plotting emission intensity from a standard against time, this period can be measured. For example, for lead this period may be 15 sec. When a sample is analyzed for lead, the first 15 sec of the discharge are not recorded. This period is called the *preburn time*, that is, the time during which the emission intensity is not stable and is not recorded.

The preburn time varies from one element to another. If the element vaporizes easily, as does tin or arsenic, for example, the preburn time is short; but if the element is, like iron, more difficult to vaporize, the preburn time is long. The preburn time should always be determined experimentally under the conditions to be used for analysis, because these also affect the volatization of the sample element.

In practice, when several different elements must be determined simultaneously, the preburn time for all elements to be determined will never be identical. If they are similar, a compromise preburn time can be used that is reasonably satisfactory for each element. If the elements are too dissimilar, a reasonable compromise may not be possible and two separate exposures and analyses may be necessary: one for volatile compounds with a short preburn time and one for nonvolatile compounds with a long preburn time.

A preburn time is always used if accurate results are required. Modern equipment can be set to incorporate the preburn time automatically into the analysis. This makes quantitative analysis easier to perform.

e. Specific Applications

Emission spectroscopy has been used for many various types of analyses. In metallurgy, for example, the presence in iron and steel of the elements nickel, chromium, silicon, manganese, molybdenum, copper, aluminum, arsenic, tin, cobalt, vanadium, lead, titanium, phosphorus, and bismuth have been determined on a routine basis. Frequently the analysis can be carried out with very little sample preparation, apart form physically shaping the sample to fit the source of discharge unit. These elements have been determined in concentration ranges varying from traces (0.001%) in iron to 30% in steels. With automatic instrumentation, numerous elements can be determined simultaneously. Other alloys that have been analyzed include aluminum alloys, magnesium alloys, zinc alloys, copper alloys (including brass and bronze), lead alloys, and tin alloys (including solders).

In the oil industry, lubricating oils have been analyzed for iron, nickel, chromium, manganese, iron, silicon, copper, aluminum, and so on. These are metals that get into the lubricant when bearings and pistons wear. Such analyses give the engine designer valuable information on the parts of bearings that need protection. The information is used in aircraft and railway engine maintenance to warn of bearing wear in the engine. The bearings may be replaced based on lube oil analysis rather than taking the engine apart for physical inspection, which is a long and expensive procedure.

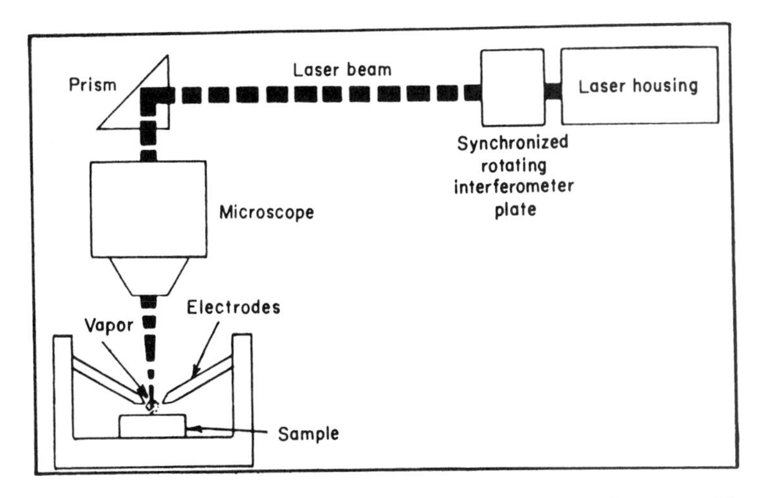

Figure 10.11 Schematic diagram of a laser microprobe. (Courtesy of Jarrell-Ash Co.)

Oil feed stocks to cat crackers (the unit in a petroleum refinery where catalytic cracking is carried out) are also analyzed for iron, nickel, and vanadium. These elements can cause a change in the product formed in the cat cracker and result in poor fuel or even hydrogen instead of good fuel. The analysis has to be rapid and sensitive.

Soil samples and animal tissue samples have been analyzed for many elements, including sodium, potassium, calcium, magnesium, zinc, copper, nickel, and iron. Similar analyses have been carried out on plant roots and plant foliage. The results have led to much information concerning animal health and its relationship to grazing grounds and food in general. Environmental samples such as air particulates, water samples, soil samples near waste dumps, and effluents from industrial plants have been analyzed by emission spectrography.

As can be seen, emission spectroscopy has many and varied uses in the qualitative and quantitative determination of the metallic elements. However, ICP has replaced emission spectrography in many of these applications.

E. THE LASER MICROPROBE

An emission method closely allied to emission spectrography involves the use of the laser microprobe. In this system, a laser beam is used to vaporize the sample, which is then excited by an electrical discharge. A schematic diagram of the instrument is shown in Fig. 10.11. The sample is loaded just beneath the two electrodes that will be used to generate the electrical discharge. A ruby laser is then focused onto the surface of the sample. The energy from the laser causes an intense local hot spot, which vaporizes a small quantity of sample. The vapor short-circuits the electrodes and electrical discharge takes place, which simultaneously excites the metals in the vapor. The latter emit typical emission spectra, which are collected and measured qualitatively in the normal fashion as on a conventional emission spectrograph.

The system is particularly useful for examining surfaces for contamination, inclusions, and other defects. The method is generally not useful for bulk analysis, because only surface atoms are vaporized. Qualitatively, the method is quite satisfactory and can provide valuable information concerning the surface that would be difficult to obtain by other techniques.

Quantitatively the method depends on the fact that the burst of laser radiation if reproducible. The amount of energy impinging on the surface is therefore reproducible and the quantity of metals vaporized from the surface is also reproducible. Based on this fact, the emission intensity can be related to the sample concentration. However, the quantitative relationship is at best tenuous and should be used only for semiquantitative analysis. However, it will provide valuable information such as the concentration levels that may be in the ultratrace, trace, percentage, or principal component level.

This method has found application in the semiconductor industry and in studies of corrosion.

F. THE USE OF RF PLASMAS AS EXCITATION SOURCES

Two types of plasma have been used to excite emission spectra: the RF plasma and the DC plasma.

1. RF Plasma

Inductively coupled plasmas (ICPs) are commercially available and are routinely used in industry and research. The process of excitation is quite different from that in an electrical discharge, but the resulting emission spectrum is similar in all respects inasmuch as the same atomic transitions are involved. A schematic diagram of the equipment is shown in Fig. 10.12.

In this system argon is directed in the form of a jet inside an induction coil. The induction coil carries an alternating current operating at radiofrequencies usually between 5 and 75 MHz. The frequency commonly used on commercial equipment is 27.1 MHz. The high-frequency electric field induces a high-frequency magnetic field. Argon, which is normally a nonconductor and would therefore not be affected by such a magnetic field, is slightly ionized using a Tesla coil. The few ions formed are immediately affected by the magnetic field. Their translational motion changes rapidly from one direction to the other, oscillating at the same frequency as the RF coil. The rapid movement in alternating directions is induced translational energy and is akin to a high tempera-

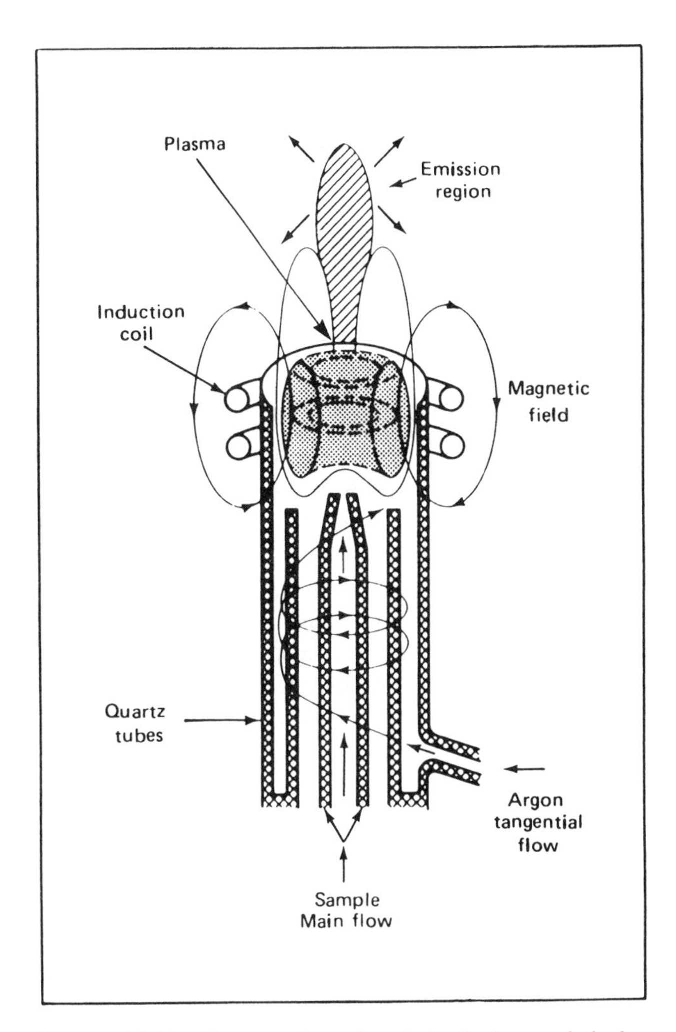

Figure 10.12 Cross section of an inductively coupled plasma torch. Note that the emission region is above the torch.

ture. The few ions formed collide with other argon atoms forming argon ions. This sets up a rapid chain reaction and rapidly forms many more argon ions. The net and final effect is that of a *plasma* in which the argon is highly ionized and each ion is oscillating in its path in sympathy with the frequency of the induction coil.

The sample is then introduced into the argon stream either as a gas, liquid, or solid. The preferred final state before entering the plasma is a gas. When the sample gas reaches the plasma, the ions collide with the gas molecules of the sample and cause intense excitation. This in turn causes intense emission available for qualitative and quantitative analysis.

The shape of a plasma in a coil is dependent on the frequency of oscillation of the RF coil. At the frequencies noted, the plasma is shaped like a donut and the sample can be directed into the interior of the plasma. If the frequency is incorrect, the plasma becomes shaped like a ball and the sample tends to float around the outside of the plasma. It does not interact with the plasma and fails to become excited in the process.

As can be seen from Fig. 10.11, there is a second stream of argon around the outside of the central argon jet. This prevents contact between the plasma and the quartz and acts as a thermal buffer, thus avoiding meltdown of the quartz in operation.

2. DC Arc Plasma

The design of another type of instrument is shown in Fig. 10.13. In practice two jets of argon issue from the anodes. These join together and form an electrical bridge with the cathode, which is made of tungsten. In operation the three electrodes are contacted together, voltage is applied, and the electrodes then separated. A plasma is generated automatically, forming a steady discharge shaped like the letter Y.

As shown in Fig. 10.13, the region that is monitored for analytical measurements is at the junction of the plasma. This small region has high intensity and a background lower than the regions in the immediate vicinity. The sample is injected into this region with a stream of argon from a separate injection system.

The system is analytically quieter in that the noise level is reduced and the background is reduced compared to the inductively coupled plasma. In addition, the power requirements and the quantity of argon required are less than that necessary to operate an RF plasma.

The temperature of the plasma is as high as 10,000 K, but in the excitation region monitored the electronic temperature is about 6000 K.

A limiting factor of the process is that desolvation, decomposition, atomization, and excitation all must take place during the brief period that the elements of interest from the sample pass through the excitation region. However, since this region is outside the main plasma, the background is reduced and the signal-to-noise ratio is improved.

Proponents of this design claim that it is simpler to operate and that the background is reduced while the sensitivities for various elements are about the same as for the RF plasma. The technique is useful for the determination of all

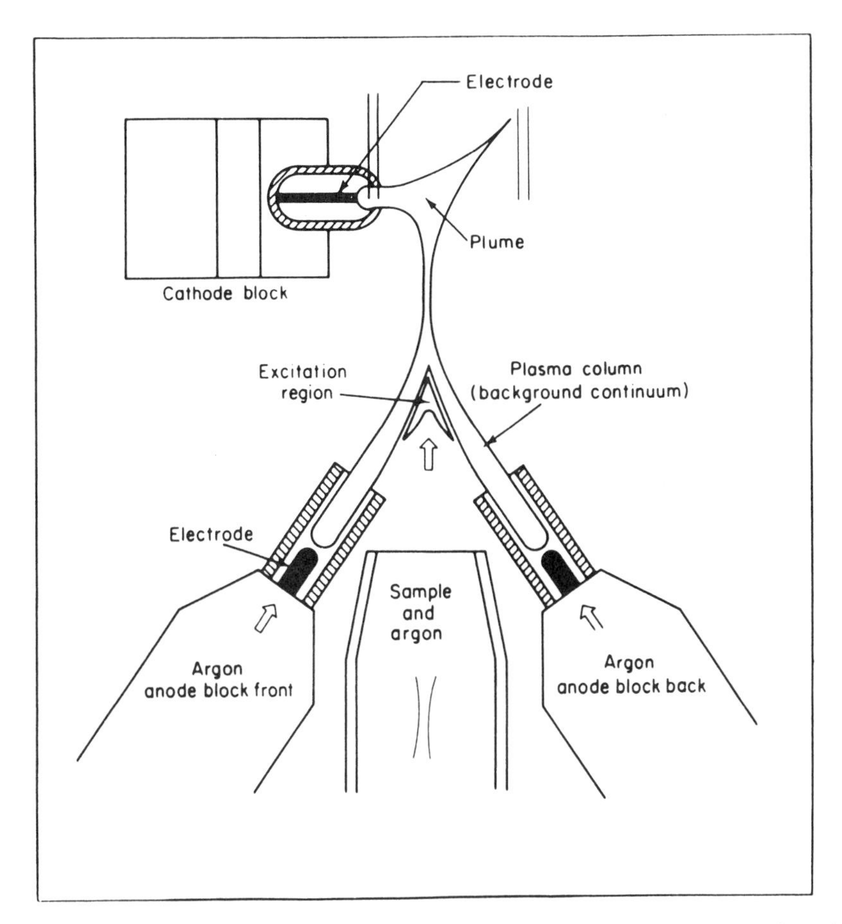

Figure 10.13 Inverted Y configuration of a plasma jet. (With permission from Spectrametrics, Inc., Andover, Mass.)

of the elements determined by the ICP. An interesting application is the determination of boron in tissue and in cells.

3. Sample Types Analyzed by RF Plasma

Samples of all three phases can be introduced into RF plasma. If the sample is a powder, it can be swept into the system; however, this will result in a loss of sensitivity because it is only when the atoms are liberated in the free state and excited that they can emit atomic spectra. The introduction of powders also tends to make the signal erratic, since the latter is dependent on the rate at which free atoms are liberated from the powder. Liquid samples can be introduced into the plasma in the form of a cloud. A nebulizer is used to break up the liquid sample into tiny droplets, which are then aspirated with the argon jets into the RF plasma. The sample rapidly breaks down, liberating free atoms, which can then be excited. Gas samples can be injected directly into the argon stream and transported directly into the RF plasma, where excitation takes place.

4. Effective Temperature of the Plasma

The effective temperature of the RF plasma is between 9000 and 10,000 K, although 6000 K is probably closer to the temperature to which the sample is actually exposed. This is an extremely high temperature compared to other emission sources. The effect of this very high temperature is to cause some ionization of the sample and the emission of ionic spectra. This effect can be reduced by the introduction of lithium salts in concentrations around 1000 ppm. The lithium salts, particularly the chloride, are easily ionized. This ionizability stabilizes the sample elements present, which remain as free atoms and are more likely to emit atomic spectra. The lithium acts as a *spectroscopic buffer*.

5. Background Spectra

The background from the continuum is very high, particularly at the tip of the induction coil in an ICP torch. However, a few centimeters above, the background drops off significantly and the sample exhibits a high signal-to-background ratio. Emission spectra are studied from this part of the plasma.

6. Sample Introduction

Liquids are the most common form of sample to be analyzed by plasma emission. These are usually introduced with a pneumatic spray, which injects aerosols into the correct argon stream entering the plasma. With low salt content solutions, the analysis is straightforward. In practice, however, it is always advisable to wash the nebulizer with distilled water after each sample in order to clear out any residue from the previous samples, thus avoiding any memory effect.

If the sample has a high salt content, such as seawater, urine, or other body fluids, it is necessary to use a spray designed to handle high salt content solutions. In practice, an argon saturator nebulizer tip enables washout of the sample and prevents clogging. One interesting method uses a carbon atomizer as a sample introduction system.

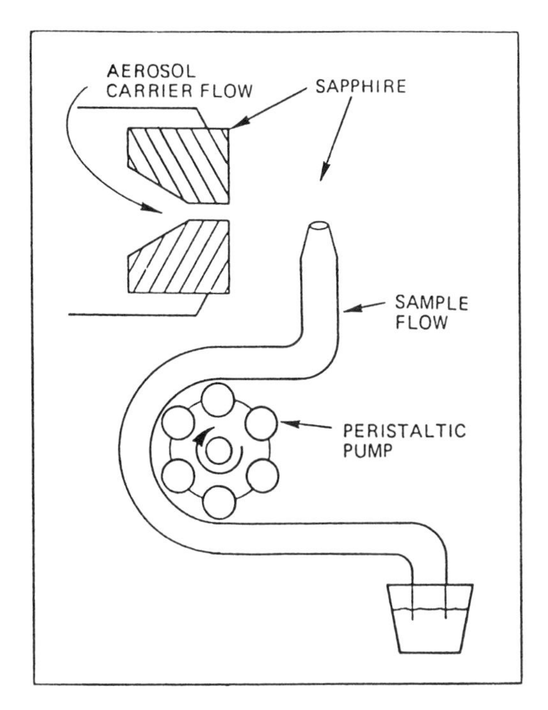

Figure 10.14 Computer-controlled cross-flow nebulizer with sapphire orifice equilibrates the torch quickly, can aspirate solutions with high dissolved solids content, resist corrosion from most samples, and minimize matrix effects.

Aqueous samples are the sample type of choice. Such samples are easy to handle and easily cleaned. Further, the water itself does not generate a very rich spectrum of its own. Organic samples are more difficult to handle since the organic material may give fragments that generate a high background. However, with suitable handling, organics can be run.

Solid samples can be analyzed, but in practice it is difficult to perform quantitative analysis when solid sample introduction systems are used that rely on floating a powder of material into the atomizer.

a. Nebulizer

Several nebulizers are available, including the pneumatic pump, cross-flow nebulizer, ultrasonic nebulizer, and high solids nebulizers, described earlier. Of these the most popular is the cross-flow nebulizer. A typical schematic diagram is shown in Fig. 10.14. A peristaltic pump ensures a steady flow rate of the sample. A cross-flow of gas nebulizes the sample and brings it to the plasma. The design can handle solutions containing high (10%) concentrations of solids and has no memory effect, and the use of a synthetic sapphire tip makes it resistant to corrosion. A comparison of the cross-flow nebulizer and the micro Babington nebulizer indicated that the latter was considerably more stable than the former.

The thermospray atomizer has been used to nebulize liquid samples both for atomic absorption and plasma emission. The advantage of this system is that the sample is reduced to vapor stage immediately rather than to a system of fine droplets. This eliminates the need for the plasma to evaporate the sample down to the residue and the problem of variable drop size causing variable evaporation efficiency. The system has the potential of significantly improving sensitivity. In preliminary studies it was found that sometimes the detection limits were worse using the thermal spray rather than the normal pneumatic nebulizer. This was because of the increase in the noise level and therefore the signal-to-noise ratio and the increase in the standard deviation of the blank sample. However, in some cases the detection limits improved, which can be explained in terms of improved nebulization efficiency. Further development of this nebulizer is needed.

Electrothermal techniques adapted from atomic absorption have been successfully applied to ICP as a means of introducing a sample into plasma. The technique is particularly useful for organic matrices since it eliminates the effects of organic material. In practice the electrothermal atomizer loosely follows the system used in atomic absorption spectroscopy, drying, ashing, and "atomizing" the sample in sequence. The atoms formed are then introduced into the ICP, where they are further atomized and excited. The method can be used fully with liquid and solid samples and needs only one microgram of material.

b. Direct Solids Elemental Analysis

Numerous attempts have been made to introduce solid samples directly into plasma. Some of these have been quite successful, and although efforts have been made to float powdered material into the atomizer, this has proved to be difficult. At this time there are no commercial models available with this capability.

Another innovation has been the use of fiber optics to convey the image of the plasma to the entrance slit of the spectrometer.

7. Laser Ablation

A technique for direct sampling of solids is laser ablation. In this system, a laser beam is directed at the solid sample. The surface material is vaporized and swept into the plasma. Eximer lasers are used as the energy source. These are gas lasers that emit strongly in the ultraviolet range. The high-energy beam has little problem in vaporizing the sample. This system is particularly useful for the analysis of metals and alloys, but has also been used for other sample types.

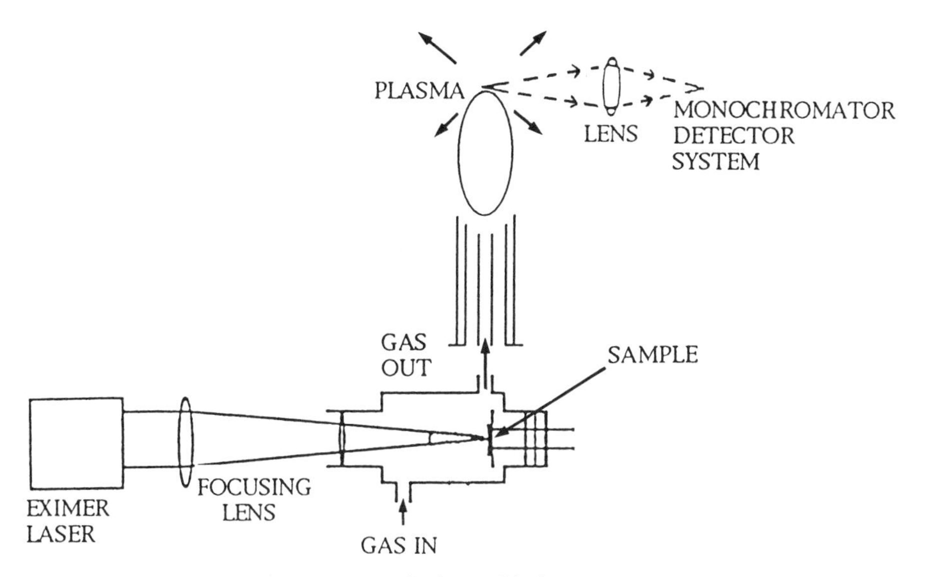

Figure 10.15 Schematic diagram of a laser ablation system.

Solid samples have also been introduced directly by forming a slurry with a suitable solvent and introducing the slurry directly into the plasma. There was little loss of sensitivity, but the noise level increased somewhat.

Organic solvents always present a problem because they tend to extinguish the plasma. Also, they emit broad-band spectra, which give rise to high background radiation. One way to reduce these problems is to remove the solvent prior to sample injection using an ultrasonic vibration and cryogenic desolvation. A schematic diagram of this system is shown in Fig. 10.15.

8. Flow Injection

Flow injection used with ICP-MS improves the system significantly. It permits using a much smaller sample size. Samples between 50 and 500 μl can be easily handled compared to the normal 3 ml used in conventional sampling. This is an important economy when only limited sample is available. Further, since the sample size is reproducibly controlled, the system lends itself to automation, an important consideration in routine laboratories. Another consideration is that the flowing solvent may be a pure liquid (water) with no dissolved solid. Only a small quantity of sample reaches the ICP, and it is washed out as soon as it passes through. The net result is that the ICP can handle much higher dissolved solid concentrations without injuring the system. The accuracy of the method is

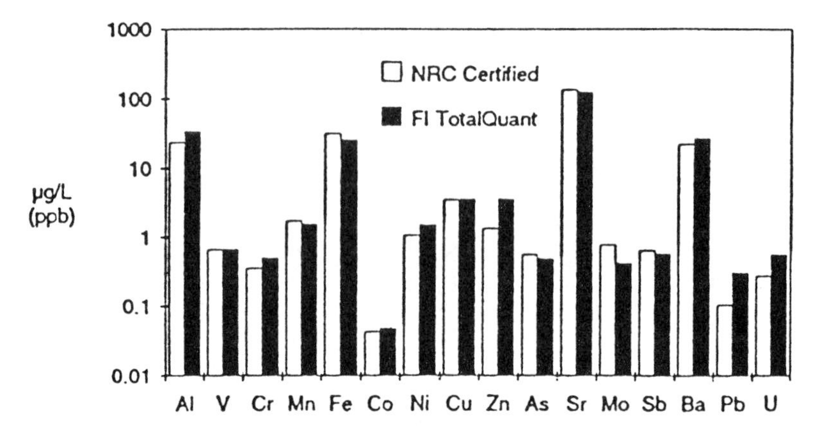

Figure 10.16 Data obtained using flow injection ICP-MS for the analysis of river water.

improved in the process as illustrated in Fig. 10.16, which shows the results obtained on a flow injection ICP-MS on Canadian National Research Council SLRS-1.

9. Analytical Use of Plasma Emission

With the use of spectroscopic buffers, the emission spectrum is similar to that of the DC arc rather than the AC spark. The system is therefore very suitable for qualitative analysis at lower levels of concentration.

Another great advantage of the ICP is that the emission intensity–versus–concentration curve is linear over ranges of up to 10^5. This is a great linear range, better than for other comparative methods of elemental analysis. It provides a means of getting quantitative data without having to dilute concentrated samples or concentrate by solvent extraction or evaporation if the concentration is too low. This extensive range excludes the use of film as a detector, because it is not possible to calibrate film over such a wide range. It is therefore normal to use photomultiplier detectors when ICP is used as the excitation source. This is not considered a disadvantage since the use of film requires a high degree of skill and patience to get good data. Photomultipliers are much more rapid and operate over a wider wavelength range and over a wide range of radiation intensities.

Another advantage of ICP is that it can be used for simultaneous multielemental analysis. Unlike emission spectrography, the induction periods for the various elements are not significantly different from each other and no preburn time is necessary. Conditions do not have to be modified to accommodate dif-

Table 10.2 Detection Limits (μg/ml) Using Plasma Emission

	ICP	DC plasma		ICP	DC plasma		ICP	DC plasma
Ag	0.004	0.004	In	0.02	0.004	Re	0.01	0.01
Al	0.008	0.002	Ir	0.02	0.020	Rh	0.01	0.001
Au	—	0.002	K	0.4	0.020	Ru	0.01	0.002
As	0.05	0.080	La	0.001	0.003	Sb	0.08	0.050
B	0.001	0.005	Li	0.02	0.002	Se	0.08	0.1
Ba	0.0004	0.0003	Mg	0.0002	0.0002	Si	0.04	0.010
Be	0.0002	0.0003	Mn	0.001	0.002	Sm	0.008	0.004
Bi	0.02	0.090	Mo	0.008	0.004	Sn	0.02	0.02
C	0.5	—						
Ca	0.0004	0.0007	Na	0.01	0.003	Sr	0.0001	0.0003
Cd	0.004	0.005	Nb	0.002	0.004	Ta	0.002	0.020
Co	0.002	0.005	Nd	0.005	0.005	Te	0.08	0.2
Cr	0.003	0.002	Ni	0.002	0.002	Ti	0.0006	0.002
Cs	2.0	0.8	Os	0.001	0.02			
Cu	0.001	0.002	P	0.004	0.090	U	0.030	0.040
Dy	0.003	0.003	Pb	0.002	0.015	V	0.0002	0.002
Er	0.004	0.002	Pd	0.001	0.002	W	0.03	0.020
Fe	0.005	0.006	Pr	0.02	0.006	Y	0.0002	0.02
Ga	0.01	0.003	Pt	0.001	0.02	Zn	0.001	0.006
Ge	0.007	0.01						
Hg	0.05	0.020	Rb	2.0	0.020	Zr	0.002	0.004

Courtesy of the Beckman Instrument Co., Irvine, California.

ferent elements. All elements can be excited at the same time in the same plasma. It is not uncommon to determine 20–30 elements at the same time using this system. However, the same problems with mechanical alignment occur with ICP as with emission spectrography. But exhaustive attention to this mechanical problem has led to designs that work satisfactorily.

The inductively coupled plasma method has not yet reached its potential, but its high sensitivity and ease of operation assures that it will become one of the most important emission methods available in elemental analysis. Typical sensitivities obtained are shown in Table 10.2.

10. Comparison Between Absorption and Emission Methods

Atomic absorption, flame photometry, and emission spectrometry all have the same goal—the qualitative and quantitative determination of elements, particularly metals. These methods have a number of things in common and a number of differences. The choice of methods for a particular analysis depends on the type of sample to be analyzed.

11. Sensitivity

The sensitivity of flame atomic absorption is better than in flame photometry and usually somewhat better than in emission spectrography; however, it has sensitivity similar to that of the ICP method. The sensitivity of the atomic absorption carbon atomizer is usually better than ICP and much better than that of flame photometry or emission spectrography.

12. Accuracy and Precision

Atomic absorption measure the absorption by ground-state atoms. The population of ground-state atoms depends on the concentration in the original sample and the efficiency of atomization. Atomization efficiency and therefore absorption is subject to errors, particularly solvent effects and chemical interferences. It is not dependent on the wavelength of absorption or the temperature of the atomizer, except that the atomizer temperature may affect atomization efficiency.

In contrast, emission methods are dependent on the number of excited atoms formed in the system. This in turn depends on the concentration of the original sample, the efficiency of atomization, and the ability to excite the free atoms. Equation (2.6) indicates that the efficiency of excitation depends on the temperature of the flame and on the energy required to cause excitation. Therefore emission intensity is dependent on the energy necessary to cause excitation. Elements such as sodium, lithium, and potassium require little energy to become excited and therefore give intense emission. However, elements such as mercury, zinc, and cadmium require significant energy and therefore are not usually observed in flames and are observed only with some difficulty in emission spectrography. The high temperature of the ICP, however, is successful in exciting these elements, and they can be quantitatively determined at low concentrations.

The effect of background radiation, temperature effects, matrix effects, and solvent effects make it more difficult to carry out quantitative analysis using emission methods and flame photometry. The precision of these methods is therefore worse than that of atomic absorption spectroscopy.

On the other hand, atomic absorption spectroscopy essentially determines one element at a time and cannot be used for qualitative analysis. In contrast, flame photometry is useful for the determination of alkali metals and alkaline-earth metal, whereas emission spectrography is useful for the determination of all the metallic elements in the periodic table. The ICP method has an extensive analytical range that supersedes those of all the methods mentioned above. However, at this point atomic absorption is the preferred method for quantitative analysis, and emission spectrography for qualitative analysis and multielemental analysis. Atomic absorption is generally more accurate, but emission spectroscopy and flame photometry are useful for qualitative analysis and simultaneously multielemental quantitative analysis.

G. INTERFACED ICP–MASS SPECTROMETER

1. Mass Spectroscopy

The conventional detector for the plasma emission spectrum is the photomultiplier. This detector is sensitive and exhibits fast response times. Unfortunately the emission spectrum of many elements is very complex, which leads to difficulties in identifying the lines to be measured for quantitative analysis and resolving lines close to each other to avoid spectral interference. Frequently spectral interference and background interference cannot be eliminated, and complicated, often error-prone sample pretreatments are necessary to correct for them.

The mass spectrometer has now been developed and used as an alternate detector to the photomultiplier. The mass spectrometer was used by J. J. Thompson in 1913 and Aston in 1919 to demonstrate that elements in the periodic table existed as various isotopes. This was a vital foundation stone in establishing the order of the elements in the periodic table. It explained why atomic weights were fractional (because elements consisted of mixtures of isotopes) rather than whole numbers. It also led to the confirmation of new elements discovered in the periodic table at the end of the nineteenth century. Since that time mass spectrometry has been devoted primarily to the analysis of organic materials.

First the large magnet mass spectrography used by Aston was developed by Dempster to be more rapid and to cover a greater molecular range. More recently the work of McLafferty has led to improved resolution using double focusing. This led to the ability to generate empirical formulas formally based on molecular weights measured to seven or eight significant figures. In more recent years the multisector mass spec has further improved resolution in this type of instrument.

Parallel to these developments has been the emergence of the time of flight mass spectrometer and the *quadropole mass spec*. The quadropole mass spec in particular has been highly exploited since it does not require that the entering ions have zero or near-zero lateral energy. This greatly facilitates the use of the quadropole mass spectrometer as a tandem instrument attached to some other system. It has been exploited particularly fruitfully in GC–mass spec combinations and more recently in HPLC–mass spec.

Table 10.3 Complementary Aspects of ICP-MS

ICP emission spectrometry	Mass spectrometry
1. Efficient but mild ionization source (produces mainly singly charged ions).	Ion source required.
2. Sample introduction for solutions is rapid and convenient.	Sample introduction can be difficult for inorganic samples. Thermal, spark-source, or secondary ion sources are generally restricted to solid samples and are time-consuming.
3. Sample introduction is at atmospheric pressure. Efficiency is poor	Often requires reduced-pressure sample introduction.
4. Matrix or from solvents interelement effects are observed and relatively large amounts of dissolved solids can be tolerated.	Limited to small quantities of sample.
5. Complicated spectra with frequent spectral overlaps.	Relatively simple spectra.
6. Detectability is limited by relatively high background continuum over much of the useful wavelength range.	Very low background level throughout a large section of the mass range.
7. Moderate sensitivity.	Excellent sensitivity.
8. Isotope ratios cannot usually be determined.	Isotope-ratio determinations are possible.

2. Advantages of Interfacing

The development of plasma mass spec using the quadropole has been a natural outcome of these developments. Some of the properties of major importance when combining these instruments include the fact that the ICP has a high ionization efficiency, which approaches 100% for most of the elements in the periodic table but has a low incidence of doubly charged ions. The mass spectra are very simple and elements are easily identified from the atomic weights and the isotope ratios exhibited. This greatly relieves the major problem of the ICP, i.e., the rich complicated optical emission spectrum. In addition the mass spectrograph is very sensitive and linear over a wide dynamic range of up to five orders of magnitude. A list of compatible properties of ICP and MS as proposed by Heifje et al. is shown in Table 10.3. These attractive features have led to considerable attention being paid to interfacing these techniques.

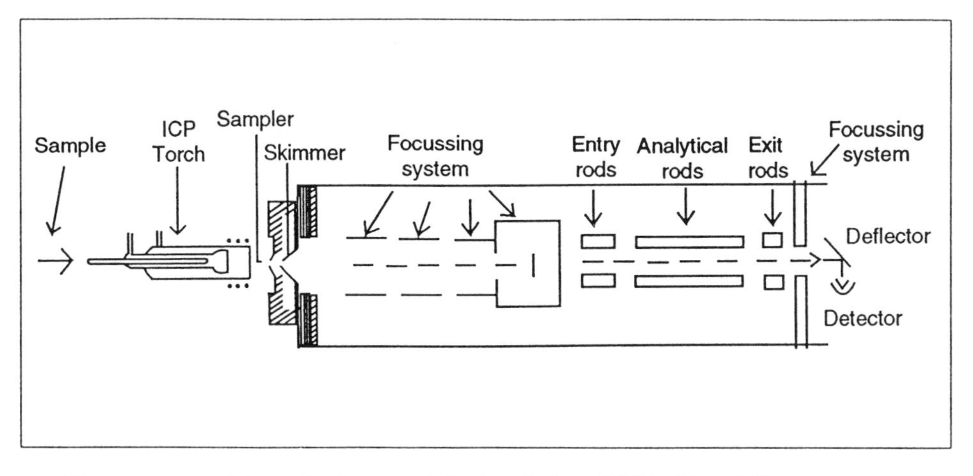

Figure 10.17 Schematic diagram of the interfacing of ICP with an MS system.

The individual components of the ICP and mass spec have already been developed considerably. However, two major problems exist. The ICP operates at itostpheric pressure and at temperatures estimated to be in the range between 5000 and 9000 K. On the other hand, the mass spectrograph operates in a high vacuum (10^{-4} to 10^{-6} torr) and at room temperature. Interfacing of the two systems is therefore the critical problem to be overcome to provide the ideal marriage of these techniques.

3. Equipment

Simplistically the instrument is an ICP interfaced with a quadropole MS. The ICP and MS systems are standard. The interfacing is the difficult step as shown in Fig. 10.17.

a. The Process of Interfacing ICP–Mass Spec

The earliest attempts to interface these systems utilized a simple metal interface (a skimmer) with a hole drilled through it with a diameter of approximately 60 μm. This was placed between the plasma and the MS. One side of the interface was exposed to the tip of the plasma and the other side to the entrance of the mass spectrometer. The interface was water cooled to prevent melting. It was anticipated that the elemental ions leaving the top of the plasma would enter the mass spectrometer, whereas much of the supporting argon gas would be discriminated against by the skimmer. Although this worked to some extent, unfortunately it became clear that it caused severe analytical problems. The water-cooled orifice created a layer of cool, slowly moving gas. Chemical

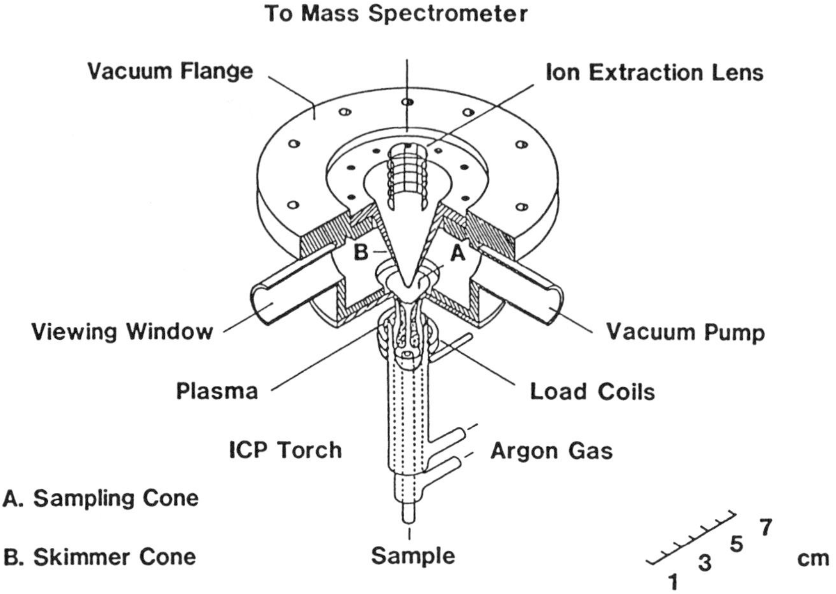

Figure 10.18 ICP and ion-sampling interface. Note that the design adopted in this laboratory in conjunction with the Perkin-Elmer Corporation has a vertical configuration, whereas a horizontal configuration between the ICP and MS is more often used. Only the first ion extraction lens is shown.

action occurred in this layer, generating metal oxide, hydroxyl ions, and other more complex ions not present in the original plasma. Further, many of these ions were not elemental in composition and complicated the mass spectrum. Their generation depended on the plasma conditions utilized and the sample injected. The presence of dilute hydrochloric, nitric, or other acids in the sample caused further interferences to the mass spectrum.

Later interfaced systems utilized two orifices. The first was the *sampler*, which was a water-cooled metal with an inverted cone shape. A circular hole with a diameter of between 0.5 and 1.0 mm was drilled through the center of the cone coaxially. A second metal cone, known as a *skimmer*, was placed immediately following the sampler, with a small gap between the two. The metal cone also contained a hole, which led to the entrance of the mass spectrometer. This system is illustrated in Figs. 10.18 and 10.19. Between the sampler and the skimmer was a region that was continuously pumped to about 1 torr. After passing through the skimmer the gases at this pressure enter the mass spectrometer at a reasonably low rate so that the mass spectrometer can be maintained at a low pressure.

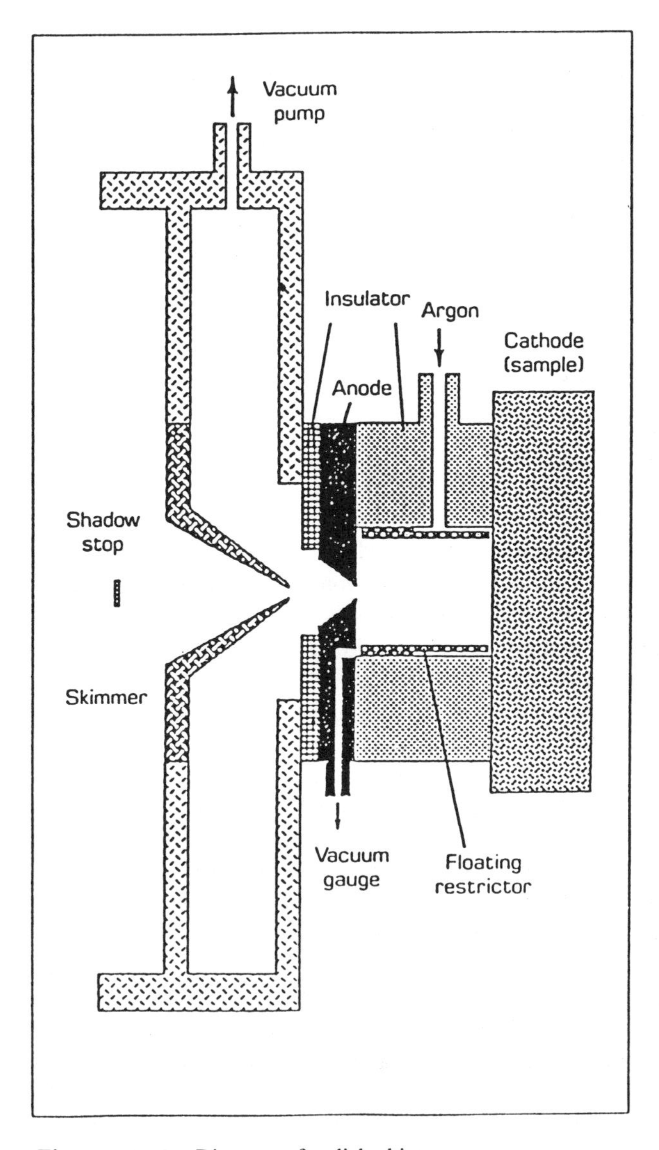

Figure 10.19 Diagram of a disk skimmer.

Location of the system in the plasma was critical and has demanded close attention. At the present time the sample is immersed in the analytical region of the plasma used for observation of the emission spectrum.

In early systems a secondary discharge occurred in the region between the two orifices. This led to the introduction of metal ions from the instrument into the mass spectrometer generating erroneous analytical data, as well as destruction of the instrument. In addition doubly charged ions were generated together with a photon background.

A series of ion-focusing elements similar to those developed in double-focusing mass spec have been utilized to introduce the ions into the quadropole. Also, photon blockers have eliminated interference effects due to the presence of large numbers of photons. Therefore, by suitable shielding and grounding, these background signals have been largely eliminated in recent models as shown in Fig. 10.17.

4. Analytical Applications

One of the most important advantages of the ICP-MS is that it can be used to *determine metals and nonmetals alike*. Also, it can detect and measure positive and negative ions. As a consequence, nonmetals are measured as the negative ion and metals as the positive ion. It can be generalized that elements on the left-hand side of the periodic table are measured as positive ions and elements on the right-hand side are measured as negative ions. This capacity is a major advantage over both atomic absorption and all forms of emission spectrography. In practice, however, it has been found that only the halides can be detected with reasonable sensitivity as shown in Tables 10.4 and 10.5. It was found that the background was higher than when positive ions were monitored, but the signal was about 50 times higher.

The use of the quadropole mass spec has quickly borne fruit since it is known to be capable of rapidly scanning the large mass ranges encountered in organic analysis. Therefore, it was not difficult to scan the range up to 250 mass units in 20 milliseconds. It should be noted that 250 mass units covers most of the elements in the periodic table. Of course, the instruments can be used in a mode that jumps from one mass range to another without covering the entire range if so desired.

The detection limits reported so far indicate that in general the system is more sensitive than the ICP optical emission spectrograph and in many cases has a sensitivity comparable with that obtained in atomic absorption using a thermal atomizer. It is of course also better than the sensitivity obtained with atomic absorption obtained using flame atomizers. Detection limits on the order of 0.1–0.2 ppb are typical (Table 10.6), whereas for ICP emission and flame atomic absorption typical results range between 1 and 40 ppb.

One interesting innovation is the determination of halo acid ions such as IO_4^-, IO_3^-, BrO_3^-, and ClO_3^-. These could not be easily separated by liquid chromatography unless long retention times were permissible. However, a modified gel permeature column worked satisfactorily. The halo anions were identified by retention time and confirmed by MS. The latter identified the halogen in ques-

Table 10.4 Comparison of the S/N Ratio for Selected Nonmetal Analytes in the Positive and Negative Ion Detection Modes

		Signal–to–background noise ratio[a]	
Element	Isotope	Positive ion	Negative ion
Boron	11	1160	ND[b]
Fluorine	19	ND	34.5
Silicon	28	95.6	ND
Phosphorus	31	507	ND
Sulfur	34[c]	23.4	ND
Chlorine	35	197	560
Arsenic	75	8090	ND
Selenium	78[c]	239	ND
Bromine	79	843	856
Tellurium	130	2300	ND
Iodine	127	3770	431

[a]Analyte concentrations are 10 μg/ml.
[b]Not detectable.
[c]Isotopes are not most abundant.
Source: Adapted from Direct determination of nonmetals in solution by atomic spectroscopy, by D. A. Melpregor, K. B. Call, J. M. Gehlhausen, A. S. Vicomi, M. Wu, L. Zhang, and J. W. Carnahan, *Anal. Chem.*, *60*(19): 1089A (1988).

tion, i.e., IO_3 and IO_4, the MS monitored the ^{127}I isotope, but the LC distinguished the original species. The halo ions are important in foods, drinking water, and body excretions such as urine.

Typical uses for ICP-MS include analysis of environmental samples for trace metals and nonmetals, the analysis of body fluids for elemental toxins, and determination of impurities in geological samples to determine age and ancient history. This method may also be useful for the pharmaceutical, chemical, and oil industries. Clearly this is an important new technique with great potential.

Table 10.5 Detection Limits for the Halogens in the Negative Ion Mode

Element	Isotope	Detection limit[a] (ng/ml)
Fluorine	19	400
Chlorine	35	80
Bromine	79	10
Iodine	127	70

[a]S/N = 3, time constant = 1 sec.
[b]F and Cl would have detection limits of 30 and 20 ng/ml, respectively, in a non–blank–limited situation.

Table 10.6 ICP-MS Detection Limits

Isotope	Detection limits	Isotope	Detection limits	Isotope	Detection limits
^{7}Li	0.07	^{59}Co	0.17	^{107}Ag	0.06
^{9}Be	0.05	^{63}Cu	0.2	^{114}Cd	0.04
^{11}B	0.3	^{64}Zn	0.4	^{120}Sn	0.08
^{24}Mg	0.06	^{69}Ga	0.03	^{121}Sb	0.02
^{27}Al	0.1	^{72}Ge	0.2	^{138}Ba	0.02
^{48}Ti	0.1	^{75}As	0.1	^{181}Ta	0.02
^{51}V	0.1	^{88}Sr	0.009	^{184}W	0.05
^{52}Cr	0.2	^{90}Zr	0.04	^{205}Tl	0.005
^{55}Mn	0.1	^{93}Nb	0.02	^{208}Pb	0.03
^{58}Ni	0.4	^{98}Mo	0.07	^{209}Bi	0.01

a. Dynamic Range

Using ion-counting measurement, a linear dynamic range of about five decades is achievable. This is much better than emission spectrography and atomic absorption spectroscopy and comparable to that obtained with ICP emission.

b. Isotope Measurements

A major new potential in the technique is the ability to measure the isotope ratios of the various elements. It is of interest to note that this was the original application of mass spectrometry both by Aston and by Thompson, and it is only now that we are again appreciating the analytical potential of this information. For example, the presence of isotopes and their relative abundance can be used to confirm the presence of the various elements. Also, the application of ICP–mass spectroscopy to geological samples for the estimation of various isotopes has many potential applications, both for dating of rocks and for identification of original point of origin.

The system has been used for measuring the relative concentrations of various lead isotopes, and this can be used to identify the source of the lead since this varies from mine to mine. It has also been used for the determination of the following trace elements in food: Na, Mg, K and Ca.

Work has also been carried out by Chong and R. S. Honk, who used argon ICP–mass spec as a GC detector for nitrogen, oxygen, phosphorus, sulfur, carbon, chlorine, bromine, boron, and iodene.

c. Interferences

Mass spectral interferences are of course observed from argon ions since these are present in high abundance in plasma. In addition, ArH^{+} and O^{2+} are present, and these may interfere with determinations of calcium and sulfur. The molecular ions from dilute solutions of reagent acids used in the original sample also cause problems. Ions such as ArN^+, ClO^+, $ArCl^+$, SO^+, and SO^{2+} are commonly observed. Based on studies of these interfering ions, nitric acid is considered to be the most attractive acid to be used in preparing sample solutions. No ions occur, however, with a mass greater than 82 from any of the common acids. Therefore, the higher part of the mass range is unaffected by these interfering ions.

Matrix Effects The introduction of the sample into the plasma suffers from the same problems in ICP–mass spec as in ICP emission. These include the complicated process of nebulization and atomization. These processes occur before introduction into the plasma–mass spec system and are not eliminated by their position.

The interface between the two systems includes the 1 torr region where deposition of products can occur. Severe suppression of signal has been observed when high concentrations (100 μg/ml) are present. This may be caused by suppression of ionization by other elements present. At this time the cause of this interference is not clear, but the fact remains that interference does take place. The problem can be overcome to some extent by limiting the concentration in the samples to less than 0.2% total solids. This can be a serious limitation, particularly when body fluids are being examined.

An example of the spectra obtained from an interfaced plasma–MS system is shown in Figure 10.20. This illustrates a valuable procedure for the difficult task of rare earth analysis.

Detection limits for the method are shown in Table 10.7. In a recent development a system was produced using plasma-MS with an electrothermal sample introduction system. Detection limits of one to two orders of magnitude greater than normal are claimed, as illustrated in Table 10.8. This must be the ultimate in interfacing.

d. Limitations

The major limitations at this point seem to be the introduction system to the ICP (a common problem with the ICP), the matrix effects of solid materials dissolved in the sample and mass ion overlap. Another serious problem is overlapping masses from different species. These arise from Ar^+, $Ar_2{}^+$, $CO_2{}^+$, CO^+, SH^+, SO^{2+}, NOH^+, ArH^+, $ArOH^+$, $ArCl^+$, ClO^+, ArS^+, etc., which derived from the plasma gas Ar and ionic compounds of Ar reacting with water and carbonaceous moities derived from the solvent or the sample.

For example, $^{40}Ar^+$ and ^{40}Ca overlap, $^{14}N_2$ overlaps with $^{12}C^{16}O$; ^{58}Ni overlaps with ^{58}Fe. Other overlapping species are ^{55}Mn and $^{40}Ar^{14}NH$, ^{51}V and $^{35}Cl^{16}O$, ^{62}Ni and $^{42}Ti^{16}O$, also ^{28}Si: $(^{14}N)_2$, ^{31}Pi $^{14}N^{16}OH$, $^{32}Si(^{16}O)_2$, ^{40}Ca: ^{40}Ar, ^{56}Fe $(^{40}Ar^{16}O)$ ^{80}Se $(^{40}Ar)_2$. Also, $MnO^+$and $MoOH^+$overlap with Cd isotopes. If the solvent is 5% HCl, as it sometimes is, overlap also occurs from

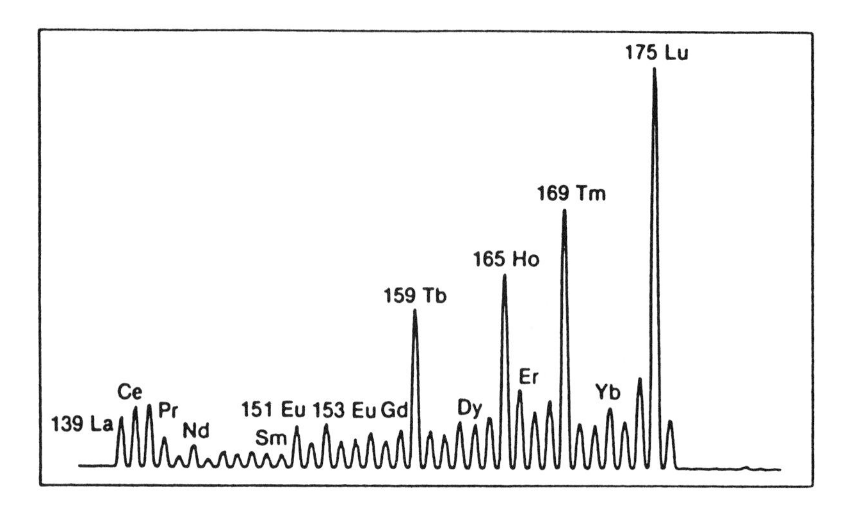

Figure 10.20 Spectra of a mixed solution of rare earths. Elemental analysis was performed using plasma torch with mass spectrometer.

^{51}V: ($^{35}Cl^{16}O$), ^{50}Ti: ($^{35}Cl^{15}N$) or ^{50}Cr ^{75}As: ($^{40}Ar^{35}Cl$). Further, Ti has five isotopes, ^{46}Ti, ^{47}Ti, ^{48}Ti, ^{49}Ti, and ^{50} Ti, which can form oxides that interfere with ^{62}Ni, ^{63}Cu, ^{64}Zn, ^{65}Cu, and ^{66}Zn.

It should be noted that the flow rate, gas mixture, and solvent used affects oxide formation and therefore the degree of isotope overlap. For reproducible results, they must be rigorously controlled.

Also, we can have doubly charged ions with the same apparent mass and a singly charged ion of another element. For example, $^{43}Ti^{+}$ and $^{86}Sr^{2+}$, $^{56}Fe^{+}$, $^{112}Cd^{2+}$, $^{66}Zn^{+}$, $^{132}Ba^{2+}$.

An inspection of the isotope tables reveals many possibilities. Of course, the problem can be assessed and perhaps corrected by using a different isotope for the analyte. The use of high resolution mass spectroscopy has been proposed, but the cost is prohibitive.

Matrix effects are caused by the solvent and other elements present in the sample. These affect atomization efficiency and therefore the strength of the MS signal. This directly affects quantitative results. The matrix also affects the formation of unstable ions such as ArO^{+}, ArH^{+}, etc. Aqueous samples act very differently from organic solvents, which in turn act differently from each other.

The problem can be overcome by matrix matching (i.e., the matrix of the sample and the standards used for calibration and matched.) This is similar to atomic absorption where the same requirement in matching solvent and predominant anion (i.e., the matrix) is required for accurate data.

Table 10.7 Relative Sensitivities Detection Limits (μg/liter)

Element	Flame AA	Hg/Hydride	Furnace	ICP emission	ICP–MS
Ag	0.9		0.005	1	0.04
Al	30		0.04	4	0.1
As	100	0.02	0.2	20	0.05
Au	6		0.1	4	0.1
B	700		20	2	0.1
Ba	8		0.1	0.1	0.02
Be	1		0.01	0.06	0.1
Bi	20	0.02	0.1	20	0.04
Br					1
C				50	50
Ca	1		0.05	0.08	5
Cd	0.5		0.003	1	0.02
Ce				10	0.01
Cl					1
Co	6		0.01	2	0.02
Cr	2		0.01	2	0.02
Cs	8		0.05		0.02
Cu	1		0.02	0.9	0.03
Dy	50				0.04
Er	40				0.02
Eu	20				0.02
F					100
Fe	3		0.02	1	0.2
Ga	50		0.1	10	0.08
Gd	1200				0.04
Ge	200		0.2	10	0.08
Hf	200				0.03
Hg	200	0.008	1	20	0.03
Ho	40				0.01
I					0.02
In	20		0.05	30	0.02
Ir	600		2	20	0.06
K	2		0.02	50	10
La	2000			1	0.01
Li	0.5		0.05	0.9	0.1
Lu	700				0.01
Mg	0.1		0.004	0.08	0.1
Mn	1		0.01	0.4	0.04
Mo	30		0.04	5	0.08
Na	0.2		0.05	4	0.06
Nb	1000			3	0.02

Table 10.7 Continued

Element	Flame AA	Hg/Hydride	Furnace	ICP emission	ICP–MS
Nd	1000				0.02
Ni	4		0.1	4	0.03
Os	80				0.02
P	50000		30	30	20
Pb	10		0.05	20	0.02
Pd	20		0.25	1	0.06
Pr	5000				0.01
Pt	40		0.5	20	0.08
Rb	2		0.05		0.02
Re	500			20	0.06
Rh	4			20	0.02
Ru	70			4	0.05
S				50	500
Sb	30	0.1	0.2	60	0.02
Sc	20			0.2	0.08
Se	70	0.02	0.2	60	0.5
Si	60		0.4	3	10
Sm	2000				0.04
Sn	100		0.2	40	0.03
Sr	2		0.02	0.05	0.02
Ta	1000			20	0.02
Tb	600				0.01
Te	20	0.02	0.1	50	0.04
Th					0.02
Ti	50		1	0.5	0.06
Tl	9		0.1	40	0.02
Tm	10				0.01
U	10000			10	0.01
V	40		0.2	2	0.03
W	1000			20	0.06
Y	50			0.2	0.02
Yb	5				0.03
Zn	0.8		0.01	1	0.08
Zr	300			0.8	0.03

All detection limits were determined using elemental standards in diluted aqueous solution.
Atomic absorption (Model 5100) and ICP emission (Plasma II) detection limits are based on a 95% confidence level (2 standard deviations) using instrumental parameters optimized for the individual element. ICP emission detection limits obtained during multielement analyses will typically be within a factor of 2 of the values shown.
Cold vapor mercury AA detection limits were determined using a MHS-20 Mercury/Hydride system with an amalgamation accessory.
Furnace AA (Zeeman/5100) detection limits were determined using STPF conditions and are based on 100-μl sample volumes.

Table 10.8 Detection Limits

	ETV-ICP-MS		Graphite furnace AAS[a]		ETV-ICP-AES(2)	
Elements	ng/ml (2 μl)	pg	ng/ml (10 μl)	pg	ng/ml (50 μl)	pg
^{107}Ag	0.08	0.16	0.01	0.1	0.1–30	1–300
^{27}Al	0.03	0.05	0.2	2	1.5–13000	7.5–26000
^{75}As	0.05	0.1	1	10	20–5000	200–10000
^{44}Ca	0.7	1.4	0.5	5	0.002–70000	0.02–139000
^{114}Cd	0.15	0.3	0.01	0.1	0.2–1640	1–3280
^{52}Cr	0.1	0.2	1	10	0.3–790	1.5–1580
^{39}K	1.5	3	0.1	1	110–83400	550–167000
^{23}Na	0.2	0.4	1	10	80–38000	400–76000
^{58}Ni	0.47	0.93	1	10	0.9–1050	4.5–2100
^{208}Pb	0.1	0.3	0.2	2	2–1400	20–2800
^{78}Se	5.7	11.4	1	10	6–600	600–3000
^{28}Si	2.7	5.4	5	50	10–500	100–2500
^{64}Zn	0.2	0.4	0.05	0.5	0.05–1700	0.25–3540

[a]JVL laboratory recent detection limits.

H. A COMPARISON OF ATOMIC SPECTROSCOPIC ANALYTICAL TECHNIQUES

Atomic spectroscopy is the most widely used technique for determining elements in almost every conceivable matrix, especially the metallic ones. The most widely used techniques are flame atomic absorption, graphite furnace AAS, and inductively coupled plasma–atomic emission spectroscopy (ICP-AES). ICP has also been used as the atomizing and ionizing device for mass spectroscopy (ICP-MS.) Table 10.9 lists several of the parameters that are important.

Table 10.9 Important Analytical Parameters

Analytical accuracy
Analytical precision
Analytical concentration range
Total analytical per sample
Instrument cost
Degree of automation
Skills required to operate
Analytical interferences and elimination
Difficulties with contamination

1. Flame Atomic Absorption

Flame AAS is the method of choice for the smaller laboratory. The equipment is relatively inexpensive and easy to operate. Interferences are few and well documented. On the negative side, the refractory elements like boron, tungsten, zirconium, and tantalum are difficult to determine because the flame is not hot enough to atomize these analytes. If more than two or three analytes must be determined in each sample, even automated AAS is a much slower process than ICP-AES.

2. Furnace AAS

Using furnace AAS as an accessory usually involves compromises that limit the quality of furnace analyses. The primary analytical advantage of furnace AAS is that its detection limits are about two orders of magnitude better than either flame AAS or ICP-AES. Like flame AAS, the refractory elements are not available and it usually determines a single element at a time. Furnace AAS is still burdened by reputation of being prone to severe interferences. However, if furnace AAS is used correctly, the technique is relatively free of interference.

3. ICP-AES

The emission signal can be recorded for all elements of interest either simultaneously or in rapid sequence, thus making ICP-AES faster than AAS techniques, especially if many elements must be determined in each sample. However, the measurement of analyte emissions at high temperature requires a monochromator of better resolution than required for AAS; thus ICP-AES is more expensive than AAS. Also, ICP-AES suffers from interferences by the large number of spectral emission lines associated with each element. This is particularly troublesome when materials to be analyzed have very complex spectra, like iron, nickel, uranium, molybdenum, and the rare earths. ICP-AES has about the same sensitivity as flame AAS and is much faster if many analytes are sought in each AAS sample. The technique requires greater operator experience than does flame AAS.

4. ICP-MS

For most elements, the ICP ionizes more than half of the elemental vapor. Coupled together, ICP-MS measures the analyte concentration by mass spectra rather than by the emission spectra. The technique is expensive but possesses the fast, multielement capabilities of ICP-AES and the great detection power of the graphite furnace. Much greater operator skills are required than for the other three techniques discussed thus far.

The ability of the mass spectrometer to measure the elements' individual isotopes with great accuracy opens up some very special applications for ICP-MS, such as isotope dilution. An isotope of the analyte of interest is added to the sample prior to sample preparation, and the altered isotope ratio is measured. If the original isotope ratio is known, as well as the altered ratio, the amount of analyte in the sample can be calculated.

A second important application is the measurement of stable isotopes added to biological systems as elemental tracers. The use of stable isotopes, measured by ICP-MS, permits these important experiments to be undertaken.

5. Detection Limits

Detection limits based on signal-to-noise ratio for these methods are listed in Table 10.7. It can be seen that ICP-MS holds a distinct advantage over the other techniques as far as sensitivity is concerned.

6. Precision, Accuracy, and Interferences

The precision of an analytical technique is generally almost constant over most of the analytical range when the analyte concentration is greater than ~50 times the detection limit. Routinely, this value is ~0.5% for flame AAS, 1.5% for ICP-AES, 3% for furnace AAS, and 2–3% for ICP-MS.

BIBLIOGRAPHY

American Society for Testing Materials, *Methods for Emission Spectrochemical Analysis*, ASTM, Philadelphia, 1964.

Anal. Chem., Application Reviews, June 1993.

Barth, K. F., Adams, D. M., Soloway, A. H., Mechelner, E. B., Adams, F., Anesuzzeman, A. K. M., *Anal. Chem.*, (1991), *63*: 880.

Brode, W. R., *Chemical Spectroscopy*, 2nd ed., Wiley, New York, 1943.

Chandler, C., *Atomic Spectra*, 2nd ed., D. Van Nostrand Co. Inc., Princeton, N.J., 1964.

Chong, N. S., and Honk, R. S., *Applied Spec.*, (1987), *41*: 66.

Denayer, E. R., and Stroh, *Am. Labs.*, (1992), Feb.: 74.

Kirkbright, G. F., and Sargent, M., *Atomic Absorption and Fluorescence Spectroscopy*, Academic, London, 1974.

Lee, Y-I., Sawan, S. P., Thiem, T. L., Teng, Y-Y., Sheddon, J. Interaction of a laser beam with metals. Part II. Space-resolved studies of laser-ablated plasma emission. *Appl. Spec.*, *46* (3): 436 (1992).
Noble, D., *Anal. Chem.*, (1994), *66*: 105.
Olesik, J. W., Elemental analysis using ICP-OES and ICP-MS, *Anal. Chem.*, *1*: 12A (1991).
Robinson, J. W., *Atomic Spectroscopy*, Marcel Dekker, New York, 1990.
Robinson, J. W., *Handbook of Spectroscopy*, CRC Press, Boca Raton, Fla., 1974.
Sacks, Richard D., Emission spectroscopy, in *Treatise on Analytical Chemistry*, Vol. 7, Wiley, New York, 1981.
Selby, M., and Heifje, G. M., *Am. Labs.*, (1987), *8*: 16.
Van born, W. A., Broekaert, J. A. C., *Anal. Chem.*, (1990), *62*: 2557.
Zamzov, D. S., Baldwin, D. A., Weeks, S. J., Bajie, S. J., and D'Silva, A. P., *Env. Sci. Tech.*, (1994), *28*: 352.
Sawyer, R. A., *Experimental Spectroscopy*, 3rd ed., Dover, New York, 1963.
Wieldera, D. A., Houk, R. S., Winge, R. K., and D'Silva, A. P., Introduction of organic solvents into ICP by ultrasonic nebulization with cryogenic desolvation, *Anal. Chem.*, *62*: 1155 (1990).

SUGGESTED EXPERIMENTS

10.1 Take several 1-g portions of dry Al_2O_3 powder to be used as the matrix. Add powdered anhydrous copper sulfate to make a powder containing 0.01% Cu. Load this into the graphite electrode of an emission spectrograph and fire under spark conditions for 3 min. Record the spectrum on the photographic plate.

10.2 Repeat Experiment 10.1, but record the spectrum after exposures of 1, 3, 5, 7, 10, 15, and 20 min. Measure the intensity of the emission lines at 327.4 and 324.7 nm. Plot the intensity of each line against the exposure time. Note that after a certain period of time, the intensity of the line does not increase, but the background does.

10.3 Using Al_2O_3 powder and anhydrous copper sulfate powder, prepare powders that contain 0.01, 0.05, 0.1, 0.2, 0.5, 0.7, and 1.0% Cu. Insert each powder separately into the emission spectrograph and expose for 3 min. Measure the intensity of the lines at 324.7 and 327.4 nm. Plot the relationship between intensity and the concentration of the copper. Note that the intensity of the lines goes through a maximum. These are reversible lines.

10.4 Repeat Experiment 10.2 using Al_2O_3 powder and stannic oxide. Note that the more volatile stannic oxide ceases to emit more rapidly than the $CuSO_4$.

10.5 Repeat Experiment 10.3, but use (a) MgO, (b) SiO_2, and (c) powdered Al metal as the matrix powder. Note that the changes in the matrix compound cause a variation in the emission from the copper, even though other conditions are constant. This is the matrix effect.

10.6 Repeat Experiment 10.3, but use (a) $CuCl_2$, (b) $Cu(NO_3)_2$, (c) Cu acetate, (d) CuO, and (e) CuS as the source of copper. Note the variation in the intensity of the copper emission when different salts are used, even though other conditions are constant. This is chemical interference.

10.7 Determine the percentage of copper in (a) brass, (b) bronze, (c) a quarter, and (d) a nickel.

10.8 Record the emission spectra of (a) Zn, (b) Sn, (c) Ni, (d) Fe, (e) Co, and (f) Cd.

10.9 Based on the emission spectra obtained in Experiments 10.7 and 10.8, what elements are present in (a) brass, (b) bronze, (c) a quarter, and (c) a nickel?

PROBLEMS

10.1 How does emission spectrography differ from flame photometry? How does this difference affect the number of elements that are detectable by this method? Explain.

10.2 Describe and illustrate two types of sample containers used for liquid samples.

10.3 A prism is able to separate the following pairs of lines. What is the required minimum dispersion of the prism at each wavelength?

Wavelengths of line pairs (nm)	Dispersion
281.7, 282.3	
350.4, 351.1	
513.5, 514.4	

10.4 When it is advantageous to use (a) film and (b) photomultipliers as a detector?

10.5 What effect does the matrix have on the intensity of the emission signal?

10.6 A calibration curve was obtained from the following solutions. The intensity of emission for each sample is shown. Prepare the calibration curve.

Sample concentration	Emission intensity (units)
0.0	15
2.0	29
4.0	43
6.0	57
8.0	71
10.0	83

Why does the curve not go through zero when the sample concentration is zero?

10.7 The emission intensity from several samples was measured using the same conditions as those used to prepare the calibration data shown in Problem 10.6. The following results were obtained. Complete the table.

Sample	Emission intensity (units)	Concentration
A	20	
B	33	
C	60	
D	41	
E	75	

10.8 What is the preburn time? Would you expect that (a) germanium and (b) titanium could be satisfactorily determined using the same preburn time? Explain your answer.

10.9 What is the resolution of a prism?

10.10 What is the resolution of a grating?

10.11 What is the resolution required to resolve the sodium D lines at 589.5 and 589.0 nm?

10.12 What is a Littrow prism?

10.13 Describe a Rowland circle.

10.14 Describe an echelle monochromator?

10.15 Describe the matrix effect.

10.16 Describe an inductively coupled plasma. What gas is used?

10.17 Describe a cross-flow nebulizer.

10.18 Describe laser ablation.

10.19 Describe interfaced ICP-MS.

10.20 What are the advantages of interfacing ICP-MS?

10.21 Describe a disk skimmer.

10.22 What are positive and negative ion detection modes?

10.23 What is the dymanic quantitative analytical range of the ICP?

10.24 What isotope overlap do you get in ICP-MS?

10.25 Compare absorption and emission methods of qualitative and quantitative analyses.

11
X-RAY SPECTROSCOPY

A. ORIGIN OF SPECTRA

An atom is composed of a nucleus and numerous electrons. The electrons are arranged in layers or shells with the valence electrons in the outer shell. The different shells, or layers of electrons, are called the *K* shell, *L* shell, *M* shell, and so on. For example, a sodium atom contains filled *K* and *L* shells and one electron in the *M* shell. An atom is shown schematically in Fig. 11.1, and a partial list of elements and their electron configurations is given in Table 11.1

When an x-ray or a fast-moving electron collides with an atom, its energy may be absorbed by the atom. If the x-ray or electron has sufficient energy, it knocks an electron out of one of the atom's inner shells (e.g., the *K* shell) and the atom becomes ionized. An electron from a higher-energy shell (e.g., the *L* shell) shell then falls into the position vacated by the dislodged inner electron and an x-ray photon is given off in the process as the electron drops from one energy level to another. The wavelength of this emitted x-ray is characteristic of the element being bombarded.

1. Energy Levels

If we plot the energy levels of the *K*, *L*,and *M* shells for a given element, we get a diagram similar to Fig. 11.2.

If an electron is dislodged from a *K* shell, it may be replaced by an electron from an *L* or an *M* shell. The latter electron emits radiation with energy

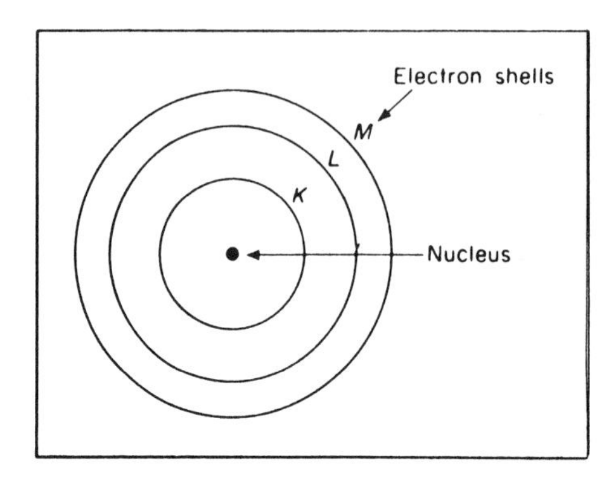

Figure 11.1 Schematic diagram of the electronic structure of an atom.

$$E(L \text{ shell}) - E(K \text{ shell})$$

where $E(L$ shell) is the energy of electrons in the L shell and $E(K$ shell) is the energy of electrons in the K shell. The energy of the radiation is given by

$$E = E_L - E_K \tag{11.1}$$

or

$$E = E_L - E_M$$

but, by Eq. (2.2),

$$E = h\nu$$

Therefore

$$h\nu = E_L - E_K$$

Hence, the frequency of the emitted x-ray is

$$\nu = \frac{E_L - E_K}{h} \tag{11.2}$$

We may also get electrons displaced from an L shell and an electron may replace it from the M shell. The energy released is E, where:

$$\begin{aligned} E &= h\nu \\ &= E\,(M \text{ shell}) - E\,(L \text{ shell}) \end{aligned}$$

Lines terminating in the K shells are called K lines, lines terminating in the L shells are called L lines, and so on. There are three L levels differing by a small amount of energy, five M levels, and so on . An electron that drops from an L shell to a K shell emits a photon with the energy difference between these

Table 11.1 Electron Configurations of Various Elements

		K	*L*		*M*			*N*	
Element	*Z*	1*s*	2*s*	2*p*	3*s*	3*p*	3*d*	4*s*	4*p*
H	1	1							
He	2	2							
Li	3	2	1						
Be	4	2	2						
B	5	2	2	1					
C	6	2	2	2					
N	7	2	2	3					
O	8	2	2	4					
F	9	2	2	5					
Ne	10	2	2	6					
Na	11				1				
Mg	12				2				
Al	13				2	1			
Si	14	Neon core			2	2			
P	15	(10)			2	3			
S	16				2	4			
Cl	17				2	5			
Ar	18				2	6			
K	19							1	
Ca	20							2	
Sc	21						1	2	
Ti	22						2	2	
V	23	Argon core					3	2	
Cr	24	(18)					5	1	
Mn	25						5	2	
Fe	26						6	2	
Co	27						7	2	
Ni	28						8	2	
Cu	29						10	1	
Zn	30							2	
Ga	31							2	1
Ge	32							2	2
As	33	Cu^+ core						2	3
Se	34	(28)						2	4
Br	35							2	5
Kr	36							2	6

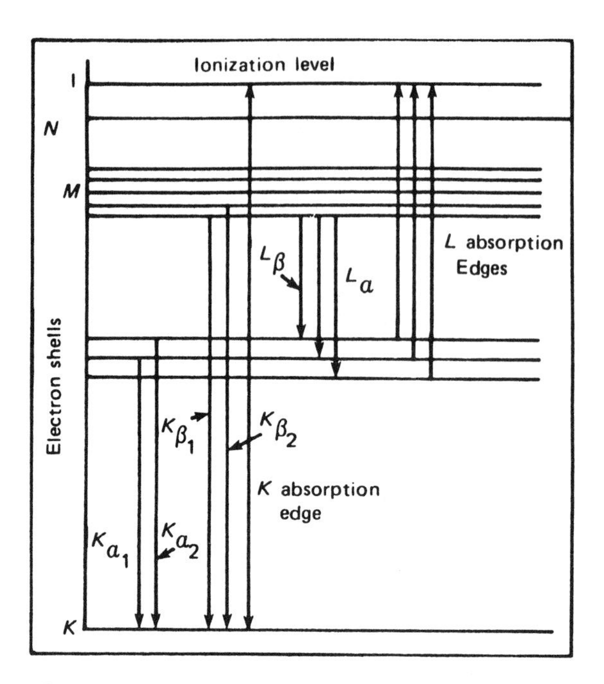

Figure 11.2 Energy level diagram for a hypothetical atom.

shells. This is called a K_α line. Since the L shell has three sublevels, we can see two K lines: $K_{\alpha 1}$, a forbidden transition, and $K_{\alpha 2}$. These are illustrated in Fig. 11.3. An electron that drops from an M shell to a K shell generates a L_β x-ray. The K_β lines have fine structure, as do the K_α lines. If an electron is ejected from an L shell, an electron from an M shell may fall into its place and emit an x-ray of characteristic wavelength with energy equivalent to the difference between the L and the M shells. These are designated the L lines. A number of L lines are possible, as indicated by Fig. 11.3. Electrons origination from an O shell and falling into the L shell generate L lines, and so on. The energy levels of the K, L, M, N, O shells, and so on, are characteristic of the element being examined. For example, the energy of an x-ray emitted when an M shell electron falls into an L shell is given by Eq. (11.1). And this is the energy of the L line. A typical emission spectrum is shown in Fig. 11.3. Values of various lines are shown in Fig. 11.4.

An x-ray emission spectrum is therefore similar for all elements, in that K_α, K_β, L_α, L_β, and so on, lines are involved. But the actual wavelengths of these lines varies from one element to another, depending on the *atomic number* of that element.

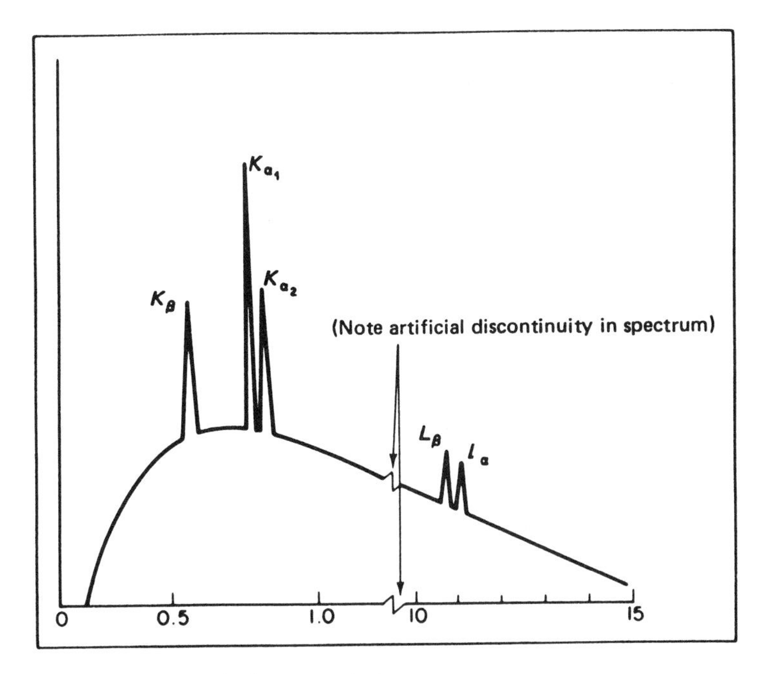

Figure 11.3 Typical x-ray emission spectrum.

A mathematical relationship was discovered between the wavelengths of the *K* series and the atomic number of the element. Similar relationships were found for the *L* lines, *M* lines, and so on.

2. Moseley's Law

The relationship between wavelength and atomic number was first discovered by Henry Moseley, a young graduate student working at Cambridge, England, during World War I. After recording the spectra from numerous elements in the periodic table, he discovered a relationship between the atomic number of the element and the wavelength of the K_α line. A similar relationship was found between the atomic number and the K_β line and the L_α line, and so on. The relationships were formulated in the *Moseley law*, which states that

$$\nu = \frac{c}{\lambda} = a(Z - \sigma)^2 \tag{11.3}$$

Where c is the speed of light, λ is the wavelength of the x-ray, a is a constant, Z is the atomic number of the element, and σ is a constant dependent on the

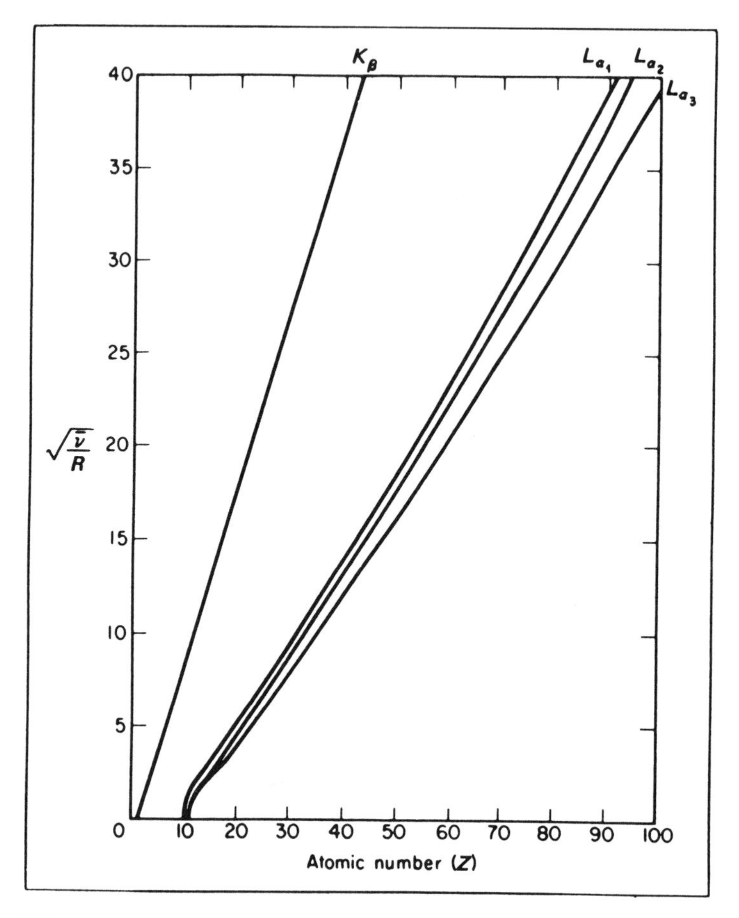

Figure 11.4 Moseley plots for x-ray levels. Here v denotes wave number and equals v/c, and R is Ryberg's constant.

series of lines (e.g., K_α or L_α lines). Shortly after this monumental discovery, Moseley was taken into the army and was killed in action. The loss to science can never be measured.

The impact of Moseley's law on chemistry was substantial, in that it provided a method of unequivocally assigning an atomic number to newly discovered elements, of which there were several at that time. Also, it clarified disputes concerning the positions of all known elements in the periodic table, some of which were in doubt at that time.

3. The Absorption Spectrum

The absorption spectrum is a plot of the absorptive power of a given element over a given x-ray wavelength range. A typical example is shown in Fig. 11.5. An abrupt change in absorptivity occurs at the wavelength of the x-ray necessary to eject an electron from an atom. The energy jump corresponds to *L*-transition *K* lines (see Fig. 11.5). These abrupt changes in x-ray absorptivity are termed *absorption edges*.

The wavelengths of the absorption edges and of the corresponding emission lines do not quite coincide. This is because the energy required to dislodge an electron from an atom (the absorbed energy) is not quite the same as the energy released when the dislodged electron is replaced by an electron from an outer shell (emitted energy). The amount of energy required to displace the electron must dislodge it from its orbital and remove it completely from the atom. This is more than the energy released by an atom falling from one energy level to another. A list of typical values is given in Table 11.2. As opposed to emission spectra, in absorption only one *K* absorption energy is involved per element. For example, in Fig. 11.2, the absorption energy of the *K* shell is the energy difference between the *K* shell and ionization energy. The K^{β} emission line is the energy difference between the *K* shell and the *M* shell. That is the energy to displace the *K* electron to ionization. Similarly, three energies are observed for the *L* levels, five for the *M* levels, and so on.

The x-ray absorption curve, such as that shown in Fig. 11.5, is a direct result of the energy levels involved in the different shells. For example, radiation with a wavelength of 1.8 Å will have a certain penetrating power. (Absorption is the inverse of penetrating power.) As the wavelengths of the x-rays decrease, the energy increases, the penetrating power increases, and the absorptivity de-

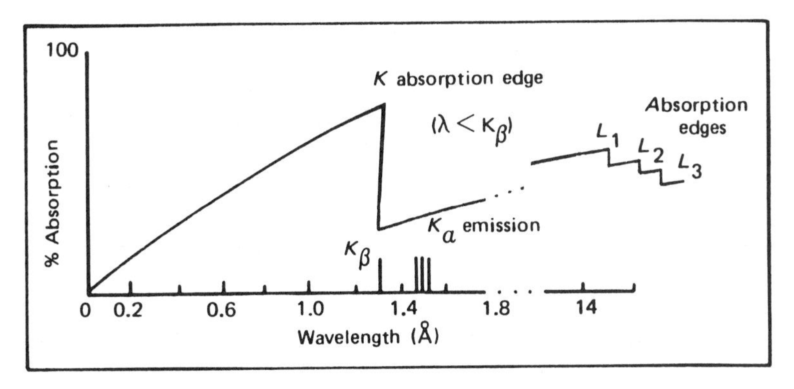

Figure 11.5 Absorption spectrum of nickel.

Table 11.2 Wavelengths of Absorption Edges and Emission Lines of Various Elements

Element	Absorption (Å)	Emission (Å)	
	K	K_β	K_α
Mg	9.54	9.56	9.9
Ti	2.50	2.514	2.748
Cr	2.07	2.085	2.290
Mn	1.895	1.910	2.106
Ni	1.487	1.500	1.658
Ag	0.484	0.496	0.560
Pt	0.158	0.164	0.186

creases. This can be seen by the downward slope of the absorption trace as it decreases from a wavelength of 1.8 Å (Fig. 11.5). As the wavelengths decrease further, the x-ray eventually has sufficient energy to displace electrons from the K shell. This results in an abrupt increase in absorption. This is manifested by the K absorption edge. After the absorption edge, the penetrating power continues to increase as the wavelength decreases further until finally the degree of absorption is extremely small at very small wavelengths. At wavelengths less than 0.2 Å, penetrating power is extremely great and we are approaching the properties of interstellar radiation such as cosmic rays, which have extremely high penetrating power.

4. Analytical Fields

There are three distinct fields of applications of x-rays to analytical chemistry: x-ray absorption, x-ray diffraction, and x-ray fluorescence.

With x-ray absorption we can tell if our feet fit into a new pair of shoes, if a bone is fractured, or is we have rocks in our head. We can also tell if there is a hole in a welded joint on the hull of a ship or if there is arthritis in an Egyptian mummy without taking the ship or the mummy apart. By using x-ray absorption we can also measure how much liquid is in a can or a storage tank without taking the liquid out of the tank or opening the container.

Through x-ray diffraction we can identify the crystal structures of various solid compounds and identify a compound from its structure. We can also determine the arrangement of molecules in a crystal. This has enabled us to obtain invaluable information on the structure of such diverse materials as chemical crystals, metals, and living tissue.

If a sample is irradiated with an x-ray beam, the sample sometimes emits x-ray beams at other wavelengths; this process is called x-ray fluorescence. The

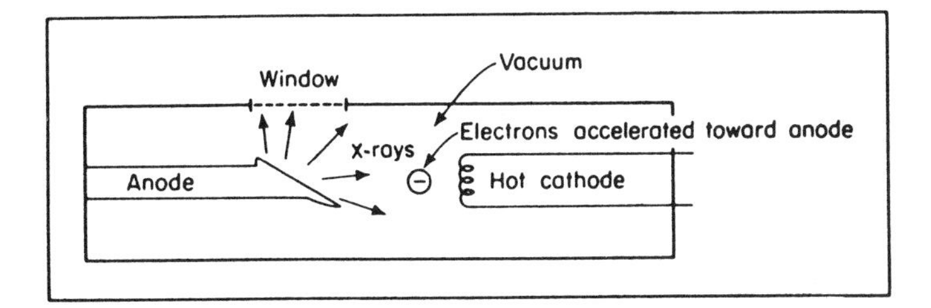

Figure 11.6 Schematic diagram of an x-ray source.

wavelength of the x-ray fluorescence enables us to determine what elements are present in a sample. The intensity of the x-rays that are emitted enables us to determine how much is present. It provides us with a method of elemental analysis that is independent of the molecular structure in which the element finds itself. The method is also nondestructive and frequently requires very little sample preparation before the analysis can be carried out.

5. X-Ray Generation

The basis of x-ray generation and analysis is as follows. When a cathode in the form of a metal wire is electrically heated, it gives off electrons. If a positive voltage, in the form of an anode, is placed near these electrons, the electrons are accelerated toward the anode. Upon striking the anode, the electrons transfer their energy to its metallic surface, which then gives off radiation. This radiation, which has a very short wavelength (0.1–100 Å), is called *x-ray radiation*. This system (see Fig. 11.6) is the basis of an x-ray source. The anode is called the *target*. Numerous metals have been used as target materials, but the most common target elements are copper, molybdenum, and tungsten. The wavelength of the radiation emitted by the target depends on the metal used and the voltage across the anode and cathode. When all the energy of the impinging electrons is turned into x-rays, the wavelength is the shortest attainable. This is termed the minimum λ or min λ. The radiation with the highest energy (and therefore the shortest wavelength) is deduced as follows. When all the energy of the electrons is converted to radiant energy, then the energy of the electrons equals the energy of the radiation. The energy of the radiation is given be $E = h\nu$, whereas the energy of the electrons is given by $E = eV$. When they are equal, $h\nu = eV$, where e is the charge of the electron, V is the applied voltage, and ν is the frequency of the radiation. But [see Eq. (2.1)]

$$\nu = \frac{\text{speed of light}}{\text{wavelength of the radiation}} = \frac{c}{\lambda}$$

Therefore

$$h\nu = \frac{hc}{\lambda} = eV \tag{11.4}$$

Rearranging, we get

$$\lambda = \frac{hc}{eV} \tag{11.5}$$

When all the energy of the electron is converted to x-radiation, the wavelength of the radiation is a minimum and we achieve minimum λ conditions. From Eq. (11.3) we get

$$\lambda_{\text{min}} = \frac{hc}{eV} \tag{11.6}$$

Inserting the values for h, c, and e, which are constants, we have the Duane–Hunt law,

$$\lambda_{\text{min}} = \frac{12{,}400}{V} \tag{11.7}$$

Where h is Planck's constant, c is the speed of light, e is the charge of an electron, V is the applied voltage across the anode and cathode (in volts), and λ is the shortest wavelength of x-rays radiated (in angstroms).

B. EQUIPMENT

We have seen that there are three distinct fields of x-ray analysis, namely, x-ray absorption, x-ray diffraction, and x-ray fluorescence. The component parts of the equipment are the same for each field, but the optical system varies for each one. The principal component parts are as follows.

1. X-Ray Source

A typical x-ray source is illustrated schematically in Fig. 11.6. Almost any solid element can be used as the source element. In choosing the element to be used for the target, it should be remembered that it is necessary for the energy of the x-rays emitted by the source to be greater than that required to excite the element being irradiated in x-ray fluorescence. As a simple rule of thumb, the target element of the source should have a greater atomic number than the elements being examined in the sample. This ensures that the energy of radiation from the source is more than sufficient to cause the sample element to fluoresce. This problem is not important is x-ray absorption or x-ray diffraction, where excitation is not necessary. In practice, the anode, or target, usually gets very hot in use, because it is exposed to a constant stream of high-energy electrons. This problem is sometimes overcome by water-cooling the anode.

One of the problems of using the x-ray source illustrated in Fig. 11.6 is that the background radiation is high compared to the x-ray lines K_α, K_β and so on. For many analytical uses, such as the selective excitations of x-ray fluorescence, this background is undesirable. It can be eliminated as follows.

If suitable radiation from a source described above falls on a metal, such as copper, it will cause the copper to fluoresce. The excited copper emits copper K and L x-ray lines, whatever the metal in the original source. The fluorescing copper can now be used as an x-ray source (or copper radiation). This source emits very little background, but does emit quite strongly at the copper K and L lines. Of course, the metal used in the target of the first source must have a higher atomic number than copper to generate fluorescence.

The background radiation emitted by the copper is very low. The intensity of radiation emitted at the K and L lines is quite high, but not as high as the radiation emitted by the x-ray sources first described (Fig. 11.6). This loss of intensity is disadvantageous, but for many purposes the disadvantage is more than offset by the low background.

The study of the radiation generated by using an electron gun (as in Fig. 11.6) is called *x-ray emission*. If the radiation is generated by exposure to x-radiation, the field is *x-ray fluorescence*. The spectrum background and the ratio of signal to background is better in x-ray fluorescence, but the absolute intensity is greater in x-ray emission.

2. Collimator

The x-rays emitted by the anode are radially directed. As a result, they form a hemisphere with the target at the center. A narrow beam of x-rays can be made by using two sets of closely packed metal plates separated by a small gap. This arrangement absorbs all the radiation except the narrow beam that passes between the gap. Decreasing the distance between the plates or increasing the total length of the gap increases resolution but decreases intensity. The system is shown in Fig. 11.7.

3. Monochromator

a. Filter

A filter is a window of material that absorbs unwanted radiation but permits radiation of the desired wavelengths to pass. A typical example is the Pd and Ag filters used for tin. The relevant spectra are shown in Fig. 11.8. In Fig. 11.8A,

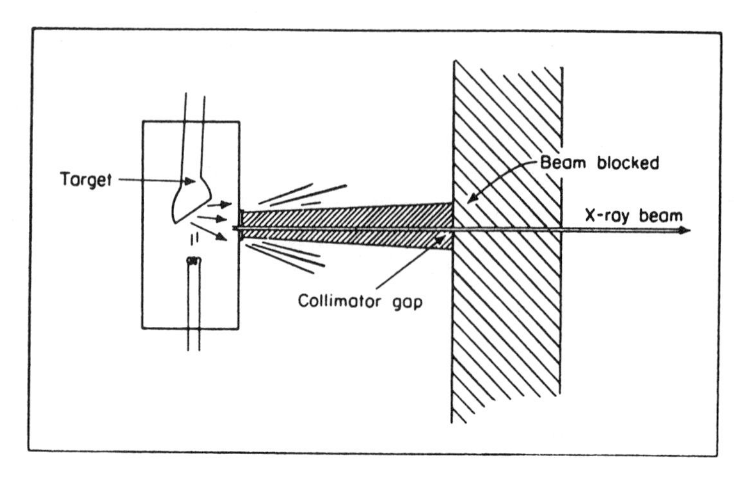

Figure 11.7 Collimator system for achieving a narrow beam of x-rays.

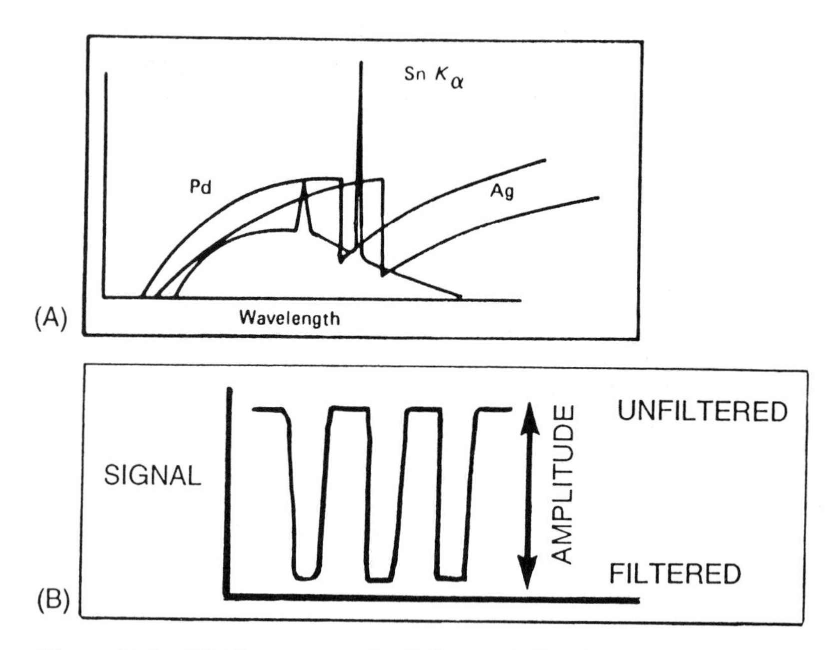

Figure 11.8 (A) Alternate use of palladium and silver filters. The concentration of tin is measured from the difference between the two readings. (B) Oscillating signal using an absorbing and a nonabsorbing filter.

it can be seen that the Pd filter strongly absorbs the short wavelength radiation but weakly absorbs the SnKα line. A silver filter absorbs both the background and the SnLα line. By rotating the two filters in front of the detector, the signal is alternately (filtered SnKα—weak) and (unfiltered SnKα—strong.) This generates an alternating signal (Fig. 11.8B), the amplitude of which is proportional to the intensity of the SnKα line and therefore the Sn concentration. In a similar fashion zirconium filters can be used for molybdenum determination. The zirconium strongly absorbs the radiation of Mo at short wavelengths, but weakly absorbs the molybdenum K_α line. The net result is that the radiation passing the zirconium filter is purer and the ratio of the molybdenum K_α line to the background is more favorable.

b. Analyzing Crystal

A crystal is made up of layers of ions or molecules arranged in a well-ordered system. An impinging beam of x-rays is reflected at each layer in the crystal (Fig. 11.9).

If we consider radiation as a wave form, the x-rays falling on the crystal are parallel waves. That is, the waves *OD* and *O'A* are in phase with each other and reinforce each other. In order for the waves to emerge as a beam, they must still be in phase with each other, that is, waves *DY* and *CX* must still be parallel or coherent. If the waves are out of phase, they destroy each other, interference occurs, and no beam is produced. In order to get reinforcement, therefore, it is necessary that the two waves stay in phase with each other after diffraction at the crystal planes.

It can be seen in Fig. 11.9 that the lower wave travels an extra distance *ABC* compared to the upper wave. It can then therefore be seen that if the extra distance *ABC* is a whole number of wavelengths, then the emerging beams *DY* and *CX* will remain in phase and reinforcement will take place. From this deduction,

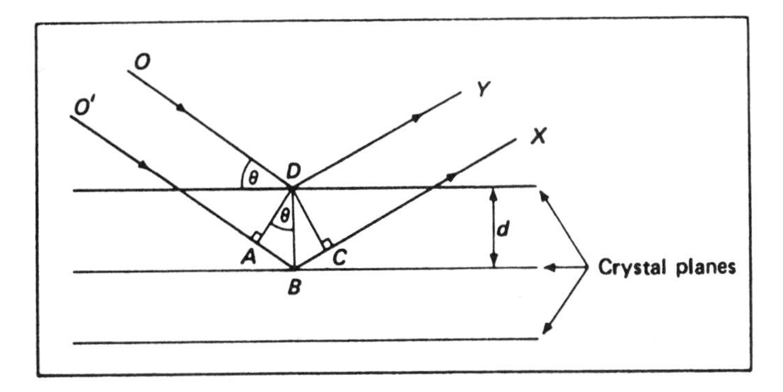

Figure 11.9 Diffraction of x-rays by crystal planes.

we can calculate the relationship between the wavelengths of x-radiation, the distance between the lattice layers, and the angle of diffraction later as follows.

X-ray waves OD and $O'B$ are parallel. The extra distance traveled by wave $O'ABCX$ in traveling through the crystal is ABC. For reinforcement, and therefore diffraction, it is necessary that this distance be a whole number of wavelengths, that is,

$$\text{distance } ABC = n\lambda \tag{11.8}$$

but

$$ABC = 2AB \tag{11.9}$$

and

$$AB = DB \sin\theta \tag{11.10}$$

where θ is the angle of incidence of the x-ray beam with the crystal; therefore

$$AB = d \sin\theta \tag{11.11}$$

where d is the distance between the crystal planes ($= DB$). Therefore

$$ABC = n\lambda = 2AB = 2d \sin\theta$$

or

$$n\lambda = 2d \sin\theta \tag{11.12}$$

The equation $n\lambda = 2d \sin\theta$ is known as the *Bragg equation*. It is named after Sir W. H. Bragg, the British scientist who first derived this formula and then proved it experimentally.

The important result of this equation is that at any given angle θ of incidence, only x-rays of a particular wavelength are diffracted by the crystal. If an x-ray beam consisting of a range of wavelengths falls on the crystal, the beam is split up by the crystal into the component wavelengths in the same way that a prism splits up white light into a rainbow. Such a crystal is called an *analyzing crystal*. The principle is illustrated in Fig. 11.10.

4. Detectors

a. Ionization Counters, Proportional Counters, and Geiger Counters

Suppose we take a cylinder of suitable material, place in its center a positively charged wire, seal the cylinder, and fill it with a filler gas, such as helium or chlorine. If an x-ray photon enters the cylinder, it will collide with and ionize

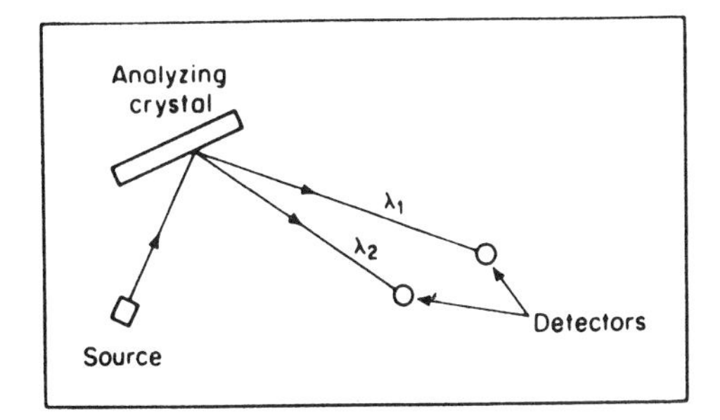

Figure 11.10 The analyzing crystal as a monochromater.

a molecule of the filler gas, creating a *primary ion pair*. With helium as a filler gas, the ion pair would be He^+ and an electron e^-, as illustrated in Fig. 11.11. The interaction

$$h\nu + He \rightarrow He^+ + e^- + h\nu$$

takes place. The electron is attracted to the center wire by its positive charge. The speed at which the electron reaches the center wire depends on the magnitude of the positive charge and the total number of pairs generated. For a given x-ray intensity, a plot of the number of electrons, or the current, reaching the center post is shown in Fig. 11.12.

With no voltage applied, the electron and the positive ion (He^+) recombine and no current flows. As the voltage is slowly increased, an increasing number of electrons reach the center post. This is depicted by part *A* of the curve in Fig. 11.12. In part *B* of the curve, all the electrons formed reach the center wire and

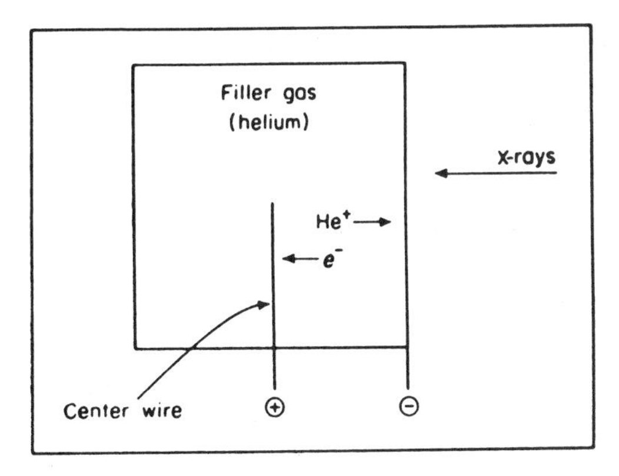

Figure 11.11 Schematic diagram of an x-ray detector.

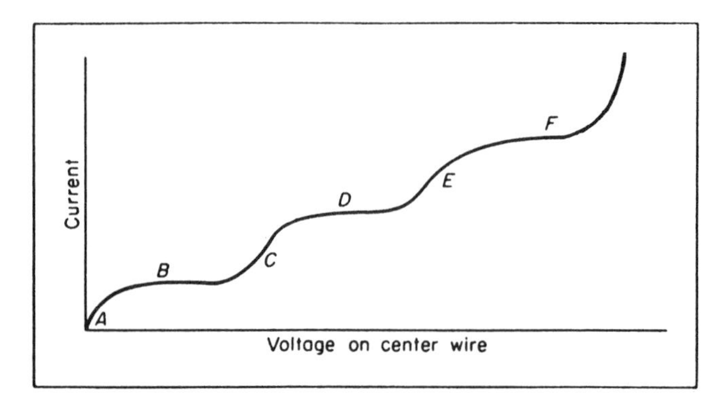

Figure 11.12 Relationship between current and voltage in an x-ray detector.

the current is independent of small changes in the voltage. However, it is dependent on the total number of ion pairs generated, which in turn is proportional to the x-ray intensity. Measurement of the signal in range *B* is therefore a measure of the x-ray intensity. A detector operating under these voltage conditions is known as an *ionization counter*.

As the voltage increases further, the electrons moving toward the center wire are increasingly accelerated until some of them have sufficient energy to collide with and ionize other atoms of the filler gas. The ionizations cause other ion pairs to be formed, as in the following interaction:

$$e^- + \text{He} \rightarrow \text{He}^+ + e^- + e^-$$

He^+ and e^- are called a *secondary ion pair*. More electrons reach the detector, which gives an increased signal. This region is represented by part *C* of the curve in Fig. 11.12. It is not desirable to work under these conditions, because a slight change in the voltage causes a significant change in signal and the detector is unstable. With a further increase in voltage, all the electrons formed in primary ion pairs produce secondary ion pairs. The current flow is independent of small voltages under these stable conditions. The signal is again proportional to the total number of ion pairs formed and therefore to the intensity of the x-ray falling on the detector. This is the basis of a *proportional counter*. In practice, proportional counters use 25% methane and 75% helium as the filler gas.

As the voltage is further increased, electrons formed in primary and secondary ion pairs are accelerated sufficiently to cause the formation of more ion pairs, producing a plasma. This results in an avalanche of electrons reaching the center wire form each x-ray photon falling on the detector. The signal, once initiated, is self-sustaining and independent of the intensity of the x-rays falling on the detector. This is the basis of the *Geiger counter*. It should be noted that a

Geiger counter gives the highest signal for an x-ray beam of given intensity. The Geiger counter can operate in two modes, one where the signal is proportional to the x-ray beam falling on it, the other where it is self-sustaining and, once started, will not extinguish. The latter conditions are very sensitive and are used only to indicate if x-rays are present or absent.

Proportional counters and Geiger counters are used extensively as x-ray detectors. Although their modes of operation are similar, they are constructed differently. It is not possible to purchase a proportional counter and change it to a Geiger counter simply by increasing the voltage. However, they are not as sensitive as *scintillation counters*, which are described next.

b. Scintillation Counters

Photomultiplier detectors (see Chap. 6, Sec. C.2.c) are very sensitive to visible and ultraviolet light, but not to x-rays, to which they are transparent. In a scintillation detector the x-radiation falls on compounds that absorb it and emit visible light. This phenomenon is called *scentillation*. The scintillating compound is generally in the form of a crystal. For some applications, however, it is more convenient to dissolve the scintillating material in a suitable solvent and form a scintillating fluid. The visible light emitted during scintillation can be detected by a photomultiplier. A schematic diagram of the equipment is shown in Fig. 11.13. The detector is useful at very short wavelengths (high energy), but its usefulness drops off at longer wavelengths. Compounds used for converting x-ray to visible light (i.e., *scintillation compounds*) include sodium iodide, anthracene, *p*-terphenyl in xylene, and naphthalene.

The optimal operating range of each detector varies. Of course, it is always best to use the detector most suitable for the work at hand. The optimal ranges for the most useful detectors are shown in Fig. 11.14.

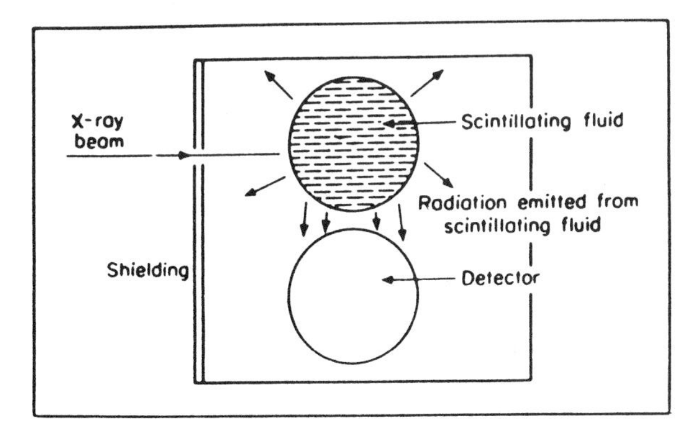

Figure 11.13 Schematic diagram of a scintillation detector.

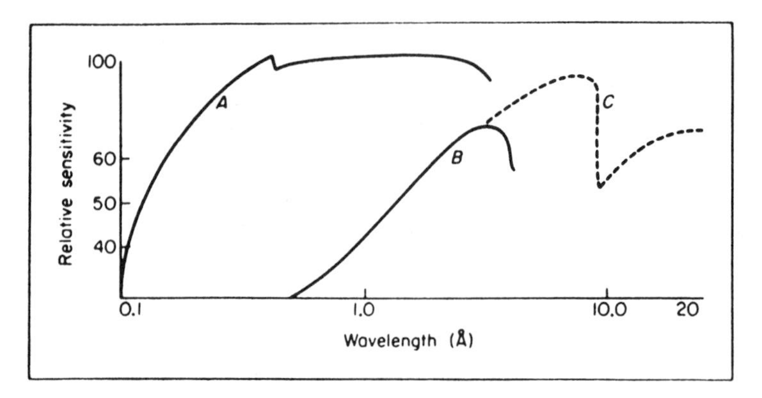

Figure 11.14 Optimal ranges for the most useful x-ray detectors: curve A, scintillation counter; curve B, Geiger counter; Curve C, proportional counter.

5. Semiconductor Detectors: Si(Li) and Ge(Li)

When an x-ray falls on a semiconductor or a silicon lithium-drifted detector, it generates an electron ($-e$) and a hole ($+e$) in a fashion analogous to the formation of a primary ion pair in a proportional counter. Based on this phenomenon, semiconductor detectors have been developed and are now of prime importance in both x-ray work and neutron activation analysis. The principle is similar to that of the gas ionization detector as used in a proportional counter, except that the materials used are in a solid state. A schematic diagram of a silicon lithium-drifted detector is shown in Fig. 11.15. In this system, a very

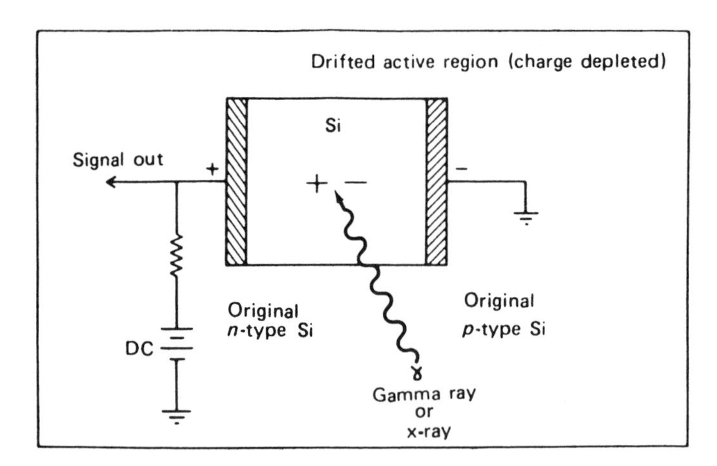

Figure 11.15 Schematic diagram of a Si(Li) detector.

pure silicon block is set up with a thin film of lithium metal plated onto one end. The density of free electrons in the silicon is very low, constituting a *p*-type semiconductor. If the density of free electrons is high, then we have an *n*-type semiconductor. Semiconductor detectors always operate with a combination of these two types (see Chap. 6, Sec. C.2.a).

The crystal is made by plating lithium onto a surface of doped silicon (*p* type) at 673 K. The temperature is then reduced in liquid N_2 and the lithium is drifted into the silicon under the influences of a strong electrostatic field. In the process, all acceptors are neutralized in the bulk material which becomes highly nonconducting. This is the "intrinsic" material. The lithium drifting is stopped before completion, leaving a region of pure Si. The temperature of the system must be maintained below 77 K, otherwise the lithium precipitates out and the detector no longer functions. In exactly the same fashion, germanium, also in group IV of the periodic table, can be used instead of silicon, making a Ge(Li) drifted detector.

The intrinsic material is light sensitive, and if it is exposed to x-rays, an electron-hole pair is formed. The energy required to make an electron-hole pair is 3.65 eV. The electron is −ve, while the "hole," which has lost an electron, is +ve. For a similar Ge lithium-drifted detector, the energy required is 2.95 eV. This is much less than the energy required for a proportional counter or a sodium iodide scintillation detector.

Under the influences of an applied voltage of about 400 V, the electrons move toward the positive charge and the holes toward the negative charge. The voltage generated is a measure of the x-ray intensity falling on the crystal. The process is analogous to that of the proportional counter. Upon arriving at the lithium coating, a pulse is generated. The voltage of the pulse equals Q/C, where Q is the total charge collected on the electrodes and C is the detector capacity. The number of pulses is a direct measure of the number of x-ray photons falling on the detector, which now provides a method of measuring x-ray intensity.

The energy necessary to elevate an electron to a conduction band in a Ge (Li)-drifted detector is only 0.66 eV and for a Si (Li)-drifted detector 1.1 eV. Therefore, at room temperature electrons tend to cross this barrier easily and become conductive even with no signal falling on the system. This causes a high noise level. Cooling liquid N_2 greatly reduces the noise level and improves the sensitivity of the detector. The is important since extremely pure germanium can now be made and Li drifting is not necessary to form intrinsic material. Using liquid N_2 for cooling is therefore not necessary but is desirable.

The sensitivity of the detector will be maximum when all x-rays falling on it are detected; however, then the detector would have no resolution and would not be able to distinguish between high-energy and low-energy x-ray beams. To overcome this problem the system was modified by applying a bias voltage across the ends of the detector. X-rays of short wavelength (high energy) penetrate deeply into the detector. But low energy (long wavelength) x-rays only penetrate close to the surface and liberate electron-hole pairs. The bias voltage enables us to distinguish between deep and shallow e^- or long- and short-wavelength x-rays. When the bisa voltage is high enough, the electrons generated will not reach the lithium surface. By varying the bias voltage it is possible to distinguish between x-rays of low energy and x-rays of high energy. This mechanism is called *sweeping the voltage* and permits us to distinguish between x-rays of various energies. This energy $E = h\nu = hc/\lambda$; therefore I, the x-ray intensity at given energy, can be plotted. This permits us to plot I versus wavelength. In practice, this is not done on a continuous basis, but the bias voltage is changed at set intervals; each interval is designated with a channel number. A typical detector may have 1024 channels, each corresponding to a different energy, and the more channels that are available, the greater the resolution of the detector. For example, if it is possible to distinguish between two energy levels on a system with 100 channels, we have a difference of 1% in the energy between each channel. However, if the detector has 500 channels, then the energy difference between each channel is 0.2%, which gives us a significant increase in resolution.

Silicon lithium-drifted detectors are widely used. Semiconductor detectors may also be made with germanium as the principal metal and doped with other elements in group 3A of the periodic table, such as gallium. The efficiency of the detector may be further increased by cooling the silicon (or germanium) to low temperatures such as that of liquid nitrogen. This increases the resistance of the silicon when unexposed to radiation, and therefore the contrast when radiation falls on a detector is greater. Low temperatures also prevent the lithium from precipitating out.

C. ANALYTICAL APPLICATIONS OF X-RAYS

There are three distinct fields of x-ray analysis: x-ray absorption, which varies with atomic weight; x-ray diffraction, which depends on the crystal properties of solids; and x-ray fluorescence, the intensity of which depends on the atomic number of the sample and the wavelength of incident radiation. Their analytical uses are described in this section.

1. X-Ray Absorption

If the wavelength of an x-ray beam is short enough (high energy), it will excite an atom that is in its path. In other words, the atom absorbs x-rays that have enough energy to cause it to become excited. As a rule of thumb, the x-rays emitted from a particular element will be absorbed by elements with a lower

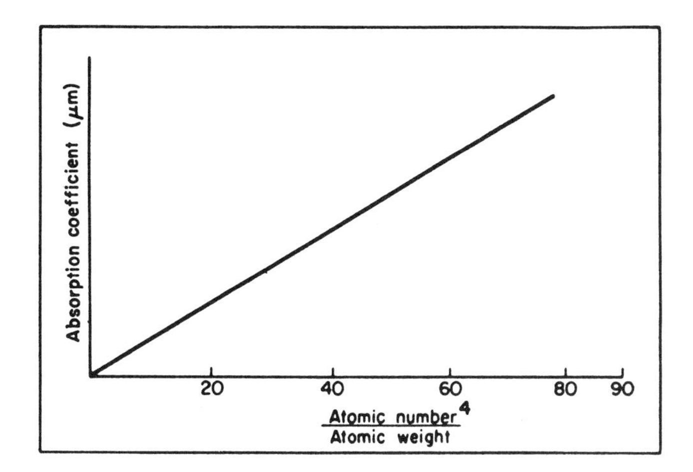

Figure 11.16 Relationship between atomic number and x-ray absorption coefficient. Note it is atomic number Z to the fourth power divided by A. This is very approximately a Z-cubed relationship.

atomic number. The ability of each element to absorb increases with atomic number. This gives a relationship as shown in Fig. 11.16.

Beer's law indicates that

$$\log\left(\frac{P_0}{P_x}\right) = \mu x \tag{11.13}$$

where

μ_x = linear absorption coefficent
x = path length through the absorbing material
P_0 = x-ray power before entering sample
P_x = x-ray power leaving sample

However,

$$\frac{\mu_x}{\rho} = \mu_m \tag{11.14}$$

where

μ_m = mass absorptive coefficient
ρ = density

But

$$\mu_m = \frac{CN_0 Z^4 \lambda^3}{A} \tag{11.15}$$

where

C = a constant
N_0 = Avogadro's number
Z = atomic number
A = atomic weight
λ = wavelength of the radiation

It can be seen that at a given wavelength, μ_m is proportional to Z^4 divided by the atomic weight. This relationship is shown in Fig. 11.16.

For example, the human arm consists of flesh, blood, and bone. The flesh or muscle is made up primarily of carbon, nitrogen, oxygen, and hydrogen. These are all low atomic number elements, and their absorptive power is very low. Similarly, blood, which is primarily water, consists of hydrogen, oxygen, plus small quantities of sodium chloride and trace materials. Again, the absorptive power of blood is quite low. In contrast, bone contains large quantities of calcium and phosphorus, primarily as calcium phosphate. The atomic numbers of these elements are considerably higher than those mentioned before, and so the absorptive power is considerably higher. When an x-ray picture is taken of an arm, the x-radiation penetrates the muscle tissue and blood quite readily, but is absorbed significantly by the bone. A photograph of this absorption indicates the location of the bone in the arm. The procedure is routinely used in medicine to detect broken bones.

Another application of x-ray absorption in medicine is to define the shapes of arteries and capillaries. Normally the blood absorbs only poorly; however, it is possible to inject a solution of strongly absorbing cesium iodide into the veins. The material is then swept along with the blood and follows the contours of the arteries. An x-ray movie picture is taken as the highly absorbing cesium iodide flows through the arteries and a film is generated that indicates the contours of the arteries. This can be used to identify breaks in the veins or arteries that could cause internal bleeding. Such internal bleeding can be the cause of a stroke. The technique may also be used to indicate a buildup of coating on the inside of the veins. This is particularly dangerous in the heart, where deposits of cholesterol restrict the flow of blood through the heart. If this is left unchecked, a heart attack will result. X-ray absorption can be used to diagnose this problem and to locate exactly the position of deposits. Surgery is made much easier by this technique, and the patient's chances of recovery are greatly improved.

Two elements in one sample will absorb the x-ray beam to different degrees. This property can be used to detect impurities, segregations, contours of different phases, and so on, in solid samples. A schematic diagram of the equipment is shown in Fig. 11.17. The photographic plate need not be removed from its wrapping, because x-ray beams will penetrate the covering paper. The presence of the different elements is indicated by patches of various intensities of gray on

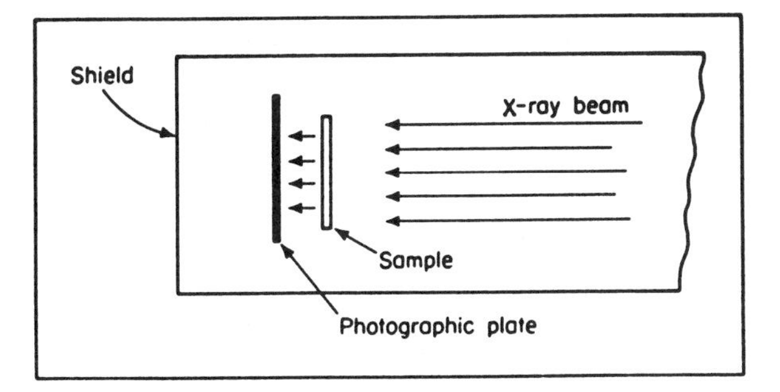

Figure 11.17 Schematic diagram of equipment for x-ray absorption.

the developed film. Where heavy elements (those with high atomic weight) have strongly absorbed the x-rays, the areas in the negative of the film are light, and where light elements (those with low atomic weight) have not absorbed the x-rays intensely, the areas are dark.

In the field of metallurgy, applications of x-ray absorption include the detection of voids or the segregation of impurities, such as oxides, in welds and other joints. Figure 11.18 shows an x-ray absorption photograph of a mechanical weld that contains voids, or blow holes. Such blow holes indicate that the weld is mechanically weak and might break in use. If the weld is weak, it must be strengthened to form a sufficiently strong joint. This is particularly important in areas such as ship building, where corrections at sea are impossible.

The technique of x-ray absorption can also be used to determine the levels of liquids in enclosed vessels or pipes without opening or breaking them. The same process can be used to detect metal supports or metal fillings inside constructed objects as diverse as buildings and small works of art.

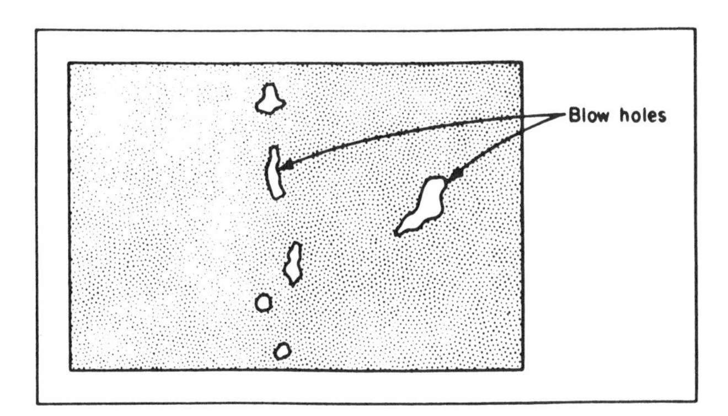

Figure 11.18 X-ray absorption photograph of a mechanical weld.

A particularly important elemental analytical application of this technique is the direct determination of lead in gasoline. In this case, however, the method is subject to interference from sulfur (which also absorbs x-rays) present in the sample. It is especially significant that x-ray absorption analysis is a nondestructive method and is independent of the chemical state of the elements concerned. Because of its low sensitivity, however, the method is more useful for major constituents than for trace metals.

One application is in the verification of works of art. A major advantage is that x-ray techniques are usually, but not always, nondestructive. Sometimes artists paint over old paintings, using the canvas for their own work and covering unrecognized masterpieces in the process. Using x-ray absorption, it is possible to reveal the covered painting without removing the top painting. When used to examine a metal horse sold for several million dollars as an ancient Greek art piece, x-ray absorption showed that the horse contained internal metal supports and was therefore a fake. This was done without destroying the art piece, in case it had been authentic.

2. X-Ray Diffraction

The ions or molecules that make up a crystal are arranged in well-defined positions. An illustration of a typical crystal structure, greatly magnified, is shown in Fig. 11.19. As we examine the structure of the crystal, we see that the ions form planes in three dimensions. If a monochromatic x-ray beam falls on such a crystal, the beam is reflected by each crystal plane. Each separate reflected beam interacts with other reflected beams. If the beams are not in phase, they

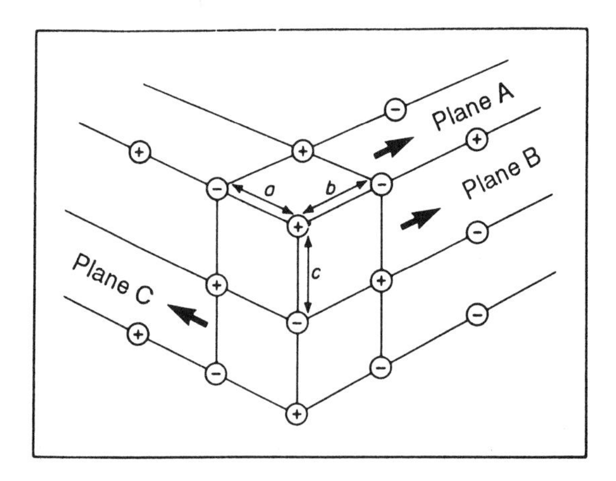

Figure 11.19 Crystal structure of sodium chloride: (+) Na ions and (−) Cl ions.

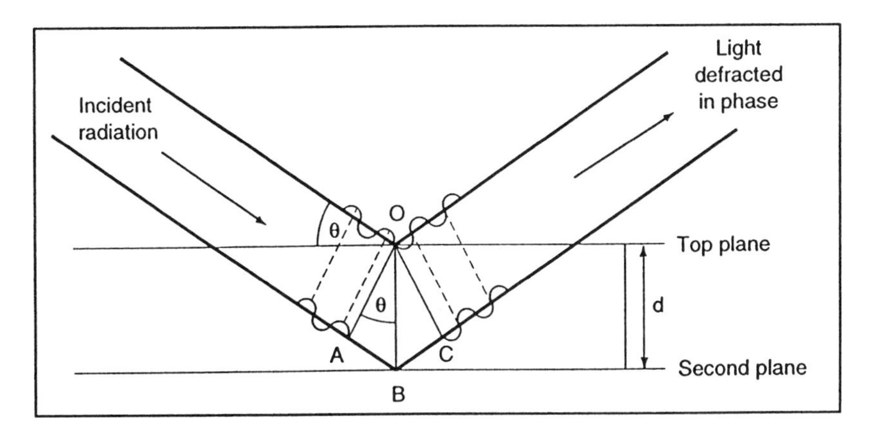

Figure 11.20 Reinforcement of light diffracted from two planes.

destroy each other and no beam emerges. The net result is a diffraction pattern of reinforced beams.

In Fig. 11.20 radiation form the source falls on the crystal, some on the top plane, some on the second plane. Since the two beams are part of the same original beam, they are in phase on reaching the crystal. However, when they leave the crystal, the part leaving the second plane has traveled an extra distance *ABC*. If *ABC* is a whole number of wavelengths, the two beams leaving the crystal will be in phase and the light coherent. If *ABC* is not a whole number of wavelengths, the two beams come together out of phase and by destructive interference the light is destroyed.

Coherence occurs when *ABC* is a whole number of wavelengths or

$$ABC = n\lambda \tag{11.16}$$

If θ is the angle of incidence of the light and d is the distance between the two planes,

$$AB = d \sin \theta$$

Also

$$BC = d \sin \theta$$

Therefore

$$ABC = 2d \sin \theta \tag{11.17}$$

From Eq. (11.16)

$$n\lambda = 2d \sin \theta$$

Table 11.3

n	λ (Å)	$n\lambda$	Order
1	60	60	First
2	30	60	Second
3	20	60	Third
4	15	60	Fourth

This is the Bragg equation, which states that coherence occurs when $n\lambda = 2d \sin \theta$. It can be used to measure d, the distance between layers in crystals, and is the basis of crystallography, an important field of science. Diffraction from a single plane is illustrated in Fig. 11.9.

For any given crystal, d is constant; hence for any given angle, θ is constant, as in $n\lambda$. Therefore, if n varies, there is a corresponding change in λ to satisfy Eq. (11.12). For a given diffraction angle, a number of diffracted lines are possible; n is known as the order. A given value $2d \sin \theta$ may equal 60, in which case each of the conditions of Table 11.3 will meet these requirements. Radiation of wavelength 60, 30, 20, or 15 Å will diffract at angle θ in first, second, third, or fourth order, respectively. This is called order overlap and sometimes creates difficulty in interpretation.

It should be noted that radiation of 30 Å will also be diffracted at a different angle in first order, as shown in Fig. 11.21. In particular, wavelengths corresponding to the low whole numbers 1, 2, 3, 4, and so on, give observable diffraction lines. Consequently, a single plane will generate several diffraction lines for each wavelength. Each of the planes in the three dimensions of the crystal will give diffraction lines. The sum total of these diffraction lines generates a diffraction pattern. From the diffraction pattern it is possible to deduce the different distances between the planes as well as the angles between these planes in each of the three dimensions. Based on the diffraction pattern, the physical dimensions of the crystal can be identified.

For any given experiment, λ is the known wavelength of the x-ray beam, θ can be measured, and from this information d can be calculated by using Eq. (11.12). By examining various sides of the crystal, the distance d_1, d_2, and d_3 between all the planes of the crystal can be calculated. These measurements are the basis for the use of x-ray diffraction.

The diffraction pattern of a single crystal of an inorganic salt is shown in Fig. 11.22. This inorganic salt always gives the same diffraction pattern, and from this pattern we can determine the dimensions of the salt crystal. Also, qualitative identification can be obtained by matching this pattern to previously identified patterns.

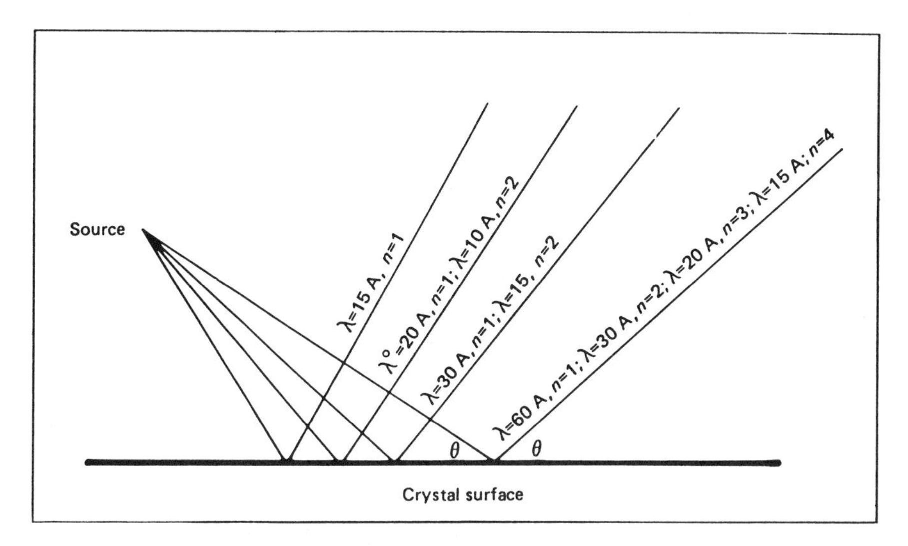

Figure 11.21 Diffraction of radiation of different wavelengths. Note that the angle of incidence and diffraction varies with wavelength.

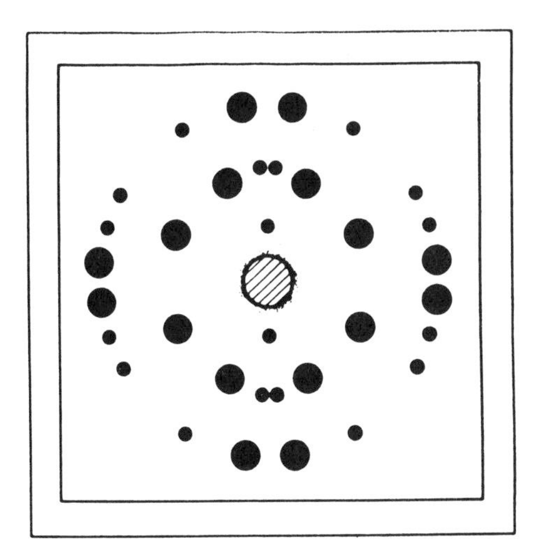

Figure 11.22 Diffraction pattern of a single crystal of an inorganic salt.

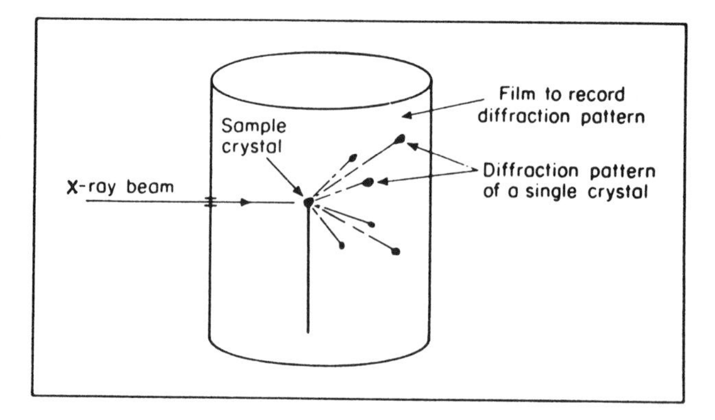

Figure 11.23 Diagram of the equipment used for x-ray diffraction studies of crystal samples.

A diagram of equipment used for x-ray diffraction studies of crystal samples is shown in Fig. 11.23. With a powder many small crystals are involved. These crystals are oriented in all possible directions relative to the beam of x-rays. Hence, instead of the sample generating only single spots, it generates a circular pattern from each crystal. These patterns overlap, and a series of diffraction lines (Fig. 11.24) evolve. A typical diffraction pattern of a powder sample is

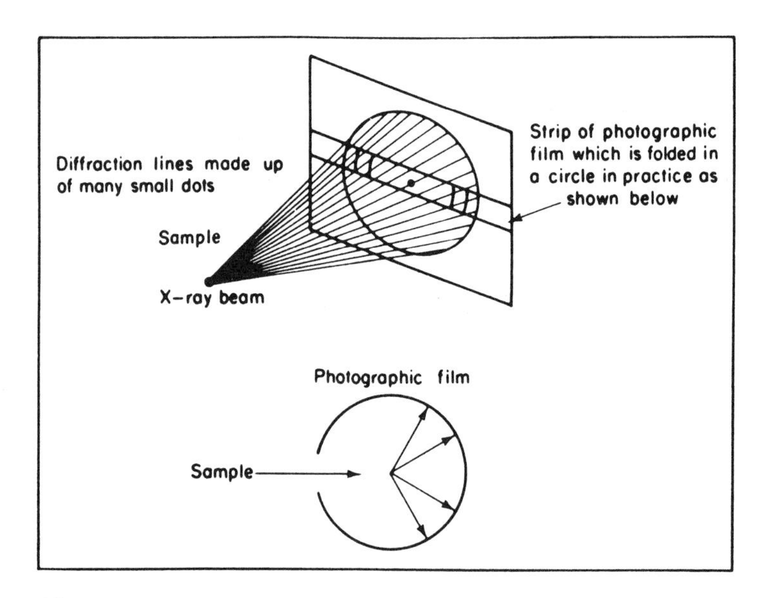

Figure 11.24 Basic diffraction geometry using the Laue technique.

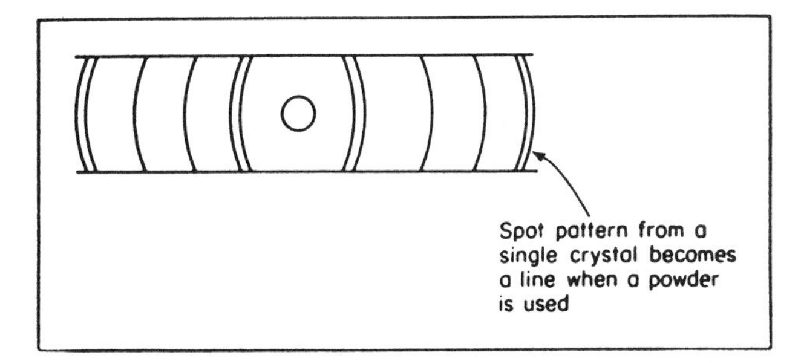
Spot pattern from a
single crystal becomes
a line when a powder
is used

Figure 11.25 Typical diffraction photograph from a powder sample. Note that when a powder is used, the spot pattern from a single crystal becomes a line which is part of a circle.

shown in Fig. 11.25. These are called Laue photographs, after von Laue, the German scientist who developed the technique. An example of the use of Laue photographs is to distinguish between crystals of different compounds.

The analytical applications of x-ray diffraction are numerous. The method is nondestructive and gives information on the molecular structure of the sample. Perhaps its most important use has been to measure the size of crystal planes. The patterns obtained are characteristic of the particular compound from which the crystal was formed. For example, as shown in Fig. 11.26, NaCl crystals and KCl crystals give different diffraction patterns. A *mixture* containing 1% KCl in NaCl would show a diffraction pattern of NaCl with a weak pattern of KCl. On the other hand, a mixture containing 1% NaCl in KCl would show the diffraction pattern of KCl with a weak pattern of NaCl.

If, on the other hand, the crystal is a *mixed crystal* of sodium potassium chloride, in which the sodium and potassium ions are in the *same crystal lattice*, there would be changes in the crystal's lattice size, as shown in Fig. 11.27. When there is a large excess of sodium over potassium, the pattern would be similar to that of sodium chloride. However, as the potassium content increases, the lattice dimensions change accordingly until they equal that of the potassium chloride when there is a very large excess of potassium chloride present. This method can be used to distinguish between a *mixture of crystals*, which would give both diffraction patterns, and a *mixed crystal*, which would give a separate diffraction pattern.

Comparing diffraction patterns from crystals of unknown composition with patterns from crystals of known compounds permits the identification of unknown crystalline compounds. It is also possible to identify a substance as being a single compound that would give a superimposed pattern for each type of crystal present.

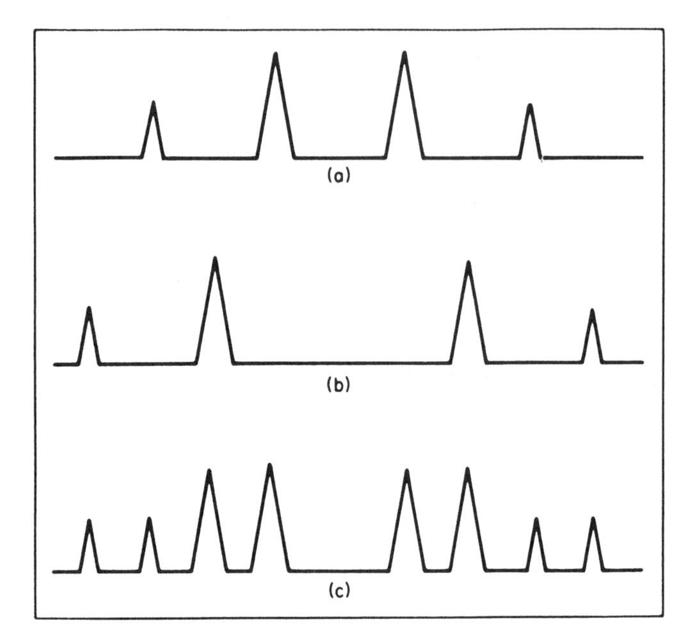

Figure 11.26 Hypothetical x-ray patterns of two salts: (a) x-ray pattern of salt A, (b) x-ray pattern of salt B, and (c) x-ray pattern of a *mixture* of salts A and B. Note that the peaks match those in (a) and (b).

Modern instrumentation utilizes computers of identifying ''unknown'' samples. An example is illustrated in Fig. 11.28. The unknown sample is put into the instrument and the x-ray diffraction pattern obtained. That pattern is then fed into a computer memory bank, which contains the diffraction patterns of a large number of unknown compounds. The computer is instructed to compare the diffraction pattern of the unknown compound with those of the known compounds and report which of these the unknown might be. The computer displays a short list of compounds. The operator may then instruct it to display the x-ray

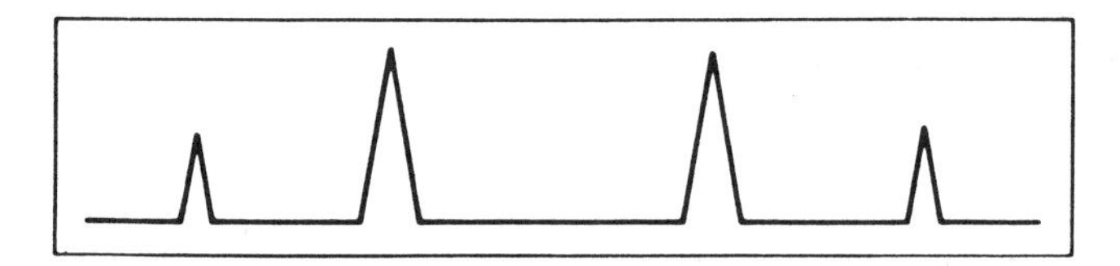

Figure 11.27 X-ray pattern of a powdered *mixed* crystal of A and B. Note that the peaks fall between those in Fig. 11.26c, showing that lattice size does not coincide with that of Fig. 11.26a or b, but is between them.

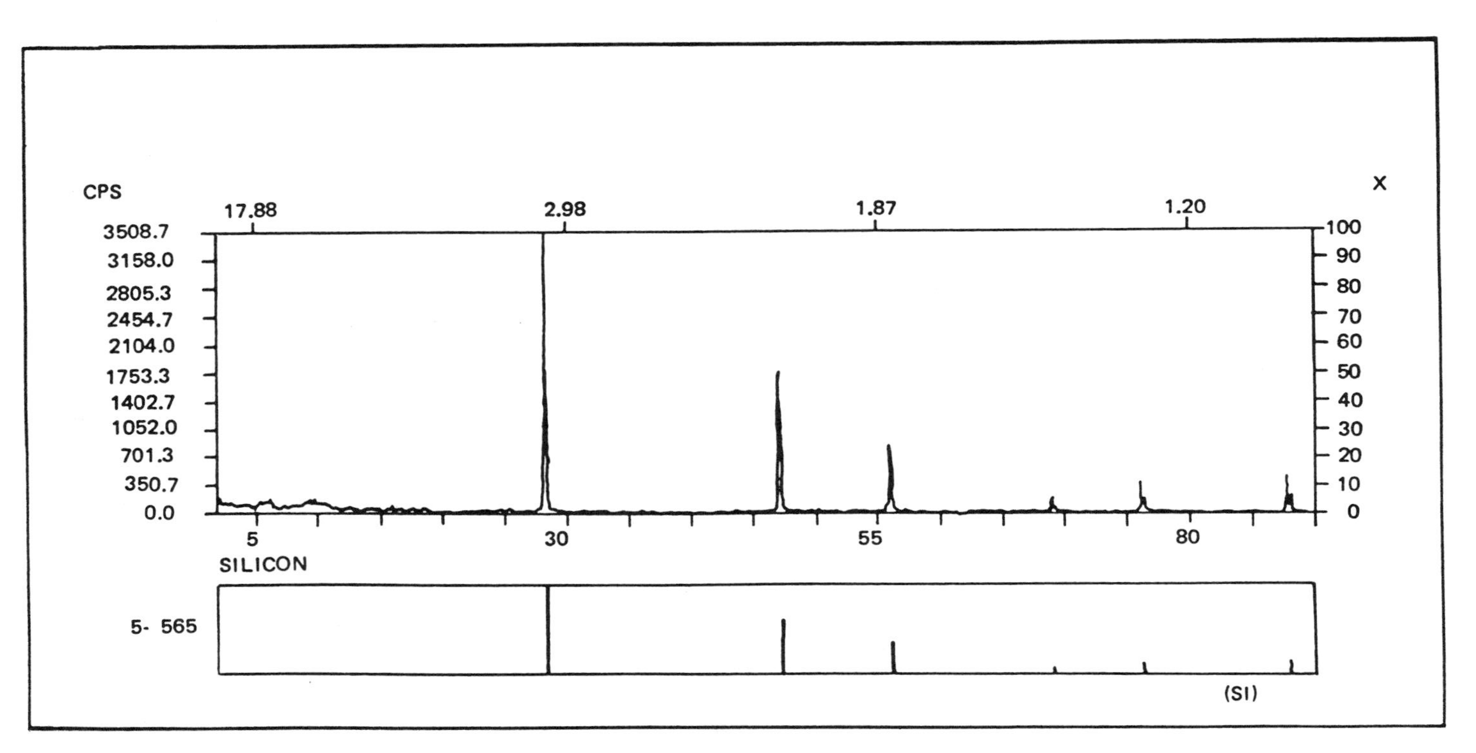

Figure 11.28 X-ray diffraction pattern of unknown material (upper spectrum). The lower spectrum was retrieved from the memory bank and is of silica.

diffraction patterns of these known compounds for visual inspection and comparison with the unknown. The operator selects the most likely candidate for the identification of the unknown compound and is at liberty to perform other confirmatory tests if necessary. Figure 11.28 shows the pattern of an unknown and that of a compound retrieved from the computer's memory bank. In this case the unknown was silica and the fit between the two spectra was good.

In polymer characterization, it is possible to determine the degree of crystallinity of the polymer. The noncrystalline portion simply scatters the x-ray beam to give a continuous background, whereas the crystalline portion creates diffraction lines. A typical diffraction spectrum of a polymer is shown in Fig. 11.29. The amorphous material in the polymer will scatter at all wavelengths and give a scattered pattern; however, the crystalline material will include crystal structures and will produce definite diffraction lines or spots. The ratio of the area of diffraction peaks to scattered radiation is proportional to the ratio of crystalline to noncrystalline material in the polymer. The ultimate quantitative analysis must be confirmed by using standard polymers with known crystallinities and basing the calculation of the known ratio of crystalline diffraction to amorphous scattering.

By measuring the intensity of diffraction patterns in several directions of a crystal, it is possible to determine if the crystals of a polymer or a metal are oriented in any particular direction. Preferred orientation can occur after the material gas been rolled out into a sheet. This is sometimes a very undesirable

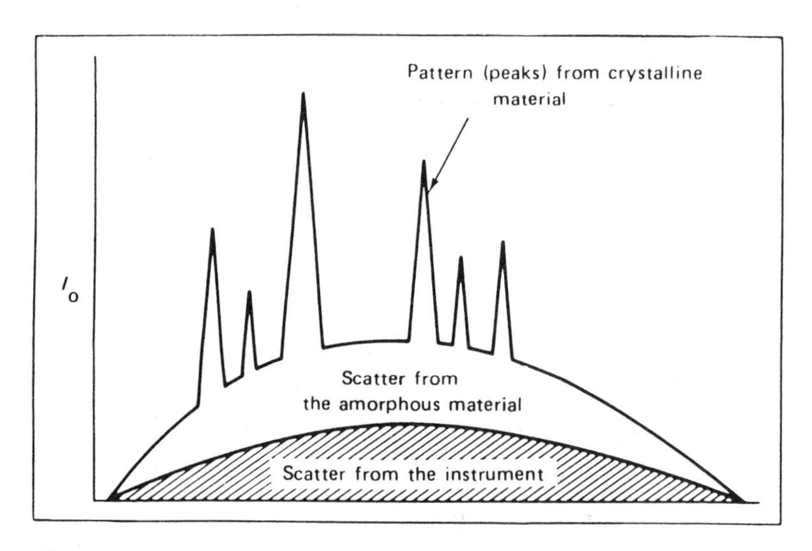

Figure 11.29 Diffraction pattern of a polymer showing crystalline, amorphous, and scattered radiation components.

property, since the material may be very weak in one direction and strong in another, with the result that it tears easily in one direction. Sometimes, however, this is a desirable property, as, for example, in wrapping material that we may wish to tear easily to open, such as cigarette package wrappings.

A property of metals that can be determined by x-ray diffraction is the state of anneal. Well-annealed metals are in well-ordered crystal form and give sharp diffraction lines. If the metal is subjected to drilling, hammering, or bending, it becomes "worked" or "fatigued," that is, its crystals become broken and the x-ray pattern more diffuse. Fatigue weakens the metal and can result in the metal breaking. It is occasionally necessary to check moving parts of metal fatigue, such as airplane wings (which move slightly during flight), combustion engine parts, and bridges. This check can be done by x-ray diffraction without removing the part from its position and without weakening it in the process of testing. These advantages of x-ray testing are frequently very important, particularly if the sample will continue to be used after testing.

Other applications of x-ray diffraction include (1) soil classification based on crystallinity. Different types of soils, such as various types of clays and sands, exhibit different types and degrees of crystallinity. Knowledge of this crystallinity gives valuable information concerning soil structure. It also tells us the effects of rain and drought and indicates the mechanism of soil erosion. (2) Analysis of industrial dusts can be effected, and their relationship to industrial disease ascertained, by means of x-ray diffraction studies. X-ray diffraction can also be used to (3) assess the weathering and degradation of natural and synthetic minerals. By designed experiments, the factors responsible for the degradation can be revealed. The same type of study can be carried out on polymers. It is possible to investigate the effects of temperature, humidity, direct sunlight, or corrosive gases on polymers. Based on the results, stable polymers have been developed that are suitable for outdoor use. (4) Corrosion products can be studied by this method. When metal samples are exposed to the atmosphere, they are susceptible to corrosion. If the corrosion is rapid, it leads to a short life for the metal product. The products of corrosion can be identified by x-ray diffraction studies; also, by research, the factors that affect the corrosion rate can be determined. Over the years, metals and alloys have been developed that are resistant to corrosion. These materials make it possible for us to build automobiles, bridges, airplanes, and trash cans that are stable over long periods of time. (5) Tooth enamel and dentine have been examined by x-ray diffraction, which reveals the chemical structure of the tooth surface. The results have enabled us to understand some of the factors that promote tooth decay and have given us insight into possible approaches to cure the problem.

In a similar fashion, x-ray diffraction has revealed (6) some of the effects of diseases on bone structure and tissue structure, as well as (7) how much crystalline material is present in a sludge or the degree of crystallinity of a polymer.

Finally, x-ray diffraction can be used for (8) the identification of crystalline compounds. These crystalline compounds may originate in the body (e.g., gallstones) or as corrosion products. They may even have been pumped from the stomach of a potential suicide victim. In each case, identification of a compound is necessary and it is essential that the results be reliable. An application of greater interest to the inorganic or physical chemist is the determination of the dimensions of crystals. This area has received much attention in the past, but is of only indirect interest to the analytical chemist, who only uses such results to identify unknown samples. It should be pointed out, however, that x-ray diffraction was a major tool in elucidating the structure of ribonucleic acid (RNA) and deoxyribonucleic acid (DNA). These are molecules intimately involved in cell reproduction.

A method of measuring x-ray diffraction is to use the equipment the optics of which are shown in Fig. 11.30. This system uses an x-ray tube, a sample specimen, and a detector that rotates in an arc described by a Rowland circle.

The sample is loaded onto the specimen holder. It may be a single crystal or a ground-up sample of the material under test packed into a small plastic tube. The latter may be used because, being made of carbon, hydrogen, and perhaps nitrogen and oxygen, all low atomic number elements, it does not significantly absorb the radiation. After mounting, the specimen is rotated relative to the x-ray source at a rate of θ°/min. Diffracted radiation comes from the sample according to the Bragg equation (11.12) and an extra θ°/min so the detector is simultaneously rotated at 2θ°/min.

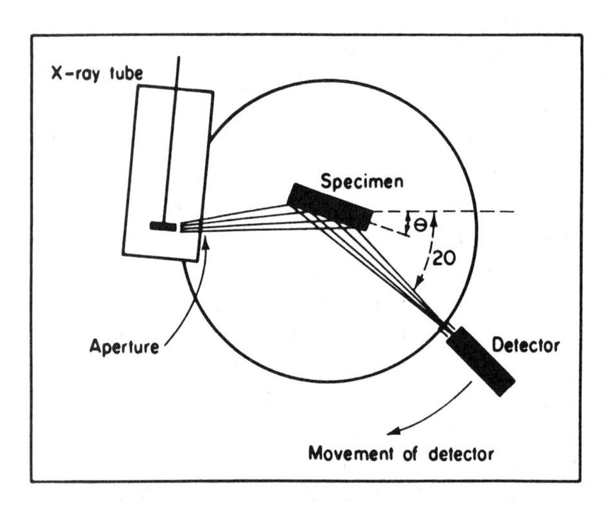

Figure 11.30 Schematic diagram of the optics of an x-ray diffractometer (goniometer).

The detector response may then be fed into a computer, which either displays the data for inspection or feeds to a memory bank for qualitative identification or quantitative analysis. This instrument is called a goniometer.

The foregoing are some of the many and varied applications of x-ray diffraction. Uses of the technique can be found in the literature or in a suitable text on the subject.

3. X-Ray Fluorescence

When an element is placed in a beam of x-rays, the x-rays are absorbed. The absorbing atoms become excited and emit x-rays of characteristic wavelength. This process is called *x-ray fluorescence*. Since the wavelength of the fluorescence is characteristic of the element being excited, measurement of this *wavelength* enables us to *identify* the fluorescing element. The *intensity* of the fluorescence depends on *how much* of that element is in the x-ray beam. Hence measurement of the fluorescence intensity makes possible the quantitative determination of an element. Figure 11.31 is a diagram of the equipment used for x-ray fluorescence.

For qualitative studies, the angle of diffraction θ is measured. From this measurement the wavelength λ of fluorescence can be calculated by using the Bragg equation,

$$n\lambda = 2d \sin \theta \qquad (11.12)$$

where d is the spacing between the crystal layers of the analyzing crystal and is therefore known. Knowing λ, the wavelength of fluorescence, we can identify the element from a chart of the fluorescence wavelengths of all the known elements.

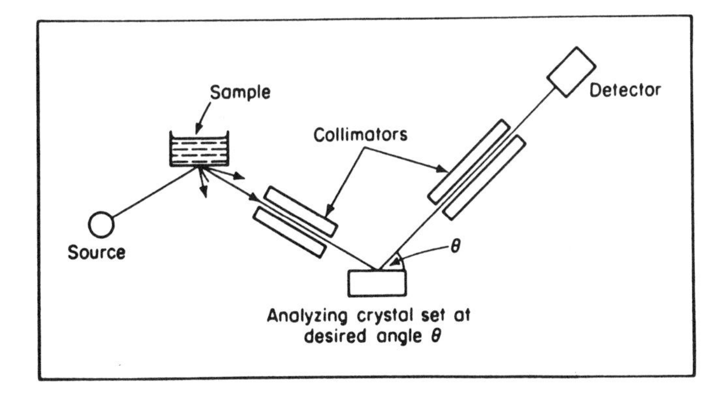

Figure 11.31 Schematic diagram of x-ray fluorescence equipment.

If the element being determined is in two chemical forms, such as calcium chloride and calcium sulfate, there is a very slight difference between the energy levels of the inner shell electrons of the calcium. This difference is too slight to be detectable by x-ray fluorescence. Consequently, x-ray fluorescence is independent of the chemical form of the element being analyzed. This is sometimes a distinct advantage over many other forms of elemental analysis.

a. Components of the Equipment

The principal problem with x-ray fluorescence is its lack of sensitivity. This factor influences the design of all the components of the equipment. The intensity of the x-ray source must be high. For this reason the voltage to the source must be high and, in fact, may be as high as 100 kV. The wavelength of the radiation from the source must be shorter than the minimum wavelength required to excite the sample, preferably less than 16% of the minimum. This requirement influences the choice of the target to be used in the source. A common target element is tungsten, which has a high atomic number and therefore a short wavelength of x-ray radiation. (Short wavelengths mean high frequency and therefore high-energy radiation.)

Selective x-ray fluorescence may be carried out by using an x-ray beam with a wavelength that excites one element but is too long (does not have enough energy) to excite another. In this system, only one element will fluoresce. The fluorescence intensity is measured when first one source and then a second source is used. The difference in intensity depends on how much element of interest is present in the sample. Background fluorescence is common to both measurements, and a correction is automatically made for this source of error.

The monochromator used is an analyzing crystal that acts with x-rays in the same way as a grating monochromator with visible light. The crystal separates x-rays of different wavelengths by diffracting them at different angles. From the Bragg equation, Eq. (11.12), we can see that when λ varies and d (crystal spacing) is constant, θ must vary. For best results the crystal should be as perfect as possible, so that d will be constant in all parts of the crystal. For best sensitivity, a curved crystal (Fig. 11.32) is used. This refocuses the x-ray fluorescence after diffraction at the crystal. Curved crystals have been made from such salts as sodium chloride, lithium fluoride, quartz, aluminum, and topaz. The choice depends on the wavelength range to be investigated (see Sec. B.4).

The sample holder is made of materials transparent to x-ray beams. Polymers such as polyethylene are composed of low atomic number elements, such as carbon and hydrogen. These make excellent sample holders, because low atomic number elements absorb very weakly. Aluminum holders are also widely used.

Liquids make the most satisfactory samples because they flow into reproducible shapes with flat surfaces. The best solvents are H_2O, HNO_3, hydrocar-

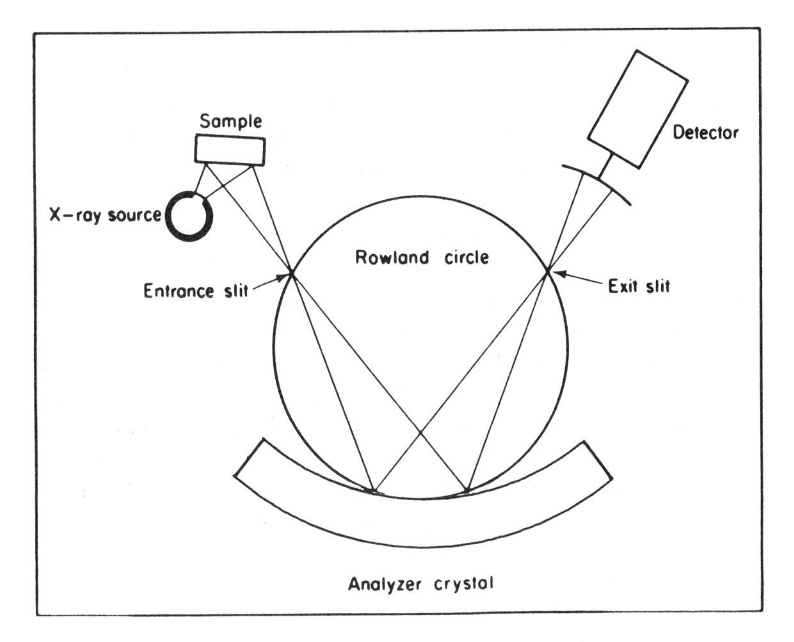

Figure 11.32 Optical system using a curved crystal analyzer to focus the fluorescence on the detector.

bons, and oxygenated carbon compounds, because these compounds contain only low atomic number elements. Solvents such as HCl, H_2SO_4, CS_2, and CCl_4 are undesirable because they contain elements with higher atomic numbers, which absorb x-rays and therefore reabsorb the fluorescence signal.

Solid samples that are not readily converted to liquids should be ground to a powder. The powder may be mixed with a borate salt or borax and formed into a briquette. This procedure provides a sample that can be easily handled and has the advantage that the borate provides a standard matrix for the sample. Furthermore, the matrix is composed of low atomic weight elements, which interact only slightly with the x-ray beam. Other solids should be ground or filed to a flat surface in order to give reproducible results.

Elements with atomic numbers between 12 and 92 can be analyzed in air. Elements with atomic numbers 5 (boron) through 11 (sodium) fluoresce at long wavelengths (low energy), and the fluorescence is absorbed by the air. Analysis of this group should be carried out in a vacuum or a helium atmosphere in order to reduce the loss of fluorescence intensity by air absorption.

b. Analytical Uses of X-Ray Fluorescence

X-ray fluorescence is a method of elemental analysis. It is most useful for the analysis of metals and nonmetals with atomic numbers greater than 12. This is a distinct advantage over emission spectrography and atomic absorption, because these methods cannot be used for the direct determination of nonmetals. With special equipment (an evacuated optical system, light-gathering crystals, and highly sensitive detectors), elements of atomic number 5 through 11 can also be analyzed. The intensity of fluorescence is independent of the chemical state of the elements. For this reason, chemical preparation prior to x-ray fluorescence measurements is frequently unnecessary. The method is nondestructive, an important feature when the sample is available in limited amounts or when it is valuable or even irreplaceable, as in the case with works of art or antiques.

The fluorescence spectra are very simple, and overlap of x-ray emission lines from different elements is unlikely; however, background emission from one element may overlap line emission from another element, as illustrated in Fig. 11.33. To measure the intensity of peak B in Fig. 11.33 a correction must be made for the background of A. This can be done by measuring the intensity of the background at a wavelength slightly greater or less than that of B (e.g., at point C) and subtracting from the intensity of B. For example,

$$\begin{aligned} \text{Intensity of B} &= 50 \text{ units} \\ \text{Intensity of C} &= 10 \text{ units} \\ \text{True intensity of peak B} &= 40 \text{ units} \end{aligned}$$

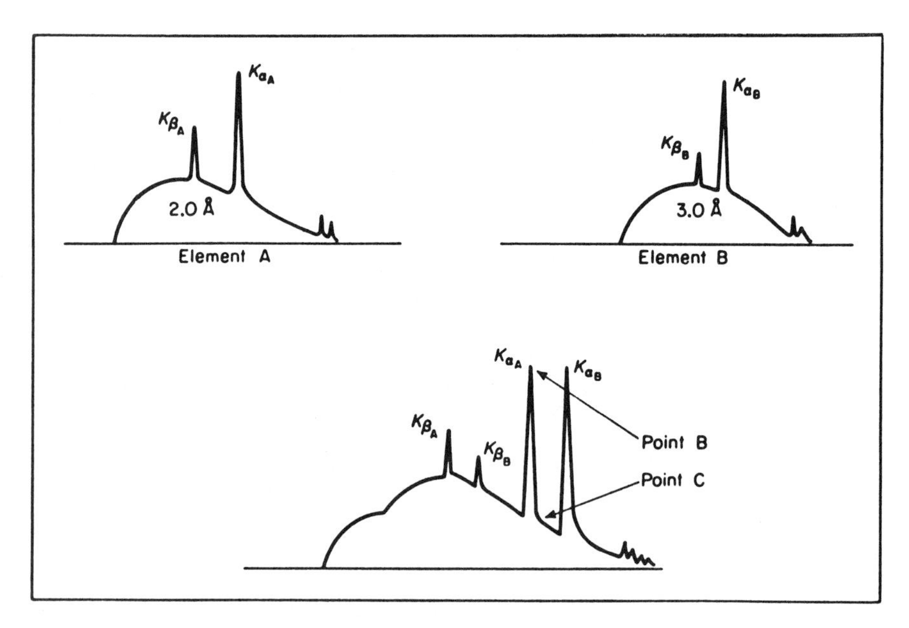

Figure 11.33 Overlap of x-ray fluorescence.

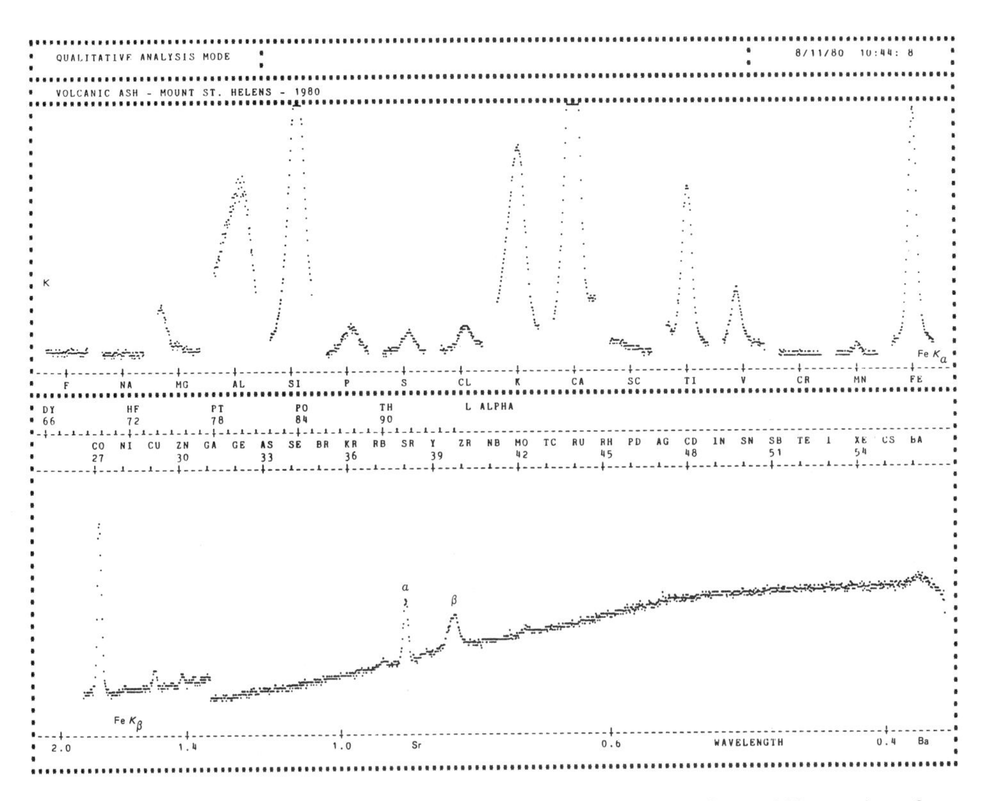

Figure 11.34 Qualitative analysis of elements in volcanic ash from 1980 eruption of Mount St. Helens.

Qualitative Analysis. Qualitative analysis can be carried out by measuring the angle of diffraction of the fluorescent x-rays. From this measurement, the wavelength of the fluorescence can be calculated. Each element fluoresces at its own characteristic wavelengths; hence the fluorescing element can be identified from a knowledge of the wavelengths of its x-ray fluorescence.

An example of qualitative analysis is given in Fig. 11.34, which shows the elemental analysis of volcanic ash from the eruption of Mount St. Helens in 1980.

Quantitative Analysis. Quantitative analysis can be carried out by measuring the intensity of fluorescence at the wavelength characteristic of the element being determined. The method has wide applications to most of the elements in the periodic table, both metals and nonmetals. It is precise, and with skilled operation it is accurate. The sensitivity limits are of the order of 10 ppm, although better sensitivity can be obtained under special and exacting conditions. The lack of sensitivity is sometimes a handicap, particularly in trace element analysis.

Applications. The fluorescence of x-rays by elements is associated with the loss of an electron from an inner electron shell. Consequently, the wavelength of fluorescence from a particular element is virtually independent of the chemical state of that element. This property allows various elements to be detected and determined with a minimum of sample preparation. Frequently no preparation whatever is required. In this respect, x-ray fluorescence enjoys a distinct advantage over many other methods of elemental analysis. As we mentioned earlier, however, it is sometimes necessary to compensate for matrix effects by fusing the sample in borax. This procedure increases the precision and accuracy of the method, but ends up diluting the sample. The following are some typical analytical uses of x-ray fluorescence.

In agriculture the technique is used for the determination of trace elements in plants and foods; the detection of insecticides on fruit and leaves; the continuous determination of phosphorus in fertilizer; the detection in fodder of elements, such as selenium, known to be harmful in large quantities; and the characterization of soils.

One of the medical uses of x-ray fluorescence is for the direct determination of sulfur in protein. The sulfur content of each of the many different forms in which protein exists in human blood varies considerably. X-ray fluorescence indicates protein distribution and provides a diagnostic link for the medical practitioner. Other medical applications include the determination of chloride in blood serum; the determination of strontium in blood serum and bone tissue (it should be remembered that the strontium content of bone and blood is affected by radioactive fallout); and the elemental analysis of tissue, bones, and body fluids.

In mining and metallurgy x-ray fluorescence is used for the analysis of ores, tailings, concentrates, and drilled cores; the continuous determination of silica in flowing slurries of ores; the continuous analysis of zinc in flowing slurries of zinc concentrates; the simple determination of lead in lead-tin alloys; the determination of chromium in stainless steels, manganese in plain steels, and tungsten in high-speed steels; the determination of copper, zinc, and tin in copper alloys; the determination of tin or zinc used as coating for steels (such as galvanized steel); the elemental analysis of slags; the classification of alloys; and the direct analysis of platinum and gold in plating solutions.

By sputtering the surface, surface analysis can be used to measure the charge in composition with depth from the surface using the Auger depth profile (Fig. 11.35). Similar data can be obtained using x-ray fluorescence, but background correction must be made.

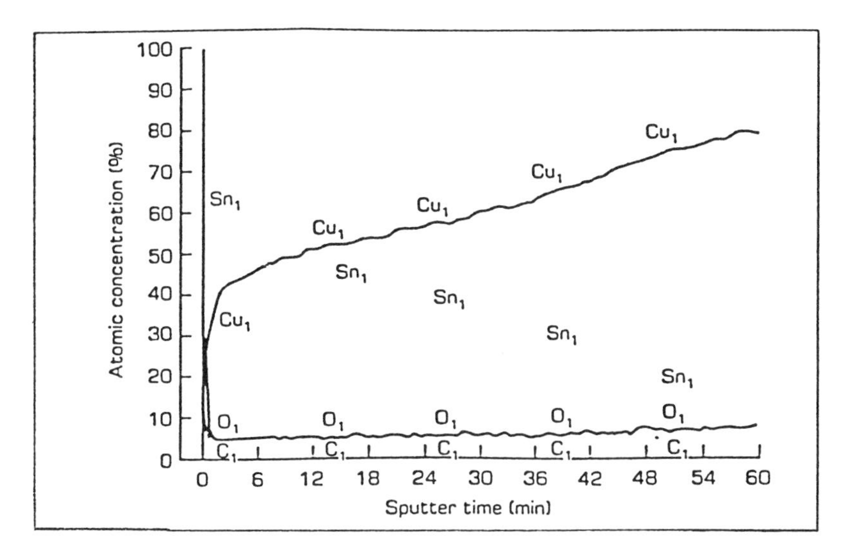

Figure 11.35 Auger depth profile of a TAB tape site showing the effect of excessive heat treatment on an Sn/Cu system.

Other applications include the determination of additives in motor oil by determining barium, zinc, phosphorus, calcium, and chloride and the determination of lead or sulfur in gasoline. Motor oil is used in automotive, airplane, and train engines, and lubricated bearings and other parts slowly wear away and enter the oil. The determination of the elements of the bearings in the used motor oil, which can be performed by x-ray fluorescence, is an indication of how badly the bearings are worn. The results reveal when the motors used in jet aircraft, locomotives, and automobiles are wearing out and in danger of failing. By identifying the metals that have worn off the motor, it is often possible to identify the actual engine component that is wearing. The latter can then be promptly replaced, thus providing a savings in time and money for the operator and increased safety for passengers.

In the rubber industry, the determination of the vulcanizing element, sulfur, can be done by x-ray fluorescence. This is a means of ensuring the production of high-quality rubber.

In space technology, the analysis of new alloys and ceramics can be carried out by x-ray fluorescence.

In all the foregoing determinations, the nondestructive nature of x-ray fluorescence is very important. This feature, coupled with the fact that sample preparation is seldom required, means that direct analysis can be performed in situ. Antiques and art objects can be characterized, and the original and copies of masterpieces distinguished from each other. Museums rely heavily on this method for examining works of art. In every case the sample is unaffected physically or chemically by the analytical process. This feature is of extreme importance to museums, aircraft manufacturers, and industrialists who use the technique for plant control.

The foregoing is a cross section of the types of samples that can be analyzed by x-ray analysis. A scan of the scientific literature will disclose many more applications of this nondestructive, multipurpose analytical technique.

BIBLIOGRAPHY

Anal. Chem., Application Reviews, June 1994.

Anderson, T. A., Evans, K. L. Carney, F., Carney, G., Berg, H., and Gregory, R., Small area x-ray fluorescence analysis of thin films, *Spectroscopy*, (1991), (Jan.): 28.

Jenkins, R., X-ray fluorescence, *Anal. Chem.*, (1984), *36*: 1009A.

Klockenkamper, R., Knoth, J., Prang, A., and Schwenke, H., Total-reflection x-ray fluorescence spectroscopy, *Anal. Chem.*, (1992), *4*(23): 1115A.

Lubhofsky, H. A., Schweikert, E. A., and Myers, E. A., *Treatise on Analytical Chemistry*, 2nd ed., Wiley Interscience, New York, 1986.

Robinson, J. W., *Practical Handbook of Spectroscopy*, CRC Press, Boca Raton, Fla., 1991.

Van Griekin, R. E., Mankowicz, A. A., *Handbook of X-ray Spectrometry*, Marcel Dekker, New York, 1992.

What's new in instrumentation, *Spectroscopy*, (1993), *8*(4): 16.

Willard, H. H., Merritt, L. L., Dean, J. A., and Settle, F. A., *Instrumental Methods of Analysis*, 7th ed., Van Nostrand, Princeton, NJ, 1988.

SUGGESTED EXPERIMENTS

X-Ray Absorption

11.1 Take several metal disks made of different pure elements, each about 1 $in.^2$ in area and ⅛ in. in thickness. It is important that the thickness of each disk be constant. The disks should be made from (a) Mg, (b) Al, (c) Fe, (d) Ni, (e) Cu, (f) Sn, and (g) Pb. Expose them to an x-ray beam simultaneously on a single sheet of photographic paper. Develop this paper. Note that the absorption of x-rays by each disk is proportional to the atomic weight of the particular element.

11.2 Place an unopened can of food in a vertical position, with x-ray film behind it. Expose the can to x-rays. Develop the film and note the level of food in the unopened can.

11.3 Take a piece of glass known to contain an entrapped bubble of air. Place it in front of x-ray film and expose it to x-rays. Note that the bubble can be detected on the film as a dark patch. Similar holes in metal castings (or other metal objects) can be detected by x-ray absorption, even though the holes are not visible to the naked eye.

X-Ray Fluorescence

11.4 Using the metal disks used in Experiment 11.1, record the fluorescence spectra of the different metals. Use a tungsten source. Identify the K_α and L_α lines. Note the relationship between the wavelengths of these lines and the atomic numbers of the metals.

11.5 Record the fluorescence spectrum of metal disks of unknown composition. By measuring the wavelength of the K_α line, identify the major components. [Use disks of brass, bronze, stainless steel (18:8), and aluminum and a quarter.]

11.6 Using silver powder and copper powder, make up powders with various ratios of these metals. Record the x-ray fluorescence spectrum of each mixture. Relate the silver/copper ratio to the ratio of the Ag K_α and Cu K_α lines. Measure the Ag/Cu ratio of various silver coins minted (a) before 1964 and (b) after 1964.

11.7 Record the fluorescence spectrum of sulfur. Locate the S K_β line. Record the fluorescence spectra of various grades of motor oil. Compare the sulfur contents of the oils by the intensities of the S K_β lines. (The oil may be held in a polyethylene bag during analysis.)

X-Ray Diffraction

11.8 Use a tungsten or copper source. Load crushed NaCl powder into the sample holder. Measure the diffraction pattern of the NaCl. Repeat with KCl. Note the difference in spectra.

11.9 Make several mixtures of NaCl and KCl powders and record the spectrum of each mixture. Note that each spectrum is that of NaCl and KCl. Measure the sodium K_β peak and the potassium K_β peak. Plot the relationship between the Na K line intensity and the percentage of Na in the powder. Repeat for potassium.

PROBLEMS

11.1 Draw a schematic diagram of an x-ray source and describe its operation.

11.2 What is the origin of the K_α x-ray lines?

11.3 State Moseley's law. What was the importance of this law to chemistry in general?

11.4 What are the three major analytical fields of x-ray spectroscopy? State three analytical applications of each field.

11.5 The intensity of x-ray fluorescence is weak compared to some other spectroscopic techniques. What instrumental changes are made to maximize the fluorescence signal?

11.6 What elements can be detected by x-ray fluorescence? What elements cannot be detected by x-ray fluorescence in the air? Explain.

11.7 Can x-rays from a tungsten target be used to excite copper atoms? Can x-rays from a copper target be used to excite tungsten atoms? Explain.

11.8 Describe an experiment you could use to calculate the percentage of crystalline material in a polymer.

11.9 What is the relationship between the wavelengths of the absorption edges and the related emission lines?

11.10 Describe an x-ray source.

11.11 Describe a filter monochromater.

11.12 Derive the Bragg equation.

11.13 (a) What is the signal-to-voltage relationship between ionization counter, proportional counter, and Geiger counter? (b) Which is most sensitive?

11.14 Describe a scintillation detector.

11.15 (a) Describe a Si (Li)-drifted detector. (b) What is an electron–hole pair?

11.16 (a) What is the mass absorption coefficient? (b) What is the relationship between the mass absorption coefficient, atomic number, and atomic weight at the same wavelength?

11.17 Describe the equipment used to take x-ray diffraction patterns.

11.18 What is an x-ray powder diffraction pattern?

11.19 Describe a goniometer.

11.20 (a) Describe a curved crystal. (b) Why is it used?

11.21 What are the main analytical uses of x-ray fluorescence?

12
Surface Analysis

Surface analysis has become increasingly important in recent years because the various methods reveal the elemental analysis, the distribution of the elements, and sometimes the chemical forms of the elements in a surface layer. This layer may vary from one molecule deep to several microns deep, depending on the technique used. Using combinations of surface techniques, such as ESCA, SIMS, and ISS-SIMS, speciation analysis is possible.

Surface analysis has found great use in understanding important fields such as corrosion, catalysis, or the functioning and breakdown of semiconductors and chips.

In medicine, it has been used to study bone structure, the surface of teeth (indicating why SnF_2 fights tooth decay), and the mechanism of toxicity of dusts and various fibers. In short, it has many valuable applications, and this list will only grow with time.

In the discussion on x-rays (Chap. 11) it will be remembered that the source of an x-ray photon is an atom that is bombarded by high-energy electrons or photons. The latter displace an inner shell electron, which is ejected from the atom, leaving an ion with a vacancy in an inner shell. An electron from an outer shell then drops into the inner shell and an x-ray photon is emitted simultaneously. The energy of the photon is equal to the difference between the energy of the orbital the electron was in originally and that of the one of which it descends. As stated in Chapter 11, Section C.3, the energy levels of these two orbitals are almost independent of the chemical form of the atom, combined or otherwise. However, we know from Chapter 6 that the energy of a valence electron varies with the chemical form and chemical environment of the combined atom and provides the basis for ultraviolet absorption analysis. This variation is reflected in the energies of the inner shell electrons, but the changes in energy involved are very small compared to the energy of the emitted x-rays themselves. The slight differences in x-ray wavelength are extremely difficult to measure, since they are such a small fraction of the nominal wavelength of the x-rays generated. It is therefore normally accepted that the energies of the emitted x-rays are independent of the chemical form of the generating atoms and cannot be observed except under very high resolution.

However, the energy E of the original electron ejected from the atom is the difference between the energy E_I of the impinging electron (or photon) and the energy E_P required to remove the electron from the atom, that is, $E = E_I - E_P$. The energy E_P will be slightly different, depending on the chemical form of the atom. We can determine this small difference by making the energy E_I of the impinging electron just slightly greater than the energy E_P required to eject the electron. The residual energy E of the emitted electron will then be small, but any variation in E_P will produce a larger relative variation in the energy E of the emitted electron, and in this way small differences in E_P can be measured. For example, if the energy of the K_α line for Al is 1487 eV, the effect of chemical environment may change this by 2 eV, resulting in Al K_α at 1485 eV. The relative shift (2/1487) is slight, and the line would be difficult to distinguish from the original 1487 eV line in x-ray emission. If the energy of the *impinging electrons* generating x-rays is 1497 eV, then the energy of the ejected electron is $1497 - 1487 = 10$ eV. But if the chemical environment changes the energy needed to eject the electron to 1485 eV, then the energy of the emitted electron is $1497 - 1485 = 12$ eV. It is easier to distinguish between electrons with 10 eV energy from electrons with 12 eV energy than electrons with energies of 1487 and 1485 eV.

Based on this phenomenon, the field of *electron spectroscopy for chemical analysis* (ESCA) was developed. A companion field, *Auger electron spectroscopy* (AES), was also developed simultaneously. Since the electrons ejected in these two techniques are necessarily of low energy so that smaller energy differences can be easily measured, the electrons cannot escape from any significant depth in the sample. The phenomenon is therefore confined to atoms, combined or otherwise, which are on the surface of the sample and provides a method of *surface analysis*. A number of techniques have been developed for surface analysis, and for convenience these will all be treated in this chapter. In order to avoid confusion, the names of these techniques are listed in Table 12.1. These techniques are frequently quite different in physical approach, but all provide information on surface atoms.

A. ESCA

When an x-ray or electron beam of precisely known energy impinges on sample atoms (combined or otherwise), inner shell electrons are ejected and the energy

Table 12.1 Techniques for Surface Analysis

Abbreviated name	Full name
ESCA	Electron spectroscopy for chemical analysis
AES	Auger electron spectroscopy
ISS	Ion scattering spectroscopy
SIMS	Secondary ion mass spectrometry
IMMS	Ion microprobe mass spectrometry

of the ejected electrons is measured (Fig. 12.1). This is the phenomenon on which ESCA is based.

The energy of the escaping electron is designated as E_L. (This proposes that the electron comes from the *L* shell.) The binding energy of this electron is given by the equation

$$E_b = h\nu - (E_L + C) \tag{12.1}$$

where

E_b = binding energy of the electron
$h\nu$ = energy of the photon (either x-ray or vacuum ultraviolet)
E_L = kinetic energy of the escaping electron
C = an instrumental constant that represents the effects of local magnetic fields on the escaping electron

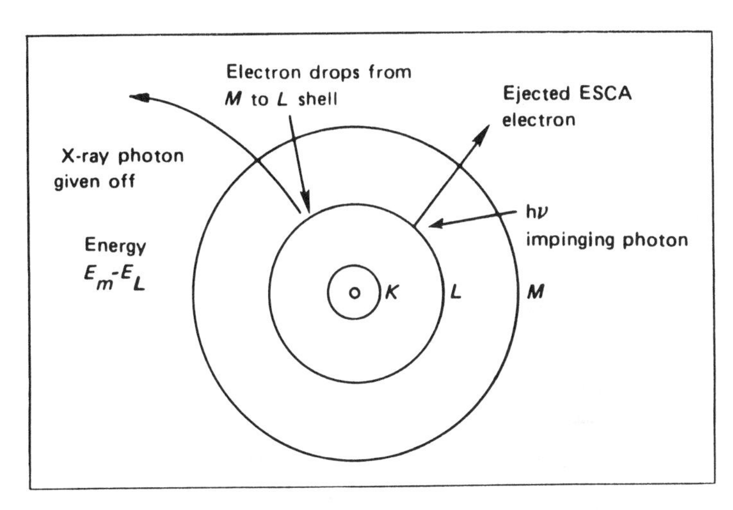

Figure 12.1 An impinging photon ejecting an inner shell electron. The residual energy of the electron is the difference between the energy of the impinging photon and the total energy required to remove the ESCA electron. An x-ray photon is also emitted in the process.

The local fields are equal to $e\phi$, where

e = electronic charge
ϕ = work function of the spectrometer

In practice, we use an x-ray source of well-defined energy; the energy can be determined by measuring the wavelength λ or frequency v of the x-rays. We also measure E_L and C, the latter being constant for a given instrument. Based on these measurements, we can calculate E_b, the binding energy of the sample electron. Changes in the chemical composition of the atom cause changes in E_b. Consequently if we measure E_b, we are able to deduce information about the chemical composition of the specimen atoms in the molecule.

So that we may get an idea of the energy levels involved in ESCA, Fig. 12.2 indicates the energies associated with valence electrons and inner core electrons for a typical atom in the middle of the period table. Superimposed on these levels are the energies of the lines emitted for He(I) and He(II), and Al K_α and Mg K_α. It will be realized, of course, that in order to eject an electron from a K

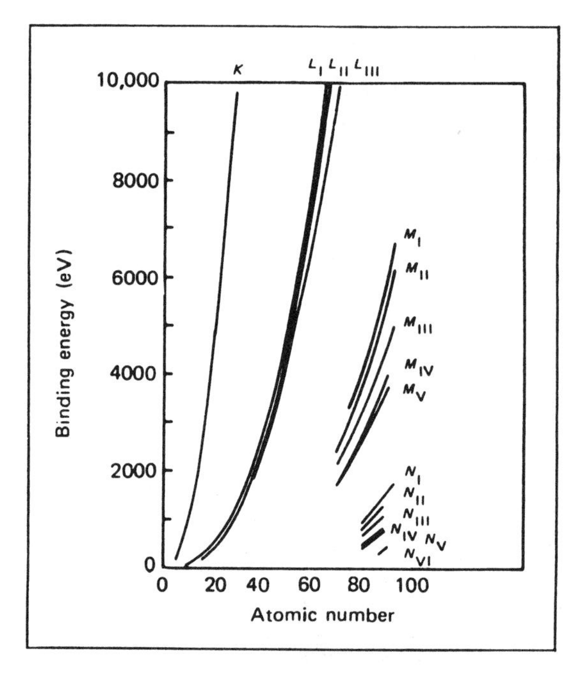

Figure 12.2 Relative energy levels of an atom with K, L, and M shells and the K_α lines of Al, Mg, and Cr. The K_α lines for Al and Mg would eject the L electrons but not the K electrons in this instance. (From *Handbook of Chemistry and Physics*, Chemical Rubber Company, 1970.)

orbital, the energy of the radiation falling on it must be greater than the energy required to displace it. The energies of the electrons in the K, L, and M orbitals of various elements, as derived from Moseley's law and based on experimental observations, are shown in Fig. 12.3. Superimposed on these levels are the energies of the lines emitted by Cr K_α Al K_α and Mg K_α. We can quickly see from Fig. 12.2 whether the energy used (e.g., Mg K_α) has sufficient energy to displace the inner shell electrons of the sample atoms.

Since E_b, the original binding energy of the emitted electron, depends on the energy of the electronic orbit and the element from which the electron is emitted, it can be used to identify the element involved. In addition, the chemical form or environment of the atom affects the energy level E_b to a much smaller but measurable extent. These minor variations give rise to the *chemical shift* and can be used to identify the valence of the atom and sometimes its exact chemical form. Quantitative measurements can be made by determining the intensity of the ESCA lines of each element.

1. Equipment

Schematic diagrams of two types of commercial ESCA apparatus are shown in Figs. 12.4 and 12.5. Each one consists of three major components: (1) the radiation source, consisting of an x-ray source and a means of providing highly monochromatic x-rays; (2) the energy analyzer, which resolves the electrons generated from the sample by energy; and (3) and detector, which measures the intensity of the resolved electrons.

a. Radiation Source

The standard x-ray source is shown in Chapter 11, Fig. 11.6; however, in ESCA it is very important that the x-ray source be monochromatic, with a linewidth

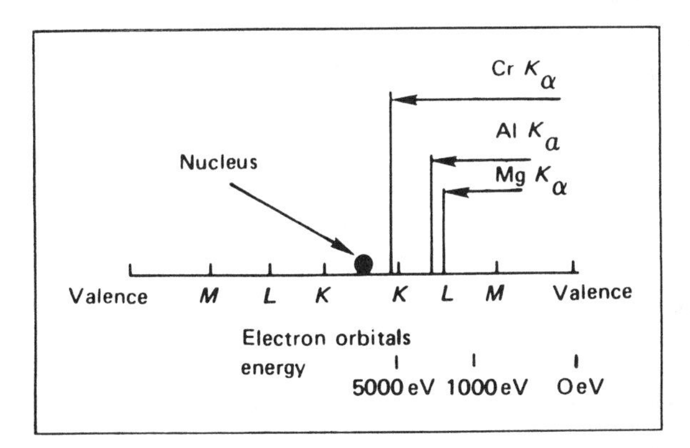

Figure 12.3 Binding energies of inner shell electrons.

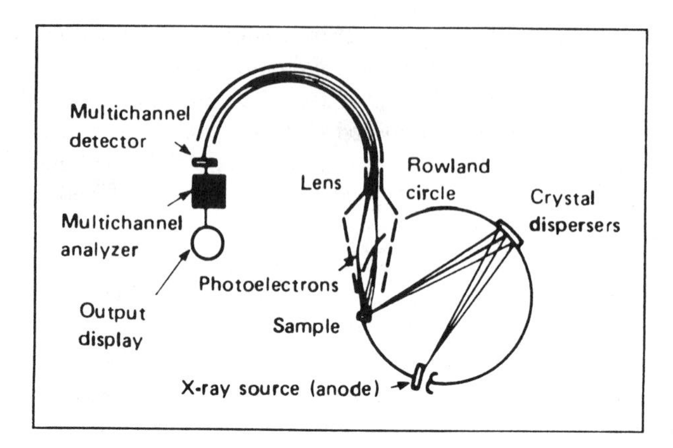

Figure 12.4 Schematic diagram of an ESCA instrument using the Rowland circle. (Courtesy of Hewlett Packard.)

extending over as narrow an energy range as possible. It can be seen from Eq. (12.1) that any variation in the energy of the impinging x-rays will produce a similar variation in the energy of the ejected ESCA electron. There are several ways to reduce the bandwidth of the source. One of the most accurate is to use a crystal monochromator and a Rowland circle, as shown in Fig. 12.4. Although

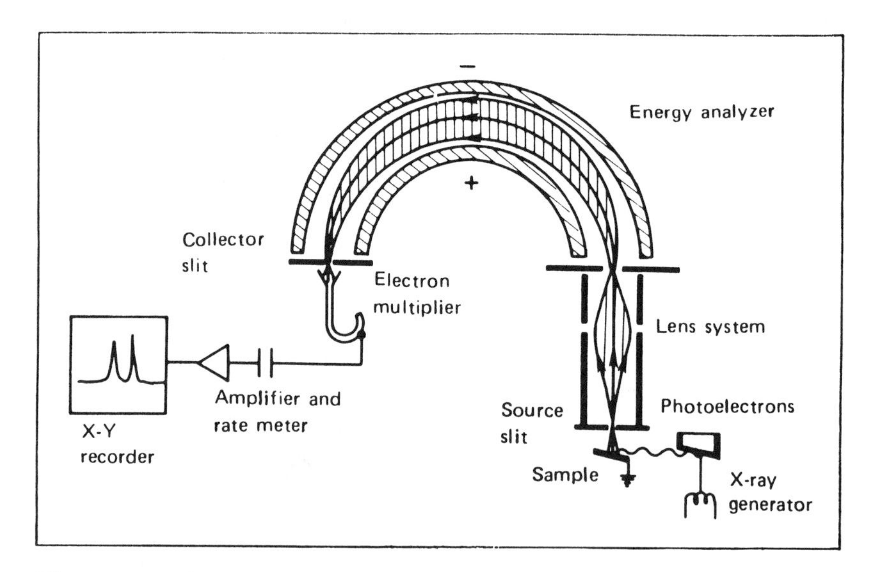

Figure 12.5 Schematic diagram of an ESCA system. (Courtesy of AEI Scientific Apparatus.)

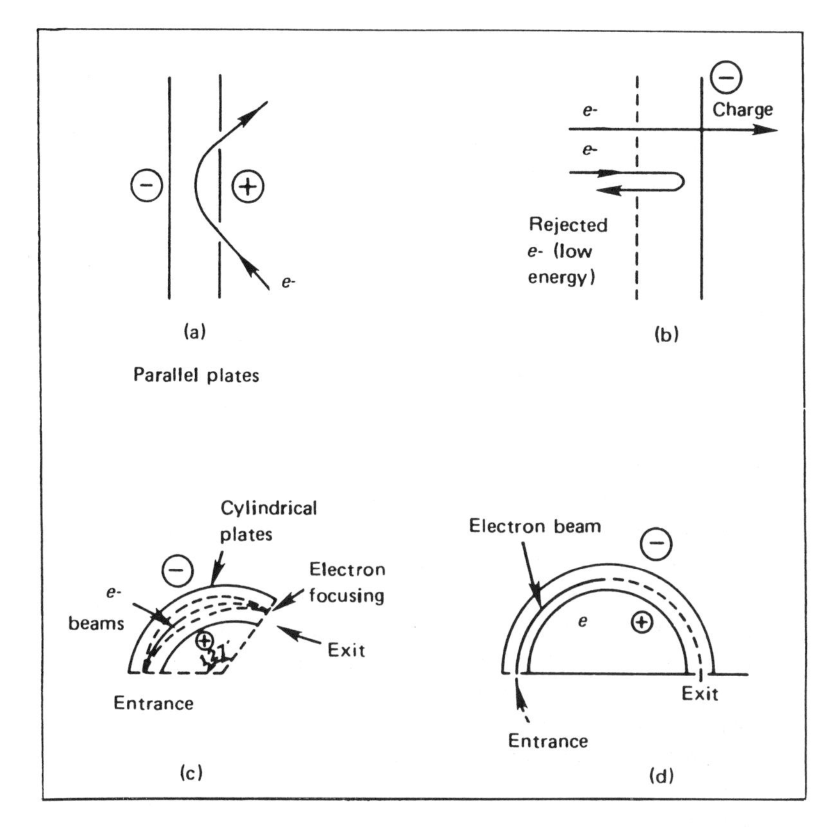

Figure 12.6 Schematic diagrams of electron energy discriminators.

this instrument gives very monochromatic radiation, unfortunately the intensity of the radiation is reduced in the process.

Other means of intensifying the radiation have been used to advantage. A common procedure is to use either aluminum or magnesium targets as the source. Each of these elements has narrow K emission lines. If an aluminum target is used, an aluminum window placed after the target removes much of the K_β and background radiation from the aluminum source. A magnesium filter can be used in a similar fashion with a magnesium source.

One problem with using x-ray beams as an energy source is that they cannot be focused; therefore, relatively large surfaces are examined. This problem can be overcome by attaching the sample to an aluminum foil, which is bombarded with an electron beam. The latter can be focused to a small point and excites the Al foil at that point. The excited Al emits Al K_α lines with a narrow wavelength range (narrow energy range). Only the sample in the immediate vicinity is exposed to this radiation. The net result is that *samples in a small region* can be excited with *narrow band radiation*, thus increasing the potential of the method.

An alternative method is to take an electron source, such as that generated by a heated cathode, and accelerate the electrons in an electrostatic field of known energy. This provides electrons of known energy. Unfortunately, there is some spread in this energy because the electrons from the heated cathode have a range of kinetic energies before acceleration, which persists after acceleration.

The energies can be filtered with energy discriminators as shown in Fig. 12.6. Figure 12.6a shows two parallel charge plates with two openings. The electrons with the required energy enter and leave the holes as in a slit system. Electrons with other energies do not penetrate the second hole. Figure 12.6b shows a second type of energy discriminator, which is a simple grid discriminator. Electrons with insufficient energy are repelled by the second grid and do not penetrate. This system only discriminates against electrons with low energies. Any electrons that have sufficient energy penetrate the second grid. A third system uses cylindical plates (Fig. 12.6c). If the angle between the planes of the entry and exit slits is 127.17° ($\pi/\sqrt{2}$ rad), a double focusing effect is obtained and the intensity of the electron beam is maintained as high as possible. Electrons with the incorrect energy are lost either to the sides of the cylindrical plates or on the sides of the exit slit. Similar modifications have led to the development of a 180° cylindrical system rather than parallel plates (Fig. 12.6d). This system does not provide such fine-tuning of the energy bandwidth, but it increases the intensity of the electron beam. This is another example of gaining in beam intensity but losing in energy discrimination, or gaining in power and losing in resolution.

b. Energy Analyzers

The energy discriminators described above are not accurate enough to be used as energy analyzers. Energy analyzers are equivalent to the monochromators used in spectroscopy. The most commonly used energy analyzers incorporate an electrostatic field which is either symmetrical or spherical. These systems are in essence an extension of the filters shown in Fig. 12.6b and d. One system is shown schematically in Fig. 12.7; it is based on the system of Fig. 12.6a. The plates form a complete cylinder, thus providing an efficient electron-trapping system and maintaining resolution. The second system is shown in Fig. 12.8 and uses a spherical electrostatic field; this is an extension of the system in Fig. 12.6c. In practice, the best results are obtained when the surface of the sample and the acceptance cross section of the energy analyzer form an angle of 42°.

Other analyzers have been designed based on the magnetic deflection of electrons, but in general these have not been very successful because of the difficulty of maintaining a uniform magnetic field.

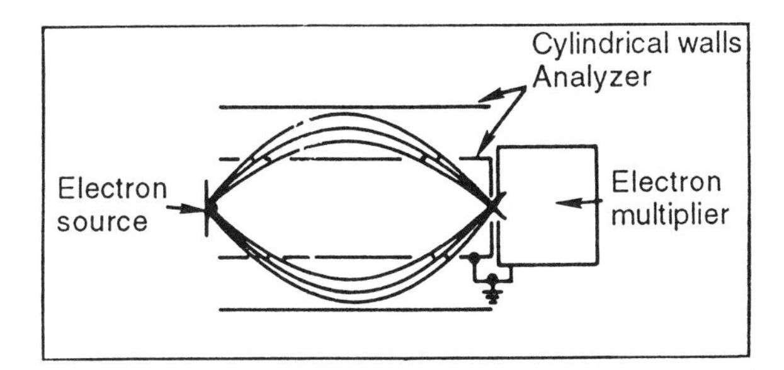

Figure 12.7 Cylindrical electrostatic electron energy analyzer.

c. Detectors

The most commonly used detector is the electron multiplier. There are several designs of electron multipliers; one is similar in all respects to the photomultiplier described in Chapter 6, Section C.2.c. The single most important difference is that the initiating impulse comes from an electron rather than a photon. A system of diodes is used as in a photomultiplier. There a number of acceptable different patterns for arranging the diodes.

Another widely used system is the channel electron multiplier (Fig. 12.9). This system utilizes a high-resistance conductor with a potential of 2–3 kV. The system may be linear or curved as in Fig. 12.9. The charged surface acts as a continuous diode and permits amplification throughout the system. Signal electrons enter one end and an amplified shower exits at the other end. Gains of up to 10^8 are possible in this system, as with the UV photomultiplier.

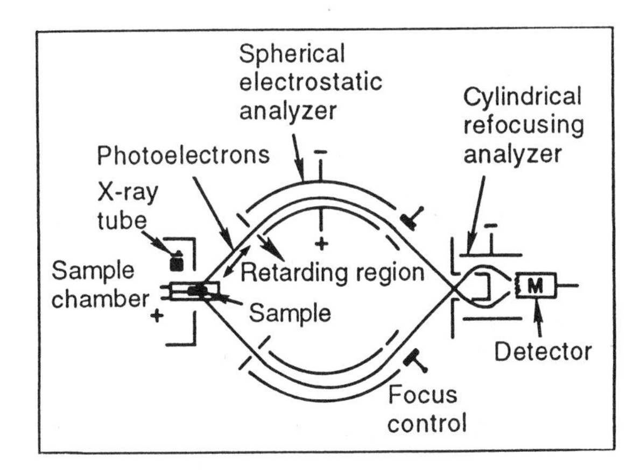

Figure 12.8 Spherical electrostatic electron energy analyzer.

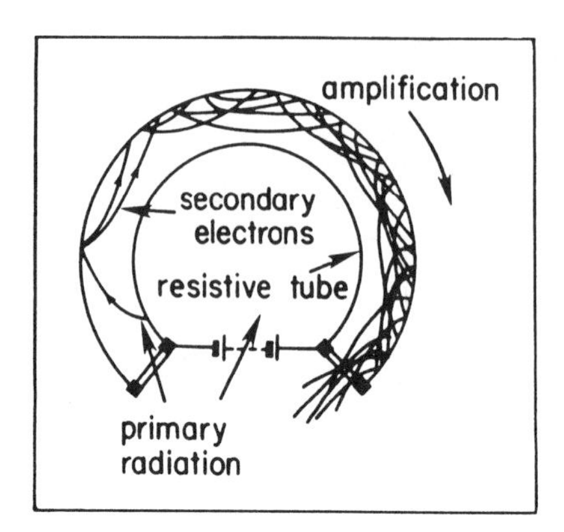

Figure 12.9 Channel electron multiplier

d. Other Necessary Components of the Equipment

The ejected electrons necessarily have low energy. They are affected significantly by local magnetic fields, including the earth's and those of any stray electrical impulses as generated by wiring to lights, equipment, elevators, and so on. These stray fields must be neutralized in the critical parts of the instrument in order to obtain useful data. One method is to enclose the critical regions with high-permeability magnetic material, which shields the sample from stray magnetic fields. Another method is to use Helmholtz coils, which produce within themselves a homogeneous field. This field may be made exactly equal and opposite to the earth's magnetic field. A feedback system to the coils can also be used; this senses variations in local magnetic fields and varies the current in the Helmholtz coils, neutralizing transient magnetic fields as they arise. This system not only neutralizes the earth's magnetic field but also local fields as they are generated.

e. Electron Flood Gun

In practice, when the sample is irradiated, electrons are ejected. An electron takes with it a negative charge, leaving the sample positively charged. Depending on the conductivity of the sample, this positive charge builds up at a steady but unpredictable rate on the surface of the sample, changing the work function of the sample itself and therefore the net energy of the ejected electrons. In practice, this would lead to a variation in the observed energy of the ejected electrons and erroneous results would be derived. It is therefore highly desirable to eliminate this variable charge.

This problem can be overcome by flooding the surface of the sample with low-energy electrons, which neutralize the positive charge built up on the surface. This is done by using an electron flood gun. Of course, these electrons in turn affect the work function of the surface to some extent, but the effect is constant and reproducible data are obtainable over an extended period of time.

f. High-Vacuum System

Since the electrons involved in ESCA are of low energy, they must not collide with other atoms or molecules in the system. This is best handled by using a very high vacuum system. Pressures of 10^{-8} Torr are typical in the analyzer sections. This also means that the sample must be kept under high vacuum. In practice, when the sample is put into the instrument, it is normally put into a sample cell, which must be then evacuated to high vacuum before it can be opened to the rest of the instrument, that is, the source, the energy discriminator, and the detector. This may present a problem when numerous samples are to be analyzed. Many instruments utilize compartments that will accommodate a number of samples, which, after loading, can be rotated in sequence into position without breaking the seal and losing vacuum.

2. Analytical Applications

The binding energies of electrons in the *K*, *L*, and *M* shells are illustrated in Fig. 12.3. These correspond to the x-ray emission and absorption spectra discussed in Chapter 11. For low atomic weight elements ($Z < 30$) two separated energy peaks are observable corresponding to the *K* and *L* shells. For elements with atomic number between 35 and 70 the L_I, L_{II}, and L_{III} lines are responsible for the observation of a triplet. For elements with atomic number greater than 70 the pattern includes *M* and *N* shell electrons. From these data it can be readily seen that qualitative identification of the elements in a sample is possible based on identification of the binding energies. In addition, there is a shift brought about by chemical environment. For example, the work of Hercules on sulfur ($Z = 16$) shows that the normal binding energy for the 2_P electrons is about 165 eV. However, changes in chemical environment cause the binding energy to range from 160 to 168 eV (Fig. 12.10). In a similar fashion, the chemical environment of carbon compounds influences the carbon ESCA spectrum as shown in Fig. 12.11.

An interesting example of ESCA analysis is lunar soil exposed to magnesium K_α radiation. The major components of the soil can be clearly identified (Fig. 12.12). It is common for samples to have absorbed oxygen and nitrogen from the earth's atmosphere on their surface. But these are not really part of the surface being examined. The contamination can lead to erroneous results if not recognized and corrected for. One method of cleaning the sample surface is to use ion bombardment with an argon ion gun. It is imperative that, once cleaned, the sample be kept in a high vacuum to prevent recontamination with oxygen and nitrogen, which are rapidly adsorbed.

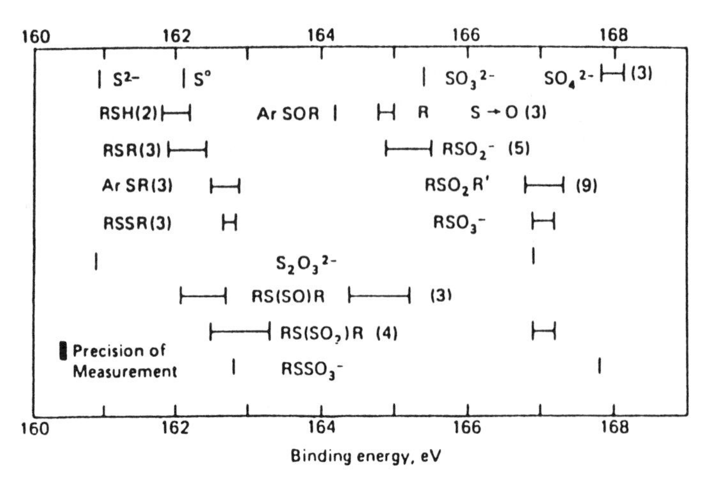

Figure 12.10 Response of sulfur 2_P electrons. Note how the different chemical forms change the binding energies. [From D. M. Hercules, *Anal. Chem.*, (1970), *42*(1): 20A.]

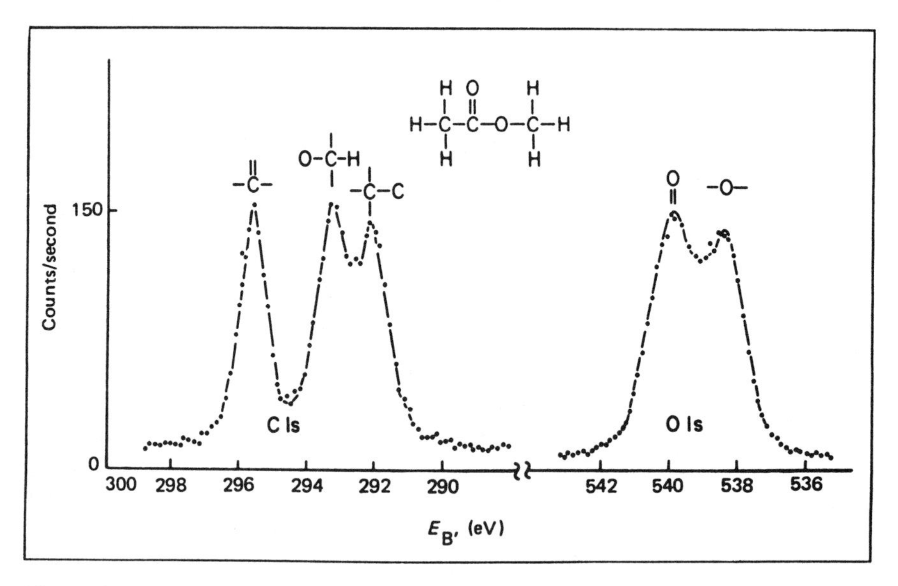

Figure 12.11 The ESCA signal for different types of carbon in methyl acetate. [From K. Siegblan, *Endeavour*, (1973), *51*: 32.]

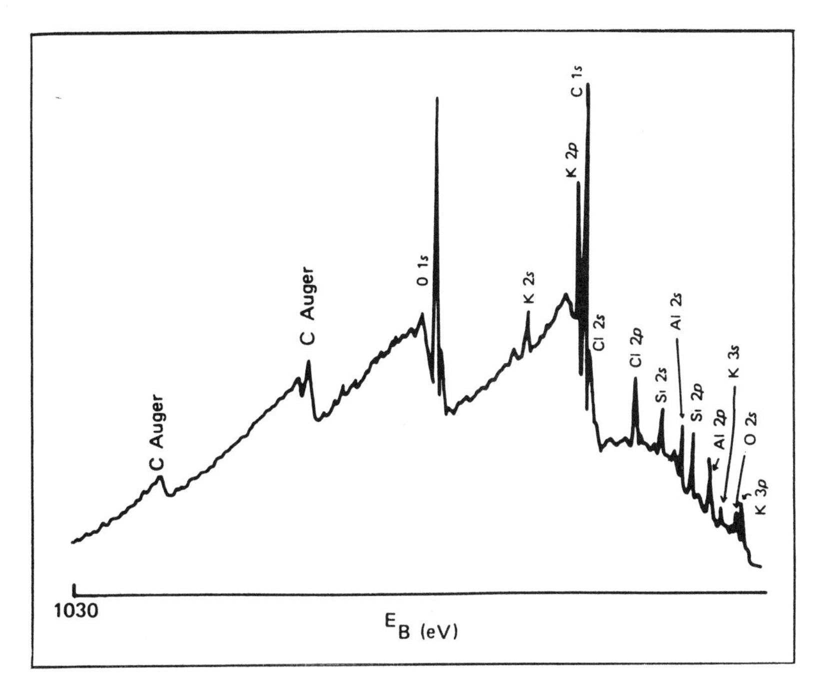

Figure 12.12 The ESCA spectrum of *lunar soil* taken on *Surveyor*. [From *Am. Lab.*, (1972), *4*(2): 7.]

Another illustration of an ESCA spectrum is shown in Fig. 12.13, which shows lead bromide and lead chloride using magnesium K_α radiation. The spectra were obtained using a McPherson instrument capable of high resolution. In this case 0.07 eV was the difference between the maxima for these two compounds. This resolution approaches the practical limitations of the procedure.

Another interesting example is shown in Fig. 12.14. A sample of pure lead was evaporated onto a surface, which was then exposed to air. The ESCA spectra were obtained at various time intervals. The spectra clearly indicate the formation of lead oxide on the surface, going through the intermediate of PbO and finally, after a number of hours, to PbO_2. Clearly this technique is capable of studying surface phenomena in situ. In summary, ESCA can be used for qualitative and quantitative characterizations of surfaces.

Electron spectroscopy for chemical analysis is capable of detecting extremely small quantities of material, for example 10^{-12} g, but the absolute sensitivities reported are somewhat misleading and must be carefully utilized. It must be remembered that ESCA is purely a surface phenomenon, and although it is common for surface areas of 1 cm^2 to be examined at one time, the sample depth exposed (see, e.g., Fig. 12.37) is only about 10^{-8} cm; therefore the total

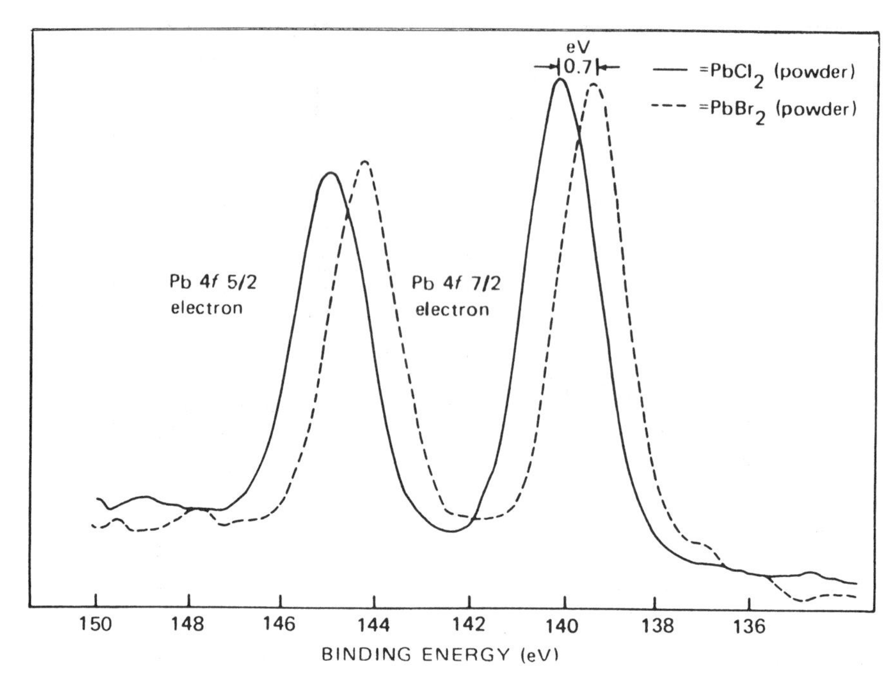

Figure 12.13 High-resolution ESCA spectrum of $PbBr_2$ and $PbCl_2$ standards obtained with a Mg K_α excitation source.

quantity of sample exposed is very small, that is, about 10^{-8} g, depending on the sample's density. Hence the concentrations of sample elements on the surface which can be detected and measured are usually in the percentage range. So, although the ESCA is capable of very high absolute sensitivity limits, its relative sensitivity limits are not high. It is not possible to detect impurities at the ppm or ppb level on the surfaces of samples without separate pretreatment.

B. AUGER SPECTROSCOPY

The process of Auger spectroscopy is similar in many respects to that of ESCA, but it differs in one fundamental step. As with ESCA, the sample is irradiated with either x-ray photons or accelerated electrons. These eject inner shell electrons, leaving a vacancy in the inner shell and an electron from the outer shell falls into the inner shell as in the ESCA process. In x-ray emission the energy balance is maintained by the emission of an x-ray photon with an energy equal to the difference between the two energy levels involved; however, in the Auger process the released energy is transferred to a second electron, which is then emitted. This is the *Auger electron*. The actual process by which the energy

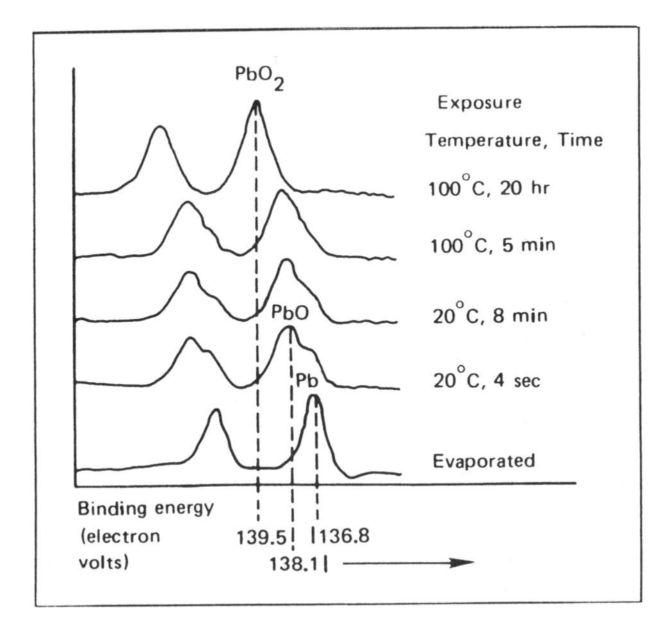

Figure 12.14 Change of the surface of a pure lead sample exposed to air and heated. [From *Endeavour*, *32*:51 (1973).]

is transferred is not clearly understood but can be represented schematically in Fig. 12.15.

In this example an electron from the *K* shell was ejected by bombardment. An electron from the *L* shell descended to the *K* shell, simultaneously transferring energy to a second electron in the *L* shell. This second electron was ejected as an Auger electron. This Auger electron is termed a *K-L-L* Auger electron. This nomenclature arises from the fact that an electron was ejected from a *K* shell followed by a transition from an *L* to a *K* shell, simultaneously liberating and *L* Auger electron. It can be seen that other Auger electrons, such as *L-M-M*, are also possible.

The Auger process and x-ray fluorescence are competitive processes for the liberation of energy from bombarded atoms. In practice, it is found that the Auger process is more likely to occur with low atomic number elements, but this probability decreases with atomic number. In contrast, x-ray fluorescence is unlikely with elements of low atomic number but increases in probability with increasing atomic number. This is illustrated in Fig. 12.16. The energies involved in Auger spectroscopy are similar in all respects to those of ESCA, since the same atomic shells are involved. A summary of Auger energies is shown in Fig. 12.17.

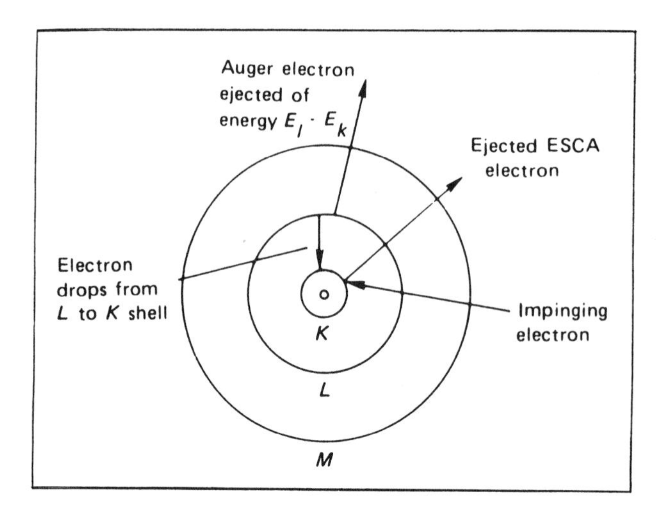

Figure 12.15 Auger process. An impinging electron ejects an inner shell electron, leaving an incomplete K shell. An electron from the L shell drops into the K shell, and simultaneously an electron is emitted from the L shell with energy $E_L - E_K$ to maintain an energy balance. This process competes for the process producing x-rays shown in Fig. 12.1.

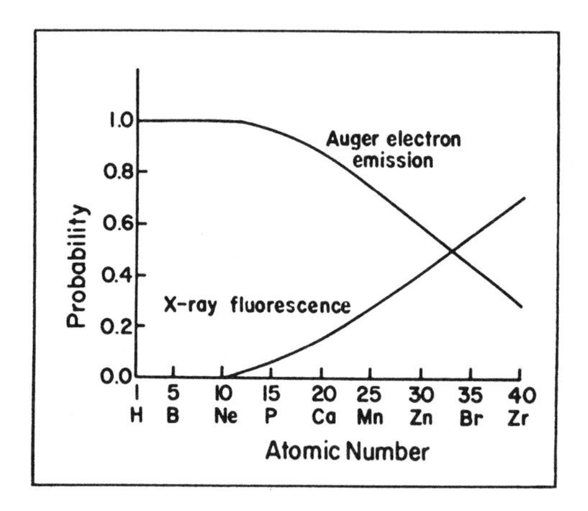

Figure 12.16 Relationship between atomic number and the probabilities of x-ray fluorescence and Auger electron ejection. [From D. M. Hercules, *Anal. Chem.*, (1970), *42*(1): 20A.]

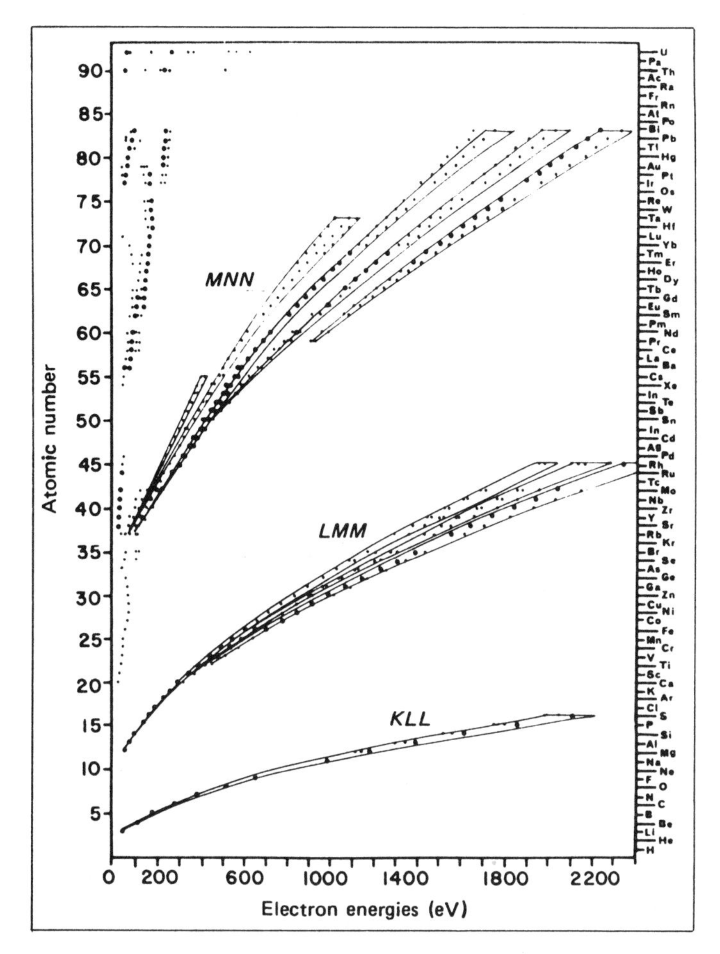

Figure 12.17 Principal Auger electron energies. (Courtesy of Physical Electronic Industries.)

1. Equipment

The equipment used in Auger spectroscopy is very similar to that used in ESCA. The same kinds of radiation sources, sample cells, and detection systems are used. A schematic diagram of a typical instrument is shown in Fig. 12.18. There is, however, one significant difference: the signal is much less intense. This results in a small signal compared to the background, which is difficult to handle. The problem is partially overcome by differentiating the signal, which is equivalent to measuring the slope of the curve. As the curve increases, the slope is

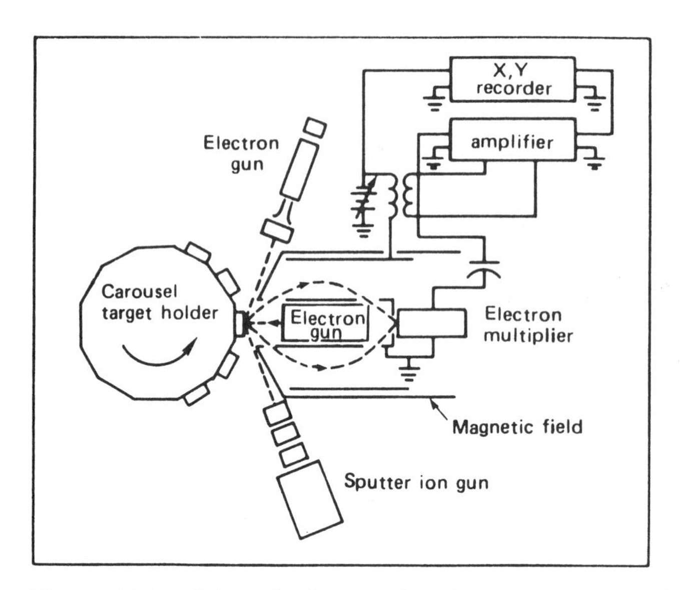

Figure 12.18 Schematic diagram of an Auger spectrometer (Phillips Electronics Industries, Inc.) with an electron flood gun and a sputter ion gun. The entire apparatus shown is enclosed in a vacuum. (Courtesy of Perkin Elmer Corporation.)

positive until the inflection point is reached. The differential curve is Al-a maximum from this point the *rate of increase* of the slope decreases until the maximum of the curve is reached. Here the slope is zero and the differential curve crosses zero. As the curve passes through the maximum, the slope becomes negative until the second inflection point, where the differential (slope) is at a minimum. After the inflection point the slope steadily approaches that of the background and the differential approaches zero. This is illustrated in Fig. 12.19. This procedure greatly facilitates the ability to locate and identify Auger signals of the associated elements. Auger spectroscopy therefore provides a method of elemental analysis of surfaces. A typical Auger signal and the differential of Heat signal is shown in Fig. 12.20.

As with ESCA, some chemical shift as experienced when elements are in different chemical environments; however, in contrast to ESCA, two shells of electrons are involved in a *K-L-L* Auger electron. If the chemical shift of both shells is the same, then no difference will be observed in the Auger electron. Although these energy shifts are usually not quite the same, the difference is very small and very difficult to detect. A further complication is the lack of sensitivity of the procedure, which demands that the x-ray source be of high intensity in order to obtain the maximum signal. This results in the use of broader-band signals, and as a consequence resolution is lost, making chemical shifts

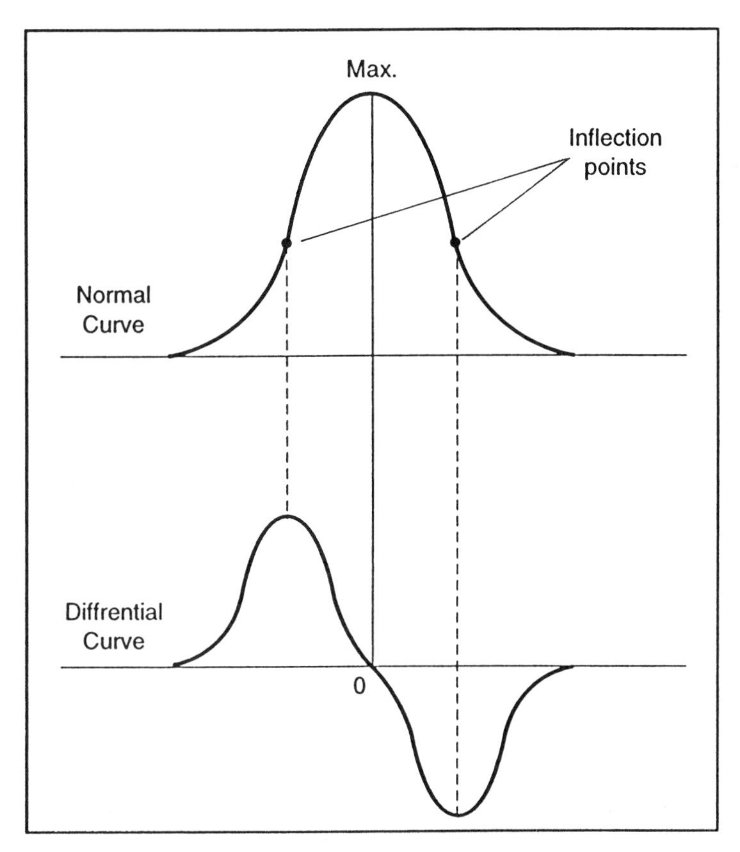

Figure 12.19 A normal curve and a differential of the curve. In the normal curve the slope is zero at the maximum. Hence at the maximum the value is zero in the differential curve. Also note that the maximum and the minimum in the differential curve are at the inflection points of the normal curve. Hence the slope changes from negative to positive or vice versa.

even more difficult to detect. It is therefore uncommon to use Auger spectroscopy for the identification of chemical form.

This problem, however, has some positive characteristics when it is realized that the Auger electron is ejected by the energy released by the neighboring electron that drops into the inner orbital. The energy released is a function of the particular atom and not of the energy used to energy used to eject the initial electron from the inner orbital. Therefore the energy of the Auger electron, is to a limited extent, independent of the energy range of the radiation source falling on the sample.

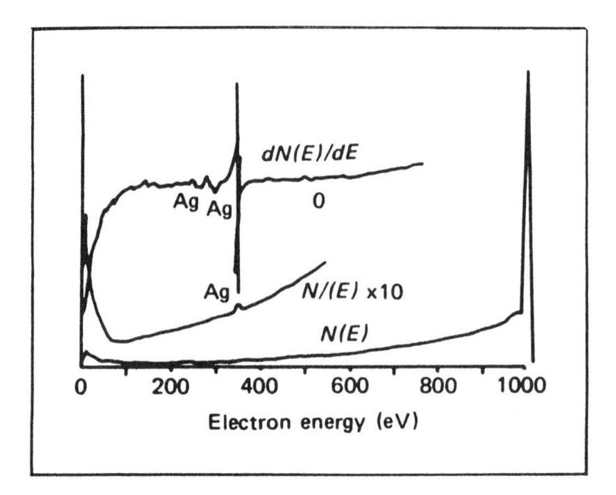

Figure 12.20 Direct Auger signal (bottom) and its differential (top). Note how the differential signal facilitates interpretation. [From R. E. Weber, *Res. Develop.*, (1972), *10*:22.]

2. Analytical Applications

Figure 12.21 shows the Auger spectrum of stainless steel after heating at 750°C for 10 min. It can be seen that the major components, ion, nickel, and chromium, are relatively unchanged by the heating. However, the carbon that was on the surface of the stainless steel before heating disappeared after heating. In addition, sulfur is present on the surface after heating. This may be due to either diffusion from inside the stainless steel to the surface during heating or adsorption from the atmosphere.

Figure 12.22 shows the surface of a spent methanation catalyst. Two areas were examined: one with a high sulfur content and the other with a low sulfur content. It can be seen that the nickel concentrations vary significantly in these areas, which may have contributed to the poisoning of the catalyst.

In more sophisticated equipment the surface examined can be exposed to an x-ray source or electron beam that systematically scans the entire surface area. By monitoring the intensity of the Auger spectrum of a particular element during scanning, it is possible to *map its distribution* on the surface examined. An example of this technique is shown in Fig. 12.23, which reveals the concentrations of sulfur and nickel on the spent methanation catalyst of Fig. 12.22. In another example, Fig. 12.24 shows the distribution of molydenum, gold, and oxygen on an integrated circuit. Using this technique it is possible to map the distributions of these three elements and their positions relative to the circuit.

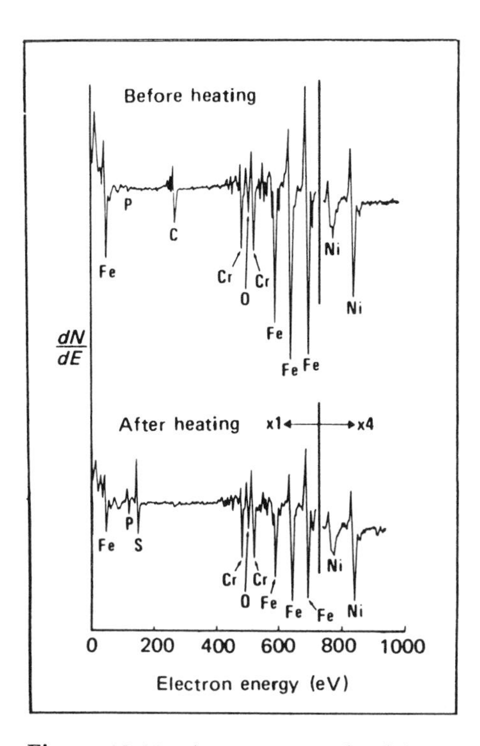

Figure 12.21 Auger spectra of stainless steel before and after heating in vacuum at 750°C for 10 min. Notice the loss of carbon and the diffusion of phosphorus and sulfur to the surface.

3. Examination of Lower Levels by Surface Stripping

Another valuable application of both ESCA and Auger spectroscoy has been developed by using an ion beam to progressively strip off the surface of a sample under controlled conditions from a sample. After each stripping, spectra can be measured and the distribution of elements recorded. The results show the change in distribution of different elements with depth. This reveals the heterogeneity of a sample, particularly near its surface.

A schematic diagram of a sample undergoing surface stripping by the sputtering technique is shown in fig. 12.25. The ion beam may be turned on for one or successive periods and the surface examined after each exposure. Figure 12.26 shows the results of such an investigation on the surface of Nichrome wire after it had been heated to 450°C in air. It can be seen that the surface was depleted on silicon until 150 Å had been stripped off. The silicon content then rose steadily. The concentration of nickel, on the other hand, was essentially zero at the outer surface, rose to a maximum at about 120 Å, and then decreased.

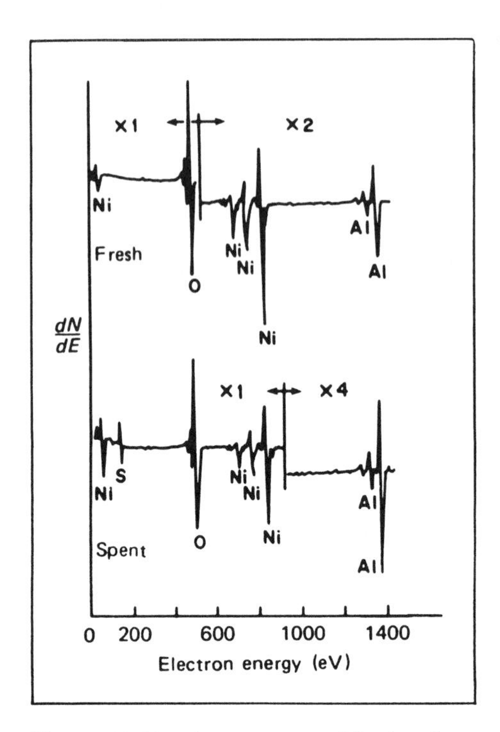

Figure 12.22 Auger spectra of fresh and spent catalyst. Note that appearance of sulfur and increased aluminum content in the latter.

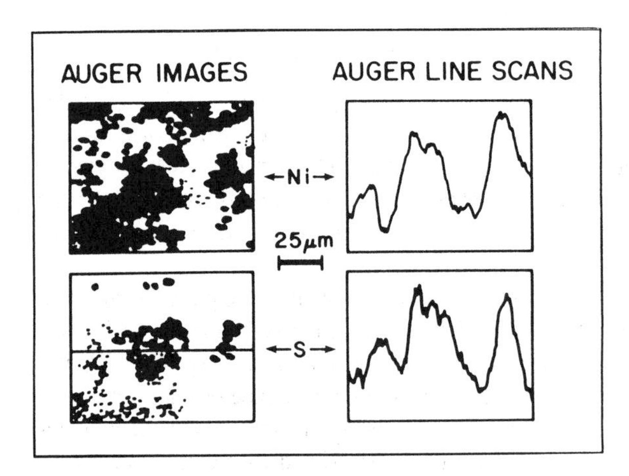

Figure 12.23 Auger distribution maps of Ni and S in a spent methanation catalyst. Note the corresponding Auger line scans.

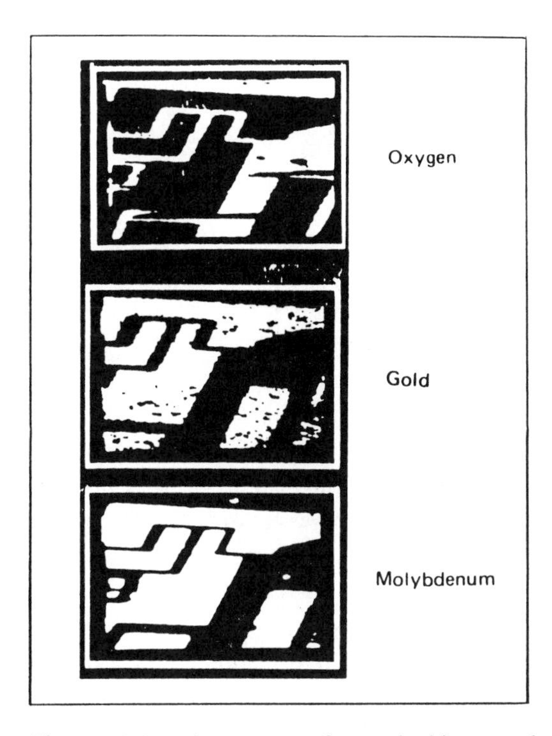

Figure 12.24 Auger maps of a standard integrated circuit, showing the distribution of oxygen, gold, and molybdenum on the surface. (Courtesy of Phillips Electronic Industries, Inc.)

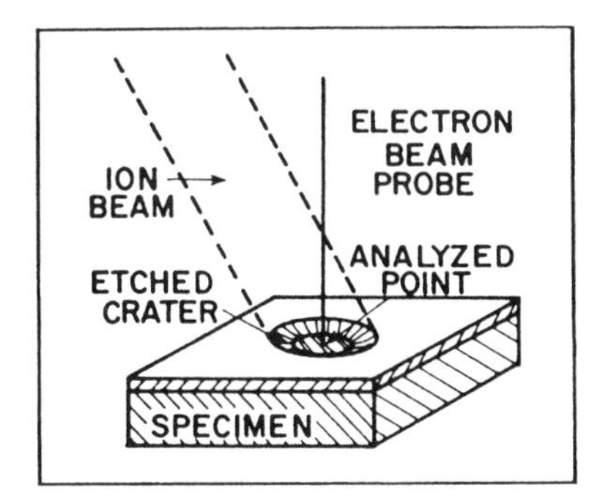

Figure 12.25 Surface stripping. An ion sputtering device removes surface layers during analysis.

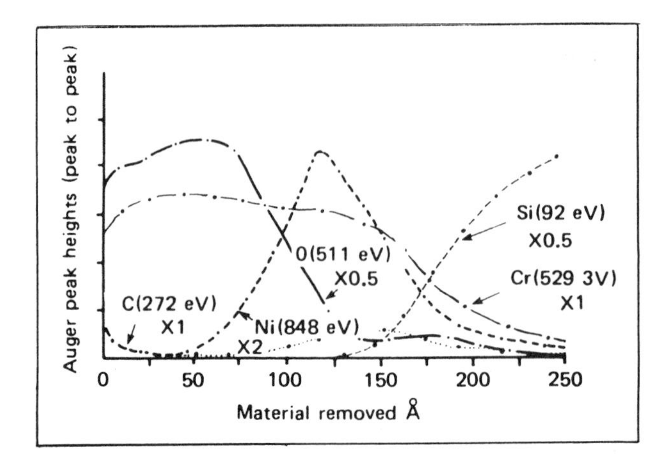

Figure 12.26 Change of composition of Nichrome film with depth. Note the wide variations in chromium, oxygen, and silicon. The sample was heated to 450°C for 30 sec. [From R. E. Weber, *Res. Develop.*, (1972), *10*: 22.]

Oxygen content was high on the surface but decreased to a low level at 150 Å and essentially disappeared at 250 Å. This type of information is very valuable in understanding the corrosion and chemical attack that take place on materials such as heating wire when heated to high temperatures.

Another example of the sputtering technique is shown in Fig. 12.27; this is the surface of silicon dioxide contaminated with chlorine. Silicon dioxide, of

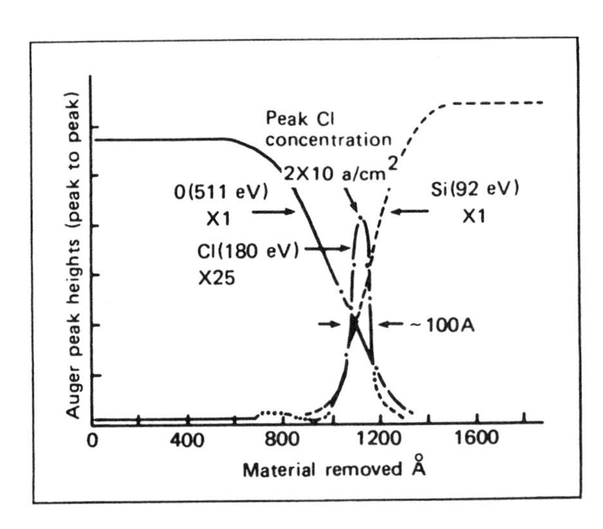

Figure 12.27 Depth profile of chlorine-contaminated silicon dioxide. Note the localized concentration of chlorine. [From R. E. Weber, *Res. Develop.*, (1972), *10*: 22.]

course, is used extensively in computer chips. The surface showed high concentrations of oxygen, which diminished with depth, and low concentrations of silicon, which rapidly increased with depth. The maximum chlorine concentrations were observed at about 1200 Å.

C. ION SCATTERING SPECTROSCOPY

In ion scattering spectroscopy (ISS) a beam of ions is directed at the sample. On arrival, the ions are scattered by the sample atoms at the surface. By measuring the energy of the *scattered* ions, one can determine the mass of the *scattering* atoms on the surface.

For a selected scattering angle, the relationship between the energy of the scattered ions and the masses of the bombarding ion and sample ion are given by the equation

$$E_s = E_0 \frac{M_s - M_{ion}}{M_s + M_{ion}} \tag{12.2}$$

or

$$\frac{E_s}{E_0} = \frac{M_s - M_{ion}}{M_s + M_{ion}}$$

where

E_s = energy of the scattered ion (after collision)
E_0 = energy of the bombarding ion (before collision)
M_s = mass of the scattering ion
M_{ion} = mass of the bombarding ion

The process is illustrated in Fig. 12.28. In this system the bombarding ion is $^3He^+$; the scattering atoms were ^{16}O and ^{18}O. Based on Eq. (12.2), the energy of the scattered helium ions can be predicted, depending on whether they were scattered by an atom or ^{16}O or ^{18}O. By measuring the relative abundance of the energies of scattered helium ions, it is possible to determine the relative amounts of ^{16}O and ^{18}O on the surface of the sample. One can readily see that other elements on the surface can also be detected and identified by the energies of the scattered helium ions. Based on the results, the elemental composition of the surface can be determined both qualitatively and quantitatively.

A schematic diagram of the equipment used for ISS is shown in Fig. 12.29. The equipment consists of an ion source, a vacuum system, an energy analyzer, and a detector. The whole system, including the sample, is under vacuum.

Several elements can be used as the ion source. The most widely used are the inert gases, particularly helium, argon, and neon. A stream of the inert gas

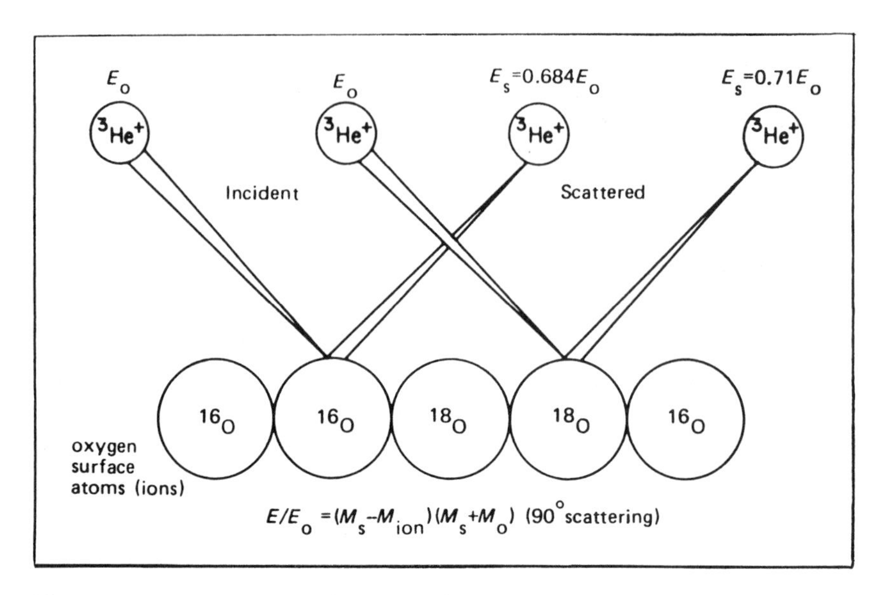

Figure 12.28 The process of ion scattering spectroscopy. The ions are scattered with different energies by ^{16}O and ^{18}O. [From A. W. Czandena, A. C. Miller, H. H. G. Jellinek, and H. Kachi, *Ind. Res.*, January (1978).]

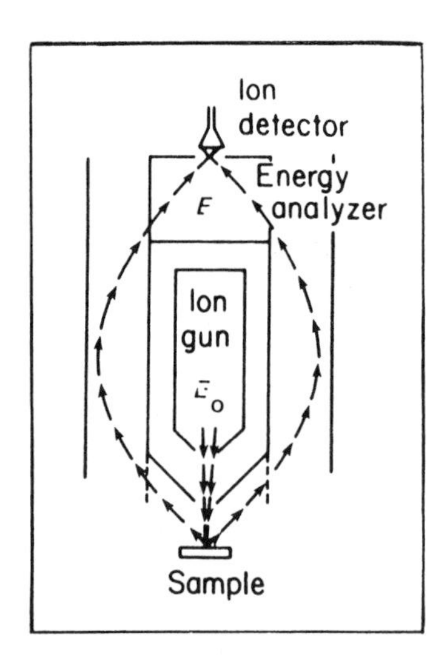

Figure 12.29 Equipment used for ion scattering spectroscopy. The total system is under vacuum.

passes through an electron grid, which ionizes the gas in much the same way as an ionizing chamber in mass spectrometry. This produces ions of the rare gas. These can then be accelerated to a known energy using an accelerating electrode. This process also is similar to that used in mass spectrometry (see Chap. 15). The accelerated ion beam is then focused onto the surface of the sample. Here the ions are scattered, and those scattered at some predetermined angle enter an energy-analyzing system, where they are separated before reaching an ion detector.

The scattering efficiencies of different rare gas ions are different for different surfaces. Some examples of the uses of ISS will be discussed later.

D. SECONDARY ION MASS SPECTROMETRY

A companion method of analysis to ISS is secondary ion mass spectrometry (SIMS). This process is slightly different from ISS, as shown in Fig. 12.30. In this instance the surface is again bombarded with a beam of an ionized inert gas, however, the inert gas ions displace ions from the sample surface. These displaced ions may be a complete molecular ion or an ionized fragment of a molecule. In addition, the displaced ion may be positively or negatively charged; in practice, both types of ions are utilized in the analytical process.

A schematic diagram of the equipment is shown in Fig. 12.31. In its most elementary form it consists of a source of ions, i.e., an ion beam, which are

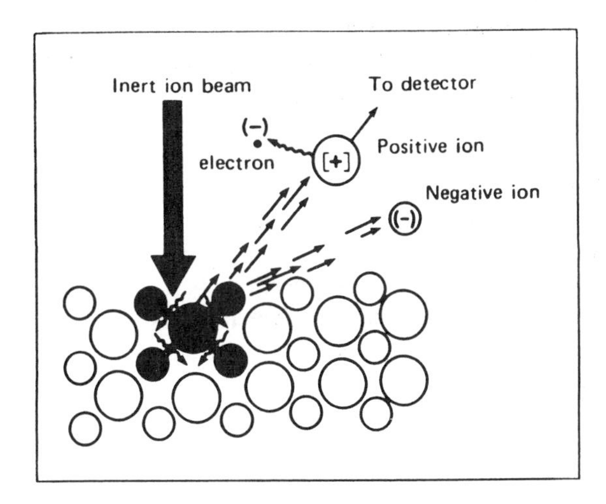

Figure 12.30 The SIMS process. Positive and negative ions are ejected from the surface by bombardment with inert ions (He^+). (Courtesy of 3M Analytical Systems, Chicago.)

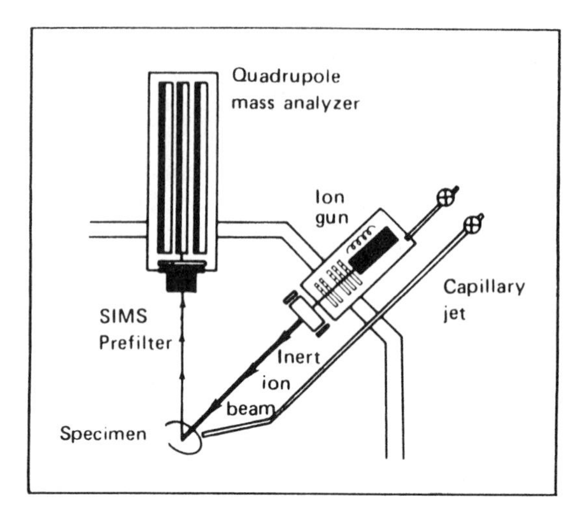

Figure 12.31 Schematic diagram of SIMS equipment. (Courtesy of 3M Analytical Systems, Chicago.)

accelerated under controlled energy conditions and focused on the sample. The ejected ions then enter a quadrupole mass spectrometer, where their masses and concentration can be determined. Both positive and negative ions are used in this same process. In practice, the positively charged ions are most valuable for examining the elements on the left side of the periodic table, whereas the negative ions are most useful for elements on the right side of the periodic table (nonmetals). Examples are shown in Fig. 12.32. With the high speed, sensitivity, and resolution of the quadrupole mass spectrometer, it is not difficult to perform a total scan of the masses emitted from a surface in a matter of seconds.

The principles of ISS and SIMS are very similar, in the method of bombarding the sample, the utilization of high vacuum, and the production of ions. In ISS the scattered ion beam is examined; in SIMS the ions ejected from the surface are examined. Generally, the particles vary considerably in mass and do not interfere with each other. The two systems can therefore be run simultaneously (ISS–SIMS), and this provides a wealth of information on elemental analysis, isotope ratios, and some information on molecular analysis, depending on the breakdown on the molecules involved.

1. Analytical Applications of ISS and SIMS

In both ISS and SIMS it is important to use a flood gun to prevent charge buildup on the sample surface. This is illustrated in Fig. 12.33, which indicates

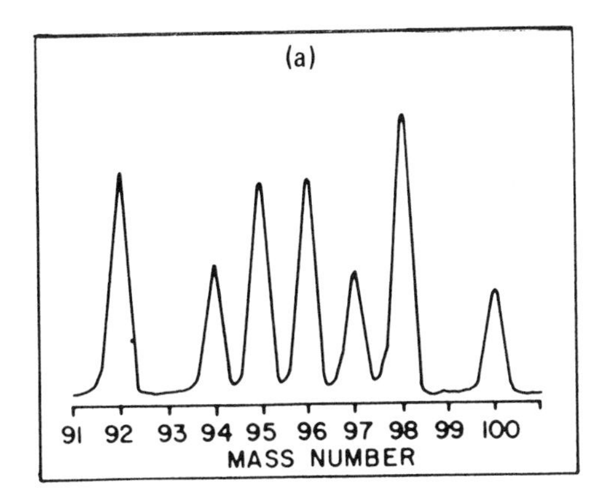

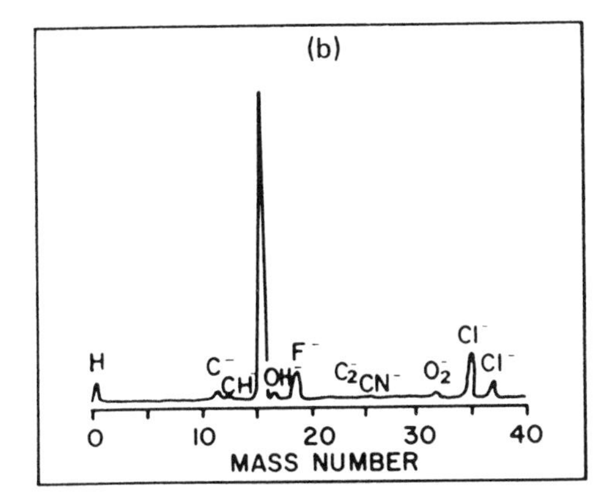

Figure 12.32 (a) Positive SIMS spectrum of molybdenum (all are isotopes of Mo) and (b) negative SIMS spectrum of stainless steel. (Courtesy of Physical Electronics Industries.)

the differences in the spectra obtained form a glass surface with and without a neutralizer or electron flood gun. It can be seen that the spectrum is much more precise when the flood gun is used. An example of the application of ISS–SIMS is shown in Fig. 12.34. This shows the elemental surface composition of treated molybdenum and gives the relative amounts of the different elements present.

The use of positively and negatively charged ions is illustrated in Fig. 12.32, which shows positive ions ejected from the high mass range and negative

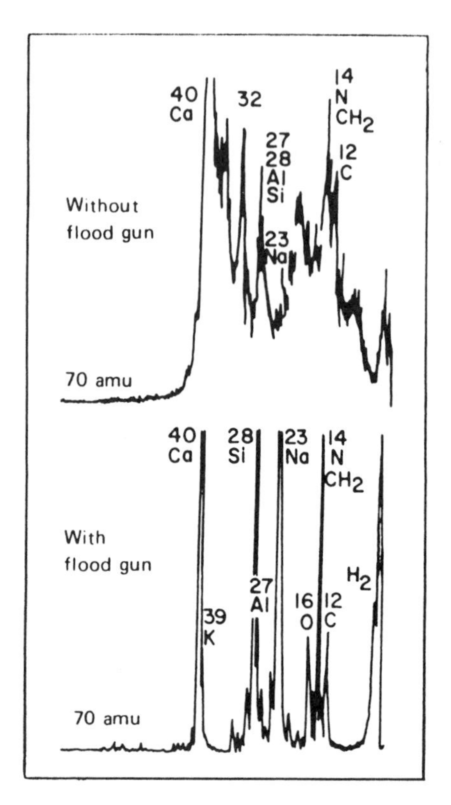

Figure 12.33 Effect of a flood gun on the SIMS spectrum of soda line glass. Note the improved resolution. (Courtesy of 3M Analytical Systems, Chicago.)

ions ejected from a stainless steel sample, indicating the presence of nonmetallic elements on the surface.

The approximate analytical sensitivity of SIMS depends to some extent on the elements being examined. Some examples are given in Fig. 12.35.

An interesting application of SIMS is the examination of aluminum surfaces being prepared for painting. It is necessary that the aluminum surface be cleaned in order to get a good painting surface. The SIMS spectra of a satisfactory and an unsatisfactory aluminum surface reveal the difference in surface elemental composition.

One application of ISS to the surface of polyamide shows the relative carbon, oxygen, and nitrogen concentrations at various depths. Ion scattering spectroscopy can show whether surface contamination is present and whether the concentrations of elements at the surface of the system reflect the bulk analysis.

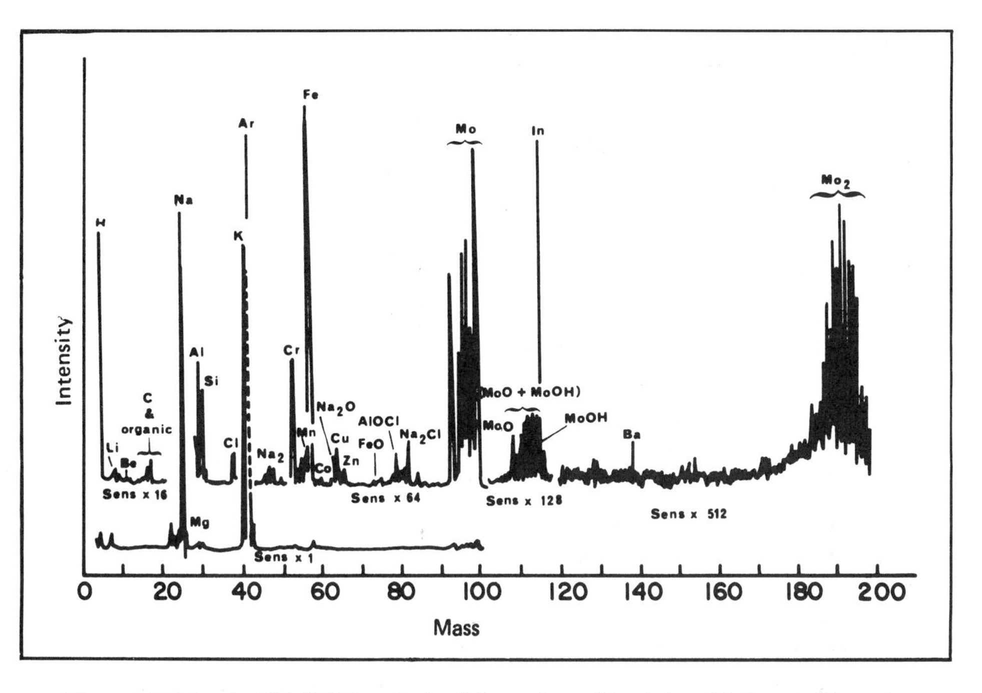

Figure 12.34 An ISS–SIMS analysis of the surface of treated molybdenum. (Courtesy of 3M Analytical Systems, Chicago.)

Changes in composition affect many surface properties, such as the tendency to crack or corrode and the ability to be chemically treated.

Figure 12.36 shows the difference in composition of aluminum sheeting and nonabrasive sheeting. A number of elements are present in the abrasive aluminum sheeting that are not present in the nonabrasive uncontaminated surface. This information is very valuable in the building industry, where aluminum sheeting is used for rapid fabrication, machining, and die stamping. Dies with abrasive surfaces wear more rapidly, and the final product is inferior. It is therefore important to control the surface of the die during use.

2. Quantitative Analysis

The analytical signal obtained in SIMS depends on a number of factors, namely, the abundance of the isotope examined on the surface, the properties of the surface, and the bombarding ion. The relationship is complex and not accurately known. Semiquantitative data are often quite informative, but quantitative data are best obtained by comparison with known materials.

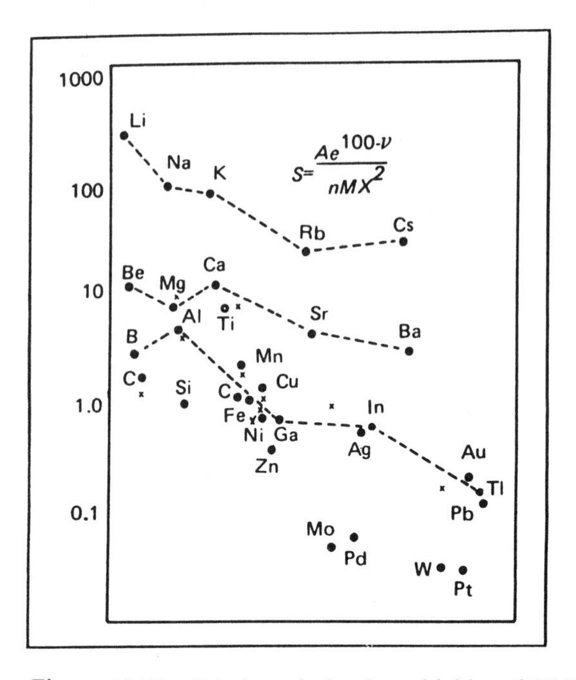

Figure 12.35 Relative calculated sensitivities of SIMS. (Courtesy of 3M Analytical Systems, Chicago.)

E. ION MICROPROBE MASS SPECTROMETRY

Ion microprobe mass spectrometry (IMMS) is an extension of SIMS but uses much more sophisticated equipment. In this system an ion beam is microfocused onto the sample so that very small points on the surface can be examined if necessary. The ion beam is generated by passing a rare gas, say, through a low-voltage, low-pressure hot cathode arc, which produces negatively and positively charged rare gas ions. The ions are then constricted and focused by electrostatic and magnetic fields. This is a *duoplasmatron*. A schematic diagram of the system is shown in Fig. 12.37.

Either inert gases, such as argon, helium, or neon, or reactive gases, such as oxygen or fluorine, can be used to produce the ion beam. The ions are accelerated to a known energy using an electrostatic field. The energies usually range between 5 and 25 keV. The beam is then constricted by two electrostatic lenses to an approximate diameter of 3–10 μm.

This ion beam may be used in any one of three modes: (1) it may be held stationary on the sample and the analysis at that particular spot obtained; (2) it may be rasted across the sample surface to obtain a distribution map of the ele-

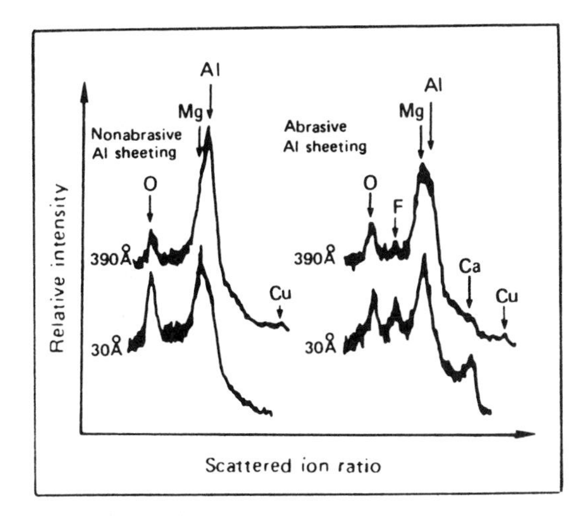

Figure 12.36 The ISS spectra for abrasive and nonabrasive aluminum surfaces. Note the slight change in composition. (Courtesy of 3M Analytical Systems, Chicago.)

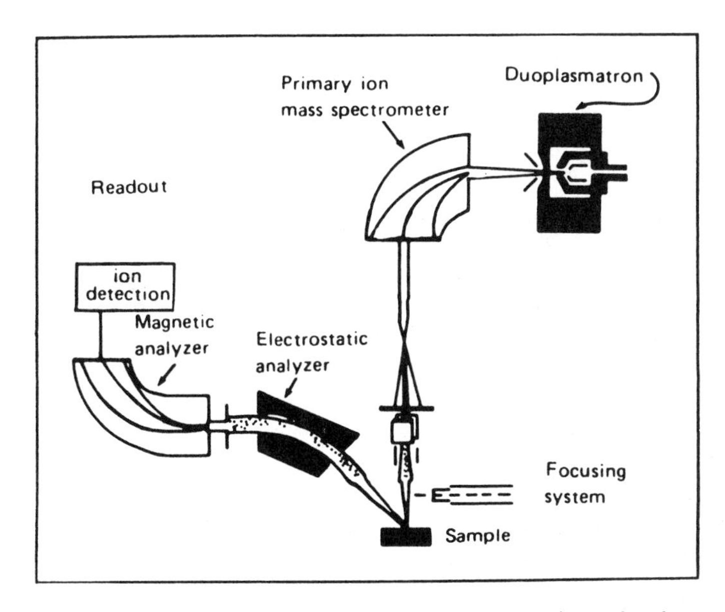

Figure 12.37 Ion microprobe mass spectrometer using a duoplasmatron.

ments on that surface; or (3) it may be focused on a single spot for an extended period of time, producing a crater that can be used to obtain a depth profile of the elements of interest.

The ions produced from the sample surface pass through an electrostatic field, which brings most of the ions to the same velocity. The ions then pass through a magnetic sector, which resolves them by mass-to-charge ratio. This is the sample principle as the two-sector mass spectrometer described in Chapter 15. With this system multielement analysis of the surface is possible. Mapping of the surface by element is also possible. In addition, the focusing potential of IMMS permits analysis of very small points on the sample surface.

One example of the use of this system is the determination of toxic elements on the surface of airborne particles. In this technique airborne particulates are gathered and the metals found adsorbed on the surface of the particulates are examined qualitatively and quantitatively. Particular attention is paid to lead, thallium, manganese, and chromium, since these are of biological interest.

Using a combination of ESCA and IMMS has proven to be very fruitful. Instruments are available that enable study of very small surface areas with both these techniques in sequence. The ESCA system permits identification of the elements present on the surface, and IMMS permits identification of the chemical forms of the elements. Use of this combination has permitted speciation analysis of thin layers of organic contaminants on the surface of polymers, functional groups on silicon wafers, and medical surgical instruments that have been electropolished. Contaminants include phosphates, sulfates, and hydrocarbons. The technique indicates the effectiveness of cleaning methods used, which is important in the manufacture and use of these products. Other uses include the detection of dimethyl silicone on SiO_2 and the adhesion of metal films on polymer optical coatings.

F. THE TRANSMISSION ELECTRON MICROSCOPE

Studies of materials being used as superconductors at elevated temperatures show that such materials exist in regions of only a few atoms. Studies are underway to extend these regions. The atomic arrangements can be studied by modified scanning transmission electron microscopy (STEM). With a light microscope, the light is focused by a lens. Structure can be determined based on the interference patterns produced by the sample. Using an electron beam, a magnet is used to focus the beam to atomic dimensions. The beam is scattered by the atoms based on their atomic number (Z). Based on the patterns of the scattered electrons, the atomic structure of the sample is revealed.

The transmission electron microscope is not usually thought of as a surface analysis technique, but since the electron beam must transmit the sample and can only transmit very small distances, only very thin samples can be examined.

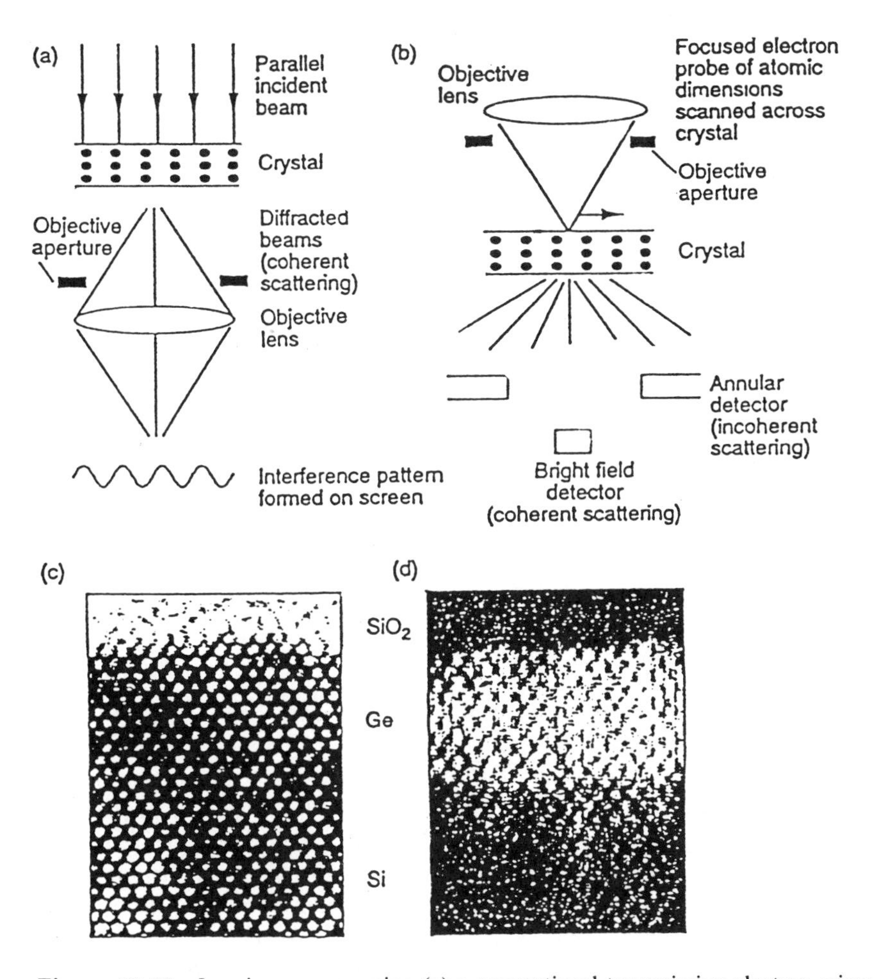

Figure 12.38 Imaging process using (a) a conventional transmission electron microscope (coherent imaging), (b) a scanning transmission electron microscope (incoherent imaging), and (c, d) atomic scattering of a SiO_2/Ge/Si system using (b).

So in a sense, it analyzes materials close to the surface. An interesting recent development is a system that studies scattering by atoms. The scattering increases with atomic number Z. A schematic diagram of a conventional electron microscope and a Z-scanning system is shown in Fig. 12.38.

G. DEPTH OF SAMPLE ANALYZED

The methods described above are useful for examining the surface of a sample to different depths. In each case the limiting factor is the ability of the electron or

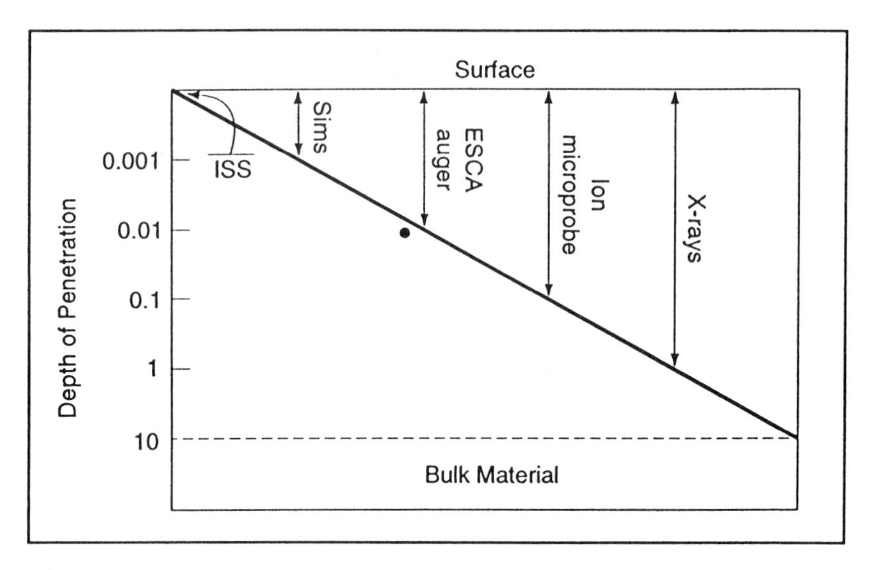

Figure 12.39 Effective surface depths of sample probe by different surface analysis techniques. [From F. W. Karasek, *Res. Develop.*, (1979), *1*: 26.]

photon to escape from the surface and still have energy representative of the chemistry or the elements involved.

The useful depths of penetration of these techniques are shown in Fig. 12.39. It can be seen that x-rays have the greatest depth of penetration, since the absorption of x-rays at these shallow depths is small, and ISS has the least penetration because the useful information obtained involves the interaction of the impinging ion with the atoms on the actual surface of the sample. In between these extremes are the techniques of SIMS, ESCA, Auger spectroscopy, and IMMS. Information obtained at depths greater than those shown in Fig. 12.38 is subject to major errors and should not be used. The techniques are not reliable in the bulk analysis of systems.

BIBLIOGRAPHY

Anal. Chem., Application Reviews, June 1994.

Carlson, T. A., *Photoelectric Auger Spectroscopy*, Plenum Press, New York, 1975.

Davis, L. E., MacDonald, N. C., Palmberg, P. W., Riach, G. E., and Weber, R. E., *Handbook of Auger Electron Spectroscopy*, 2nd ed., Physical Electronics Industries, Eden Prairie, Minn., 1976.

Hercules, D. M., *Anal. Chem.*, (1970), 42, 28A.

Lee, L. H., Ed. *Characterization of Metal and Polymer Surfaces*, Vol. 1, Academic Press, 1974.

Pennycook, S. J., Atomic scale imaging of materials by Z contrast scanning transmission electron microscope, (1992), *Anal. Chem., 64*: 263A.

Perry, S. S., and Somorjai, G. A., Characterization of organic surfaces, (1994), *Anal. Chem., 66*(7): 403A.

PROBLEMS

12.1 The wavelength of x-ray fluorescence radiation of an element is virtually independent of the chemical form of the element, but in ESCA the reverse is true. Explain.

12.2 Why is a flood gun used in ESCA-Auger equipment?

12.3 ESCA-Auger equipment is useful for the analysis of elements present in concentrations greater than 1% although the absolute sensitivity of the method is about 10^{10} g. Explain.

12.4 Draw a schematic diagram of an ESCA instrument.

12.5 Why is ESCA sensitive to the chemical form of an element, but Auger is not used for obtaining this information?

12.6 Why is an ESCA signal recorded directly, but an Auger signal is integrated before recording?

12.7 X-ray fluorescence is most sensitive for elements with a high atomic number. What is the relationship between Auger sensitivities and atomic number?

12.8 The concentration of nickel on the surface of a Nichrome wire depends on the depth below the wire surface. How may this variation be measured?

12.9 Why is it necessary to analyze surface samples under vacuum? How may several samples be examined consecutively?

12.10 Describe ISS. What depth of surface can be measured by this method? How does that compare with ESCA?

12.11 How can the concentration and distribution of a selected metal on a plane surface be measured and presented?

12.12 What sources are commonly used for generating ESCA and Auger spectra?

12.13 How can the effects of rapid variation in local magnetic fields on ESCA signals be eliminated? Why is it necessary to do that?

12.14 What are derivative curves? Explain their use.

12.15 What is SIMS? Describe its analytical use.

12.16 What are the effective surface depths of (a) ISS, (b) SIMS, (c) x-ray fluorescence, and (d) ESCA?

12.17 Describe, using a schematic diagram, an Auger instrument.

12.18 Describe the detectors used in ESCA.

12.19 Describe two types of energy discriminators. What are their functions?

12.20 Describe two types of energy analyzers. What is their function?

13
Chromatography

The field of chromatography has changed dramatically in the past few years. Early commercial equipment involved simple gas chromatography with universal or specific detectors. Now great progress has been made in liquid chromatography and in interfacing gas or liquid chromatography and mass spectrometry, infrared, diode array systems, and even nuclear magnetic resonance.

Whole new fields of endeavor in biological and natural science are opening up and great strides are being made in agriculture, medicine, pharmaceuticals, etc., based on these new sources of information. Many of these innovations are discussed in this chapter.

Chromatography is a method of separating a mixture of compounds into its component. It enables us to separate trace impurities and/or major fractions from each other. Based on the time required for elution (retention time), we can obtain considerable information about the identity of each component. However, final identification usually requires confirmation by some other analytical procedure, such as infrared (IR) spectroscopy, nuclear magnetic resonance (NMR), or mass spectrometry (MS). Quantitative analysis can be carried out by measuring the area of the chromatographic peak. Hence chromatography can be used for qualitative and quantitative analysis.

Chromatography was first demonstrated experimentally in 1906 by Michael Tswett, a Russian botanist who was interested in separating the pigments in biological materials like leaves and grass. He made up a column of crushed chalk and then passed down an extract of the leaves, continually washing them with the original solvent. The solution separated into colored fractions, which we now know to be compounds such as chlorophyll and carotene. These brightly colored compounds were responsible for his naming the procedure *chromatography*, after the Greek words *chroma* and *graphos*, meaning "color" and "writing," respectively.

The procedure was largely ignored until 1944, when Martin and Synge published a paper that contained much of the theory of chromatography. Their procedure was for liquid chromatography. As with Tswett, the publication was largely ignored by the scientific world. In 1952, Martin and Synge reinforced their theoretical concepts and demonstrated their procedure by using it to separate amino acids. This was a major breakthrough to biochemistry, since prior chromatographic separation, identification and purification of these very important biological compounds were extremely tedious, time-consuming, and unreliable. Their procedure was a very important contribution to biochemistry in particular and to science in general. The field of chromatography has flourished since those days and has spread to many areas of endeavor. Initially it was strictly a method of separation usually followed by some method of identification and measurement. However, procedures are now available that couple chromatography with other major analytical fields such as infrared GC-IR and mass spectrometry GC-MS to permit automatic qualitative and quantitative analysis. Techniques such as these comparing two or more instruments are called *hyphonated techniques*.

Chromatography is probably the most important single analytical technique used today and will probably continue to be so for the foreseeable future. It is a cornerstone of molecular analytical chemistry in particular. Recently its coupling with atomic absorption spectroscopy has extended its application to elemental analysis and to speciation of the metal. Speciation is an increasingly important field since it tells us not only the concentration of the metal but the various forms in which it exists.

The information obtainable by gas chromatography is particularly useful to the research organic chemist or biochemist who wants to know what materials he has synthesized in the laboratory or separated from living tissue. In addition to its application to pure research, it is valuable to the industrial scientist who wants to know the composition of her own or her competitor's products, as well as to many other people involved in the characterization of matter. It can also be used as a method of preparing very pure compounds, such as in the pharmaceutical industry or in the manufacture of pure chemicals, and is now used commercially to do so.

A. PRINCIPLES OF CHROMATOGRAPHY

The principles of chromatography can be explained in terms of the following experiments.

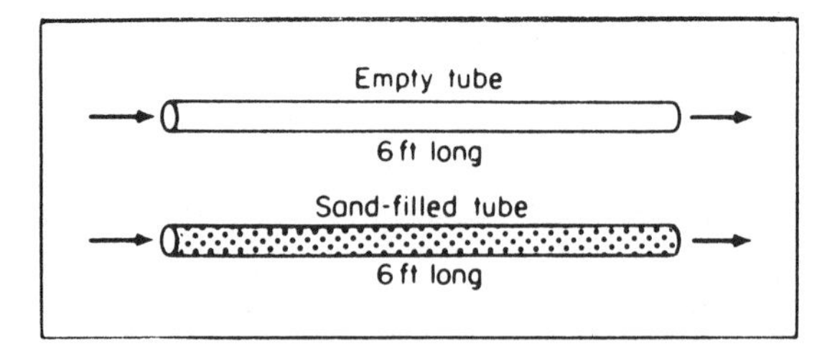

Figure 13.1 (Top) Empty 6-ft-long tube. The flow rate of a gas in the empty tube is 3 ft/min; therefore the time needed for the gas to traverse the empty tube is 2 min. (Bottom) Sand-filled 6-ft-long tube. The flow rate of the gas in the sand-filled tube is 2 ft/min; therefore the time needed for the gas to traverse this tube is 3 min.

A gas, e.g., N_2, that is flowing smoothly at the rate of 3 ft/min down an empty tube that is 6 ft long takes 6/3 = 2 min to flow from one end of the tube to the other (Fig. 13.1, top). If such a tube were filled with sand (Fig. 13.1, bottom), the gas would flow through it more slowly. If the rate at which the gas flows in the sand-filled tube is 2 ft/min, it will take the gas 6/2 = 3 min to traverse the tube. The sand-filled tube in this example has some properties of a gas chromatography column. The gas is the moving or *mobile phase*. The sand is the *stationary phase*. The gas that emerges after it has passed through the column is called the *effluent*. In practice the mobile phase should be relatively insoluble in the stationary phase; otherwise the stationary phase becomes overloaded.

To further our analogy, consider the following property of soluble gases. In a vessel containing a nonvolatile liquid A we place a soluble gas compound B (Fig. 13.2). A short time after the vessel has been sealed, the compound B in the

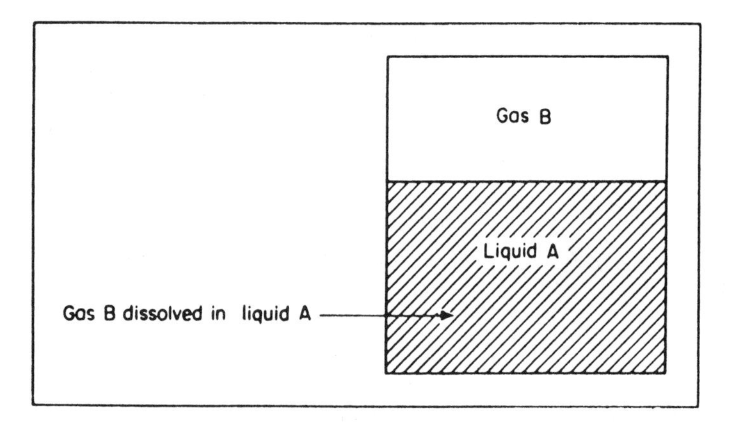

Figure 13.2 Gas in equilibrium with a liquid.

gas phase comes to equilibrium with the compound B dissolved in the liquid phases, and the distribution of B between the two phases remains constant. Although the total number of gas molecules above the liquid and in the liquid stays the same, a rapid interchange takes place between the molecules in the two states; that is, molecules from the gas pass into the liquid at the same rate that dissolved gas molecules leave the liquid and becomes gaseous. The molecules are said to be in *dynamic equilibrium*. This is constant if the temperature is kept constant. It can be shown that, on the average, each molecule of B spends a constant fraction of the time in the liquid phase and the remainder of the time in the gas phase. For the purposes of our discussion, we hypothesize that compound B spends 40% of its time in liquid A and 60% of its time as a gas. Also, the equilibrium will be established more quickly if the thickness of the liquid is small—otherwise, dissolved molecules B wander from the interface and the equilibrium is disturbed.

To proceed with our illustration, we remove the sand from the tube shown in Fig. 13.1, coat it with a thin film of liquid A, and replace the coated sand in the tube. Liquid A is now known as the stationary phase, or substrate, and gas N_2 is known as the carrier gas. The sand is the support. We inject gas B into the packed tube. We saw earlier that it took 3 min for gas N_2 to flow through the tube filled with uncoated sand; moreover, we have hypothesized that when gas B is in equilibrium with liquid A, it spends 40% of its time in the liquid. Therefore for 40% of the time compound B spends in the tube, it does not travel down the tube. During the remaining 60% of the time (i.e., during the time in which it is in the gaseous state), gas B moves down the tube at the normal flow rate of 2 ft/min. On the average the gas flows at $(60/100) \times 2$ ft/min $= 1.2$ ft/min. The time taken to pass through the column at this rate is

$$\frac{6 \text{ ft}}{1.2 \text{ ft/min}} = 5 \text{ min}$$

Alternatively, we can say that if gas B were flowing all the time, it would take 3 min to flow through the tube. Therefore it must spend 3 min flowing to reach the end of the tube. But it spends only 60% of its time in the gas phase, and 3 min is 60% of the total time in the column. Therefore the total time in the tube is $3 \times (100/60) = 5$ min.

The distribution coefficient for the gas in Fig. 13.2 is given by

$$K = \frac{\text{concentration in liquid}}{\text{concentration in gas}} = \frac{40}{60}$$

and this is reflected directly in the time it takes for the gas to elute from the chromatography column.

Based on this relationship, we can state that the time spent in the gas phase, or mobile phase, is the same as the time it takes for the carrier gas to pass through the column, that is, t_0. The time spent in the solvent phase, or stationary phase, is the *extra* time it takes for the sample to pass through the column, that is, $t_r - t_0$, where t_r is the total time for the sample to pass through the column, or the retention time. We can therefore present K in terms of the chromatographic experiment as

$$K = \frac{t_r - t_0}{t_0} \tag{13.1}$$

or

$$K = \frac{t_r}{t_0} - 1 \tag{13.2}$$

where

t_0 = time spent by sample in mobile phase
t_r = total time for sample to elute
$t_r - t_0$ = time spent in stationary phase

This equation is described more explicitly in Section G.

In practice, we can inject air into a column and it will emerge after time t_0. This assumes that it does not interact with the stationary phase. If we inject compound B and a small air sample together, the chromatograms will appear as in Fig. 13.3a. If we substitute a second compound C for compound B in Fig. 13.3b and it is found that C spends only 20% of its time in liquid A and 80% of its time as a gas, then when C is permitted to flow through the tube packed with coated sand, it moves at the rate of 2 ft/min for 80% of the time and is at rest for 20% of the time. The average flow rate should be $2 \times (80/100) = 1.6$ ft/min. The total time needed for gas C to flow through the tube is $6/1.6 = 3.75$ min.

If we were to pass a mixture of gases B and C down the tube, gas C would emerge after 3.75 min and gas B after 5 min. Passage of a mixture of the two gases down the tube results in separation of the mixture.

Inert gases, such as helium, flow through chromatographic columns at a constant velocity; that is, such gases do not interact with and thus spend no time in the stationary phase. For this reason inert gases are often used as carrier gases. In practice, the carrier gas flows steadily through the column. The sample is injected into the carrier gas and is swept onto the chromatographic column. The sample then shares its time between the stationary phase and the moving carrier gas.

We have seen that with a liquid substrate, the process can be used to separate gases from each other (gas–liquid chromatography). By using solid substrates, we can separate liquids from each other (liquid–solid chromatography). Different species of molecules spend different periods of time in the stationary

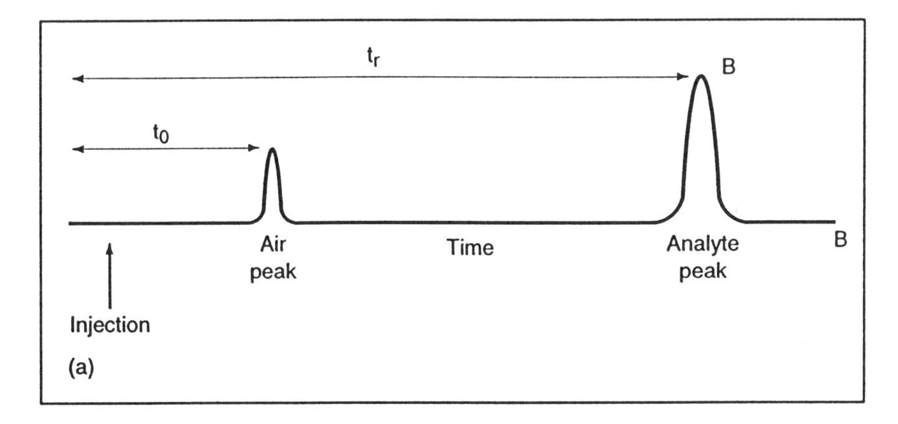

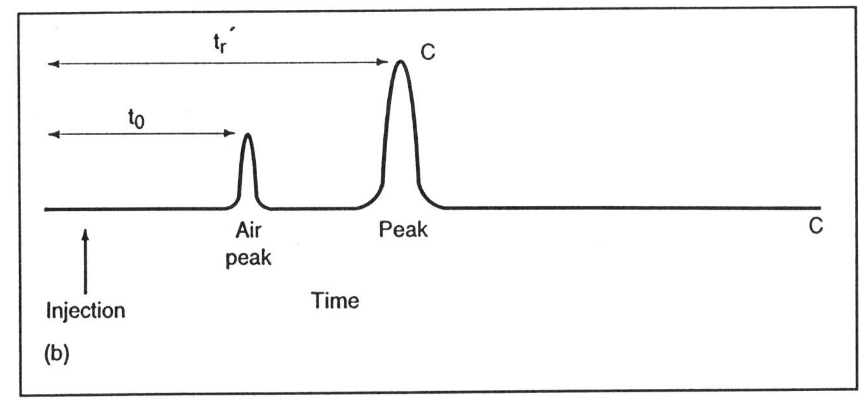

Figure 13.3 (a) Typical chromatogram with air peak and emergence of analyte B. (b) Chromatogram showing emergence of air peak after t_0 and compound C after t'_r.

phase (or substrate) and therefore different percentages of time in the mobile (or moving) phase. The variation in time spent in the moving phase, in turn, affects the length of time a gas takes to pass through and emerge from a chromatographic column. A record of the emergence of various compounds plotted against time is called a *chromatogram*.

This process of separation is the basis of all forms of chromatography. The factors that affect the time distribution between the mobile phase and the stationary phase differ from one branch of chromatography to another. The most important branch is gas–liquid chromatography, which will be discussed later. The next three most important branches are gas–solid, liquid–liquid, and liquid–solid chromatography, where the first phase is the mobile phase and the second is the stationary phase. In the following paragraphs, we consider some properties of the substances involved in chromatography, as well as some hypothetical data illustrating chromatographic analysis.

In the example described previously, compound B distributed itself between the liquid phase A and the gas phase in the ratio of 40% (liquid) to 60% (gas). If we add a second gas such as helium that can be used as a carrier gas, the distribution of B between the gas phase and the liquid phase will not be disturbed. The distribution can then be described by

$$K_B = \frac{\text{concentration of B in liquid (stationary) phase}}{\text{concentration B in gas (mobile) phase}} = \frac{X_s}{X_m} = \frac{40}{60} \tag{13.3}$$

In this case K is the partition coefficient. Frequently it is expressed in terms of moles per liter, in which case

$$k'_B = \frac{\text{moles in stationary phase}}{\text{moles in mobile phase}} = K_B \frac{V_s}{V_m} \tag{13.4}$$

where V_m and V_s are the volumes of the mobile (gas) and stationary (liquid) phases, respectively, and k'_B is the solute partition ratio. Also, more correctly, $k' = (t_r - t_0)/t_0$.

Suppose that we have a second gas C, which spends 80% of its time in the gas phase and 20% in the liquid phase. The partition coefficient for gas C is given by

$$k'_{(C)} = \frac{\text{moles of C in stationary phase}}{\text{moles of C in mobile phase}} = \frac{20}{80}$$

When the partition coefficients are widely different, we can expect separation. On the other hand, if the partition coefficients are similar, it is less likely that separation will be achieved. The separation factor or *selectivity factor* α can be given as

$$\alpha = \frac{K_B}{K_C} \text{ or } \frac{k'_B}{k'_C} \tag{13.5}$$

A chromatogram illustrating this separation is shown in Fig. 13.4.

A guide to how much time difference there will be between the emergence of two compounds such as B and C is the time difference each spends in the stationary phase $t_r - t_0$. They both spend the same time in the mobile phase. This is reflected by α, the selectivity factor. If α equals or approaches unity (1), separation will be difficult because the two peaks emerge at about the same time.

Unfortunately, selectivity factors are only a guide to the actual ability of a column to separate two components. Other factors influence the final result. For example, the two gases are in dynamic equilibrium between the liquid phase and the mobile phase. If the mobile phase is flowing too fast, then this dynamic equilibrium is approached at all times but never quite achieved. This results in

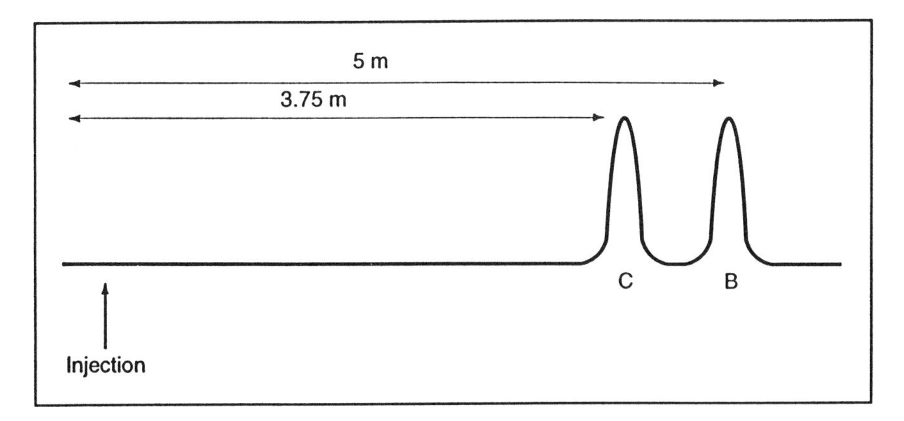

Figure 13.4 Separation of B and C by retention time.

a spreading of the components B and C along the column. Sometimes it results in a lack of separation, even though the selectivity factors indicate that the two components should be separable. Further relationships are discussed in Section G. Here the capacity fraction K is deduced in terms of the retention time of the sample and the flow rate of the solvent.

1. rf Value

A more accurate way to present the progress of the sample down the column is the *rf value*. This is the ratio of the flow rate of the sample to the flow rate of the solvent over the stationary phase:

$$\frac{t_r}{t_0}$$

Alternatively, B and C may diffuse extensively in the mobile phase and even overlap, resulting in poor separation. This is illustrated by Fig. 13.5.

B. EFFICIENCY OF THE CHROMATOGRAPHIC PROCESS

1. The Separation of Multicomponent Mixtures

The efficiency of a chromatographic column is a measure of its ability to separate the components of a given mixture. To separate the components, it is necessary that they not spread out in the column and overlap each other. In Fig. 13.4, which represents good separation, the two components emerge within 1.25 min of each other and are completely separated. In Fig. 13.5, an example of bad separation, the components are not separated, even though the centers of the peak emerge with the same time difference as in Fig. 13.4.

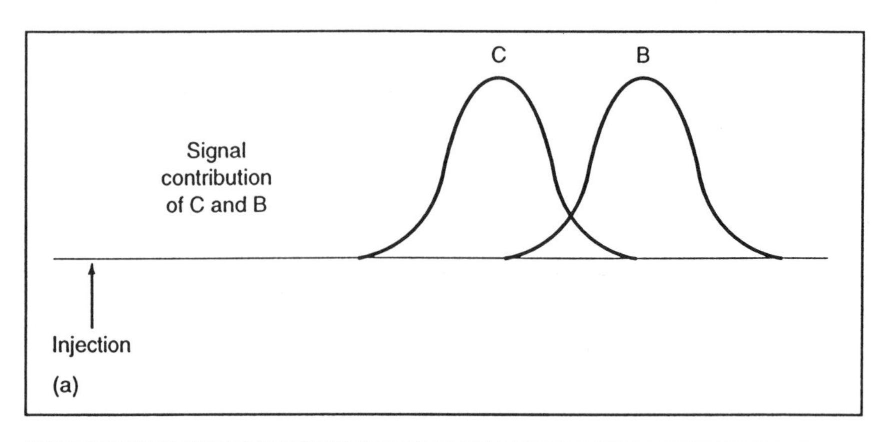

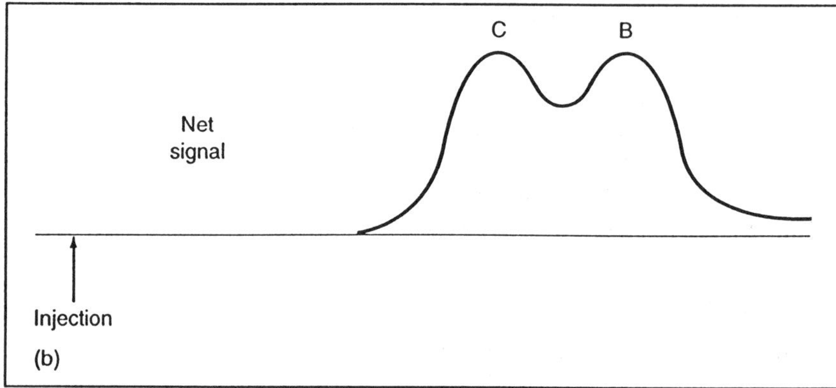

Figure 13.5 (a) Poor separation is achieved even though time separation (α) was good. (b) Total signal from unresolved peaks.

2. Measuring Efficiency

a. The Theoretical Plate

The unit in which the efficiency of a chromatographic column is expressed is the *theoretical plate*, a term borrowed from distillation processes. If a mixture of two components is heated, the lower-boiling material boils over first. However, the higher-boiling substance has a finite vapor pressure at the temperature at which the lower-boiling material distills over. Some of the higher-boiling substance vaporizes with the lower-boiling one. The simple distillation conditions, with no refluxing, described here as referred to as a single-plate distillation. The product of such a distillation would be enriched in the volatile component as compared to the undistilled mixture. If this product were distilled again, a further enrichment in the volatile component would take place. This would be equivalent to a two-plate distillation. Repeated distillations would produce an increasingly pure product. By changing the design of the distillation equipment (e.g., by permitting some refluxing), the number of plates of a single distillation can be increased and better separation of the components can be made from one efficient distillation.

A chromatographic separation is in many ways similar to a distillation separation. The efficiency of chromatographic separations was first measured by A. J. P. Martin in *theoretical plates*, a term he proposed. He measured the total number of theoretical plates by injecting a single compound together with a small bubble of air into the column. A typical resulting chromatograph is shown in Fig. 13.6.

Time for air to emerge = T min (from injection to maximum of air peak)
Time for sample to emerge = $T + X$ min (from injection to maximum of sample peak)
Time for sample peak to pass detector = Y min

The total number N of theoretical plates in the column is given by the equation

$$N = 16 \left(\frac{X}{Y}\right)^2 \tag{13.6}$$

which is an empirical formula first developed by A. J. P. Martin. It is assumed that air does not interact with the liquid substrate. It therefore spends no time in the substrate but flows down the column in a minimum time. This period, called the *dead time*, is a measure of how long a sample takes to flow through the column if it does not interact with the substrate. Time X is a measure of the extra

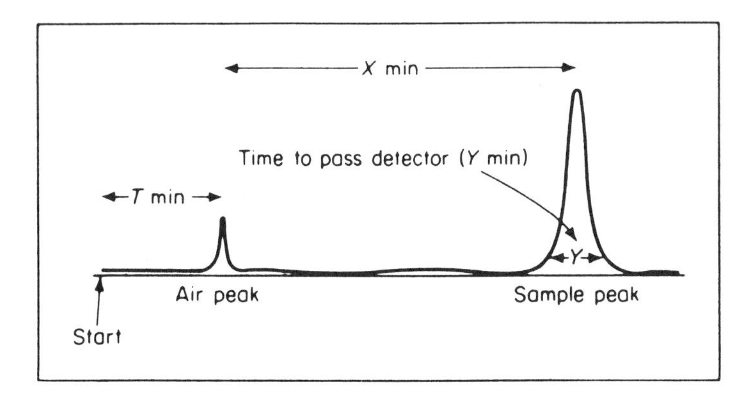

Figure 13.6 Chromatogram of a single compound. The difference in the time needed for the sample and the air to reach their respective peaks is X min.

time the sample spends in the column. This time is spent in the stationary phase. Time Y is a measure of the length of time taken by the sample to pass the detector; this is a direct measure of how much the sample has spread out or diffused in the column during its journey down the column. Note that the dead time is t_0 and X is $t_r - t_0$.

No two columns can be physically packed with substrate in exactly the same manner. Hence r must be determined experimentally for each column. For example, for a column 6 ft long, these data were obtained:

Time for air peak maximum to emerge = 1 min
Time for sample peak maximum to emerge = 5 min
Time for sample peak to pass detector = 1/2 min

$$\text{Number of theoretical plates } N = 16\left(\frac{X}{Y}\right)^2 = 16\left(\frac{5-1}{1/2}\right)^2$$

$$= 16\left(\frac{4}{1/2}\right)^2 = 16 \times 64$$

$$= 1024$$

Resolution R. The resolution or separation of two adjacent peaks is given by

$$R = \frac{t_A - t_B}{0.5\ (\text{width A} + \text{width B})}$$

where

t_A = time for compound A to emerge
t_B = time for compound B to emerge
width A = width of Peak A
width B = width of Peak B

Fig. 13.7 illustrates this. The resolution of the peaks in Fig. 13.7 is 5 m − 4 m = 1 m. Divided by the average of 15 and 25 sec: i.e. 20 sec

$$R = \frac{60\ \text{sec}}{20\ \text{sec}} = 3$$

If the peaks are not skewed or distorted, when $R = 1$, there is 3% overlap between the peaks. If $R = 1.5$, overlap is 0.2% and can be ignored. If $R < 1$, overlap is significant. If $R > 1.5$, overlap is insignificant. The fundamental equation defining resolution is given by

$$R_s = \sqrt{\frac{N}{16}}\left(\frac{\alpha - 1}{\alpha}\right)\frac{k'}{k' + 1}$$

where $\sqrt{N/16}$ is dependent on column efficiency, $(\alpha - 1)/\alpha$ is dependent on selectivity, and $k'/(k' + 1)$ is dependent on the rate of migration.

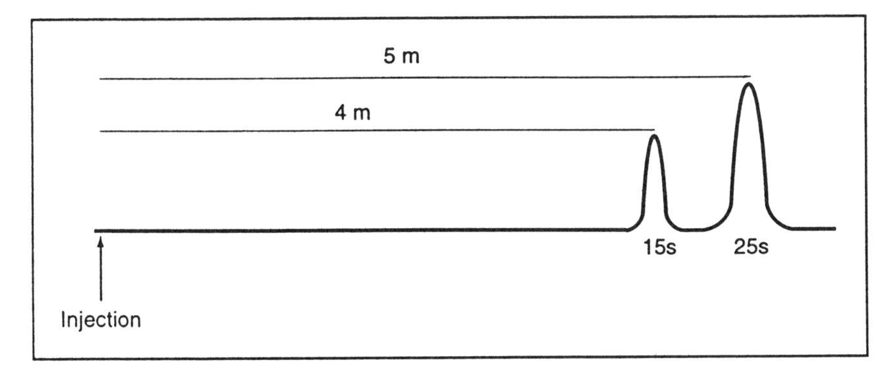

Figure 13.7 Resolution of the peaks emerging after 4 and 5 minutes, respectively, with peak width of 15 and 25 sec, respectively.

With older equipment it was common to observe an air peak any time a sample was injected. The time for the air peak to emerge is the dead time and is a measure of the flow rate of the mobile phase. This assumes that the air does not interact with the stationary phase. With modern equipment, unless air is injected intentionally, it is unusual to observe an air peak in a gas chromatogram. It is quite common, therefore, but less accurate, to take the time for the sample peak to emerge (in this case, 5 min) as X.

In practice, it is more meaningful to speak of the height of each theoretical plate, that is, the *theoretical plate height* H, which is given as

$$H = \frac{L}{N} \tag{13.7}$$

where

L = length of the column
N = number of theoretical plates

Thus, in our example

$$H = \frac{6 \text{ ft}}{1024} = \frac{72}{1042} \text{ in.}$$

In earlier discussions on the dynamic equilibrium of a gas between the stationary liquid phase and a mobile gas phase, it was seen that a certain period of time was necessary before equilibrium took place. This time period can be interpreted as the distance traveled down a column before equilibrium takes place. This distance is essentially the theoretical plate height. It is important that this distance be kept to a minimum. A *low plate height* indicates that the sample *equilibrates rapidly* and goes through many periods of equilibrium before emerging from the end of the tube. However, if the plate height is great, then there are only a few equilibrations and the sample becomes smeared inside the column and resolution between the two components becomes poor. The factors that affect the plate height or the theoretical plate height H are discussed below.

b. The Van Deemter Equation

By comparing the theoretical plate heights of two columns, we can obtain a direct measure of their relative efficiencies. It will be noted that the total length of the column was not included in Eq. (13.6). Column length is a separate variable and should not be allowed to confuse the issue of the most efficient operating conditions of a given column.

Factors that affect the length of time a component takes to pass through a column include the ratio of the times spent in the flowing gas phase and in the stationary phase. This ratio is affected by the affinity between the component and the substrate, the molecular weight of the component, the flow rate of the carrier gas, and the length of the column.

At first sight, it appears that increasing column length sufficiently would enable us to separate any two components, no matter how similar they might be. Unfortunately, this is not true because the components continue to diffuse and overlap in the extra column length. Up to a point, an increase in column length gives an increase in total theoretical plates. With great increases in column length, however, the diffusion of the sample becomes more important. In the limit, no increase in efficiency results from an increase in column length.

Factors that affect how much a component spreads as it flows down the column include *diffusion* in the gas phase and in the liquid phase, *eddy currents* in the gas phase, the *geometry of the packing*, how fast the sample comes to equilibrium between the mobile and the stationary phase (*mass transfer coefficient*), and the *carrier gas flow rate*. Each of these factors increases the degree to which the component spreads out in its passage down the column. The relationship between these variables and column efficiency is given by the *Van Deemter equation*, which states that

$$H = A + \frac{B}{V} + CV \tag{13.8}$$

where

H = theoretical plate height

A = a constant dependent on eddy currents in the flowing gas. This is affected by the geometry of the packing and the column walls, as well as how smoothly the gas flows. Also, A depends on some molecules traveling farther in a more circuitous path than others, causing them to arrive at various times as shown in Fig. 13.8.

B = a constant dependent of the sample in the gas phase and liquid phase. The diffusion rate is affected by temperature and how fast the sample

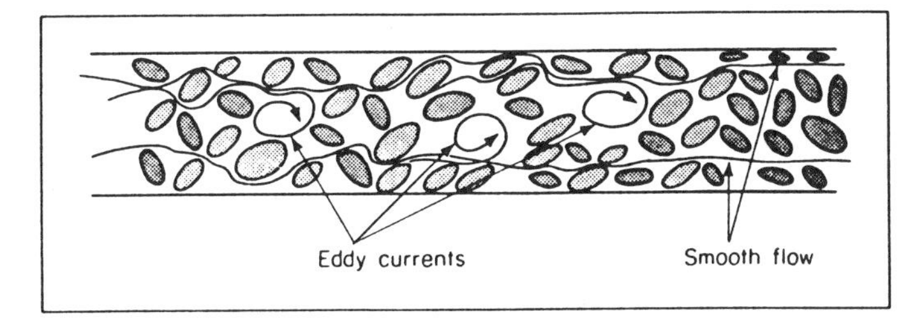

Figure 13.8 Eddy currents in a flowing gas.

molecules diffuse into the carrier gas or substrate. Even if a sample is not flowing, it will diffuse and spread out along the column as shown in Fig. 13.9.

C = a constant dependent on the mass transfer between the sample molecules in the liquid phase and gas phase. This constant is dependent on how fast the dynamic equilibrium is set up between a gas in the mobile phase and in the stationary phase.

V = flow rate of the carrier gas down the column.

This basic equation has been refined by numerous workers and is still the subject of considerable research. The constants A, B, and C are characteristic of any individual column and vary from one column to another.

The constant A is given by $A = 2\lambda d_p$ where λ is the geometry constant and d_p is the average particle diameter of the support. The value of A also depends on the eddy currents that form in the flowing gas. If the column were perfectly packed and the gas flow rate were slow enough, there would be no eddy currents, but A would never be zero, because in practice there are irregularities in the packing (Fig. 13.8).

Small eddy currents cause the sample to be spread out along the column. This leads to an overlap between two species that have similar separation prop-

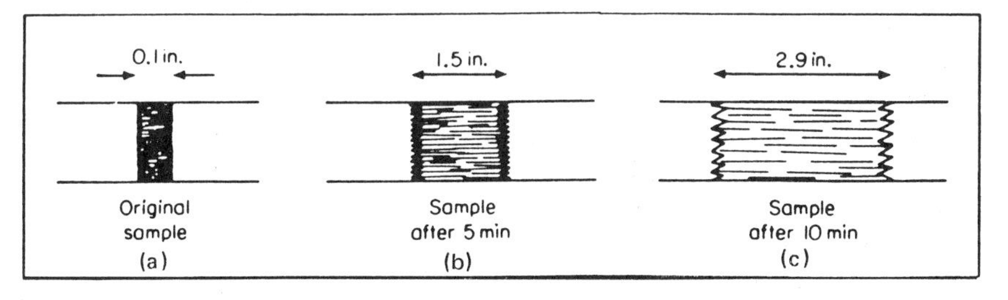

Figure 13.9 Diffusion of a sample into a carrier gas within a column: (a) when first injected, (b) after 5 min, and (c) after 10 min.

erties. If the column is unevenly packed, the eddy currents will be numerous and separation on the column will be poor. Good packing is therefore essential.

The packing also affects the physical process of moving down the column itself and therefore the total distance traveled. Two individual molecules may start off from the same point to move down the column, but one may follow a path around numerous particles, and that path will be affected directly by the particle size and the particular location of the particles relative to each other. A second molecule may go through a different path and the actual physical distance traveled may be less because the path was straighter. This is akin to two people walking across a canyon with many boulders on the terrain. The actual path that each one chooses may vary considerably in length, and therefore the time it takes to go from one side to the other may vary considerably, even though both people walked at the same speed. The degree of variation is dependent on the size of the boulders. If they are very large, then it is conceivable that the time difference may be considerable. If they are small, however, it is likely that the time difference will also be small. In the practical case of the column, we have perhaps 10^{20} molecules progressing down the column, and each conceivable pathway will be used by some of the molecules. The result is that some molecules will arrive in a minimum of time and some will take a maximum time, and a Gaussian distribution of times will be taken for all the molecules. The geometry of the system directly affects the shape of the distribution curve of the molecules in the columns. The greater the particle size, the flatter and broader the curve will be and, consequently, the time distribution will be at a maximum. This physical problem would be encountered even if other conditions were perfect. Improving packing in the column will only reduce the problem; it cannot eliminate it. It is therefore essential that the packing be good and the particle size be reasonably small to obtain optimal conditions. If the flow rate is very high, turbulence occurs and with it increased eddy currents and loss of resolution. In practice, an optimum flow rate is determined.

The constant B depends on the physical diffusion of the sample in the carrier gas stream. Suppose that a sample is injected into the carrier gas and that its original displacement in the column is 0.1 in. Furthermore, let us suppose that the carrier gas is not flowing down the column. The sample slowly diffuses into the carrier gas. After 5 min its displacement in the column is 1.5 in., after 10 min 2.9 in., and so on. Clearly, no eddy currents are formed. Nevertheless, the sample slowly spreads out into the carrier gas (Fig. 13.9). (But diffusion of the sample in a carrier gas will take place even if the carrier gas is flowing down the column. The longer the sample is in the column, the greater the diffusion that will take place.

If the system shown in Fig. 13.9 is maintained and the sample is in the column for 5 min, it spreads out to at least 1.5 in. along the column. The diffusion rate is independent of eddy currents but is dependent on the physical properties of the sample. Highly diffused molecules, such as hydrogen or methane, diffuse rapidly in the column, and separation becomes more difficult. However, molecules that move more slowly in the gas phase, such as high molecular weight materials, diffuse slowly during their flow down the column. It should be remembered that the diffusion referred to is the diffusion rate of the sample at the temperature of the column. High molecular weight samples often need to be analyzed in a hot column before they will flow at all. Under these conditions their diffusion rates are sometimes comparable with those of lower molecular weight samples at lower temperature. Similarly, molecules with very low molecular weights may need to be cooled before they can be separated. This reduces the diffusion rate of the molecules.

Similar properties affect the diffusion of the sample in the liquid phase. If the flow rate is too slow, the sample stays in the column too long and diffusion is too high. The total diffusion depends on the total length of time spent passing through the column. This time is inversely proportional to V, the flow rate. Hence the diffusion constant is divided by V to assess its effect on the theoretical plate height.

The constant C is proportional to the rate at which the dynamic equilibrium is set up between molecules in the stationary phase and in the mobile phase. When the equilibrium is set up quickly, the molecules rapidly come to equilibrium and progress down the tube as the carrier gas flows. Molecules in the stationary phase tend to come out quickly and are not left behind, so equilibrium can be rapidly established. On the other hand, if the equilibrium is slow in developing, then molecules that flow down the tube with the carrier gas will be swept away from molecules remaining in the liquid phase, and the equilibrium will be only partially obtained. This results in a considerable broadening of the band, because molecules that statistically spend an extended period of time in the mobile phase will travel faster than molecules that statistically spend most of their time in the stationary phase. The problem is accentuated by an increase in the flow rate of the carrier gas. Consequently, in the Van Deemter equation the term C is always multiplied by V, the carrier gas flow rate.

It can be seen from these considerations, as illustrated by the Van Deemter equation, that increasing V decreases the time in the column and decreases the effect of B. But, at the same time, the effect of C, the mass transfer coeficient, is increased.

For each column, a practical compromise must be reached to determine the best flow rate for minimum H and, therefore, best resolution for each column. This is performed experimentally as follows.

c. Relationship Between H and V: Experimental Determination

The carrier gas flow rate V refers to the *linear velocity* of the carrier gas down the chromatographic tube. In practice, it is quite difficult to measure the linear flow rate. On the other hand, it is not difficult to measure the volume flow rate of the carrier gas (i.e., the rate in volume per unit time at which the carrier gas leaves the column). This can be measured directly with the use of a bubbler as shown in Fig. 13.10. The bubbler is attached to the end of the column and allowed to come to equilibrium. Soap solution is then squeezed into the bottom of the new tube and soap bubbles form across the tubing. The soap bubbles then travel at the velocity of the carrier gas. Markers are etched on the side of the vertical tube, and the volume between the markers is known. The time taken for a bubble to pass from one marker to the next is then a direct measure of the volume flow rate. In practice, several bubbles have to be injected onto the tube before one reaches the top marker, which occurs only after the inside of the tube is completely wet and the bubble is not broken en route.

The volume flow rate is a reasonable measure of the linear flow rate, and for a given column the relationship shown in Fig. 13.11 between the theoretical plate height and the carrier gas velocity holds. In this instance the volume flow rate is substituted for the velocity of the carrier gas. The data cannot be transferred between different columns, since the volume flow rate in a column is related to the linear flow rate and is dependent on the effective cross-sectional area inside the column. This will not be a constant between different columns; hence the data obtained on one column do not pertain to a second column and cannot be used to assess the optimum gas flow rate to produce the best theoretical plate height on any given column.

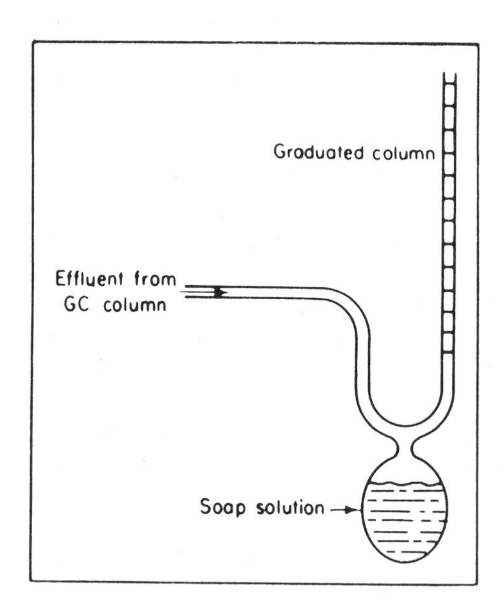

Figure 13.10 Bubbler to measure the volume flow rate of a carrier gas in gas chromatography.

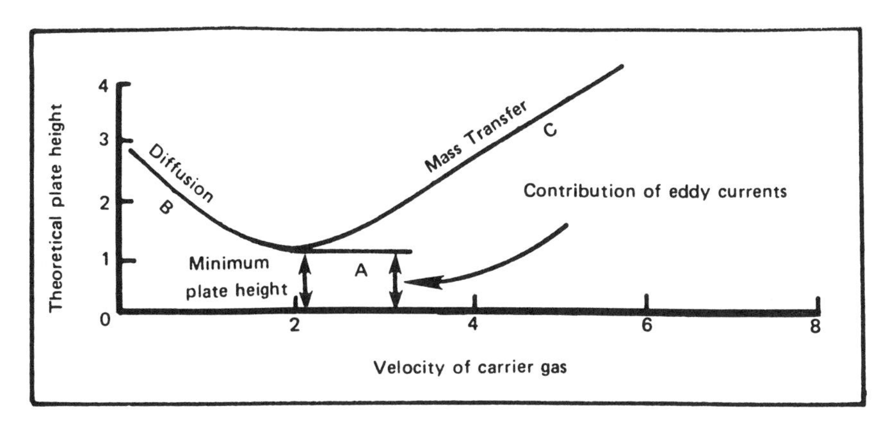

Figure 13.11 Relationship between the theoretical plate height H and the carrier gas flow rate V.

Furthermore, in practice, it is not convenient to measure continuously either the volume flow rate or the linear flow rate. Normally, under routine operations, the *gas pressure* is measured and controlled. It is assumed that the conditions in the column do not change in use, so that a given pressure will result in a given linear flow rate.

The optimal conditions must be determined experimentally. These are best derived by utilizing the Van Deemter relationship. The term C expresses the resistance to transfer of the sample molecules between the gas phase and the liquid phase. If the velocity of the carrier gas is changed, then, as predicted by the Van Deemter equation, the theoretical plate height changes (Fig. 13.11).

The column is most efficient when H is a minimum. The optimum value of V, the carrier gas flow rate to obtain the minimum H, can be determined experimentally. The determination is done by injecting an aliquot of a sample into a column with the carrier gas flowing at a given rate V and determining H from Eqs. (13.4) and (13.5) as follows.

The chromatogram is traced as in Fig. 13.6. By measuring X and Y we can calculate N, the number of theoretical plates in the column:

$$N = 16\left(\frac{X}{Y}\right)^2 \tag{13.6}$$

If the column length is L, then

$$H = \frac{L}{N}$$

By repeating the determination several times with the carrier gas set at different flow rates, we can plot the curve relating H and V as in Fig. 13.11. From this the optimum flow rate V can be determined and the minimum value for H can be determined experimentally.

C. EQUIPMENT

The equipment used for gas chromatography is basically composed of a sample inlet at one end of a metal column packed with stationary phase material, followed by a detector system (Fig. 13.12). The components include a device for regulating the carrier gas flow, the sample injection system, a separation column at a controlled temperature, and a detector and readout system.

1. Carrier Gas

The carrier gas functions as the mobile gas phase, which propels the sample down the tube. For this to be reproducible, it is vital that the rate of flow of the gas remain constant. The flow rate must therefore be carefully controlled. This control is usually achieved with pressure gauges and flow meters. These devices make it possible for the flow rate to be kept constant for periods up to a day.

The carrier gas itself must be inert toward the sample and the substrate and must be thermally stable. The most commonly used carrier gases are helium, argon, nitrogen, and hydrogen. Of these, helium is used most often because it is inert, can be operated at high flow rates, and is the most convenient for a number of detectors, such as the thermal conductivity detector. Argon or nitrogen can be used if cost is a factor.

Hydrogen shares most of these advantages, but it is chemically reactive and may destroy the sample. Also, its flammability makes it more difficult to handle safely in the laboratory.

2. Sample Injection System

The sample must be introduced onto the column in the form of a compact "plug." The normal way is to inject the sample with a syringe into an injection

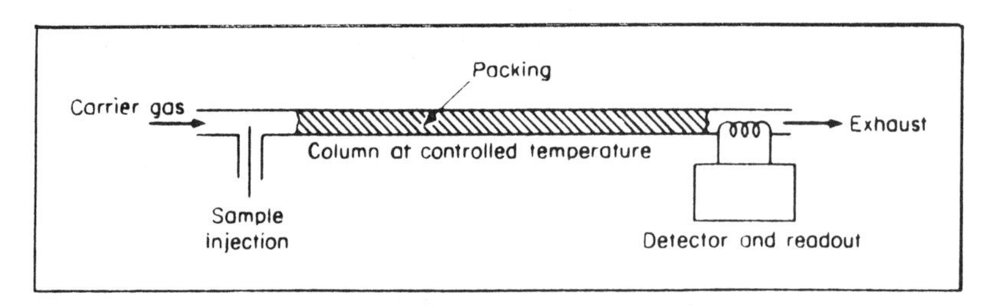

Figure 13.12 Schematic diagram of a gas chromatograph.

chamber, as shown in Fig. 13.13. The sample is best injected in the form of a liquid or a gas. If the sample vaporizes slowly, the nonvolatile components enter the carrier gas over a relatively long time and emerge from the column over a correspondingly extended time period. Poor resolution results. In order to ensure that the sample stays in a single plug (plug flow), it must be vaporized rapidly after injection. Rapid sample vaporization is brought about by heating the injection cell to temperatures up to 350°C. When the sample is injected into the injection cell, it immediately vaporizes, and the vapors are swept onto the substrate by the carrier gas. Ideally, the sample spreads evenly over the cross section of the column and so remains as a plug. Care must be taken not to overheat the injection cell or the sample may decompose. If prior knowledge indicates that the sample may decompose or isomerize upon heating, it may be necessary to reduce the temperature of the inlet system to a safe temperature. Of course, the column temperature must not exceed the inlet temperature; or rapid sample evaporation takes place. The extra temperature achieves nothing, but the risk of decomposition is increased.

The sample may be introduced into the injection port with a syringe. This is a very convenient way of introducing an accurately known volume of material into the column. The syringe enters the injection port by way of a replaceable *septum*. The latter may be made of neoprene rubber or silicone rubber if very high temperatures are expected. The septum self-seals after withdrawal of the syringe needle, thus preventing the sample from blowing back out after it expands rapidly upon introduction into the high-temperature injection port. Inevitably, after numerous injections the septum's ability to self-seal is lost and the sample may escape. At this time a new septum should be put into place.

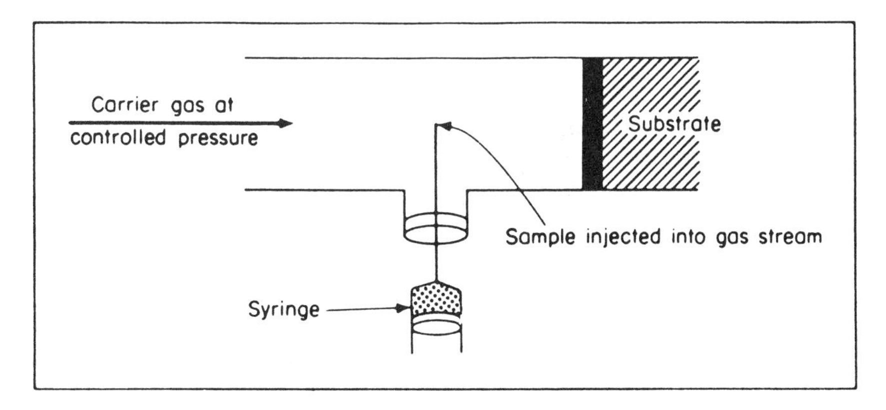

Figure 13.13 Schematic diagram of a sample injection system.

3. Packed Column

The column is a metal tube packed with the *support*, which is coated with the *substrate*. The support should have a high surface area in order to promote maximum contact between the stationary phase and the mobile phase, thus assuring rapid equilibrium, and a low value for C, the mass transfer coefficient. It should also be chemically inert so that it cannot interact with either the sample or the substrate. A material commonly used as a support is crushed, used firebrick, which combines the properties of being inert and possessing a high surface area. Furthermore, because it has been used in high-temperature furnaces for extended time periods, it contains no volatile compounds that might generate extraneous peaks.

The surface of the support is coated with a liquid substrate. Interaction between the sample components and the liquid substrate is the basis of chromatographic separation. The substrate must be chemically inert even at high temperatures. Its vapor pressure must be very low over the entire temperature range of operation. As a solvent, it should provide differential partitioning between the components of sample. Typical examples of substrates are shown in Table 13.1. Many more substrates are readily available in the commercial literature.

Column packings are selected based on the type of sample to be analyzed. Different packings are more suitable than others for analysis of various sample types. For example, a Carbowax 20M column is preferred for the separations of alcohols, esters, pesticides, and essential oils. On the other hand, the diethylene glycol adipate column is used for the separation of fatty acids, esters, and pesticides. Squalane is useful for the separation of hydrocarbons, and silicone gum can be used for the separation of alcohols, fatty acids, esters, and aromatics. These are many other columns are commercially available. A wealth of information is available from the manufacturers, who will readily recommend what columns should be used for separating different sample types. A great deal of

Table 13.1 Substrates Used in Packed Columns

Material	Maximum temperature (°C)
Squalane	150
Apiezon-L grease	300
Methyl silicone gum (SE 30)	350
Polyalkylene glycol (Ucon L. B550x)	200
Methyl silicone oil (DC 200)	200
Carbowax 20 M	250
Diethylene glycol adipate	200

research has gone into this phase of chromatography, and most of the information available from the instrument manufacturers is reliable and can be used to advantage.

Filling the column with packing is an exacting step. It is important that the packing be evenly distributed and contain no voids. (This keeps the constant A low.) In practice, the metal tube from which the column is to be made is left uncoiled. One holds it vertically in a suitable place, such as a stairwell—being sure to plug up the bottom of the column. Substrate is slowly added to the top. The column is shaken well after each addition. An ultrasonic resonator can be used to advantage to vibrate the column at this stage. After packing, the top end is plugged to prevent the substrate from moving while in use and creating voids. The column is then curled into a coil so that it will be more compact for handling. At this point, it is ready for use in the instrument. The more direct way is to buy one from a commercial manufacturer.

4. Detectors

The function of the detector is to detect and measure the different components of the sample as they emerge from the column. The choice of detector depends on the type of analysis being performed. High sensitivity can be achieved by using a flame detector, but the sample is destroyed in the process. High selectivity can be achieved with electron capture detectors. Other detectors, such as the thermal conductivity detector, are nonselective and nondestructive. The most common gas chromatographic detectors are described in the following paragraphs.

a. Argon Ionization Detectors

In the argon ionization detector (Fig. 13.14), argon must be used as the carrier gas in the chromatographic separation. The basis of the detector's operation is

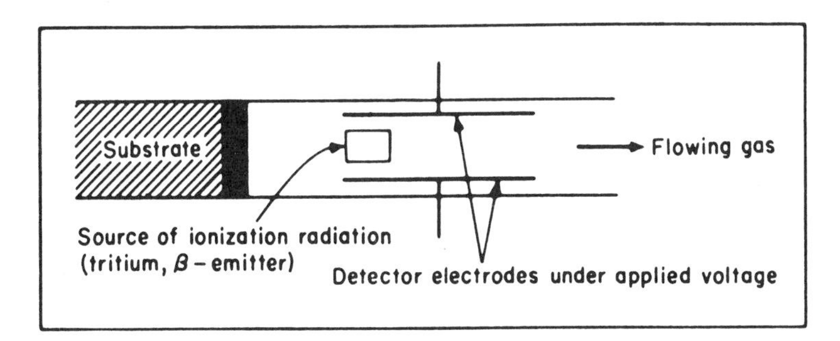

Figure 13.14 Schematic diagram of an argon ionization detector.

as follows. Two electrodes inside the detector cavity are separated by the carrier gas. A potential difference (voltage) is applied across the electrodes. Under normal circumstances, the carrier gas (argon) acts as an insulator, and no current flows between the electrodes. If the sample gas becomes ionized, it becomes conducting and its resistance is changed greatly. In turn, this changes the current flowing between the electrodes. The flow of current can be measured and provides a method of detection.

The ionization step is carried out as follows. A radioactive source, in this case tritium, emits β-rays (electrons) and causes the argon to be *excited* to the metastable state Ar* The excited atoms are *not ionized* and do *not* conduct electricity; however, they have sufficient energy to ionize most organic molecules; consequently, when excited metastable argon atoms collide with organic molecules from the sample, the organic molecules become ionized and conduct electricity. This causes current flow between the electrodes. The current creates a signal proportional to the amount of sample passing through the detector. The signal can be amplified, and very sensitive gas chromatographic detectors have been developed based on this technique.

The steps in the ionization are

$$(1)\ \mathrm{Ar} + e^- \ (\beta \text{ emitted from tritium}) \rightarrow \mathrm{Ar}^*(\text{excited}) + e^- \qquad (13.9)$$

$$(2)\ \mathrm{Ar}^* + \mathrm{org} \rightarrow \mathrm{org}^+ + \mathrm{Ar} + e^- \ (\mathrm{org} = \text{organic molecule})$$

The argon ionization detector is sensitive over a wide range of concentrations. Such detectors respond to most inorganic and organic compounds but do not detect water vapor, air, and some halogen compounds. They do not detect H_2O, CH_4, C_2H_6, O_2, N_2, CO, CO_2, or fluorinated compounds C_xF_y.

As indicated previously, a noble gas must be used as the carrier gas. The most popular are argon and helium.

b. Flame Ionization Detectors

In the flame ionization detector the effluent from the gas chromatographic column is fed into an air-hydrogen flame (Fig. 13.15). Two wire electrodes are also placed in the flame and an electrical potential is applied across them.

When only the inert carrier gas enters the flame, the current across the electrodes is small and constant. When an organic compound from the end of the chromatographic column enters the flame, the compound is broken up into fragments that are highly conducting. These are easily detected by the change in the current flowing across the electrodes. This detector is capable of detecting 10^{-12} g of organic material. It is one of the most sensitive detections available for gas chromatography (GC) and can be used for conventional GC and capillary column GC (see Sec. D.2). It is sensitive to all organic compounds, including

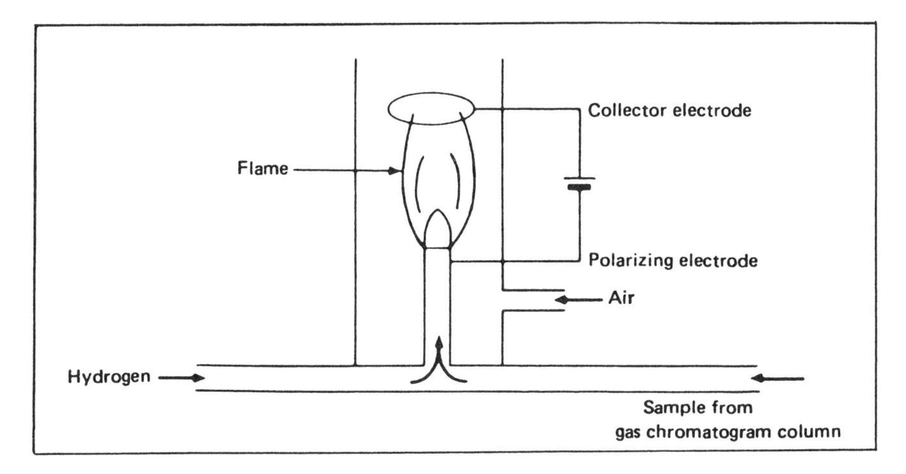

Figure 13.15 Schematic diagram of a flame ionization detector.

hydrocarbons and heteroorganic compounds. It has several disadvantages, however. It is not sensitive to many inorganic compounds. The degree of change in conductivity depends on the particular molecules. Hence its response to a given quantity of an organic compound depends on the particular compound. The relationship of response to sample weight must be determined for each compound to be determined. The flame ionization detector is not sensitive to H_20, CO, CO_2, N_2, or O_2, which are normal flame products, or to the inert gases, which can therefore be used as carrier gases without affecting the signal.

c. Double Flame Detectors

Insecticides, fungicides, and pesticides frequently are organic compounds containing either phosphorus or halides. Consequently, attention has been paid to developing detectors that are many times more sensitive to these latter elements than to other elements. This enables us to detect trace of these materials in the presence of many other compounds.

The first such detector used two flames as shown in Fig. 13.16. The effluent from the GC column enters the lower flame, which functions as a simple flame detector to burn and detect virtually all compounds that leave the column. The products of combustion pass through the alkali metal halide screen into the upper compartment. The combustion products are drawn into the upper flame. The gases CO, CO_2, H_2O, and N_2, which are the combustion products of hydrocarbons, carbohydrates, amines, and so on, do not change the flame conductivity and are not detected, but compounds containing halides or phosphorus are selectively detected by the second flame.

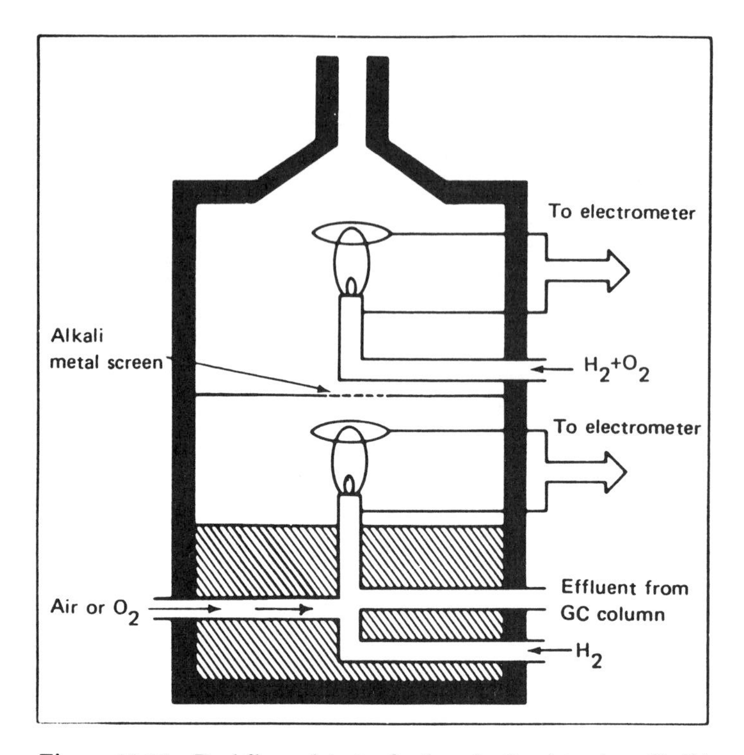

Figure 13.16 Dual flame detector for the selective detection of halide and organophosphorous compounds.

The advantage of this detector is that the signal from the first flame gives the total chromatogram, whereas the signal from the upper flame selectively detects the halide and phosphorus compounds. Using this combination, it is possible to obtain both a selective and a nonselective chromatogram. The sensitivity of both sections of this detector is very high, in the picogram range.

d. Flame Emission Detectors

A different approach has also been used for the selective detection of sulfur or phosphorus compounds. In this technique the flame emission in the upper part of the flame is monitored. The emission lines of sulfur and phosphorus are monitored and their presence or absence detected.

These detectors have been widely used for the detection and measurement of insecticides, such as DDT, because they contain halides. Identification of the actual compound has been achieved based on the retention time of the peak, so this method readily distinguishes between compounds with different retention times, but not between compounds with the same retention time. Confirmation of identification should be carried out with a separate technique, such as mass spectrometry. Unfortunately, such confirmation is often difficult at the low con centrations of insecticides, for example, encountered in our environment. Much of the early work that led to the banning of DDT was in error because the compounds detected were in many instances PCBs (polychlorinated biphenyls), which are also used as an electrical insulator and which have overlapping retention times with DDT. This error has cost many lives through disease and millions of dollars in lost productivity. Nowadays it is simple to avoid this error by using hyphenated techniques such as GC–MS rather than simple retention time.

e. The Cross-Section Detector

The cross-section detector (Fig. 13.17) consists of two metal electrodes across which a voltage is applied. The effluent from the gas chromatography column passes between the electrodes. Radioactive ^{90}Sr ionizes the organic sample, creating an ion pair according to the equation

$$\text{organic compounds} + e^- \ (\beta \text{ emitted from } ^{90}\text{Sr}) \rightarrow \text{org}^+ + 2e^- \qquad (13.10)$$

The generated ion and electrons cause a change in the current flowing between the electrodes. The quantity of electrical current flowing is a measure of the quantity of the components passing through the column. The detector is sensitive to all organic compounds; however, its sensitivity is low. It has a wide analytical dynamic range, in that it can measure the concentrations of compounds up to 100%. It is not destructive. The carrier gas used with these detectors is hydrogen or nitrogen. In summary, it is a nondestructive, universal detector used for the determination of major components.

f. Thermal Conductivity Detector (Katharometer)

A common GC detector is the *katharometer* or *thermal conductivity detector*; a schematic diagram of this device is shown in Fig. 13.18. As the carrier gas from

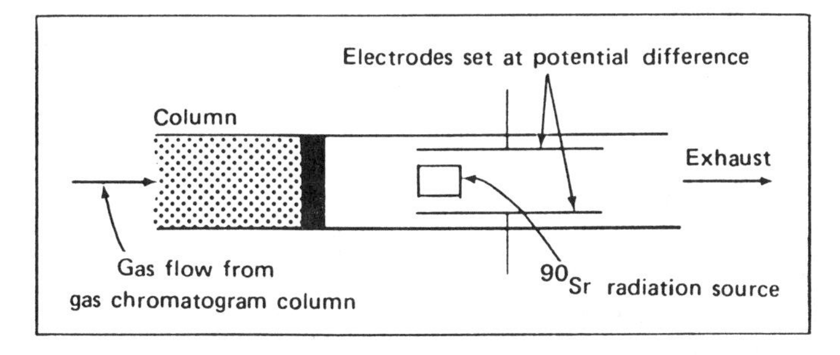

Figure 13.17 Schematic diagram of a cross-section detector.

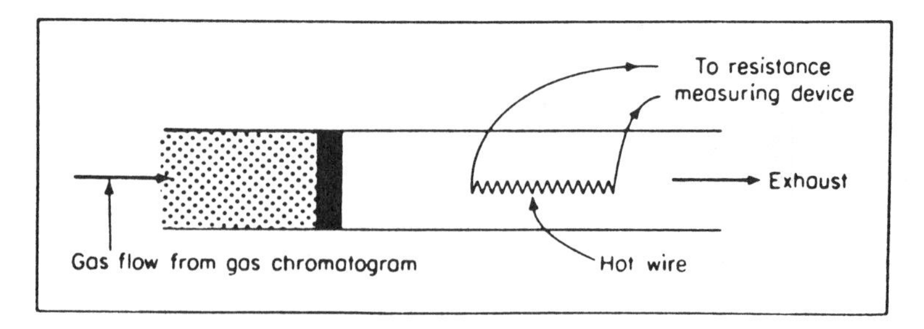

Figure 13.18 Schematic diagram of a thermal conductivity detector.

the gas chromatograph emerges from the column, it flows through the detector. The detector consists of a hollow tube with a wire situated in the central axis. The wire is heated electrically and reaches a steady temperature when the heat gained from electrical energy is equal to the heat lost to the surrounding gas (i.e., the carrier gas). Under these conditions of steady temperature, the electrical resistance of the wire is steady.

When a different gas (e.g., a sample component) flows past the wire, there is a change in the thermal conductivity of the surrounding gas. Heat is conducted away from the wire at a different rate. The temperature of the wire changes and therefore the electrical resistance of the wire changes. This change in resistance is used to detect the presence of different gases flowing past the conductor wire.

The basis of the thermal conductivity detector is the change in thermal conductivity when a sample flows past the conducting wire. For maximum sensitivity, this change should be a maximum. It is therefore advantageous to use a carrier gas with a thermal conductivity significantly different from that of organic gases (sample components). For this reason, helium or hydrogen is the preferred carrier gas.

It should be noted that hydrogen is very reactive chemically. If it is used as a carrier gas, care should be taken to avoid reaction with the sample during an analysis. Such reaction would completely vitiate the analytical results.

Hydrogen has a much higher thermal conductivity than any other gas. For this reason, it can be used as the carrier gas and all other gases can be detected using the thermal conductivity detector. The sensitivity of helium is almost as good as that of hydrogen, since it has a better thermal conductivity than all other gases except hydrogen. It is chemically unreactive and is therefore usually the carrier gas of choice. This is particularly important if low molecular weight gases such as CO_2, O_2, and H_2O are to be detected and measured. However, when the analyses are confined to higher molecular weight gases, nitrogen and argon can be used effectively with some savings in cost.

The hot wire indicated in Fig. 13.18 is part of a *Wheatstone bridge* arrangement as shown in Fig. 13.19 (see Chap. 1). This system can be used in either of two ways. In the first, if there is a change in conductivity of the gas passing over the resistance wire, there is a change in its temperature and a change in its resistance. This causes the Wheatstone bridge to become unbalanced, and a current passes between points A and B. The current flowing can be used to drive a recorder which provides a trace of the current change, which in itself is a record of the gas chromatographic trace.

The second procedure is to use the current flowing between A and B as a negative feedback system to drive a mechanism for varying resistance R_2. When zero current flows between A and B, the feedback system goes to zero, but when R_1 changes, a current flows between A and B. The feedback system changes R_2, which stabilizes at a new resistance R'_1. At this point R'_2, the resistance of the hot wire divided by R'_2, is again equal to R_3 and R_4. In short, the variation in resistance of the sample detector is responsible for varying R_2 to bring the ratio R'_1/R'_2 back to balance. The actual resistance of R'_2 is recorded, and this provides us with a gas chromatogram as before. This system is the *null point system* and is often preferred to one that measures the current across the bridge directly.

The thermal conductivity detector is not the most sensitive detector, but it is satisfactory for a wide variety of analytical applications. It detects all components with a thermal conductivity different from that of the carrier gas. Also, it is nondestructive. This is a distinct advantage over other detectors if the sample

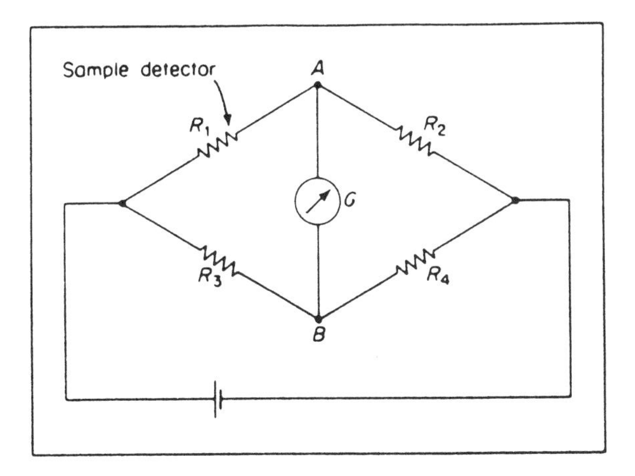

Figure 13.19 Wheatstone bridge arrangement.

must be trapped after separation and used for other purposes. It is commonly used in conventional GC, but it is not sufficiently sensitive for capillary column GC (see Sec. D.2).

g. Electron Capture Detector

The electron capture detector has been used to advantage when high sensitivity and selectivity toward halide, phosphorus, or nitrogen compounds are required. The schematic diagram of the detector (Fig. 13.20) is very similar to that shown in Fig. 13.17. The source of electrons, or β-rays, is $^{3}H_2$ (tritium) or ^{63}Ni foil rather than ^{90}Sr. Tritium is a radioactive gas that emits β-rays. It is contained (absorbed) on platinum foil. The β-rays, or electrons, traverse the effluent from the gas chromatogram. Many compounds, such as paraffins and simple hydrocarbons, are virtually transparent to the electrons, but organic halide, phosphorus, and nitrogen compounds are not, and they capture the electrons. There is an abrupt change in the number of electrons reaching the collector, and this provides the detection signal.

The best carrier gas for use with this detection system is nitrogen or hydrogen. Noble gases such as argon are unsatisfactory because they become excited (not ionized) and act as in the argon ionization detector. As noted for the

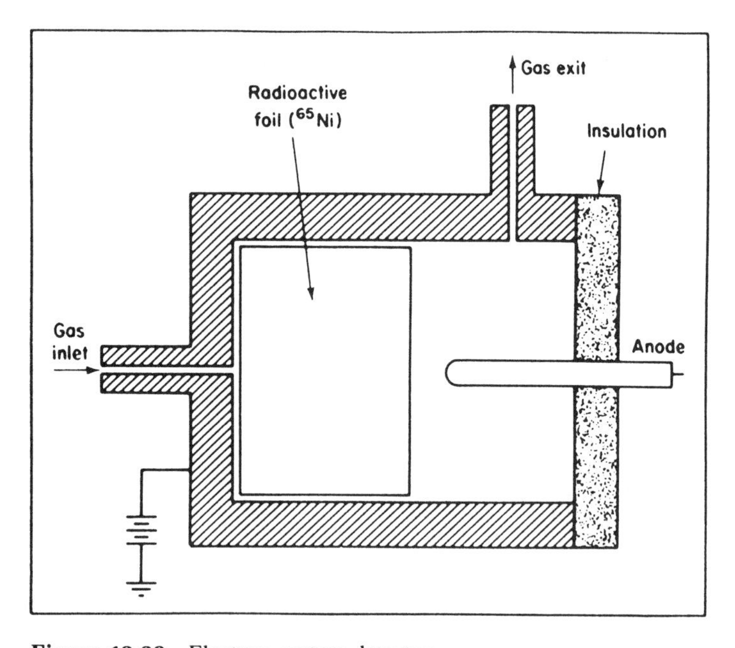

Figure 13.20 Electron capture detector.

katharometer, care must be taken to avoid chemical reaction when hydrogen is used as the carrier gas.

h. Sensitivity

The sensitivities of the detectors described above vary significantly. They are given in Table 13.2.

An examination of the mode of operation and these detectors emphasizes the importance of maintaining a steady (carrier gas) flow rate. Any change causes a signal in the detector that could be mistaken for a component. It might also cause a drifting baseline, which would introduce errors in quantitative analyses.

Most detectors measure a change in the electrical resistance, or the conductivity, of some part of the detector. The common method of measuring a change in resistance of the detector is to make it one arm of a Wheatstone bridge (Fig. 13.19).

The signal from the detector is made to operate a pen. This enables us to record the signal and therefore the amount of sample passing the detector. The recorder moves steadily with time. The result is a chromatogram. A typical chromatogram is shown in Fig. 13.21. If an AC current is used instead of a DC current, the current flowing through the galvanometer wire is AC. This current can be amplified electronically, and the resultant equipment is significantly more sensitive than when a DC Wheatstone bridge is used. Alternately, the read out may be recorded on a tape and presented on a screen for further use. There are literally thousands of other illustrations which could have been chosen since

Table 13.2 Sensitivity of GC Detectors

Detector	Sensitivity (g)	Linear range	Comments
Thermal conductivity	10^{-8}	10^{4}	Universal sensitivity; nondestructive
Flame ionization	10^{-13}	10^{6}	Detects all organic compounds; the most widely used GC detector; destructive
Electron capture	10^{-13}	10^{2}	Detects halo-, nitro-, and phosphorus compounds; response varies significantly; nondestructive
Flame emission	10^{-11}	10^{3}	Sulfur and phosphorus compounds; response varies widely with compound; destructive
Gas density balance	10^{-5}	10^{5}	Universal; low sensitivity; nondestructive
Argon ionization	10^{-12}	10^{5}	Universal; argon carrier gas necessary; nondestructive
Cross section	10^{-6}	10^{5}	Universal; detects major components

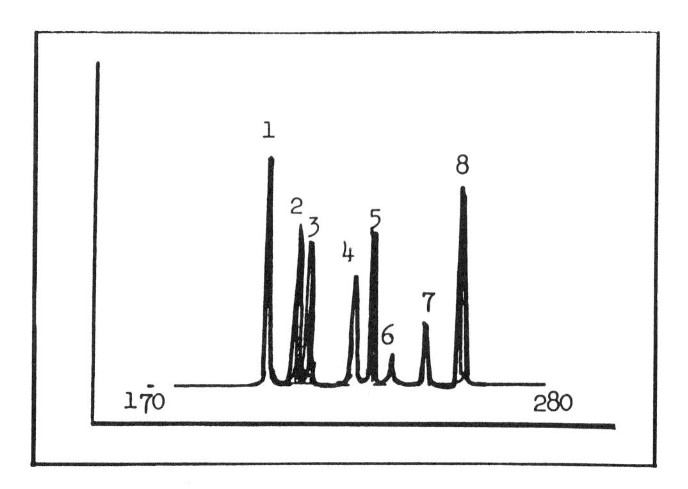

Figure 13.21 Gas chromatographic trace of street drugs; (1) procaine, (2) methadone, (3) cocaine, (4) codeine, (5) morphine, (6) unknown, (7) heroin, and (8) quinine.

chromatography has now become a cornerstone in the analysis of biochemical, medical, chemical, pharmaceuticals, environmental, clinical, and many other endeavors.

D. BRANCHES OF GAS CHROMATOGRAPHY

There are a number of different analytical requirements that elicit different aspects of gas chromatography. For this reason, modification of the equipment has led to the development of the distinct branches described in the following sections.

1. Packed-Column Gas Chromatography

Packed-column gas chromatography, as the name implies, involves the use of packed columns with internal diameters of about 0.25 in. The tube may vary in length from 3 to 20 ft. This method, the workhorse of gas–liquid chromatography, is used for all forms of conventional organic molecular analysis, including both qualitative and quantitative analysis. It is most satisfactory for quantitative analysis because the reasonably large sample size allows accurate measurement. Samples of the order of 0.01 ml are usually analyzed. The resolution is reasonable, but not excellent. The injection of a larger quantity of sample permits the trapping of components as they emerge from the column. The trapped fractions can be positively identified by other methods, such as IR absorption spectroscopy, UV absorption, NMR, or mass spectrometry, or they may be interfaced directly with MS-FTIR or diode array detectors. The large sample size allows the use of all conventional detectors. Highly sensitive detectors are not required and may be overloaded by these large quantities. To prevent overloading it may be necessary to split the sample into two streams before the components reaches the detector. A large portion is thrown away and a small portion goes through the detector. The splitting is performed by a sample splitter (Fig. 13.22).

The splitter separates off the bulk of the sample. In doing so, it allows a constant fraction of all samples to reach the detector. It is important that the fraction that reaches the detector be a constant fraction of the sample; otherwise quantitative interpretation is impossible. Detectors that need the protection of sample splitters in packed-column gas chromatography include flame and electron capture detectors.

The use of packed columns permits many variations of substrate and support. The substrates can therefore be selected for use at high or low temperatures or for programmed temperature work (see Sec. D.4). This type of column is therefore widely used for routine analysis; however, it has neither the high resolution of capillary columns nor the high sample capacity of preparative-scale columns, each of which is described in the following sections.

2. Capillary Column Gas Chromatography

Very narrow columns (less than 0.02 in. in diameter) are used in capillary column gas chromatography. The capillary columns, which may be several hundred feet long, are too thin and too long to pack with a support coated with a substrate. Usually, therefore, they are not packed; instead, a liquid substrate coats the inside walls of the column. The smooth unpacked interior considerably reduces the *A* (geometric) factor in the Van Deemter relationship and contributes to the high resolving power of the technique. In addition, the low pressure drop per unit length allows very long columns to be used.

The inside walls of the capillary are coated with substrate by forcing the latter through the column under high pressure, a difficult operation that requires special equipment. In general, capillary columns are prepared by the manufacturer and are sold ready to use. Samples of the order of 1 μg are analyzed, but even these may require size reduction with a sample splitter to avoid swamping

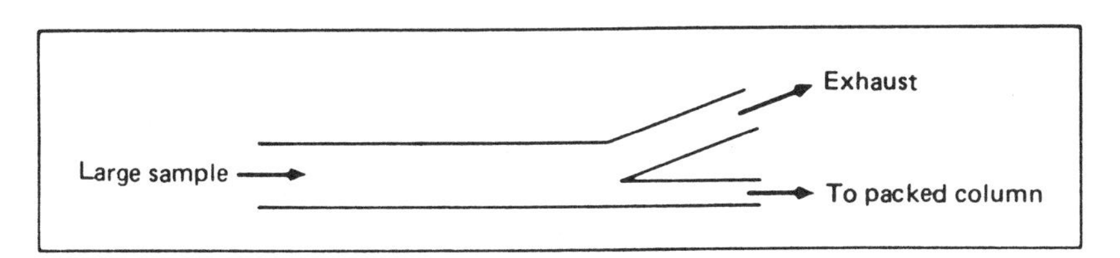

Figure 13.22 Schematic diagram of a sample splitter.

the detector. Because of the small sample size, highly sensitive detectors such as flame ionization electron capture detectors or argon ionization detectors must be used for this branch of gas chromatography. The method is capable of very high resolution: 10^5 theoretical plates on a single column is not uncommon. It is therefore excellent for multicomponent mixtures, particularly if these are present in small concentrations that might not be easily resolved or noticed in the presence of the large chromatographic peaks generated by major components in packed-column chromatography. An example is the trace developed by Milos Novotny on the volatile components of a body fluid (Fig. 13.23). The sample cannot be recovered after injection into these columns, particularly if destructive flame ionization detectors are used. The column is therefore unsuitable for isolating pure compounds.

3. Preparative-Scale Gas Chromatography

Preparative-scale gas chromatography is more important to the organic chemist than to the analytical chemist. Large-bore packed columns are used with any suitable nondestructive (thermal conductivity) detector. The most-common size

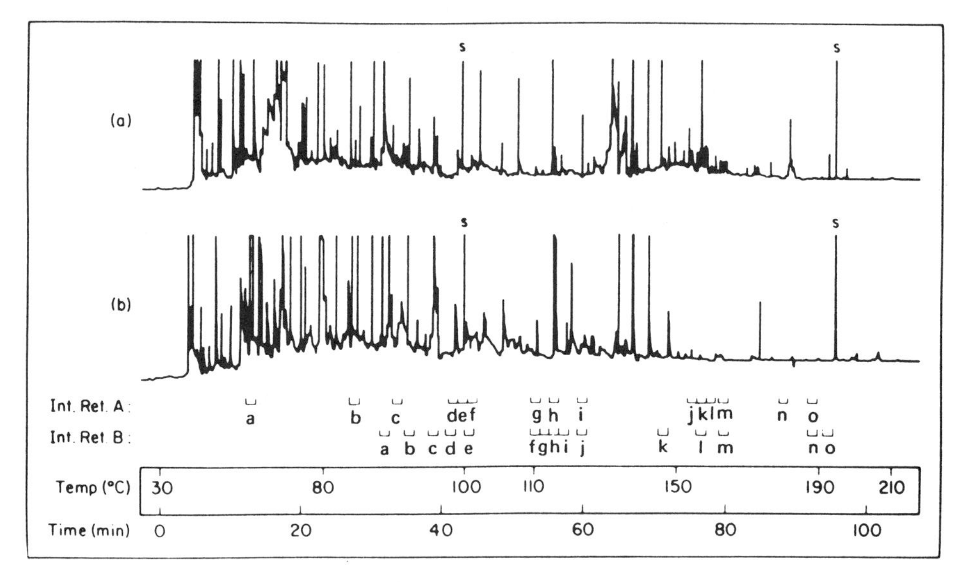

Figure 13.23 (a) Chromatogram of the urinary volatiles of a diabetic male. (b) Chromatogram of the urinary volatiles of a normal female. Scale markings indicate intervals selected as significant in the distinction of normal and diabetic profiles. (Courtesy of M. Novotny, Indiana University.)

is 5 cm in diameter. This system lacks high resolution, but progress has been made in improving the resolution of prep scale chromatography. This has revolved around improving the *A* factor in the Van Deemter equation. This was done by decreasing the particle size of the substrate by using microparticles of silica and optimizing the flow rate. Some column designs have used compression of the columns to reduce voids in the packing. Mixtures of components that are not difficult to separate can be separated on such columns. Relatively large samples (up to 100 ml) can be handled with comparative ease. The components are usually trapped as they leave the column, at which stage they are very pure. The analytical chemist may use the trapped fractions for identification when the mixture's components are unknown.

One of the most important uses of preparative-scale gas chromatography is in the preparation of very pure organic compounds. By trapping the required fraction but not the other fractions, it is possible to obtain a compound in a very pure form. Thus chemists need not go through the tedious steps of multidistillation or multicrystallization. Conversely, the other components, which include the impurities, may also be trapped and identified. Identification of side products in an organic reaction provides important information to the organic chemist regarding the mechanism of the reaction.

In analytical chemistry, preparative-scale gas chromatography is used for trapping the components of a mixture of organic materials and identifying each of them by methods such as IR absorption spectroscopy, NMR, and mass spectrometry. This analytical information is of great service to the chemist, chemical engineer, and pharmacist, and to all concerned with the manufacture or research synthesis of organic and some inorganic compounds. Also, preparative-scale gas chromatography can be used to reveal the source of unwanted impurities. It has been used to collect and purify pollutants in air or water. It has also disclosed impurities in many forms of samples.

Commercial operations based on preparative-scale gas chromatography have been developed as a means of making large quantities of highly pure compounds. Large columns up to several feet wide have been investigated, with encouraging results indicating that gross samples of many liters can be purified this way. This technique can be used to provide a commercial source of very pure compounds that hitherto have had to be prepared individually by the bench chemist. Compounds such as pharmaceutical drugs and fine chemicals produced by preparative-scale gas chromatography are purer, cheaper, and more readily available than those purified by other means, such as distillation.

As a commercial means of purification, preparative-scale gas chromatography competes with distillation, the cost of which increases rapidly as increased purity is required. Figure 13.24 indicates the relative cost per pound of a given compound, at various degrees of purity, produced by preparative-scale gas chromatography and by distillation.

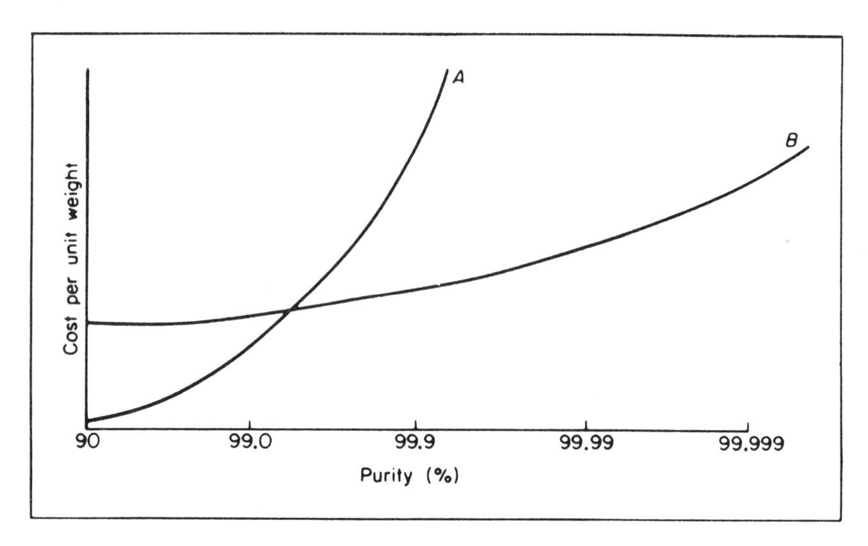

Figure 13.24 Relative cost per pound of a given compound at various degrees of purity: (A) distillation and (B) preparative-scale gas chromatography.

Specific commercial applications of preparative-scale gas chromatography include (1) the identification of flavors from coffee, tea, wines, beer, whiskey, and citrus fruits; (2) the production of research chemicals; (3) the reduction of the toxicity of pharmaceuticals along with increased purity; and (4) the preparation of pure drugs for medical research.

4. Programmed-Temperature Gas Chromatography

In order to obtain reproducible analytical results, it is necessary to control the temperature of the column very carefully. The temperature directly affects the tendency of organic compounds to enter the gas phase and therefore affects K, the distribution coefficient. At low temperatures, if the boiling point is high, the compound will spend most of its time in the stationary phase and emerge from the column only after a prolonged time period. Under these circumstances, the GC peak is very much broadened and the data not very useful. However, the temperature may be controlled such that the vapor pressure of each component is reasonably high. It should not be so high that the components are volatile and boil. If compounds boil, they spend their time in the gas phase and pass through the column without separating. In general, however, a suitable temperature can be found at which all compounds spend appreciable but different times in the gas phase. If the sample contains compounds with widely different boiling points, however, a suitable single temperature cannot be found at which each component spends a reasonable fraction of time in the gas phase.

This problem was overcome by using a programmed-temperature procedure, which was developed by Steve dal Nogare of du Pont. In this method, the sample is injected into the column in the normal way. The temperature of the column is maintained at some suitable low temperature, such as 50°C, during injection. The column temperature is then increased at a controlled rate (e.g., 20°C/min) up to a maximum temperature as high as 300°C. Higher temperatures are not commonly used because at such temperatures the substrate may be lost by vaporization, and in the process it may coat and destroy the detector. To avoid this problem, capillary columns have been developed using a deactivated metal surface as the stationary phase. Normally, metals catalyze decomposition or reaction, but deactivation prevents that problem. Temperatures as high as 430°C have been successfully used for the analysis of waxes. It is also necessary to check that the sample components are stable at these temperatures. If the components decompose or isomerize, the analyses will be unreliable because any such fractions collected will have originated in the column and not the sample. It is important that the column be heated uniformly during the analysis; otherwise the sample components will be diffused and column efficiency decreased. It should be noted that although the column may be at a low temperature initially, the inlet port must be maintained at a high temperature (350°C) to ensure rapid vaporization of the sample after injection and subsequent transport onto the front of the column.

At the beginning of the chromatogram, the temperature of the column is low. The low-boiling components emerge in an orderly fashion and can be resolved. As the temperature increases, the vapor pressure of the middle- and higher-boiling components increases and they in turn emerge from the column and are resolved and analyzed.

This technique has extended the use of gas chromatography to the analysis of mixtures containing components with a wide range of molecular weights. Typical examples include (1) alcohols from CH_3OH to $C_{20}H_{41}OH$, (2) paraffins from CH_4 to $C_{40}H_{82}$, (3) olefins from C_2H_4 to $C_{40}H_{80}$, and (4) natural fatty acids (up to $C_{15}H_{31}COOH$). There are numerous examples in other molecular species, such as amines and halides. Also, mixtures of these compounds can be successfully separated and analyzed via this technique. A typical chromatogram is shown in Fig. 13.25. It is not difficult to imagine that the separation of such a mixture by any other means would be virtually impossible.

A serious problem with programmed temperature GC is that at the elevated temperatures used, the sample compounds may decompose or isomerize. This leads to a direct error in the analyses obtained. This problem may be overcome by using liquid chromatography, usually run at room temperature where sample molecules are stable.

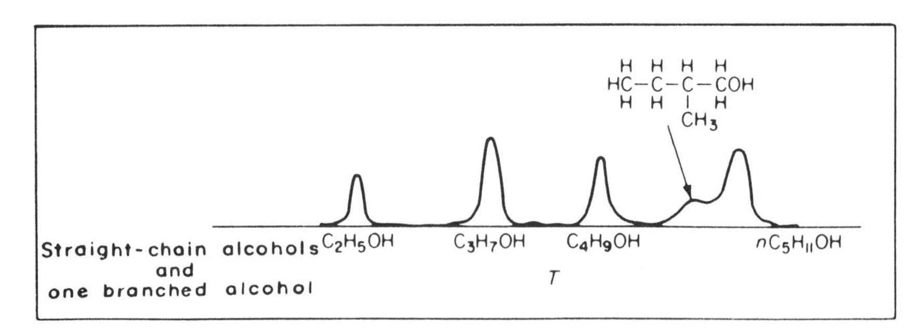

Figure 13.25 Typical chromatogram of alcohols.

E. ANALYTICAL APPLICATIONS OF GAS CHROMATOGRAPHY

The principal applications of gas chromatography are the qualitative and quantitative analysis of liquids, gases, and vapors, particularly of organic compounds. Any stable compound that can be vaporized below 300°C can be determined by this method. It should be noted that the compound must be stable with respect to isomerization and decomposition at these temperatures, or the method gives erroneous results. Compounds that are unstable at these temperatures or that are not volatile can be analyzed by liquid chromatography. This will be discussed later.

1. Qualitative Analysis

The most widely used method for qualitative identification is based on the retention time of the sample. This is the time difference between insertion of the sample in the injection port and the appearance of the maximum of the chromatographic peak on the recorder. This is the *retention time* t_r. A more accurate time is *adjusted* retention time, which is corrected for the retention time of unretained material such as air. Consequently,

$$\text{adjusted } t_r = \text{unadjusted } t_r - T_{air} \text{ or } t_r - t_0 \qquad (13.11)$$

If the conditions of the column are kept constant, the retention time of a particular compound will be constant. Consequently, to identify two compounds that may or may not be the same, we may inject them into a gas chromatograph. If their adjusted retention times are not equal, they are not the same compound. On the other hand, if the adjusted retention times are equal, there is a much increased possibility that they are the same compound. But this confirmation is not complete, since it is possible for two compounds to have identical retention times. Even if the retention times are very similar but not identical, it is difficult to distinguish between them if they are analyzed at different times. Consequently, a more certain method of identification is to follow the following sequence.

Suppose we suspect that an unknown material is naphthalene. We run the sample in the normal way and identify the suspected naphthalene peak on the gas chromatogram. We then take a second aliquot of the sample and add naphthalene to it. The mixture is then analyzed as before. If the compound is indeed naphthalene, then the suspected chromographic peak will increase in size. If, however, the compound has a similar but not identical retention time, then we will find two overlapping peaks that provide one unsymmetrical peak. This method of identification is fairly but not completely reliable, because it is possible, though unlikely, that compounds with identical retention times may be confused.

Another method is to run the sample through a second column with a different stationary phase. If the retention times still come out the same, the probability of correctly identifying the unknown peak is again greatly increased. If the sample is run through a third column and the same retention time obtained, the chance that the sample and standard are not the same is vanishingly small. This technique, although time-consuming, is very valuable in environmental analysis, where position identification may be vital. The technique was developed by Ed Overton, Department of Environmental Studies, L.S.U.

Qualitative analysis of the components of a mixture of organic compounds can be carried out by separating the components by gas chromatography and then identifying the pure components. This can be done by trapping the components as they emerge from the column (Fig. 13.26). A separate trap is used for each component.

The material caught in the trap may be identified by IR spectroscopy, NMR, or mass spectrometry. This procedure is effective but time-consuming.

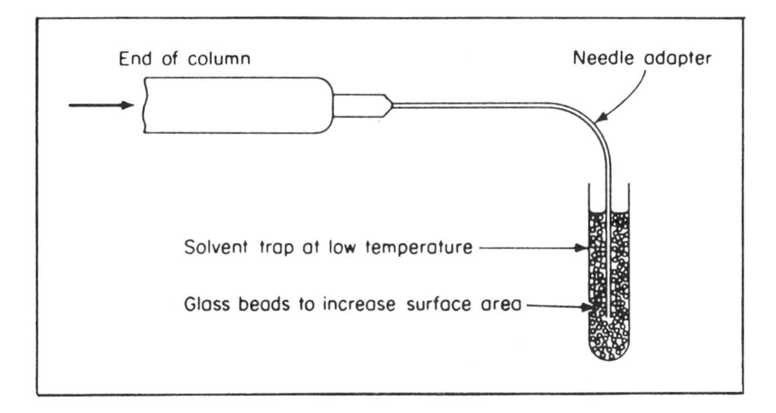

Figure 13.26 Schematic diagram of a trap for a component emerging from a gas chromatography column.

Recently, equipment has been developed to carry out the identification directly. This can be done by attaching a mass spectrograph to the end of the gas chromatographic column. The components emerging from the column are fed directly into the mass spectrometer (MS). As each different component enters the mass spectrograph, it is fractured and passes down the magnetic field in the normal manner. Each component may be identified from its cracking pattern. (When a molecule enters the ionization chamber of a mass spectrometer, it "cracks" into fragments. The type and number of fragments constitute that molecule's cracking pattern. This and related phenomena are discussed more fully in Chap. 15.)

The mass spectrograph must be able to record the cracking pattern of one component before the next component emerges from the gas chromatographic column. With a very fast response instrument, the minimum separation time is about 1 sec. With this GC–MS combination, any two compounds that can be separated by 1 sec on a gas chromatographic column can be identified by MS without prior separation and collection. This procedure saves a great deal of time and enables us to identify compounds that could not be collected in a trap with such a short time separation. The most common mass spectrometers used for the work are the quadrupole MS and the fast scan sector MS.

As an alternative, a rapid-scanning FTIR absorption spectrometer can be attached to the end of a gas chromatographic column. Again, the emerging components are fed directly into the sample cell of the instrument. The IR absorption curve can be obtained within 3 sec, allowing immediate analysis of successive peaks. Organic functional groups of the components may be identified this way. Because the components have recently emerged from the gas chromatographic column, they are generally pure compounds. This enables us to make positive identification of the compound with an increased degree of confidence, because no other compounds are present to complicate the spectra. By using two sets of techniques, gas chromatography plus mass spectrometry and gas chromatography plus IR spectroscopy, we can obtain the molecular weight, cracking pattern, and IR absorption trace for the different compounds of a mixture. This information is frequently sufficient to enable us to identify each compound present in the mixture. It is normal to use FTIR for this step.

2. Interfacing Between Gas Chromatography and Mass Spectrometry

In order to interface GC and MS successfully, several requirements must be met. First, the GC peak may pass the GC detector within a few seconds. It is therefore necessary that the mass spectrum be obtained within an extremely short period of time to relate the GC and the mass spectrum. Second, the effluent from a gas chromatograph contains a very high preponderance of the carrier gas. If this were injected into the mass spectrometer, the latter would be flooded and the mass lines of the sample would be greatly diminished. It is therefore necessary to eliminate much of the carrier gas before injection into the mass spectrometer. Third, when the data are provided by the mass spectrometer, it is necessary to record this at extremely high speed. These problems have all been overcome and GC–MS interfacing, which is now routine in many laboratories across the country.

The mass spectrometer of choice is the quadrupole, which has the advantage of being able to accept samples over a wide angle. This is particularly useful in GC–MS interfacing. It is also easy to program a quadrupole mass spectrometer to complete an analysis within a few seconds. Unfortunately, the resolution of the quadrupole mass spectrometer is not equal to that of the conventional magnetic mass spectrometer, but for qualitative identification, which is all that is required in this instance, the quadrupole mass spectrometer is usually quite satisfactory. Recent modifications in the drift chamber and magnet have led to the development of high-speed mass spectrometers capable of resolutions of 10,000 with a recording time of 1 sec per decade. Other systems use the ion trap MS, which also has a fast response time.

The elimination of the carrier gas is carried out by using a *jet separator* as shown in Fig. 13.27 The effluent from the gas chromatograph proceeds through a fine capillary and is directed onto a larger-bore capillary, which leads to the inlet of the mass spectrometer. During the short transition from the GC effluent to the mass spectrometer inlet, the gases diffuse in all directions.

To take advantage of this diffusion, helium or hydrogen is used as the carrier gas. Each of these gases has a very high diffusion rate. Upon leaving the GC capillary column, most of the helium or hydrogen diffuses out of the path of the jet stream and does not enter the inlet of the mass spectrometer. The jet separator is continuously pumped out with a vacuum pump.

The sample compounds issuing from the GC also diffuse, but at a much slower rate. To allow for some diffusion, the capillary column entering the mass spectrometer has a wider diameter than the exit from the GC.

This technique permits a great concentration of the components of the gas chromatograph and elimination of much of the carrier gas. It is necessary, of course, to use either hydrogen or helium to achieve the high sample concentration.

The third problem involves the recording of the data issuing from the mass spectrometer. Clearly a pen-and-paper type of recorder would not be suitable for such high rates of recordings. This problem is overcome by recording directly onto magnetic tape. The magnetic tape is quite capable of recording all the data that can be generated by the mass spectrometer within 1 or 2 sec. It can then be computerized and the signal printed out at leisure. In practice, mass spectra can be taken automatically every 15 sec of a GC run and recorded directly onto mag-

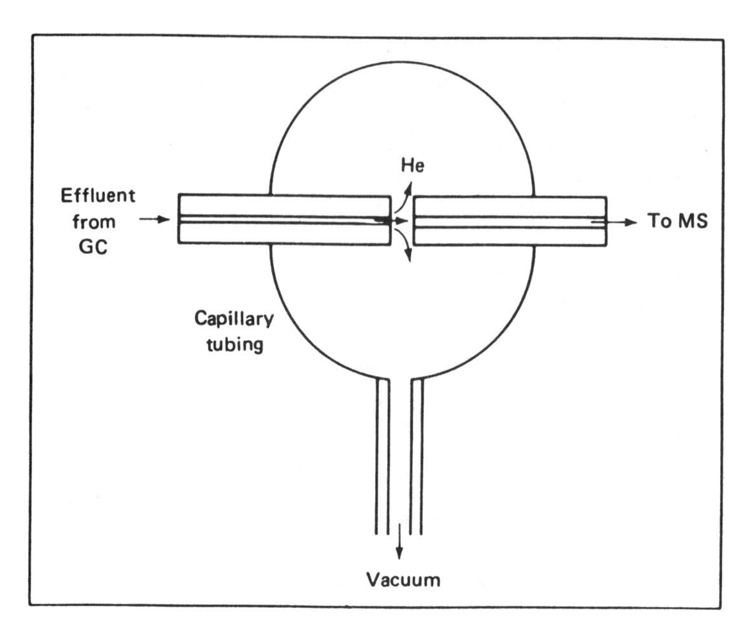

Figure 13.27 The GC-MS jet separator. Diffusion of the carrier gas (He) leads to a loss of helium during the transfer from the GC to the MS. The sample is concentrated in the process. Note that the capillary to the MS entrance is wider than the capillary from the GC to accommodate some diffusion of the sample.

netic tape. After completion of the gas chromatographic separation, the operator can demand the computer to record any peak at will or, better still, record a peak and take away the background at the beginning of the peak and display the net signal. This is a more precise representation of the mass spectrum of the material in the compound. In addition, with the use of computers, the mass spectra can be interpreted with some degree of competence from MS memory banks. It is possible to identify minor components even in the presence of major quantities of unresolvable background material. Figure 13.28 illustrates a GC trace of a sludge material found in the bottom of a commercial reaction vessel. By obtaining the mass spectra at positions 4 and 3, the mass spectrum of compound 4 could be deduced as the difference between these two, that is, the mass spectrum of 4 minus the background 3 at that point. Similar identification could be made for compounds corresponding to peaks 2–6. The information obtained was pertinent to identifying the source of the sludge.

Techniques have now been developed to attach FTIR systems directly to the effluent from the GC column. This is more difficult to do than the GC–MS interface, but it has been achieved by using a gold-plated capillary light pipe. The latter is heated above the column temperatures to avoid sample plating out. It is

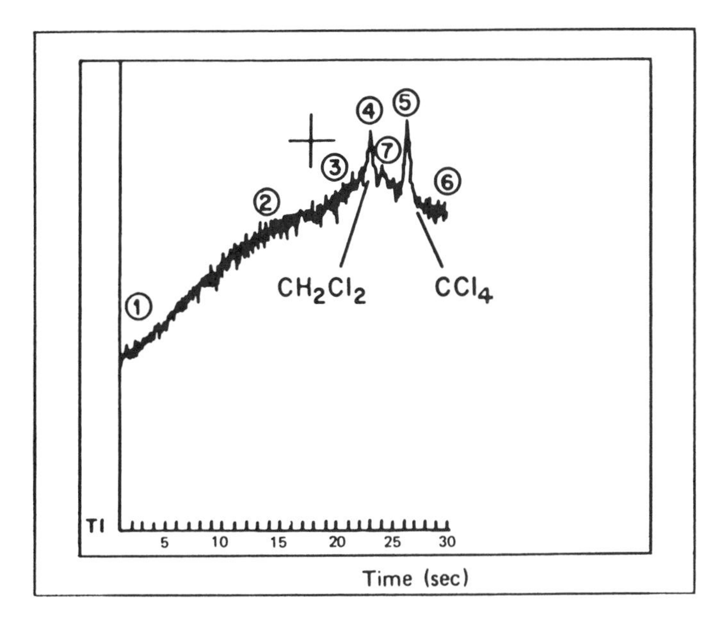

Figure 13.28 GC trace of sludge obtained from commercial process. Mass spectra of peaks 4 and 5 can be obtained by correcting for background MS at positions 3, 7, and 6.

important that the mobile phase not absorb in the infrared. The ends of the capillary have transparent windows attached.

The net result is that the IR spectra can be obtained as effluents are emerging from the column. This is extremely valuable in characterizing complex mixtures. Used in conjunction with a quantitative detector such as thermal conductivity, quantitative and qualitative analysis can be rapidly obtained. The use of GC–MS and GC–FTIR provides a reliable approach to solving analytical problems of complicated, relatively unstable samples.

3. Quantitative Analysis

The *area* of a chromatographic peak is proportional to the number of gas molecules (and, therefore, moles) of the component reaching the detector. Unfortunately, none of the detectors available respond to the same extent for an equal number of molecules of different compounds. For this reason, a *response factor* must be ascertained experimentally for each compound being quantitatively determined. The conditions under which the response factor is determined must be the same as those in which the samples are analyzed. We can obtain the response factor experimentally by injecting a known amount of the compound into the gas chromatograph and then measuring the areas of the relevant response peaks.

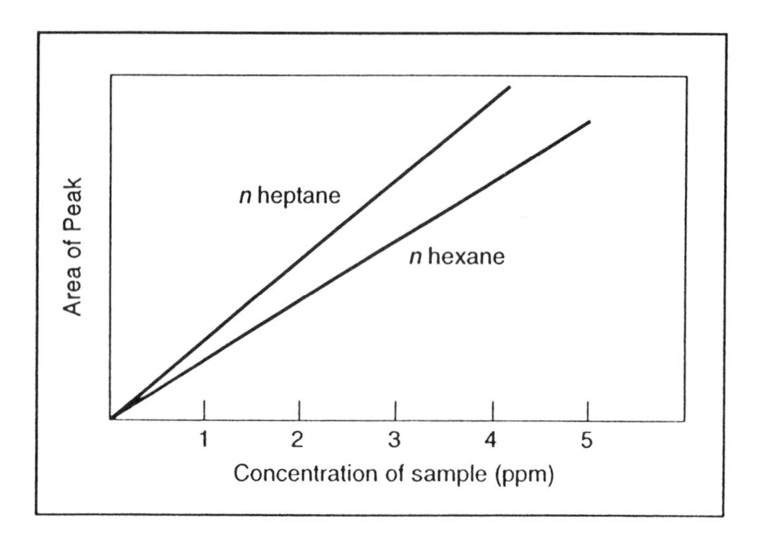

Figure 13.29 Calibration curves for *n*-heptane and *n*-hexane showing a difference in response per weight of sample.

For example, using a thermal conductivity detector, we may set up a calibration curve by injecting a series of hexane samples of volumes 0.01 to 0.1 m and preparing a calibration curve such as that shown in Fig. 13.29. The response is dependent on how much sample was injected and the difference in thermal conductivity between *n*-hexane and the mobile phase carrier gas, e.g., argon.

If we repeat the procedure for n-heptane using the same volumes and the same carrier gas, the differences in thermal conductivity are greater and a different calibration curve is obtained, as shown in Fig. 13.29.

This type of difference in response for different compounds holds true for all types of detectors, so quantitative analysis must take this into effect.

The response factor is given by the relationship

$$\text{response factor} = \frac{\text{area of peak}}{\text{weight or volume of component A}}$$

$$= \text{area per unit weight (or volume) component A} \qquad (13.12)$$

For example, if the volume of component A injected equaled 0.11 ml and the peak area equaled 3.3 cm^2, the response factor would be 3.3/0.11 = 30 cm^2/ml of component A.

For 0.35 ml of an actual sample containing an unknown concentration of component A analyzed under the same conditions, it was found that the area of the relevant peak was 2.1 cm^2. The calculations of how much compound A was present in the original sample are as follows:

$$\text{total volume of sample injected} = 0.35 \text{ ml}$$

$$\text{area of component A peak} = 2.1 \text{ cm}^2$$

$$\text{response factor for component A} = 30 \text{ cm}^2/\text{ml of component A}$$

$$\text{volume of component A} = \frac{2.1 \text{ g cm}^2}{30 \text{ cm}^2/\text{ml}} = 0.07 \text{ ml}$$

$$\text{volume \% of component A in sample} = \frac{0.07}{0.35} \times 100\% = 20\%$$

With the same approach, the response factor for all components of a mixture can be obtained and a complete quantitative analysis performed, provided that conditions are kept constant and the response factors remain constant. Once these factors are obtained, routine analysis is rapid and reliable. It should be noted that the response factor for component A will not be the same as the response factor for some other component, even though the components may be chemically similar.

It is incorrect to calculate the concentration of the sample based on the area of the sample peak and the total area of the peaks eluted. It is quite possible that some components of the sample never eluted, so the total area is not a measure of the total weight or volume of sample injected. In addition, the area of each peak must be correlated with a response curve to the actual quantity present.

If the peaks are very narrow, peak height can be used as an affirmate measurement of peak area. This relationship does not hold if the peak is not narrow or is distorted, and the method should be avoided if possible.

F. LIQUID–SOLID CHROMATOGRAPHY

A mixture of liquids can be separated by being passed over a solid substrate in the same manner as gases are separated by being passed over a liquid substrate. The liquid solvent is the mobile phase and the solid is the stationary phase. As with gas chromatography, the difference in the flow rate of the components of the mixture depends on how much time is spent by each component in the stationary phase and how much is spent in the mobile phase. A suitable inert solvent is used as the mobile phase in the same way an inert carrier gas is used in gas chromatography. The ratio of time spent by the sample in the solvent and on the surface of the stationary phase depends on the solubility of the sample in the solvent and the absorption effects of the stationary surface on the sample molecules. Each of the factors in the Van Deemter equation is equally pertinent in liquid–solid chromatography; that is, geometry, diffusion, mass transfer, and flow rate. i.e., $H = A + B/V + CV$

However, there are some important differences between liquid–solid and gas–liquid chromatography that should be noted. In gas–liquid chromatography, the proportion of time the sample spends in the stationary liquid phase and in the mobile gas phase is governed primarily by the affinity (or solubility) of the sample molecules for the stationary liquid phase. Provided that the stationary phase does not become saturated with sample molecules, the capacity of the column stays reasonably constant and will operate with acceptable efficiency for extended periods of time. Also, different gases will have different volatilities. If two gases have equal solubilities in the stationary phase, but different volatilities, they will have different retention times. The liquid mobile gas phase is inert and does not significantly affect the equilibrium. It is generally considered that all gases are soluble in all other gases, provided that no chemical reaction takes place.

This situation is modified in liquid–solid chromatography. As before, the sample molecules distribute their time between the stationary solid phase and the mobile liquid phase. As before, the sample molecules progress down the column when they are in the mobile phase. The equilibrium is disturbed by two other variables, however. First, the sample molecules have a finite solubility in the mobile liquid phase (solvent). The lower the solubility, the less time the molecules will spend in the mobile phase. The solubility changes greatly from one solvent to another. So, in contrast to gas–liquid chromatography, the choice of the mobile phase greatly affects the rate of progression of the sample along the column.

As with gas chromatography, the theoretical plate height H is dependent on A, the geometry of the system. This relationship is illustrated in Fig. 13.30, which shows the relationship between H and V using substrate with different particle size S.

It can be seen that as A decreased from 37 μm to 10 μm, H decreased considerably. It is also noteworthy that the negative effect of C with increasing flow rate was greatly reduced when A was as small as 10 μg. Presumably, the equilibrium of the analyte between the mobile phase and the stationary phase was established more rapidly with the increased surface area per gram of the substrate. Analyte molecules were therefore not separated so much at high flow rate (V), and resolution was maintained.

As a direct consequence, higher flow rates can be used without sacrificing resolution. Faster analyses can therefore be carried out with the many advantages that follow faster throughput.

The interaction between the sample molecules and the stationary solid phase is also different from the simple solubility-controlled phenomenon encountered in gas–liquid chromatography. In liquid–solid chromatography, the sample molecules do not dissolve in the solid stationary phase; rather, they are adsorbed on its surface.

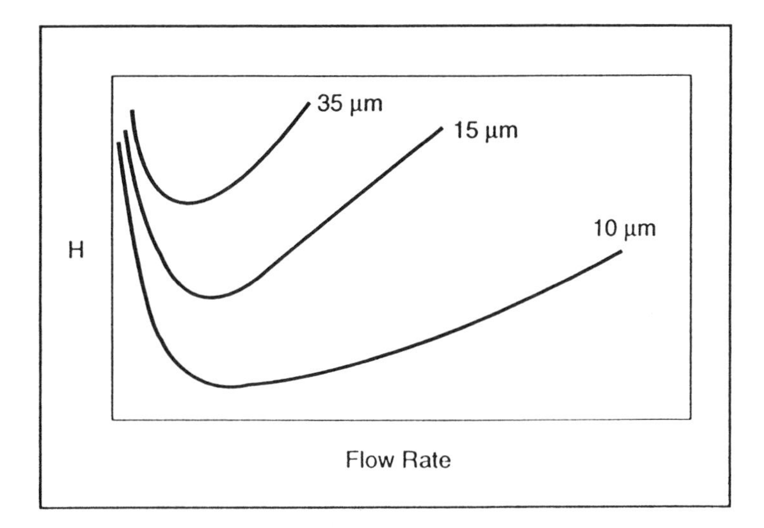

Figure 13.30 Effect of particle size on the relationship between *H* and flow rate. Note improved efficiency (lower *H*) with decreased particle size.

The phenomenon of adsorption is not simple or haphazard. The sample molecules are not adsorbed by all parts of the surface, but only at "active sites." The true nature of active sites is not clearly understood. We do know that they are dependent on local electrical charges created by the surface molecules themselves. They are also affected by the dimensions of the crystal lattices that make up the surface. In general, the closer the fit between the crystal lattice size and the dimensions of the sample molecule, the greater will be the affinity between the two and the adsorption effects will be increased.

The active sites, however, attract not only the sample molecules, but also the solvent (mobile phase) molecules. Consequently, there is competition between the solvent molecules and the sample molecules for the active sites. Generally, there is a large excess of solvent molecules present, and because of the mass action effect, the solvent molecules are able to displace the sample molecules from the active sites. By choosing a solvent with low affinity for the active sites, the sample molecules are permitted to remain on the surface of the stationary phase for an acceptable length of time; but if the solvent has a high affinity for the active sites, the sample molecules are quickly displaced and flow rapidly through the column. As can be imagined, the correct choice of solvent and solid substrate is critical in liquid–solid chromatography.

Another problem peculiar to solid substrates is their loss of capacity with time to handle samples of suitable size. The capacity of the stationary phase is directly proportional to the number of active sites on its surface rather than the gross surface area. When the number is high, the capacity is high, and vice versa.

In practice there are often low concentrations of impurities present in the solvent or in the samples analyzed that have a very high affinity for the active sites. They are not easily displaced once they become adsorbed. The active sites become lost to the system and, gradually, as all the active sites become covered with impurities, the surface becomes ineffective as a stationary phase. As a result, chromatographic separations are no longer possible.

The active sites may be released or regenerated by removing the strongly adsorbed species. This can be done by heating the solid to some recommended temperature to volatilize the adsorbed molecules or in certain circumstances by burning them off. After cooling under prescribed conditions, the material should again be ready for use.

A very useful application of liquid–solid chromatography is for the separation of high molecular weight compounds. Typical samples include organic reaction residues, distillation residues, high molecular weight compounds and compounds that may isomerize or decompose in gas–liquid chromatographic columns. Such samples cannot be analyzed by gas chromatography because their boiling points are too high for them to become vapors, even in programmed-temperature gas chromatography, or because the compounds are unstable.

Another problem with reaction residues is that they frequently contain a very large number of different compounds. In such instances, the chromatographic peaks overlap and adequate separation cannot be made. The compounds may be solid at normal temperature, but they may be dissolved in a suitable solvent. Typical samples include high molecular weight esters, acids, olefins, paraffins, vitamins, and biological compounds. They may be separated by functional groups by using a suitable combination of solvents and stationary phase material which allow, for example, the separation of esters from olefins. The esters may then be separated on a different column.

The more important branches of liquid–solid chromatography are described in the following sections.

1. Paper Chromatography

Cellulose filter paper is often used as the stationary phase in paper chromatography. Since it is hydrophylic, it is usually covered with a thin film of water. The procedure is often regarded as liquid-liquid chromatography. A typical setup for paper chromatography is shown in Fig. 13.31. The stationary phase is paper, and the mobile phase is a solvent that flows over the paper. The separation of different compounds is brought about by the difference in time spent in the moving solvent and on the stationary paper.

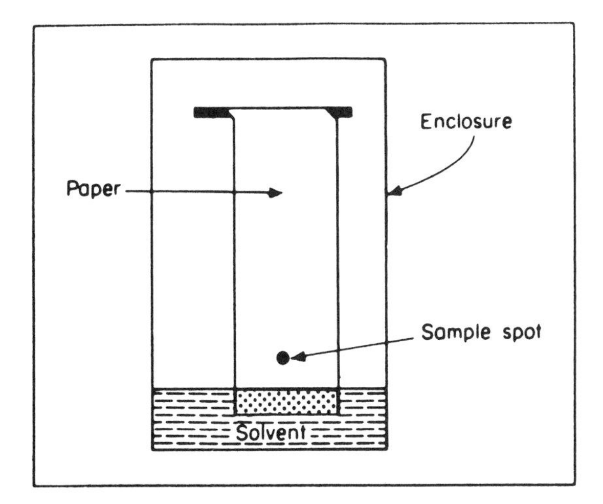

Figure 13.31 Separation by paper chromatography.

In practice, a strip of paper of uniform thickness is suspended with one end dipping into a solvent. The latter is drawn up the paper at a steady rate by the capillary action of the paper fibers. This movement causes the solvent to flow past the sample. As it flows past the sample, the components of the sample dissolve and are displaced along the paper, their rate of movement being proportional to the flow rate of the solvent over the paper and the time each component spends on the stationary phase. Different components flow at different rates and are separated in the process. Of course, it is also possible to arrange the equipment so that the solvent is in a trough at the top of the paper and therefore flows downward. The solvent is helped by gravity in this case and flows faster, but sometimes more unevenly. This separation brings about "development" of the chromatogram.

It is important to prevent air currents from drying out the mobile phase. This would lead to an uneven flow rate of the mobile phase and a distorted chromatogram. This can be achieved by placing the entire system inside an inverted glass jar.

The position of each component can be detected by treating the components with a reagent that colors them and thus reveals their positions. The identity of each component can be determined by comparing the flow rate with that of known samples. For example, if a sample were known to contain three components A, B, and C, the three components could be located by adding separate spots of A, B, and C to the paper and developing the chromatogram until the separation is achieved (Fig. 13.32). When this is done, component A of the sample will have traveled the same distance as the spot known to be compound A.

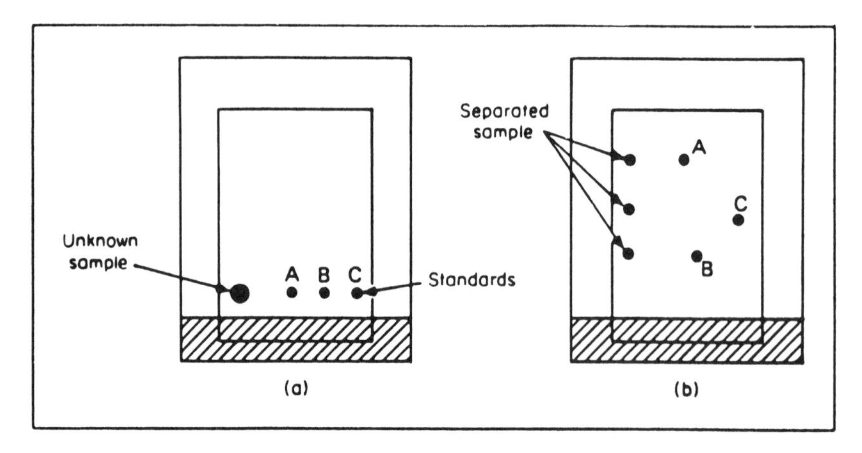

Figure 13.32 Schematic diagram of the location of components (a) before separation and (b) after separation by paper chromatography.

This identifies component A. The other components are located and identified in the same fashion.

If the identity of the components is not known, then after separation in the chromatogram they must be identified by other methods, such as mass spectrometry, IR absorption, NMR, or UV absorption.

2. Thin-Layer Chromatography

Thin-layer chromatography is similar to paper chromatography, except that a thin (0.25 mm) layer of some inert material, such as Al_2O_3, MgO, or SiO_2, is used as the substrate instead of paper. A layer of any one of these oxides is made from a slurry of powder in a suitable inert solvent. The slurry is spread evenly over a flat surface (glass) and dried. It may be spread manually or mechanically. The advantage of using inert substrates instead of paper is that more reactive developing reagents, such as strong acids, can be used to detect the compounds without destroying the substrate.

The detection of molecules or ions in paper or thin-layer chromatography can present a problem. If the molecules are colored, they can be detected visually; however, very few organic molecules are colored intensely enough to be detected with the eye. For some compounds, specific reagents are available that react with compound to form a colored compound that can then be seen. Unfortunately, only relatively few compounds can be detected this way. One important example is lead and lead compounds. After exposure to iodine vapor, the location of the lead is revealed by a brown stain. In the majority of instances, however, the sample compound is transparent and cannot be located visually.

Many large molecules fluoresce under UV light. This is particularly true of polynuclear aromatic compounds, such as benz(a)pyrene. Many of these compounds are significant in medicine, biochemistry, and pollution control. This characteristic provides a convenient method of analysis for such compounds. The sample is separated in the normal fashion and then located by placing it under UV light. A special box can be used that contains short-wavelength UV light, which causes some compounds to fluoresce; the box also contains a second lamp, with longer-wavelength radiation, which causes other compounds to fluoresce. Using two such lamps affords some specificity. The enclosed box contains a viewer so that even faint fluorescence from the sample can be seen by the analyst.

Some high molecular weight compounds cannot be detected with specific reagents and are not fluorescent. These may be located by impregnating the entire chromatographic plate with a fluorescent material. Under the UV lamp the whole plate fluoresces; however, spots occupied by a nonfluorescing sample on a fluorescing layer appear dark. A commercial fluorescent material used for this work is ninhydrin, a commercial name for 1,2,3-nidantrione monohydrate.

Thin-layer chromatography is used extensively for the quantitative determination of high molecular weight compounds, particularly in medical and biological research. Qualitative identification can be based on the rate at which a compound flows along the plate. A compound can be identified by direct comparison with a known compound, as shown above. This requires guessing the identity of the components in order to use pertinent "known compounds." A better method involves measuring the *Rf value* of the sample. This is the ratio of the flow rates of the sample and the solvent over the plate. For any particular system of substrate, solvent, and sample, the Rf value is constant (at a constant temperature). An unknown compound is then compared to compounds with the same Rf value, which are relatively few in number, and subsequent identification is then much easier.

Quantitative analysis of compounds separated by thin-layer chromatography can be carried out by estimating the area of the spot, and, based on prior calibration, an estimation of the amount of material in the spot can be made. More accurate determination can be made by carefully removing the entire sample spot (and the substrate) from the plate. The sample is then removed from the substrate by dissolving it in a known volume of a suitable solvent. It can then be determined by UV absorption, IR absorption, or some other suitable analytical method.

a. Determination of Best Mobile Phase and Best Stationary Phase

Another valuable use of thin layer chromatography is for the correct selection of mobile and stationary phases in columns to be used for the analysis of unfamiliar samples. If an analytical chemist is confronted with an unusual sample to be analyzed by HPLC, a common approach is to guess the best stationary phase and mobile phase to be used. However, it is not uncommon for some components of the sample to remain stationary at the injection point and never emerge. The operator has no way of anticipating this event. However, if the same stationary phase and mobile phase are used on a thin layer chromatograph and the sample tested, it is immediately apparent if all components move down the plate or not.

The technique can be used to test various combinations of stationary phases and mobile phases very quickly. From this preliminary study, the optimum stationary and mobile phases can be used in HPLC to give best resolution and leave no component on the column.

3. Column Chromatography

Larger samples can be separated using a column of substrate rather than a thin layer. The equipment is illustrated in Fig. 13.33. It is the procedure originally used by Tswett. The size of the column is usually about 3 ft × 1/2 in., but it varies considerably according to requirements. The sample (about 1 g) is first dissolved in a suitable solvent and the solution is loaded onto the substrate at the top of the column. More solvent is then added. The components of the sample flow down the column at different rates, again depending on the distribution of time spent in the mobile and stationary phases. They separate and are collected in fractions as they emerge. The large quantities of solvent used may be evaporated from the collected fractions if it is necessary to obtain the components in a pure state. The solvent and the dissolved sample are collected as they drip off the end of the column.

There are several ways this can be done. One way is to use a fraction collector. In this instrument a large number (up to 100) of small tubes are arranged

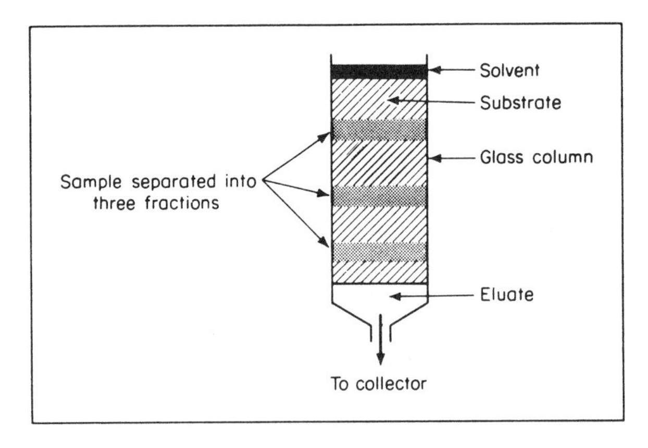

Figure 13.33 Schematic diagram of the equipment used for the separation of larger samples by column chromatography.

on a circular plate. One tube is positioned beneath the column and collects the effluent. After about 5 ml has been collected in the first tube (this volume can be varied at will), the disk rotates, moving the next tube into position to collect the next 5 ml of effluent from the column. The process is repeated many times until the chromatogram is complete and the effluent has been collected in numerous small fractions. Each fraction can then be examined separately and the various components of the original sample identified. If convenient, all the fractions containing a particular compound can be poured together, thus isolating all of that particular component from the original sample. The process can be repeated for all the fractions.

As an alternative method, the components in the effluent from the column can be detected using either their properties of UV absorption or refractive indices. This enables one to monitor the effluent and split it into different portions, each containing one component of the sample. The portions can then be examined in any convenient way.

In column chromatography, numerous materials have been used as substrate, including Al_2O_3, SiO_2, MgO, $CaCO_3$, and resins. Some selectivity can be obtained by using different solvents to separate the components. Typical solvents include heptane, ether, benzene, esters, and alcohol. Their selectivity depends on matching the dielectric constant of the solvent to that of the desired fraction. The solvent selected should be inert toward the sample. The solubility of the different components of the sample should vary somewhat in order to allow good separation of the components on the column. The method is very time-consuming and labor-intensive. It has largely been displaced by HPLC, which is faster and more efficient.

4. Successive Elution

For complex examples it may be necessary to use a series of solvents. For example, a particular solvent may dissolve and elute only one group of compounds, such as hydrocarbons, within a reasonable time, leaving the rest of the compounds containing other functional groups undissolved and close to the top of the column. After the group of compounds that are soluble in this first solvent has been eluted from the column, a new solvent may be used to dissolve and separate a second group of compounds on the column. For instance, the first solvent used in the analysis of a complex mixture of compounds might be heptane. The latter removes paraffins from the sample. Having been removed from the original sample, the paraffins can be further separated by means of a second chromatographic column or by gas chromatography (or other suitable techniques).

A different solvent, such as benzene, which will elute aromatics from the sample, may now be used on the paraffin-free original sample. After all the aromatics have been removed, the benzene may be replaced by a third solvent,

Table 13.3 Solvent Strengths of Common Solvents

Solvent	Solvent strength parameter
n-Pentane	0.00
Cyclohexane	0.04
Carbon tetrachloride	0.18
Toluene	0.29
Benzene	0.32
Ethyl ether	0.38
Chloroform	0.40
Tetrahydrofuran	0.45
Acetone	0.56
Methyl acetate	0.60
Acetonitrile	0.65
1-Propanol	0.82
Ethanol	0.88
Methanol	0.95
Formic acid	Large

such as diethyl ether. Each change in solvent brings about the elution from the sample of a different set of organic functional groups. This makes possible complete separation of the sample into compounds with common functional groups. Each set of fractions can be studied separately by further chromatography with a different column or by IR, UV absorption spectroscopy, or NMR. This makes possible qualitative and quantitative analysis of the components of the original sample. It is important that the solvents be in sequence in increasing polarity or dielectric constant, which dissolves compounds with increasing polarity. The polarity may be expressed as solvent strength. A list of common solvents and their solvent strengths on alumina is shown in Table 13.3. If solvents are used out of order, e.g., acetone was used following heptane, then all the types of compounds with solvent strength less than acetone will be eluted together and nonseparation obtained.

5. Gradient Elution

The abrupt change in solvents described above sometimes leads to problems, particularly in collecting fractions near the interface between the two succeeding mobile phases. This technique has been improved upon using a process of *gradient elution*. In this system, two solvents may be used that are completely miscible in each other, but with different dielectric constants. After the sample is loaded on the top of the column, the first solvent, which has a lower solvent strength, is used as the mobile phase. After a suitable period of time a small amount of the second solvent is introduced. With time, its concentration in the first solvent progressively increases until finally the mobile phase consists entirely of the second solvent.

Throughout the operation there is a gradient in the concentrations of the two solvents. This leads to a progressive change in solvent strength and in the solubility of some of the components of the sample. Any two solvents may be used, provided that they are miscible. Frequently, solvents that are somewhat different in their properties are selected in order to separate a greater range of components. Popular pairs of solvents include water-methanol, water-acetonitrile, and dioxane-tetrahydrofuran.

The identity of an unknown compound can be revealed based on its retention time, which can be confirmed by comparison with a known standard. This identification process is not completely reliable, because it is possible for the standard and reference to have the same retention time even though they are different compounds. Further confirmation can be obtained by running the same sample and following the same procedure using a different column and different conditions. The likelihood of two different compounds maintaining the same retention time under different conditions for gradient elution is very unlikely, but not impossible. To obtain complete confirmation, it is necessary to trap out the sample and identify it by other analytical methods.

6. Displacement Chromatography

It is not always convenient or necessary to perform a chromatographic separation as just described. Sometimes all the necessary information can be obtained using displacement analysis. In this method, the sample is completely dissolved in a suitable solvent and loaded onto the chromatographic column. At this point, the sample and a minimum amount of the first solvent are at the top of the column. This solvent is chosen because it easily dissolves the sample. As the solution percolates through the column, a second solvent, such as methanol or formic acid, is introduced as the mobile phase at the top of the column. This second solvent is chosen so that it has a high affinity for the stationary phase. Such a solvent completely displaces all the components of the sample from the surface of the substrate. Each component, however, has a somewhat different affinity for the surface of the substrate. Those with the least affinity move down the column most quickly and proceed to the front of the sample fraction. The compounds with the most affinity for the substrate are the most difficult to displace and so become the last of the fraction. The whole sample is linearly displaced but arranged in order, with the components emerging in the reverse order of their affinity for the substrate. This method is rapid and makes possible the handling of large samples. It avoids the use of large quantities of solvent and ensures a separation in a short period of time. The separation into components is not complete but allows a rapid semiquantitative analysis of the major components of the sample.

7. Affinity Chromatography and Chiral Analysis

In affinity chromatography, a substrate is used that interacts specifically with a particular compound. The substrate has a strong affinity with that compound, hence the name. Other compounds elute rapidly from the column, leaving the compound of interest to elute after a reasonable time delay.

Many of these substrates or liquids are naturally occurring compounds such as enzyme inhibitors, antibodies, or dyes. Such systems are expensive to operate and can only be justified for analysis where they are necessary. Some reduction in cost can be achieved by using synthetic liquids rather than those separated experimentally using expensive labor.

The types of samples analyzed include amino acids, amino alcohols, and proteins used in pharmaceuticals, the chemical industry, foods and drugs, and biotechnology. The normal UV or refractive index detectors can be used, but very high sensitivity (10^{-18} g) has been claimed using laser-induced fluorescence to excite the molecules and measuring the fluorescence.

Recently, separation of *chiral compounds* with the same empirical formula has been achieved. This is important because the effect of drugs on patients depends on the chirality of the drug. In the case of β-blockers used to treat heart patients, one optical isomer acts as desired, but the other optical isomer is not a β-blocker but does reduce blood pressure.

Since many drugs are optically active, it is becoming more and more important to distinguish between the effects of the different isomers and, if necessary, treat the patient with the correct isomer.

It should be noted that racemic mixtures can also be separated using bonded silica as support. These have in turn been replaced by cross-linked synthetic polymer supports with hydroxyl functionality. Hypercarb D.L. uses a standard 100 mm × 4.6 mm column together with chiral selectors as the mobile phase. The latter include beta-cyclodextrin, *N*-carbobenzoxy-glycyl-L-proline (L-ZGP), or di-cyclo-hexyl (RR) tostrate (DCHT.) Using these combinations, a single column can be used to separate many different racemic mixtures, thus greatly reducing the cost of the analysis. This is illustrated in Fig. 13.34.

8. Detectors Used in Liquid Chromatography

Two types of detectors are commonly used in liquid chromatography. These are based on the UV absorption and the refractive index of the sample.

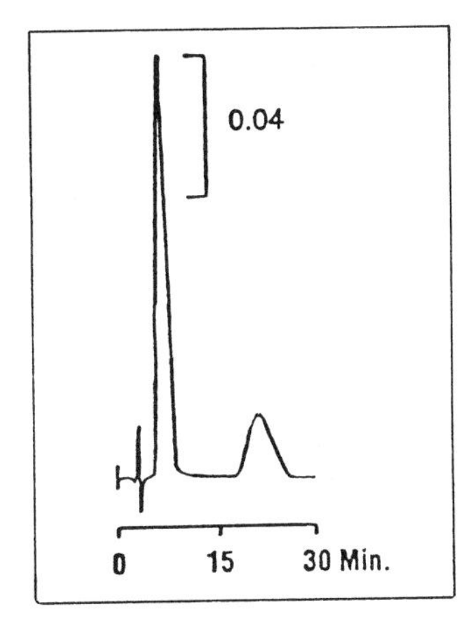

Figure 13.34 Separation of optical isomers using specialized columns. Column: Hypercarb DL, Eluant: 3 mM β-cyclodextrin ACN/0.05M phosphate (pH 12).

a. UV Adsorption Detectors

The basis of these detectors is that a sample compound absorbs UV radiation more than the mobile phase solvent. For that reason, it is an advantage to choose a mobile phase that absorbs little of the UV, such as heptane. A schematic diagram of such a detector is shown in Fig. 13.35. The source may be a deuterium

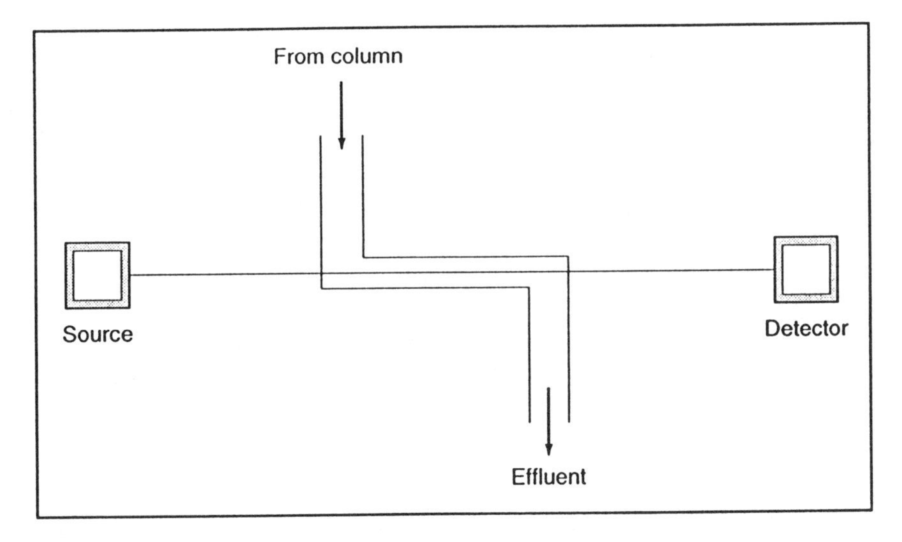

Figure 13.35 UV absorption detector.

lamp or a Hg lamp. The detector is a photomultiplier.

There are three variations of this detector: (1) fixed wavelength, (2) variable wavelength, and (3) scanning wavelength detectors. The fixed wavelength detector is the least expensive. It monitors a single wavelength such as 220, 280, 313, 334 or 365 nm. These wavelengths can be selected by putting a suitable filter in the light path.

A problem with this method is that, although most compounds that absorb UV will absorb at these wavelengths, the degree of absorption, and therefore the sensitivity, varies significantly between different types of compounds. For example, aromatics will absorb strongly, but paraffins will not.

The *variable wavelength* allows the operator to choose the wavelength to monitor based on the absorption spectra of the compound of interest. This extra degree of freedom is clearly an advantage but entails knowing the identity of the compounds of interest. It does not help if this information is unavailable.

Photo Diode Array. The scanning wavelength detector utilizes a *photo diode array* (PDA) as a detector. This enables the complete UV absorption spectrum to be obtained in a fraction of a second. Consequently, as each component emerges from the column, its complete UV spectrum can be acquired without trapping. This goes a long way to identify the compound.

The spectra are stored on a tape and retrieved as required. Since this is a nondestructive detector, the effluent can be connected to a thermospray interface to eliminate most of the solvent; the remaining sample is then injected directly into a quadrapole MS and the mass spectrum is obtained.

The UV and MS spectra provide quite reliable confirmation of the identification of the component. Since no trapping or purification is necessary, reaction intermediates can sometimes be identified.

Biological metabolites have been studied this way. This technique is particularly valuable since biological metabolites are usually present in very small quantities. Trapping and purification would be particularly difficult. But combined HPLC-PDA-MS provides enough information for confirmed identification or for strong indication of the metabolites involved.

The UV absorption method is very sensitive and is compatible with gradient elution. It is also nondestructive. Unfortunately, it is not universal. Only those compounds that absorb in the ultraviolet region are detected, and although this includes many compounds, such as aromatics, unsaturated compounds, and halogens, it does not include compounds such as paraffins.

b. Refractive Index Detectors

A more universal detector is based on a difference in refractive index between the mobile phase and sample components. When light passes from one phase to another, its path is bent. If we place a stick in water it appears to be bent at the interface between the water and the air (the water surface). The degree of bending depends on the differences in refractive index between the two phases. This is air and water in this case, but is the mobile phase and the sample component in the case of the HPLC detector.

A schematic diagram of a commercial refractive index detector is shown in Fig. 13.36. Radiation from a suitable light source is split by a mask into two beams A and B. These pass through a focusing quartz lens, which directs the two beams to (A) the upper half (B) the lower half of a complex prism. This is a system made up of two partial prisms fitted into each other. Beam A passes through the thick part of one prism, which contains the HPLC effluents. Beam B passes through the thick part of the other prism containing only mobile phase. The total volume of the cell is about 3 μl.

When there is no component present, there is mobile phase only in both half-prisms, and the radiation falls on the detector. The focusing lens is used to adjust the beams if necessary.

When a sample component passes through the upper half-prism, there is a change in refractive index. Beam A is distorted relative to beam B and is deflected from the detector. This changes the light intensity falling on the detector and a signal is generated. These detectors have a wide range of linearity and can be used over the entire refractive index range.

A reflection-type refractometer is shown in Fig. 13.37. The system is based on the fact that the amount of light reflected by a surface depends on the angle incidence of the light (Fresnel's law) and the refractive index of the solution.

Light from the source falls on the solvent and the reference, and this is focused on two detectors. When a component emerges from the HPLC reaches the detector, that beam of light is deflected and does not fall on the detector. A signal is therefore generated.

Other HPLC detectors include a *fluorescence detector*. These detect only those compounds that fluoresce. The selectivity can be an advantage. The sensitivity of these detectors if very high (10^{-12}), especially when lasers are used as the light source. Laser-induced fluorescence is a common observation, and

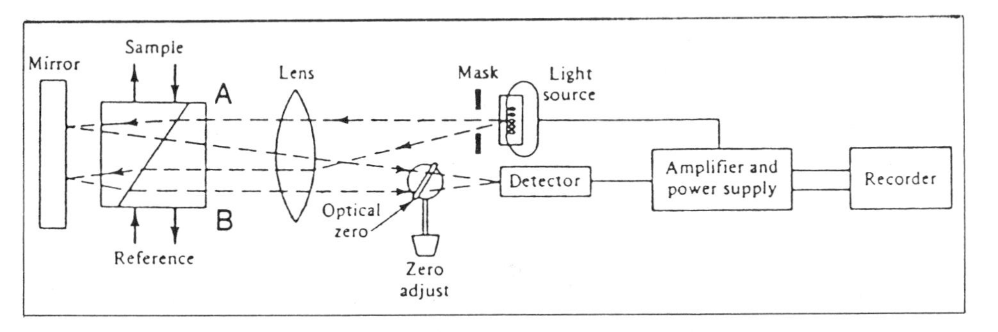

Figure 13.36 Refractive index detector (Courtesy Waters Assoc.)

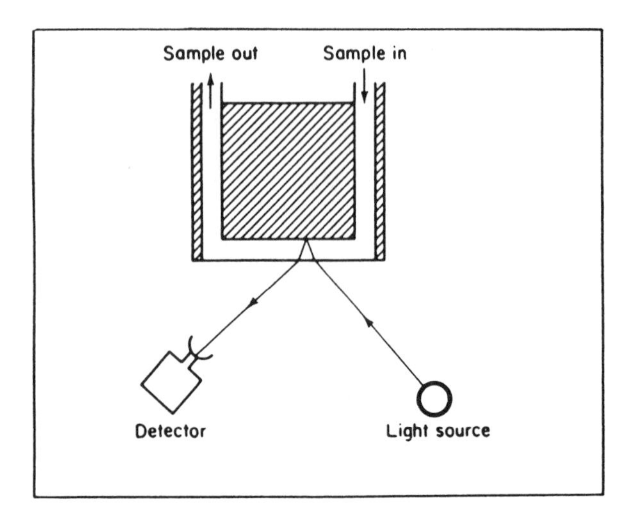

Figure 13.37 Reflection refactometer detector.

many compounds can be detected this way. The widespread applicability of the method has been extended to nonfluorescing compounds by adding a fluorescent material to the mobile phase. When a nonfluorescing material passes the detector, there is a drop in fluorescence intensity and the component is detected. Sensitivities of 10^{-13} g are claimed.

Electrochemical detectors are very valuable when polar mobile phases are used (water or methanol). The method is very sensitive, since often picograms can be detected. Many biologically active compounds can be detected this way.

9. HPLC Detectors

In general these detectors are nondestructive and sensitive to 10^{-7} g. The choice of detector depends on the information type. The fixed wavelength UV detector is the least informative, and the diode array UV detector is the most informative. With increasing attention to the life sciences, this field will continue to grow strongly in the future.

The refractive index detector can accommodate any suitable solvent but cannot be used for gradient elution. The UV detector has improved detection limits (10^{-9} g) and can be used with gradient elution but can only detect those compounds that absorb UV.

HPLC-MS. Sophisticated equipment using HPLC interfaced with MS is now commercially available. The interface involves a system to evaporate the solvent, which should boil at relatively low temperatures. The remaining effluent is then introduced into a mass spectrometer, usually a quadropole or, in some cases, a time-of-flight MS.

This system has the advantage of providing the mass spectrum of the compound without prior separation or trapping factions from the HPLC. Although these systems are commercially available, they are not routine in operation and require operator attention.

HPLC-FTIR. Because of the speed in which analyses can be obtained, HPLC has been usefully interfaced with FTIR. This enables a complete IR scan to be obtained in a fraction of a second, i.e., while a component is passing through the detector. This permits the IR spectrum to be obtained without trapping the sample, and results are available immediately.

In summary, HPLC detectors are now available to provide simple component detection and/or to provide the complete UV absorption spectrum, a mass spectrum, or an IR spectrum. This permits rapid qualitative identification and quantitative determination to be available with a minimum of handling or time delay.

G. LIQUID–LIQUID CHROMATOGRAPHY

Liquid–liquid chromatography was first developed by Martin and Synge in 1941. They developed and illustrated the use of this technique for the separation of amino acids. This was a major step forward in biochemistry and earned them the Nobel Prize. Their basic premise was to use liquid phases that were mutually insoluble as the mobile and stationary phase. The stationary liquid phase was adsorbed on the surface of a substrate. The amount of material involved was very small, so that a thin layer of adsorbed liquid was held in place on the surface of the substrate. A mobile liquid phase could flow over this stationary phase without removing it from the column.

A sample was then injected into the column in the normal fashion and distributed itself between the stationary phase and mobile phase according to the distribution coefficient. When the sample was in the mobile phase, it progressed down the column and eventually reached the detector. Separation of two samples was achieved when their distribution coefficients were different, and therefore the times spent in the mobile phase were different. Consequently, the elution times would be different and separation would be effected.

As described earlier, the total number of theoretical plates in the column is given by Eq. (13.4), and the theoretical plate height is given by Eq. (13.5).

The factors that affect peak broadening are similar to those encountered in gas–liquid chromatography and liquid–solid chromatography. In gas–liquid chromatography, the sample in the mobile phase is in the gas phase. It is considered that gases are mutually soluble in each other (unless chemical reaction takes place); however, this is not true with liquids; the solubility of one liquid in another is a physical property of that liquid–liquid system. The solubility is usually limited; it may be low or high, but is not usually infinite. This is particularly true of solvents selected for liquid–liquid chromatography, where a limited solubility leads to control of chromatographic separation.

If a sample is partially soluble in two immiscible solvents, it will come to equilibrium and distribute itself between these solvents with X_m in the mobile phase and X_s in the stationary phase, where X_s and X_m are the quantities of sample compound involved, from which we can derive the distribution coefficient K_x as

$$K_x = \frac{X_s}{X_m} \tag{13.13}$$

If K_x is high, the sample is mostly in the stationary phase, and vice versa.

If we include the volume of mobile and stationary phase in the calculation of distribution, we have a new constant, the capacity factor k'_x, where

$$k'_x = \frac{\text{total moles of } X \text{ in stationary phase}}{\text{total moles of } X \text{ in mobile phase}} \tag{13.14}$$

$$= \frac{V_s}{V_m}\frac{X_s}{X_m} = \frac{V_s}{V_m} K_x \tag{13.15}$$

where

V_s = volume of stationary phase in column
V_m = volume of mobile phase in column

When a peak maximum reaches the detector, half of the component has eluted in the retention volume V_r. Half is left in the column distributed between the stationary and mobile phases. Therefore

$$V_r X_m = V_m X_m + V_s X_s \tag{13.16}$$

$$V_r = V_m + V_s \frac{X_s}{X_m}$$

$$= V_m + V_s K_x \tag{13.17}$$

$$= V_m \left(1 + \frac{V_s}{V_m} K_x\right)$$

$$= V_m (1 + k'_x) \tag{13.18}$$

$$= \frac{V_r}{V_m} - 1 = k'_x \tag{13.19}$$

But

$$V_r = \text{(flow rate of sample)} \times \text{(retention time)}$$
$$= ft_r$$

t_r = retention time of sample

Also,

$$V_m = \text{volume of mobile phase in the column}$$
$$= ft_0$$

where

ft_0 = flow rate of solvent

t_0 = time for solvent molecule to pass through column

From Eq. (13.2) we can deduce that

$$k'_x = \frac{t_r - t_0}{t_0} = \frac{t_r}{t_0} - 1 \tag{13.20}$$

The selectivity factor K determines if two compounds will be resolved by a particular column. We can therefore state that the selectivity factor

$$K = \frac{k'_1}{k'_2} \tag{13.21}$$

This equation is fundamental to both gas and liquid chromatography. It illustrates the dependence of the retention time on the relative solubility of the sample in the stationary phase and the mobile phase. This basic relationship holds in liquid–liquid and gas–liquid chromatography.

Going back to first principles, t_0 is the time the sample is moving in the mobile phase and $t_r - t_0$ is the extra time, spent in the stationary phase; hence $(t_r - t_0)/t_0$ is the distribution of time spent by the sample between the stationary phase and the mobile phase.

1. Selection of Mobile and Stationary Phases

It is of paramount importance that the two liquids chosen for liquid–liquid chromatography have virtually no solubility in each other. Two suitable liquids would, therefore, be water and hexane. If water is selected as the stationary phase and hexane as the mobile phase, then the sample is injected in the hexane and separation takes place in the normal fashion. However, it should be noted that water is very slightly soluble in hexane and therefore, with time, the hexane would dissolve the stationary water phase and it would be lost from the column. In practice, this problem is overcome by presaturating the hexane with water before use. The presaturated hexane will not dissolve water from the stationary phase, which is therefore left intact throughout the chromatographic separation. This process is successful only if the mutual solubility is very low.

This places a limitation on liquid–liquid chromatography, since it is important that the stationary phase be a reasonable solvent for the sample but not be soluble in the mobile phase. Samples that can be separated by liquid–liquid chromatography are therefore limited to materials with comparatively low capacitance values. A change of mobile phase is difficult under these conditions, and great care must be made in selecting solvents that will fill the requirements necessary for separation.

In practice, loss of stationary phase occurs even if its solubility in the mobile phase is slight. With continued use the stationary phase is slowly dissolved off the column. This results in loss of substrate material and therefore a breakdown of the procedure.

2. Reversed-Phase Chromatography

In normal chromatography the mobile phase and the sample are organic, and usually the stationary phase is polar. Under these circumstances organic samples can be analyzed. If this relationship is reversed, that is, the mobile phase is polar, such as water, and the stationary phase is organic, then we have *reversed-phase chromatography*.

In this system, the support which may be glass or silica, is used as a base onto which organic material is bonded. This provides an organic surface and acts as an organic substrate of stationary phase. The mobile phase can then be aqueous in nature or any other solvent with a high dielectric constant. The most common stationary phases used to date include silica gel esterified with alcohols to form silicate esters. Also, the SiOH groups on silica gel can be silanized with organochlorosilanes, forming surfaces onto which the substrate has been chemically bonded. This method has been used successfully for reversed-phase chromatography.

When bonded organic surfaces are used as the stationary phase, it is possible to change the mobile phase. One method of doing this is to use water as the initial mobile phase and after a suitable period change to methanol as the mobile phase. A more popular and efficient method is to use gradient elution (p. 628). In this system, water may be used as the initial solvent, but the composition of the solvent varies with time as increasing percentages of methanol are included. The final solvent is methanol. Other pairs is solvents include water–acetonitrile and tetrahydrofuran–dioxane.

Reversed-phase chromatography has been very widely exploited in biological analysis, particularly for body fluids, such as blood, urine, saliva, and other fluids. These samples are essentially aqueous in nature, with organic compounds either dissolved or mixed into the water. For example, blood contains many organic materials dissolved in a salty aqueous solvent. The advent of reversed-phase chromatography was a major breakthrough in the analysis of these types of samples.

3. High-Performance Liquid Chromatography

One of the early problems with liquid chromatography was the slow rate at which the analysis took place. Early methods used gravity feed, and it was not uncommon for an analysis to take several days to complete. This led to great delay, but also the excessive time on the column inevitably led to loss of resolution by diffusion, and so on. Consequently, for a number of years liquid chromatography was not widely used as a means of separating organic compounds. This problem was largely overcome by the advent of high-performance liquid chromatography (HPLC). In this system pressure is applied to the column, forcing the mobile phase through at a much higher rate. The pressure is applied using a pumping system. The action of the pump is critical, since it must not pulsate and mix up the sample being separated in the solvent, causing it to lose resolution. Development of pumps has proceeded quite quickly over the last several years, and now it is possible to achieve good resolution under the conditions required for HPLC.

All of the factors affecting separation in liquid chromatography apply to HPLC. The factors affecting plate height, the sample distribution between the stationary and mobile phases, and the selection of stationary and mobile phases still pertain even under the conditions of HPLC. The principal advantage of the system is the speed at which separations take place. Because of the decrease in time, diffusion in the column is reduced and resolution improved.

Emphasis has been placed on the size of the particles making up the substrate. It has been found that the smaller the size, the better the resolution. This is because of the decrease in the geometry factor A in the Van Deemter relationship.

Pressures used normally range from 30 to 200 atm, depending on the type of column used. The pressure is varied to provide the optimum linear flow rate of the mobile phase. It is that pressure which gives the smallest theoretical plate height. It is derived experimentally as in gas chromatography where the relationship H/V is developed.

Although GC is fast and simple to operate, HPLC has now become the workhorse of chromatography (see Table 4). At this time about 5% of all chromatographic analysis are carried out by GC, while 60–70% are performed by HPLC. In contrast to GC, especially programmed-temperature GC, nonvolatile thermally labile and polar samples can be analyzed by HPLC. These include acids, bases, ionic compounds, surfactants, macromolecular compounds, phtha-

Table 13.4 Comparison of GC and HPLC

Method	Characteristics
GC	Inexpensive
	Fast
	High resolution
	Interfaced with MS
HPLC	Analyzes nonvolatile and thermally unstable organic compounds
	Used for inorganic ionic samples
	Interfaced with MS with difficulty
GC + HPLC	Small sample required
	Highly sensitive
	Nondestructive
	Used for quantitative analysis based on signal area
	Used for qualitative analysis based on RF value (not conclusive)

late azo dyes, and polyaromatic hydrocarbons (PAH). These types of sample include very many industrial, research, and environmental samples, such as drinking water, air samples, solid wastes, and pesticides. An example of the use of HPLC in drug analysis is shown in Fig. 13.38.

Sensitivities of 10^{-9} g can be obtained using UV absorption or refractive index detectors. Sensitivities of 10^{-12} g can be detected by UV fluorescence.

4. Capillary Column Liquid Chromatography

Recently, capillary column liquid chromatography has been developed to the point where it is useful and is providing resolutions hitherto unavailable in liquid–liquid chromatographic procedures. The method is based on the use of glass capillary columns coated with reversed-phase substrates. This provides a stable organic stationary phase. The very high resolution encountered in capillary gas chromatography is also found in capillary liquid chromatography and for the same reasons. The method has been used successfully to characterize body fluids such as blood and urine. These systems are essentially aqueous in nature, containing organic compounds that can be separated on the bonded capillary column. The use of a pump forces the solvent through at a relatively high rate, so analysis can be completed in a reasonable time. The method finds many applications in the natural sciences because of its high speed, high resolution, and the fact that it can be used for high molecular weight materials such as proteins, peptides, and polynucleotides.

In conclusion, it can be seen that liquid–liquid chromatography has a major advantage over gas chromatography in that it can be used for the determination

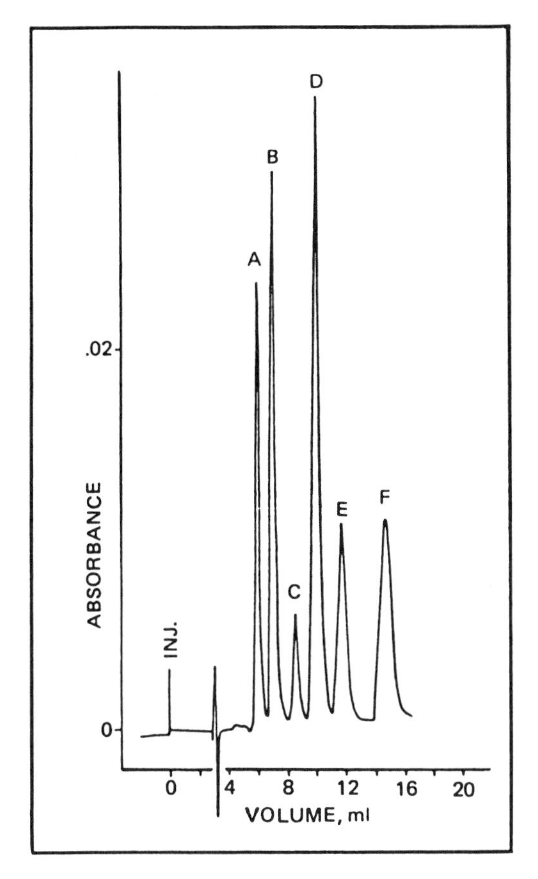

Figure 13.38 Separation of barbiturates by HPLC: (A) phenobarbital; (B) butabarbital; (C) mephobarbital; (D) pentabarbital; (E) secobarbital; (F) methohexital.

of high molecular weight materials and for materials that are unstable and cannot be separated using programmed-temperature GC. Liquid chromatography can therefore be used for many samples encountered in the natural sciences, such as biochemistry, biology, medicine, and clinical science. However, one advantage that gas chromatography still holds over liquid chromatography is that it is much faster in the separation and determination of low molecular weight materials and for materials that are similarly stable and can be readily vaporized using programmed-temperature GC.

Some excellent work in the field of capillary liquid chromatography has been carried out by Milos Novotny; high resolution was achieved and extremely complicated mixtures separated out.

5. Ion Exchange Resin Columns

For many years it has been known that clays include metals ions in their composition and that these ions can change places with other metal ions in their vicinity. This observation was important to the understanding of soil chemistry. Zeolites (sodium aluminum silicates) have some properties similar to those of clays. Using this phenomenon, natural and artificial zeolites became useful in removing Ca^{2+} and Mg^{2+} from water. Ca^{2+} and Mg^{2+} disolved in water cause it to be "hard." Hard water reacts with some kinds of soap—wasting the soap and forming a scum. Also these ions deposit on boilers and kettles when water is boiled in them on a long-time basis. This deposit is called "fur" and causes a waste of fuel needed to boil this water. The Ca^{2+} and Mg^{2+} ions dissolved in the water are exchanged for Na^+ ions from the zeolite, thereby providing soft water. Ion exchanges have been created by combining a polymer and a functional group. Such ion exchangers have properties very similar to those of zeolites. Two types are available, cationic and anionic exchangers. Both use a resin as the insoluble inert support, but they use different functional groups in order to provide different types of exchangers used for different purposes.

a. Resin Structure

A typical resin may be prepared by polymerizing styrene and divinyl benzine:

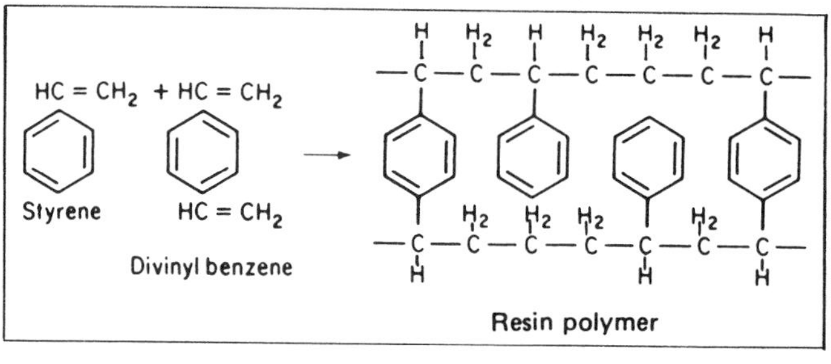

$HC = CH_2$ + $HC = CH_2$

Styrene

$HC = CH_2$

Divinyl benzene

Resin polymer

The polymer contains many hundreds of carbon atoms, which provide a strong three-dimensional molecular matrix, or framework, for the resin. Onto this matrix the functional groups are added as described below.

b. Cation Exchange Resins

The functional group in cation exchange resins is usually an acid. By sulfonation, a sulfonic acid functional group, SO_3H, may be added to essentially all the benzene rings of the resin. This provides an ion exchange resin of the formula Res—$(SO_3H)_n$, where Res represents the resin polymer or matrix and $(SO_3H)_n$ the numerous attached sulfonic acid groups. Similar resins can be made by using carboxylic acid. The formula for this resin is represented as Res—COOH.

In each case, it can be seen that there is an acidic hydrogen attached by way of the functional group, such as the carboxylic acid, which is chemically bound to the resin. When a solution containing another ion, such as Na^+, passes over the resin, exchange takes place. The H^+ from the resin is very loosely held and drifts away from the functional group into the water. Its place on the resin is taken by a Na^+ from the solution. In short, the affinity of the resin for Na^+ is greater than for H^+ and an exchange takes place. The net effect is that the reaction

$$\text{Res—COOH} + Na^+ \rightarrow \text{Res—COONa} + H^+ \quad (13.22)$$

takes place. If a sulfonic acid group is used instead of carboxylic acid, the exchange is

$$\text{Res—}SO_3H + Na^+ \rightarrow \text{Res—}SO_3Na + H^+ \quad (13.23)$$

Almost any metal ion will exchange with a hydrogen ion in these circumstances. This provides a method of removing metal ions from aqueous solutions. Of course, the aqueous solution becomes acidic if only the metal is exchanged with H^+. A resin with H^+ on the functional group is said to be in the *acid form*. If Na^+ instead of H^+ is attached to the functional group, the resin is said to be in the *sodium form*.

The exchange between ions on the resin and ions in solution is a general phenomenon in that other resin forms can be used. Selective exchange of other resin forms can be performed by the correct choice of resin form. For example, if a resin in the sodium form is used, it can be used for exchanging other metal ions, but not sodium. It can therefore be used for extracting other metal ions from sea water without interference from the Na^+ already present in the sea water. The exchange of copper ion follows the equation

$$\text{Res—}(SO_3)_2^{2-}2Na^+ + Cu^{2+} \rightarrow \text{Res—}(SO_3)_2\,Cu + 2Na^+$$

The copper ions are removed from the solution and are replaced by sodium ions.

A cation column was used to obtain the chromatogram shown in Fig. 13.39, which shows the metals present in orange juice. Such analyses are important because the metal content affects the product's taste, shelf life, and contributions to health in the form of trace metals.

c. Anionic Resins

Anions can be removed from solution using the same principle as that used to exchange cations. The functional groups are generally amines or quaternary ammonium salts. The resin form can be presented as Res—NH_3OH. It exchanges with anions in the following manner:

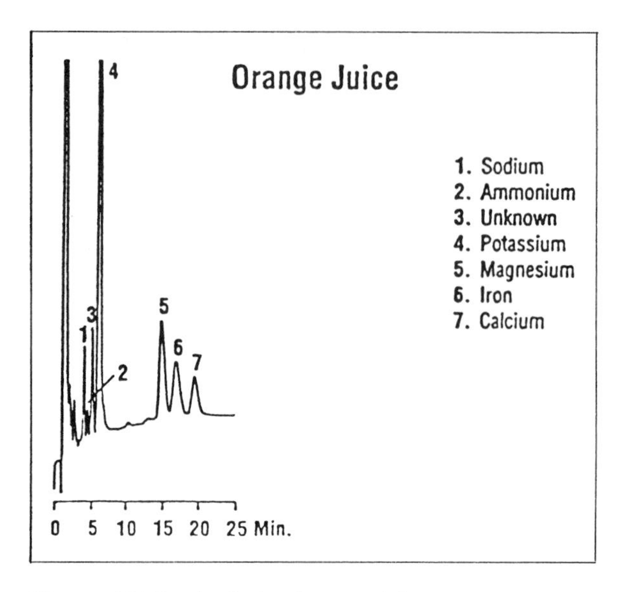

Figure 13.39 Analysis of orange juice.

$$\text{Res—NH}_3\text{OH} + \text{Cl}^- \rightarrow \text{Res—NH}_3\text{Cl} + \text{OH}^-$$

This provides a means of removing anions from aqueous solution by exchanging them with an OH^- ion from the resin. By using a mixture of anion and cation exchange resins, it is possible to exchange both the anions and cations in the water and replace these with H^+ and OH^-. This removes all the metal ion impurities and enables us to obtain water that may often be purer than distilled water.

d. Principles Governing Ion Exchange

The rate of ion exchange is controlled by the law of mass action. For example, in the case illustrated above, Na^+ from solution will replace H^+ from the resin. However, if very strong acid is washed through the resin, the high concentration of H^+ in the acid will by mass action reverse the process and replace the Na^+ from the resin.

At all times the metal ions present are in competition for the functional groups. In general, at equal concentrations the functional group sites are taken by the metal with the greatest affinity for that group. The affinity is controlled as follows.

1. The ion with the highest charge has the highest affinity for the functional group; that is,

   $$Na^+ < Ca^{2+} < Al^{3+}$$

and

$$Cl^- < SO_4^{2-}$$

2. The ion with the greatest size (and charge) has the highest affinity, that is,

$$Li^+ < Na^+ < K^+ < Cs^+ < Be^{2+} < Mg^{2+} < Cu^{2+} < Sn^{2+}$$

and

$$F^- < Cl^- < Br^- < I^-$$

If the ion becomes too large, distortion of the resin takes place and the rule breaks down.

e. Ion Exchange Columns as a Means of Chromatographic Separation

Separation of ions on a column of ion exchange resin can take place according to rules similar to those that apply to chromatography in general. The principal difference between the two processes is that in chromatography there is an equilibrium in the distribution of the sample between the substrate and the mobile phase. This distribution is controlled by the physical properties of the components. On an ion exchange column the distribution between the two phases is affected by properties that can be considered chemical.

The separation of two ionic components on a resin column follows the Van Deemter equation [Eq. (13.6)] very closely. Band broadening is caused by effects such as diffusion and eddy currents similar to those encountered in gas chromatography.

f. Analytical Applications of Ion Exchange Chromatography

We have seen that by using a mixture of anion and cation exchange resins, both anions and cations can be removed from the water. This process, which is inexpensive, rapid, and requires little operator attention, provides very pure water and is widely used in analytical laboratories. One shortcoming of the method is that it does not remove organic (nonionic) impurities. This must be eliminated by a different process if very high purity is desired.

Ion exchange columns can be used as chromatographic columns for the separation of ions in aqueous solutions. They can also be used to concentrate desired ionic components of a sample. For example, a water specimen may contain 1 part in 10^7 of gold ion. By passing the mixture down a suitable column, the gold is removed from the water. If a large volume of water is available, several gallons (or more) may be processed on one column. The gold may then be removed from the column by passing down a little (100 ml) strong acid such as HCl. The gold that was originally in several gallons of water is now concentrated in 100 ml of acid. After being concentrated in this fashion, the gold may be determined by a suitable analytical technique. Interfering ions may be removed from the samples prior to analysis. For example, phosphate may be removed prior to the determination of barium or calcium.

The total cation content of a sample may be determined by using a cation exchanger and titrating the H^+, which exchanges with the metal ions present. By using an anion exchanger, the total anion content may be similarly determined, after titration of the OH^- liberated. The anions in acid rain may be determined by ion chromatography.

In organic chemistry amino acids can be separated into groups by using a suitable resin column. The pH of the solution containing each group can then be adjusted by the addition of acid or base. The change of pH has the effect of changing the charge on the molecule. As a result, the affinity of the amino acids for the resin is changed. Further separation of the groups of amino acids into subgroups is therefore possible, leading to the isolation of individual amino acids from mixtures that may have been quite complicated originally.

If we make up a column of anion exchange resin and cation exchange resin, we will remove both anions and cation from solution. For example, if a water sample contained NaCl as an impurity, the Na^+ would be replaced by H^+ and the Cl^- by OH^-. The H^+ and OH^- generated would form water and the result would be purified water. Similarly, if we had $FeSO_4$ as an impurity in the water, each Fe^{2+} would be replaced by $2H^+$ and each SO_4^{2-} by $2OH^-$, and again pure water would result.

Commercially, ion exchange resins are used extensively for bulk water softening for both industry and private residences. In this case, Ca^{2+} and Mg^{2+}, the principal causes of water hardness, are exchanged with Na^+, which does not cause hardness. Ion exchange has also been used to extract trace elements from the sea.

In practice, the resin is easily regenerated. For example, a cation (acid) resin may become saturated with other metal ions as follows:

$$\text{Res—SO}_3\text{H} + \text{M}^+ \rightarrow \text{Res—SO}_3\text{M} + \text{H}^+$$

By pouring strong acid over the resin, the exchange process is reversed by the mass action effect of the hydrogen ion. The exchanged metal ions are then washed away from the column in the acid solution according to the equation

$$\text{Res—SO}_3\text{M} + \text{H}^+ \rightarrow \text{Res—SO}_3\text{H} + \text{M}^+$$

Anionic resins may be regenerated by passing NaOH down the column, thus forcing the following reaction to take place:

$$\text{Res—NH}_3\text{X} + \text{OH}^- \rightarrow \text{Res—NH}_3\text{OH} + \text{X}^-$$

As before, the anions X^- are washed away from the resin.

If the resin is a mixture of cationic and anionic resins, these must be separated (e.g., by flotation) before regeneration is possible. If this is not done, half of the resin mixture becomes ineffective. For example, if the mixture is treated with a strong acid, the cationic resin will be regenerated but the anionic resin will be saturated. After separation by flotation or some other suitable process, the resins are treated separately, washed clear with water, and then recombined for further use.

H. ELECTROPHORESIS AND ELECTROCHROMATOGRAPHY

The movement of ions under the influence of an applied potential is electrophoresis. The velocity-charged particles, including ions, move in under the influence of an applied potential in a solvent as follows

$$v_{epi} = \mu_{epi} \times E$$

where

v_{epi} = electrophoretic velocity
E = applied voltage
μ_{epi} = electrophoretic mobility

Also

$$\mu_{epi} = \frac{q}{6\pi a \eta}$$

where

q = charge on the electron cloud
a = effective radius of the particles
η = viscosity of the solvent

Hence

$$v_{epi} = \frac{q}{6\pi a \eta} \times E$$

Using simple electrophoresis, columns have been developed with as many as 10^5 theoretical plates for separating large molecules. Even better resolution has been obtained with small ions.

Electrochromatography is a combination of electrophoresis and liquid–solid chromatography. A sheet of paper or other suitable substrate is used to support the sample and the solvent. The components of the sample are separated by downward displacement in the normal chromatographic fashion. Meanwhile, a horizontal electrical potential is applied, perpendicular to the path of downward displacement. The net effect is that the components of the sample move downward and sideways simultaneously. A diagram of the equipment used is shown in Fig. 13.40.

The direction of the path of any particular component of the sample depends on its rate of downward displacement and its rate of sideways displacement. For example, in the equipment in Fig. 13.38, the distance traveled downward by all components was 20 in. Each component took a different period of time to reach

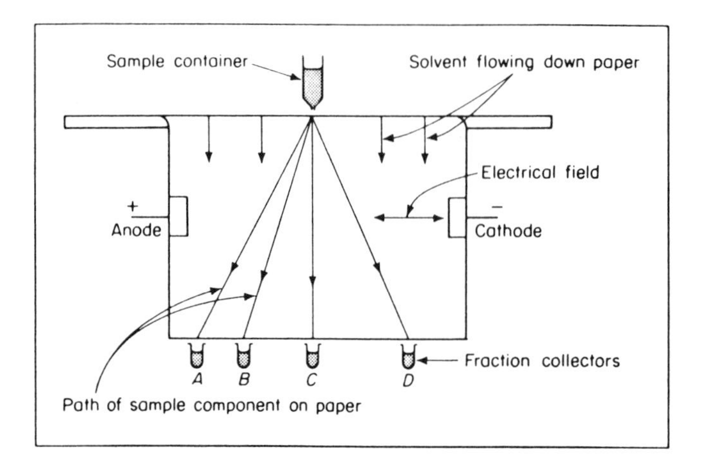

Figure 13.40 Schematic diagram of the equipment used for electrochromatography.

the collector, depending on its downward flow rate. Component A traveled 4 in. toward the anode while it traveled 20 in. downward. Component B traveled 2½ in. toward the anode while it traveled 20 in. downward. Component C did not travel toward the anode or cathode, whereas component D traveled 4 in. toward the cathode while flowing downward. The net result was that each component was widely separated from the other components upon arrival at the base of the paper.

1. Continuous Electrochromatographic Analysis

The downward flow rate of each component of a sample being analyzed by electrochromatography is controlled by all the factors that lead to separation in liquid–solid chromatography. These factors, which are discussed in the earlier parts of this chapter, include (1) the solubility of the sample molecule in the displacing solvent and (2) the component's tendency to be held stationary on the surface of the substrate. If there are any significant attractive forces (such as adsorption forces) between the component and the substrate, the component tends to stay on the surface and therefore flow slowly down the paper. The same effect is observed if the surface area of the substrate is increased.

The sideways motion of the components is controlled by the applied voltage. First, any component that is charged positively will move toward the cathode because of the attractive force of the applied voltage. Similarly, any molecule that is negatively charged will move toward the anode, whereas uncharged components will not be attracted either way. The charge therefore controls the horizontal direction and speed at which the component moves. How far the components move depends on how fast they move and how long they travel. The total time taken is the time for the component to reach the base of the paper. How fast a component travels laterally depends on several factors, such as the applied voltage and the ion mobility. The greater the applied voltage, the faster a component moves, because the attractive force is greater. The greater the charge on the component, the faster it moves in the applied field, again because the electrical attractive force is greater.

The larger the molecule, the slower it moves. This is because large components are less mobile and spend more time on the solid surface. When the component is on the solid surface, it is virtually stationary. Since a component adhering to the surface of the substrate is virtually immobile, the net effect of this is to slow down the sample.

Use of this procedure is not widespread, but for special applications it is superior to many other methods.

2. Capillary Electrochromatography

Electrophoresis is defined as the movement of charged particles in a magnetic field. Originally, these particles were colloids, but now studies have included charged molecules and simple ions.

The rate at which the charged system moves in an applied field depends on the ion mobility. This in turn depends on the charge (increased charge results in increased mobility), the size of the particle (the smaller, the faster), and the solvent. For a given solvent, small highly charged ions (PO_4^{-3}) move faster than large, low-charge molecules (I^-) or colloids.

If we place a mixture of charged particles in a capillary and apply a voltage to the ends of the capillary, the ions will flow toward the electrode with a charge their own. They will flow at different rates, and this difference can be used as a method of separation.

This type of separation has been invaluable in the clinical analysis field where a great variety of biological samples have been analyzed. These biological samples include blood serum, urine, stomach acids, spinal fluids, and saliva. It has been used to separate proteins, alkaloids, steroids, amino acids, carbohydrates, nucleic acids, antibiotics, and drugs; and is therefore, valuable for the diagnosis of disease or drug addiction.

The method is simple to apply. Identification is based on ion mobility and is most useful when the analyte is already known. It is less useful for qualitative identification except as a useful separation step. Electrochromatography is also known as zone electrophoresis, ionophoresis, and electromigration.

3. Isodalt Chromatography

Molecules such as carboxylic acids have a negative charge when ionized and therefore a negative potential. Similarly, amines have a positive charge and a positive potential in solution. Molecules that contain both amines and acids, such as amino acids, have a net potential somewhere between that of the amines and carboxylic acid.

This may be utilized by putting the molecules onto an LC chromatography (paper) system. If they are now put into an electrical potential gradient that is negative at one and positive at the other, they will flow in the direction in which they are attracted. If they are net negative in charge, they flow toward the positive charge until they reach a point on the paper where the potential is equal to their own: This is the *isoelectric point*. The molecules are no longer moved by the potential gradient. Because the charge and the ionization of the acid and amine are closely related to the hydrogen ion concentration, the potential gradient is often referred to as a pH gradient. If an amino acid is in a basic solution, it behaves like an acid; in an acid solution it behaves like a base. At some pH value it is relatively neutral and acts like a zwitterion. This point is also the isoelectric point and there is no mobility caused by the potential; that is, there is no electrophoretic mobility. This fact has been utilized as a means of separating complex mixtures of molecules, such as proteins and lipids.

The procedure is to take a paper dampened with a conducting solvent and place a positive charge at one end and a negative charge at the other end. The charge distribution is shown in Fig. 13.41. A spot of sample can then be placed between the charges. The components of the sample are mobilized by the action

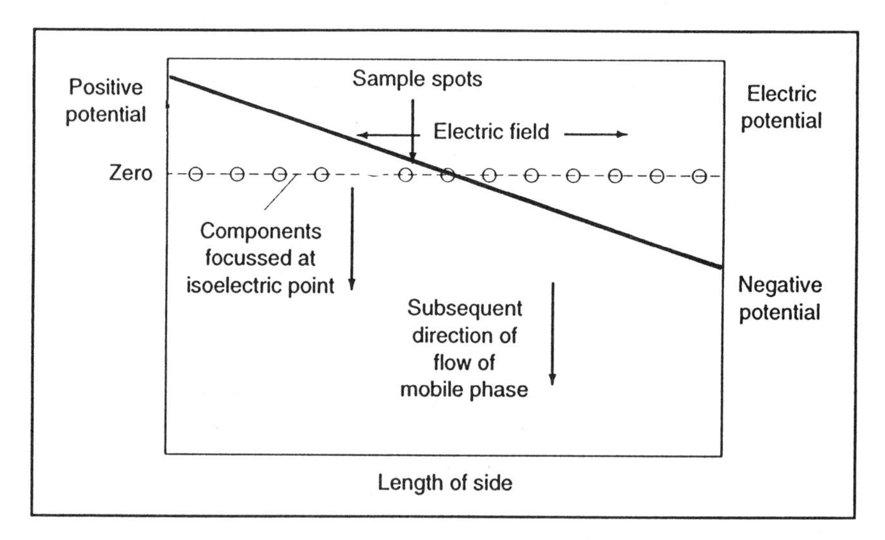

Figure 13.41 Charge distribution along the length of one side of the two-dimensional isodalt chromatogram. Local variation may occur because of local variations in resistance.

of the potential gradient on their own net charge. Since many of these components have a different net charge, they flow differently. Their flow will stop when they reach their own particular isoelectric points. By this means the sample can be separated into fractions according to the isoelectric points of the components. The fractions will lie on the line joining the positive and the negative charge applied to the paper.

After the components come to rest and are in equilibrium with the solution, the potential may then be removed and the sample treated as on a paper chromatogram. A solvent is added at one end of the paper and the sample components may then be further separated by a flow of solvent (mobile phase) perpendicular to the direction of the original potential applied. The components separated by differences in isoelectric point in the y direction will be further separated by normal chromatographic procedures similar to those encountered in paper chromatography.

The result is a two-dimensional separation of the components. They are separated in one direction by their isoelectric points and in the second direction by chromatography. This is illustrated in Fig. 13.42. The process is called isodalt chromatography.

Isodalt chromatography has been used by N. G. Anderson at the Argonne National Laboratory and by E. Jellum in Oslo, Norway. With this technique they were able to separate the multitude of components in blood and thereby characterize blood samples. Jellum used the method to try to diagnose cancer. He had 2000 patients from which he took and stored a blood sample every year for many years. With time some of these patients developed cancer. He then ran an isodalt chromatogram on their blood samples and located a spot that was not present in the blood of normal patients. This spot may be indicative of cancer. Jellum then took the stored blood samples of the cancer patients from previous years and noted how many years previously it was first possible to detect the telltale spot. By developing this technique he hopes to detect cancer in patients long before it becomes medically evident and diagnosable. Although much remains to be done to verify and understand the procedure, it has opened the door to possible early diagnosis. This, of course, is the essence of preventive medicine, since early detection is critical in curing cancer. Work is continuing in the computerized detection and location of unusual spots, since these are extremely complex procedures. It would seem that isodalt chromatography has unlimited possibilities in medicine and other branches of life science, as well as in many phases of industry.

4. Analytical Applications

The principal use of electrochromatography, like all forms of chromatography, is to separate the components of a sample. These may then be identified and/or

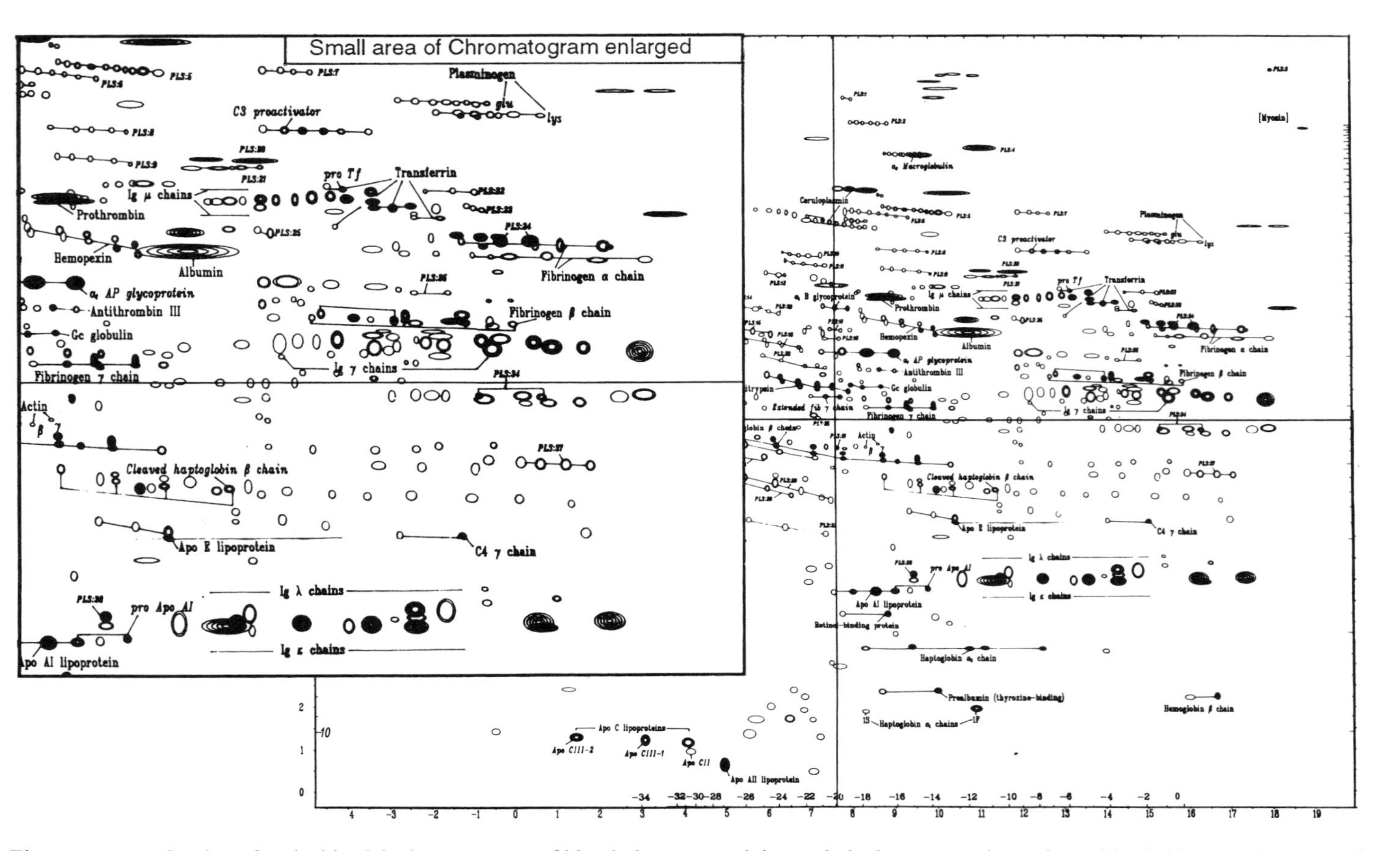

Figure 13.42 Section of typical isodalt chromatogram of blood plasma containing carbohydrates, protein, amino acids, lipids, etc. (Courtesy of Dr. Norman G. Anderson and Dr. Leigh Anderson.)

determined by any suitable means, such as IR or atomic absorption spectroscopy. In general, electrochromatography is more useful for aqueous samples, but it has also found applications with organic solvents. The method is continuous; that is, once the equipment is operating, sample can be added continuously to the top of the paper and collected at the base. It is also nondestructive. The components of the sample are usually unaltered by the process.

Changes in the composition of human serum (which appear when a healthy person becomes unhealthy) can be detected by using this method. It can also be used to separate dyes from each other or to remove impurities from them.

Metals can be separated from each other and from anions by electrochromatography. It is readily understandable that changes in pH greatly affect the effective change on a metal ion, so careful control is necessary. Furthermore, the addition of complexing agents greatly changes all the flow characteristics of the components. Two metals that are normally difficult to separate can be separated by selectively complexing one metal and leaving the other in its normal state.

5. Gel Permeation Chromatography

Gel permeation chromatography (GPC), or exclusion chromatography, uses a porous material as the stationary phase and a liquid as a mobile phase. The diameters of the pores of the porous material are on the order of 5–300 Å, which is similar to the size of many molecules. The latter penetrate the pores according to their size. Small molecules penetrate more rapidly than large molecules, which frequently are excluded from the smaller pores present. This results in a difference in the rates at which the molecules pass down the column, the smaller molecules traveling faster than the larger molecules.

There are generally two types of columns packings: porous glasses or silicas and porous cross-linked organic gels such as dextrans, methacrylate-based gels, polyvinyl alcohol–based gels, and hydroxyethyl cellulose gels. The detectors used are based on UV fluorescence, UV absorption, or changes in refractive index.

The chromatogram obtained results from a separation based on the sizes of the sample molecules, which are directly related to the molecular weights of the components. If the sample is a single polymer, the chromatogram represents the molecular weight distribution. This method is very valuable for determining the molecular weight distributions of polymers up to very high molecular weights. The actual calculation is usually based on a comparison with a standard polymer material of known molecular weight distribution. The procedure is also used to assess the molecular weights of biological compounds, such as proteins. Gel permeation chromatography is also capable of separating different polymers from each other and, under the correct conditions, mixtures of polymers can be characterized with respect to percentage and weight range of each polymer present.

Gel permeation has been used for such important analyses as the determination of aldehydes and ketones in auto exhaust. Both of these types of compounds are considered to be somewhat toxic. It has also been used to separate C_{60} and C_{70} fullerenes (Buckeyballs), as can be seen in Fig. 13.43.

Care must be taken in interpreting the results since the separation is based on the size of the molecule rather than the actual molecular weight. This means that the shape of molecule has a significant effect on the results. For example, two molecules of the same molecular weight, one straight chained and the other highly branched, will give somewhat different results because each has a different penetrating power through the column. This must be taken into account when selecting standards for calibrating the results.

An advantage of gel permeation chromatography is that it can be carried out at room temperature and the samples are not decomposed because of exposure to high temperature. This is especially important when dealing with very high molecular weight compounds, which are easily broken into two or three fragments, resulting in great changes in apparent molecular weight.

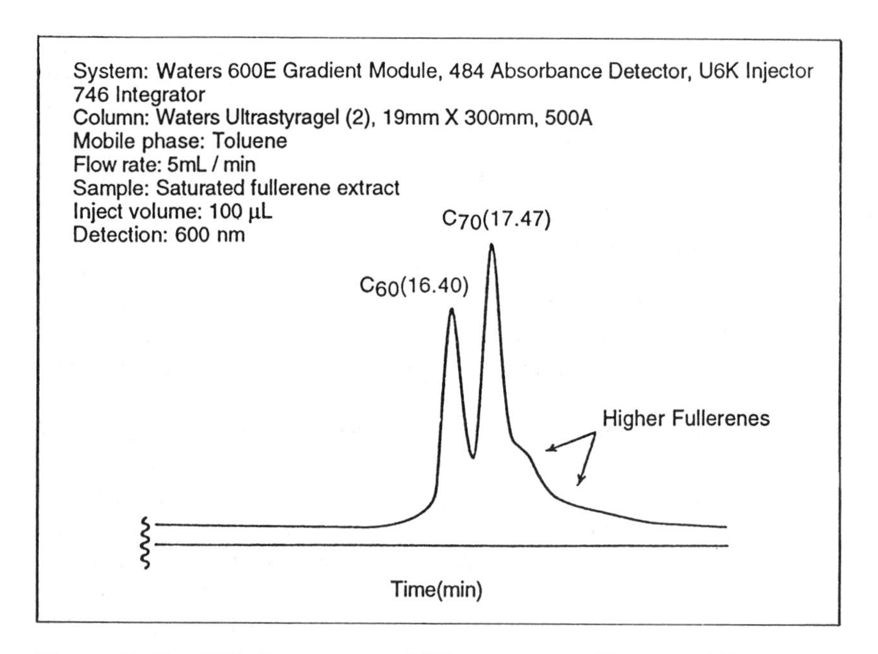

Figure 13.43 GPC chromatogram of fullerene extract. (Courtesy of Waters Assoc.)

6. Supercritical Fluid Chromatography

A compound such as CO_2 is a gas at normal temperature and pressure (NTP) and, like all gases above a certain critical temperature, increasing the pressure to any value will not result in the formation of a liquid. However, if the temperature is dropped below this critical temperature, increasing pressure will eventually lead to the gas becoming a liquid.

At a particular combination of critical temperature and pressure, the density of the gas phase and the liquid phase is the same. This state is neither a true liquid nor a true gas but is a *supercritical fluid* (SCF). This state has different properties from either a gas or a liquid, as shown in Table 13.5.

This system has been used for chromatography, and its use is expanding.

To date, most attention has been focused on CO_2, C_2H_6, and N_2O, with critical temperatures of 31, 32, and 37°C, respectively. The temperatures and the necessary pressures can be accommodated in conventional GC.

An important property of supercritical fluids is their ability to dissolve nonvolatile molecules. Certain important industrial processes are based upon the high solubility of organic species in supercritical CO_2. For example, CO_2 has been employed for extracting caffeine from coffee beans to produce decaffeinated coffee and for extracting nicotine from cigarette tobacco. Supercritical CO_2 readily dissolved n-alkanes containing up to 22 carbon atoms, di-n-akylphthalates with the alkyl groups containing up to 16 carbon atoms, and various polycyclic aromatic hydrocarbons.

a. Operating Conditions

To adapt equipment to supercritical applications, it is only necessary to provide an independent means for controlling the internal pressure of the system. Such systems are available commercially.

b. Effect of Pressure

Pressure changes in supercritical chromatography affect k'. For example, increasing the CO_2 pressure in a packed column from about 70 to 90 atm reduced the elution time for hexadecane from about 25 to 5 min. This effect is general

Table 13.5 Typical Values for Gases, Liquids, and Supercritical Fluids

Property	Gas (NTP)	SCF	Liquid
Density	1×10^{-2}	0.3	1.0
Diffusion coefficient (cm^2/sec)	3×10^{-1}	10^{-3}	1×10^{-5}
Viscosity (g/cm/sec)	2×10^{-4}	2×10^{-4}	2×10^{-2}

and has led to the type of gradient elution in which the column pressure is increased linearly as the elution proceeds. The results are analogous to those obtained with programmed-temperature gas chromatography and solvent-gradient elution in liquid chromatography.

c. Stationary Phases

So far, most supercritical fluid chromatography have utilized column packings commonly used in liquid chromatography. Further, capillary supercritical fluid chromatography has been performed with stationary phases of organic films bonded to capillary tubing. Because of the low viscosity of supercritical fluids, long columns (50 m or more) can be used. This results in very high resolutions in a reasonable elapsed time.

d. Mobile Phases

The most commonly used mobile phase for supercritical fluid chromatography is CO_2. It is an excellent solvent for many organic molecules, and it is transparent in the ultraviolet range. It is odorless, nontoxic, readily available, and inexpensive when compared with other chromatographic solvents. Carbon dioxide's critical temperature of 31°C and its pressure of 72.9 atm at the critical temperature permits a wide selection of temperatures and pressures without exceeding the operating limits of modern HPLC equipment.

Other substances that have served as mobile phases for supercritical chromatography include ethane, pentane, dichlorodifluoromethane, diethyl ether, and tetrahydrofuran.

e. Detectors

The most widely used detectors are those found in LC, i.e., ultraviolet absorption, refractive index, flame ionization, and mass spectrometry.

f. Supercritical Fluid Chromatography Versus Other Column Methods

As shown in Table 4, several physical properties of supercritical fluids are intermediate between gases and liquids. Hence, SCF chromatography combines some of the characteristics of both gas and liquid chromatography. For example, like gas chromatography, supercritical fluid chromatography is inherently faster than liquid chromatography because of the lower viscosity and higher diffusion rates in supercritical fluids. High diffusivity, however, leads to band spreading, a significant factor with gas chromatography but not liquid chromatography. The intermediate diffusivities and viscosities of supercritical fluids result in faster separations than are achieved with liquid chromatography together with lower zone spreading then encountered in gas chromatography.

Fig. 13.44 compares the performance characteristics of a packed column when elution is performed with supercritical CO_2 and a conventional mobile phase.

The rate of elution is approximately four times faster using SCF. The roles of the mobile phase in gas, liquid, and supercritical fluid chromatography are somewhat different. In gas chromatography, this mobile phase serves but one function—zone movement. In liquid chromatography, the mobile phase provides not only transport of solute molecules but also influences selectivity factors (α). When a molecule dissolves in a supercritical medium, the process resembles volatilization but at a much lower temperature than under normal circumstances. Thus, at a given temperature, the vapor pressure for a large molecule in a supercritical fluid may be orders of magnitude greater than in the absence of that fluid. As a result, important compounds such as high molecular

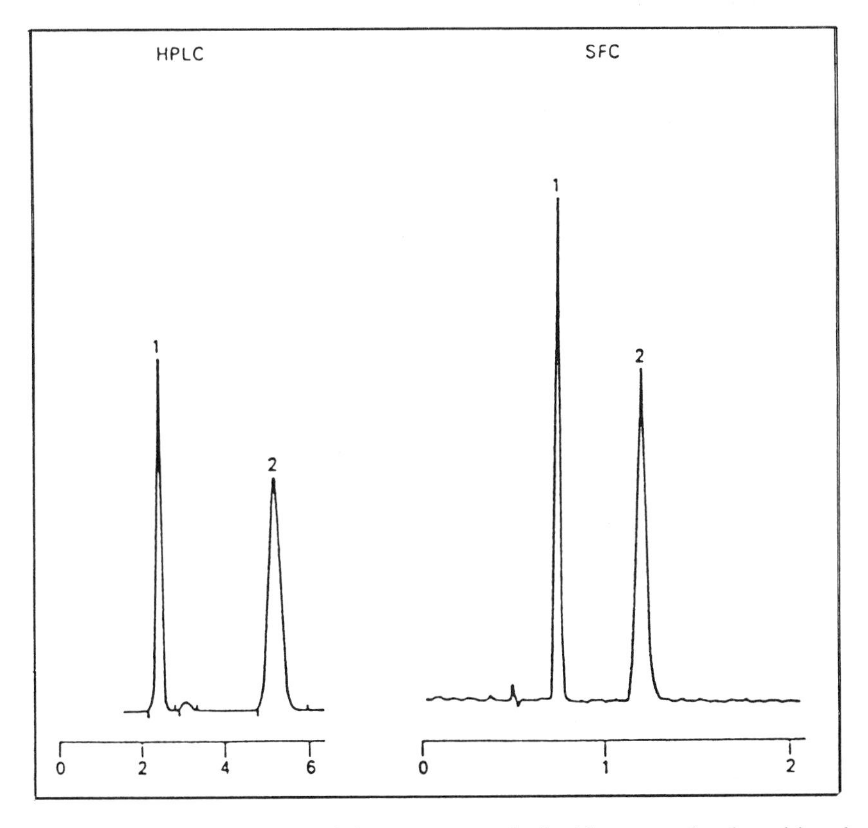

Figure 13.44 Comparison of chromatograms obtained by conventional partition chromatography (HPLC) and supercritical fluid chromatography.

weight compounds, thermally unstable species, polymers, and biological molecules can be brought into a much more fluid state than a normal liquid solution of these same molecules. Interactions between solute molecules and the molecules of a supercritical fluid must occur to account for their solubility in these media. The solvent power is thus a function of the chemical composition of the fluid. Therefore, in contrast to gas chromatography, there is the possibility of varying α by changing the mobile phase.

The solvent power of a supercritical fluid is also related indirectly to the gas density and thus the gas pressure. One manifestation of this relationship is the general existence of a threshold density below which no solution of solute occurs. Beyond this critical value, the solubility increases rapidly and then levels off.

g. Applications

Supercritical fluid chromatography can handle larger molecules than gas chromatography with higher efficiencies than liquid chromatography. It is also considerably easier to interface with mass spectrometers than is its liquid counterpart. Its potential applications based upon the high resolution of coated capillary columns may be particularly significant.

BIBLIOGRAPHY

Anal. Chem., Application Reviews, June 1994.

Bauer, H. H., Christian, G. D., and O'Reilly, J. E., *Instrumental Analysis*, Allyn and Bacon, Boston, 1978.

Brown, P. R., High performance liquid chromatography: past developments, present status, future trends, *Anal. Chem.*, (1990), *62*: 995A.

Buyter, L., Duvekot, J., Peene, J., and Mussche, P. L., *Am. Lab.*, (1991), *Aug*: 13.

Cox, G. B., *Column Watch*, (1992), *6*(9): 691.

Dal Nogare, S., and Juvet, R. S., *Gas Liquid Chromatography*, Wiley, New York, 1959.

Dietz, M. L., and Horwitz, E. P., Novel chromatography materials based on nuclear waste processing chemicals *LC-GC*, (1993), *11*(6): 424.

Ettre, L. S., *Basic Relationships of Gas Chromatography*, Perkin-Elmer, Norwalk, Conn., 1977.

Ettre, L. S., *Introduction to Open Tubular Columns*, Perkin-Elmer, Norwalk, Conn., 1979.

Fujimoto, C., and Jinimo, K., GC-MS and GC-FTIR interfacing, *Anal. Chem.*, (1992), *64*(8): 476A.

Glanz, J., Laser tweezers grab cells, viruses, DNA—and biologists, *R&D Mag.*, (1993), (January): 20.

Griffiths, P., and de Haseth, J., GC FTIR. *Anal Chem.*, (1983), *55*: 1273.

Jinimo, K., Fujimoto, C., and Uematsu, G., HPLC-FTIR, *Am. Lab.*, (1984), February 39.

McNair, H., and Bonelli, E. J., *Basic Gas Chromatography*, Varian Associates, Palo Alto, Calif., 1969.

Majors, R. E., Thin-layer chromatography—a survey of the experts, *LC-GC*, (1990), *8*(1): 760.

Martin, A. J. P., and Synge, R. L. M., *Biochem. J.*, (1941), 35:91.

Morris, C. J., and Morris, P., *Separation Methods in Biochemistry*, Wiley, New York, 1963.

Novotny, M., Recent advances in microcolumn liquid chromatography. *Anal. Chem.*, (1988), *60*: 500A.

Parikh, I., and Cuatrecasas, P., Affinity chromatography, *C&EN*, (1985), August 26: 17.

Snyder, L. R., and Kirkland, J. J., *Introduction to Modern Liquid Chromatography*, 2nd ed., Wiley, New York, 1979.

Spraul, M., Hofmann, M., Dvortsak, P., Nicholson, J. K., and Wilson, I. D., High-performance liquid chromatography coupled to high-field proton nuclear magnetic resonance spectroscopy: applications to the urinary metabolites of ibuprofen, *Anal. Chem.*, (1993), *65*: 327.

Takao, T., and Novotny, M., Packed microcapillary columns in high performance liquid chromatography, *Am. Chem.*, (1978), *50*: 271.

Wehr, T., Zhu, M., Rodriguez, R., Burke, D., and Duncan, K., High performance isoelectric focusing using capillary electrophoresis instrumentation. *Chem. Br.*, (1990), Sept. 23.

Willard, H. H., Merritt, L. L., Dean, J. A., and Settle, F. A., *Instrumental Methods of Analysis*, Van Nostrand, New York, 1981.

Yound, E., Kuhn, W. G., *Anal. Chem.*, (1991), *6*(5): 276A.

SUGGESTED EXPERIMENTS

13.1 Set a chromatograph at a fixed temperature and carrier flow rate. Inject a sample of hexane and record the chromatogram. measure the number of theoretical plates in the column. Repeat at different carrier gas flow rates. Correlate the number of theoretical plates with the flow rates. Set the gas chromatograph at the optimum carrier gas flow rate.

13.2 Using the conditions set in Experiment 13.1, inject a mixture of hexane and heptane. Note the separation of the two paraffins. Add heptane to the mixture and reinject. Note the increase in the peak caused by the heptane. This is one method of identification. Compare with Experiment 5.2.

13.3 Make up mixtures of hexane and heptane of various proportions. Inject the mixtures into the gas chromatograph operating under the optimal conditions found in Experiment 13.1. Measure the areas of the peaks in each chromatogram. Relate the relative areas of the peaks to the relative percentages of the compounds in the mixtures.

13.4 Repeat Experiment 13.2 using normal, secondary, and tertiary butanol. Note the long time taken for elution. Increase the column temperature by 10°C and repeat the experiment. Continue to increase the column temperature by intervals of 10°C until the elution time is acceptable.

13.5 Repeat Experiment 13.4 using a mixture of methanol, propanol, butanol, hexanol, and decyl alcohol. Note that no single column temperature is satisfactory for all the compounds.

13.6 Repeat Experiment 13.5 with a programmed-temperature gas chromatograph. Experimentally determine the optimum temperature program rate.

13.7 Inject a mixture of unknown compounds into a gas chromatograph equipped with trapping equipment. Trap out the components separately as they emerge from the column. Identify them by IR spectroscopy or NMR.

13.8 Brominate a sample of benzene. Inject it into a g as chromatograph. Identify the products of the reaction by trapping the components and identifying them by IR spectroscopy, NMR, or MS.

PROBLEMS

13.1 Complete the following table (time for air peak = 10 sec):

Time for sample peak to emerge (sec)	Time for sample peak to pass detector (sec)	Number of theoretical plates
170	10	
250	10	
250	20	
610	10	
610	30	

13.2 What is the Van Deemter equation? In a column 6 ft long, $A = 0.1$, $B = 0.05$, $C = 0.2$, and $V = 0.1$ ft/sec. What is the theoretical plate height H? How many theoretical plates are in the column?

13.3 Illustrate a gas chromatographic instrument and describe the principal component.

13.4 Describe three detectors used in gas chromatography.

13.5 What are the differences between packed-column chromatography and capillary column chromatography? What detectors are used in capillary column chromatography?

13.6 What are the applications of programmed-temperature gas chromatography? Explain.

13.7 The following data were obtained in a quantitative determination:

Weight of pure compound A injected into a column	0.0121 g
Area of relevant peak	2.42 in.^{2}
Weight of sample injected into column	0.115 g
Area of peak of compound A in sample	4.6 in.^{2}

What percentage of compound A is present in the sample?

13.8 How many different components can be separated and identified by thin-layer chromatography?

13.9 What are the principal analytical uses of column chromatography?

14
Thermal Analysis

When matter is heated, it undergoes certain physical and chemical changes. These physical and chemical changes take place over a wide temperature range. Physical changes such as melting or boiling may occur at widely varying temperatures, depending on the material involved. For example, helium is a liquid at 6 K, whereas carbon is stable in an inert atmosphere up to 5000 K. Chemical changes, such as decomposition or reaction, may also take place at very different temperatures.

The physical and chemical reactions a sample undergoes when heated are characteristic of the material being examined. By measuring the temperature at which such reactions occur and the heat involved in the reaction, we can characterize the compounds present in the material. The majority of known inorganic compounds have been so characterized. The physical and chemical changes that take place when an unknown sample is heated provide us with information that enables us to identify the material. These changes also indicate the temperature at which the material in question ceases to be stable under normal conditions. This information is very useful to industrial chemists, such as those who make varnishes, paints, and polymers, since it allows them to predict the service lifetime of such compounds.

The analytical procedures that take advantage of these temperature- and heat-related properties are thermogravimetric analysis (TGA), differential thermal analysis (DTA), differential scanning calorimetry (DSC), thermometric titrations (TT), and direct injection enthalpy measurements. These are discussed below.

A. THERMOGRAVIMETRY

Thermogravimetry was born out of problems encountered in gravimetric analysis. For example, a common precipitating and weighing form for the determination of calcium was calcium oxalate. In practice, the calcium oxalate was precipitated, filtered, and the filtrate dried and weighed. The drying step was difficult to reproduce. For example, a particular routine procedure might recommend drying at 110°C for 20 min and cooling prior to weighing. The analyst would then obtain very reproducible results derived from his or her own work. A second worker might recommend drying at 125°C for 10 min, cooling, and weighing. As before, the analyst would obtain very reproducible results, but these results might not agree with those of the first research worker. The thermogram of calcium oxalate is shown in Fig. 14.1. This problem is more easily understood when we examine the thermogram and find that water is driven off at temperatures up to 225°C. The water that is driven off includes not only uncombined but absorbed water from the precipitating solution, but also water of crystallization that is bound to the calcium oxalate. It is only when both types of water are driven off that reproducible results are achievable. If the sample were heated to 110°C, the absorbed water would be driven off, but a small amount of combined water would also be lost. Similarly, if the drying temperature were 125°C, the absorbed water would be driven off and a different small amount of combined water would be lost. As long as a drying temperature was rigidly adhered to, the results would be reproducible, but the results using a different drying temperature would show differences because the amount of water crystallization obtained would vary from one set of conditions to another.

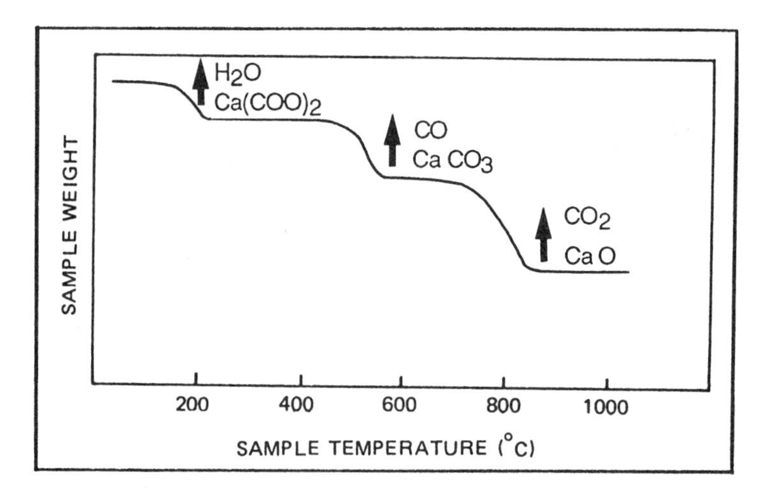

Figure 14.1 Thermogram of calcium oxalate.

The problem was solved when Claude Duval, working at the Sorbonne in Paris, produced thermogravimetric curves for many types of precipitates and deduced in their correct drying temperatures for gravimetric analysis. In the case of calcium oxalate a temperature in excess of 200°C was necessary to ensure that all water was driven off. The system rapidly became more important when it was realized that it could also be used to characterize compounds in order to study chemical reactions and decompositions at elevated temperatures.

1. Equipment for Thermogravimetry

Thermogravimetry (TGA) is concerned with the change of weight of a material as its temperature changes. First, this determines the temperature at which the material loses weight. This loss indicates decomposition or evaporation of the sample. Second, the temperature at which no weight loss takes place is revealed, which indicates stability of the material. These temperature ranges are physical properties of chemical compounds and can be used for their identification.

Knowledge of the temperatures at which a sample is unstable and subject to decomposition or chemical change is important to the engineer because it reveals the temperature range over which such materials as polymers, alloys, building materials, packing materials, and high-pressure valves may *not* be used, as well as the temperatures at which they may be used safely. It is also important to the analytical chemist, because it helps distinguish samples such as two different polymers or differentiates between a polymer that would fail in service and one that would be satisfactory.

Another important piece of information that can be obtained by TGA is the *weight lost* by a sample heated to a given temperature. This helps the inorganic or analytical chemist determine the composition of a compound and follow the reactions involved in its decomposition. It also enables the analytical chemist to identify crystals of unknown composition or determine the percentage of a given compound in a mixture of compounds. For example, if calcium carbonate ($CaCO_3$) is heated to 850°C, it loses 44% of its weight (Fig. 14.1). Also, the gas evolved can be collected and identified as CO_2. This observation virtually confirms that the reaction

$$CaCO_3 \rightarrow CaO + CO_2$$

takes place at this temperature. The loss of CO_2 would also lead to a loss in weight of 44%, and the products of the reaction can be identified as CaO and CO_2. The formation of CO_2 can be verified by injecting the offcoming gases into a mass spectrometer and confirming by the spectrum observed.

a. The Chevanard Balance

The equipment used for TGA consists of a furnace for measuring the temperature of the sample, a balance for continuously weighing the sample, and a re-

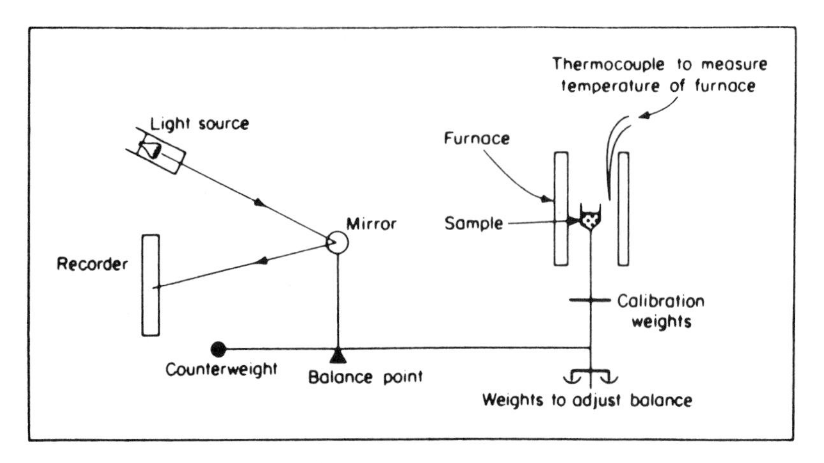

Figure 14.2 Chevanard balance.

corder. The first balance to find widespread use in pioneering work by Duval and other workers in this method of analysis was the Chevanard balance (Fig. 14.2). In practice, the sample was loaded into the crucible and a counterweight added to restore the arms of the balance to equilibrium. The furnace was heated at a controlled rate, starting from the lowest temperature (usually room temperature) and ending at temperatures as high as 1000°C. The rate at which the temperature was increased was slow enough to allow the sample to remain close to chemical and thermal equilibrium with its surroundings throughout the heat treatment. If the sample was heated too rapidly, the apparent temperatures at which chemical or physical changes took place would appear to be too high. If the rate at which the sample was heated was too slow, the experiment would take too long. Moreover, the Chevanard balance was a delicate instrument. Shaking the table on which it stood disturbed it so that the readings it gave were suspect. In addition, the furnace required cooling after each sample before the next one could be tested. This cooling took an inordinate amount of time (up to 2 hr). At most, two analyses per day could be carried out. The system is no longer in use.

b. Other Thermogravimetry Instruments

Numerous types of TGA instruments are commercially available at the present time. Basically, they all measure the same physical parameter, that is, the change in physical weight against the temperature increase. It should be noted that there appears to be no advantage to measuring weight changes against temperature decrease, but there are several disadvantages, which is why this approach has never become popular in the practicing laboratory.

The methods for weighing the sample on a continuous basis have used both single-pan and double-pan balances. The actual weight change has been monitored either by a torsion balance, spring balance, or self-balancing devices or by using a Cahn electrobalance.

The balance shown in Fig. 14.3 monitors weight change by the current generated in a transducer coil. In this system the sample is loaded into the sample holder. The temperature of the sample is steadily raised by increasing the tem-

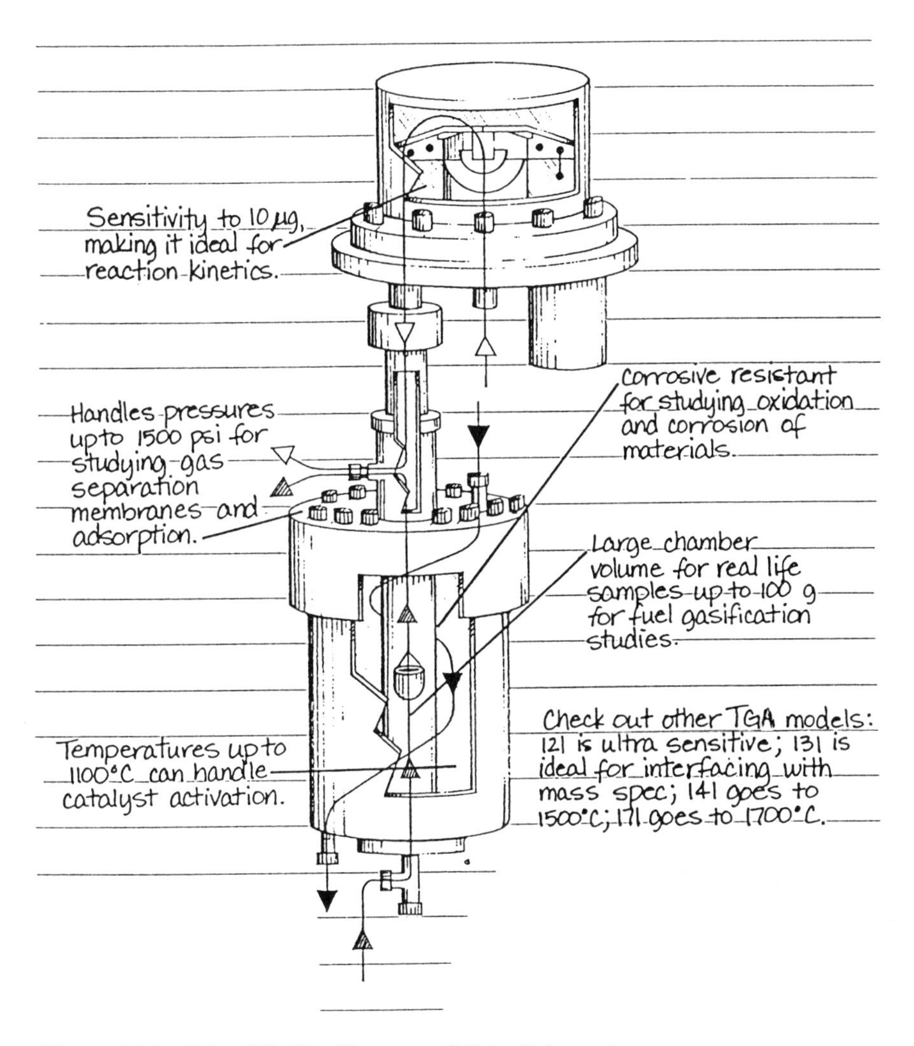
Sensitivity to 10 μg, making it ideal for reaction kinetics.
Corrosive resistant for studying oxidation and corrosion of materials.
Handles pressures upto 1500 psi for studying gas separation membranes and adsorption.
Large chamber volume for real life samples up to 100 g for fuel gasification studies.
Check out other TGA models: 121 is ultra sensitive; 131 is ideal for interfacing with mass spec; 141 goes to 1500°C; 171 goes to 1700°C.
Temperatures up to 1100°C can handle catalyst activation.

Figure 14.3 Cahn TG 151. (Courtesy of Cahn Balances.)

perature of the furnace. Any decrease in the sample's weight is registered on the electrostatic balance. This movement generates an electrical signal proportional to the weight loss of the sample. The electrical signal is used to move the recorder in the *Y* direction. The temperature change in the oven also generates a signal in a thermocouple, which is relayed to the recorder in the *X* direction. The combination of the two signals is recorded on an *X–Y* recorder to present a plot of the sample's weight against the temperature.

2. Thermograms

The plot of weight against temperature produced by a thermogravimeter is called a *thermogram* (Fig. 14.4). The gases given off by a sample when it loses weight during a TGA run can be collected and identified, as can the compound remaining in the sample cell at each step in the process. The thermogram in Fig. 14.1 clearly shows that calcium oxalate forms stable compounds in the temperature ranges 240–400, 500–650, and 860–1000°C.

As discussed earlier, calcium oxalate was frequently used as a gravimetric weighing form of calcium. For reproducible results, the thermogram in Fig. 14.1

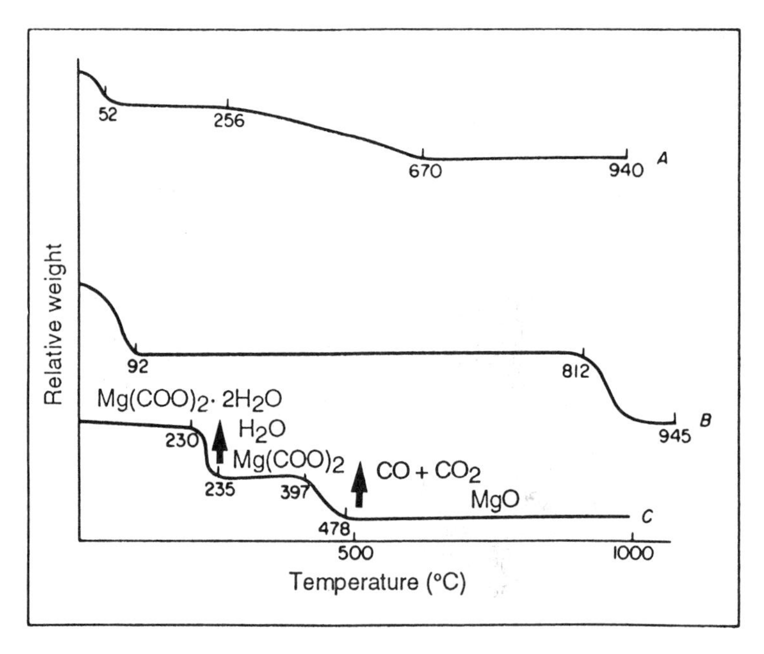

Figure 14.4 Typical thermograms of (*A*) mercurous chlorate, (*B*) silver chromate, and (*C*) magnesium oxalate.

shows that it is necessary to heat the sample to at least 228°C before the precipitate used in this gravimetric analysis relates to the weight of calcium in the sample.

3. Analytical Applications of Thermogravimetry

One of the first important applications of thermogravimetry was the determination of correct drying temperatures for precipitates used in gravimetric analysis. This knowledge was of vital importance if accurate and reproducible results were to be obtained from gravimetric analysis. A second important application was the identification of the gases given off while a sample's temperature is increased. In addition, the composition of the residue can be determined. This information reveals the chemical decomposition process of heated materials and permits identification of the formulas of the residue.

For example, from Fig. 14.1 it can be determined that when calcium oxalate [$Ca(COO)_2H_2O$)] is heated, it first loses water of crystallization and forms $Ca(COO)_2$. Upon further heating (up to about 400°C), CO is given off and the reaction $Ca(COO)_2 \rightarrow CaCO_3 + CO \uparrow$ occurs. Finally, at even higher temperatures (about 800°C), the $CaCO_3$ formed at 400°C decomposes to form $CaO + CO_2 \uparrow$.

A third application of TGA is the identification of the compounds present in mixtures of materials. When such mixtures are heated using a thermogravimeter, a thermogram is produced for each component. These thermograms are super-imposed on each other to provide a single composite thermogram for the sample. Interpretation of the complete thermogram requires that the individual thermograms be separated and identified. Not only can the components of the mixture be identified, but a quantitative determination of each is possible from the thermogram.

An illustration of the application of TGA to quantitative analysis is the determination of the magnesium oxalate content in a mixture of magnesium oxalate [$Mg(COO)_2 \cdot 2H_2O$] and MgO. The TGA curve of magnesium oxalate is shown in Fig. 14.4. Magnesium oxide is stable at the temperatures indicated and does not lose any weight.

It can be seen from Fig. 14.4 that when $Mg(COO)_2 \cdot 2H_2O$ is heated to a temperature above 478–500°C, it forms MgO according to the reaction

$$Mg(COO)_2 \cdot 2H_2O \rightarrow MgO + CO + CO_2 + 2H_2O$$

The final weight is therefore

$$\frac{\text{mol wt MgO}}{\text{mol wt } Mg(COO)_2 \cdot 2H_2O} \text{ per gram of } Mg(COO)_2 \cdot 2H_2O$$

We can calculate, therefore, that 1 g of dry magnesium oxalate would lose 0.73 g upon heating to 500°C.

From Fig. 14.4 the following data can be obtained:

Original weight of sample	2.50 g
Weight of sample after heating to 500°C	1.04 g
Loss in weight	1.46 g

But we have already calculated that 1 g of $Mg(COO)_2 \cdot 2H_2O$ loses 0.73 g upon heating to 500°C; therefore the weight of $Mg(COO)_2 \cdot 2H_2O$ in the original sample was

$$\frac{1.46}{0.73} = 2.0 \text{ g}$$

$$\text{percent in original sample} = \frac{Mg(COO)_2 \cdot 2H_2O \times 100}{\text{total weight of sample}}$$

$$= \frac{2.00}{2.50} \times 100 = 80\%$$

If we ran a mixture of magnesium oxalate [$Mg(COO)_2 \cdot 2H_2O$] and calcium oxalate [$Ca(COO)_2 \cdot H_2O$], we could calculate the percentage of each present from the thermogram, assuming the mixture contains no other compounds.

The thermogram obtained is shown in Fig. 14.5. By examining Figs. 14.1 and 14.5, it is evident that the CO_2 evolving at 700°C is from $Ca(COO)_2 \cdot H_2O$. We know one molecule of $Ca(COO)_2 \cdot H_2O$ generates one molecule of CO_2. If the weight loss due to CO_2 was 22%, then the weight of $Ca(COO)_2 \cdot H_2O$ is

$$\frac{\text{mol wt } Ca(COO)_2 \cdot H_2O}{\text{mol wt } CO_2} \times 22\% = \frac{146}{44} \times 22 = 73\%$$

Assuming that the weights of calcium oxalate and magnesium oxalate total 100%, then it follows that the percentage of $Mg(COO)_2 \cdot 2H_2O$ = 100 − 73 = 27%.

4. Differential Thermogravimetry

Examination of a TGA curve will show that a sample's weight loss associated with a particular decomposition occurs over a considerable temperature range.

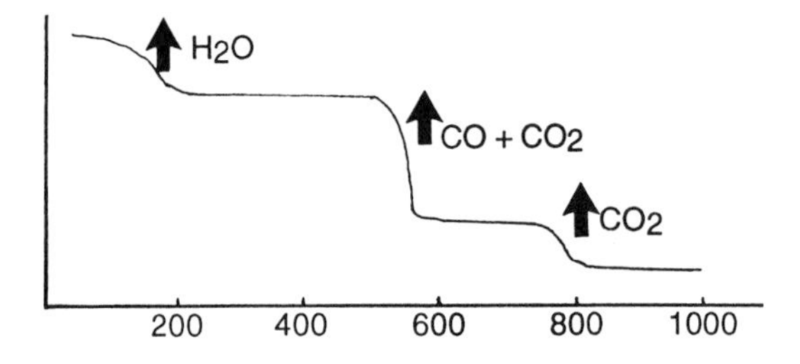

Figure 14.5 Thermogram of a mixture of Ca oxalate and Mg oxalate.

When TGA is used to identify an unknown compound, this wide range is a handicap because the uncertainty of identification is increased. This problem can be partially overcome by derivative thermogravimetry (DTG). In this instance a derivative of the TGA curve is taken. Figure 14.6 shows a thermogram of nylon and its DTG curve. There is a minimum in the DTG curve at the point of inflection of the TGA curve. This makes identification simpler. Consequently DTG is a valuable method of data presentation for qualitative analysis.

Another example of the application of thermogravimetry is in the characterization of coal. A typical curve is shown in Fig. 14.7. This curve indicates, in a single analysis, the percentage of volatiles, the fixed carbon, and the amount of ash that would be expected from the coal examined. This is very valuable information in the characterization of the coal and for its handling and subsequent use. For example, Fig. 14.7 indicates a considerable amount of volatile material. The composition of this volatile material is variable and may contain valuable chemicals that can be used in the pharmaceutical industry, the dyestuff industry, and the chemical industry, among others. Of course it can be used as a fuel as coal gas.

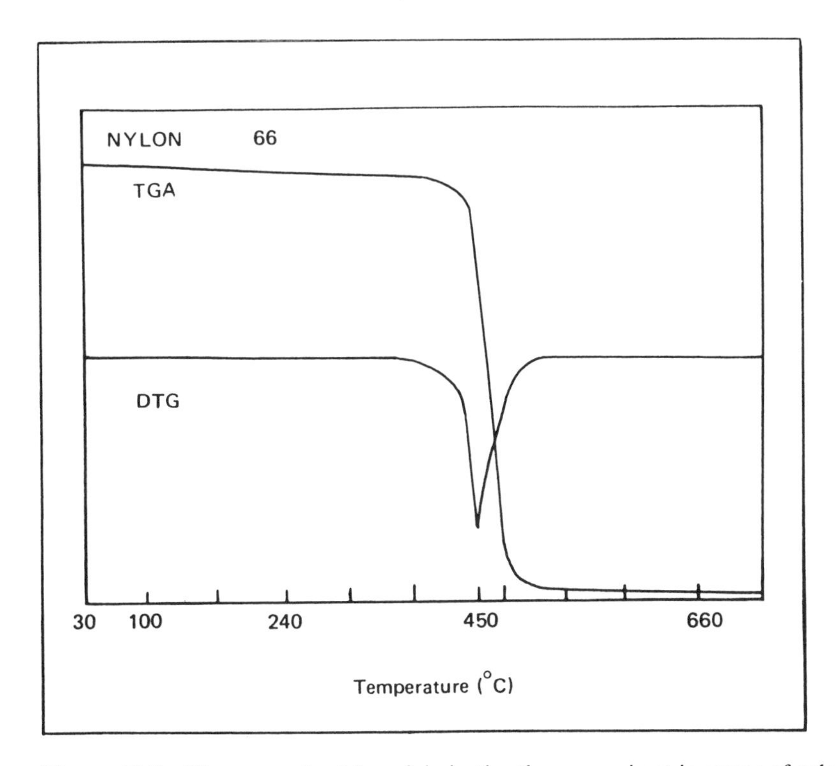

Figure 14.6 Thermogravimetric and derivative thermogravimetric curves of nylon 66.

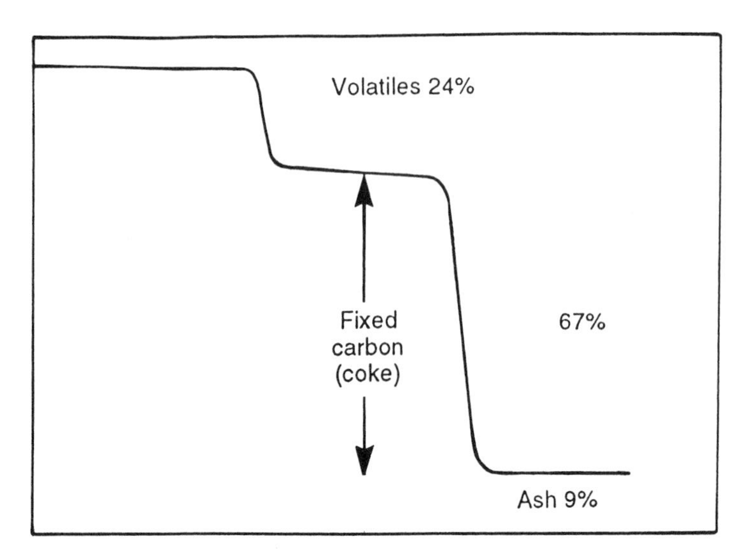

Figure 14.7 Thermogravimetric curve of coal showing the percentage of volatiles, fixed carbon, and ash. These data can be used to characterize the quality of the coal.

The composition of the evolved gas is determined using *evolved gas detection* (EGD) or, if quantitative analysis is required, *evolved gas analysis* (EGA). These techniques are essentially a combination of TGA and mass spectrometry (MS) or tandem mass spectrometry (MS–MS). The evolved gases from the thermogram are sampled with a capillary or jet separator and enter the mass spectrometer. Usually a quadrupole mass spectrometer using chemical ionization is the preferred instrument.

An important application of EGD and EGA is in the characterization of shale. Huge reserves of shale are available that contain large quantities of combustible, volatile organic materials. Unfortunately the percentage of volatiles is small. Theoretically, the recoverable organic material can be used as fuel. Based on EGA results, an estimate can be made of how much fuel is recoverable and whether it is profitable to process the shale as a major fuel source. So far no economic way has been found to recover these huge quantities of low-grade fuel.

a. Calibration of Temperature

For sample characterizations it is important to know the precise temperatures at which transitions occur. This entails calibrating the temperature of the sample while the analysis is carried out. It is particularly important to measure the temperature of the sample rather than that of the furnace, and this is difficult because the detectors are near but not in the sample. The problem is compounded by the fact that at temperatures below 500°C, most of the heat transferred from the furnace to the sample takes place by convection and conduction, but at temperatures above 500°C, where the furnace is red hot, most of the energy is transferred by radiant energy.

The time-honored device for measuring temperature is the thermocouple. A number of places have been suggested for the thermocouple so that accurate measurements may be obtained, but the switch from conduction–convection to radiative energy transfer makes this quite a complicated problem. However, the thermocouple is still widely used.

Another method is to use standards that have known transition temperatures and either include these in a sample or run their thermogram under standardized conditions from which calibration data can be obtained. A problem with this technique is that black samples such as coal behave differently from white samples such as calcium phosphate under the influence of radiant energy, and the sample temperatures will therefore be different under the same furnace conditions.

An accurate technique has been developed based on the Curie points of standard ferromagnetic materials. A ferromagnetic material is magnetic under normal conditions, but at a characteristic temperature (the Curie point) its atoms become disoriented and paramagnetic and the material loses its magnetism. To take advantage of this phenomenon, we weigh the ferromagnetic material continuously, with a small magnet placed above the balance pan. The standard's apparent weight is then its gravitational weight minus the magnetic force it experiences due to the magnet immediately above. At the Curie point, the standard loses its magnetism and the effect of the magnet is lost. There is a change in the apparent weight of the standard, and this can be recorded. The temperatures of the Curie point transitions of ferromagnetic materials are well known and therefore can be used for calibration purposes. To increase the accuracy of the data, the readings may be presented as derivative thermograms as shown in Fig. 14.8. This method is accurate and is convenient for temperature calibration.

B. DIFFERENTIAL THERMAL ANALYSIS

Differential thermal analysis (DTA) is concerned with the rate of change of the temperature of a sample as it is heated at a constant rate of heat input. When a sample is heated slowly, its temperature increases at a steady rate. However, if an endothermic reaction, such as melting, takes place, the sampler's temperature remains constant even though the sample continues to absorb heat from its surroundings. The heat is used to melt the sample, and the sample's temperature will not change until it has completely melted. When the temperature remains constant, its rate of change is zero. Losing CO_2 to the atmosphere, driving off water, changing crystal structure, and decomposing are other endothermic reactions.

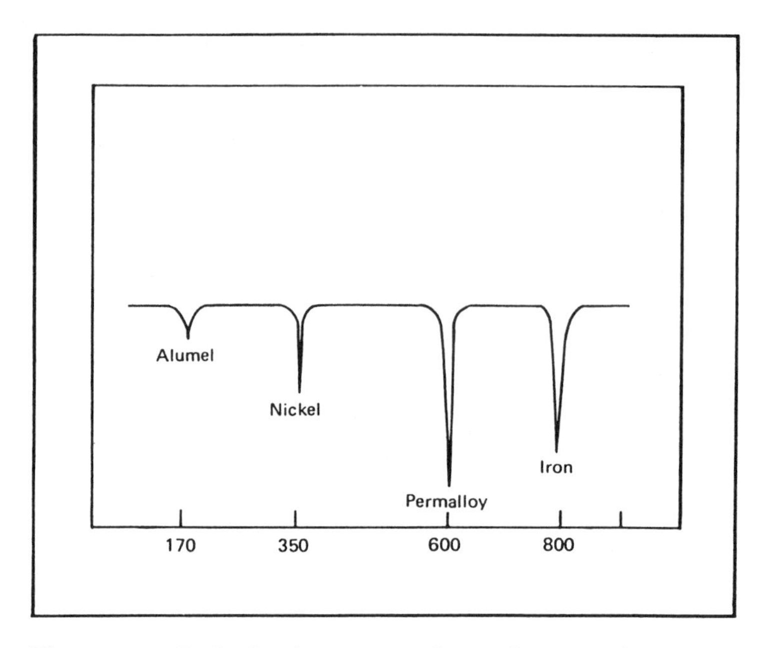

Figure 14.8 Derivative thermograms of some ferromagnetic materials showing their Curie points.

A change of phase, such as from solid to liquid or from one solid to another, will also produce a similar effect. It is not necessary that the sample's weight change in order to produce a DTA response. However, if a weight change does take place, as in drying, a signal will invariably be generated in the process. Phase changes can be exothermic or endothermic. Endothermic reactions absorb heat and the peak is negative, but exothermic reactions release heat and the peak is in the opposite direction (see Fig. 14.15).

1. Equipment

The equipment used in DTA studies is shown in Fig. 14.9. The sample is loaded into the crucible, which is then inserted into the well at the right. A reference sample is made by placing a similar quantity of inert material (such as Al_2O_3) in a second crucible. This crucible is inserted in the well at the left. The third (center) well is used to house a thermocouple to measure the temperature of the block. The dimensions of the two crucibles and of the cell wells are as nearly identical as possible; furthermore, the weights of the sample and the reference are also virtually equal. The metal block surrounding the wells acts as a heat sink. By mixing equal amounts of reference material and sample material, the

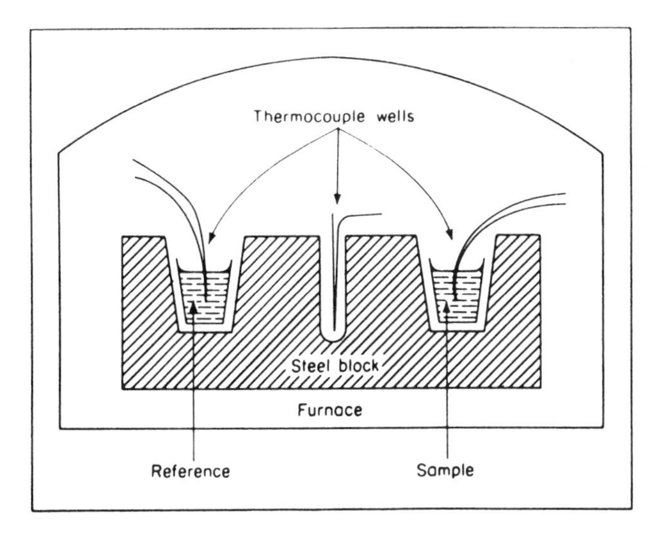

Figure 14.9 Equipment for DTA.

heat conductivities of the two cells become more similar to each other and more accurate results are obtained.

The temperature of the heat sink is slowly increased using an internal heater. The sink in turn simultaneously heats the sample and reference material. A thermocouple is placed in each of the crucibles in such a fashion that their voltages generate opposing electrical signals. This allows the difference in temperature between the sample and the reference to be recorded. If there is no difference in temperature, the electrical signals generated by the thermocouples cancel each other out and no signal is generated, even though the actual temperatures of the sample and reference are both increasing.

When a physical change takes place in the sample, heat is absorbed or generated. For example, when a metal carbonate decomposes, CO_2 is evolved and heat is absorbed from the crucible and the sample, which in turn remains at a lower temperature than the reference. The temperature difference between the sample and reference generates a net signal, which is recorded. A plot of the temperature difference against the heat sink temperature is the DTA thermogram. A typical example of a DTA thermogram is shown in Fig. 14.10.

If, in the course of heating, the sample undergoes a phase transition or a weight loss that results in the generation of heat, that is, an exothermic reaction, the sample becomes hotter than the reference material. In this case, the sample heats up to a temperature higher than that of the reference material until the reaction is completed. The sample then cools down or the temperature of the reference cell catches up until its temperature and that of the heat cell once again

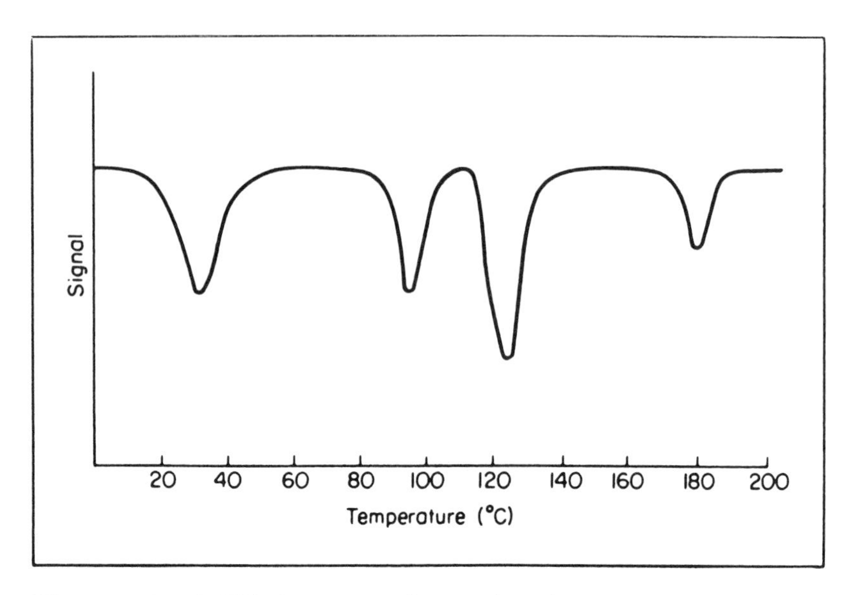

Figure 14.10 A DTA thermogram of ammonium nitrate.

become equal. Such an effect is shown on the thermogram as a peak that moves in a positive (upward) direction rather than in the negative direction, which allows us to distinguish between exothermic and endothermic reactions. This is another characteristic that enables us to identify the chemical under examination, if it is an unknown material, or to measure the amount of heat lost or gained during a phase transition.

2. Analytical Applications

Differential thermal analysis is based on changes of heat flow into the sample. Using DTA, we can detect the decomposition or volatilization of the sample, just as we can with TGA. In addition, however, physical changes that do not involve weight changes can be detected by DTA. Such changes include crystallization, changes in phase, homogeneous reactions in the solid state, and degradation. In each of these changes there is a flow of heat between the sample and its surroundings caused by endothermic or exothermic transitions.

It should be noted that a change in the sample's weight does not necessarily take place at a transition temperature detected by DTA. An instance of the use of transitions where no change in weight occurs is the DTA characterization of semicrystalline materials, for example, polymers. The physical properties, such as strength and flexibility, of a polymer depend (among other things) on its crystallinity. A polymer is made up of high molecular weight compounds that are

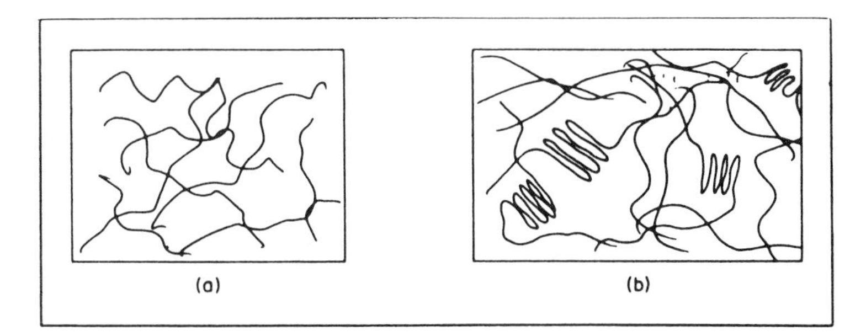

Figure 14.11 The crystal structures of polymers: (a) amorphous arrangement (molecules are intertwined) and (b) crystalline and amorphous arrangement (some molecules are folded over in an orderly manner).

shaped somewhat like long chains. When these chains are stuck together like a rumpled-up piece of string, the polymer is said to be amorphous. With proper treatment, however, some of these long chains can be made to form crystal-like zones that are regularly oriented (Fig. 14.11). The DTA curves of the two samples in Fig. 14.11 would appear as in Fig. 14.12.

in Fig. 14.12a no transition took place until the sample decomposed. However, the softening of the polymer absorbed some heat; hence the curve is not linear. This nonlinearity is typical of an amorphous sample. It can be seen from Fig. 14.12b (crystalline sample) that a transition occurred starting at about 200°C, whereas decomposition commenced at 480°C. The transition resulted from the melting of the crystals in the polymer. As in the case for any other crystal, heat is required to break down the crystal structure. This heat requirement is indicated by the transition points in the curve.

We can further deduce from Fig. 14.12b that the area of the transition peak at 180°C is directly proportional to the weight of crystalline material in the sam-

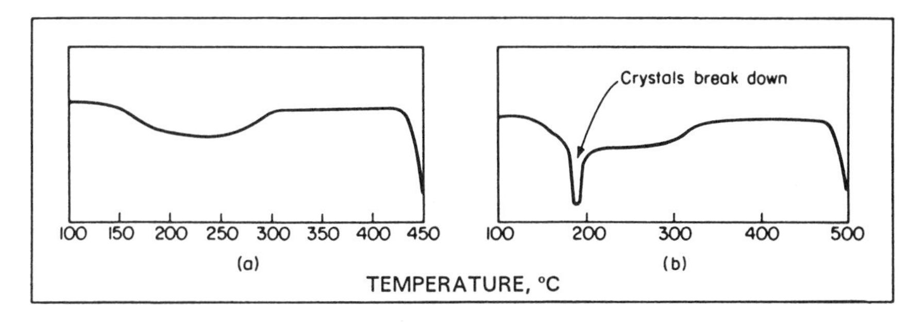

Figure 14.12 The DTA curves of two polymers: (a) amorphous and (b) crystalline.

ple. The relationship between area and percentage of crystallinity can be calibrated by obtaining the DTA curves of samples of known crystallinity. From these calibration data we can calculate the percentage of crystallinity of a polymer of unknown crystalline content.

Figure 14.12b also indicates the temperature range over which melting took place. This information can be used to get an idea of the size of the crystals. In fact, the curve provides two important pieces of information; an engineer could ascertain that the polymer loses strength at temperatures above 150°C, and a chemist or chemical engineer could ascertain the material's crystallinity and the temperature at which breakdown occurs.

Qualitative DTA can be carried out by comparing a sample's DTA curve with those of known standards. This comparison makes it possible to identify the materials present in the sample. Mixtures of compounds can be analyzed quantitatively by DTA by measuring the amounts of heat involved in the various transitions. These are calibrated by comparison with known amounts of the same compounds. From the data, an estimate of the quantity of each compound present can be made.

Differential thermal analysis can be used for industrial controls, such as for the determination of the structural and chemical changes occurring during sintering, fusing, and other heat treatments; identification of different types of synthetic rubbers; identification of the components of alloys; and determination of structural changes in the production of metals sheets and wire. It is also used to study the mechanism of fatigue and other sources of metal failure.

C. DIFFERENTIAL SCANNING CALORIMETRY

The most useful applications of both TGA and DTA are for qualitative and semiquantitative analysis. Identification of transition temperatures and phase changes and the determination of exothermic or endothermic transitions are readily obtained. Only semiquantitative data can be obtained with these techniques, because accurate information on the specific heats of mixtures of unknown composition is not available. Also, the thermal conductivity for mixtures is not known, and therefore there are variations in heat transfer rates between the heat sink and the reference and sample cells. Consequently, analytical errors are introduced.

Better quantitative information can be obtained by using differential scanning calorimetry (DSC). The experimental setup is similar to that used for DTA, but the analysis block is replaced by a differential calorimeter. The sample and reference are heated directly with separate heating coils.

A heating coil makes the temperature of the reference material increase at a constant rate. A second heating coil is placed in the sample. The sample and the reference are kept at equal temperatures. When a phase change or weight loss occurs, the sample and reference temperatures become slightly different, which generates a current in the thermocouple system measuring the temperature difference between the two cells. This current activates a relay, causing extra power to be directed to the cell at the lower temperature. In this manner, the temperatures of the reference and sample cells are kept virtually equal throughout. The quantity of electrical energy used in heating the sample and the reference is measured accurately and continuously. In turn, the electrical energy is an exact measure of the number of calories used in heating the cells.

The resultant thermogram is similar to a DTA trace, but more accurate and reliable. Endothermic changes are recorded as heat input into the sample, and exothermic changes as heat input into the reference. The area of the peaks is an exact measure of heat input involved. Differences in heat capacity or thermal conductivity do not affect the results. From the data, accurate quantitative analytical results can be obtained. This is a very valuable technique, for it enables the determination of enthalpies of reactions. It should also be noted that if the equipment is built to take measurements under adiabatic conditions, it is also possible to measure specific heats accurately.

The use of DSC can be extended by running samples under different pressures. Changing pressure has several effects on chemical systems:

1. Vaporization occurs at different temperatures.
2. Reactions involving offcoming vapors are modified.
3. Heterogeneous reactions involving the offcoming vapors are accelerated or retarded.
4. Decomposition generating volatile compounds is modified.
5. Reactions with gaseous toxins can be studied.
6. Absorption and desorption effects can be studied.

Figure 14.13 shows the absorption and desorption of H_2 by palladium. This reaction has important implications in catalytic effects.

It is also possible to examine the sample in an atmosphere of a reactive gas, thus studying the reaction between the sample and the gas. The system shown in Fig. 14.3 would accommodate such work.

Another important application is the curing of polymers used for thermosetting that must be treated before use. A thermogram of uncured polyester is shown in Fig. 14.14. This figure includes the normal DSC curve and the first differential to the curve. It shows that most of the glass transition occurred before 50°C was reached but accelerated at higher temperatures. The shoulder at 120°C indicates the formation of some glassy material.

In summary, TGA can be used to detect changes in weight with changes of temperature. In addition, DTA can be used to detect phase changes. However, good quantitative data on phase changes (and decomposition) can best be obtained using DSC.

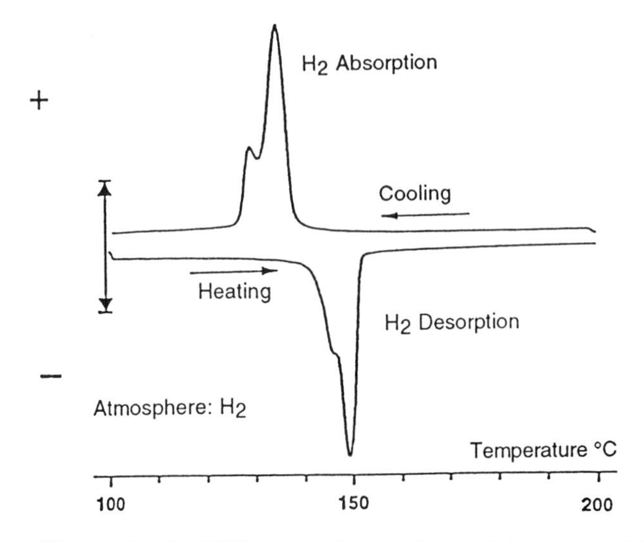

Figure 14.13 DSC curve of adsorption and desorption of hydrogen under constant pressure.

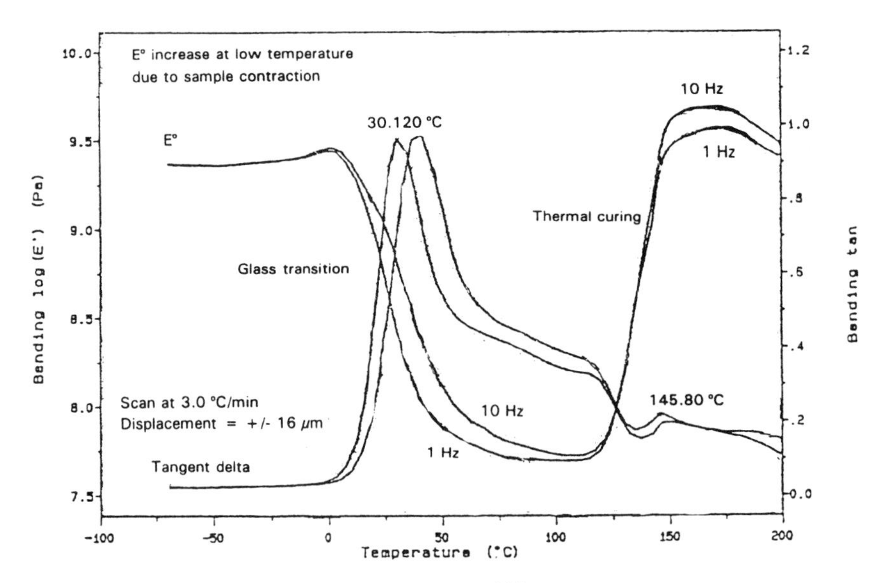

Figure 14.14 Thermal scan of uncured Kevlar™ polyester.

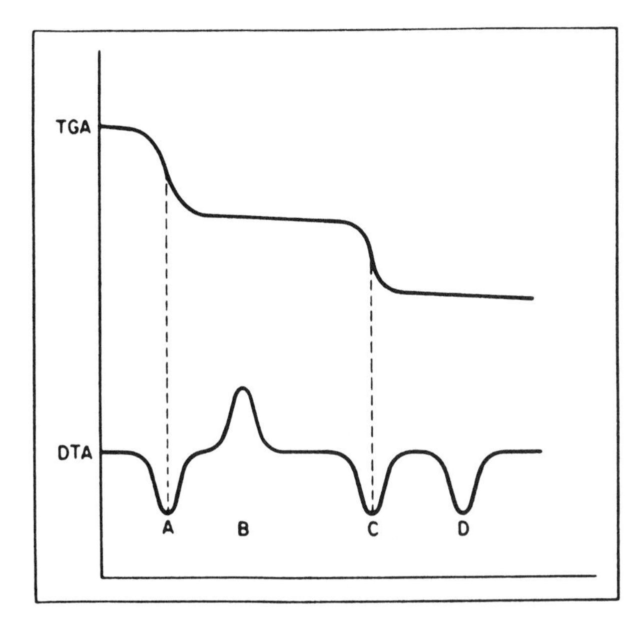

Figure 14.15 The TG and DTA curves of a hypothetical sample. The two steps in the TG curve result from weight losses. They correspond to peaks A and C in the DTA curves, Peaks B and D are, respectively, an exothermic and an endothermic phase change.

D. COMBINATION TECHNIQUES

The interpretation of thermograms is facilitated when both a TGA thermogram and a DTA thermogram are run on the sample. The appearance of a peak on both curves indicates that the sample has undergone a weight loss; the appearance of a peak in the DTA curve but not the TGA curve discloses that the change did not involve a weight loss but must have been caused by a phase change, such as melting. This is illustrated in Figs. 14.15 and 14.16.

The gases evolved during TGA, DTA, or DSC can be trapped and identified. This step provides important information for interpreting the reactions of the sample at elevated temperatures. Moreover, as we mentioned earlier, the remaining sample can be retrieved at any time during TGA or DTA studies and checked for composition changes or changes in crystallinity by techniques such as x-ray diffraction.

Equipment has now been developed in which the offcoming gases are fed directly into a mass spectrometer. This permits immediate identification of the offcoming gases and greatly facilitates interpretation. The procedure is also much easier to handle since labor-intensive trapping steps are eliminated. A schematic diagram is shown in Fig. 14.17.

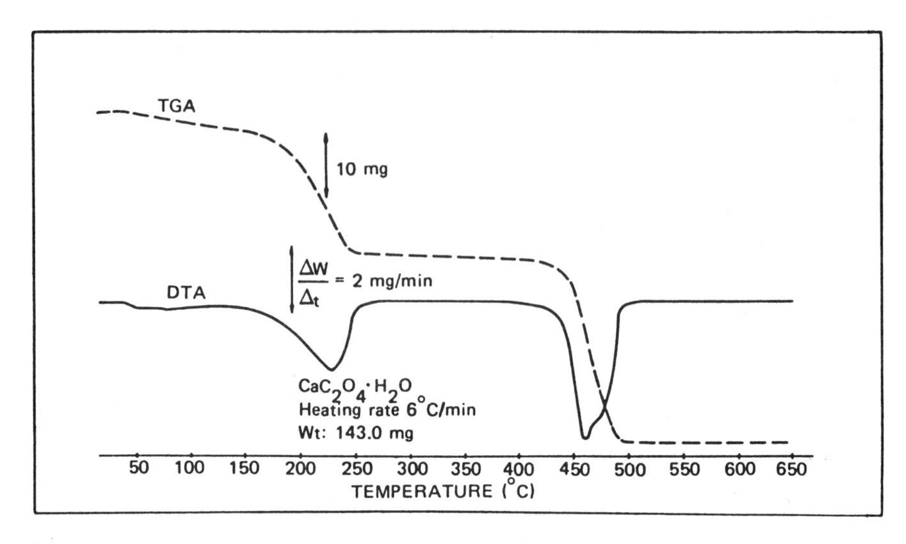

Figure 14.16 TG and DTA curves of $CaC_2O_4{\cdot}H_2O$.

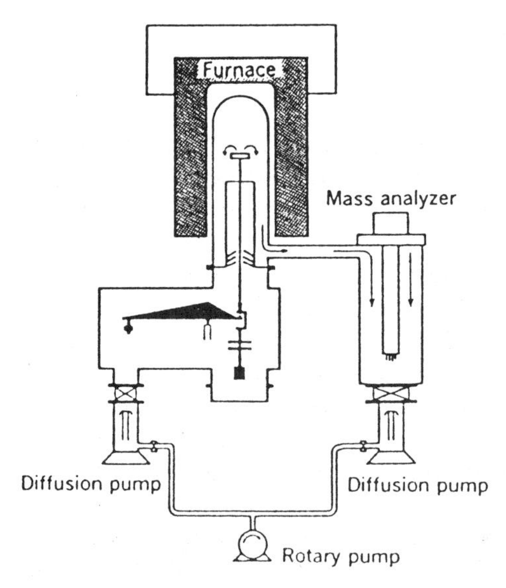

Figure 14.17 Schematic diagram of interfaced thermal analyzer–MS (Courtesy of Mettler Instrument Corp.)

Table 14.1 Partial List of Sample Types and Properties Examined by Thermal Analysis

	Samples						
Properties	Chemicals	Elastomers	Explosives	Soils	Plastics	Textiles	Metals
Identification	x	x	x	x	x	x	x
Quantitative composition	x	x	x	x	x	x	
Phase diagram	x	x	x		x		x
Thermal stability	x	x	x		x	x	
Polymerization	x	x			x	x	x
Catalytic activity	x						x
Reactivity	x	x			x	x	x
Thermochemical constants	x	x	x	x	x		

x = sample type that has been characterized for this property.

Experimental work using TGA–DSA mass spectroscopy on calcium oxalate showed the presence of some CO_2 at the 470°C step. This may be caused by the presence of air oxidizing the CO to CO_2.

In short, TGA and DTA provide a means of studying the chemical and physical changes that take place in a sample when it is heated, and these changes can be used to characterize the material. Thermal analysis has found wide application in such areas as ceramics, polymer sciences, metallurgy, and gravimetry. A brief summary of these applications is presented in Table 14.1.

Some other applications of TGA are as follows:

Determinations of environmental stability
Thermal stability in inert atmospheres
Oxidative stability
Analysis of coals and coal products
Lubricants and lube oils
Polyethylene/Carbon black
Moisture determinations
Residual solvent determinations
Ash and inert filler determinations
Characterization and analysis of clays and minerals, oil shales, tar, sands, and so on
Microdistillation of petroleum and syncrude fractions
Sublimation temperatures
Determination of the plasticizer content of polymers
Estimation of catalytic activity
Determination of flame-retardant properties of polymer additives

Determination of loss on ignition of industrial raw materials
Characterization of cements
Establishment of stoichiometries of dehydration

E. THERMOMETRIC TITRATIONS

Thermometric titration depends on measuring the heat generated during a chemical reaction. Usually the reactions take place at room temperature under adiabatic conditions. In this procedure a titrant of known concentration is added to a known volume of sample. The titration reaction follows the chemical equation

$$\text{A (sample)} + \text{B (titrant)} \rightarrow \text{C (products)} + \text{heat}$$

For any particular reaction, a mole of A titrated with a mole of B will generate a fixed quantity of heat, which is the heat of reaction. If there is half a mole of A present, half as much heat is generated, even if there is an excess of B. Heat is generated only while A and B react with each other; an excess of either one does not cause generation of heat. The reaction is usually set up in an insulated beaker, a Dewar flask, or a Christiansen adiabatic cell. This ensures a minimum of heat loss from the system during titration.

In practice, the temperature of the system is measured as the titrant is added to the sample. A typical plot is shown in Fig. 14.18. Typical pH responses for the titration of boric acid and HCl with NaOH are shown in Fig. 14.19. It is clear that the process used to derive the plot in Fig. 14.19 is easy to use for the titration of boric acid.

It can be seen that the temperature of the mixture increases until approximately 4.5 ml of titrant have been added to the sample. This is the end point of the reaction. Further addition of titrant causes no further reaction, because all of the sample A has been consumed. The temperature of the mixture steadily decreases as the cool titrant is added to the warm mixture. The endpoint can therefore be determined as the abrupt change in slope that occurs at the equivalence point.

For simple titrations the equipment necessary is not particularly sophisticated; however, if one wants to measure specific heats or the heat of reaction, better control is required. For example, temperature control and insulation are necessary in order to ensure that the temperatures of the sample and the titrant are equal before the titration. It is possible to set these temperatures to within 0.001°C. A constant-delivery pump may be used to advantage. Provided that the volume delivered per unit time is accurate and reproducible, it is possible to record the temperature of the mixture on a strip-chart recorder and use this as a means of measuring the volume of titrant added. In these circumstances it is not necessary to add exactly the required amount of titrant. Any excess added does

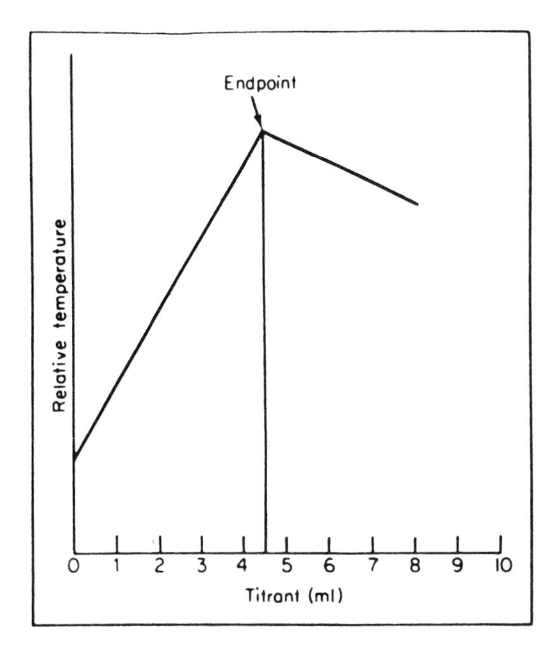

Figure 14.18 Typical thermometric titration plot for boric acid. The endpoint is denoted by an abrupt change in the slope of the curve.

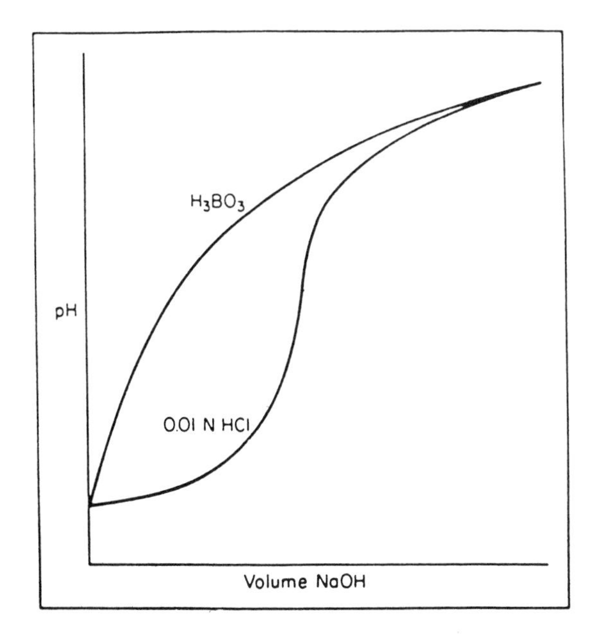

Figure 14.19 Responses for the thermometric titration of boric acid and HCl with NaOH.

not result in an increase in temperature, but merely in a change of slope in the curve of temperature versus volume of titrant. This is therefore not a source of error as it is in conventional volumetric analysis.

In thermometric titration it is assumed that the rate of reaction is relatively fast and that the endpoint will occur as soon as an excess of titrant is added. This assumption is not valid if the reaction is slow; the position of the endpoint will be distorted and errors will be involved.

F. DIRECT INJECTION ENTHALPIMETRY

When a chemical reaction occurs, heat is liberated or absorbed. This is the heat of reaction ΔH, or enthalpy. It is a reproducible physical property for a given reaction $A + B \rightarrow AB + \Delta H$. Therefore the magnitude of ΔH depends on the quantity of reactants involved in the reactions. Excess of any of the reactants does not take part in the generation of energy.

Direct injection enthalpimetry is similar in many respects to thermometric titrations; however, one essential difference is that an excess of titrant is added very rapidly to the sample and the reactants mix vigorously. The temperature is then measured against time, as shown in Fig. 14.20. We may suppose that the following reaction takes place:

$$\text{A (sample)} + \text{B (reactant)} \rightarrow \text{C (products)} + \text{heat}$$

The quantity of heat generated is a function of the number of moles that take part in the reaction. This, in turn, is controlled by the amounts of A and B. Any excess of either A or B does not react. In this instance, we have added an excess of B; therefore the amount of heat generated must be a function of the amount of A present in the mixture. If the amount of A present were halved, then the amount of heat generated would be halved. Since there is an excess of B present in both cases, this would not affect the amount of heat generated.

The method is quite useful, particularly if the rate of chemical reaction between A and B is slow. The final quantity of heat evolved is not a function of time, but a function of the concentration of sample. This is a distinct advantage over conventional volumetric analysis and, in some instances, thermometric titrations. In the latter case, it is assumed that the rate of reaction is relatively fast and that the endpoint will occur as soon as an excess of titrant is added. This assumption is not valid if the reaction is slow; the position of the end point will be distorted and errors will be involved.

Direct injection enthalpimetry and thermometric titrations are very useful for the volumetric analysis of materials, such as boric acid, which are virtually impossible to titrate using endpoint indicators or pH indicators. The methods can also be used in biological studies where the reaction rates may be slow. For ex-

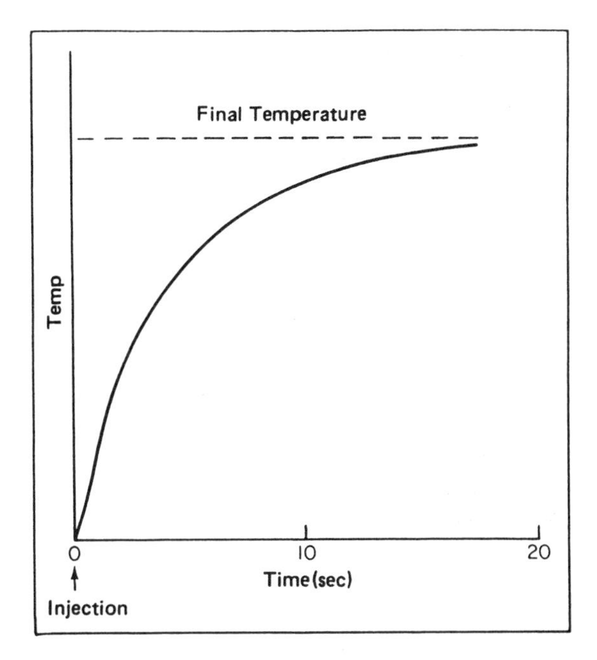

Figure 14.20 Direct injection titration. An excess of reactant B is added. The amount of sample A present is then calculated from the final temperature. The method is valid even if the reaction is slow.

ample, proteins have been titrated with acid or base, antibodies have been titrated with antigen, and enzyme–coenzyme systems have been studied.

A word of caution should be mentioned about possible interferences, particularly if two simultaneous reactions take place. Since both reactions may generate heat, a direct error will be involved with injection enthalpimetry methods. However, with thermometric titrations, frequently consecutive reactions take place and can be determined consecutively without prior separation. This has been demonstrated in the titration of calcium and magnesium with oxalate.

G. CONCLUSIONS

Thermometric methods of analysis have wide analytical applications. They can be used not only for the identification of chemical compounds, but also for the quantitative analysis of mixtures of compounds and for characterizing such materials as polymers. In polymer characterization not only can the degree of crystallinity be measured, but different polymers give different traces in thermal analysis and can be distinguished from each other. The fields of thermometric titrations and direct injection enthalpimetry are distinct advances over conventional volumetric analysis and can be used to advantage in the analysis of biological systems.

BIBLIOGRAPHY

Anal. Chem., Application Reviews, June 1994.

Connolly, M., and Tobias, B., *Am. Lab.*, (1992), *1*: 38.

Duval, C., *Inorganic Thermogravimetric Analysis*, 2nd ed., American Elsevier, New York, 1963.

Earnest, C., Modern Thermogravimetry, *Anal. Chem.*, (November):1471A (1984).

Ewing, G. W., Instrumental Methods of Chemical Analysis, 4th ed., McGraw-Hill, New York, 1975.

Washall, J. W., and Wampler, T. P., Direct-pyrolysis Fourier transform-infrared spectroscopy for polymer analysis. *Spectroscopy*, *6* (4):38 (1992).

Wendlandt, W. W., *Thermal Methods of Analysis*, 3rd ed., Wiley, New York, 1986.

SUGGESTED EXPERIMENTS

14.1 Using a TGA instrument, determine the TGA curve of sodium sulfate. Note the high stability of the salt. Repeat with sodium zinc uranyl acetate. Note that the composition varies up to 360°C.

14.2 Repeat Experiment 14.1 using KCl. Note the initial loss of free water and then the loss of bound interstitial water up to temperatures above 200°C. Could temperatures below 200°C be used to dry KCl crystals successfully?

14.3 Using a solution of $AlCl_3$ (in acid), neutralize and precipitate the Al as $Al(OH)_3$ by the addition of NH_4OH. Filter the precipitate. Obtain the TG curve of the precipitate. Note the steady loss of weight up to 1000°C. Could this precipitate be dried to a constant weight at any temperature below 1000°C?

14.4 Take a fresh piece of ham. Obtain the TG curve up to 150°C. Note the loss in weight due to (a) water evaporation and (b) fat loss. Compare with other hams.

14.5 Obtain the TGA curve of butanol. Heat slowly. Note the slow evaporation of the butanol throughout and then the rapid loss at the boiling point.

14.6 Using DTA equipment, obtain the DTA curve of benzoic acid. In the first instance, heat the acid rapidly; in the second, heat it slowly. Note the difference in the apparent melting and boiling ranges of the benzoic acid when the different heating rates are used. Which rate gives the more accurate data?

14.7 Obtain the DTA and TGA curves of benzoic acid. Note that the TGA curve does not indicate the melting point but does indicate the boiling point.

14.8 Obtain the DTA curve for a sample of commercial varnish or paint. Note (a) the loss of thinners at low temperatures and the actual amount of varnish in the sample, (b) the steady loss of adsorbed thinner at elevated temperatures, and (c) the temperature at which the varnish decomposes in air. Compare these results with those you obtained for other commercial samples.

PROBLEMS

14.1 Describe the Aminco thermogravimeter with a diagram.

14.2 A sample of mixed calcium oxalate monohydrate and silica weighed 7.02 g. After heating to 600°C, the weight of the mixture was 6.560 g. What was the weight of calcium oxalate in the original sample?

14.3 What information can be obtained by DTA that cannot be obtained by TGA?

14.4 Describe a DTA instrument.

14.5 What are the advantages of using a combination of DTA and TGA?

15

Mass Spectrometry

Mass spectrometry (MS) is currently one of the most rapidly advancing fields of instrumental analysis. During its brief history, it has developed from an inorganic method proving that simple elements existed as various isotopes to one of the cornerstone techniques used to elucidate the structure of the human genome.

In the past few years extraordinary innovations and developments in mass spectrometry have revolutionized the applications of the field. New procedures, such as fast atom bombardment (FAB), secondary ion mass spectrometry (SIMS), thermospray, electrospray, laser desorption and plasma desorption, ion trap, and ICP–MS have been extensively developed and are in routine use. These and other techniques will be discussed in this chapter.

Mass spectrometry is an analytical technique that provides qualitative and quantitative information concerning the molecular weight and molecular structure of organic and inorganic compounds. This procedure can be used as a qualitative analytical tool to identify and characterize different materials of interest to the chemist or biochemist. With it one can also analyze quantitatively mixtures of gases or liquids and, in some cases, solids.

The analytical chemist is attracted to mass spectrometry mainly because of its speed, sensitivity, and reliability. It is one of the few methods that can be used to determine directly molecular weights accurately up to 10,000 and approximately up to 100,000. The method also provides a quantitative analysis in concentrations as low as the ppm level.

A. EARLY DEVELOPMENTS AND EQUIPMENT

J. J. Thompson in 1913 first used mass spectrometry to demonstrate that neon consisted of two nonradioactive isotopes, neon 20 and neon 22. This was revolutionary because it demonstrated that elements existed as isotopes with different atomic weights and simultaneously explained why the apparent atomic weight of an element based on chemical reactions was not a whole number, but a fraction made up of a mixture of isotopes. In 1923, Francis W. Aston used a more elaborate and more accurate system to determine the atomic weights of all the elements and the isotope ratios of each particular element. This was extremely useful to inorganic chemists and helped solve many of the problems concerning the position of elements in the periodic table at that time.

Equipment used in the early development of mass spectrometry maintained a constant accelerating voltage V and a constant magnetic field B. The paths of particles with different mass numbers were different. The distribution of masses was determined with a photographic plate as shown in Fig. 15.1a.

Unfortunately, the resolution of this equipment was not good. The highest mass numbers that could be separated were about 150. Also, the use of a photographic plate limited the sensitivity of the instrument. This would be a severe handicap to the modern analytical chemist, who must deal with samples as small as a milligram or less. Furthermore, photographic plates are very poor vehicles to use for quantitative analysis. Their response is not linear with exposure, and poor photographic development can cause major errors when several different plates are compared. As a result, the instrument has been greatly refined over the years. Other MS designs are shown in Fig. 15.1.

The first instrument to use an electrical detection system rather than a photographic plate was designed in 1918 by A. J. Dempster. This instrument was more accurate in measuring the *abundance* of various ions rather than their actual masses as was required in isotope work.

It was not until 1942 that research in the oil industry resulted in the development of commercial mass spectrometers (by the Atlantic Refining Corp.) for petroleum analysis. Subsequent developments have led to the development and extension of mass spectrometry to many branches of organic and biological chemistry.

B. THE DEMPSTER MASS SPECTROMETER

This instrument played a most important role in the history of mass spectrometry. Special attention will be given to it because understanding this instrument leads to an understanding of all later developments. It is shown in Fig 15.2.

The mass spectrometer is an instrument that separates individual particles, such as ionized atoms or molecules, by the difference in their masses. The molecules or atoms are bombarded by electrons and become ionized by the process, as shown in the following equation:

$$\text{mol} + e^- \rightarrow \text{mol}^+ + 2e^- \tag{15.1}$$

where

mol = molecule (or atom)
e^- = electron
mol^+ = ionized molecule (or atom)

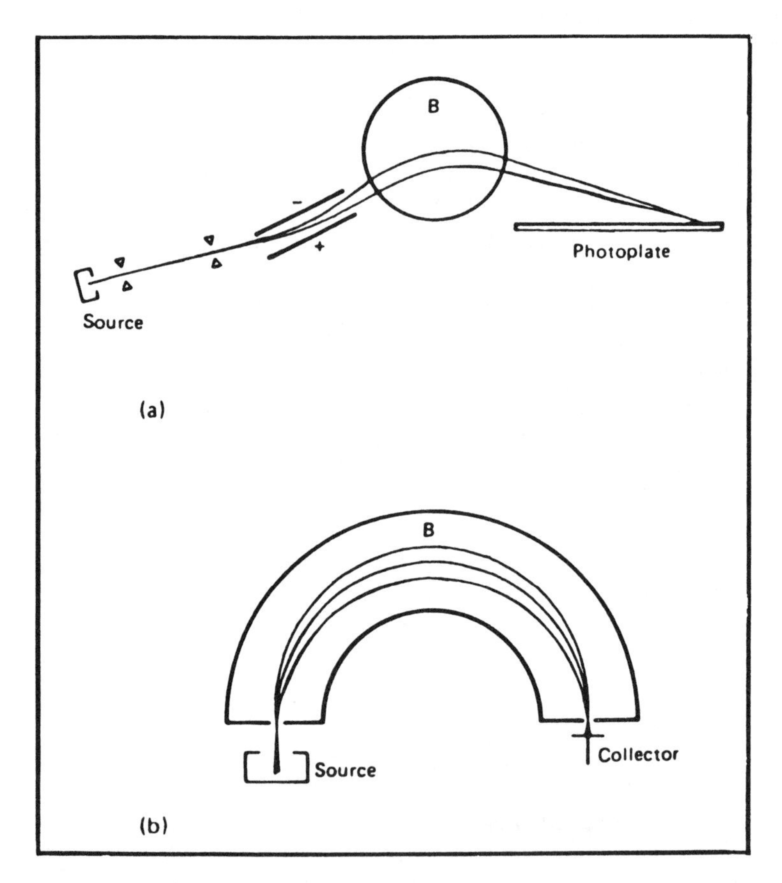

Figure 15.1 Early mass spectrometer designs: (a) Aston, 1919; (b) Dempster, 1918; (c) Mattauch-Herzog, 1935; (d) Bainbridge, 1933. In each case, B signifies the magnetic field.

The ions are then accelerated in an electric field at a voltage V. If the particles all have a single positive charge, the intensity of the charge is equal but opposite to the charge of an electron. The energy of each particle is equal to the charge times the acceleraty voltage, eV. But a particle's energy can also be measured as its kinetic energy. In this case, the kinetic energy is $\frac{1}{2}mv^2$. The two formulas

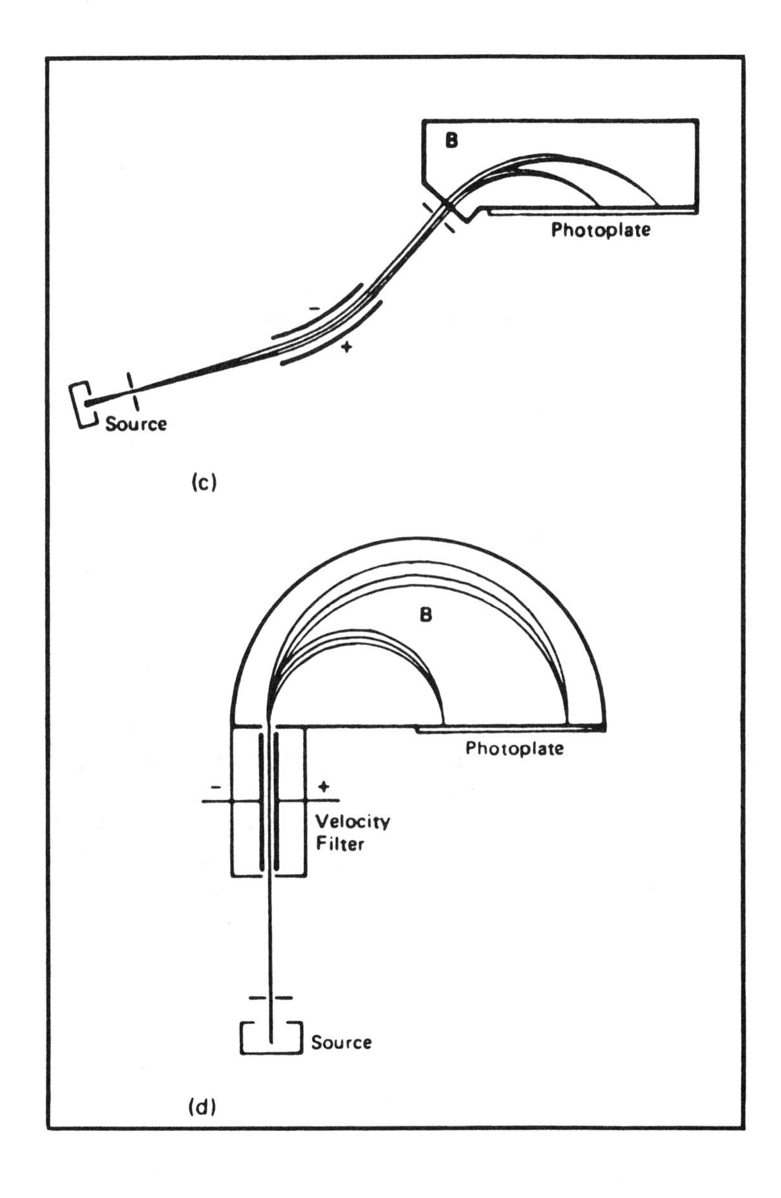

are mathematical expressions of the energy of the same particle and are therefore equal to each other:

$$1/2mv^2 = eV \tag{15.2}$$

hence

$$v = \left(\frac{2eV}{m}\right)^{1/2} \tag{15.3}$$

where

m = mass of the particle
v = velocity of the particle
e = charge of an electron
V = accelerating voltage

All particles have the same energy eV. However, when we apply the kinetic energy equation $\frac{1}{2}mv^2$, m varies from particle to particle as the mass of the particles varies. Simultaneously, therefore, the velocity v changes such that $\frac{1}{2}mv^2$ remains a constant. This relation can be expressed for two different particles as follows:

$$eV = 1/2m_1v_1^2 \text{ (particle 1)}$$

$$eV = 1/2m_2v_2^2 \text{ (particle 2)} \tag{15.4}$$

$$1/2m_1v_1^2 = 1/2m_2v_2^2$$

where

m_1 = mass of particle 1
v_1 = velocity of particle 1
m_2 = mass of particle 2
v_2 = velocity of particle 2

We can see, therefore, that the velocity of a particle depends on its mass.

After the charged particles have been accelerated by an applied voltage, they enter at right angles to a homogeneous magnetic field B. This field acts on the particles, making them move in a circle. The attractive force of the magnet equals Bev, but the balancing centrifugal force of the particle is equal to mv^2/r, where r is the radius of the circular path, these two forces are equal:

$$\frac{mv^2}{r} = Bev \tag{15.5}$$

or

$$\frac{1}{r} = \frac{Bev}{mv^2}$$

and

$$r = \frac{mv}{eB} \tag{15.6}$$

Substituting for v [Eq. (15.3)], we get

$$r = \frac{m}{eB}\left(\frac{2eV}{m}\right)^{1/2}$$

Squaring both sides, we have

$$r^2 = \frac{m^2}{e^2B^2}\frac{2eV}{m} = \frac{m}{e}\frac{2V}{B^2} \tag{15.7}$$

$$r = \left(\frac{2V}{B^2}\frac{m}{e}\right)^{1/2}$$

or

$$\frac{m}{e} = \frac{B^2r^2}{2V} \tag{15.8}$$

Also, for a given instrument

$$2mV = B^2r^2e = \text{constant} \tag{15.8A}$$

That is, the radius of the circular path of the particle depends on the accelerating voltage V, the magnetic field B, and the ratio m/e. When V and B are kept constant, the radius of the circular path depends on the mass m of the ionized molecule. The relation between m, the radius r of the circular path of the ion, V, and B is the basis of the separation of particles according to their mass. This was the basis of the system used by Aston.

For ions with a single charge, the charge is e and is a constant. A few ions are doubly charged, and with two positive charges their path is affected by the magnetic field accordingly. These ions are ignored in this treatment but are taken into account in rigorous analytical treatments.

In the large magnet mass spectrometer the accelerating voltage V is varied while the applied magnetic field B is held constant. The drift chamber allows only ions with a trajectory of radius r to pass through. For a particular voltage, then, only ions with masses that satisfy Eq. (15.7) will exit the chamber. Consequently, for different values of V, ions with different masses will pass through the drift chamber (the lower the voltage, the greater the mass of the ion passing through to the detector), and by varying V we can scan the mass range of the sample ions.

1. Resolution of a Mass Spectrometer

The ultimate value of an MS instrument depends on its *resolution*, that is, its ability to separate two particles of different masses. Numerically, the resolution is equal to the mean mass of the two particles divided by the difference in mass. For example, to separate two particles of mass 999 and 1001 requires a resolution of 500, that is, 1000 (average of 999 and 1001) divided by 2 (difference between 999 and 1001). The greater the resolution of the instrument, the greater its analytical usefulness. So we may say

$$\text{resolution} = \frac{\text{average mass}}{\text{difference in mass}} \tag{15.9}$$

In practice, it is found that if we wish to distinguish between particles of mass 600 and mass 599, the resolution required is 599.5, that is, essentially 600. As a rule of thumb, if we wish to distinguish between particles in the 600 mass number range, we need a resolution of 600. If we need to distinguish between particles in the 1200 range, we need a resolution of at least 1200. Obviously, an increase in resolution will make it easier to distinguish between two particles in the same mass range, and a decrease in resolution will make it more difficult to resolve them.

A schematic diagram of a conventional (large magnet) mass spectrometer is shown in Fig. 15.2a. This equipment focuses the ion beam on a narrow slit, which restricts the system to a well-defined radius r. The ions that travel in a circle of this radius can be selected according to their mass by changing the accelerating voltage V. The latter determines the speed of the particle. Alternatively, the magnetic field B may be varied, which adjusts the circular motion of the particle. In practice, it is found to be more convenient to maintain B constant and vary V, the accelerating voltage [see Eq. (15.7)]. The components of the system are described below.

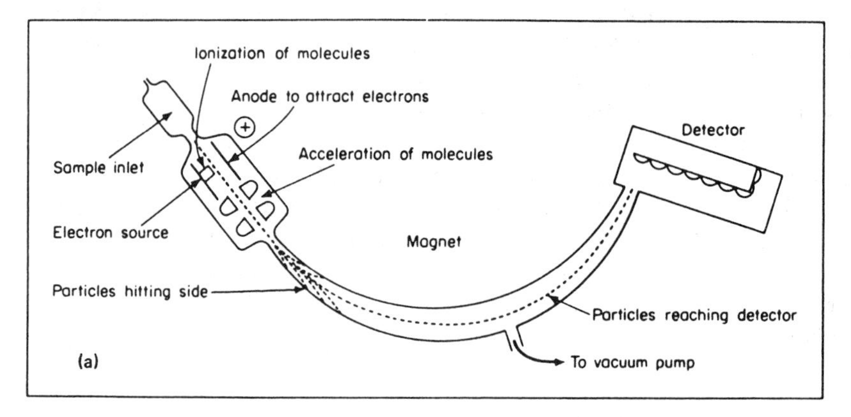

Figure 15.2 (a) Schematic diagram of a mass spectrometer. (b) Magnetic mass spectrometer. (c) Forces and energies associated with the magnetic mass spectrometer to reach the detector: $Bev = mv^2/r$.

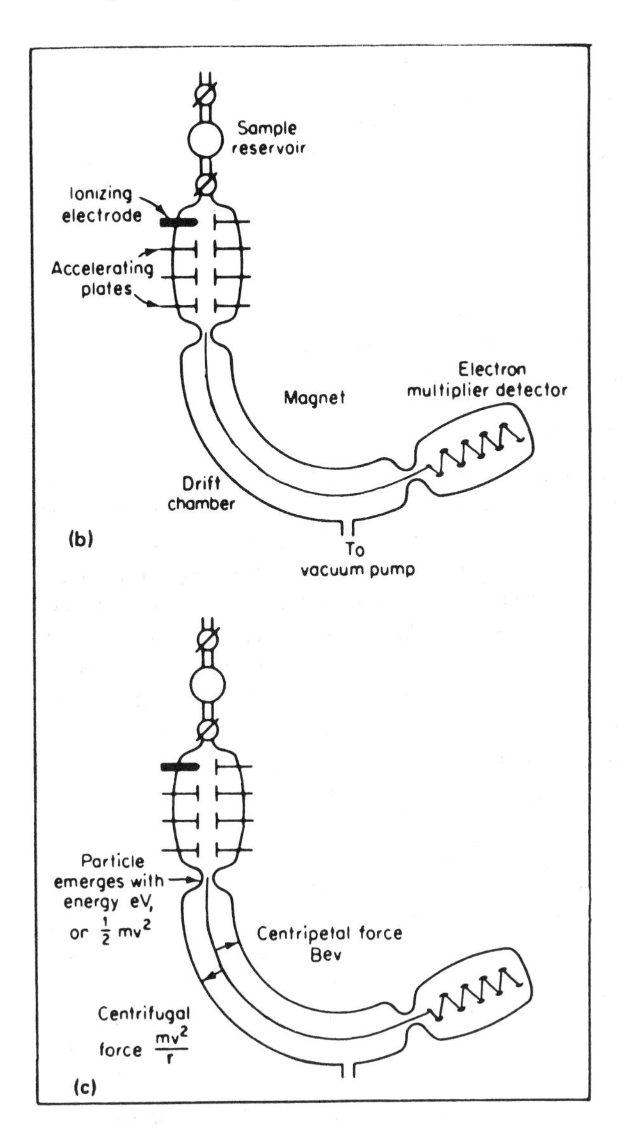

2. Inlet System

For successful ionization and acceleration, all samples must first be converted to the gaseous state. For this purpose the inlet system is usually heated to temperatures up to 400°C, depending on the volatility of the sample. Unstable samples should not be exposed to excessively high temperatures, otherwise decomposition will result before the sample enters the mass spectrometer. It is also necessary that the rate at which the sample enters the ionization chamber remains constant throughout the analysis so that the relative abundances measured at the beginning and the end of the analysis are constant. This is very important for the quantitative analysis of mixtures and the qualitative identification of unknown compounds.

3. Ionization

Upon leaving the inlet system, the sample enters the ionization cell. Here a beam of electrons is accelerated across the molecules of the sample. The molecules become ionized according to the reaction

$$\text{mol} + e^- \rightarrow \text{mol}^+ + 2e^-$$

If the electrons in the beams are accelerated at only a low voltage, they move slowly and the molecules tend to remain intact. As the voltage is increased, some molecules will become ionized. This voltage is a direct measure of the strength of the molecular bond being broken and is called the *appearance potential*. If the electrons are accelerated by a high voltage (up to 100 V), their energy is increased. Upon collision with the sample molecules, the latter are reproducibly shattered into well-defined fragments. Each fragment is charged and proceeds through the instrument as a simple charged particle. Care must be taken at this stage to prevent the fragments from recombining to form molecules that were not present in the original sample. This can best be done by maintaining the system at a low pressure, thus decreasing the number of collisions between fragments or molecules inside the cell.

A compromise is necessary in selecting the ionization voltages used. If the voltage is too low, few molecules will be ionized and many of them will pass through the chamber without becoming ionized at all. If the voltage is too great, the molecules will be shattered and reduced to small fragments. If it is important to measure the mass of the whole molecule, it is desirable to maintain as many molecules as possible in the ionized but unfragmented condition. This gives information on the molecular weight of the sample but little information on its structure. On the other hand, if information on the structure of the molecule is necessary, the voltage should be increased in order to break off fragments that can be identified separately. As a compromise, many commercial mass spectrometers operate with the ionization chamber at approximately 70 V. Knowledge of the structure of the fragments can be used to reconstruct the original molecule before fragmentation. This process is in many ways similar to putting together a jigsaw puzzle. Great skill and art are required on the part of the operator to (successfully) elucidate molecular structures from fragmentation patterns. Computer programs and memory banks are now being used to perform this operation with some degree of success.

4. Accelerating Chamber

The ionized particles enter the accelerating cell, where they come under the influence of the accelerating voltage V. The speed of the particles increases as described in Eq. (15.3).

Under fixed voltage conditions, only one mass would satisfy Eq. (15.8) and reach the detector. In order to scan the entire mass range of fragments, the voltage is varied at a constant rate. This introduces the fragments into the magnetic field at a range of velocities and a range of mass numbers that satisfy Eq. (15.8). The accelerating grids are shaped like slits so that a parallel beam of fragments is formed. Any fragments traveling at an angle to the beam collide with the side of the slit and are removed. This is desirable because any such particles would have velocity components at right angles to the desired flow. That would complicate Eq. (15.3) and hence Eq. (15.8) and lead to an incorrect assignment of the masses reaching the detector and to mass overlap, i.e., loss of resolution.

5. Drift Chamber (Magnetic Field Path)

The accelerated fragmented ions enter the drift chamber and come under the influence of the magnetic field. The forces that determine their motion are described in Eq. (15.8). Figure 15.2b illustrates a simple machine with a circular magnetic field path; other designs with cycloidal paths and sector paths are also in commercial use. The drift chamber is under a high vacuum to prevent fragments from colliding with each other or with extraneous molecules. These chambers are glass lined. The length of the circular drift chamber is usually about 3 ft but varies considerably with design. Some instruments include an electrostatic analyzer at the exit of the drift chamber to improve the resolution of the instrument.

6. Detector

The most common detector is the electron multiplier (Fig. 15.3). The fragments from the drift chamber arrive at the electron-emissive surface of a dynode amplifier, which acts in a similar fashion to a photomultiplier or electron multiplier but is initiated by the impact of a fragment. A collision releases several electrons from this surface. These electrons are then accelerated to a second such surface, which, in turn, generates several electrons for each electron that bombards it. This process is continued until a cascade of electrons arrives at the collector. In this fashion, a single fragment can generate thousands of electrons. The detection system is very sensitive and has a very fast response.

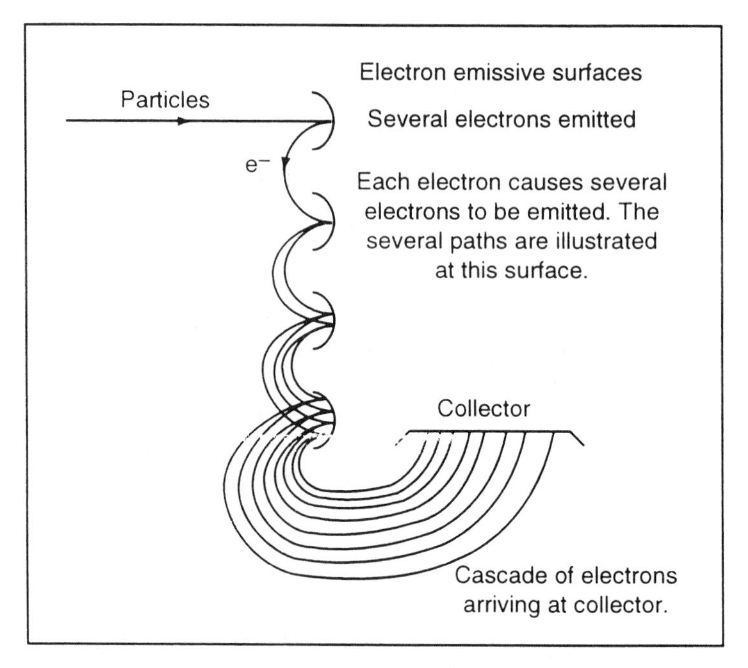

Figure 15.3 Diagram of the dynode amplifier.

7. Analytical Sequence

In a typical analysis, the gas sample is bled very slowly at a constant rate into the ionization chamber, where it is ionized and fragmented. The accelerating voltage V is set at a high value, and the particles are accelerated to a high velocity. Under these conditions, only particles with a small mass can be deflected sufficiently by the magnet; hence only articles with low mass numbers (H, D, etc.) hit the collector and are counted. The accelerating voltage V is then progressively decreased and simultaneously particles of increasing mass number reach the collector. When V reaches zero, the record of the distribution of masses in the sample is completed and is presented as a *mass spectrum*. The sequence in which the particles hit the detector follows from Eq. (15.8). When V is a maximum, m must be a minimum. As V decreases, m increases. All other variables in Eq. (15.8) are constant under experimental conditions. A typical mass spectrogram is shown in Fig. 15.4.

The time required to obtain a complete mass spectrum depends on the particular instrument. For many instruments, it is in the neighborhood of 20 min. For some applications this presents a problem because conditions such as sample

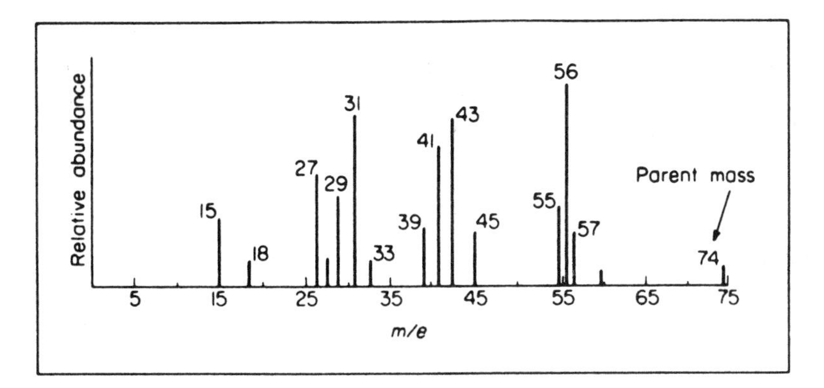

Figure 15.4 Typical mass spectrogram of *n*-butanol.

feed rate, ionizing voltage, and applied magnetic field must remain constant over this extended period of time.

In recent years, the time required to complete a mass spectrum of a single compound has been reduced to as little as 1 sec. This reduction has allowed the instrument to be used in conjunction with a gas chromatograph. Emerging compounds are fed directly into the mass spectrometer and a mass spectrum is obtained for each separate peak, provided that there is 1-sec difference in the retention times of the compounds. High-speed scanning is always performed with some loss of resolution; however, this loss is not too important if the instrument is used in a qualitative sense, that is, to identify the compound only. In the example cited, quantitative measurements are carried out by measuring the areas of the peaks on the gas chromatogram. With the GC–MS combination, qualitative and quantitative analyses of complex mixtures can be performed.

Instruments are now commonly manufactured that can handle compounds with molecular weights up to 4000. Special equipment has an even higher range (up to 7000). The resolution of such equipment varies between 10,000 and 30,000, depending on the manufacturer. With higher molecular weights it becomes a challenge to prevent all the molecules from fragmenting during the operation. With careful control, however, this problem can be overcome, and mass spectrometry is useful in characterizing high molecular weight compounds.

C. IMPROVEMENT IN ITEMS OF EQUIPMENT: IONIZATION PROCESSES

1. Electron Impact

This technique, described in Section B.3, worked very well for small molecules. However, it was readily apparent that even with small molecules it was often difficult to obtain the parent mass of the analyte. Since identification of the molecular weight of a compound is a very attractive feature of mass spectroscopy, other ionization processes were developed that greatly enhanced the parent mass abundance but still provided information on the cracking pattern.

2. Electron Spray Ionization (ESI)

When large molecules are ionized to M^+, the mass-to-charge ratio is very high, and under the conditions of T.O.F. or large magnet MS, the velocity of the molecular ion leaving the acceleration chamber is small. Different large molecules have similar low velocities, and it is difficult to resolve them This is one of the important difficulties in the use of mass spectroscopy for the study of large molecules.

The problem was greatly reduced by J. B. Fenn et al., who contrived to generate molecules and fragments with multiple charges M^{n+}. A given population of sample molecule may have a number of charges such as M^{10+}, M^{11+}, M^{12+}, M^{13+}, etc. Each ion would be detected given a regular spectrum for the same molecule, differing by the charge on each. Examples of such spectra are shown in Figs. 15.5 and 15.6. A schematic diagram of the equipment is shown in Fig. 15.7. It can be seen that the apparent mass numbers vary regularly as the charge varies. The actual mass can be calculated as:

$$\text{apparent mass} \times \text{charge} = \text{actual mass}$$

For example, in Fig 15.6A, if the apparent mass = 1430 and the charge is 10 as in M^{10+}, then the actual mass is 1430 × 10 = 14,300. This is confirmed by a second peak, which is apparently 15.89 and the charge 9, giving 15.89 × 9 = 14,300.

In practice, the charge is not known, but it is assumed that successive charges vary by 1. The apparent masses of the fragments are known. A series of equations can be set up as follows:

$$\begin{aligned} M_1 \times X &= Y \\ M_2 \times (X + 1) &= Y \\ M_3 \times (X + 3) &= Y, \text{ etc.} \end{aligned}$$

where M_1, M_2, and M_3 are the "apparent" masses of the particles.

Solving for Y gives the true molecular weight of the molecule. An example of a raw spectrum and a deconvoluted spectrum is shown in Fig. 15.5. The top figure shows the raw data, the bottom figure shows the true spectrum after mathematically treating the raw data.

ESI has been used in conjunction with most conventional forms of mass spectromety, such as large magnet, multisector TOF, quadropole, and FTMS.

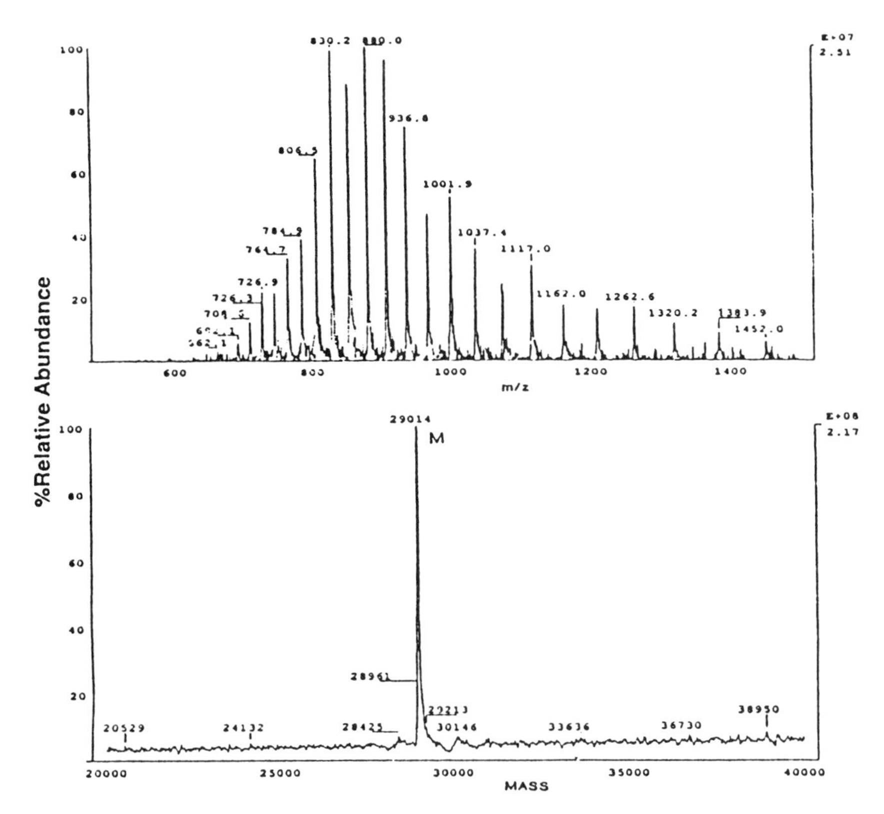

Figure 15.5 Original spectrum (top) and deconvoluted spectrum (bottom) of carbonic anhydrase. (Reprinted with permission from Finnigan, M., *Analytical News*, (1990) Feb.)

Masses up to 100,000 have been measured with reasonable accuracy. ESI has also been used to study synthetic polymers providing information on molecular weights and molecular weight distributions.

Using ESI–FTMS, only extremely small samples are required. Spectra have been recorded from samples as small as 30 picomols.

The electron spray ionization source is somewhat complicated. The sample is presented with a capillary through a series of skimmers, each at a different voltage, causing multiple ionization. The ions then enter the mass spectrometer instrument for detection.

ESI has already greatly extended the molecular weight range of mass spectrometry and will continue to do so.

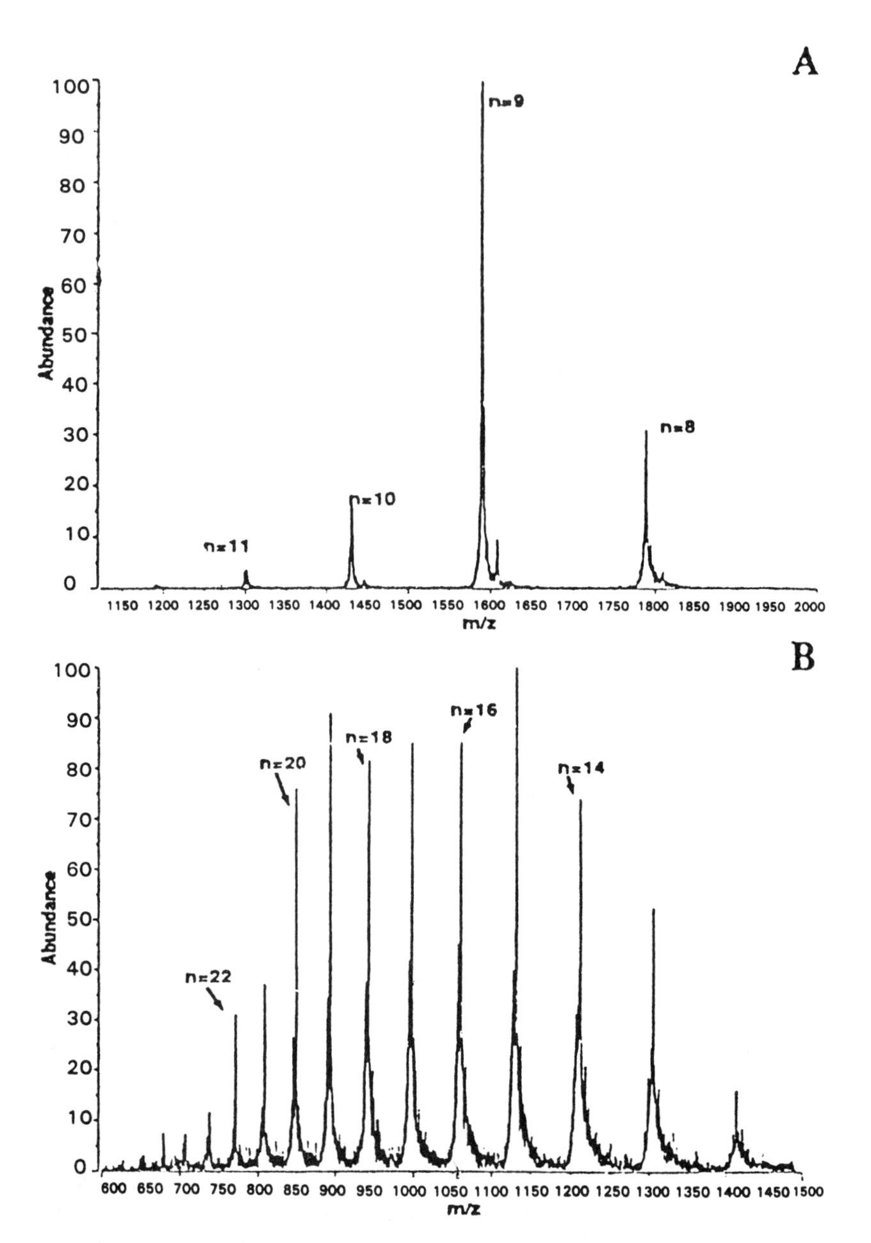

Figure 15.6 ESI mass spectra of proteins. (A) Hen egg-white lysozyme; (B) equine myoglobin.

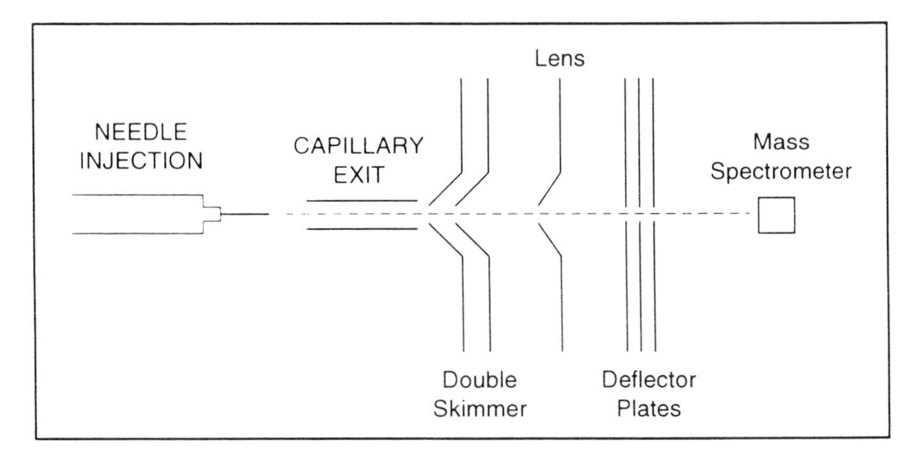

Figure 15.7 A schematic diagram of the electrospray ionization.

3. Fast Atom Bombardment

It is difficult to volatilize and ionize large molecules using conventional techniques. Further, if they are volatilized, they are normally shattered by the electron beam, reducing the value of the analytical data. This problem has been somewhat overcome by using fast atom bombardment (FAB). In this procedure, the sample is "dissolved" in a high-boiling solvent, preferably glycerol, although carbowax monothioglycerol and 2,4-pibentylphenol have also been used. The solution in placed on a metal strip, which is the target.

Argon ions are produced in a heat filament ion source. These are accelerated under an electrostatic field of 3–8 kV towards the target. The fast-moving ions exchange their charge with slow-moving argon atoms but lose no kinetic energy in the process. This results in fast-moving argon atoms and slow-moving argon ions. The latter are removed from the system using a deflection plate. The fast-moving atoms now strike the target, liberating molecular ions, which may be positively charged M^+ or negatively charged M^-. If the mass spectrometer is set up for positive ions, M^+ and most are, the negative ions M^- are rejected and take no further part in the process. Besides argon, xenon and cesium have been used as bombarding atoms.

Sample fragmentation is greatly reduced in the process, resulting in a large parent mass with somewhat unstable molecules. This provides information on the molecular weight of the molecule, which is particularly important in biological samples such as proteins. Numerous workers have recorded spectra from molecules with molecular weights up to 15,000.

Although a strong parent mass is obtained with FAB, fragmentation patterns are also obtained (Fig. 15.8). Each peak indicates the loss of an amino

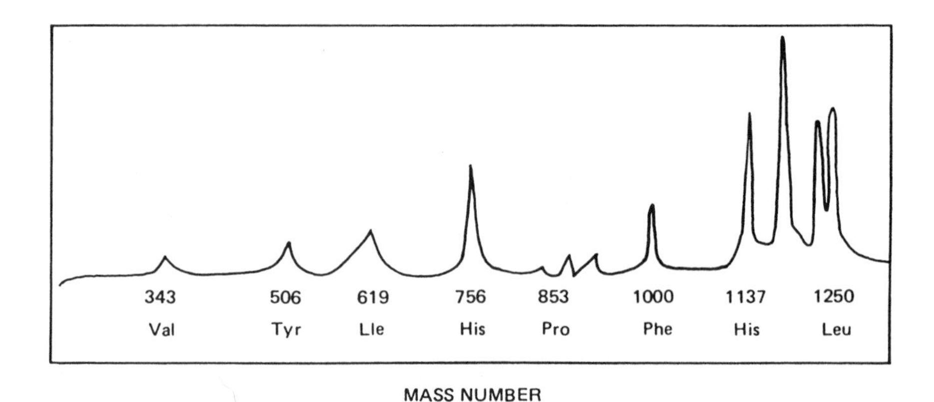

Figure 15.8 Typical fast atom bombardment: mass spectrum of a simple protein (protonated argiotensin). Each peak represents loss of an amino acid as indicated.

acid. This information can be used to obtain the amino acid sequence in the original protein. An illustration is shown in Fig 15.9.

An important advantage of FAB is that only the atoms or ions vaporized (i.e., sputtered) from the glycerol are lost; the remainder can be recovered. Consequently, samples as small as 1 μg can be loaded into the instrument, and after analysis 90% of this can be recovered.

4. Continuous Flow FAB

Despite its attractiveness, FAB has several drawbacks, such as the required use of a nonvolatile, viscous solvent (glycerol). This results in ion-suppression effects, decreased sensitivity, and a high background. This technique was therefore improved by using continuous flow FAB (CFFAB). In this system, the sample is introduced into the mass spectrometer through a fused silica capillary. The tip of the capillary is the target. Hence the solution is bombarded by fast atoms produced as described above. Solvent is flowing continuously and the liquid sample is introduced by continuous flow injection (Fig. 15.10). The bombarded sample produces fragmentation patterns and a large parent mass.

Typically, the solvent is 95% water and 5% glycerol. The ability to inject aqueous samples is an enormous advantage in biological studies. Very frequently, the samples are aqueous in nature, such as blood, urine, and other body fluids.

The sensitivity at the lower molecular weight range (1500) is increased by two orders of magnitude. Further, the background is reduced because of the reduced amount of glycerol present. In addition, when the solvent alone is injected, a background signal can be recorded, (1) and then when the sample is

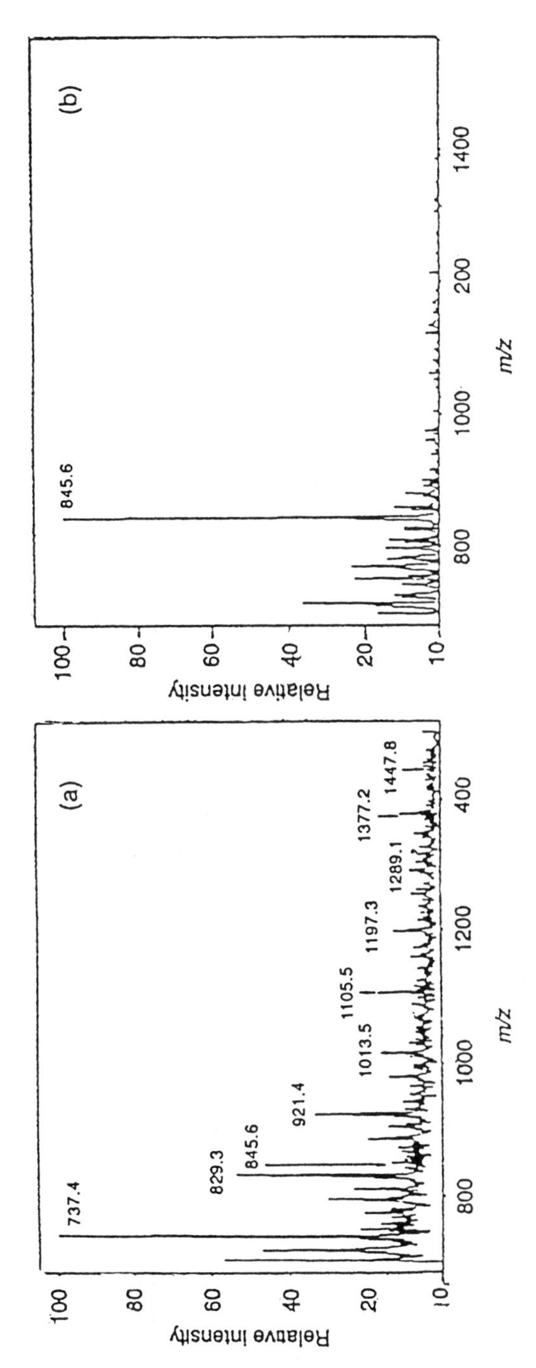

Figure 15.9 Mass spectra of approximately 200 pmol of an oligopeptide having an $(M+B)^+$ at m/z 845.6. (a) Spectrum obtained with standard FAB techniques. (b) Spectrum obtained with CF-FAB by flow injection with background subtracted.

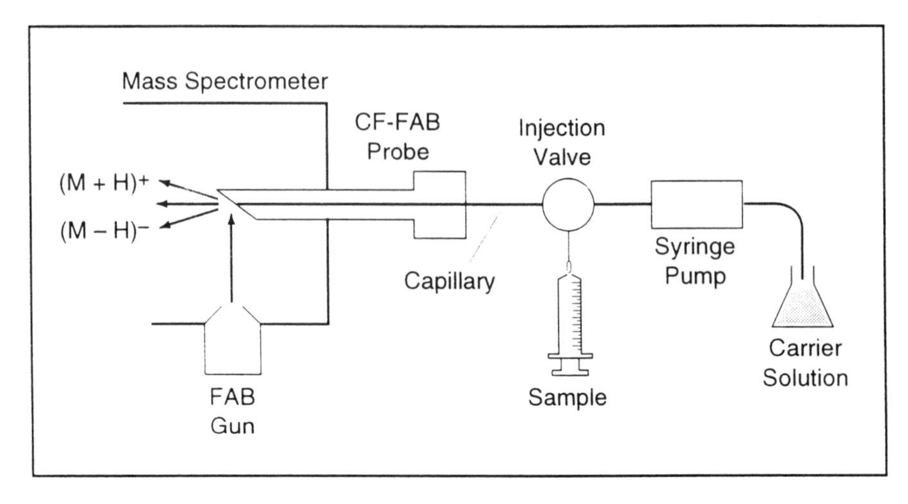

Figure 15.10 Schematic diagram of CF–FAB MS operated in the flow injection mode.

injected a signal compound of background plus sample is obtained (2) by subtracting 1 from 2, the net signal of the sample is obtained. This procedure is especially valuable for trace analysis, where it is claimed that concentrations as low as 10^{-12} g can be detected.

The system can be incorporated into LC–MS systems. The mobile phase is the solvent used. The effluent from the LC is transported directly into the mass spectrometer and the MS obtained by CFFAB. This provides real-time analysis of each peak if necessary and is capable of analyzing a wide range of components from trace levels to major components.

In summary, FAB and CFFAB have greatly increased the potential of mass spectrometry by increasing the molecular weight range of molecules whose parent mass can be determined. The system can be directly attached to LC, permitting identification of the components of a solution. Also, trace analysis is possible. Importantly, the method can be applied to biological samples, a research area of great importance to the health sciences, environmental science, and other aspects of human life.

5. Field Ionization

In this system the electron impact source is replaced by a field ionization source. This consists of a small wire or a sharp-edged piece of metal maintained at a very high electrostatic field, between 10^5 and 10^8 V/cm. The molecules entering the high electric field lose electrons owing to the tunnel effect, producing molecular ions. The molecular ions then proceed into the acceleration chamber and drift chamber to the detector in the normal fashion. The advantage of this system is that there is very little fragmentation of the molecule and therefore a significant enhancement of the abundance of the molecular ion, or parent mass. The disadvantage of the system is that its efficiency is orders of magnitude lower than that of the electron impact source, with a corresponding loss in sensitivity. However, the source's ability to increase the relative abundance of the parent mass and the corresponding isotope masses greatly facilitates the elucidation of the sample's molecular weight and empirical formula.

6. Chemical Ionization Mass Spectrometry

One of the difficulties of mass spectrometry is that the sample molecule is usually fragmented in the ionization chamber and very little of the peak generated by the parent mass is observed. This is increasingly the case as the molecular weight of the sample molecule increases. For high molecular weight compounds the parent peak often is not observed at all, and qualitative identification of the sample from its mass spectrum is very difficult. This problem can be greatly reduced by using chemical ionization as a means of ionizing the sample.

In this system, we modify the ionization chamber by decreasing the size of the entrance and exit ports to enable operation at increased pressure (1 Torr). By increasing the vacuum pumping capacity, we can maintain the pressure in the drift chamber at 10^{-5} Torr to prevent molecular collisions during the analysis. We then mix the sample with a low molecular weight gas such as methane (CH_4), with the methane in great excess (e.g., 1000:1).

When the gas mixture enters the ionization chamber, the methane becomes ionized almost exclusively, creating ions such as CH_4^+, CH_3^+, CH_2^+, and so on. These recombine in the ionization chamber to form secondary reactant ions:

$$CH_4^+ + CH_4^+ \rightarrow CH_5^+ + CH_3^+$$

It is these reactant ions, in turn, that cause ionization of the sample molecule. The latter frequently loses or gains a proton in the process.

Fragmentation of the sample molecule is greatly reduced, and a large peak at a mass number one less or one greater than the parent mass (PM + 1 or PM − 1) is observed. Of course, very large peaks generated by the methane fragments are observed, but these are at the low-mass end of the spectrum and do not interfere with observations of the sample peaks. Typical reactions are

$$CH_5^+ + C_2H_4OH \rightarrow CH_4 + C_2H_6OH^+ \text{ (PM + 1)}$$

$$CH_5^+ + C_2H_6 \rightarrow CH_4 + C_2H_5^+ \text{ (PM − 1)} + H_2$$

With different low molecular weight reactant gases, such as N_2, propane, isobutane, water, or methanol, chemical ionization with different degrees of fragmentation of the sample can be achieved. The balance between the abundance of the parent mass and the fragmentation pattern can be varied considerably, depending on what information is sought from the mass spectrogram. For molecular weight information, the parent peak would be emphasized; for structural information, the fragmentation pattern would be needed. Chemical ionization mass spectrometry has been very useful in the study of high molecular weight samples encountered in biochemistry, such as drugs and body fluids.

7. Laser Desorption and Matrix-Assisted Laser Desorption Ionization (MALDI)

Laser desorption has been used to ionize large molecules with some success. Lasers of many different wavelengths have been used with wavelengths from the far UV to the far IR. By focusing the laser, spots smaller than 1μm can be analyzed. This is an important application, because it permits analysis of surfaces and gives spatial resolution to examine strains, etc.

More importantly, laser desorption permitted the analysis of large biomolecules such as protein and carbohydrates. However, the process was erratic until certain guidelines were discussed and followed. These are described below.

First, controlled energy transfer from the laser to the molecule is best achieved with resonant absorption, i.e., the laser wavelength must be at an absorption wavelength of the molecule. Examples are lasers in the far UV, which couple with electronic transitions or lasers in the far IR, which couple with rotational-vibrational levels.

Second, in order to avoid decomposition, the absorbed energy must be quickly dispersed within the molecule. However, in practice, molecules with molecular weights greater than 1000 for biopolymers and 9000 for synthetic polymers decomposed, and molecules of greater molecular weight could not be studied.

This problem was relieved by using matrix-assisted laser desorption ionization. In this system, the sample is suspended in a matrix. The function of the matrix is to disperse the large amounts of energy absorbed from the laser. This prevents decomposition of the molecule.

In practice, a small quantity of the sample is suspended in a suitable solid or liquid matrix. The matrix acts as a strong absorber of the laser radiation. This system transfers energy to the analyte but reduces decomposition. Typical matrices and optimum laser wavelengths are shown in Table 15.1. A matrix is chosen that absorbs the laser radiation but at a wavelength at which the analyte absorbs moderately. This diminishes the likelihood of shattering the molecule.

Table 15.1

Matrix	Wavelength		
Nicotinic acid	266 nm	2.94 nm	10.6 μm
2,5-Dihydroxy	266 nm	337 nm	355 nm
benzoic acid	2.8 μm	10.6 μm	
Succinic acid	2.94 μm	10.6 μm	
Glycerol (liquid)	2.79 μm	2.94 μm	10.6 μm
Urea (solid)	2.79 μm	2.94 μm	10.6 μm

A schematic diagram of the equipment is shown in Fig. 15.11. It can be seen that the system is simple in design. This ionization process has been used in conjunction with many types of mass spectrometers.

Analytes with molecular weights up to 300,000 have been studied. Figure 15.12 shows the mass spectrum of a monoclonal antibody from the mouse. This spectrum reveals how two or three analyte molecules may combine together during ionization; also, multiple ionization is possible.

There is no question that matrix-assisted laser ablation has opened new areas of research and will be invaluable in studies of large biomolecules, such as proteins and carbohydrates, and of synthetic polymers, assisting in molecular weight determination, molecular weight distribution, and structural information for each.

D. NONMAGNETIC MASS SPECTROMETERS

1. Quadrupole Mass Spectrometry

The quadrupole mass spectrometer does not involve the use of a heavy magnet; rather, it uses four electromagnetic poles (Fig. 15.13). The opposite pairs of poles A and B, and C and D, are each connected to the opposite ends of a DC source, such that when C and D are positive, A and B are negative. The pairs of electrodes are then connected to an electrical source oscillating at radiofrequencies. They are connected in such a way that the potentials of the pairs are continuously 180° out of phase with each other. The magnitude of the oscillating voltage is greater than that of the DC source, resulting in a rapidly oscillating field. The magnetic poles would ideally be hyperbolic in cross section to provide a uniform field. Under these conditions, the potential at any point between the four poles is a function of the DC voltage and the amplitude and frequency of the RF voltage. The shape of the poles varies with different manufacturers.

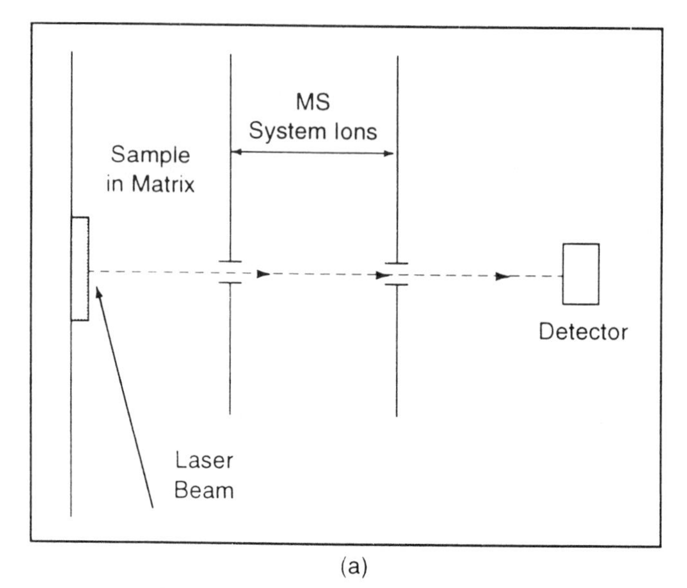

(a)

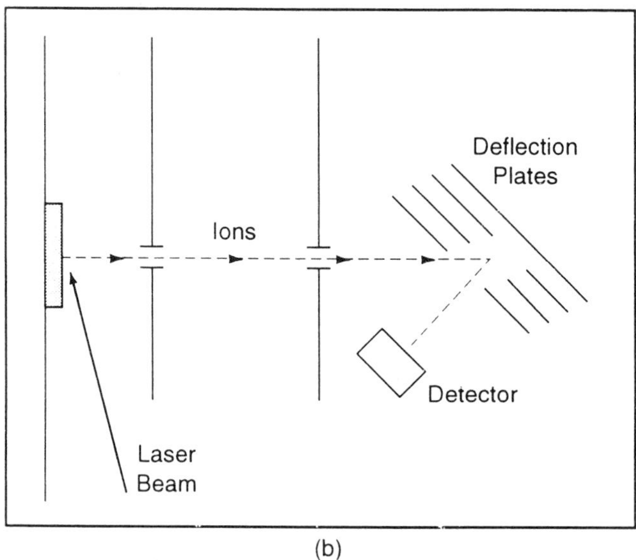

(b)

Figure 15.11 Schematic diagram of matrix-assisted laser desorption ionization (MALDI) with deflector plates. Ions are deflected but neutrals are not and are removed from the beam hitting the deflector.

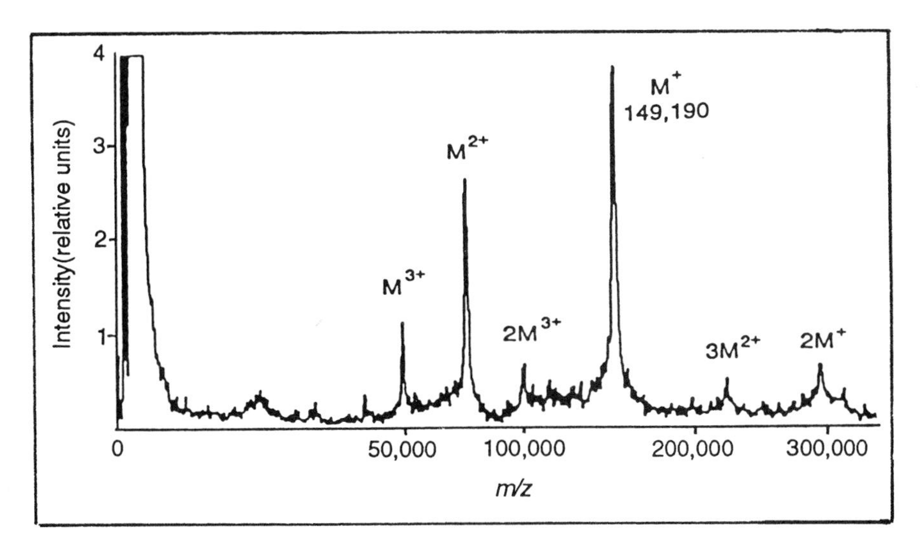

Figure 15.12 LDI mass spectrum of a monoclonal antibody from the mouse.

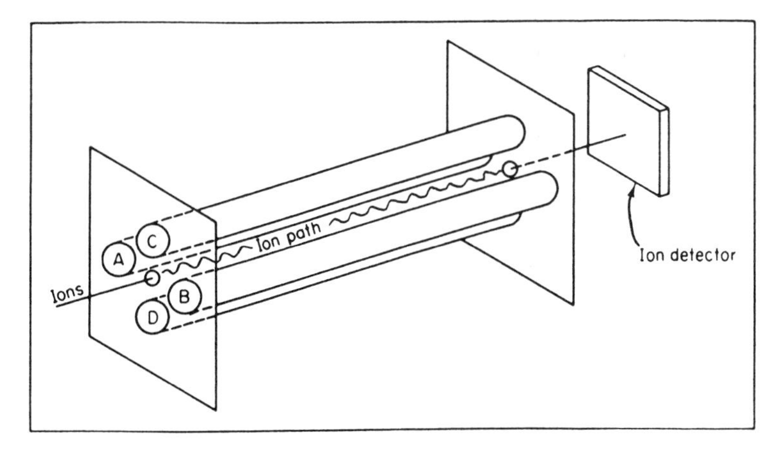

Figure 15.13 Quadrupole mass spectrometer. Poles AB and CD are connected.

If an ion is introduced into this space, the force on the ion is given by the following equations:

$$F_x = e\frac{\delta\phi}{\delta x} = -e\frac{(V_{DC} + V_0 \cos \omega t)2x}{r^2} \tag{15.10}$$

$$F_y = -e\frac{\delta\phi}{\delta y} = +e\frac{(V_{DC} + V_0 \cos \omega t)2y}{r^2} \tag{15.11}$$

where ϕ is the potential. This gives the following equations of motion:

$$\frac{d^2x}{dt^2} + \frac{2}{r^2}\frac{e}{m}(V_{DC} + V_0 \cos \omega t)x = 0 \tag{15.12}$$

$$\frac{d^2y}{dt^2} + \frac{2}{r^2}\frac{e}{m}(V_{DC} + V_0 \cos \omega t)y = 0 \tag{15.13}$$

where

V_{DC} = voltage of the DC signal
V_0 = amplitude of the voltage of the RF field
ω = frequency of oscillation of the RF field (rad/sec)
r = half the distance between the inner edges of opposing poles such as A and B

In practice, a sample is introduced into an ionization chamber similar to that used in conventional magnetic mass spectrometry. The ions produced are then introduced into the space between the quadrupoles and driven under their own impetus and the applied DC field from the point of access between the quadrupoles to the detector at the other end.

The RF field does not accelerate the ions toward the detector; instead it produces an oscillating magnetic field. If the mass of the ion and the frequency of oscillation are compatible, the ion will oscillate toward the detector and eventually reach it; however, if they are not compatible, that is, the conditions required by Eqs. (15.12) and (15.13) are not met, the ions will oscillate with an increasingly wide path until they collide with the quadrupoles or are pulled out by the pumping system. In any case, the ions will not progress to the detector.

Using this technique, the separation of ions of different masses can be achieved by several methods. The most commonly used method is to keep the frequency of oscillation of the RF field constant while varying the potentials of the DC and RF fields in such a manner that their ratio is kept constant. It can be shown mathematically that the best resolution is obtained when the ratio V_{DC}/V_0 is equal to 0.168. If the ratio is greater than this number, a stable path cannot be achieved for any mass number; if the ratio is lower than this number, resolution is progressively lost (the ions collide with the quadrupole or with each other and are lost from the system). To keep this ratio constant, the DC changes as the frequency of the RF changes.

The resolution of the system is dependent on the number of oscillations an ion undergoes in the drift chamber. Increasing the pole lengths, therefore, increases resolution and extends the use of the system to higher molecular weight compounds. This same improvement can be brought about by increasing the frequency of the RF field. The pole diameter is also important. If the diameter is increased, the sensitivity is greatly increased, but then the mass range of the system is decreased. The manufacturer must come to a compromise with these factors when designing an instrument for analytical use. The resolution achievable with the quadrupole mass spectrometer is of the order of 20,000 units. As with the conventional mass spectrometer, the sample must be available in the gas phase and must be ionized.

One major advantage of the quadrupole mass spectrometer over magnetic spectrometers is that it has a wide angle of acceptance. Even though ions may enter the quadrupole field with considerable lateral energy, they will nevertheless come under the influence of the applied magnetic field and will oscillate normally. This is in contrast to the large magnet mass spectrometer, which will reject ions with any noticeable lateral kinetic energy. The latter causes loss of resolution in the conventional mass spectrometer, but not in the quadrupole mass spectrometer.

Although the quadrupole mass spectrometer does not have the accuracy and precision of magnetic analyzers, it is very fast while maintaining some degree of accuracy. It can provide a complete mass spectrum within a few seconds. This property and its wide angle of acceptance make it suitable for coupling to a gas chromatograph.

The combination technique of gas chromatography (GC–MS) has been very fruitful in many fields of research and routine analysis. It has permitted mass spectra of individual peaks to be recorded from which the compound can frequently be identified. In this combination, quantitative analysis is carried out by measuring the area of the gas chromatographic peak, and qualitative identification is achieved as necessary using the mass spectrum.

The advent of computers has further increased the potential of this technique. For example, the GC trace of a very impure sample is shown in Figs. 15.14 and 13.25. The GC peak is found to be on top of a lot of background material emerging from the GC column. This is not an uncommon occurrence, but generally any mass spectra taken under these conditions would be of little value, since they would not distinguish between the background material and the sample material. However, this problem can now be overcome with the assistance of computers and memory storage.

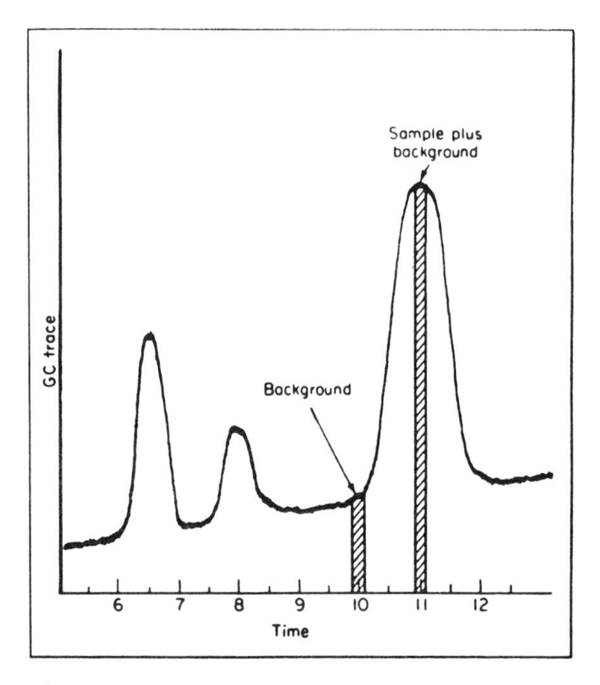

Figure 15.14 A GC trace of a very impure sample. Mass spectrometer data are taken at various points on the curve. The background spectrum can then be subtracted from the sample and background spectrum, leaving the net sample spectrum.

In the example cited, it is possible to direct the instrument to record the mass spectrum of a sample every 15 sec. At the end of the run the computer is instructed to retrieve run number 10 (which is all background) and run number 11 (which is sample plus background), subtract spectrum 10 from spectrum 11, and read out the difference. This difference should be very close to the mass spectrum of pure compound present in the peak. The information can be displayed either as a permanent record on paper or on an electronic display.

It is possible to add a memory bank to the system. The computer can then be instructed to search through the memory bank to see whether it has a mass spectrum similar to that of the compound just analyzed. Often it will display several possibilities. The memory approach is not foolproof, but it is very valuable in the hands of an expert who can distinguish between likely and unlikely answers.

In summary, the quadrupole mass spectrometer has provided a ready means for identifying effluents from gas chromatographic columns and will undoubtedly find increased use in this and other areas in the future.

2. Time-of-Flight Mass Spectrometry

Like the quadrupole mass spectrometer, the time-of-flight mass spectrometer (TOF) does not use a heavy permanent magnet to separate the charged particles. It separates them by the differences in their velocities as they leave the accelerating chamber and pass through the drift chamber. The energy relationship of the particles remains the same as that of the conventional mass spectrometer, that is [Eq. (15.2)],

$$1/2mv^2 = eV$$

If the accelerating voltage V is kept constant, then $\frac{1}{2}mv^2$ is constant. From Eq. (15.2) we can deduce that

$$\frac{m}{e} = \frac{2V}{v^2}$$

$$= \frac{1}{v^2} \times \text{constant}$$

This means that for a constant accelerating voltage, light particles (m is small) move very fast, but heavy particles move slowly. This difference in velocity allows the separation of the particles. More exactly, the kinetic energy is $zeVs$, where e is the charge on the electron, z is the number of charges for $M^{+}{}_{1}z = 1$ for $M^{2+}z = 2$, etc., V is the accelerating voltage, and s is the length of the accelerating chamber. For the same instrument s is a constant.

However, the kinetic energy also equals $\frac{1}{2}mv^2$, where v is the velocity of the emerging ions, therefore

$$v^2 = \frac{2zeVs}{m}$$

$$v = \left(\frac{2zeVs}{m}\right)\frac{1}{2}$$

$$v = \sqrt{\frac{2zeVs}{m}}$$

The time taken t to pass through the drift chamber (time of flight) is given by

$$t = \frac{L}{v} \tag{15.14}$$

where L is the length of the drift chamber. Substituting for v,

$$t = \sqrt{\frac{m}{2zeVs}}L$$

Squaring both sides,

$$t^2 = \frac{mL^2}{2zeVs}$$

or

$$\frac{m}{e} = \frac{t^2 zeVs}{L^2}$$

with a constant accelerating voltage. For a given instrument L, e, s, and v are constant and can be equated to K. Hence,

$$\frac{m}{e} = Kt^2 z$$

and m is proportional to (time of flight)^2.

a. Equipment

A schematic diagram of a time-of-flight spectrometer is shown in Fig. 15.15 In such mass spectrometers, the sample is bled into the ionizing chamber at a reduced pressure and becomes ionized by a pulse of electrons flowing across its path. This pulse of electrons is of very short duration and constitutes the time of ionization. It is known as pulse desorption (PD), and its short duration (10^{-6} sec) greatly assists in measuring t. The energy pulse may be from radioactive ^{252}Cf. The charged particles are accelerated by the voltage V applied across the grids. At discrete time intervals, very short bursts of charged particles are allowed into the drift chamber. Fast-moving particles separate from those that are moving more slowly and form bunches of particles. The fragments in each bunch are moving at the same speed, and it can be deduced that they have the same mass. The arrival of these bunches is detected and recovered and stored on a tape. The recorded data can then be retrieved and displayed. The fragmentation pattern of each burst of molecules entering the ionization chamber is displayed simultaneously.

In recent years, the value of TOF has been greatly extended by first using matrix-assisted laser desorption as a vaporizing technique. This has greatly ex-

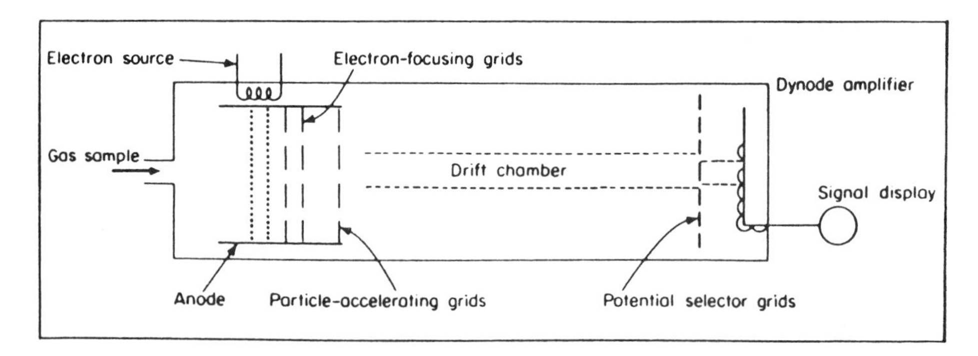

Figure 15.15 Time-of-flight mass spectrometer.

tended the range of molecular weight that can be vaporized. Also, much greater accelerating voltages (up to 30 kV have been used. This greatly increases *V*, which is particularly important for very large molecules. Protein with molecular weights up to 100,000 have been studied using this method.

The technique has been used for sequencing studies of compounds such as methionine enkephalen and from digests of staphylococcus nuclease. Like all analytical techniques, TOF has particular uses for which it is most suitable.

b. Analytical Uses

The resolution of a TOF is considerably less than for large magnet MS. Therefore, resolution of particles of different mass numbers is more difficult with TOF, so it is less useful than conventional MS for the measurement of exact mass numbers. For example, it would not be possible to separate the particles C_2ON ($m/e =$ 53.9980) and C_3H_2O (m/e 54.0106) using TOF. A high-resolution MS would have to be used. On the other hand, a complete scan may only take a few seconds. This is a distinct advantage over conventional MS, which may take as long as 20 min. This speed makes possible the "continuous" analysis of samples with rapidly changing composition.

One important use of the TOF is as a chromatographic detector. The effluent from the gas chromatographic column is fed into the TOF, and each peak of the chromatogram can by partially identified from the mass spectrum obtained. Passing the sample through a gas chromatographic column separates the components from each other. The size of the chromatographic peaks indicates how much of each component is present (quantitative data.) The separated components may be fed directly into a TOF as they leave the chromatographic column. The cracking pattern obtained can be used to identify the molecule (qualitative data.) This combination of gas chromatography and TOF MS makes it possible to perform qualitative and quantitative analyses of organic mixtures in a comparatively short time.

For many years, the maximum mass that could by accommodated by the TOF was about 300. Time-of-flight MS was therefore considered lower in performance than conventional MS. Recent advances have made quadrupole MS suitable for use with gas chromatography, thus diminishing the value of time-of-flight MS in this application. However, in recent years, the capabilities of the instruments has been improved. The accuracy of measuring t, the time taken to reach the detector, and hence v, the velocity of the ion, has been improved.

This has led to the use of TOF for "measuring" the molecular weights of large molecules such as small proteins or fragments of large proteins. It has been used to measure molecular weights up to 100,000 with an accuracy of ± 100. Measuring the molecular weight of a protein or a protein fraction to within ± 100 daltons is a major step forward in characterizing polymers, either synthetic or natural, such as protein or carbohydrates.

E. TRAPPED ION MASS SPECTROMETRY (FOURIER TRANSFORM MS)

Another form of mass spectrometry, Fourier transform mass spectroscopy (FTMS), has been introduced, which permits the use of the Fourier transform for the collection and presentation of data. In this method the sample is introduced and trapped in a cubic cell, shown schematically in Fig. 15.16. The sample enters through a small hole in the trap plate. It is then ionized and the ions come under the influence of a magnetic field lined up along the path traveling from the entrance hole to the collector trap. The ions generated from the sample react to the magnetic field by going through a circular motion at right angles to the magnetic field.

The frequency of this motion is given by

$$\omega = \frac{e}{m}H \tag{15.16}$$

where

ω = frequency of rotation of the ions in radians per second
e = the charge on the ion
m = mass of the ion
H = applied magnetic field

As can be seen, the frequency of rotation depends on the mass of the ion.

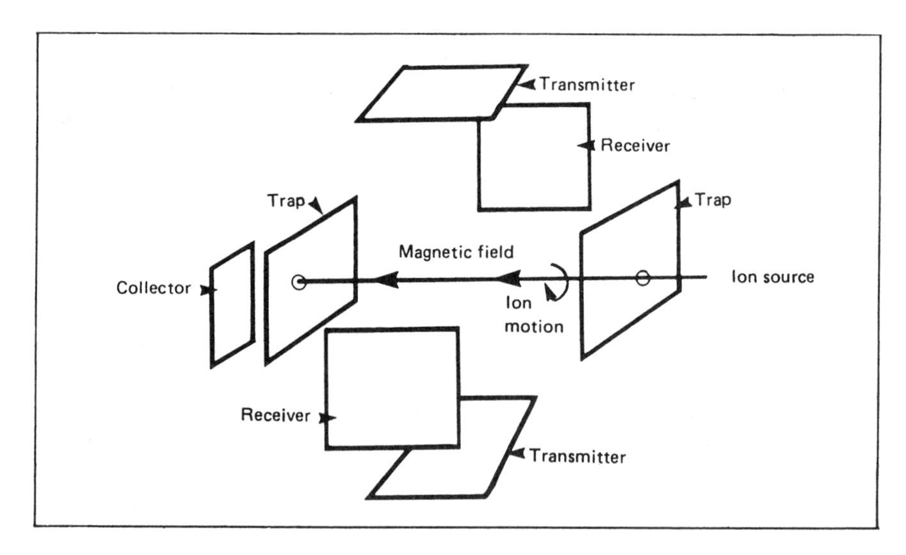

Figure 15.16 Diagram of exploded ion trap mass spectrograph. With Fourier transform, this provides FTMS.

A radio transmitter making up two sides of the trap is then turned on at some predetermined frequency. If the frequency coincides with the frequency of rotation of the ion, the ion picks up energy by absorbing the radiation. There is no change in the frequency of rotation of the ion in this process, but its amplitude of rotation increases.

At some fixed time interval the radio frequency is turned off and the rotating ion emits radiation at its rotating frequency, giving a decaying signal when plotted against time.

A group of identical ions rotates at the same frequency and emits a coherent signal at this frequency. As the group of positive ions approaches the receiver plate, its charge attracts electrons to the inside surface of the plate. As the group recedes, the electrons are released. This process is repeated with each cycle and an ''image current'' is produced in the receiver plates with a frequency equal to that of the oscillating ions. The signal perceived by the receiver plate may then be transformed using a Fourier transform to provide the mass spectrum signal. This provides qualitative information on the mass or molecular weight of the ion, whether it is a fragment or a parent mass. The magnitude of the image current depends on the number of ions in the group and provides a quantitative measure of their concentration.

To scan various mass ranges the frequency of the RF transmitter can be varied between controlled frequencies (e.g., 0.05 to 5.0 mHz) and over a controlled time. At some time during this variation, different ions will be in resonance with the radio frequency and will subsequently emit a signal as they go out of resonance. The collective signals are accumulated and using Fourier transform can be transformed into the total mass spectrum of the sample.

The technique is particularly useful when used in conjunction with a gas chromatograph. It is usually necessary to use helium as a carrier gas so that low pressures can be more easily obtained in the cell by pumping off the volatile and easily defusible helium atoms.

Several methods of ionization have been used successfully; these include electron impact and more recently bombardment with cesium, potassium, or sodium ions. The use of cesium ions to ionize the samples has been very productive. Cesium aluminosilicate filaments heated to 1000°C act as the ion source. The ions enter through the ion source hole and appear to be very efficient in ionizing the sample. Sensitivities of the order of 100 times greater than would be expected have been realized.

Sodium ions have been used in similar fashion; in this case the parent ion of the sample was not observed but the ion of formula $(M + Na)^+$ was observed, suggesting a complex forming between the sample and the sodium ion. Laser desorption processes have been used that provide a more gentle ionization process for larger molecules, reducing fragmentation and allowing easier interpretation.

Each of the techniques gives strong parent peaks as well as abundant fragmentation patterns. Both of these properties are useful for structural interpretation of the sample.

The pressure in the ion trap is preferably kept at about 10^{-8} Torr. This is necessary for high resolution. Cesium iodide ion clusters with mass numbers up to 12,000 have been observed. In addition, polymers of polyhexafluoropropylene oxide up to mass numbers 6841 have been detected. However, when the molecular weight of organic compounds exceeds 400–500, difficulties in resolution are experienced.

Workers in the field have claimed resolutions in the order of one part per 100,000. High resolution such as this is very valuable in deducing exact empirical formulas.

One problem with this system is that when used in conjunction with a gas chromatograph (under normal operating conditions) the effluent from the GC may overload the trap. As a consequence it is necessary to throw away as much as 99% of the sample using a stream splitter and permit only 1% to proceed to the mass spectrometer. This splitting can introduce quantitative errors, particularly when the molecular weights of the components are high.

The quad ion trap consists of three electrodes: two end cap electrodes and a ring electrode to provide an RF potential. The system is a hybrid of the quadrupole MS and the conventional ion trap. The former discriminates between different masses, while the latter stores them under controlled conditions.

The instrument has high sensitivity (10^{-18} mol), high mass range for molecular weight determination, structural elucidation from fragmentation patterns, is capable of analyzing mixtures, and can scan at the rate of 5000 daltons per sec.

The quad ion trap was originally used as a GC detector for environmental analyses by monitoring ambient air and water by purging the water with helium. The ion trap has been interfaced with super critical fluid GC and LC. This system has also been used in protein sequencing, a procedure of rapidly increasing importance in biopolymer analyses such as protein sequencing.

F. HIGH-RESOLUTION MASS SPECTROMETRY

The atomic weights of the isotopes of the elements are nominally whole numbers, but actually they are only close to being whole numbers. If we use ^{12}C as the standard, the atomic weights of some common isotopes are as shown in Table 15.2. As a consequence, two empirical formulas may have the same nominal molecular weight but actually differ slightly but significantly. For example, $C_{20}H_{26}N_2O$ and $C_{16}H_{28}N_3O_3$ have the same molecular weight (310), but if we take into account the exact atomic weights of the nuclei involved, the molecular weights are actually 310.204513 and 310.213067, respectively. Other empirical

Table 15.2 Atomic Weights of Common Isotopes

Isotope	Atomic weight
^{1}H	1.007825
^{2}H	2.004102
^{12}C	12.000000
^{14}N	14.003074
^{16}O	15.994915

formulas with this nominal mass also have slightly different actual molecular weights. With sufficient resolution it is possible to distinguish between all possible empirical formulas. More important, knowing the exact molecular weight allows us to identify the empirical formula of the sample. This is true not only of the parent mass, but also of fragments of the molecule, and this provides valuable information on large fragments of very large molecules. To distinguish different empirical formulas, high resolution is necessary and instruments have been designed for this specific purpose. Much of the pioneer work was carried out by F. W. McLafferty and Cornell University using double focusing.

The loss of resolution is largely caused by the narrow beam of ions spreading out while traveling down the drift chamber. Similar masses may spread out enough to actually overlap each other and therefore be unresolved. The method McLafferty used to reduce this problem was basically to refocus the ions in the drift chamber (Fig. 15.17).

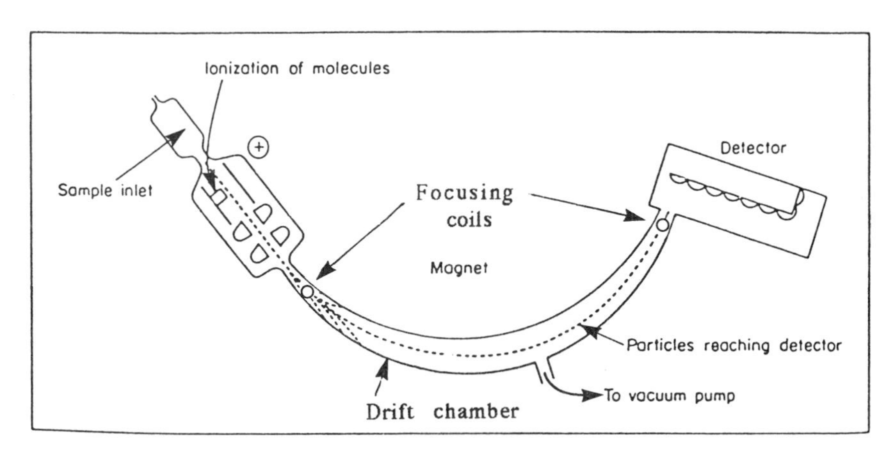

Figure 15.17 Schematic diagram of double focusing mass spectrometer designed by F. W. McLafferty.

A small coil was placed at the exit of the accelerating chamber. By passing a small DC current through this coil, a magnetic field was set up that focused the emerging ions to a finite spot at the center of the coil. Simultaneously the lateral energy of the ions was reduced.

A similar focusing coil was placed at the end of the drift chamber immediately in front of the detector. This again focused the ions to a point in the center of the coil. This system was called *double focusing*. It greatly improved resolution but did not reach the objective of a resolution of 150,000 that McLafferty had set for himself. His work, however, opened up a whole new dimension in mass spectrometry, which was pursued by others. Better resolution is now obtained using *multisector MS*, described below.

High-resolution MS can be used to identify the empirical formulas of fragments of large molecules such as proteins. It can also be used to assign empirical formulas to newly synthesized organic compounds or complex molecules isolated from natural products.

G. MULTISECTOR MASS SPECTROMETRY

When a molecule or fragment is ionized, it has a small amount of lateral energy before it enters the accelerating chamber. In a given population, some ions have more lateral energy than others; the distribution of this residual energy in Gaussian. After the ion passes through the accelerating chamber, its energy becomes $eV \pm \delta E$, where δE is small compared to eV but not zero. This leads to a spreading of the ions in the drift chamber and a corresponding loss of resolution in conventional mass spectrometers.

The problem can be overcome if the magnet is shaped so that its field focuses ions in the drift chamber onto an exit slit, thereby eliminating their spread. A conical magnet is most commonly used. Another method is to have the ion beam enter and leave the drift chamber at an angle (10–30°) as shown in Fig. 15.18.

By using a similar approach with an electrostatic field and a magnetic field, double focusing can be achieved and high resolution obtained. The process also allows fast scanning to be carried out. A complete scan can be achieved in 1–10 sec with reasonable resolution. A decade of mass numbers can be scanned in 1 sec. A multisector MS is shown in Fig. 15.19.

H. SECONDARY ION MASS SPECTROMETRY

Secondary ion mass spectrometry (SIMS) is technically a field of mass spectrometry, and almost any type of mass spectrometer can be used for its application. However, it is essentially a method of surface analysis because only the surface layers of the sample are exposed to the process, and so only an analysis

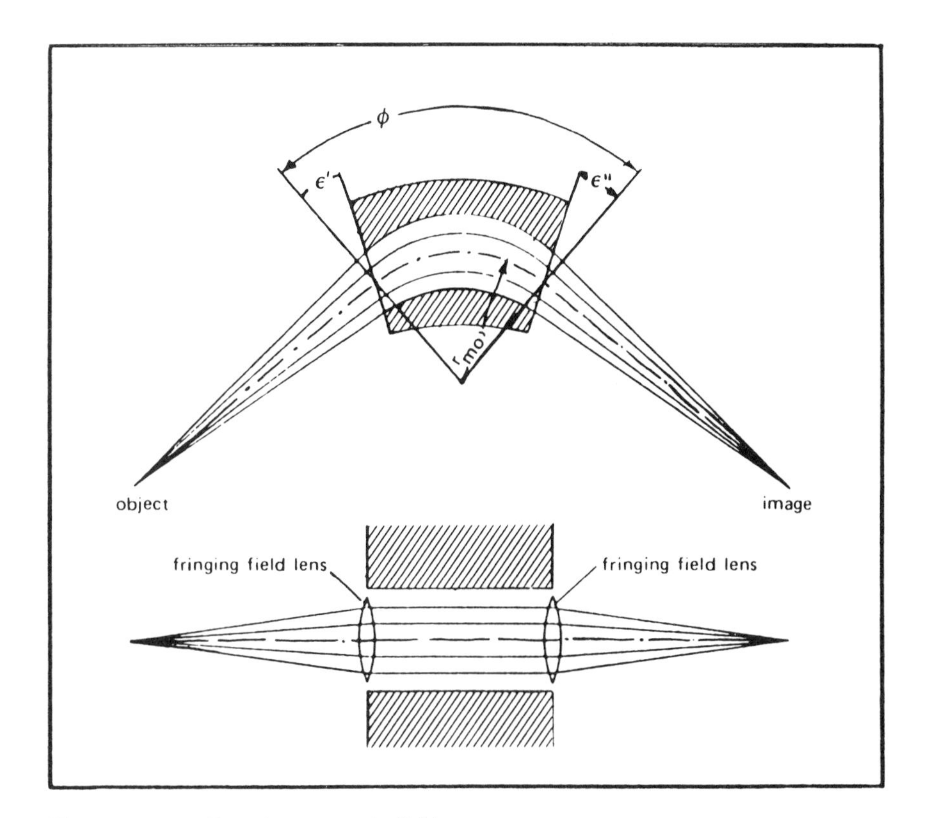

Figure 15.18 Focusing magnetic fields.

of the surface is obtained. This is of itself very valuable, but it should be noted that the composition of the surface of a sample is a very unreliable indicator of the bulk analysis of that sample. The technique is described in more detail in Chapter 12.

I. ANALYTICAL USES OF MASS SPECTROMETRY

1. Elemental Analysis

The original use of mass spectrometry was for the detection and determination of elements. The elements have different mass numbers, so MS also provided a method of determining atomic weights. The various elements as well as their isotopes could be separated from each other with this technique. This made it possible to obtain the isotope distribution of pure elements. The application of mass spectrometry to the determination of atomic weights and isotope distribution was crucial in the development of atomic chemistry and physics.

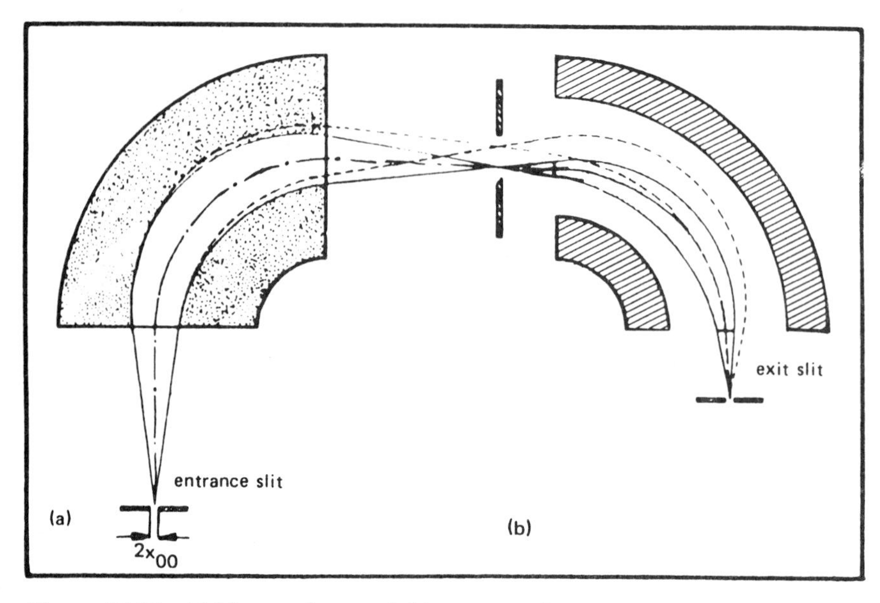

Figure 15.19 (a) Mass analyzer and (b) energy analyzer.

Recently this field has resurfaced in the form of inductively coupled plasma–MS (ICP–MS), which permits determination of all of the elements in the periodic table at very low concentration levels. As many as 60 different elements can be determined simultaneously over a very wide dynamic range.

This technique has great advantages over all other techniques for elemental analysis. Some techniques can determine metals, some nonmetals, some all atomic weights greater then 27, but only ICP–MS can determine all elements at high sensitivity and over a wide range of concentrations.

In addition, isotope ratios can be obtained, providing geological information and information on the mining source of origin of many elements such as lead. This information is valuable in environmental studies where the source of lead pollution can be determined with increased certainty by labeling the lead and distinguishing it from lead from other sources.

A high-temperature torch decomposes the sample into elemental form, which becomes ionized. Therefore the products of the plasma torch contain as elements all the components present in the original sample. A high percentage of these elements are in the form of ions and therefore do not need to pass through an ionization chamber. The method is particularly useful for simultaneous multielemental analysis of metals and nonmetals at the ppm level.

Although the plasma itself is very hot (10,000 K), it has a very low thermal capacity and can be cooled quickly. This can be facilitated by using a water-

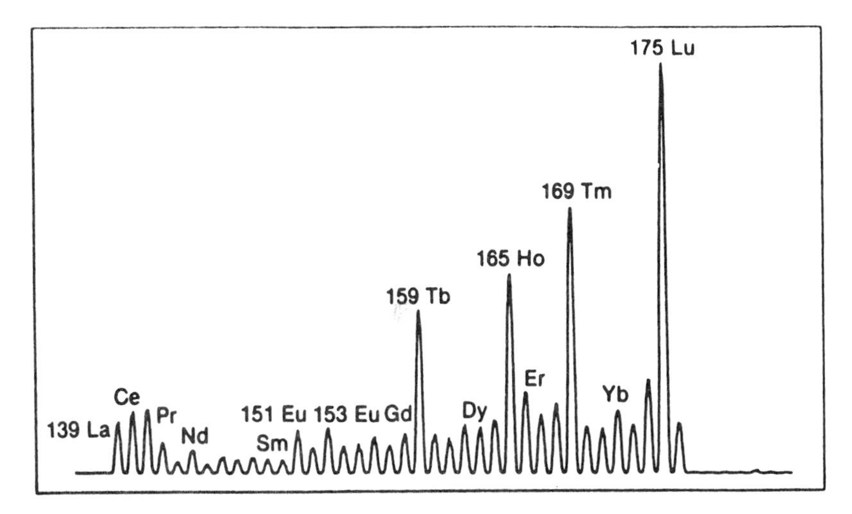

Figure 15.20 Spectra of a mixed solution of rare earths. Elemental analysis was performed using plasma torch with mass spectrometer.

cooled ring before the entrance to the mass spectrometer, which also helps cool the sample ions as they enter the instrument. Only low resolution is required to discriminate between different elements because their masses are their atomic weights. The abundance of the mass number of each element is a quantitative measure of that element's concentration in the original sample. In Fig. 15.20 an illustration of an elemental analysis using this technique is given. Figure 15.21 shows the detection limits of the method in parts per billion.

2. Molecular Analysis

In recent years the most important application of mass spectrometry has been to molecular analysis, particularly that of organic molecules. It is perhaps the only method that can be used to find the empirical formulas of compounds that are not completely pure. Impurities interfere with other methods, such as combustion techniques, and cause them to give incorrect results; however, in MS, by calculating and subtracting the contribution of the impurity to the total spectrum, the composition of the sample molecule can be obtained.

One of the more interesting direct applications of mass spectrometry is in blood analysis. The speed of the analysis allows surgeons to monitor the blood of patients during operations. The concentration in the blood of compounds such as CO_2, CO, O_2, N_2, and anesthetics (such as N_2O) can be controlled in this way.

I A	II A	III B	IV B	V B	VI B	VII B	VIII	VIII	VIII	I B	II B	III A	IV A	V A	VI A	VII A	O
H																H	He
2 Li	Be											2 B	50 C	N	O	30* F	Ne
2 Na	1 Mg											3 Al	Si	300 P	1* S	Cl	Ar
K	Ca	3 Sc	2 Ti	2 V	1 Cr	1 Mn	4 Fe	0.8 Co	2 Ni	1.5 Cu	2 Zn	Ga	Ge	50 As	80 Se	100 Br	Kr
0.8 Rb	0.8 Sr	0.8 Y	1 Zr	1 Nb	2 Mo	Tc	4 Ru	1 Rh	0.4 Pd	1 Ag	2 Cd	1 In	2 Sn	Sb	0.4 Te	2 I	Xe
1 Cs	1 Ba	1 La	Hf	Ta	2 W	2 Re	0.8 Os	2 Ir	1 Pt	0.4 Au	1.3 Hg	0.6 Tl	0.8 Pb	0.6 Bi	Po	At	Rn
Fr	Ra	Ac															

1 Ce	1 Pr	2 Nd	Pm	2 Sm	2 Eu	2 Gd	2 Tb	1 Dy	0.4 Ho	2 Er	0.5 Tm	0.9 Yb	1 Lu
0.7 Th	Pa	1 U	Np	Pu	Am	Cm	Bk	Cf	Es	Fm	Md	No	Lw

*$\mu g\ ml^{-1}$.

Figure 15.21 Detection limits using plasma MS analysis (in nanograms per milliliter).

Mass spectrometry has many industrial applications. For example, in petroleum chemistry it is used to identify chemicals in hydrocarbon fractions, to detect and identify impurities (such as ethane) in polymer feeds (such as ethylene), and to determine CO, CO_2, H_2, and N_2 in feed streams used for making alcohols. It is also used for gas analysis and the analysis of jet and *automobile exhausts*. In natural products complicated molecules such as steroids can be determined in the presence of each other, and compounds occurring in biological organisms can be identified by MS.

In organic chemistry reaction products and byproducts can be identified by MS, particularly with the help of gas chromatography. Impurities in concentrations as low as a few parts per billion can be detected; MS is widely used for this purpose. In inorganic chemistry, with special inlet techniques, the elemental composition of materials as diverse as crystals and semiconductors can be determined.

One special application of MS is for the characterization of *polymers*. In this case, the polymer is pyrolized (i.e., decomposed by heating) under controlled conditions. The pyrolysis products are fed into the inlet of a mass spectrometer and identified. From the results, much information concerning the structure of the polymer can be obtained. For example, in chlorinated polymers, MS can distinguish between polymers in which chlorination occurs in blocks in the structure of the polymer and those in which chlorination is random.

In summary, mass spectrometry can be used for atomic and molecular weight determinations. Its most important application to the analysis of organic compounds is for the identification of pure compounds and the quantitative analysis of mixtures of compounds. The method is rapid, reliable, and, with suitable computers, subject only to minor interferences.

3. Protein-Sequencing Analysis

Proteins make up some of the most important components of the human body and indeed all animal life. They are major components in all living cells and are therefore of major importance in all chemical studies of life sciences.

Proteins are made up of amino acids, which, as the name implies, include an amino group and an acid, usually a carboxylic acid. There are 28 amino acids (Table 3), and the many hundreds of thousands of proteins are made up of these building blocks. Some proteins contain many different kinds of amino acids, some only a few. DNA, for instance, contains only four different amino acids, but many of each. What distinguishes one protein from another is the sequence in which these are arranged in the molecule. Sequencing has become a cornerstone of the research into protein chemistry and understanding how they function.

Recombinant DNA has led to the use of protein-based pharmaceuticals and agricultural products, with large-scale production being developed. To ensure that the products are correctly formulated, protein sequencing is routinely required. The amino acids are listed in Table 15.3. In order to write down a struc-

Table 15.3 Amino Acids and Their Abbreviations

Name	Abbreviation	Formula
Glycine	Gly	$NH_2CH_2CO_2H$
Alanine	Ala	CH_3CHCO_2H NH_2
Valine[a]	Val	$(CH_3)CHCHCO_2H$ NH_2
Leucine[a]	Leu	$(CH_3)_2CHCH_2CHCO_2H$ NH_2
Isoleucine[a]	Ileu	$CH_3CH_2CH—CHCO_2H$ CH_3 NH_2
Phenylalanine[a]	Phe	$C_6H_5CH_2CHCO_2H$ NH_2
Serine	Ser	$HOCH_2CHCO_2H$ NH_2
Threonine[a]	Thr	$CH_3CH—CHCO_2H$ OH NH_2
Lysine[a]	Lys	$NH_2(CH_2)_4CHCO_2H$ NH_2
δ-Hydroxylysine	Lys-OH	$NH_2CH_2CH(CH_2)_2CHCO_2H$ OH NH_2
Arginine	Arg	NH $C—NH(CH_2)_3CHCO_2H$ NH_2 NH_2
Aspartic acid	Asp	$HO_2CCH_2CHCO_2H$ NH_2
Asparagine	Asp-NH_2	$NH_2COCH_2CHCO_2H$ NH_2
Glutamic acid	Glu	$HO_2C(CH_2)_2CHCO_2H$ NH_2

Table 15.3 Continued

Name	Abbreviation	Formula
Glutamine	Glu-NH_2	$NH_2CO(CH_2)_2CH(NH_2)CO_2H$
Cysteine	CySH	$HSCH_2CH(NH_2)CO_2H$ [H] ⇅ [O]
Cystine	CyS—CyS	$S-CH_2CH(NH_2)CO_2H$ $\vert$ $S-CH_2CH(NH_2)CO_2H$
Methionine[a]	Met	$CH_3S(CH_2)_2CH(NH_2)CO_2H$
Tyrosine	Tyr	$HO-C_6H_4-CH_2CH(NH_2)CO_2H$
Thyroxine	Thy	$HO-C_6H_2I_2-O-C_6H_2I_2-CH_2CH(NH_2)CO_2H$
Proline	Pro	$CH_2-CH_2-CH_2-NH-CHCO_2H$ (ring)
Hydroxyproline	Hypro	$HOCH-CH_2-CH_2-NH-CHCO_2H$ (ring)
Tryptophan[a]	Try	indol-3-yl $-CH_2CH(NH_2)CO_2H$
Histidine	His	$CH=CCH_2CH(NH_2)CO_2H$, ring N=CH–NH

[a]Amino acids essential in human diets.

ture more easily these have been given shortened signatures such as Ag = Argenine and Pro = Proline.

Traditionally, a degradation procedure has been used for this analysis. The terminal NH_2 group (all proteins have a terminal NH_2 group) is coupled to phenylisocyanate under alkaline conditions. The terminal amino acid is then cleared from the molecule, derivatized, and analyzed by HPLC. The procedure is repeated until the sequence of the entire protein is obtained. Say that a fragment of a molecule has the sequence Arg-Pro-Lys-Pro-Gln-Phe-Phe-Gly-Leu-Met-NH_2. Using the above procedure to sequence this relatively short protein would take about 12 hours of skilled lab work. Clearly, faster methods would be very attractive.

Much research into the use of mass spectrometry n this endeavor has been performed. The basic idea is that the parent mass (PM) is the whole molecule. Then the cracking pattern would reveal primarily the successive losses of amino acids.

The group of masses (PM $-$ X) next to the PM would be caused by the loss of an amino acid, which is identified as having a mass of PM $-$ (PM $-$ X).

The third highest group in the cracking pattern with mass PM $-$ X undergoes loss of a second amino acid. Its mass would be (PM $-$ X) $-$ (PM $-$ Y).

For example, if the PM is 10,000 and PM $-$ X is 9925, this corresponds to a mass loss of X, which is 75 (Gly). If the next mass occurs at 9836, this is caused by a further loss of the second amino acid Y. Its mass is determined as (PM $-$ X) $-$ (PM $-$ X + Y), i.e., 9925 $-$ 9836, which denotes that Y = 89 (Ala).

The sequence, therefore, starts off Gly, Ala, etc. Since the complete cracking pattern is available, it is theoretically possible to get the entire sequence in one analysis. In practice, the cracking pattern is complicated by the molecule cracking in several places simultaneously, leaving fragments of the molecule. Interpretation is performed by piecing together the information in the same way that jigsaw puzzles are solved.

In large molecules, several mass spectrometries may be performed in sequence. The first one fractures the molecule. The second one isolates one fragment or mass and sequences that. For very large molecules, three mass spectrometers may be put in tandem to sequence fragments of the fragments.

These systems are termed MS–MS and MS–MS–MS, respectively, and are increasingly uncommon in studies of large molecules such as natural polymers, proteins, etc. A schematic diagram of such an instrument is shown in Fig. 15.22.

An ultimate formidable task has been initiated in sequencing the human genome, which has a known molecular weight of several million. To give an idea of the magnitude of this problem, when the entire sequence is known and written down using the three-letter symbols for amino acids, and using standard typesetting, the name will be 5 miles long.

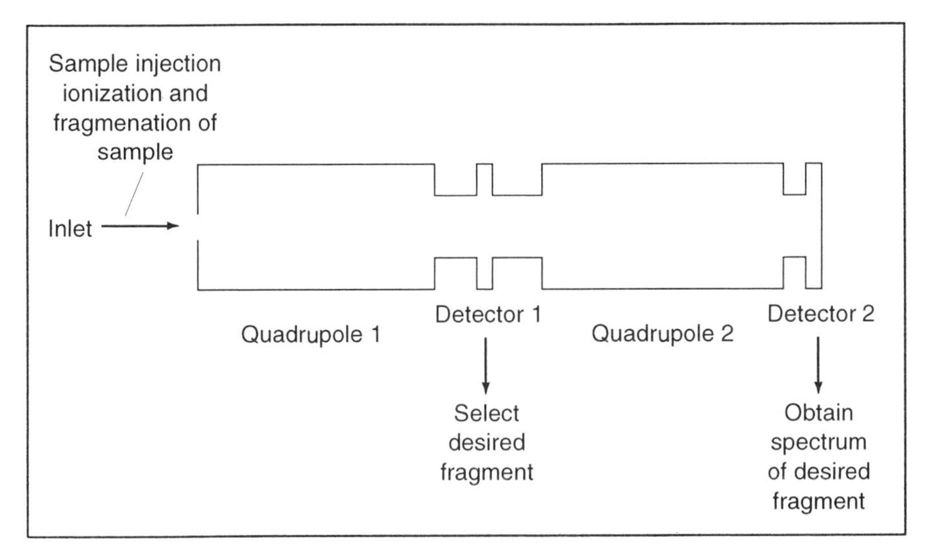

Figure 15.22 Schematic diagram of MS-MS. This system can be extended to obtain the mass spectrum of fragments of the "desired fragment."

4. Special Equipment for GC–MS Combinations

Both the quadrupole mass spectrometer and the time-of-flight mass spectrometer can be used in conjunction with a gas chromatograph, primarily because of the high speed at which mass spectra can be obtained. Although the signal in the form of electrical impulses is readily obtained in these systems, the problem of recording the data in a readily interpretable form was difficult to overcome. Clearly, it is not possible to use a pen and paper, because the pen simply cannot move at a sufficient velocity. Early attempts utilized a light spot moving over light-sensitive paper. The system was reasonably successful at recording the data, but failed in practice because with extended inspection the signal on the light-sensitive paper slowly faded under the influence of room light. Data recorded under this system was therefore not permanent. This is a severe disadvantage in practice.

The problem was overcome by recording the electrical impulses directly onto magnetic tape. This tape system is very convenient to store and is permanent and easily reproducible. If desired, the tape can be used to drive a pen recorder at low speeds and provide a permanent visual record of the mass spectrum. The recordings obtained in this leisurely fashion are much easier to handle and interpret. In addition, data can be recorded digitally, which is a great advantage when elucidating formulas based on isotope ratios.

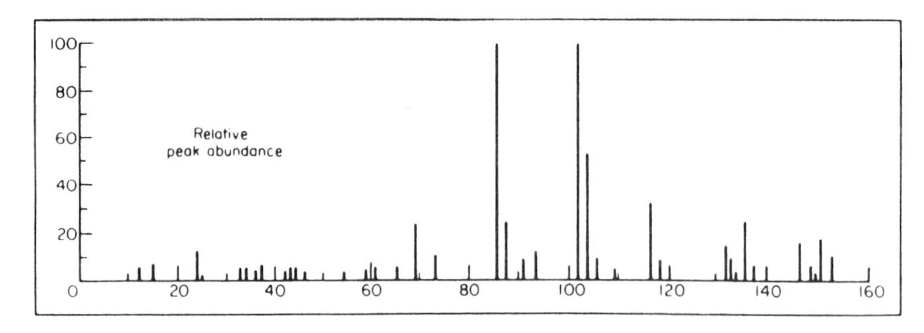

Figure 15.23 Digital readout of a typical recorded mass spectrum derived as the difference between the mass spectrum of a GC peak and the nearby background.

Another advantage is that the mass spectrum recorded on the tape can be connected to a memory bank of mass spectra also recorded on tape. A computer can then compare the mass spectrum of the sample with all the mass spectra in its memory bank and retrieve possible formulas for the sample. Memory banks for various types of samples (e.g., environmental, industrial) are commercially available. The computer searches the memory bank for the most likely compounds in the sample based on the recorded spectrum. The operator makes the final choice. The use of memory banks is increasing at a great rate and will clearly become a very valuable means of identification in the future. However, it is still frequently necessary to interpret spectra from first principles if only to confirm memory bank identification.

A typical recorded mass spectrum and digital readout are shown in Fig. 15.23

5. GC–FTIR–MS Systems

Combination techniques already commercially available have certain limitations. For example, the GC–MS combination separates the compounds and provides the mass spectrum of any desired peak. The mass spectrum may provide the parent mass and hence the molecular weight of the compound. Also, the cracking pattern of the molecule can be identified by comparison with known spectra or from first principles. But it does not directly confirm the functional groups present.

On the other hand, GC–FTIR separates the compounds, and the FTIR identifies the functional groups present in any desired peak but provides no data on molecular weight. A combination of the three techniques has been developed (Fig. 15.24).

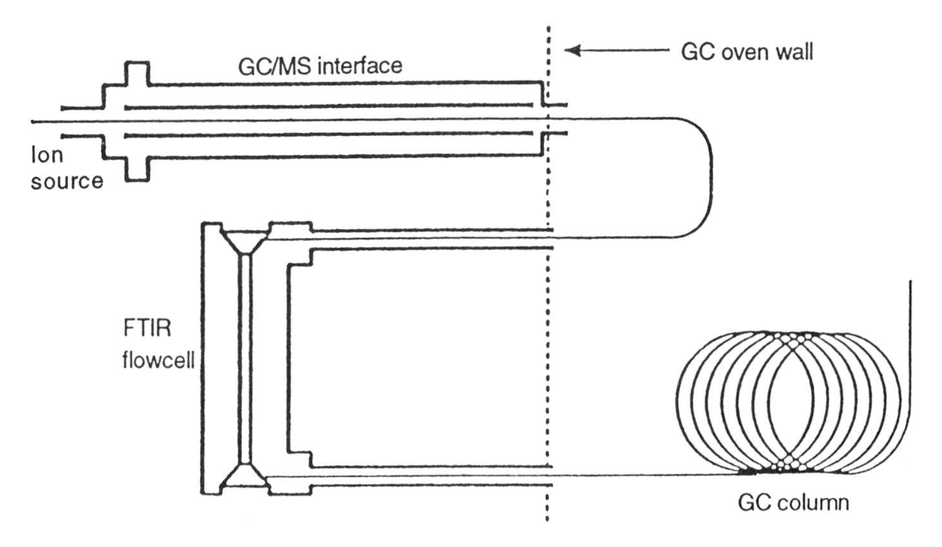

Figure 15.24 Schematic diagram of a GC-FTIR-MS system which operates in continuous flow mode.

The effluent from the GC enters the FTIR, where rapid repetitive IR scans can be performed, thus generating the IR spectrum of the GC peak. The effluent from the FTIR then enters the GC by way of the GC–MS interface.

The data provided by each technique are complementary to each other. For example, homogeneous series of alcohols or ketones have similar IR spectra but different MS spectra. In analyses involving isomers, often the MS is similar, but the IR distinguishes them.

The system has been used as process control in infant formula manufacturing, nutritional sports drinks, flavoring compounds, and oil additives such as tricresyl phosphates (some of which have been shown to be toxic). Gasoline samples can be screened for compounds such as alcohols, aromatics, olefins, or carbonyls.

This system can save much time in the hands of a skilled operator, who can use the area of the GC peak for quantitative analysis and the FTIR–MS data or qualitative identification.

6. Thermal Analysis–MS Interface

Thermal analysis techniques, TGA, and DSC provide physical measurements with temperature changes from which the molecular weight, and therefore the formula of the offcoming gases, can be deduced. However, this is not proof of the formula. Evolved gas analysis (EGA) is best provided by interacting with

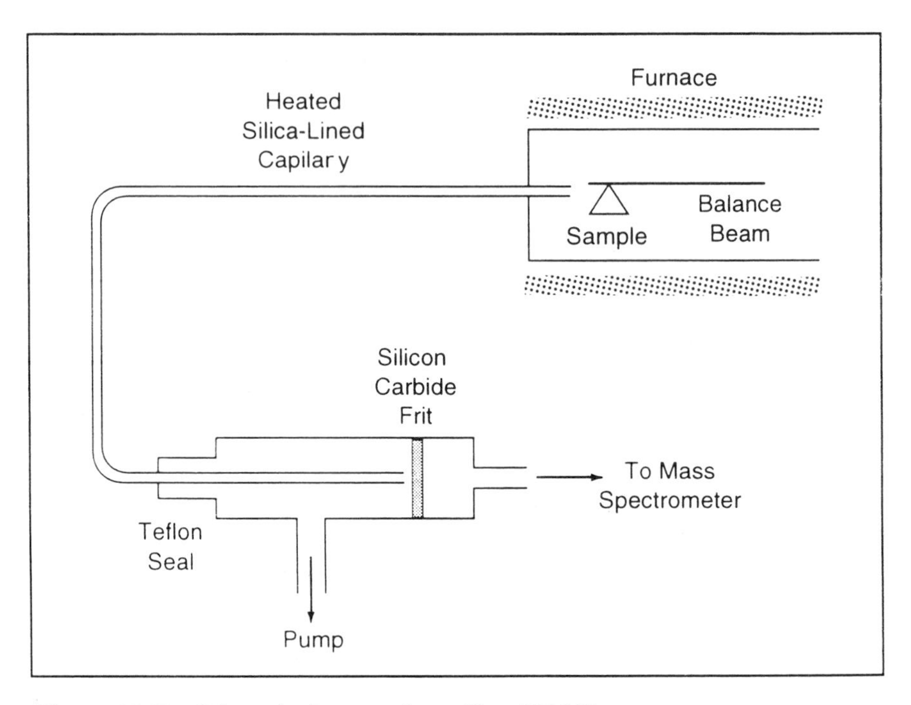

Figure 15.25 Schematic diagram of a capillary TG-MS.

MS. But, suffice it to say that MS can reveal the molecular weight of a molecule and information on its structure. The mass spectrum used is a quadrupole MS, which can provide a spectrum in a few seconds. Schematic diagrams of the instrument are shown in Fig. 15.25. The MS and thermal analyzers are standard but the interface is not.

The problem with interfacing is that the thermal anayzer is at atmospheric pressure, but the mass spectrometer runs at very low pressure. A schematic diagram of the interface is shown in Fig. 15.26.

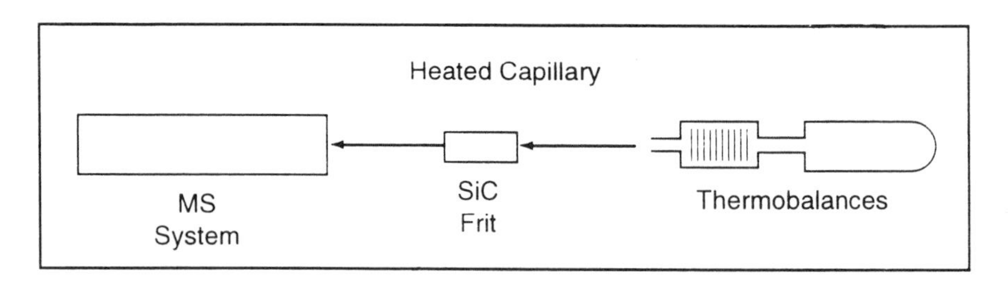

Figure 15.26 Interface between thermobalance and MS.

The pressure drop is accomplished in two stages. The first stage utilizes a flexible stainless steel capillary (0.3 mm × 6 ft) lined with silica to make it inert. One end is in the thermobalance; the other end is attached to the second stage of the MS pump. The transfer capillary is heated to prevent condensation. A silicon carbide frit is placed in line and provides a second means of reducing the pressure before the sample enters the mass spectrometer.

To cover a wide molecular weight range, logrithmic histogram scanning (LGM) can be used. This permits slow scanning in the low mass range for inorganic gases such as CO, CO_2, and H_2O, but more rapid scanning in the high mass range encountered in polymer characterization.

After a trial run using LGM scanning, the preliminary data allow pertinent masses to be scanned with greater care and attention.

7. Typical Applications

Mass spectral data taken of the thermogram of calcium oxalate verify the loss of H_2O, CO, and CO_2 at various stages. In polymer analysis, it is possible to determine the original monomer. Also in the case of co-polymers, i.e., polymers made up of two or more monomers, it is possible to determine the approximate percentage of each in the final polymer.

J. INTERPRETATION OF MASS SPECTRA FROM FIRST PRINCIPLES

1. Cracking Patterns and the Parent Mass

When large organic molecules enter the ionization chamber of a mass spectrometer, they break into ionized fragments. If the ionization conditions are kept constant, a given molecule will always produce the same fragments. The mass and abundance of these fragments are called the molecule's *cracking pattern*. Each molecule has its own characteristic cracking pattern under particular ionization conditions. An unknown compound can be identified by comparing its cracking pattern with those of known pure compounds. Also, by measuring the relative intensities of the patterns, quantitative determinations can be made. The interpretation of cracking patterns permits the quantitative analysis of organic mixtures and even of isomers of the same compound. The cracking patterns of larger organic molecules are quite complicated, and the composition of mixtures of such compounds can be determined only by involved calculations. These calculations are tedious, lengthy, and always subject to error. The use of computers has greatly eased this problem.

Recall that in the cracking pattern of a particular molecule, the greatest mass detected (the parent mass, or PM) is frequently that of the unfragmented but ionized molecule. The PM is usually of low abundance and in fact is not always observed with large molecules under normal ionization conditions. This fact must be remembered when setting up the conditions for analysis. If information concerning the parent mass is required, the ionizing voltage should be kept low enough to reduce fragmentation of the molecules. In short, the parent mass indicates the molecular weight of the molecule; the cracking pattern is used to reveal the structure of the molecule.

When pure compounds are examined, the detection of the parent mass enables us to determine the molecular weight of a compound. The determination of the parent mass by mass spectrometry is one of the most rapid and reliable methods of determining the molecular weight of organic compounds. The techniques used for identifying the parent mass are not discussed in this book. More information on the subject can be obtained from more advanced textbooks.*

2. Quantitative Analysis of Mixtures

Suppose that a sample is known to contain only the isomers of butanol. It can be seen from Table 15.4 that mass 33 is derived from *n*-butanol, but not from the other isomers. A measurement of mass 33 would therefore provide a basis for measuring the *n*-butanol content of the mixture. Also, we can see that the abundances of masses 45, 56, and 59 vary greatly among the isomers. By measuring the actual abundances of these three masses in the sample and applying the ratio of the abundances from pure compounds, three simultaneous equations with three unknowns (i.e., the percentages of *n*-butanol, *sec*-butanol, and *tert*-butanol) can be obtained. The three equations can be solved and the composition of the sample determined. It is customary to perform this step on a computer, which gives rapid and reliable results. Programs can be set up to handle multicomponent systems, make all necessary corrections, and present the results.

It is of interest to examine the cracking patterns of these isomers of *n*-butanol further (Table 15.4). First it should be noted that 74 is the PM and is of low or zero abundance. To enhance this peak, a different ionization process would be necessary.

In the spectrum of *n*-butanol, the most abundant peak is 31. This corresponds to $-CH_2OH$ and indicates that *n*-butanol is most likely to fragment by breaking off the CH_2OH fragment. The abundant peak at 56 indicates the loss of an OH from the parent molecule.

In *sec*-butanol, the most abundant peak is at 45. This corresponds to CH_3CHOH, indicating a loss of C_2H_5.

---

*See, for example, Fred W. McLafferty, *Interpretation of Mass Spectra*, 2nd ed., Benjamin, New York, 1973.

Table 15.4 Mass Distribution of the Fragments of Isomers of Butanol[a] using e^- impact at 70 eV

m/e	*n*-Butanol $CH_3CH_2CH_2CH_2OH$	*sec*-Butanol CH_3CH_3COH (H, CH_3)	*tert*-Butanol CH_3COH (CH_3, CH_3)
15	8.4	6.80	13.3
27	50.9	15.9	9.9
28	16.2	3.0	1.7
29	29.9	13.9	12.7
31	*100	20.31	35.5
33	8.5	0	0
39	15.6	3.4	7.7
41	61.6	10.1	20.8
42	32.4	1.7	3.3
43	61.4	9.8	14.5
45	6.6	*100	0.6
55	12.3	2.0	1.6
56	90.6	1.0	1.5
57	6.7	2.7	9.0
59	0.3	17.7	*100
60	0	0.7	3.2
74	0.8	0.3	0

[a]Parent mass = 74. Asterisks in table body denote the most abundant fragment. The abundance is equated to 100 for each pattern, but the abundances of these peaks probably differ from each other.

In *tert*-butanol, the most abundant peak is 59, corresponding to a loss of 74 − 59, i.e., 15. This is the mass of a CH_3 fragment, leaving

$$\begin{matrix} CH_3 \\ CH_3 \end{matrix} > C{-}OH$$

Therefore, *tert*-butanol is most likely to fragment by first breaking off CH_3.

3. Structural Analysis

The technique of identifying a pure compound by comparing its spectrum with known spectra works well if we already know a great deal about the compound and if the compound has been documented in terms of its fragmentation pattern. This enables us to make an educated guess as to the identity of the compound and confirm it by direct comparison with known spectra. Unfortunately, this is not always the case; more frequently, the nature of the sample is not known.

Furthermore, the sample may not be pure, and therefore direct comparison of spectra will not constitute a valid confirmation of the compound's identity. In these circumstances it is necessary to take the spectrum and deduce the information from it.

One of the most important pieces of information that can be obtained is the molecular weight of the compound. From it we can immediately conclude a great deal about the compound. For example, for an organic compound if the molecular weight were 54, we would be able to conclude that the compound can only have empirical formula C_2NO, $C_2H_2N_2$, C_3H_2O, C_3H_4N, or C_4H_6. Some of these possibilities can be eliminated by inspection, because it is not possible to have a molecule with certain empirical formulas. For example, C_2NO cannot exist. Thus, if we know the molecular weight, we have considerable information about the empirical formula of the molecule.

We said earlier that the identification of the parent mass, which corresponds to the molecular weight, is not easy and is outside the scope of this book. We can safely assume, however, that the identification can be made. Let us therefore suppose that in the compound mentioned above, we had identified the PM as 54. We would then find a peak at 54 and a small peak at 55. This latter peak exists because carbon exists as the isotopes ^{12}C and ^{13}C.

In every natural compound the ^{13}C will result in a peak that is one mass number greater than the PM, which assumes that the atomic weight of carbon is 12. This ^{13}C peak is generally designated as PM + 1. For example, with methane, CH_4, the molecular weight is 16, assuming that carbon has an atomic weight of 12 (i.e., $^{12}C^1H_4 = 16$). Let the abundance of this peak be 100. But the abundance of ^{13}C is 1.1% of ^{12}C, therefore $^{13}C^1H_4 = 17$, and its abundance is 1.1% of $^{12}C^1H_4$. This designated as:

	MW	Abundance
$^{12}CH_4$	16	100
$^{13}C^1H_4$	17	1.1

Similarly, for ethane, C_2H_6, we can designate the MS pattern as:

			MW	Abundance
	P	$^{12}CH_3{}^{12}CH_3$	26	100
	P + 1	$^{12}CH_3{}^{13}CH_3$	27	1.1
	P + 1	$^{13}CH_3{}^{12}CH_3$	27	1.1
Total	P + 1		27	2.2
	P + 2	$^{13}CH_3{}^{13}CH_3$	28	0.01
				(ie., 1% of 1% of 100)

The P + 2 peak at 28 can usually be ignored in small molecules, but not large molecules. We can see, therefore, that if there is 1.1% natural abundance of ^{13}C compared to ^{12}C. only one carbon is present in the molecule, then the ratio (PM + 1)/PM = 1.1%. If two carbons are present in the molecule, the probability of ^{13}C being present is twice as great and (PM + 1)/PM = 2.2%.

In fact, there is a definite relationship between PM + 1 and PM that is directly related to the number of carbons present in the molecule:

$$\frac{\text{PM} + 1}{\text{PM}} = \text{at least } 1.1\% \times \text{number of C atoms in the molecule} \qquad (15.17)$$

This relationship is very valuable in characterizing an unknown compound.

After identifying the parent mass and measuring the ratio (PM + 1)/PM, it is possible to calculate the number of carbons in the molecule. For example, in the molecule of molecular weight 54, if (PM + 1)/PM is 2.36%, only two carbons are present, and the formula can only be $C_2H_2N_2$ (if three carbons were present, the ratio (PM + 1)/PM would be at least 3.3%). It should also be noted that if the ratio (PM + 1)/PM = 2.36%, there cannot be more than two carbons; however, there is a possibility that there are less than two if extraneous fragments are present. If the ratio were 3.3, three carbons would be present and the formula would be C_3H_2O; similarly, if the ratio were 4.42, the formula would be C_4H_6. Using the isotope ratio of carbon, it is possible to deduce the number of carbons present, and from this information the empirical formula of the molecule can often be deduced. Note that this simplified treatment ignores the isotopes of nitrogen or hydrogen (deuterium) that may be present. A correction must be made in more rigorous treatment.

What would be the formula of the compound with the mass spectrum shown in Table 15.5? Let us guess which is the parent mass. If it is 17, the possible

Table 15.5 Mass Spectrum of an Unknown Compound

m/e	Abundance
1	3.1
2	0.17
12	1.0
13	3.6
14	9.2
15	85
16	100
17	1.11

Table 15.6 Identity of Fragments in Methane Cracking Pattern[a]

m/e	Predominant fragment
1	H
2	D/H_2
12	C
13	CH
14	CH_2
15	CH_3
16	CH_4
17	$^{13}CH_4$

[a]Other isotope mixtures such as $^{13}CH_3$ for *m/e* 16 can be imagined, but their abundance is minor.

formulas are OH, NH_3, and CH_5; if it is 16, the possible formulas are O, NH_2, and CH_4. Of all these possibilities, only CH_4 is likely. The ratio (PM + 1)/PM = the ratio of the masses at 17 and at 16 = 1.11/100, meaning that one carbon is present in the molecule. This number fits with CH_4. If we look at the spectrum again, we can identify each fragment as shown in Table 15.6. Such a fit is not possible with NH_3; therefore we conclude that the compound is CH_4.

Similar isotope effects are found with other elements in the periodic table. For example, oxygen contains isotopes ^{16}O, ^{17}O, and ^{18}O, all of which exist in a constant proportion to each other in nature. Since the natural abundance ratio of ^{16}O to ^{18}O is 100:0.2, the number of oxygens can be calculated from Eq. (15.18):

$$\frac{\text{PM} + 2}{\text{PM}} = 0.20\% \times \text{no. of O atoms} + \left(\frac{1.1\% \times \text{no. of C atoms}}{100}\right)^2 \tag{15.18}$$

For example, the mass spectrum of methanol is shown in Table 15.7. It can be deduced with practice that the parent mass is 32 and that the ratio of 33 to 32 is slightly over 1.1%, indicating not more than one carbon. Also, the ratio of 34 to 32 is approximately 0.21%, indicating not more than one oxygen.

There are some other interesting points in the spectrum. For instance, the most abundant peak is at 31, indicating that methanol very easily loses one hydrogen and forms CH_3O. Another abundant peak is at 29, indicating that methanol loses three hydrogens to form COH. The next most abundant peak is at 15, which is equivalent to CH_3. The fragment 17 is not particularly abundant, in-

Table 15.7 Mass Spectrum of Methanol

m/e	Relative abundance
12	0.3
13	0.7
14	2.4
15	13.0
16	0.2
17	1.0
28	6.3
29	64.0
30	3.8
31	100
32	66
33	1.0
34	0.14

dicating that although OH fragments are formed, most are not ionized positively and do not reach the detector.

A list of common isotopes and their relative abundance in nature is given in Table 15.8. It can be seen that the ratio of ^{35}Cl to ^{37}Cl is 100/33. For a compound such as CH_3Cl the PM is 50, but a peak will exist at PM + 2 because there will be ^{37}Cl present. The ratio (PM + 2)/PM will be the same as the natural ratio of ^{37}Cl to ^{35}Cl, that is, 32.7%.

Table 15.8 Relative Natural Abundance of Common Isotopes

Isotope	Mass	Relative abundance	Mass	Relative Abundance	Mass	Relative abundance
H	1	100	2	0.016		
C	12	100	13	1.08		
N	14	100	15	0.36		
O	16	100	17	0.04	18	0.20
F	19	100				
Si	28	100	29	5.07	30	3.31
P	31	100				
S	32	100	33	0.78	34	4.39
Cl	35	100			37	32.7
Br	79	100			81	97.5
I	127	100				

Based on knowledge of the parent mass and the isotope ratios, it is possible to come to well-founded conclusions on the empirical formula and the molecular weight of an unknown compound. When used in conjunction with other methods, such as IR spectroscopy and NMR, this method makes it possible to identify the compound.

K. INTERPRETATION OF SPECTRA

The spectra involved are usually obtained by electron impact since other techniques enhance the parent mass and greatly reduce the fragmentation pattern. Consequently, the following discussion is concerned mostly with electron impact spectra.

There are two ways to interpret such spectra. The first is to compare the spectrum you have with those in a memory bank. Since there are approximately 200,000 known chemicals, no bank contains the spectra of all known compounds. Rather it is typical for a memory bank to contain around 5000 spectra of compounds pertinent to a particular topic. Such topics include environmental compounds, pharmaceuticals, natural products, oil industry compounds, etc.

In practice, the analyte spectrum is entered into the memory bank, which then, using a computer, compares it to the spectra in its memory. There may be six compounds with spectra similar to the analyte. The names of these components are presented in the screen.

The operator may request to see these components' spectra and manually compare them to that of the analyte. Based on his or her best judgment, the analyst the identity of the unknown.

This method requires little training on the part of the analyst to identify the compound. However, a more challenging problem is to identify the compound from first principles. These techniques are discussed below.

1. Identification of Mass Spectra from First Principles

Interpretation of mass spectra is much like solving a jigsaw puzzle: all the pieces are there, but putting them together involves a lot of guesswork followed by confirmation.

Three major steps are involved:

1. Identification of the parent mass
2. Using isotope ratios to count the number of (a) carbons, (b) oxygens, (c) other elements present
3. Identifying fragments of the molecule from the fragmentation pattern

Table 15.9 Mass Spectrum of Methane

m/e	Relative abundance
1	3.1
2	0.17
12	1.0
13	3.9
14	9.2
15	85.
16	100.
17	1.11
18	0.04

2. Identification of the Parent Mass

This is often the most difficult step to take, particularly with larger molecules, since the abundance of the PM may be very small or zero. In this text, we shall emphasize compounds with PMs that can be identified or deduced with reasonable certainty. If the parent mass is present, it must be the highest *m/e* in the spectrum, excluding the effects of the isotope. Examples are shown for methanol (Table 15.7), methane (Table 15.9), and benzene (Table 15.10). In each case, the parent mass was very abundant and not difficult to identify. This is not always the case, as will be seen in later examples.

When a molecule is ionized by electron impact, it undergoes the reaction.

$$M + e^- \rightarrow M^+ + e^- + e^-$$

Thus, a PM is always a +*ve* ion. It is also evident that if an organic molecule loses an electron, it must be left with an unpaired electron, i.e., it is a free radical. This ion with an unpaired electron is termed an *odd electron ion. The parent mass or molecule ion is always an odd electron ion.* For example, in the reaction $C_2H_6 + e^- \rightarrow C_2H_6^+ + e^- + e^-$, the $C_2H_6^+$ is the molecular ion (or parent mass).

However, the molecule may fragment rather than ionize, leaving a pair of electrons behind on one ion. This is an *even electron ion*. Such an ion arises by *fragmentation* and *cannot* be the parent mass. Hence, the parent mass is *never* an even electron ion. As an example, C_2H_6 may react as follows:

$$C_2H_6 + e^- \rightarrow C_2H_5^+ + H + e^- + e^-$$

The $C_2H_5^+$ has lost a H and cannot be the PM.

In practice, both reactions will occur and masses equivalent to C_2H_6 (30) and C_2H_5 (29) will be present. The 30 peak is the parent mass.

Table 15.10 Mass Spectrum of Benzene

m/e	Relative abundance	m/e	Relative abundance
37	4.0	53	0.80
37.5	1.2	63	2.9
38	5.4	64	0.17
38.5	0.35	73	1.5
39	13.	74	4.3
39.5	0.19	75	1.7
40	0.37	76	6.0
48	0.29	77	14.
49	2.7	78	100.
50	16.	79	6.4
51	18.	80	0.18
52	19.		

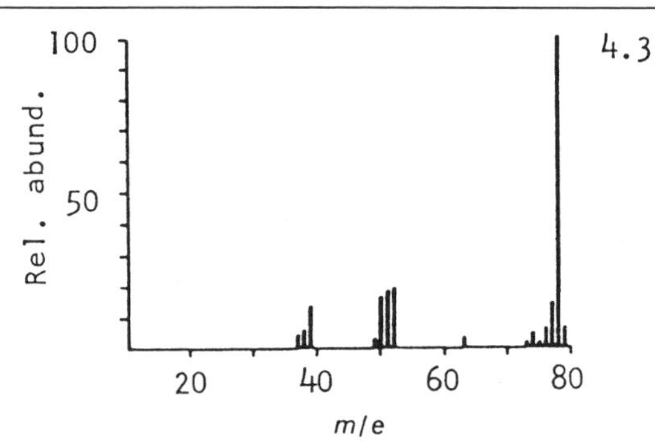

Because of the atomic weights and valences, it transpires that the *PM is always an even number, unless an odd number of N's are present*. This is a most important rule to remember. Examples are as follows:

Compound	PM
CH_4	16
C_2H_5OH	46
C_6H_6	78
$C_2H_5NH_2$	45
$C_6H_5NH_2$	93
C_6H_{14}	86
C_6H_{12}	84

The next step is to use the relative abundance of natural isotopes (see Table 15.8) to identify the elements present. Notice that tritium ^3H and ^{14}C are not listed. This is because the abundances of these radioactive isotopes are so small that they can be disregarded.

However, it can be seen that the ratio $^{12}C/^{13}C$ is 100/1.08. Hence, for a simple compound like methane, if the parent mass $^{12}C^{1}H_4$ has a mass of 16 and an abundance of 100, then $^{13}C^{1}H_4$ must have an abundance of 1.08, i.e., mass 17/16 = 1.08%.

With C_2H_6, we have $^{12}C_A{}^{12}C_B{}^{1}H_6$ for a mass of 30. But we also will have $^{13}C_A{}^{12}C_B{}^{1}H_6$ mass 31 *and* $^{12}C_A{}^{13}C_B{}^{1}H_6$ mass 31. Both combinations are equally possible. If the abundance of mass 30 is 100, then $^{13}C^{12}C^{1}H_6$ 31 is 1.08 and $^{12}C^{13}C^{1}H_6$31 is also 1.08, so the total abundance of mass 31 is 2.16% of mass 30.

Using a similar process for C_3H_8 mass 44, the abundance of 45 will be 3 × 1.08% of the 44 (1.08% for each carbon present, i.e., 3.3).

As a general rule the relative abundance off the P + 1 peak compared to the P peak depends on the number of carbons present. For simple hydrocarbons:

$$\frac{(P + 1^+)}{(P^+)} = 1.08\% \times \text{number of carbons present}$$

If N is present in the molecule, the ratio $^{15}N/^{14}N$ is 0.36% (Table 15.11).

To correct for this possibility, the general expression for compounds containing only C, H, N, and O (and F, P, and I since these have only one mass.) is

$$\frac{(P + 1^+)}{(P^+)} = 1.08\% \times N^0 \text{ of C present} + 0.36\% \times N^0 \text{ of N present} \qquad (15.19A)$$

The ratio $(P + 1^+)/(P^+)$ enables us to count carbons.

For example, in the case of benzene, the PM is 78, P + 1 is 79, and the relative abundance is 6.6%, confirming 6 carbons. Also, in the case of methane, the ratio 17/16 is 1.1%, confirming 1 carbon.

Using this technique it is possible not only to count carbons in the parent peak but also in fragments from the peak. For example, in the spectrum for *tert*-butanol, the PM is 74 but is not present. However, the ratio 60/59 is 3.2%, indicating three carbons. This would support the formula $C_3H_6OH^+$ and would arise from the loss of a CH_3 from the original molecule.

3. Counting Oxygens

Examination of the Table 15.11 shows that oxygen has two important isotopes, ^{16}O and ^{18}O, with a relative abundance $^{18}O/^{16}O$ of 0.2%. The spectrum of water is shown in Table 15.11.

Table 15.11 Spectrum of Water

m/e	Relative abundance
1	<0.1
16	1.0
17	21.
18	100.
19	0.08
20	0.22

To calculate the number of carbons and oxygens in a molecule, we assume that only C, H, N, O, F, P, and I present, and there are no interferences from other ions.

$$\frac{(P + 1^+)}{(P^+)} = 1.1\% \times \text{no. of C atoms} + 0.36\% \times \text{no. of N atoms} \qquad (15.19B)$$

$$\frac{(P + 2^+)}{(P^+)} = 0.2\% \times \text{no. of O atoms} + \frac{(1.1 \times \text{no. of C atoms})^{2\%}}{200} \times \text{no. of O atoms} \qquad (15.20B)$$

These formulas apply to fragments and to PMs. The PM $^1H_2{}^{16}O$ is 18 but there is also a mass at 20 (P + 2) caused by $^1H_2{}^{18}O$, and its abundance is 0.2% of the abundance of 18 ($^1H_2{}^{16}O$).

We can generate another general rule that

$$\frac{(P + 2)^+}{(P)^+} = 0.2\% \times \text{no. oxygen present} \qquad (15.20A)$$

This ignores molecules large enough to contain two ^{13}C isotopes. The general equation is, therefore,

$$\frac{(P + 2)^+}{(P)^+} = 0.2\% \times \text{no. oxygens} + (1.1\% \text{ no. C})^2 \qquad (15.21)$$

Note the squared term, i.e., the correction is $0.01 \times (\text{no. C})^2$.

Looking at spectrum 15.7 the PM is probably 32. (If it were 31, where would 32 come from?)

The number of carbons is calculated as:

$$\frac{P + 1}{P} = \frac{0.98}{66} = 1.2\% = 1 \text{ carbon}$$

If there were 2 carbons, this ratio would have to be at least 2.2%. Hence, there cannot be more than 1 carbon. Also,

$$\frac{(P + 2)}{(P)} = \frac{0.14}{66} = 0.21$$

indicating 1 oxygen. We have

$$\begin{aligned} 1\ C &= 12 \\ 1\ O &= \underline{16} \\ &\ \ \ 28 \end{aligned}$$

But the PM is 32. Therefore, *four* mass numbers are missing. These can be either one helium or four hydrogens. Clearly it is four hydrogens, resulting in the formula CH_4O. But 15.7 also indicates *m/e* 15 (CH_3) and *m/e* 17(OH), i.e., CH_3OH. Therefore, there must be 4 (i.e., 32 − 28) hydrogens present to give a formula COH_4 or CH_3OH.

Using these equations we shall look at some other spectra.

Looking at Table 15.12.if

PM = 44
Ratio 45/44 = 1.2% = 1 carbon = 12
46/44 = 0.42% = 2 oxygens = 32

Therefore, the compound is CO_2.

Looking at Table 15.13, if

PM = 27
Ratio 28/27 = 1.5 = 1 carbon
29/27 = 0.06 = 0 oxygen

If the PM is 27, there must be an odd number of nitrogens present. There cannot be three, because the mass would be too high. Therefore, There must be 1 nitrogen. The mass accounted for is 1 C (12) + 1 N (14), i.e., 26. The PM is 27, indicating 1 hydrogen present, to give a formula HCN.

Table 15.14 shows the mass spectrum of acetylene. It can be deduced as follows. If the parent mass P is 26, then P + 1 is 2.2, indicating 2 carbons. Their mass is 24. The missing mass of 2 can only be two hydrogens. Hence, the formula is C_2H_2. Note that the P + 2 peak is too small to indicate the presence of oxygen, and the even number PM indicates zero (or two) nitrogens. Also, *m/e* = 25 = C_2H; *m/e* 24 = C_2. There is no *m/e* 23, indicating 3 hydrogens.

PM = 26
Ratio 27/26 = 2.2% = 2 C
No H = 26–24
Formula H_2C_2 or HC ≡ CH, i.e., acetylene

Table 15.12 Spectrum of CO_2

m/e	Relative abundance
12	8.7
16	9.6
22	1.9
28	9.8
29	0.13
30	0.02
44	100.
45	1.2
46	0.42

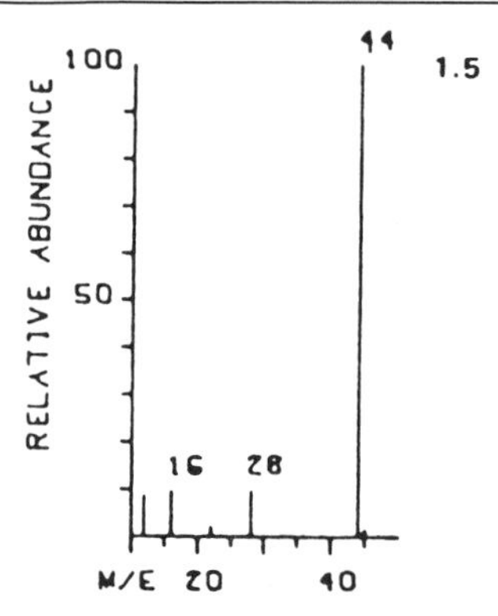

Table 15.13 Spectrum of HCN

m/e	Relative abundance
12	4.2
18	1.7
13.5	0.88
14	1.6
15	0.12
26	17.
27	100.
28	1.6
29	0.06

Table 15.14 Spectrum of Acetylene

m/e	Relative abundance
12	0.91
13	3.6
14	0.10
24	6.1
25	23.
26	100.
27	2.2
28	0.02

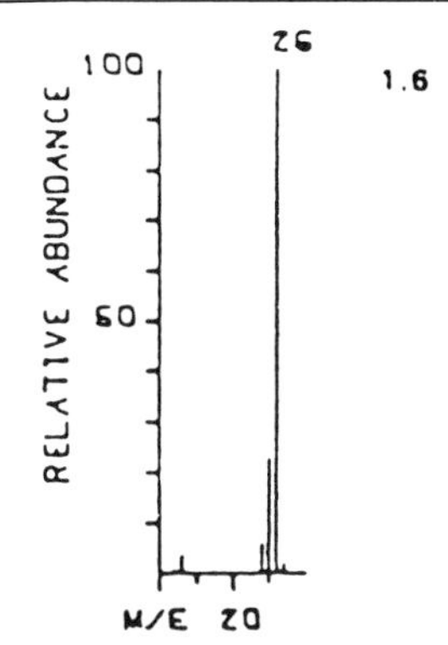

4. Compounds with Hetero Elements

Using the isotope ratios given in Table 15.11, we can examine other elements, particularly the halides.

Table 15. shows two strong peaks at 36 and 38 in an abundance ratio of 100/33. This is typical of Cl. If Cl is present, then mass 36 is ^{35}Cl plus 1 H (only H could increase the mass from 35 to 36). Also, mass 38 would be ^{37}Cl plus 1 H. This would be the spectrum of hydrochloric acid, HCl. Note that the ratio 38/39 indicates zero carbons and 38/40 no oxygens, supporting the formula HCl. The ratio 36/37 cannot be used to count carbons because 37 is an isotope of Cl.

5. Clusters of Halogens

If we have two Cls present in a molecule, the distribution of masses in each molecule may be ^{35}Cl ^{35}Cl, ^{37}Cl ^{35}Cl, ^{35}Cl ^{37}Cl, or ^{37}Cl ^{37}Cl. But, ^{35}Cl:^{37}Cl

Table 15.15 Spectrum of HCl

m/e	Relative abundance	
35	12.	
36	100.	
37	4.1	Cl cluster
38	33.	
39	0.01	

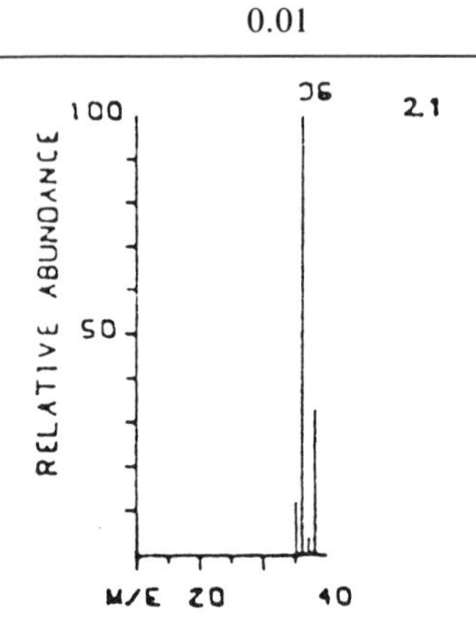

abundance is 100:33, so the probability of these mixed isotopes distribution occurring is 100:66:11.

Consequently, any molecule RCl_2 containing two Cls will exhibit masses of R + 70, R + 72, and R + 74 with a relative abundance due to Cl isotopes of 100:66:33. This is a Cl_2 *isotope cluster*.

Similarly, if three chlorines are present (RCl_3), a cluster will occur with masses R + 105, R^- + 107, R^- + 109, and R^- + 111 with approximate abundances of 100:100:33:4.

Similar sets of clusters can be determined for bromine compounds. Br has two isotopes, ^{79}Br and ^{81}Br, and they are approximately equal in abundance.

A compound RBr containing 1 Br will have parent masses at R + 79 and R + 81, and they will be about equal in abundance. If 2 Brs are present, as in RBr_2, there will be masses of R + 158, R + 160, and R + 162 from R $^{79}Br^{79}Br$, $R^{79}Br^{81}Br$, $R^{81}Br^{79}Br$, $R^{81}Br^{81}Br$, and their relative abundances will be 55:100:45 based purely on probability.

In Table 15.16, the masses at 130, 132, 134, 136 indicate a cluster of 3 chlorines and a PM of 130. If true, $3^{35}Cl$ = 105, leaving a mass of

Table 15.16 Spectrum of Trichlorethylene

m/e	Relative abundance	
129	30.	
130	100.	Cl_3 cluster
131	31.	
132	98.	Cl_3 cluster
133	12.	
134	32.	Cl_3 cluster
135	1.7	
136	3.5	Cl_3 cluster
137	0.07	

130 − 105 = 25 to be accounted for. This is probably C_2H, giving a total formula C_2HCl_3 or HClC═C—Cl_2. This compound is trichloroethylene, a commercial solvent.

In Table 15.17, the abundance of the two peaks at 94 and 96 indicates 1 bromide. This is confirmed by masses at 79 and 81 in equal abundance. If the PM is 94 and ^{79}Br is present, a group of mass 15 must be present. This is confirmed by an abundant mass at 15. Based on the 15:16 ratio, it contains 1 C and is probably CH_3. Consecutive masses of 13, 14, 13, 12 confirm 3 hydrogens, ending in C with no H. Hence, the total molecules is Br + CH_3 or CH_3Br—methyl bromide.

6. General Appearance of the Spectrum Using Electron Impact

The general appearance of the mass spectra depends on the types of compounds analyzed. Figure 15.27 shows the spectra of n-C_{15} and n-C_{30} paraffins. It can be seen that these spectra are very similar and would be hard to distinguish even though one molecule is twice as big as the other. The parent mass is very small using electron impact ionization. (To get the PM, FAB or chemical ionization could be used.)

It can also be noted that with each compound, there are groups of fragments differing by 14 from each other. This is caused by the loss of successive numbers of CH_2 (methylene) groups as the molecules are fragmented.

For the homologous series of paraffin, masses 15, 29,43, 57, etc. will be noted starting at 15 (CH_3). However, for other homologous series, the same general appearance is observed by the numbers displaced. Table 15.18 shows the terminal group and the first several masses of the series.

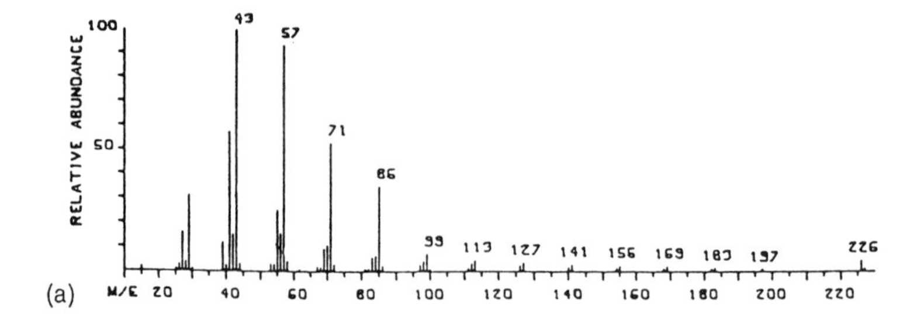

(a)

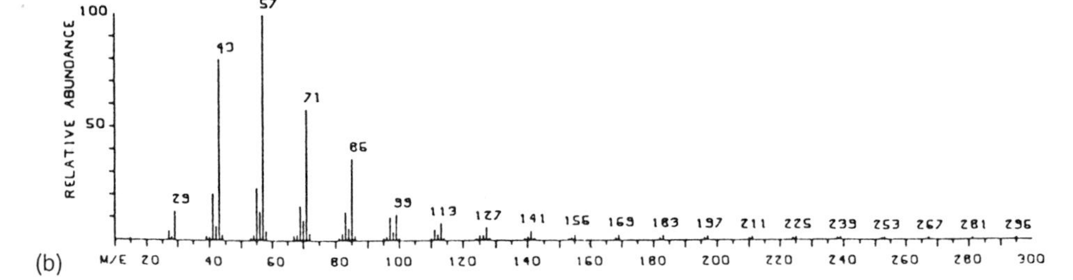

(b)

Figure 15.27 (a) Mass spectrum of *n*-hexadecane and n-tridecane. The relative abundances of the *m/e* 506 and 507 peaks are 0.46% and 0.18%, respectively. The relative abundances of the C_nH_{2n+1} peaks decrease regularly from *m/e* 309 = 0.7% to *m/e* 505 = 0.1%.

Such regular grouping separated by 14 indicates a paraffinic-type homologous series, and the actual number reveals the type of compound (amino acid, alcohol, etc.) On the other hand, aromatic compounds present quite different spectra, as can be seen in Table 15.19.

In this spectrum, the PM is 128. The ratio P+1/P is 11/100, indicating 10 carbons. Their mass is 10 × 12 − 120. The molecule, therefore, has 128 − 120 i.e., 8 hydrogens. The formula indicated is $C_{10}H_8$ and is probably naphthalene.

Naphthalene

It will also be remembered that the MS for benzene (Table 15.10) showed a very strong peak at 78, the parent mass. These spectra are typical of aromatics. The molecules are very stable and do not easily fragment. Hence, the PMs are very strong.

Table 15.17 Spectrum of Methyl Bromide

m/e	Relative abundance	*m/e*	Relative abundance	
12	1.2	48	0.95	
13	1.4	79	10.	Bromide cluster
14	3.8	81	10.	
15	59.	91	4.2	
16	0.62	92	2.4	
39.5	0.19	93	6.8	
40.5	0.20	94	100.	Bromide cluster
46	1.3	95	3.6	
46.5	0.30	96	96.	
47	2.3	97	1.1	
47.5	0.28			

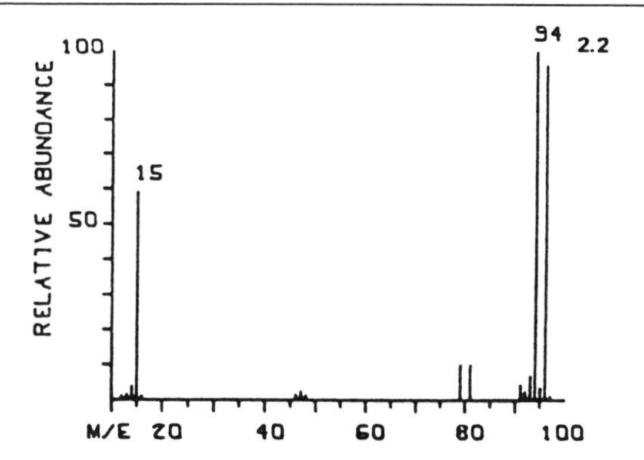

Table 15.18 Homologous Series Separated by CH_2

Functional group	End group	Series
Amine	$-CH_2-NH_2$	30 44 58 . . .
Alcohol	$-CH_2-OH$	31 45 59
Aldehyde	$-CHO$	29 43 57 71
Ketone	$-C(=O)-CH_3$	43 57 71

Table 15.19

m/e	Relative abundance	m/e	Relative abundance
38	1.8	64.5	1.1
39	3.9	65	0.24
40	0.19	66	0.02
41	0.01	74	4.7
42	0.06	75	4.9
42.67	0.10	76	3.3
43	0.16	77	4.1
43.5	0.02	78	2.7
49.5	0.24	79	0.19
50	6.4	87	1.4
50.5	0.09	88	0.22
51	12.	89	0.73
51.5	0.37	90	0.06
52	1.6	101	2.7
53	0.27	102	7.1
54	0.04	103	0.64
55	0.14	104	0.15
55.5	0.03	110	0.06
56	0.12	111	0.09
56.5	0.06	112	0.02
61	1.4	113	0.18
61.5	0.08	125	0.85
62	2.7	126	6.1
62.5	0.20	127	9.8
63	7.4	128	100.
63.5	0.95	129	11.
64	10.	130	0.52

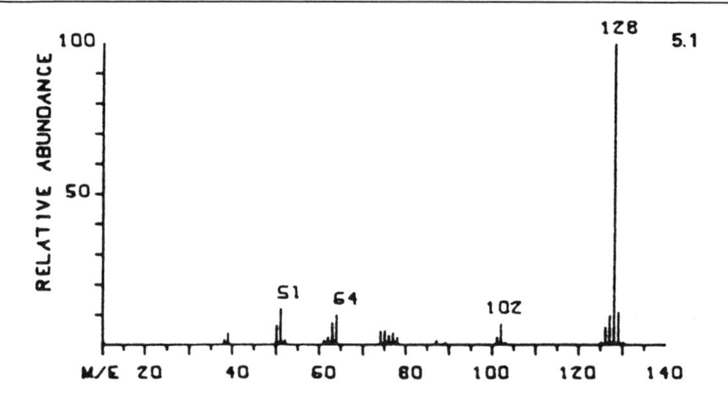

It will also be noted that with naphthalene, masses at 64.5, 63.4, 62.5 are observed. These are not fragments with fractional masses. Rather, they are doubly charged ions. It will be remembered that

$$m/e = \frac{B^2r^2}{2V} \tag{15.8}$$

and it is normally assumed that $e = 1$. But if the ion is doubly charged to m^{++}, it appears to have half the mass to satisfy Eq. (15.8). Therefore, these ions really have masses of $64.5 \times 2 = 129$, $63.5 \times 2 = 127$, $62.5 \times 2 = 125$. The apparent mass of 42.67 is triply changed, i.e., M^{+++}, and its true mass is $52.67 \times 3 = 128$.

Again, this illustrates how stable the aromatic molecule is. As a general rule, increased unsaturation increases stability (double bonds must be broken rather than single bonds). Also, rings are more stable than straight chains. In summary, aromatic and aliphatic molecules can be distinguished at a glance.

Aromatics with side chains will have characteristics of both. This in itself is informative.

7. Rings Plus Double Bonds

The formula for *n*-hexane is C_6H_{14}. The formula for cyclohexane is C_6H_{12}, the same as the formula for hexene. The presence of a ring or double bond resulted in a change in the ratio of carbons to hydrogens. This change is a general property and based on it the following formula is derived.

The number of rings plus double bonds (db) in a molecule of formula $C_xH_yN_zO_m$ is

$$x - \frac{1}{2}y + \frac{1}{2}z + 1 \tag{15.22}$$

For example, for *n*-hexane, x = 6, y = 14, so the number of rings is $6 - (\frac{1}{2} \times 14) + 0 + 1 = 0$. For cyclohexane, x = 6, y = 12, therefore rings + db $= 6 - \frac{1}{2}12 + 1 = 1$. For benzene, C_6H_6, rings and db $= 6 - (\frac{1}{2}6) + 1 = 4$, i.e., 3 double bonds plus 1 ring. Referring to the spectrum in Table 15.20,

PM	= 72	
P + 1	= 3.5	Indicating 3 C = 36
P + 2	= 0.48	Indicating 2.0 = 32

3C + 2.0 = mass 68
Indicating 72–68 = 4H
Formula is C_3H_4O

$$\text{Rings + d.b.} = 3 - \frac{1}{2}(4) + 1 = 2$$

Table 15.20

m/e	Relative abundance	m/e	Relative abundance
25	5.2	45	32.
26	38.	46	2.5
26.5	0.15	47	0.14
27	74.	52	1.4
27.5	0.26	53	6.0
28	12.	54	2.3
29	4.3	55	74.
30	0.13	56	2.6
31	0.48	57	0.19
41	1.2	71	4.3
42	1.3	72	100.
43	5.8	73	3.5
44	14.	74	0.48

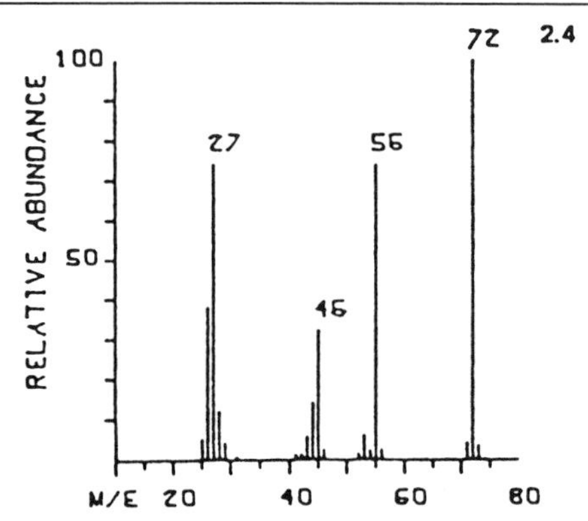

Also, mass 45/47 ratio indicates 2 oxygens = 32 45 − 32 = 13, which is probably CH. So 45 is probably 2 oxygens + CH or COOH, the remainder of the molecule has mass 72 − 45 = 27, including 2 C and a double bond.

Probable formula $H_2C{=}CH - COOH$.

When atoms other than CHNO are present, the atoms are listed according to the lowest valence, i.e., halides, valence 1 acts as H. For even electron ions (fragments) the true value will be increased by 1/2.

Hence the general formula is:

$$\text{Rings and double bonds} = x - \frac{1}{2}y + \frac{1}{2}z + 1$$

where x = C, Si, etc.
y = H, F, Cl, Br, I
z = N, P

8. Important Neutral Species

When a molecule ionizes it follows the reaction $M + e^- \rightarrow M^+ + e^- + e$, as described earlier. However, if the molecular ion fragments, it can do so in two ways

$$M^+ \rightarrow A^+ + B$$

$$M^+ \rightarrow A + B^+$$

In the first case we would detect the A^+ ion and in the second case the B^+ ion. Since the probabilities of these two reactions are not equal, the observed abundances of A^+ and B^+ are not equal. Further, we should remember that in 10^{-6} moles there are more than 10^{17} molecules, so both reactions will take place to some extent.

If we look the the first reaction, $M^+ \rightarrow A^+ + B$, we can get some information on molecular formula because $M^+ = A^+ + B$. If we know their masses, we can get information on the groups present in the original molecule.

Rearranging, $M^+ - A^+ = B$.

B is a neutral species that is identified by differences in mass numbers. It is instructive to reexamine Table 15.4. In the *n*-butanol spectrum, the PM is 74, but the most abundant fragment is 31, corresponding to CH_2OH^+. The neutral fragment involved was 74 − 31 = 43, probably C_3H_7. This indicates that the most likely place to fragment is shown below.

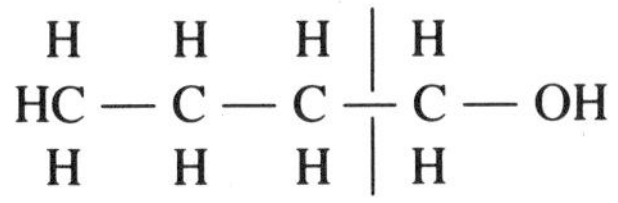

```
 H    H    H | H
HC — C — C —|— C — OH
 H    H    H | H
      Fragmentation
```

The peak at 57 shows a loss of 74 − 57 = neutral fragment 17 (OH), which appears to take a H with it, probably to form H_2O leaving mass 56. The mass at 59 indicates loss of mass 75 − 59 = 15 (i.e., CH_3).

With *sec*-butanol,

```
            H
           HCH
  H    H    |
HC — C — C — OH
  H    H    H
```

the most abundant peak is at 45, indicating loss of a neutral fragment, 74 − 45 = 29, probably C_2H_5. Favored fragmentation is

```
            H
           HCH
  H    H    |
HC — C — C — OH
  H    H    |
            H
```

Note that the 59 peak (loss of CH_3) is more abundant than with *n*-butanol. There are two CH_3s in the molecule.

With *tert*-butanol

```
        H
       HCH
    H   |
  H     |
HC — C — OH
  H     |
       HCH
        H
```

the most abundant peak is 59, showing a loss of 74 − 59 = 15. Looking at the structure, it is easy to understand, since any one of the CH_3 fragments may break off.

Important neutral fragments are 1 (H), 15 (CH_3), 17 (OH), 29 (C_2H_5), 16 (O) or 16 (NH_2), 35-37 Cl, 19 F, 79-81 Br, 129 I, 28 CO, 27 HCN, 30 NO, 44 CO_2, 17 (NH_3), 77 phenyl. Although these fragments can be identified as described, the relative + ion will probably also be present, but perhaps at a low abundance.

Based on this information, let us look at the spectrum in Table 15.21. The general appearance indicates an aromatic. PM is probably 182. $\frac{P+1}{P}$ shows 12 or 13 C. $\frac{P+2}{P}$ is inconclusive because we have many Cs, therefore there is a possibility of 2 ^{13}Cs in the molecule.

The most abundant peak is 105, showing the loss of a neutral fragment of 182 − 105 = 77, probably phenyl (C_6H_5). $\frac{P+1}{P}$ of this fragment suggests 7 car-

Table 15.21

m/e	Relative abundance	m/e	Relative abundance	m/e	Relative abundance	m/e	Relative abundance
27	1.3	74	2.0	105	100	153	1.8
28	1.0	75	1.7	106	7.8	154	1.4
38	0.37	75.5	0.21	107	0.51	155	0.16
39	1.1	76	4.3	119	0.02	164	0.06
50	6.2	76.5	0.33	126	0.63	165	0.44
51	19.	77	62.	127	0.40	166	0.06
52	1.4	78	4.2	128	0.13	181	8.2
53	0.29	79	0.16	139	0.17	182	60.
63	1.3	91	0.08	151	1.1	183	8.5
64	0.61	104	0.48	152	3.4	184	0.71
65	0.09						

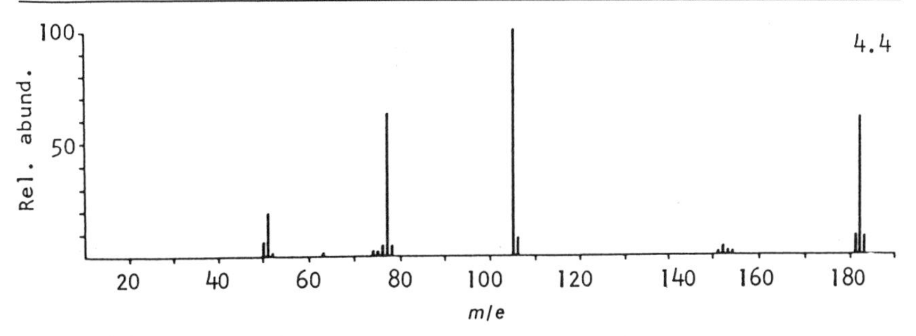

bons. A second strong peak is at 77, showing a further loss of 105 − 77 = 28 (CO). The suggested structure is, therefore,

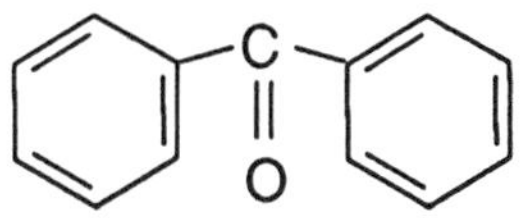

9. Conclusions

Interpreting mass spectra is like filling in crossword puzzles. Practice is necessary. Confirmation can rarely be achieved by MS alone. Corroborative evidence is often necessary from IR, which confirms the presence of functional groups, and NMR, which confirms functional groups and (by spin-spin splitting) the placement of these groups.

BIBLIOGRAPHY

Alexander, M. L., Hemberger, P. H., Cesper, M. E., and Nogar, N. S., Laser desorption in a quadrupole ion trap: mixture analysis using positive and negative ions, *Anal. Chem.*, (1993), *65*: 1600.

Anal. Chem., Application Reviews, June 1993.

Biemann, K., *Mass Spectrometry: Organic Chemical Applications*, McGraw-Hill, New York, 1962.

Buchanan, M. V., and Hettich, R. L., Fourier transform mass spectrometry of high-mass biomolecules. *Anal. Chem.*, (1993), *65*(5): 245.

Caprioli, R. M., Continuous flow FAB MS., *Anal. Chem.*, (1990), *62*(8): 177A.

Cody, R. N., Tamura, L., and Musselman, B. D., Electrospray ionization/magnetic sector MS calibration and accuracy, *Anal. Chem.*, (1992), *64*: 1561.

Cooks, R. G., Mchuckey, S. A., and Kaiser, R. E., Ion trap mass spectroscopy, *Canad. En.*, (1991), Mar. 25: 26.

Creswell, C. J., and Runquist, O., *Spectral Analysis of Organic Compounds*, Burgess, Minneapolis, 1979.

Duncan, W. P., *Research and Development*, (1991), Apr.: 57.

Fenn, J. B., Mann, M., Ming, C. K., Wong, S. E., and Whitehouse, C., *Science*, (1989), *64*: 246.

Goldner, H. J., Electrospray excites scientists with its amazing potential, *R&D Mag.*, (1993), *10*: 43.

Karas, F. H., Beavis, R. C., and Chait, B. T., Matrix-assisted laser desorption ionization mass spectrometry of biopolymers, *Anal. Chem.*, (1991), *63*(24): 1193A.

Kiser, K. W., *Introduction to Mass Spectrometry and Its Applications*, Prentice-Hall, Englewood Cliffs, N.J., 1965.

McLafferty, F. W., *Interpretation of Mass Spectra*, 2nd ed. Benjamin, New York, 1975.

Mahoney, J., Perel, J., and Taylor, S., Primary ion source for FAB MS. *Am. Lab.*, (1984), *March*: 92.

Silverstein, R. M., Bassler, G. C., and Morrill, T. C., *Spectrometric Identification of Organic Compounds*, 4th ed., Wiley, New York, 1981.

Smith, R. D., Wahl, J. H., Goodlett, D. R., and Hofstadler, S. A., *Anal. Chem.*, (1993), *65*(13): 574 A.

Spectra, Finnigan MAT, Vol. 9, Nos. 2 and 3, Fall 1983.

Voreos, L., Electrospray M.S., *Anal. Chem.*, (1994), *66*(8): 481A.

What's new in instrumentation, *Spectroscopy*, (1993), *8*(4).

SUGGESTED EXPERIMENT

15.1 Run the mass spectrometer with no sample injected. It should give no spectrum. Any spectrum found is caused by contamination from previous samples or by air leakage. Before further work can be performed, either the instrument should be cleaned or account should be taken of the contamination spectrum.

PROBLEMS

15.1 What are the functions of the ionizing and accelerating chambers of a mass spectrometer?

15.2 Complete the following table:

$r = 1$ m	$B = 0.1$ Tessla
$V = 10{,}000$ V	$m =$
$V = 5{,}000$ V	$m =$
$V = 1{,}000$ V	$m = 500$
$V = 500$ V	$m =$
$V = 100$ V	$m =$
$V =$	$m = 10$
$V =$	$m = 200$

15.3 What is the parent mass? What is the importance of identifying it to an organic chemist?

15.4 Briefly describe a time-of-flight mass spectrometer.

15.5 Name the principal analytical uses of mass spectrometry.

15.6 Identify the compound giving the following mass spectrum:

m/e	Relative abundance
1	<0.1
16	1.0
17	21
18	100
19	0.15
20	0.22

Identify the composition of the compound and the isotopic weights of each fragment.

15.7 Identify the compounds giving the following mass spectra:

(a)

m/e	Relative abundance
12	8.7
16	9.6
28	9.8
29	0.13
30	0.02
44	100
45	1.2
46	0.42

(b)

m/e	Relative abundance
35	12
36	100
37	4.1
38	33
39	0.01

15.8 Describe a Dempster MS.
15.9 What is the resolution of a mass spectrograph?
15.10 a) Describe electron impact ionization.
b) Describe electron spray ionization.
15.11 Describe FAB.
15.12 Describe chemical ionization.
15.13 Describe laser desorption.
15.14 Describe with a diagram a quadrupole MS.
15.15 Describe a time-of-flight MS.
15.16 Describe Fourier transform MS.
15.17 What is high-resolution MS? Why is it used?
15.18 How is MS used for elemental analyses?
15.19 How is MS interfaced with GC?

Identify the following compounds from their mass spectra:

15.20

m/e	Relative abundance
28	6.3
29	64.
30	3.8
31	100.
32	66.
33	0.98
34	0.14

15.21

m/e	Relative abundance
14	17.
15	100.
16	1.0
19	2.0
31	10.
32	9.3
33	89.
34	95.
35	1.1

15.22

m/e	Relative abundance
12	3.3
13	4.3
14	4.4
15	0.07
16	1.7
28	31.
29	100.
30	89.
31	1.3
32	0.21

15.23

m/e	Relative abundance
16	5.2
24	0.8
25	0.04
32	11.
34	0.42
48	49.
50	2.3
64	100.
65	0.88
66	4.9
67	0.04

15.24

m/e	Relative abundance
14	5.2
19	8.4
33	42.
52	100.
53	0.39
71	30.
72	0.11

15.25

m/e	Relative abundance	m/e	Relative abundance
12	2.7	49	4.2
13	3.0	50	1.5
14	0.63	51	0.31
15	0.05	59	2.6
24	4.0	60	24.
25	15.	61	100.
26	34.	62	9.9
27	1.2	63	32.
30	0.06	64	0.67
31	0.32	74	0.02
35	7.0	95	3.0
36	1.9	96	67.
37	2.3	97	3.3
38	0.68	98	43.
47	6.5	99	1.2
47.5	0.22	100	7.0
48	5.9	101	0.40
48.5	0.19		

15.26

m/e	Relative abundance	m/e	Relative abundance
15	6.3	51	1.0
16	0.20	52	0.27
26	1.4	53	1.2
27	15.	54	0.20
28	2.4	55	2.8
29	38.	56	4.3
38	1.4	57	100.
39	13.	58	4.4
40	1.4	59	0.08
41	41.	71	0.04
42	2.3	72	0.01
43	1.6		

15.27

m/e	Relative abundance	*m/e*	Relative abundance
15	2.8	42	4.2
16	0.03	43	19.
26	1.4	44	3.9
27	10.	45	100.
28	5.2	46	2.5
29	6.0	47	0.19
30	0.33	57	0.41
31	4.5	58	0.20
39	5.8	59	4.3
40	1.0	60	0.51
41	7.2	61	0.03

15.28

m/e	Relative abundance	*m/e*	Relative abundance
12	13.	38	6.2
14	2.1	50	25.
19	2.0	69	100.
24	2.7	70	1.08
26	11.	76	46.
27	0.11	77	1.0
31	22.	95	2.4
32	0.28	96	0.06

16

Electrochemistry

R. J. Gale and James W. Robinson

Department of Chemistry
Louisiana State University
Baton Rouge, Louisiana

The original analytical applications of electrochemistry, electrogravimetry, and polarography were for the determination of trace metals in aqueous solutions. The latter method was reliable and sensitive enough to detect concentrations as low as 1 ppm of many metals. Since that time, several different types of electrochemical techniques have evolved, each useful for particular applications.

A species that undergoes reduction or oxidation is known as an electroactive species. Electroactive species in general may be solvated or complexed, ions or molecules, in aqueous or nonaqueous solvents. Electrochemical methods are now used not only for trace metal ion analyses, but also for the analysis of organic compounds and for continuous analysis. Applications have been developed that are suited for quality control of product streams in industry. Under normal conditions, concentrations as low as 1 ppm can be determined without much difficulty. By using electrodeposition and then reversing the current, it is possible to extend the sensitivity limits for many electroactive species by three or four orders of magnitude, thus providing a means of analysis at the ppb level.

In practice, electrochemistry not only provides means of elemental and molecular analysis, but also can be used to acquire information about equilibria and reaction mechanisms from research using polarography, amperometry, conductometric analysis, and potentiometric titrations. The analytical calculation is usually based on the determination of current or voltage or on the resistance developed in a cell under conditions such that these are dependent on the concentration of the species under study. Electrochemical measurements are easy to

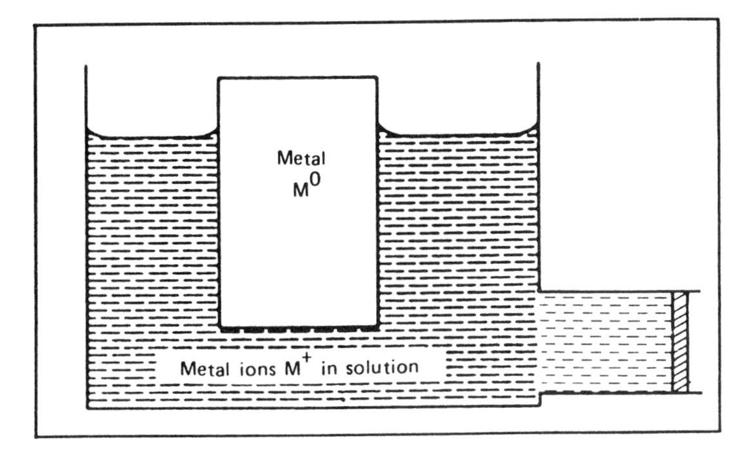

Figure 16.1 A half-cell.

automate because they are electrical signals. The equipment is often far less expensive than spectroscopy apparatus.

A. ELECTROCHEMICAL CELLS

At the heart of electrochemistry is the electrochemical cell. When a metal is immersed in a solution of its own ions, a potential difference (voltage) is created between the metal and the solution. The potential difference arises from the metal atoms going into solution as ions, liberating electrons in the process. This system constitutes a half-cell (Fig. 16.1). The metal strip in the solution is called an electrode and the ionic solution is called an electrolyte. The reaction between the metal strip and the ionic solution can be represented as

$$M^0 \rightarrow M^+ + e^- \tag{16.1}$$

where M^0 is an uncharged metal atom, M^+ is a positive ion, and e^- is an electron. This is an oxidation reaction, because the metal has been oxidized from an uncharged atom to a positively charged ion. In the reaction, the metal ions enter the solution (dissolve) and the electrode is called an anode. Expressing this differently, we say that at the anode, oxidation of the metal occurs according to the reaction shown in Eq. (16.1).

It is important to note that the reaction is often reversible. In practice, a state of equilibrium is reached by the system's components M^0, M^+, and e^-. Furthermore, the reaction is not restricted to systems where only one electron is involved. More properly, the reaction should be represented as

$$M^0 \rightarrow M^{n+} + ne^- \tag{16.2}$$

where n is a whole number. It has been found that with some metals the spontaneous reaction is in the opposite direction (to the left) and the metal ions tend to become metal atoms, taking up electrons in the process. This reaction can be represented as

$$M^0 \leftarrow M^+ + e^-$$

or

$$M^+ + e^- \rightarrow M^0 \quad (16.3)$$

This is a reduction reaction because the positively charged metal ions lose their charge and become neutral atoms. The neutral atoms deposit on the electrode. This is *electrodeposition*. In the course of this reaction, the electrode loses free electrons and is termed a *cathode*. At the cathode, reduction of an electroactive species always take place. As with oxidation reactions, the components are in equilibrium if there is negligible current; furthermore, more than one electron per atom may be involved. The reaction is better represented as

$$M^{n+} + ne^- \rightarrow M^0 \quad (16.4)$$

Either the oxidation or the reduction system can constitute a half-cell. It is not possible to measure directly the potential difference between the metal and the solution in either case. In other words, direct observation of a half-cell potential has never been achieved practically. We can join two half-cells to form a complete cell (Fig. 16.2) with a potential difference manifested by a current flowing across the connections. The potential developed is the difference between the potentials of the two half-cells. This is the basis of a potentiometric cell, or battery (Fig. 16.3).

In the typical cell shown in Fig. 16.3, the two half-cells follow the reactions

$$Cu^{2+} + 2e^- \rightarrow Cu^0 \quad \text{and} \quad Zn^0 \rightarrow Zn^{2+} + 2e^- \quad (16.5)$$

but no reaction will take place, and no current will flow, unless the electrical circuit is complete. In the electrolyte the current flow is ionic (ion motion), and in the external circuit the current flow is electronic (electron motion).

1. Measuring the Relative Voltage of Standard Half-Cells

Completion of the circuit in Fig. 16.2 is effected through the voltmeter and the glass frit. The glass frit is a porous membrane that prevents the solutions in the two half-cells from mixing rapidly but permits ionic conductance. Alternatively, a salt bridge can be used instead of the glass frit (Fig. 16.3). A salt bridge is usually a gel in which an inert electrolyte is dissolved. For example, a salt bridge can be made by dissolving potassium chloride (KCl) in hot agar gel. The

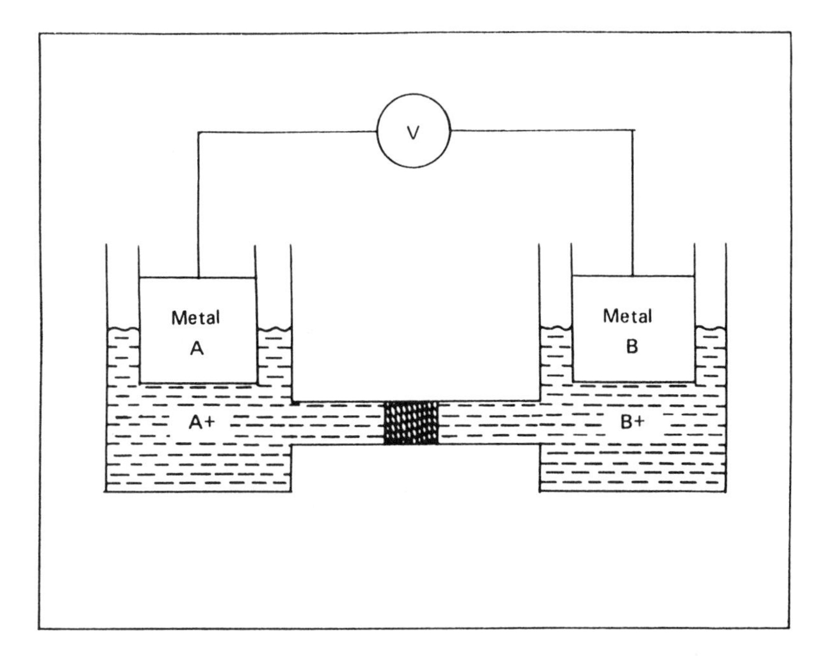

Figure 16.2 A complete cell.

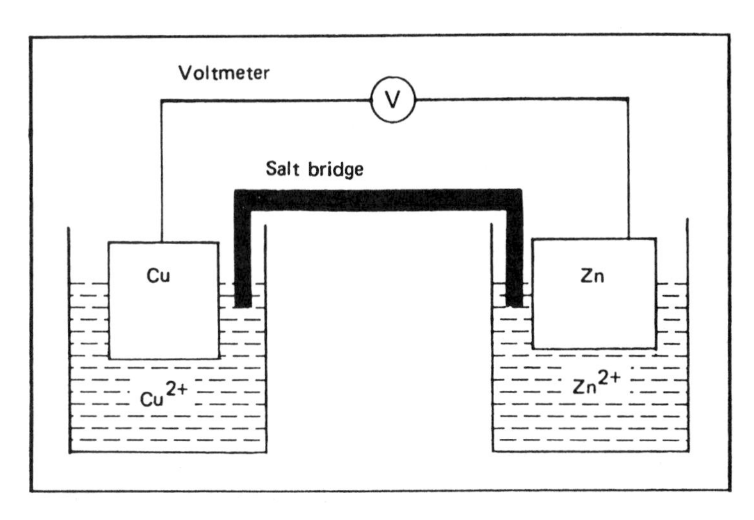

Figure 16.3 A copper/zinc cell.

solution is then transferred to a glass tube, and after it has cooled, the bridge is physically strong enough to provide a stable electrical (ionic) connection. If there is a difference in the concentration or types of ions of the two half-cells, a small potential is created at the junction of the membrane or salt bridge and the solution, but because the rates of diffusion of K^+ and Cl^- ions are similar, when a salt bridge is used the potential is very small.

An inert salt bridge must be used to electrically connect the two solutions of a cell. If we were to use a metal wire, the wire itself would constitute a pair of half-cells, one at each end, and this would complicate our calculations. Another role of the gelatinous salt bridge is to prevent the mixing of the two solutions in the two half-cells.

We explained earlier that it is not possible to measure the absolute voltage difference between the metal strip and a solution of its ions; however, we can compare the voltage difference between two half-cells. In practice, it has been arbitrarily decided and agreed upon that the voltage difference between hydrogen ion and hydrogen gas under standard conditions shall be defined as zero. In other words, the cell utilizing the reaction

$$1/2H_2 \rightarrow H^+ + e^- \qquad E_0 = 0.0 \tag{16.6}$$

has zero voltage. By connecting this standard half-cell with any other half-cell and measuring the voltage difference developed, we can determine the relative voltage developed by the second half-cell.

Consider, for example, a cell made up of the two half-cells

$$2H^+ + 2e^- \rightarrow H_2 \quad \text{and} \quad Zn^0 \rightarrow Zn^{2+} + 2e^-$$

The total voltage developed under standard conditions is +0.763 V. But the voltage of the hydrogen half-cell is zero by definition; therefore the voltage of the zinc half-cell is −0.763 V.

By substituting other half-cells, we can determine their voltages (actually, their relative voltages). For instance, for the reaction and voltage of each half-cell of a copper/zinc cell we have

$$Zn^0 \rightarrow Zn^{2+} + 2e^- \qquad E_0 = -0.763 \text{ V}$$

$$Cu^{2+} + 2e^- + Cu^0 \qquad E_0 = +0.337 \text{ V}$$

The total voltage developed is the difference between the voltages of the two half-cells written as reductions, that is,

$$+0.337 - (-0.763 \text{ V}) = +1.1000 \text{ V}$$

In tables of standard potentials, all of the half-cell reactions are expressed as reductions. The sign is changed if the reaction is reversed to become an oxidation. In a spontaneous reaction, when both half-cells are written as reductions, the half-cell with the more negative potential will be the one that oxidizes (note, however, that we can never have a real cell with two reduction half-cells). The negative sign in the difference equation reverses one of the reduction processes to an oxidation.

In the reaction of our example, zinc metal dissolves, forming zinc ions and liberating electrons. Meanwhile, an equal number of electrons are consumed by copper ions, which plate out as copper metal. The net reaction is summarized as

$$Zn^0 + Cu^{2+} \rightarrow Cu^0 + Zn^{2+} + 1.1000 \text{ V} \tag{16.7}$$

Zinc metal is oxidized to zinc ions, and copper ions are reduced to copper metal. The reaction comes to an end when either all the copper ions are exhausted from the system or all the zinc metal is dissolved, or an equilibrium situation is reached when both half-cell potentials are equal.

The zinc metal forms the anode and the copper metal the cathode. The copper cathode becomes depleted of electrons because these are taken up by the copper ions in solution. At the same time the zinc anode has an excess of electrons because the neutral zinc atoms are becoming ionic and liberating electrons in the process. The excess electrons from the anode flow to the cathode. The flow of electrons is the source of external current; the buildup of electrons at the anode and the depletion at the cathode constitute a potential difference that persists until the reaction ceases. If the process is used as a battery, such as a flashlight battery, the battery becomes dead when the reaction ceases.

When a voltage is spontaneously generated by a cell, the electrode that is *negatively charged* is the *anode* and the *positively charged* electrode is the *cathode*.

2. Types of Half-Cells

a. Metal/Metal Ion Cell

A metal/metal ion cell is a common type of half-cell consisting of a metal electrode in contact with its own ions. An example is the silver cell represented as $Ag \mid Ag^+ \parallel$. In the silver cell the silver metal is in the reduced state and the silver ion is the oxidized form. As described earlier, a potential difference develops between the two. The *vertical stroke* (or a comma) between Ag and Ag^+ indicates a difference in phases (e.g., solid | liquid) in the constituents of the half-cells that are in contact with each other. The *double vertical stroke* after Ag^+ indicates the *termination of the half-cell* and the point where it is connected to a second half-cell by a membrane junction.

b. Redox Cell

In this important type of half-cell, known as a redox couple or redox cell, typically two ions of the same element act as the oxidizing and the reducing form of the ion. In the Pt | Fe^{2+}, Fe^{3+} || cell, for example, a solution containing both Fe^{2+} and Fe^{3+} acts as the electrolyte, the Fe^{2+} being the reduced form and the Fe^{3+} being the oxidized form. An inert material, in this case a platinum wire, acts as the electrode and provides a vehicle for the electrons involved. The potential produced by the half-cell depends on the concentrations of Fe^{2+} and Fe^{3+} present in the solution. The potential is described by the Nernst equation [Eq. (16.9)], which is used for reduction–oxidation reactions (see Sec. B). Glassy carbon and platinum are frequently used as inert materials for electrodes in organic electrochemistry and for detector electrodes.

c. Gas Ion Cell

The hydrogen half-cell, which is written as Pt | H_2 | H^+ ||, is an example of a gas ion half-cell. The platinum wire is inert and does not undergo a net reaction with either the H_2 gas or the H^+ ion; it does, however, provide a vehicle, or catalyst, for the addition or subtraction of electrons in the system. The electrical potential is developed between the hydrogen ion H^+ and the hydrogen gas H_2.

d. Concentration Cell

In a concentration cell the reaction occurring at one electrode is exactly reversed at the other. Thus there is no resultant chemical change upon discharge of the cell, and the electrical energy is due to the transfer of material from one concentration to another.

For example, in a hydrogen half-cell, hydrogen gas is bubbled over an inert electrode (e.g., a platinum wire) immersed in an acid solution. When the conventionally defined standard conditions prevail—that is, when the pressure of the hydrogen is 1 atm, the temperature is 25°C, and the concentration of the acid solution is 1 molar (abbreviated 1 *M*)—the potential E_0 of this half-cell is zero. [Strictly speaking, the activity of the hydrogen ion should be 1, and the substitution of concentrations instead of activities is valid only at low concentrations (see below).] The reaction of the hydrogen half-cell is

$$\tfrac{1}{2}H_2 \rightarrow H^+ + e^- \qquad E_0 = 0.0000\ \text{V}$$

When a second half-cell, say, of metal A, is joined to a hydrogen half-cell, the total voltage developed is the sum of the voltages of the half-cells. The net potential *E* developed by the two half-cells is numerically equal to the potential developed by the metal A half-cell. That is,

$$\begin{aligned}\text{net potential } E \text{ developed by complete cell} &= E(\text{hydrogen} - E(\text{metal A})\\ &= 0 - E(\text{metal A})\\ &= -E(\text{metal A})\end{aligned}$$

Using this convention, by measuring the voltage developed by the metal and the hydrogen half-cells, we can measure the potential of the metal half-cell directly.

The term E(metal A) is the reduction potential. The sign of the voltage of the metal half-cell is opposite to that measured.

We said earlier that the potential, or electromotive force (emf), of a half-cell can be calculated using the Nernst equation [Eq. (16.9) when standard conditions do not hold. We can recall that by definition the potential of a standard hydrogen half-cell (i.e., one having a 1 M hydrogen ion solution) is zero. As we will see from Eq. (16.10), the emf of a hydrogen half-cell having a 0.1 M hydrogen ion concentration is +0.0591 V. If this half-cell were joined to a standard hydrogen half-cell, the potential developed by the resulting complete cell would be the difference in potential of the two half-cells:

$$E(\text{total}) = E([\text{H}^+] = 1\ M) - E([\text{H}^+] = 0.1\ M) = 0 - (-0.0591\ \text{V})$$

The complete cell is driven by the difference in concentration between the H^+ solutions. Such combinations are called *concentration cells*.

3. Sign Conventions

The actual sign of the half-cells has been defined in a number of different ways, and this variety has led to considerable confusion. The convention used here is in accord with the recommendations of the International Union of Pure and Applied Chemistry meeting in Stockholm in 1953. In this convention the standard half-cell reactions are written as reductions. Elements that are more powerful reducing agents than hydrogen show negative potentials, and elements that are less powerful reducing agents show positive potentials. For example, the $Zn \mid Zn^{2+}$ half-cell is negative and the $Cu \mid Cu^{2+}$ half-cell is positive.

4. Activity Series

The tendency for a molecular species to become oxidized or reduced in a half-cell determines the sign and potential of the half-cell. The tendency is strongly related to the chemical reactivity of the species concerned in aqueous systems. Based on the potential developed in a half-cell under controlled conditions, the elements may be arranged in an order known as the activity series (Table 16.1).

In general, the metals at the top of the activity series are most chemically reactive and tend to give up electrons easily, following the reaction of $M^0 \rightarrow M^{n+} + ne^-$. The metals at the bottom of the series are more noble and therefore less active. They do not give up electrons easily; in fact, their cations will accept electrons from metals above them in the activity series. In the process, the cations become neutral metal atoms and plate out of solution, while the more active metals become ionic and dissolve. This is illustrated as follows:

$$\begin{aligned} \text{metal A (active)} &\rightarrow \text{metal ion A}^+ + e^- \\ \text{metal ion B}^+ \text{ (noble)} + e^- &\rightarrow \text{metal B} \\ \text{net result: A} + \text{B}^+ &\rightarrow \text{A}^+ + \text{B} \end{aligned} \tag{16.8}$$

In short, the more active metals displace the less active metals from solution. As an example, if an iron strip is immersed in a solution of copper sulfate, some of the iron dissolves, forming iron ions, while the copper ions become metallic and copper metal plates out on the remaining iron strip.

This reaction is particularly important when B^+ refers to a hydrogen ion solution (acid). Based on this principle, all elements above hydrogen in the activity series are capable of displacing hydrogen from solution; that is, it is possible for them to be dissolved in an acid solution. The metal ionizes and enters the

Table 16.1 The Activity Series of Metals

Metal	E(half-cell) (V)	Chemical reactivity
Li	−3.05	These metals displace hydrogen from acids and dissolve in all acids, including water
K	−2.92	
Ba	−2.91	
Sr	−2.89	
Ca	−2.87	
Na	−2.71	
Mg	−2.37	These metals react with acids or steam
Al	−1.66	
Mn	−1.19	
Zn	−0.762	
Cr	−0.74	
As	−0.60	
Sb	−0.51	
Cd	−0.043	
Co	−0.28	These metals react slowly with all acids
Ni	−0.257	
Sn	−0.138	
Pb	−0.126	
Fe	−0.037	
H_2	0.000	
Bi	+0.32	These metals react with oxidizing acids; e.g., HNO_3
Cu	+0.342	
Ag	+0.7996	
Hg	+0.851	
Pt	+1.12	These metals react with aqua regia (3 HCl, 1 HNO_3)
Au	+1.50	

solution while the H^+(acid) forms H_2 and usually bubbles off. In contrast, noble metals, such as platinum and gold, cannot be dissolved by all acids, but only by oxidizing acids such as aqua regia, a mixture of HCl and HNO_3.

B. THE NERNST EQUATION

It has been found experimentally that certain variables affect the potential of a half-cell. These variables include the temperature and concentration of the solution and the number of electrons transferred. Mathematically, the relationship between the potential of a half-cell consisting of a metal in contact with its ions and the variables involved is represented by the Nernst equation:

$$E = E_0 + \frac{RT}{nF} \ln \left(\frac{\text{molar concentration of ions}}{\text{metal concentration} = 1 \text{ (by def.)}} \right) \tag{16.9}$$

where

E = potential (emf) of the half-cell
E_0 = emf of half-cell under standard conditions
R = constant (8.314 J/°C)
T = absolute temperature
n = number of electrons transferred during reaction (also equal to the *valence change* of the metal)
F = Faraday number (96,495 C)
ln = log to base e

If the values of R, T (25°C = 298 K), and F are inserted into the equation and the natural log is converted to log to the base 10, the equation reduces to

$$E = E_0 + \frac{8.314 \times 298 \times 2.303}{n \times 96{,}495} \log \text{(molar concentration of ions)}$$

$$= E_0 + \frac{0.0591}{n} \log\text{(molar concentration of ions)}$$

$$= E_0 + \frac{0.0591}{n} \log[\text{ions}]$$

It should be noted that the brackets following the "log" literally mean "the molar concentration of." For example, $[Fe]^{2+}$ means "the molar concentration of ferrous ion."

The equilibrium $M^0 \rightarrow M^{n+} + ne^-$ can be rewritten as

reduced form $\rightarrow$ oxidized form + ne^-

or

red $\rightarrow$ ox + ne^-

where n is the number of electrons transferred.

More accurately, the potential developed is proportional to the logarithm of the activity of the ion rather than to the logarithm of the molar concentration of ions. The activity is the activity coefficient γ times the molar concentration of the ion. The Nernst equation can therefore be written as

$$E = E_0 + \frac{RT}{nF} \ln(\gamma \times \text{molar concentration of ions})$$

The activity coefficient approaches unity when the solution is very dilute, and under these conditions Eq. (16.9) is accurate. At higher concentrations, however, the activity coefficient usually decreases. For accurate work, a correction should be made for this decrease.

In the Nernst equation, the metal is in the reduced form and the ions are in the oxidized form. If the equilibrium is written in the form of an oxidation–reduction reaction, then the Nernst equation is written in the general form

$$E = E_0 + \frac{RT}{nF} \ln \left(\frac{[\text{ox}]}{[\text{red}]}\right) \tag{16.10}$$

or

$$E = E_0 - \frac{RT}{nF} \ln \left(\frac{[\text{red}]}{[\text{ox}]}\right) \quad \text{(note the sign change) and inversion to} \left(\frac{[\text{red}]}{[\text{ox}]}\right)$$

where

[ox] = concentration of oxidized form of metal ions
[red] = concentration of reduced form of metal ions

By inserting the values for the constants, Eq. (16.10) reduces to

$$E = E_0 - \frac{0.0591}{n} \log \left(\frac{[\text{red}]}{[\text{ox}]}\right) \tag{16.11}$$

This brings us to the very important relationship between E, the emf of the half-cell, and the concentration of the oxidized and reduced forms of the components of the solution. Both pH measurements and ion-selective electrode techniques are based on this relationship.

1. Standard Potentials

In the Nernst equation, the expression E_0 is the emf of the half-cell under standard conditions. A half-cell is said to be under standard conditions when it is one of the following at a temperature of 25°C:

1. A pure liquid or pure solid (e.g., a metal electrode in the standard state)
2. A gas at a pressure at 1 atm (760 mmHg)
3. A solute at 1 M concentration (more accurately, at unit activity)

If we compare a 1 M solution and a 0.01 M solution, the more concentrated solution acts as though it is less than 100 times more concentrated than the dilute solution. It is then said that the activity of the 1 M solution is less than unity. The effective concentration of the solution is equal to molarity × activity coefficient. In practice, the activity is deduced by comparing it with very dilute solutions where the activity coefficient is considered to be unity.

A half-cell is also under standard conditions when it is either of the following:

4. A saturated solution of a slightly soluble solute (e.g., AgCl)
5. A saturated solution of a dissolved gas at 1 atm pressure and 25°C

When the reactants are at unit activity, by applying the simplified form of the Nernst equation [eq. (16.11)], we get

$$E = E_0 - \frac{0.0591}{n} \log\left(\frac{[\text{red}]}{[\text{ox}]}\right)$$

But [ox] = [red] and also both ox and red are at unit activity; hence

$$E = E_0 - \frac{0.0591}{n} \log 1$$

But since log 1 = 0,

$$E = E_0 - \frac{0.0591}{n} \times 0$$

or

$$E = E_0$$

This is the emf of the half-cell under standard conditions.

2. Reference Cells

In order to measure the emf of a half-cell, it is necessary to compare it with a second half-cell and measure the voltage produced by the complete cell. The second half-cell serves as a reference cell. Although the hydrogen cell serves as the standard reference in the activity series, in practice it is not always convenient to use a hydrogen cell as a reference cell; hence other, more stable cells have been developed. In principle, any metal ion system could be used under controlled conditions to provide a standard half-cell, but in practice, many metals are unsatisfactory materials. Active metals, such as sodium and potassium, are subject to chemical attack by the electrolyte. Other metals, such as iron, are difficult to obtain in the pure form. With some metals, the ionic forms are unstable to heat or to exposure to the air. Also, it is frequently difficult to control

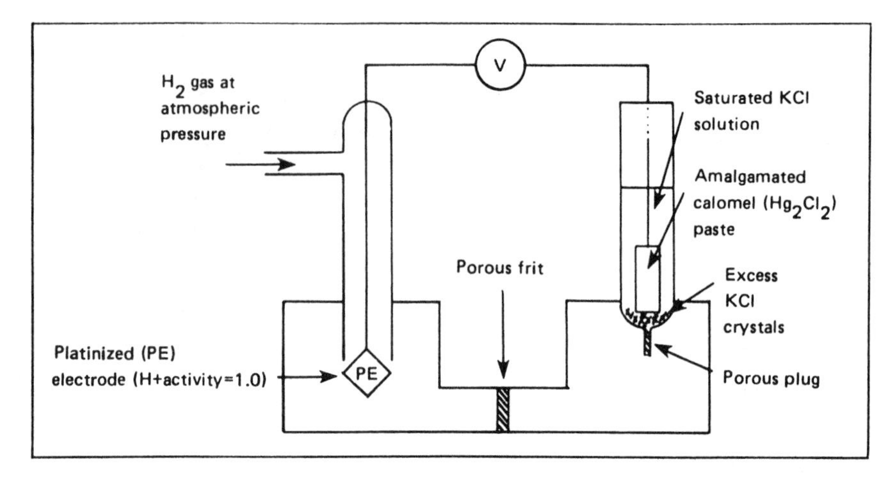

Figure 16.4 Saturated calomel half-cell and standard hydrogen electrode (25°C). (Note: A slight vapor pressure correction is necessary for the H_2 gas.)

the concentration of the electrolytes accurately. As a result, only a few systems provide satisfactory stable half-cells. Some of the more common reference cells are described below.

a. The Saturated Calomel Electrode

The saturated calomel electrode (SCE) is composed of metallic mercury in contact with a saturated solution of mercurous chloride, or *calomel* (Hg_2Cl_2). The mercurous ion concentration of the solution is controlled through the solubility product by placing the calomel in contact with a potassium chloride solution of known concentration. A typical calomel electrode is shown in Fig. 16.4. The half-cell reaction is

$$Hg_2Cl_2 + 2e^- \rightarrow 2Hg^0 + 2Cl^-$$

The oxidized form is Hg_2Cl_2, the reduced form is $2Hg^0 + 2Cl^-$. Applying Eq. (16.11), we obtain

$$E = E_0 - \frac{0.0591}{2} \log \left(\frac{[\text{red}]}{[\text{ox}]} \right)$$

$$= E_0 - \frac{0.0591}{2} \log \left(\frac{[Hg^0]^2 \, [Cl^-]^2}{[Hg_2Cl_2]} \right)$$

But Hg_2Cl_2 and Hg are in standard states at unit activity; therefore

$$E = E_0 - \frac{0.0591}{2} \log \left(\frac{1 \times [Cl^-]^2}{1} \right)$$

$$= E_0 - \frac{0.0591}{2} \log ([Cl^-]^2)$$

When the KCl solution is saturated and its temperature is 25°C, the concentration of chloride ion, $[Cl^-]$, is known and

$$E = +0.2412 \text{ V}$$

For a 1 *N* solution of KCl, the electrode is called a normal calomel electrode (NCE) and *E*(NCE) = +0.2801 V. The notation for the half-cell describes the electrode, the reduced form of the metal (these two may or may not be the same), and the oxidized form of the metal. The concentration of the electrolyte is also noted. For example, the calomel half-cell is symbolized as

$$\text{Hg} \mid \text{Hg}_2\text{Cl}_2, \text{ satd. KCl} \parallel$$

The double vertical stroke indicates the point at which the half-cell is joined to a second half-cell. A slight contamination of KCl always occurs because of leakage of the saturated KCl solution into the second half-cell through the fine porous plug or cracked glass bead junction. This may be prevented, if desired, with a second (remote) junction. A complete cell, such as a calomel and a hydrogen cell, is symbolized as

$$\ominus \text{ Pt, H}_2 \mid \text{H}^+ \parallel \text{satd. KCl, Hg}_2\text{Cl}_2 \mid \text{Hg} \oplus$$

It should be noted that in the second half-cell (the hydrogen cell), the electrode Pt is not the same as the reduced "metal" H_2. Hence both are shown abbreviated in the symbolized description. A practical saturated calomel electrode should always have some crystals of KCl visible in the bottom of the cell.

b. The Silver/Silver Chloride Cell

Another common reference cell found in the chemical-processing industry and useful in organic electrochemistry is the silver/silver chloride half-cell. The cell consists of silver metal coated with a layer of silver chloride. Again, the same principle applies as in the SCE, but in this case the silver ion activity depends on the solubility product of AgCl, which is in contact with a solution of known chloride ion activity. Reference electrode manufacturers will supply 3.5 *M* KCl (saturated with AgCl) for filling this reference half-cell. It provides a reproducible and reliable electrode free from mercury salts.

C. ELECTROANALYTICAL METHODS

Diverse analytical techniques, some highly sensitive, have been developed based on potentiometry and measurements of current (or charge) arising from various controlled voltage perturbations. Those widely encountered are defined briefly here and explained in detail in the following sections.

The Nernst equation indicates the relationship between the concentration of a solution and the emf produced by a half-cell involving that solution. The electrochemical technique called *potentiometry* utilizes the potential developed by the half-cell as a measure of the concentration of the components of the solution. Only a negligible current should be allowed to flow throughout the determination of the potential to avoid significant changes in the concentration of the component being measured. This, in turn, would cause the potential to change.

In electrodeposition, metallic elements are plated out onto an electrode and weighed (electrogravimetry). The weight of the metal deposited is a measure of the concentration of the metal originally in solution. This is a form of coulometry, which is based on Faraday's laws of electrolysis. It is known that 1 faraday (96,495 C) of electricity is required to reduce 1 gram-equivalent weight of an electrolyte. By measuring the quantity of electricity required to reduce (or oxidize) a given sample exhaustively, the quantity of electrolyte reduced can be determined, provided the reaction is 100% efficient (or of known efficiency). Mass, or charge i (A) $\times$ t (sec), can be used as a measure of the extent of the electrochemical reaction.

In *polarography* a controlled potential is applied to one electrode known as the working electrode. The term *polarography* is used when the working electrode is a dropping mercury electrode or a static (hanging) mercury drop. The auxiliary electrode (or counterelectrode) is normally a Pt wire or foil. A third (reference) electrode is used as a basis for control of the potential at the working electrode. The current of analytical interest flows between the working and auxiliary electrodes, and the reference acts only as a high-impedance probe. As the voltage is progressively increased or decreased with time (sweep voltammetry), resultant changes in the anodic or cathodic currents occur whenever an electroactive species is oxidized or reduced, respectively.

A powerful family of techniques is available using voltammetry. Polarography is a special case of *linear sweep voltammetry* because the electrode area increases with time as each drop grows and falls every 4 sec or so; that is, the voltage is changing as the electrode area increases. Polarography is especially useful for analyzing and studying metal ion and metal complex reductions and solution equilibria. Another technique for studying electrochemical reaction rates and mechanisms involves reversing the potential sweep direction to reveal the electroactive products formed in the forward sweep. In this way it is possible to see if the products undergo reaction with other species present or with the solvent. This technique is called *cyclic voltammetry* and is usually carried out at a solid electrode. In a special case of voltammetry called *anodic stripping voltammetry* the electrolyzed product is preconcentrated at an electrode by deposition for a reasonable period at a fixed potential. The product is then stripped off with a rapid reverse potential sweep in the positive direction. Peak currents on the reverse sweep are used to determine the analyte concentration from a standard additions calibration. In principle, this method is applicable to both anodic and cathodic stripping.

In *conductometry* an alternating (AC) voltage is applied across two electrodes immersed in the same solution. The applied voltage causes a current to flow. The magnitude of the current depends on the conductivity of the solution, and the composition of the sample is deduced from the measurement of the conductivity. This method makes it possible to detect changes of composition in a sample during chemical reactions (as, e.g., during a titration). Conductivity cells are finding application as chromatography detectors.

1. Potentiometry

Potentiometry is concerned with the measurement of the potential, or voltage, of an electrochemical cell. Accurate determination of the potential developed by a cell requires a negligible current flow during measurement. A flow of current would mean that a faradaic reaction is taking place, which would change the potential from that existing when no current is flowing.

Measurement of the potential of a cell can be useful in itself, but it is particularly valuable if it can be used to measure the potential of a half-cell. This can be accomplished by connecting the half-cell being measured to a reference half-cell to form a complete cell. When the total voltage of the two half-cells is measured, the difference between this value and the voltage of the reference cell is the voltage of the other half-cell. This can be expressed as

$$E(\text{total}) = E(\text{sample}) - E(\text{reference}) \tag{16.12}$$

By convention, in abbreviated descriptions of cells, the half-cell on the right is measured with respect to the half-cell on the left. When a standard hydrogen electrode is on the left, the potentials will be consistent with the standard reduction potentials, for example, $H_2, Pt \mid H^+ \parallel Fe^{3+} \mid Pt$. The total cell potential is often expressed in the form

$$E(\text{total}) = E_{\text{right}}(\text{written as reduction}) - E_{\text{left}}(\text{written as reduction})$$

Because one must always have one oxidation and one reduction occurring, the minus sign takes care of this fact.

The potential of the total cell is measured experimentally. The potentials of common standard cells can be found in reference tables. When these two pieces of information are available, the potential of the sample cell can be calculated. Most commercially available instruments are standardized to a particular reference cell. Common standard half-cells include the hydrogen cell and the saturated calomel cell. The hydrogen cell is not easily handled, but the saturated calomel half-cell is robust and reliable and is commonly used for such applications.

Having determined the potential of the sample half-cell versus a reference half-cell, it is possible to calculate the concentration of the ions in the solution. The relationship between ion concentration and the potential of the half-cell is given by the Nernst equation [Eq. (16.10)], which at 25°C can be stated for a reduction as

$$E = E_0 + \frac{RT}{nF} \ln \left(\frac{[M^{m+}]}{[M]} \right)$$

but since the [M] is the pure metal electrode, it is under standard conditions, and the equation can be rewritten as

$$E = E_0 + \frac{0.0591}{n} \log[M^{m+}] \tag{16.13}$$

where $[M^{m+}]$ is the concentration of metal ion in solution. This equation states that the relationship between the emf (E) produced by a half-cell and the concentration of the metal ions in the solution is logarithmic. For most metal systems we know the values of E_0 and n. Hence, by measuring E, we can calculate the concentrations of metal in the solution.

a. Equipment: The Measuring Potential

The potential of a complete cell can be measured with a solid-state circuit that requires negligible input current but can output the voltage to a meter, digital readout, or recorder. Figure 16.5 illustrates an operational amplifier (op-amp)

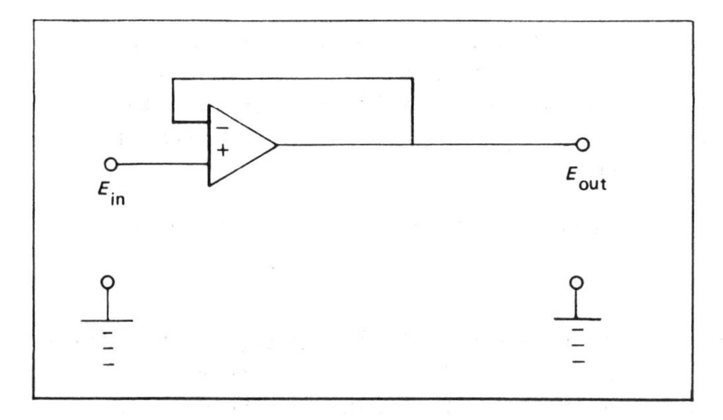

Figure 16.5 Schematic of an operational amplifier with voltage feedback (the voltage follower).

set up as a voltage follower. The triangle represents the amplifier with two inputs on the left and an output on the right. It may contain a large number of discrete components, for example, transistors. Usually such devices require ±15 V power lines, which are omitted in circuit diagrams. The voltage input marked + is called a *noninverting input*, and the one marked − is the *inverting input*. The principle of operation is quite simple. If A is the gain (typically 10^5), in equation form we can write

$$\begin{aligned} E_{\text{out}} &= -A(E^- - E^+) \\ &= -A(E_{\text{out}} - E_{\text{in}}) \end{aligned} \tag{16.14}$$

where E^- and E^+ are the voltage values at the inverting and noninverting inputs, respectively. The minus sign means that the polarity of the voltage difference is reversed. Rearranging, we have

$$E_{\text{out}} = \frac{E_{\text{in}}}{1 + 1/A} \tag{16.15}$$

and because A is very large,

$$E_{\text{out}} \approx E_{\text{in}} \tag{16.16}$$

Notice that E_{out} is fed back into the noninverting input (E^-). This is termed voltage feedback.

The function of this type of circuit is to offer a high-input impedance which can be typically 10^5–10^{13} ohms (Ω). This means that the + and − inputs require negligible current and can be used to monitor the voltage of a highly resistive source such as a glass electrode. However, the output current and voltage capabilities are considerable, depending on the device properties and the power supply limitations. It seems odd, at first, that the circuit just reproduces the voltage of the cell but does so in such a way as to make available much larger currents than are available from the electrochemical cell.

A second important use of operational amplifiers and feedback circuits is to control the potential at a working electrode. The circuit which performs this task is called a *potentiostat*. Potentiostats and modern polarography equipment are called three-electrode devices because they attach to a working electrode, a reference electrode, and a counterelectrode (Fig. 16.6). The principle of operation is that the potential between the working electrode (WE) and the reference electrode (REF) is maintained by a feedback circuit. The applied voltage E_{appl} may be a constant voltage or some signal generator voltage, such as a linear ramp. To understand how this circuit works, consider starting with no potential at the amplifier output. The output voltage E_{out} will be given, as before, by

$$E_{\text{out}} = -A(E^- - E^+) \tag{16.17}$$

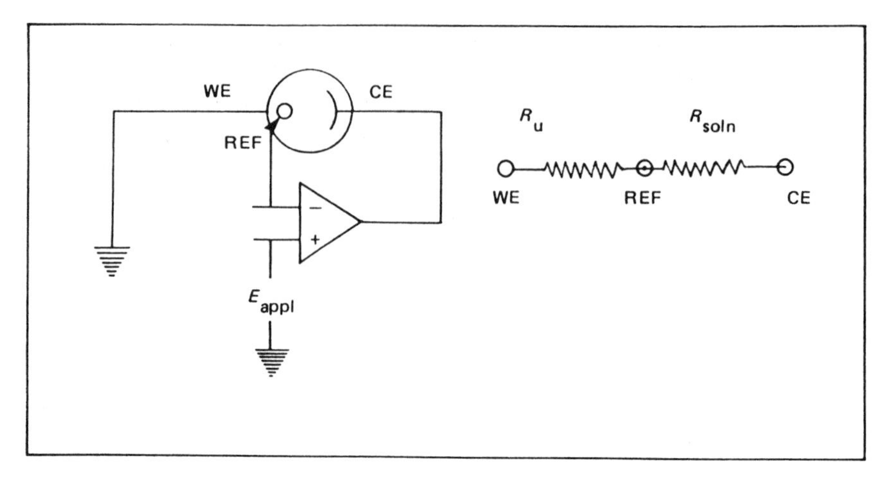

Figure 16.6 Schematic of a simple potentiostat and representation of the cell as an electrical (equivalent) circuit.

Because at time $t = 0$, $E^- = E_{appl}$, the voltage E_{out} will rise, with the same sign, until $E_{out} = E_{appl}$ at the point REF. Assuming that the uncompensated resistance R_u between WE and REF can be essentially ignored, this ensures that WE is held at E_{appl} and the potential of the counterelectrode can "float" to maintain zero differential input at the − and + inputs. Note that negligible current flows to the inverting (−) input through the reference electrode. The measured cell current flows between the working electrode (WE) and the counterelectrode (CE). In practice, the uncompensated resistance R_u is kept small by placing the reference electrode close to WE, by using an excess of supporting electrolyte, and by keeping the net cell current small by using microelectrodes, for example, a dropping mercury electrode. Then the voltage drop in the electrolyte between WE and REF, iR_u, need not be compensated for.

On modern potentiostats, the applied voltage is given as the *actual* sign of the voltage on the WE, and there is no need to assume a potential reversal. In other words, if you set the potentiostat at −0.500 V, WE is at −0.500 V versus whatever reference electrode you have in the actual cell. If, for example, this electrode was a saturated calomel electrode, then on the standard hydrogen electrode (SHE) scale at 25°C,

$$\begin{aligned} E_{WE} &= -0.500 + 0.241 \\ &= -0.259 \text{ V vs. SHE} \end{aligned}$$

If the cell had a silver/silver chloride reference (3.5 *M*) and the equipment were set at −0.500 V vs. REF, then

$$E_{WE} = -0.500 + 0.205$$
$$= -0.295 \text{ V vs. SHE}$$

In each of these cases, when we state our values on the hydrogen scale, the potential at the working electrode is numerically less. On the other hand, for an applied potential of +0.500 V vs. SCE on the potentiostat,

$$E_{WE} = +0.500 + 0.241 = 0.742 \text{ V vs. SHE}$$

This raises an important point. It does not matter which reference electrode we use; it is best to use the one most suited for the chemical system being studied. We should always, however, quote our potentials versus the particular reference and be sure to label the axes of current–voltage curves with the appropriate reference electrode. Electrode potentials are relative values.

b. Analytical Applications

Potentiometry is used in the study of metal ion concentrations, changes in ion concentration and pH, and even in the analysis of gases and organic compounds. Some of the analytical techniques in which potentiometry is applied are discussed in the following paragraphs. (To derive full value from the following discussions, you should be conversant with the fundamentals of volumetric analysis.*)

Direct Measurement of a Metal Concentration. The concentration of a metal ion in solution may be measured directly via the potential developed by the half-cell involved and by applying the Nernst relationship [Eq. (16.18)]. For example, a silver/silver ion half-cell plus a calomel electrode gives the following relationship:

$$E(\text{complete cell}) = E(\text{Ag}) - E(\text{reference})$$

where E(Ag) is the emf of the silver half-cell and E(standard) is the emf of the calomel cell. But

$$E(\text{Ag}) = E_0(\text{Ag}) + \frac{RT}{nF} \ln[\text{Ag}^+]$$

Therefore

$$E(\text{complete cell}) = -E(\text{reference}) + E_0[\text{Ag}] + \frac{RT}{nF} \ln\,[\text{Ag}^+]$$

and

$$\log[\text{Ag}^+] = \frac{E(\text{complete cell}) - E_0(\text{Ag}) + E(\text{reference})}{2.303\ RT/nF} \qquad (16.18)$$

*See, for example, G. H. Ayres, *Quantitative Chemical Analysis*, 2nd ed., Harper and Row, New York, 1968.

In an experiment using this combination of half-cells, it was observed that E(complete cell) = 0.400 V. Under the conditions used, E(reference) = 0.246 V (for a calomel electrode). Also, it is known that $E_0(Ag)$ = 0.799 V (potential of silver half-cell under standard conditions). Therefore, substituting in Eq. (16.18), one obtains the following results:

$$\log[Ag^+] = \frac{0.400 - 0.799 + 0.246}{0.0591}$$

$$= -\frac{0.153}{0.0591}$$

$$\log[Ag^+] = -2.59$$

Therefore

$$\text{concentration of } Ag^+ = \text{antilog}(-2.59)$$
$$= 2.57 \times 10^{-3}\ M$$

Titration of Ag^+ with Cl^-. The concentration of the Ag^+ ion in solution can be used to determine the equivalence point in the titration of Ag^+ with Cl^-. In this titration the following reaction takes place:

$$Ag^+ + Cl^- \rightarrow AgCl \downarrow \text{ (precipitation)}$$

The concentration of the Ag^+ in solution steadily decreases as Cl^- is added.

At the equivalence point $[Ag^+] = [Cl^-]$. But, for the sparingly soluble salt, $[Ag^+]\,[Cl^-]$ is a constant called the *solubility product* K_{sp}. In the case of the AgCl precipitation reaction $K_{sp} = 1 \times 10^{-10}$ at 25°C. Therefore,

$$[Ag^+]\,[Cl^-] = 1 \times 10^{-10}$$

But, as we have seen, $[Ag^+] = [Cl^-]$ at the equivalence point; therefore

$$[Ag^+]^2 = 1 \times 10^{-10}$$
$$[Ag^+] = \sqrt{1 \times 10^{-10}} = 10^{-5}\ M$$

From this calculation, at the equivalence point $[Ag]^+ = 10^{-5}M$. But by applying the Nernst equation to this solution, we get

$$E(Ag) = E_0 + \frac{0.0591}{1}\log[Ag^+]$$
$$E_0(Ag) = +0.799$$

Therefore

$$E(Ag) = +0.799 + \frac{0.0591}{1} \times (-5)$$

$$= +0.799 - 0.295$$

$$= +0.504$$

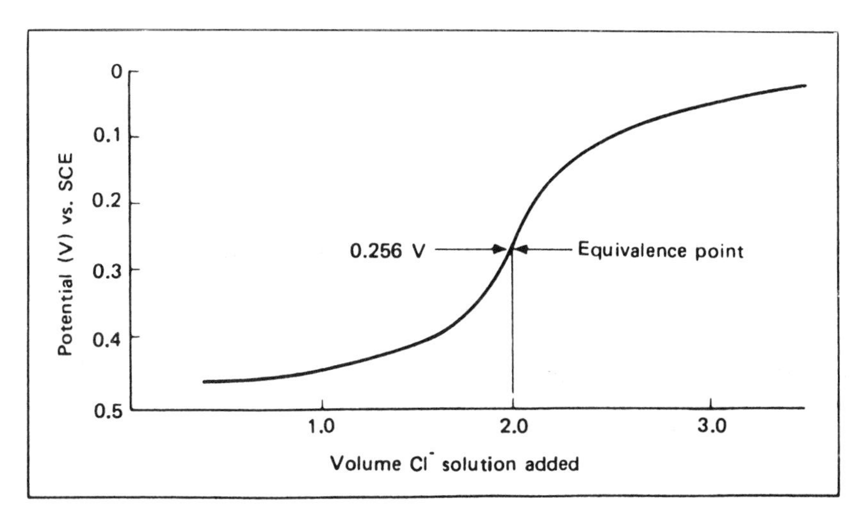

Figure 16.7 Relationship between Ag^+ cell potential and the volume of Cl^- solution added.

At the equivalence point the emf of the half-cell is calculated to be 0.504 V vs. SHE. If we make a complete cell by using the silver half-cell and a SCE half-cell (+0.241 V), the observed voltage at the equivalence point will be given by the relationship

$$\begin{aligned} E(\text{observed}) &= E(\text{Ag}) - E(\text{reference}) \\ &= +0.504 - 0.241 \\ &= +0.263 \text{ V vs. SCE} \end{aligned}$$

This information tells us that at the equivalence point the emf of the complete cell is +0.263 V vs. SCE.

In the titration of silver ion, the chloride solution should be added in small portions and the emf of the cell measured after each addition. By plotting the relationship between the volume of Cl^- solution added and the voltage of the cell, we can determine the volume of Cl^- solution necessary to reach the equivalence point. A typical curve relating the volume of chloride to the voltage for titration with 0.1 N Cl^- solution is shown in Fig. 16.7.

In the example below, the equivalence point is reached after adding 2.00 ml of Cl^- solution to 5.00 ml of silver solution. The normality of the silver can be determined by applying the equations

$$\text{volume} \times \text{normality of chloride solution} = \text{volume} \times \text{normality of } Ag^+ \text{ solution}$$

$$\text{volume of } Cl^- \text{ solution} = 2.00 \text{ ml}$$

but

$$\text{normality of Cl}^- \text{ solution} = 0.1\ N$$

and

$$\text{volume of silver solution} = 5.00 \text{ ml}$$

Therefore $2.00 \times 0.1 = 5.00 \times$ the normality of the silver solution, that is,

$$\text{normality of silver solution} = \frac{2.00 \times 0.1}{5.00} = 0.040\ N$$

From this, the weight of Ag^+ present in the solution can be calculated as follows:

$$1 \text{ liter of } 1\ N\ Ag^+ \text{ contains } 108 \text{ g } Ag^+$$

Therefore

$$5 \text{ ml of } 1\ N\ Ag^+ \text{ contains } \frac{108 \times 5.00}{1000} \text{ g}$$

Thus

$$5 \text{ ml of } 0.04\ N\ Ag^+ \text{ contains } \frac{108 \times 5.00 \times 0.040}{1000} \text{ g} = 0.0216 \text{ g}$$

Differential Curves. In Fig. 16.7 the titration curve was simple and the equivalence point easily detected; however, in dilute solutions and titrations involving slower reactions the titration curve may be flatter and difficult to interpret. The problem can be simplified by using equipment that records the first and/or second differential of the titration curve. The curves obtained for a relationship such as that in Fig. 16.7 are shown in Fig. 16.8.

When the first differential of the curve is taken, the slope is measured and plotted against the volume added. As the potential changes, the slope changes. At the endpoint, the slope of the curve ceases to increase and therefore begins to decrease. The slope goes through a maximum and starts to decrease. This can be seen in the curve of Fig. 16.8b.

The second differential measures the slope of the first differential curve and plots this against the volume added. At the endpoint the curve in Fig. 16.8b goes through a maximum. At this point the slope of the curve is horizontal; that is, it has zero slope. The second differential curve is shown in Fig. 16.8c.

As can be seen in Fig. 16.8c, there is a rapid change in the second differential curve, whose values go from positive to negative, and at the equivalence point the value equals zero. This feature has been used to construct automatic titrating machines, which cut off the supply of titrant immediately after the equivalence point has been reached. The volume of the titrant can be read off accurately and quickly.

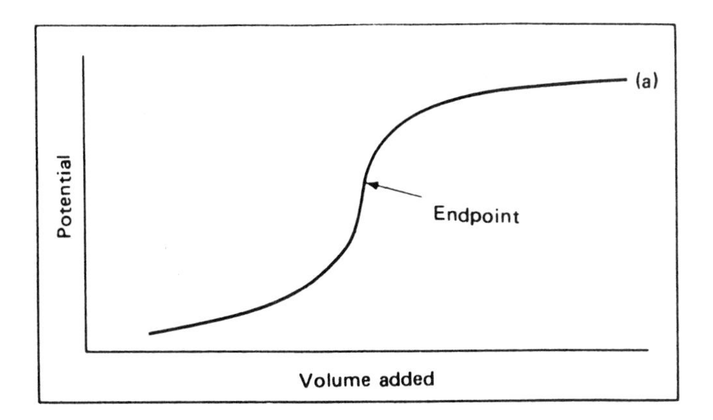

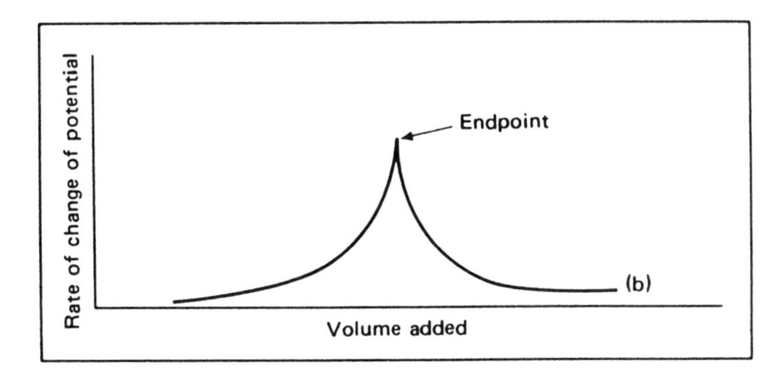

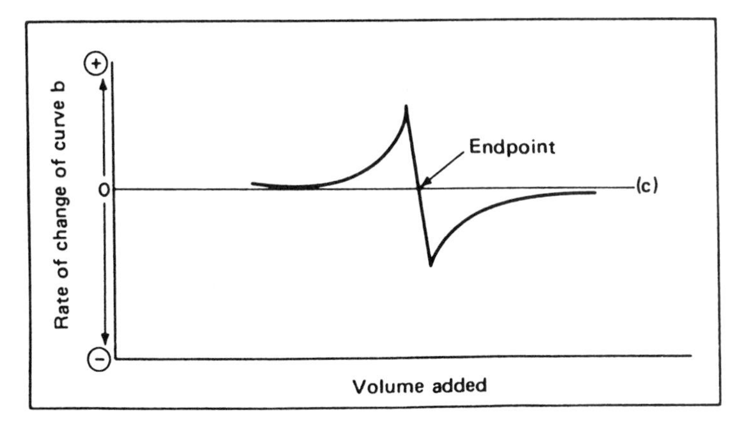

Figure 16.8 Relationship between (a) a simple potentiometric titration curve and its (b) first and (c) second differentials.

Determination of K_{sp}. It was shown in the previous example that the equivalence point of the titration of silver with chloride can be determined by a knowledge of K_{sp} for AgCl. From the value of K_{sp} we were able to calculate the emf of the half-cell at the equivalence point. By the same token, if the emf of a half-cell at the equivalence point is known, then K_{sp} can be calculated.

For example, precipitated AgCl is added to distilled water and shaken until a saturated solution is formed. Special care is taken to purify the AgCl and the water. During the process a little of the precipitated AgCl dissolves in the water to form Ag^+ and Cl^-. This solution is used as the electrolyte in a Ag–Ag^+ half-cell and the emf is measured. (By definition, for a sparingly soluble salt, $K_{sp} = [Ag^+]\,[Cl^-]$.)

In an actual experiment it was found that E(Ag), the emf of a half-cell made with a saturated solution of AgCl, was 0.504 V vs. SHE. Applying the Nernst equation gives us

$$E(\mathrm{Ag}) = E_0 + \frac{0.0591}{1}\log\,[\mathrm{Ag}^+] - 0.0$$

$$0.504 = 0.799 + \frac{0.0591}{1}\log[\mathrm{Ag}^+]$$

from which it can be calculated that $\log\,[Ag^+] = -5$; therefore

$$[\mathrm{Ag}^+] = 10^{-5}$$

But

$$[\mathrm{Ag}^+] = [\mathrm{Cl}^-] \quad \text{(both are dissolved in equal concentrations)}$$

and

$$K_{sp} = [\mathrm{Ag}^+][\mathrm{Cl}^-]$$

Therefore

$$\begin{aligned} K_{sp} &= 10^{-5} \times 10^{-5} \\ &= 1 \times 10^{-10} \end{aligned}$$

Procedures such as this can be used to determine the solubility product K_{sp} of other sparingly soluble salts.

pH Measurements. One of the most important applications of potentiometry is for the determination of $[H]^+$ or the pH of the solution, where pH is defined as the negative log of the hydrogen ion concentration. The procedure can be used for the determination of the pH of an ordinary solution, which is important if the solution is drinking water, the water in a swimming pool, or a chemical in an industrial process. Frequently, industrial reactions and other chemical reactions carried out by the research chemist are very sensitive to pH, which therefore must be controlled.

The determination of $[H^+]$ in a solution is a special case of the determination of metal ion concentration. The importance of this determination has merited special equipment. Hydrogen ion concentration can be determined directly by employing the Nernst equation. The most common hydrogen cell is the glass electrode (see Sec. C.1.c), which, for convenience, is generally preferred to electrodes involving H_2 gas. A saturated calomel electrode is commonly used as the reference electrode. The cell may be formulated as

$$\text{SCE} \;||\; \underset{\text{(unknown)}}{[H_3O^+]} \;|\; \text{glass membrane} \;|\; \underset{\text{(internal reference)}}{[H_3O^+] \;||\; 0.1\,M \;|\; \text{AgCl, Ag}}$$

The potential developed by the cell is given by (if we ignore the liquid junction potential)

$$E(\text{cell}) = E(H^+) - E(\text{reference})$$

Also,

$$E(H^+) = E_0(H) + \frac{RT}{nF}\ln[H^+]$$

$$E(\text{cell}) = -E(\text{reference}) + E_0(H) + \frac{RT}{nF}\ln[H^+]$$

$$= -E(\text{reference}) + E_0(H) + \frac{0.0591}{1}\log[H^+] \quad \text{at } 25°\text{C}$$

But $E_0(H)$, the emf of a standard hydrogen cell, is zero by definition. Hence

$$E(\text{cell}) = -E(\text{reference}) + \frac{0.0591}{1}\log[H^+]$$

The emf of a standard calomel cell = +0.241 V vs. SHE; therefore

$$E(\text{cell}) = -0.241 + 0.0591\log[H^+]$$

By rearranging we get

$$-\log[H^+] = \frac{[E(\text{cell}) + E(\text{reference})]}{0.0591}$$

$$= \frac{[E(\text{cell}) + 0.241]}{0.0591}$$

If in a particular experiment the observed voltage of the cell is −0.6549 V vs. SCE, then

$$-\log[H^+] = -\frac{(-0.6549 + 0.2412)}{0.0591} = 7 \quad \text{or} \quad [H^+] = 10^{-7}\,M$$

In order to avoid using very small numbers involving 10 raised to a negative power, it is more convenient to express the hydrogen ion concentration as the negative log of the hydrogen ion concentration using base 10. This is called, by definition, pH; that is, the *pH* of a solution is the *negative log of the hydrogen ion concentration* in that solution. For example, if $[H^+] = 10^{-1}$ *M*, then pH = 7.0. In the foregoing example, therefore,

$$-\log[H^+] = 7, \text{ or pH} = 7.$$

In commercial equipment, the potential of the cell is measured but the dial readings give pH directly. The instrument is calibrated at one point by using a buffer of known $[H^+]$, and the units of pH on the scale correspond to various voltage readings at 25°C.

pH Titrations. Direct measurement of pH can be used for acid–base titration. For example, if NaOH is added to HCl, the following reaction takes place:

Original solution: $HCl \rightarrow H^+ + Cl^-$
Addition of NaOH: $H^+ + Cl^- + Na^+ + OH^- \rightarrow H_2O + Na^+ + Cl^-$

In the process H^+ is removed and the potential of the hydrogen half-cell changes according to the equation

$$E(\mathrm{H}) = E_0(\mathrm{H}) - \frac{0.0591}{1} \mathrm{pH}$$

that is, the voltage changes by −0.0591 V for each unit change in pH. Near the neutralization point the pH changes very rapidly with the addition of NaOH. A plot of the potential of the hydrogen half-cell versus the volume (in milliliters) of NaOH added gives the relationship in Fig. 16.9. The equivalence point of the reaction is the inflection point of the curve.

Use of this type of pH titration allows analyses to be carried out in colored solutions. Also, it solves the problem of selecting the correct indicator for a particular acid and base titration. The endpoint can be determined more accurately by using a first or second differential curve as described earlier.

Titrations of Weak Acid and Base Mixtures. Most weak acids or bases are too weak to be titrated in aqueous solutions. It is common practice, therefore, to choose a nonaqueous solution for these determinations. The nonaqueous solutions are chosen based on their acidity, dielectric constant, and the solubility of the sample in the solvent. It is important that the acidity of the solvent not be too great; otherwise the titrant is used up in titrating the solvent rather than the sample. The same principles apply to the titration of bases, in that the solvent must not be too basic; otherwise reaction between the sample base and the titrant is not observed. The dielectric constant affects the relative strength of dissolved acids. For example, negatively charged acids become relatively stronger compared to uncharged acids (such as formic acid) as the dielectric constant of the solvent is decreased. It is also important that the sample be soluble in the solvent, otherwise the reactions will be incomplete and unobservable.

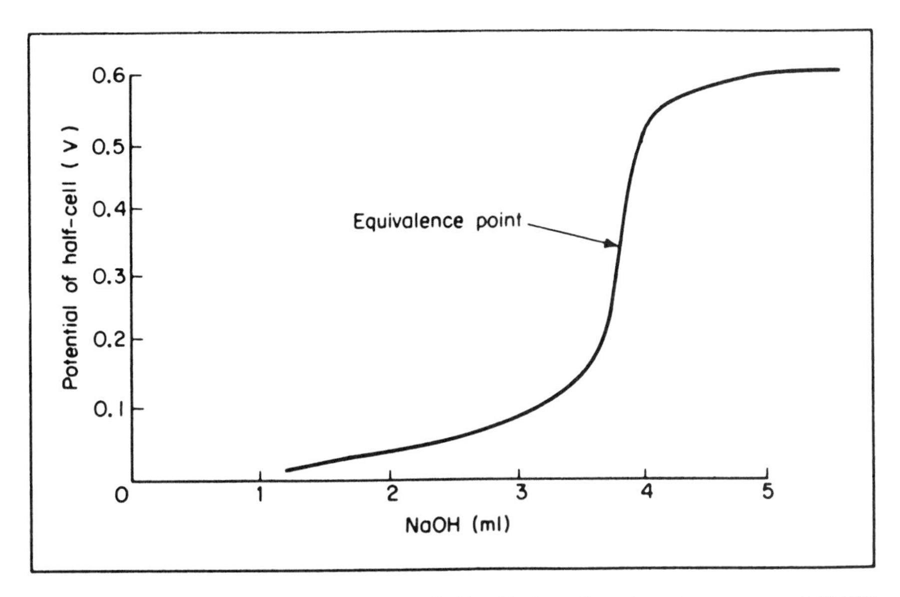

Figure 16.9 Potential of the hydrogen half-cell plotted against the amount of NaOH added.

The types of cells used vary with the solvent employed. Glass electrodes may be used if the solvent is an alcohol, ketone, or acetonitrile. Platinum electrodes may be used for titrations with tetrabutylammonium hydroxide, $(C_4H_9)_4NOH$, dissolved in a benzene–methanol mixture.

When the base is added to the mixture, it first reacts with the strongest acid until reaction is virtually complete. It then commences reaction with the second strongest acid until the second reaction is complete. Of course, there will be some overlap between the two, but if the emf is plotted against the volume of titrant added, two equivalence points can be observed. These two acids may be either a dibasic acid or a mixture of acids. This system holds up with weak acid–strong base titrations. A typical illustration is shown in Fig. 16.10. This process has found very valuable applications in the titration of products, such as amino acids.

c. Ion-Selective Electrodes

The analytical value of potentiometry has been greatly enhanced by the development of ion-selective electrodes. As their name implies, these electrodes are very sensitive to the concentration (or, more accurately, the activity) of a particular ion in solution and less sensitive to the other ions present. Because the electrode is most sensitive to one particular ion, a different electrode is needed

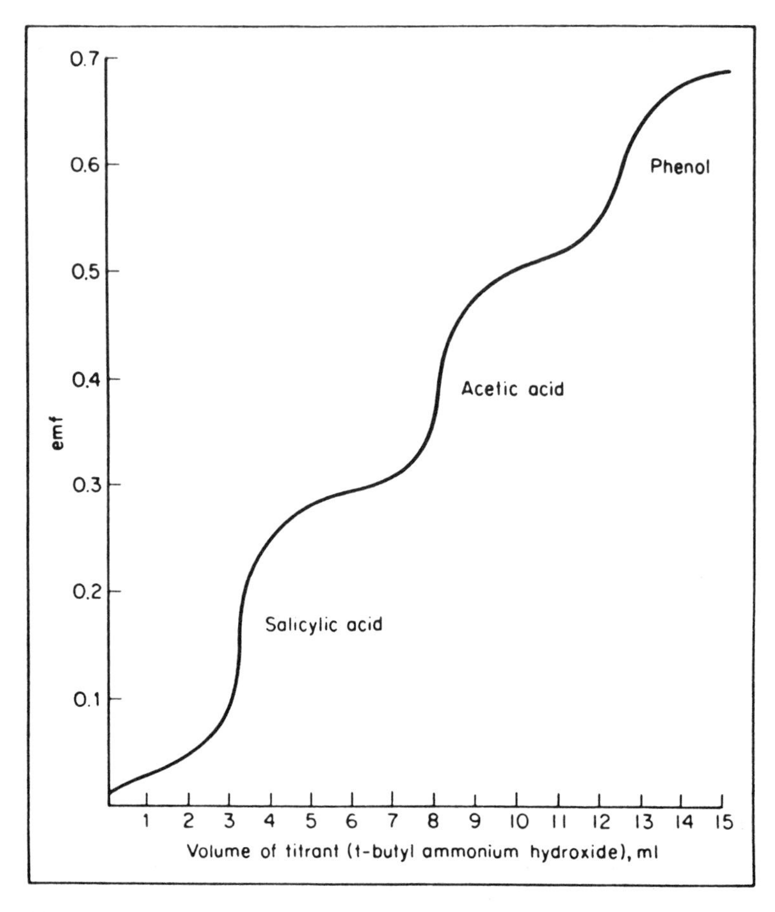

Figure 16.10 Weak acid–strong base titration. Note the equivalence points after the addition of 3.5, 8.5, and 13.0 ml of titrant.

for each ion to be studied. Furthermore, ion-selective electrodes are often slightly sensitive to other ions; consequently, corrections must be made for such interferences if very accurate results are needed.

The first ion-selective electrode to be used was the glass electrode. This electrode is sensitive to the hydrogen ion concentration and almost insensitive to the concentration of other ions; it is therefore a true ion-selective electrode. Based on this design, other electrodes have been developed. In general, these electrodes fall into three categories: (1) glass electrodes, (2) anion-selective electrodes, and (3) liquid–liquid membrane electrodes.

Glass Electrodes. When a thin glass membrane is placed between two solutions of different acid strength, a potential difference develops across the glass membrane, which can be used to measure the pH of solutions. The functioning of the glass electrode seems to depend on the permeability of hydrogen ions through the glass.

In practice, the pH on one side of the glass is fixed by using 0.1 *M* HCl solution saturated with AgCl. The internal electrode is a silver unit coated with AgCl and provides a reference electrode sealed inside the container (Fig. 16.11a). A second reference half-cell, usually a saturated calomel electrode, may be employed to complete the electrochemical cell. Sometimes, however, the glass electrode has this reference incorporated into its structure and provides a small porous plug near the glass bulb to complete the circuit; these electrodes are termed *combination electrodes* (Fig. 16.11b, c).

Glass electrodes are calibrated with standard buffer solutions. Buffer solutions are those in which the pH remains relatively constant (1) over wide ranges of concentration and (2) after the addition of small quantities of acid or base. For example, a solution of ammonium acetate is a good buffer for pH 7. If a little acetic acid is added, the acetate ion generated displaces the equilibrium between the ammonium ion and the acetate ion. This in turn adjusts the equilibrium between the $[OH^-]$ and $[H^+]$ present in the solution, and they tend to return to their original concentrations. The net result is that $[H^+]$ changes very little upon addition of an acid. In similar fashion, $[OH^-]$ changes very little upon addition of a base. The solution is said to be buffered. Standard buffer solutions to calibrate pH meters are commercially available or may be formulated from mixtures described in chemical handbooks.

The glass electrode is subject to asymmetry potentials. These potentials, which arise from strain in the glass, can cause errors in the reading of the cell. The electrical resistance of the glass electrode is enormous (50–10,000 MΩ). An ordinary potentiometer would indicate practically no observable voltage, which is why a high-impedance solid-state measuring circuit is required and the connection wire is shielded to minimize electrical pickup. Modern instruments drain no more than 10^{-10}–10^{-12}A from the cell being measured. The pH meter is calibrated (to compensate for asymmetry errors) by immersing the glass electrode in a buffer solution of known pH (e.g., pH 7). It is best to calibrate the instrument to a pH near to that at which the instrument will be used. If necessary, calibrations can be performed at other acid levels, such as pH 4 or 10, by using other buffer solutions to check the linearity over any range of study.

The exact way the pH-sensitive glass electrode functions is still not clearly understood, but certain facts have been ascertained. For example, when the glass is in contact with the internal solution of the cell and the sample solution,

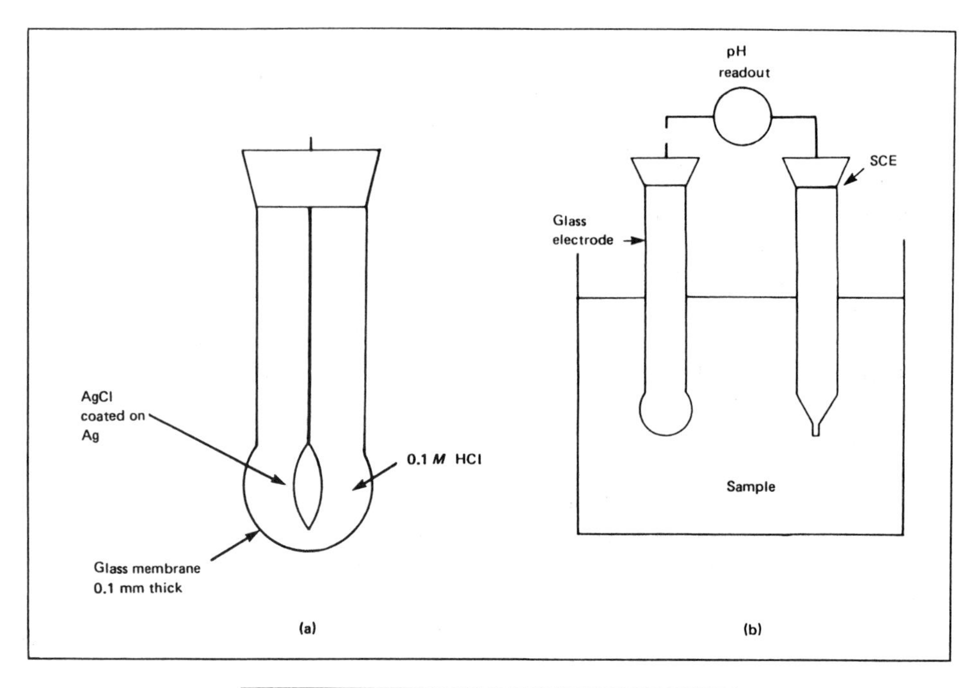

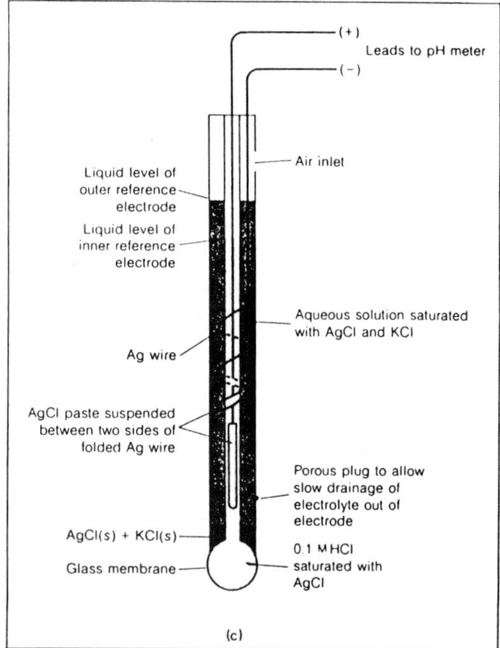

Figure 16.11 (a) Glass electrode, (b) glass electrode shown with external standard calomel electrode to complete the measuring circuit, and (c) glass electrode with internal standard Ag/AgCl pH electrode.

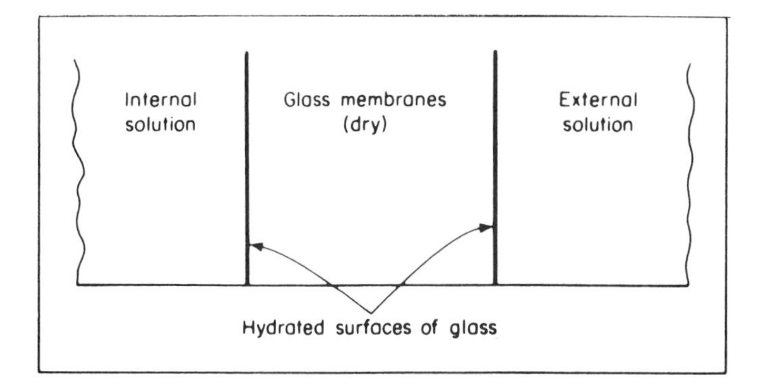

Figure 16.12 Glass membrane between two solutions.

several layers of hydrated silica are formed. Each exposed surface of the glass becomes hydrated, forming a thin gel layer. Between the surfaces the glass is dry, as illustrated in Fig. 16.12. At each surface the sodium ions and hydrogen ions are in equilibrium according to the equation

$$Na^+(\text{solution}) + H^+(\text{glass}) \rightleftharpoons Na^+(\text{glass}) + H^+(\text{solution})$$

which represents an ion exchange mechanism.

The electrical charge is relayed by the charge-bearing ion from the surface of the glass exposed to the external solution to the surface exposed to the internal surface. No individual ion penetrates the glass; rather, a charged ion moves a short distance and transfers its energy to the next ion. This process is continued through many charge-bearing ions, and ultimately the charge reaches the other surface of the glass.

Glass Electrodes Selective to Cations Other Than Hydrogen. It has been found that the current is carried entirely by the cation with the lowest charge that is available in the glass used. By changing the chemical composition of the glass, the ion to which the electrode is sensitive is changed. For example, glasses containing less than 1% Al_2O_3 are sensitive to hydrogen ions but almost insensitive to other ions present. On the other hand, a glass whose composition is 11% Na_2O, 18% Al_2O_3, 71% SiO_2 is highly selective toward sodium, even in the presence of other alkali metals. For example, at pH 11 this electrode is approximately 3000 times more sensitive to Na^+ than to K^+. The ratio of the sensitivity of the electrode to a solution of sodium to its sensitivity to a solution of potassium (of equal strength) is called the *selectivity factor* (in this case it is 3×10^{-3}). A sodium electrode has different selectivity factors for each element. In practice, it is sometimes necessary to correct for this low sensitivity to other ions. The electrode is also sensitive to H^+concentration, and this sensitivity must be controlled in order to obtain reproducible results.

At present the relationship between glass composition and ion selectivity is empirical. Studies are being made, however, to provide a better understanding of this process. Meanwhile, glass electrodes have been made that are selectively sensitive to H^+, Li^+, Na^+, K^+, and Ag^+. It will be noted that these electrodes are all selective toward cations.

Anion-Selective Electrodes. The ion-selective glass electrodes described earlier contain anionic sites that attract certain cations. Such cations ''fit'' the anionic site with respect to electrical charge and physical size. In a similar fashion, materials can be found that contain cationic sites suitable for anions with the correct charge and size. As with the glass electrode, these electrodes must also exchange ions (in this case, anions).

Sparingly soluble inorganic salts, such as silver iodide, have been found satisfactory. Unfortunately, these salts are not physically strong. They must therefore be used in conjunction with a support. Silicone rubber has been found effective for this purpose. The precipitate is suspended in the rubber in such a way that there is physical contact between the particles of the precipitate distributed throughout the rubber.

Electrodes have been developed in which selectivities are claimed for iodide, bromide, chloride, perchlorate, sulfate, and phosphate ions. As we said before, a separate electrode must be used for each ion. It has been shown that these electrodes are virtually insensitive to other ions present. For example, the response of the iodide-selective electrode to iodide ions is about 200 times as great as its response to bromide ions and at least a million times greater than its response to sulfate ions. Furthermore, the iodide-selective electrodes are insensitive to any cations present. In practice, solutions of KI, ZnI_2, BaI_2, and CeI_3 give responses that are dependent on the iodide concentration but independent of the predominant metal ion.

Solid-state electrodes have also been developed, similar in principle to the use of precipitate electrodes. These electrodes use single crystals of inorganic materials doped with rare earths. Such electrodes are particularly useful for fluoride ion analysis and sulfide analysis. Similar solid-state electrodes have also been developed for chloride, bromide, and iodide ions.

Liquid–Liquid Membrane Electrodes. In the past extensive work has been performed on the study of liquid ion exchangers. The use of such ion exchangers in ion-specific electrodes has been made difficult, however, by the mechanical problems involved in the cell. The liquid ion exchanger must be in contact with the liquid sample, but mixing of the liquids must be kept minimal. This problem was solved by using a semipermeable membrane. This membrane allowed the liquids to come in contact with each other but not mix. A schematic diagram of a liquid–liquid membrane cell is shown in Fig. 16.13. Based on this

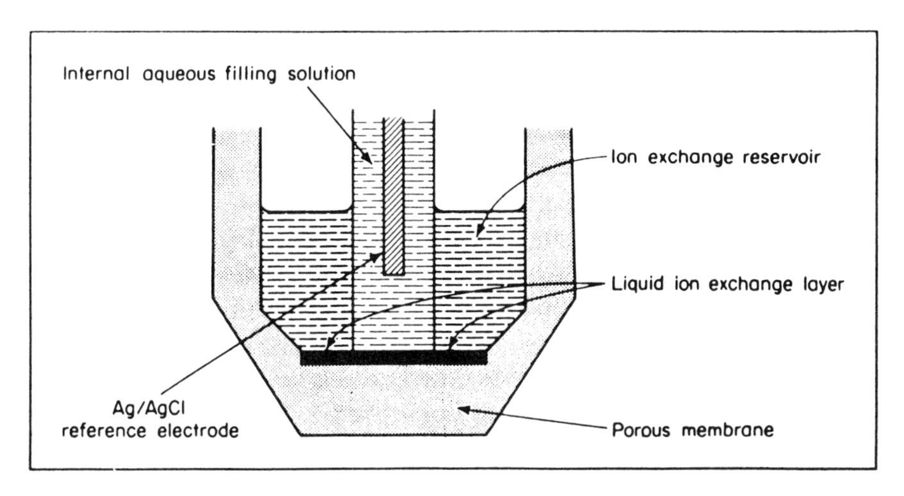

Figure 16.13 Liquid–liquid ion-selective electrode.

principle, cells have been developed that are selective to calcium and calcium plus magnesium concentrations; these cells are used for measuring water hardness. Similar cells have been used for copper determination.

Selectivity factors for calcium-sensitive electrodes are approximately as indicated in Table 16.2. Similar anionic liquid cells have been developed that are sensitive to chloride, perchlorate, and nitrate.

Quantitative Analytical Applications. The relationship between the cell potential and the concentration of the sample ion in solution is given by the Nernst equation as

$$E = E_0 + \frac{RT}{nF} \ln[M^+]$$

Table 16.2 Selectivity Factors for Calcium-Selective Electrodes

Cation	K
Sr^{2+}	1.5×10^{-2}
Mg^{2+}	0.5×10^{-2}
Ba^{2+}	1.7×10^{-3}
Na^+	3×10^{-4}
K^+	3×10^{-4}

$$\text{Selectivity factor } K = \frac{\text{sensitivity to metal ion activity}}{\text{sensitivity to calcium ion activity}}$$

where $[M^+]$ is the activity of the metal ion in solution. In the case of the sodium-selective electrode, the potential is

$$E = E_0 + \frac{RT}{nF} \ln[Na^+] \text{ where } n = 1$$

It must be remembered, however, that the cell is *selective* but *not specific* toward sodium. For example, if potassium were present in the sample, a small increase in potential would be observed. This increase can be deduced from the selectivity factor. In the case of potassium this factor is 3×10^{-3}, and the potential developed due to potassium would be given by

$$E = E_0 + \frac{RT}{F} \ln[3 \times 10^{-3}][K^+]$$

where $[K^+]$ is the activity of the potassium ions in the sample and $n = 1$. The total potential of the cell is therefore

$$E = E_0 + \frac{RT}{F} \ln([Na^+] + (3 \times 10^{-3})[K^+])$$

It must also be remembered that the cell potential is proportional to the activity of the metal salt rather than its concentration. The activity is a measure of the extent of thermodynamic nonideality in the solution. The activity coefficient is usually less than unity, so the activity of a solution is generally lower than the total concentration, but the values of activity and concentration approach each other with increasing dilution. If the metal salt is not completely ionized, the activity is further decreased. Decreased ionization can be brought about because of a weak ionization constant, chemical complexation, or a high salt concentration in the solution. Any of these factors will cause a change in the potential of the cell, even if the metal concentration is constant. In practice, it is better to determine the relationship between the cell potential and metal ion *concentrations* experimentally. A calibration curve relating potential and concentration of metal in mixtures similar to the sample must be prepared. This curve eliminates many unsuspected sources of error and is therefore much more reliable than a theoretical curve. Either a calibration curve or the standard additions method may be used to determine experimental unknowns.

In the construction of standard addition curves for calibration it must be remembered that the potential is proportional to the log of the concentration. Let the original metal ion concentration be C_0 and the solution volume be V_0 (in liters). The initial potential

$$E_1 = \text{const.} + 2.303 \frac{RT}{nF} \log C_0 \qquad (16.19)$$

It is usual to add a small volume V_s of a standard solution of concentration C_s from a microliter pipette such that the concentration becomes $(C_0V_0 + C_sV_s)/(V_o + V_s)$ and the new potential E_2 is given by

$$E_2 = \text{const.} + 2.303 \frac{RT}{nF} \log\left(\frac{C_0V_0 + C_sV_s}{V_0 + V_s}\right) \tag{16.20}$$

Subtracting Eq. (16.19) from Eq. (16.20), we have

$$\Delta E = 2.303 \frac{RT}{nF} \log\left(\frac{C_0V_0 + C_sV_s}{C_0(V_0 + V_s)}\right)$$

If we note that $V_0 \gg V_s$,

$$\Delta E \approx 2.303 \frac{RT}{nF} \log\left(\frac{C_0V_0 + C_sV_s}{C_0V_0}\right) \approx 2.303 \frac{RT}{nF} \log\left(1 + \frac{C_sV_s}{C_0V_0}\right)$$

This expression, in turn, can be rearranged as

$$1 + \frac{C_sV_s}{C_0V_0} = 10^{nF\,\Delta E/2.303RT} \tag{16.21}$$

From which we can obtain the concentration of the unknown by plotting $10^{nF\Delta E/2.303RT} - 1$ versus V_s for several additions of the standard. Then the slope m will have the value C_s/C_0V_0 and the unknown concentration is calculated from

$$C_0 = C_s/mV_0 \tag{16.22}$$

The method is usually very sensitive but depends upon the particular ion-selective electrode and type. Also, special buffers may be required to avoid problems with potentially interfering ions.

Qualitative Applications. Ion-selective electrodes can be used for the analysis of aqueous and nonaqueous solutions alike and have therefore found increasing applications in organic chemistry, biochemistry, and medicine. Furthermore, in many circumstances no sample preparation is necessary. As a result, the use of ion-selective electrodes is particularly valuable for obtaining rapid results with no loss of sample. These electrodes can even be used for continuous analysis of waters and industrial plant streams simply by inserting the electrodes into the sample (such as the river or plant stream) and measuring the voltages generated. The voltages may be calibrated in terms of sodium concentration, if a sodium ion-specific electrode is used, and the sodium concentration read off the dial directly. The concentration of other ions can be determined in the same fashion by using the pertinent ion-selective electrodes.

Some of the specific analyses include the determination of sodium and potassium in bile, nerve and muscle tissue, kidneys, blood plasma, urine, and other body fluids. This method has also been used for the analysis of marine muds, sea water, river water, and industrial waters (such as boiled water), as well as for the determination of Na^+, K^+, Mg^{2+}, Ca^{2+}, Ag^+, Cr^{2+}, Rb^+, NH_4^+, I^-, F^-, Cl^-, Cd^{2+}, Pb^{2+}, and CNS^-in organic samples. In aqueous samples, the ions Br^-, Cd^{2+}, Ca^{2+}, Cl^-, CN^-, Cu^{2+}, F^-, I^-, Pb^{2+}, NO_3^-, ClO_4^-, Na^+, CNS^-, S^{2-}, and Ag^+ have been determined at very low levels of concentration with ion-selective electrodes.

d. Oxidation–Reduction Reaction

The value of the emf of a half-cell can be used to follow reduction–oxidation (redox) titrations. For example, the oxidation of stannous ions by ceric ions follows the chemical reaction

$$Sn^{2+} + 2Ce^{4+} \rightarrow Sn^{4+} + 2Ce^{3+}$$

It is usually best to first identify the two half-cell reactions and to find the total number of electrons n in the reaction (n is the same for each half-cell reaction):

$$Sn^{2+} - 2e^- \rightarrow Sn^{4+}$$

$$2Ce^{4+} + 2e^- \rightarrow 2Ce^{3+}$$

The emf $E(A)$ of the half-cell is given by the Nernst equation for a stannous/stannic half-cell:

$$E(A) = E_0(Sn^{2+}/Sn^{4+}) + \frac{0.0591}{2}\log\left(\frac{[ox]}{[red]}\right) \tag{16.23}$$

For the ceric/cerous half-cell

$$E(B) = E_0(Ce^{3+}/Ce^{4+}) + \frac{0.0591}{2}\log\left(\frac{[Ce^{4+}]^2}{[Ce^{2+}]^2}\right) \tag{16.24}$$

At all times the emf produced by each cell must be equal to that of the other because the mixture cannot exist at two potentials. Therefore at chemical equilibrium

$$E(A) = E(B)$$

Thus, from Eqs. (16.23) and (16.24), we have

$$E_0(Sn^{2+}/Sn^{4+}) - E_0(Ce^{3+}/Ce^{4+}) = \frac{0.0591}{2}\left[\log\left(\frac{[Sn^{4+}]}{[Sn^{2+}]}\right) - 2\log\left(\frac{[Ce^{4+}]}{[Ce^{3+}]}\right)\right]$$

$$= \frac{0.0591}{2}\left[\log\left(\frac{[Sn^{4+}]\,[Ce^{3+}]^2}{[Sn^{2+}]\,[Ce^{4+}]^2}\right)\right] \tag{16.25}$$

However, it can be shown that the equilibrium constant for reaction is given by

$$K_{eq} = \frac{[Sn^{4+}]\,[Ce^{3+}]^2}{[Sn^{2+}]\,[Ce^{4+}]^2} \tag{16.26}$$

Therefore

$$E_0(\mathrm{Sn}^{2+}/\mathrm{Sn}^{4+}) - E_0(\mathrm{Ce}^{3+}/\mathrm{Ce}^{4+}) = \frac{0.0591}{2} \log K_{eq} \tag{16.27}$$

A similar relationship is true for all equilibrium redox reactions. With knowledge of $E_0(Sn^{2+}/Sn^{4+})$ and $E_0(Ce^{3+}/Ce^{4+})$ it is possible to calculate the equilibrium constant for the reaction. This principle can be used in redox reactions and is valuable for determining equilibrium constants.

It is possible to calculate the *equivalence point potential* by noting that we cannot ignore the small concentrations of Sn^{2+} and Ce^{4+} remaining, even though the overall reaction may be close to completion. The stoichiometry at the equivalence point demands that

$$2[\mathrm{Sn}^{4+}] = [\mathrm{Ce}^{3+}]$$

and

$$2[\mathrm{Sn}^{2+}] = [\mathrm{Ce}^{4+}]$$

If we let the potential at equivalence be E_{eq},

$$E_{eq} = E_0(\mathrm{Ce}^{3+}/\mathrm{Ce}^{4+}) + \frac{0.0591}{2} \log\left(\frac{[\mathrm{Ce}^{4+}]^2}{[\mathrm{Ce}^{3+}]^2}\right)$$

$$2E_{eq} = 2E_0(\mathrm{Sn}^{3+}/\mathrm{Sn}^{4+}) + \frac{0.0591}{2} \log\left(\frac{[\mathrm{Sn}^{4+}]^2}{[\mathrm{Sn}^{2+}]^2}\right)$$

Adding these two expressions gives

$$3E_{eq} = E_0(\mathrm{Ce}^{3+}/\mathrm{Ce}^{4+}) + 2E_0(\mathrm{Sn}^{2+}/\mathrm{Sn}^{4+}) + \frac{0.0591}{2} \log\left(\frac{[\mathrm{Ce}^{4+}]^2\,[\mathrm{Sn}^{4+}]^2}{[\mathrm{Ce}^{3+}]^2\,[\mathrm{Sn}^{2+}]^2}\right)$$

If we now substitute in this expression the values for the Ce ion species, the log term becomes equal to zero:

$$0.0591 \log\left(\frac{2[\mathrm{Sn}^{2+}]^2\,[\mathrm{Sn}^{4+}]^2}{2[\mathrm{Sn}^{4+}]^2\,[\mathrm{Sn}^{2+}]^2}\right) = 0.0591 \log 1 = 0$$

Thus $3E_{eq} = E_0(Ce^{3+}/Ce^{4+}) + 2E_0(Sn^{2+}/Sn^{4+})$. From a table of standard reduction potentials,

$$E_0(\mathrm{Ce}^{3+}/\mathrm{Ce}^{4+}) = 1.28 \text{ V} \qquad (1\ M\ \mathrm{HCl})$$

$$E_0(\mathrm{Sn}^{2+}/\mathrm{Sn}^{4+}) = 0.14 \text{ V} \qquad (1\ M\ \mathrm{HCl})$$

$$E_{eq} = \frac{2(0.14) + 1.28}{3} = +0.52 \text{ V vs. SHE}$$

In practice, it is often easier to observe the equivalence point than to calculate it!

The relationship between the emf of the cell and the number of milliliters of ceric salt added to the stannous solution is of the same form as Fig. 16.7. The equivalence point is denoted by a rapid change in potential as the ceric salt is added. If necessary, the first and second differentials of the curve may be taken to detect the endpoint as shown in Fig. 16.8. From the data obtained, calculations of solution concentrations and other variables are done in the same manner as in conventional volumetric analysis.

The same principle can be used for many oxidation–reduction reactions. The potential of the cell depends on the concentration of the oxidized and reduced forms of ions present. During a titration these concentrations vary as the chemical reaction proceeds:

$$\mathrm{red(A)} + \mathrm{ox(B)} \rightarrow \mathrm{ox(A)} + \mathrm{red(B)}$$

where red(A) is the reduced form of ion A, ox(B) is the oxidized form of ion B, and so on. At the equivalence point there is a rapid change in potential with the addition of the titrating solution. This change makes it possible to detect the equivalence point of the reaction. Typical oxidation–reduction reactions include

$$2Fe^{2+} + Sn^{4+} \rightarrow 2Fe^{3+} + Sn^{2+}$$
$$5V^{2+} + 3MnO_4^- + 4H^+ \rightarrow 5VO_2^+ + 3Mn^{2+} + 2H_2O$$
$$IO_3^- + 5I^- + 6H^+ \rightarrow 3H_2O + 3I_2$$
$$I_3^- + 2S_2O_3^{2-} \rightarrow 3I^- + S_4O_6^{2-}$$

It may be necessary to correct for interferences to the method. For example, the presence of traces of other oxidizing or reducing compounds may displace the equilibrium and give an incorrect voltage for the equivalence points. Many equilibria are pH sensitive, and this too may generate inaccurate results. Care must be taken to correct for these interference effects.

Similarly, by use of potentiometry, the potentials of species in saturated solutions of insoluble salts may be measured, and based on the data the solubility products may be calculated.

Electrolysis. It was shown earlier that when a standard cell is made from a Zn/Zn^{2+} half-cell and a Cu/Cu^{2+} half-cell, Zn dissolves, Cu^{2+} is deposited as Cu metal, and a potential of 1.100 V is developed from the spontaneous cell thermodynamics. If a sufficient voltage is applied in the opposite direction, from a power source or battery, there will be a tendency for these reactions to reverse, provided that no new reactions occur. In this case, if we try to electrodeposit Zn by the reaction

$$Zn^{2+} + 2e^- \rightarrow Zn$$

the evolution of hydrogen will take precedence by the reaction

$$2H^+ + 2e^- \rightarrow H_2$$

and we say that the Zn/Zn^{2+} half-cell is *chemically irreversible*. On the other hand, the copper will dissolve if made an anode in an acidic solution, and the reaction

$$Cu \rightarrow Cu^{2+} + 2e^-$$

is said to be *chemically reversible*. The process of causing reduction and oxidation reactions at electrodes is called *electrolysis*.

The differences and similarities between an electrochemical cell and an electrolysis cell are summarized in Fig. 16.14. Oxidation always occurs at the anode and reduction always occurs at the cathode; notice, however, that in a spontaneous cell the *positive* electrode is the one at which reduction takes place, whereas in an electrolysis cell the *negative* electrode is the one at which reduction takes place. Negatively charged anions, such as Cl^-, NO_3^-, and SO_4^{2-},

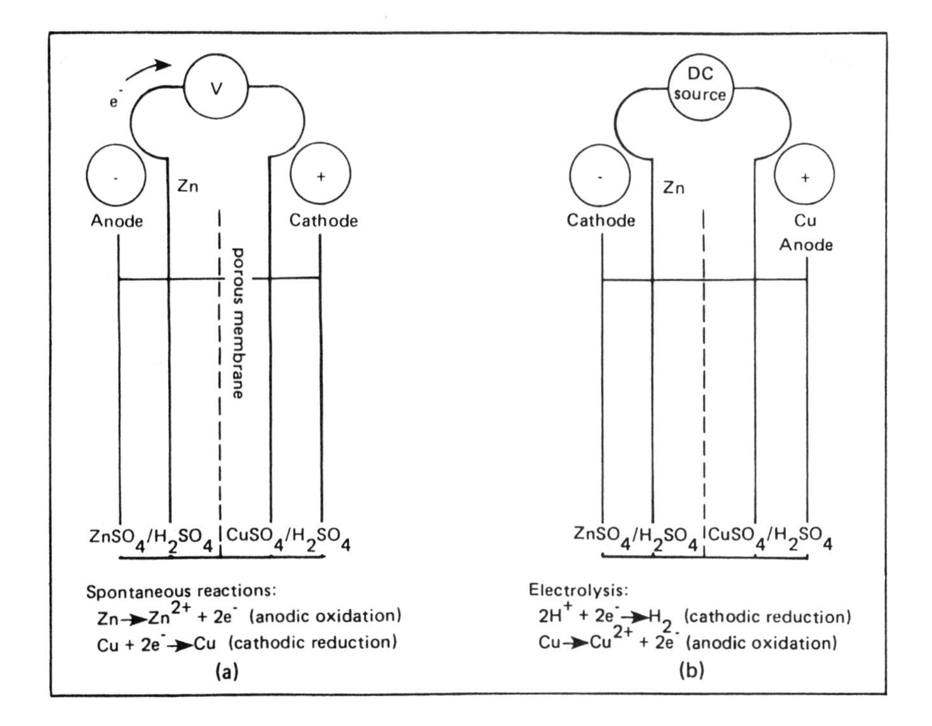

Figure 16.14 Comparison of (a) an electrochemical cell and (b) an electrolysis cell.

are attracted to a positive electrode. Positively charged cations, such as Na^+, Ca^{2+}, and Mg^{2+}, are attracted to a negative electrode. Under the influence of the applied voltage in an electrolysis cell the H^+ and Zn^{2+} ions are attracted to the negative electrode, the cathode; the SO_4^{2-} and OH^- ions are attracted to the positive electrode, the anode. Electrons are consumed at the cathode by H^+ ions to evolve hydrogen gas in the electrolysis cell, but are consumed by Cu^{2+} ions in the spontaneous cell to deposit Cu atoms. For continuous operation of either type of cell, it is necessary for both the anode and cathode reactions to proceed with equal numbers of electrons.

Frequently the reaction that proceeds at either the anode or cathode is the decomposition of the solvent; if water is the solvent, it may electrolyze. Note that water electrolysis depends on the pH:

$$\begin{aligned}
\text{For reductions: } & 2H^+ + 2e^- \rightarrow H_2 \\
& 2H_2O + 2e^- \rightarrow H_2 + 2OH^- \\
\text{For oxidations: } & 2H_2O - 4e^- \rightarrow O_2 + 4H^+ \\
& 4OH^- - 4e^- \rightarrow O_2 + 2H_2O
\end{aligned}$$

Electrogravimetry. If a solution of a metallic ion, such as copper, is electrolyzed between electrodes of the same metal (i.e., copper), the following reaction takes place:

$$\begin{aligned}
& Cu^0 \rightarrow Cu^{2+} + 2e^- \qquad \text{(anode)} \\
& Cu^{2+} + 2e^- \rightarrow Cu^0 \qquad \text{(cathode)}
\end{aligned}$$

The net result is that metal dissolves from the anode and deposits on the cathode. The phenomenon is the basis of electroplating (e.g., chromium plating of steel), electrowinning, and electrorefining. Also, it is the analytical basis of an electrodeposition method known as *electrogravimetry*. This involves the separation and weighing of selected components of a sample. Some of the elements that can be determined in this manner include Ag, Bi, Cd, Co, Cu, In, Ni, Sb, Sn, and Zn.

For example, a metal alloy may contain nickel and copper. The alloy can be dissolved and the solution electrolyzed, with the result that the copper is selectively and exhaustively deposited on the cathode. The basic apparatus required is a power supply, an inert cathode and anode (usually platinum foil or mesh), and arrangements for stirring. Sometimes a heater is used to facilitate the processes. The electrodeposited copper should form an adherent coating so that it can be washed, dried, and weighed. Instead of dissolution of the platinum metal taking place at the platinum anode, oxygen gas is liberated. As a consequence, platinum is a useful electrode material for redox purposes. The initial reactions are as follows:

cathode	$Cu^{2+} + 2e^- \rightarrow Cu$	$E_0 = +0.337$ V
anode	$2H_2O - 4e^- \rightarrow O_2 + 4H^+$	$E_0 = -1.229$ V
Net reaction	$2H_2O + 2Cu^{2+} \rightarrow 2Cu + O_2 + 4H^+$	$E_{net} = -0.892$ V

We may now consider on a more quantitative basis what influence the applied potential has. If the cell could behave as a thermodynamic cell and the reactants were under standard conditions, we might expect that the voltage needed would be that given by the net reaction above, $E_{net} = -0.892$ V. But this would be for an infinitesimal, reversible change and, similar to any other chemical reaction, an electrochemical reaction has a reaction rate constant. In fact, an additional voltage termed the *overpotential* η is required to drive the electrode reactions. About 2 V has to be applied to reach the current onset of the cell described above, and this threshold is sometimes called the *decomposition voltage*. Thus the total potential that we must apply across the cell is

$$E_{decomp} = E_{net} + IR + \eta$$

An overpotential contribution is required for both the anodic and the cathodic reaction and, in this case, it is mainly to oxidize water at the Pt anode. As copper is deposited and the concentration of Cu^{2+} ions falls, both the ohmic potential drop IR across the electrolyte and the overpotential decrease. If we assume that the solution resistance R remains fairly constant, the ohmic potential drop is directly proportional to the net cell current. The overpotential increases exponentially with the rate of the electrode reaction. Once the Cu^{2+} ions cannot reach the electrode fast enough, we say that *concentration polarization* has set in. An ideally polarized electrode is one at which no faradaic reactions ensue; that is, there is no flow of electrons in either direction across the electrode–solution interface. When the potential of the cathode falls sufficiently to reduce the next available species in the solution (H^+ ions, or nitrate depolarizer), the copper deposition reaction is no longer 100% efficient.

Vigorous stirring, therefore, is necessary for several purposes. It helps to dislodge the gas bubbles at the electrodes. More importantly, it provides convective transport of the Cu^{2+} ions to the cathode as the solution becomes increasingly dilute in Cu^{2+} ion. At this point, if the applied voltage is raised to increase the current density, the potential could become sufficiently negative to cause the reduction of the next easiest reduced species present, in this case H^+ ions:

$$2H^+ + 2e^- \rightarrow H_2 \qquad E_0 = 0.00 \text{ V}$$

Such evolution of H_2 gas can produce unsatisfactory copper deposits, which may flake off and lead to a poor analytical determination. To prevent this, it is usual to add a cathodic depolarizer of NO_3^- ion to the solution. A *depolarizer* is a species more readily reducible than H^+ ion but one that does not complicate the cathodic deposit by occlusion or gas entrapment. In this instance, the product is ammonium ions:

$$NO_3^- + 10H^+ + 8e^- \rightarrow NH_4^+ + 3H_2O$$

If the potential becomes too negative at the cathode, the Ni^{2+} ion might commence to deposit and form an alloy deposit. Note that the order of selectivity is not always that predicted by the standard potentials in the activity series. Control of the pH and the chelates or complexing agents in a solution may alter the reduction–oxidation potentials from those of an uncomplexed species.

Perhaps the most common example of overvoltage encountered in electrochemistry is that needed to reduce H^+ ions at a mercury electrode. On a catalytic Pt surface (platinized Pt, which is a large-area, black Pt deposit) the H_2/H^+ ion couple is said to behave reversibly. This means that one can oxidize hydrogen gas, or reduce H^+ ions, at the standard reduction potential 0.0 V under standard conditions. At a mercury cathode, however, it takes about −1 V vs. SHE to reduce H^+ ions, because hydrogen evolution is kinetically slow. This phenomenon has great importance in polarography, because it enables the analysis of many of the metals whose standard potentials are negative of the standard hydrogen electrode.

Controlled potential electrolysis (potentiostatic control) is an alternative method. This requires a three-electrode cell, so as not to polarize the reference electrode. Controlled potential methods enable one to be even more selective, and if two components have electrochemical potentials which differ by no more than several hundred millivolts, it may still be possible to shift these potentials by complexing one of the species. One disadvantage of exhaustive electrolysis is the time required for analysis, and faster methods of electrochemical analysis are described below.

In summary, important practical considerations in precise electrodeposition are (1) rapid stirring, (2) optimum temperature, (3) correct current density (usually expressed in amperes per square decimeter), and (4) control of pH. The proper conditions have all been worked out for hundreds of systems; when these conditions prevail, deposits are bright and adherent and no unusual precautions are necessary in handling (washing, drying, weighing, etc.) the electrode. In commercial electroplating, additives (glue, gelatin, thiourea, etc.) are used as brighteners and for good adherence. They are less useful in analytical deposition, because they would cause the weight of deposit to be too high and would create an interference in the method. Electrodeposits improve the appearance, wear, and/or corrosion resistance. Electrolysis is also used to recharge batteries, study electrode reactions, form pure metals from solutions and create shapes that cannot be machined (electroforming), and eliminate metallic impurities from solutions.

2. Coulometry

a. Analytical Determinations Using Faraday's Law

The analytical methods described in the foregoing section depend on weighing the electrodes before and after plating out the element under test. As an alternative to weighing the deposit, the quantity of electricity used to deposit the metal can be measured. From this measurement the quantity of metal deposited or the amount of ions reduced or oxidized can be calculated. The calculation is based on Faraday's law, which states that *equal quantities of electricity cause chemical changes of equivalent amounts of the various substances electrolyzed.*

Example. (a) A sample of stannic chloride was reduced completely to stannous chloride according to the reaction

$$Sn^{4+} + 2e^- + Sn^{2+}$$

The applied current was 9.65 A and the time taken for reduction was 16 min 40 sec. What was the initial weight of stannic ion present?

$$\begin{aligned} \text{number of coulombs} &= \text{current } i \text{ (A)} \times \text{time } t \text{ (sec)} \\ &= 9.65 \times 1{,}000 \text{ C} \\ &= 9{,}650 \text{ C} \\ \text{number of faradays} &= \frac{9{,}650}{96{,}500} = 0.1 \text{ F} \end{aligned}$$

But 1 faraday will reduce 1 g-eq weight of stannic ion:

$$1 \text{ g-eq wt.} = \frac{\text{atomic wt. of tin}}{\text{valence change}} = \frac{118.69}{2}$$

that is, 1 F will reduce 59.35 g of stannic ion, or 0.1 F will reduce 5.935 g of stannic ion.

(b) If the original volume of the solution was 250 ml, what was the molarity of the solution?

A *molar solution* is one that contains in 1 liter of solution the same number of grams of solute as the molecular or atomic weight of the solute. The atomic weight of tin is 118.69. A molar solution of tin therefore contains 118.69 g/liter of solution. The volume of the solution is known to be 250 ml. A molar solution with this volume would contain (118.69 × 250 g)/1000 = 29.67 g. It was shown that 0.1 F was used. This amount was sufficient to reduce 5.935 g of stannic ion. Therefore the molarity of the solution was 5.935/29.67 = 0.2 *M* stannic chloride. The weight deposited can be calculated from the equation

$$w = \frac{CM}{nF} \tag{16.28}$$

where

M = molecular weight or formula weight of the sample
w = weight of sample oxidized or reduced
C = number of coulombs used
n = number of electrons involved, or the valence change during the reaction period

Utilizing this technique, it is possible to electrolyze standard solutions such as silver nitrate into which two electrodes are dipped. The quantity of electricity passing between the electrodes determines how much silver is deposited. Conversely, by weighing the quantity of silver deposited, it is possible to calculate the quantity of electricity that is passed through the system. This system can be used as a means of measuring the quantity of electricity that has passed; the system is a *coulometer*.

A gas coulometer is more accurate than a silver chloride coulometer. In a gas coulometer two platinum electrodes are immersed in an aqueous solution. During electrolysis H_2 and O_2 are evolved. The volume of H_2 and O_2 evolved is a direct measure of the number of coulombs of electricity that pass between the platinum electrodes. This volume can be measured quite accurately and is more convenient than weighing a solid material, such as silver, which must be dried prior to weighing. With careful technique, it is possible to evaluate the Faraday constant to seven significant figures.

More recently, electronic integrators have been used that are capable of measuring the number of coulombs by measuring the product of current and time and integrating over the total time period. These electronic integrators are accurate to 0.01% and are capable of giving results as accurate as chemical coulometers.

b. Controlled Potential Coulometry

Coulometry is frequently carried out under conditions of controlled potential. This is achieved by using a third, reference electrode in the system, as described earlier. The reference electrode maintains the potential at the working electrode at a constant value or permits an applied voltage pulse or ramp to be added to the working electrode. The reason for using controlled potential can be seen when we examine the Nernst equation. As a metal is deposited under experimental conditions, the concentration of the remaining metal ions in solutions steadily decreases; therefore, in order to continue the deposition, the potential applied to the system must be steadily increased. When the potential is increased, different elements may begin to deposit and interfere with the results. Their deposition results in an increased weight of metal deposited and an increased number of coulombs passing through the cell. Controlled potential is therefore used to eliminate interferences from other reactions, which will take place at different potentials.

In theory, the equivalence point is never reached, because the current decays exponentially. Since a small amount of material always remains in solution, a correction must be made for it by measuring the current flowing at the end of the analysis. This quantity should be subtracted from the integrated signal in order to give an accurate measure of the material that has deposited during the experiment.

Controlled potential coulometry is usually used to determine the number of electrons involved in a reaction when studies are being carried out on new inorganic or organic compounds. It is also valuable in generating unstable or highly reactive substances in situ with good quantitative control. With the use of Eq. (16.28), it is possible to measure C, w, and M and then deduce n, the number of electrons involved.

c. Coulometric Titrations

An aqueous iodide sample may be titrated with mercurous ion by anodizing a mercury pool electrode. When metallic mercury is oxidized to mercurous ion by a current passing through the system, the mercurous ion reacts directly with the iodide ion to precipitate yellow Hg_2I_2:

$$Hg - e^- \rightarrow Hg^+$$
$$2Hg^+ + 2I^- \rightarrow Hg_2I_2$$

The reaction continues and current passes until all the iodide is used up. At this point some means of endpoint detection is needed. Two methods are commonly adopted. The first uses an amperometric circuit with a small imposed voltage that is insufficient to electrolyze any of the solutes. When the mercury ion concentration suddenly increases, the current will rise because of the increase in the concentration of the conducting species and, possibly, the faradaic reactions. The second method involves using a suitable indicator electrode. An indicator electrode may be a metal electrode in contact with its own ions or an inert electrode in contact with a redox couple in solution. The signal recorded is potentiometric (a cell voltage versus a stable reference electrode). For mercury or silver we may use the elemental electrodes, because they are at positive standard reduction potentials to the hydrogen/hydrogen ion couple.

Figure 16.15 illustrates an apparatus suitable for the coulometric titration described above. The anode and cathode compartments are separated with a fine glass membrane to prevent the anode products from reacting at the cathode, and vice versa. The porosity should be such as to allow minimal loss of titrant in the course of the experiment. The electrolysis circuit is distinct from the endpoint detector circuit. A constant current source may be used, but it is not mandatory, provided that the current is integrated over the time of the reaction, that is

$$q = it \qquad \text{(constant current)}$$

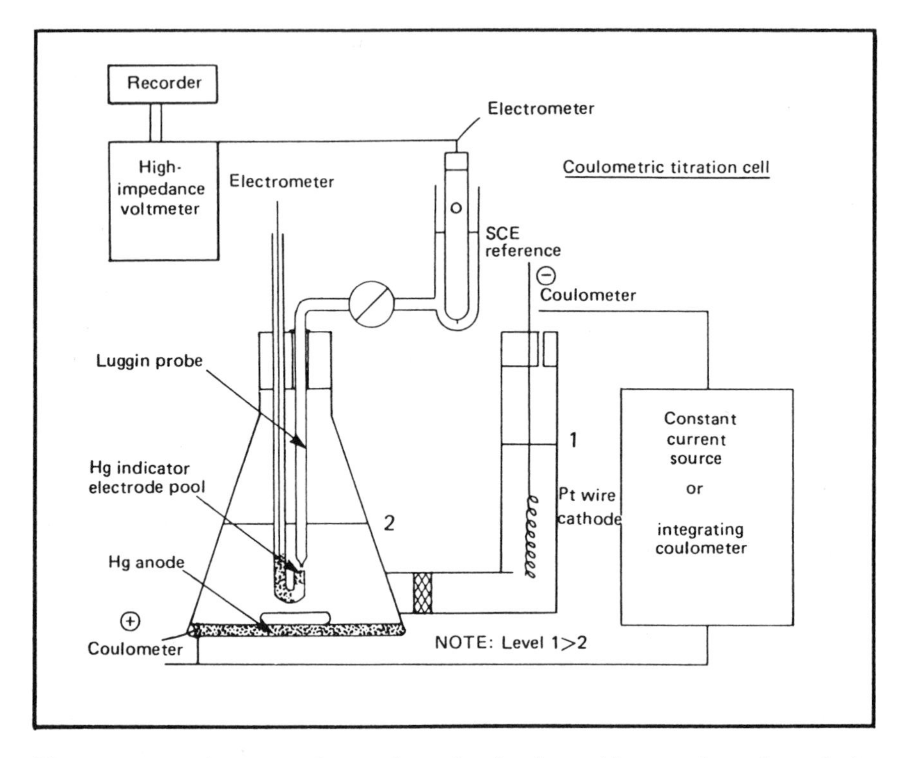

Figure 16.15 Apparatus for coulometric titration with potentiometric endpoint detection.

or

$$q = \int_0^t i(t)\, dt \qquad \text{(integrator)}$$

The endpoint is sensed by recording the voltage across the potentiometric cell, which can be written as

$$\text{Hg} \mid \text{Hg}^+,\ 0.2\ M\ \text{NaNO}_3 \mid 0.1\ M\ \text{HClO}_4 \parallel \text{SCE}$$

The Luggin probe used to monitor the potential of the Hg indicator electrode is spaced close (~1 mm) to the mercury droplet to reduce ohmic resistance. The top barrel is cleaned, wetted with concentrated $NaNO_3$ solution, and closed. Sufficient conductivity exists in this solution layer to permit the small currents necessary for potentiometric determination. When an excess of mercurous ion is generated, the potential of the Hg/Hg^+ ion couple varies in accordance with the Nernst equation, and a voltage follower may be used to output the voltage curve to a Y–time recorder. For precise work, dissolved oxygen should be removed from the supporting electrolyte and efficient stirring is needed.

This type of reaction has been used for the determination of mercaptan, halide, and phosphorus compounds by using a silver electrode. The principle has also been used for many other types of titrations. Another example is the determination of ferrous ions, Fe^{2+}, in the presence of ferric ions, Fe^{3+}. This reaction can be controlled directly by coulometric analysis, but cannot be carried out to completion. As the concentration of Fe^{2+} decreases, the Nernst equation indicates that the potential necessary to continue oxidation will steadily increase until water is oxidized and oxygen is evolved. This will take place before completion of the oxidation of ferrous to ferric ions.

The problem can be overcome by adding an excess of cerous ions, Ce^{3+}, to the solution. The Ce^{3+} ion acts as a *mediator* and itself undergoes no net reaction. With coulometry, the cerous ions can be oxidized to ceric ions, which are then immediately reduced by the ferrous ions present back to cerous ions, generating ferric ions. The reaction continues until all the ferrous ions have been oxidized. At this point any new ceric ions that are formed are not reduced but remain stable in solution, and there is a change in the current flow that signals the endpoint. The endpoint can also be detected using an indicator sensitive to ceric ions.

d. Analytical Applications

Some typical important industrial applications of coulometry include the continuous monitoring of mercaptan concentration in the materials used in rubber manufacture. The sample continuously reacts with bromine, which is reduced to bromide. A third electrode measures the potential of the Br_2 versus the Br^- and, based on the measurement, automatically regulates the coulometric generation of the bromine.

A second example is the continuous analysis and process control of the production of chlorinated hydrocarbons. The chlorinated hydrocarbons are passed through a hot furnace, which converts the organic chloride to HCl. The latter is dissolved in water and the Cl^- titrated with Ag^+. The Ag^+ is generated coulometrically. It is necessary for the sample flow rate to be constant at all times. The coulometric current is proportional to the Cl^- concentration, which activates a $Ag^+/AgCl$ cell throughout the operation.

Another interesting application of coulometry is in the generation of a chemical reagent in solution. Such reagents as acids, bases, reducing agents, and oxidizing agents can all be generated in solution and allowed to react in the manners used in volumetric analysis. In effect, volumetric titrations can be carried out by generating the titrant electrochemically; the amount of titrant used is then measured directly by the amount of electricity used to generate it. This method is very exact, frequently more so than weighing the reagent, because of interfering traces of impurities in the reactant. With coulometry we can obtain results that are more accurate than those possible with the purest reagents. The accuracy depends on our knowledge of the value of the faraday. The method may provide us with the ultimate method of making primary standards for volumetric analysis.

3. Conductometric Analysis

Ohm's law states that the resistance of metal wire is given by the equation

$$i = \frac{E}{R} \tag{16.29}$$

where

E = voltage applied to the wire
i = current of electrons flowing through the wire
R = resistance of the wire

The resistance R depends on the dimensions of the conductor:

$$R = \frac{\rho L}{A}$$

where

ρ = resistivity
L = length
A = cross-sectional area

Another valuable parameter, especially when we consider the mechanisms of current flow in solutions, is conductivity, where

$$\text{conductivity } \sigma = \frac{1}{\rho}$$

The units of conductivity are written $\Omega^{-1}m^{-1}$(reciprocal ohm m^{-1}, mho m^{-1}, or S m^{-1}, where S is the siemen). The conductance of a solution is a measure of how well it carries a current, in this instance by ionic rather than electronic carriers, and it refers to a volume 1 cm long and with 1 cm^2 cross-sectional area.

The charge carriers are ions in electrolyte solutions, fused salts, and colloid systems. The positive ions M^+ migrate through the solution toward the cathode, where they may or may not react faradaically to pick up electrons. Anions, symbolized as A^-, migrate toward the anode, where they may or may not deliver electrons. The net result is a flow of electrons across the solution, but the electron flow itself stops at each electrode. Faradaic reaction of the easiest reduced and oxidized species present may occur, and hence compositional changes (reduction and oxidation) may accompany ionic conductance.

Electrolyte conductivity depends on three factors: the ion charges, mobilities, and concentrations. First, the number of electrons each ion carries is important, because A^{2-}, for example, carries twice as much charge as A^-. Second, the speed with which each ion can travel is termed its *mobility*. The mobility of an ion is the limiting velocity of the ion in an electric field of unit strength. Factors that affect the mobility of the ion include (1) the solvent (e.g., water or organic), (2) the applied voltage, (3) the size of the ion (the larger it is, the less mobile it will be), and (4) the nature of the ion (if it becomes hydrated, its effective size is increased). The mobility is also affected by the viscosity and temperature of the solvent. Under standard conditions the mobility is a reproducible physical property of the ion. Because in electrolytes the ion concentration is an important variable, it is usual to relate the electrolytic conductivity to a unit equivalent conductivity. This is defined by

$$\Lambda = \frac{1000K}{C_{eq}}$$

where

Λ = equivalent conductivity ($\Omega^{-1}cm^2$ per gram-equivalent)
K = electrical conductivity per cubic centimeter
C_{eq} = equivalent concentration

Electrolyte solutions only behave ideally as infinite dilution is approached. This is because of the electrostatic interactions between ions, which increase with increasing concentration. As infinite dilution is approached,

$$\Lambda^0 = F(U^0_+ + U^0_-) = \lambda^0_+ + \lambda^0_-$$

where

U^0_+ and U^0_- = cation and anion mobilities
λ^0_+ and λ^0_- = cation and anion equivalent ion conductivities at limiting dilution

The λ^0's are not accessible to direct measurement, but they may be calculated from transport numbers. Kohlrausch's law of independent ionic conductivities states that at low electrolyte concentrations the conductivity is directly proportional to the sum of the n individual ion contributions, that is,

$$\Lambda^0 = \sum_{i=1}^{n} \lambda^0_i$$

Table 16.3 shows some typical equivalent conductances. From this we can deduce that the equivalent conductance of potassium nitrate is 74 (K^+) + 71 (NO_3^-) = 145, and that of nitric acid is 350 (H^+) + 71 (NO_3^-) = 421. The

Table 16.3 Equivalent Conductances of Various Ions at Infinite Dilution at 25°C

Anions	Conductance (Ω^{-1} cm^2 eq^{-1})	Cations	Conductance (Ω^{-1} cm^2 eq^{-1})
OH^-	198	H^+	350
Cl^-	76	Na^+	50
Br^-	78	K^+	74
ClO_4^-	74	Ag^+	62
NO_3^-	71	Cu^{2+}	55
CH^3COO^-	41	Zn^{2+}	53
$(SO_4^{2-})_2$	80	Fe^{3+}	58

change in conductivity of a solution upon dilution or replacement of one ion by another by titration is the basis of conductometric analysis.

The electrolytes above (KNO_3, HNO_3) are strong and dissociate completely in water. Weak electrolytes do not dissociate completely and many of the ions exist in a combined form as un-ionized molecules. Now the effects of inter ionic forces are less important because the ion concentrations are lower and, in fact, the degree of ionization (α) is readily obtainable from conductance measurements:

$$\alpha \approx \frac{\Lambda}{\Lambda^0}$$

For example, for the weak acid HA,

$$\underset{(1-\alpha)c}{\mathrm{HA}} \rightleftharpoons \underset{\alpha c}{H^+} + \underset{\alpha c}{A^-} \quad \text{molar concentrations}$$

We can express an apparent dissociation constant in terms of conductivities by substitution.

$$K' = \frac{\alpha^2 c^2}{(1-\alpha)c} \approx \frac{\Lambda^2 c}{\Lambda^0(\Lambda^0 - \Lambda)}$$

This is an expression of Ostwald's dilution law, and the equation can be used to determine both K' and Λ^0. Conductance measurements may also be used to find the solubility of sparingly soluble salts.

a. Equipment

The equipment is basically a Wheatstone bridge (see p. 26) and conductivity cell, as illustrated in Fig. 16.16. Resistance A is made up of the cell containing

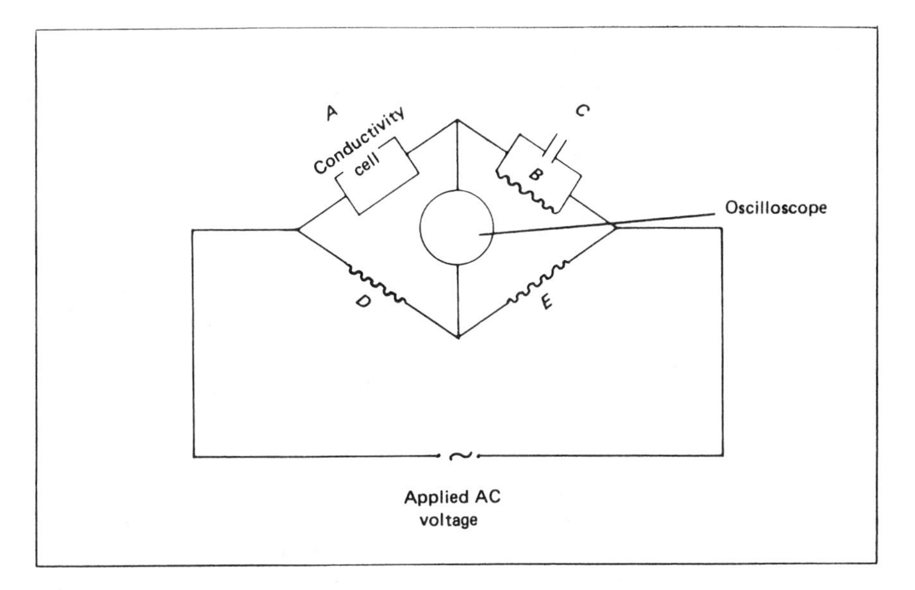

Figure 16.16 Wheatstone bridge arrangement for conductometric analysis.

the sample; *B* is a variable resistance; resistances *D* and *E* are fixed. Resistor *B* and variable capacitor *C* may be adjusted so that the balance point can be reached. At this point

$$\frac{R_A}{R_B} = \frac{R_D}{R_E}$$

The resistances *B*, *D*, and *E* can be measured, and from these measurements the resistance and hence the conductance of the cell can be calculated. A *small* superimposed AC voltage (~20 mV peak to peak) at 1000 Hz is best as a signal, because then faradaic polarization at the electrodes is minimized. The null detector may be a sensitive oscilloscope or a tuned amplifier and meter.

The electrochemical cell usually is constructed with two parallel Pt foil electrodes, each about 1 cm^2 in area. Because the dimensions of a constructed cell are not ideal, one must calibrate a particular cell using KCl solutions of known strength. For example, at 25°C, 7.419 g of KCl in 1000 g of solution has a specific conductivity of 0.01286 $\Omega^{-1}cm^{-1}$. One further requirement is that the Pt electrodes be platinized. This is achieved by cleaning the Pt in hot concentrated HNO_3 and electrodepositing a thin film of Pt black from a 2% solution of platinic chloride in 2 *N* HCl. The Pt black is a porous Pt film, which increases the surface area of the electrodes and further reduces faradaic effects.

b. Analytical Applications

When the ions of one element are replaced in solution by ions of another element, a change in conductivity occurs. For example, when NaOH is added to HCl, the following reaction occurs:

$$\text{Chemical reaction: } HCl + NaOH \rightarrow NaCl + H_2O$$
$$\text{Ionic reaction: } H^+ + Cl^- + Na^+ + OH^- \rightarrow Na^+ + Cl^- + H_2O$$

The original solution contains $H^+ + Cl^-$ in water; the final solution contains $Na^+ + Cl^-$ in water. It can be seen that Na^+ replaces H^+. If we plot the conductivity while NaOH is being added, we observe the relationship depicted in Fig. 16.17. In part *A* of the curve, H^+ ions are being removed by the OH^- of the NaOH added to the solution. The conductivity slowly decreases until the neutralization point *C* is reached. In part *B* of the curve, the H^+ ions have been effectively removed from solution. Further addition of NaOH merely adds Na^+ and OH^- to the solution. An increase in conductivity results. The contribution of the Cl^-, Na^+, H^+, and OH^- ions to the total conductivity of the solution can be seen from Fig. 16.17. The volume of NaOH required to titrate the HCl can be measured by projecting lines *A* and *B* to the intersection *C*. Point *C* indicates the required volume of NaOH.

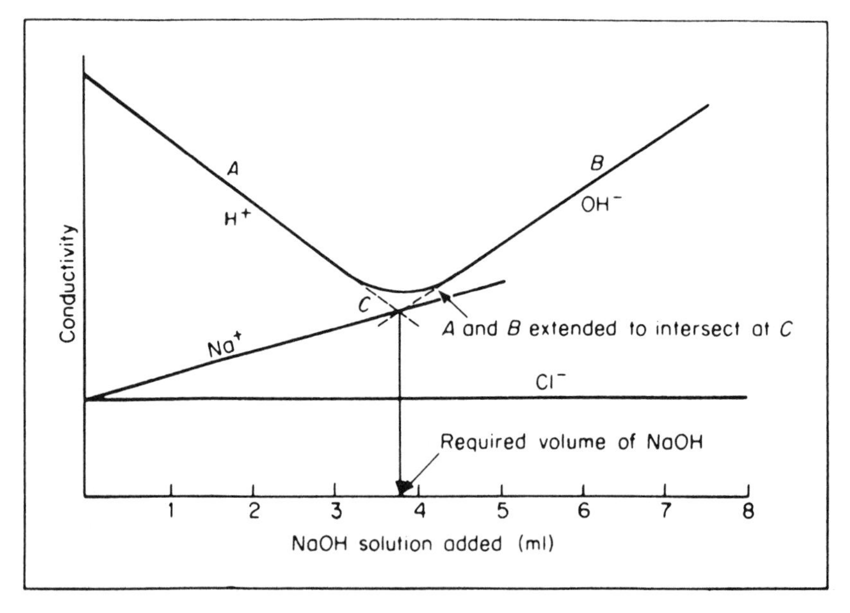

Figure 16.17 Solution conductivity during a HCl–NaOH titration (*C* is the neutralization point).

Conductometric analysis can be used for many titration procedures where ionic solutions are involved. Although it is quite accurate, it has the disadvantage of sometimes requiring sample preparation (dissolving the sample and dilution) and may require samples of significant size (e.g., several milliliters).

Weak acids can also be titrated with NaOH and the endpoint detected by conductometric analysis. A typical curve is shown in Fig. 16.18. As with all weak acids, the H^+ concentration is low and in equilibrium with the acetate ion. At the equivalence point, however, all the H^+ has been neutralized. Any further addition of NaOH has the effect of adding Na^+ and OH^- to the solution. A sharp increase in conductivity occurs, as shown by point C in Fig. 16.18.

Mixtures of weak and strong acids can be titrated with NaOH and the endpoints detected by conductometric analysis. In such mixtures the strong acid is neutralized before the weak acid. Titrations involving mixtures of acids such as these are difficult to perform using indicators because of the problem involved in detecting both endpoints separately. With conductometric analysis the problem is simplified. A typical conductometric curve is shown in Fig. 16.19. It can be seen that two abrupt changes in conductivity take place (C' and C''); these correspond to the titration endpoints. The contributions from the Na^+, Cl^-, acetate$^-$, OH^-, and H^+ to the total conductivity are noted.

The method can be used to show how many acids are in a mixture and to indicate whether they are weak or strong acids. It can also be used to find the endpoints of reactions where precipitation takes place. An example is the titration of silver solution with chloride solution. Again the endpoint is manifested by a sharp change in conductivity. The procedure has also been used for mon-

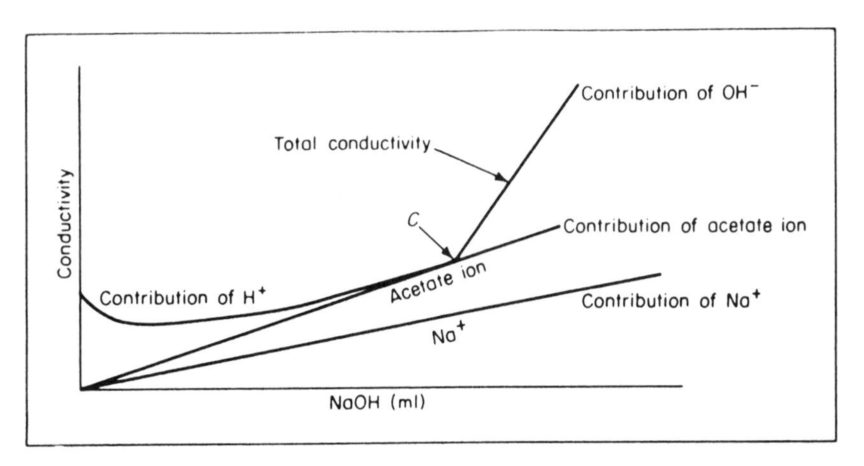

Figure 16.18 Solution conductivity during an acetic acid–NaOH titration (C is the neutralization point).

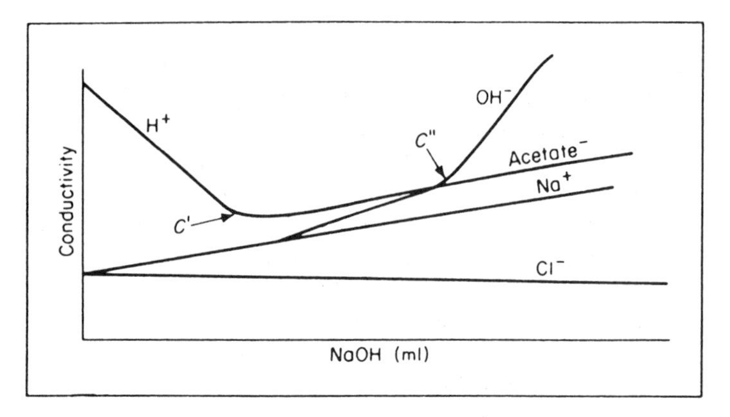

Figure 16.19 Solution conductivity during titration of HCl and CH_3COOH with NaOH. Here C' and C'' designate the titration endpoints.

itoring process streams, and more recently has become popular as a means of detection in liquid chromatography.

c. Nonaqueous Titrations

In nonaqueous solvents, such as alcohol, it is possible to titrate Lewis acids that cannot be titrated in aqueous solutions. For example, it is possible to titrate phenols in alcohol solutions. Phenols act as Lewis acids, releasing hydrogen ion, which can be titrated with a suitable basic material. Since the titration is carried out in a nonaqueous medium, it is necessary to use a base that is soluble in the solvent. A base commonly used is tetramethylammonium hydroxide. This material is basic but soluble in organic solvents and will neutralize Lewis acids.

The procedure can also be used to investigate new molecular species that may be titrated with acids or bases soluble in nonaqueous solvents, such as those found in natural products. The principal problem with the method is that the change of conductance at the endpoint may not be as sharp as in aqueous media, but it is nevertheless identifiable and measurable.

4. Polarography

When a potential difference is applied across two electrodes immersed in a solution, even in the absence of an electroactive species of interest a small current arises due to background reactions and dissolved impurities. If the solution contains various metal ions, these do not electrolyze until the applied negative potential exceeds the reduction potential of the metal ion, that is, becomes more negative. The difference in the current flowing through a solution under two conditions—(1) with the potential less negative than the reduction potential and (2) with the potential more negative than the metal ion reduction potential—is the basis of polarography. Polarography is not restricted to reductions (negative potential sweep), although these are more common. It is also possible to sweep to positive potentials and obtain oxidation curves. Metal cations, anions, complexes, and organic compounds all can be analyzed using polarography. The plot of current against applied potential for a sample solution is called a polarogram.

Polarography is the study of the relationship between the current flowing through a conducting solution and the voltage applied to a dropping mercury electrode (DME). It was discovered by Jaroslav Heyrovsky more than 60 years ago and has since resulted in tens of thousands of research studies. Heyrovsky's pioneering work in the field earned him the Nobel Prize in 1959. In the past two decades many variations of Heyrovsky's classical polarographic method have evolved, principally to improve the sensitivity and resolution of this analytical method. Two important variations are described below, namely, normal pulse polarography and differential pulse polarography.

Classical DC polarography uses a linear potential ramp, i.e., a linearly increasing voltage. It is, in fact, one subdivision of a broader class of electrochemical methods called *voltammetry*. The term *polarography* is usually restricted to electrochemical analyses at the dropping mercury electrode. Figure 16.20 illustrates the potential excitation used in DC polarography and the net current wave form relationship, usually shown as an i–E rather than an i–t curve. The net S-shaped, or sigmoidal, curve obtained when a substance is reduced (or oxidized) is called a polarographic wave and is the basis for both qualitative and quantitative analyses. The *half-wave potential* $E_{1/2}$ is the potential value at a current one-half of the limiting current i_L of the species being reduced (or oxidized). The magnitude of the limiting diffusion current i_L, obtained once the half-wave potential is passed, is a measure of the species concentration. It is standard to extrapolate the background residual current to correct for impurities which are reduced (or oxidized) before the $E_{1/2}$ potential of the species of interest is reached. Although DC polarography is no longer commonly used for analytical purposes, its study is important because it provides the basis of the newer, more powerful variants of classical polarography.

a. Classical or DC Polarography

A modern apparatus for DC polarography is shown in Fig. 16.21. The three-electrode system for aqueous work consists of the DME as the working electrode, a Pt wire or foil auxiliary electrode, and a saturated calomel electrode (SCE) for reference. The potential of the working electrode (DME) is changed by imposition of a slow voltage ramp versus a stable reference (Fig. 16.20). At

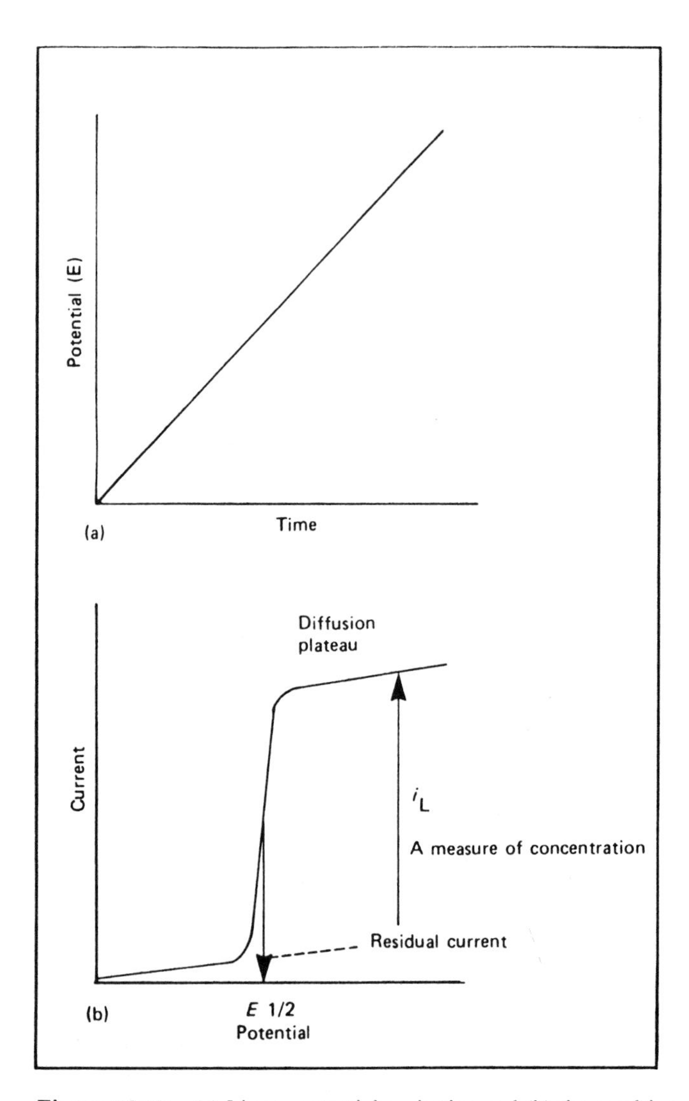

Figure 16.20 (a) Linear potential excitation and (b) the resulting polarographic wave.

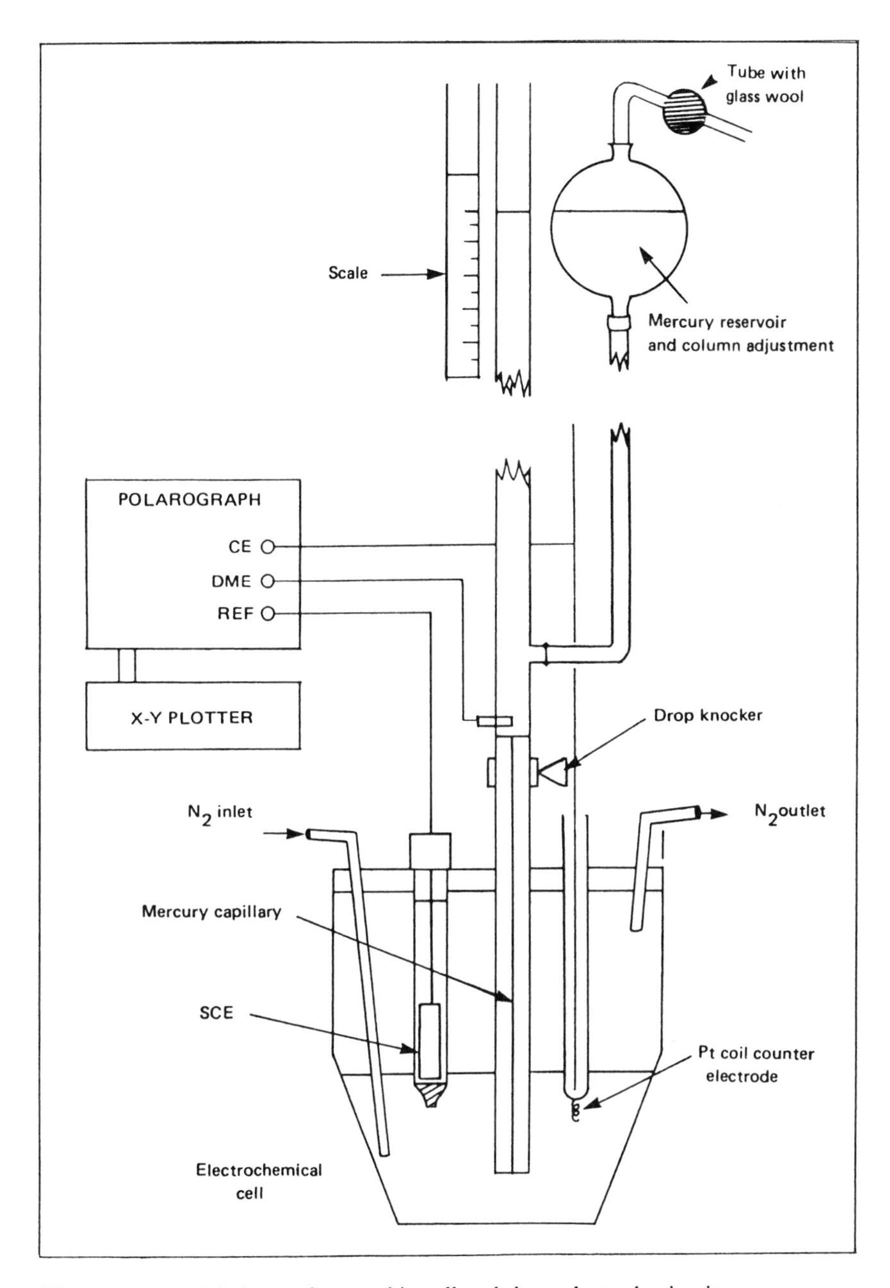

Figure 16.21 Modern polarographic cell and three-electrode circuit.

the heart of polarography is the dropping mercury electrode. A very narrow capillary is connected to a mercury column which has a pressure head that can be raised and lowered. Typically, the level of the mercury column above the tip of the capillary is about 60 cm and the natural interval between drops is 2–6 sec. The choice of mercury for the electrode is important for several reasons:

1. Each fresh drop exposes a new Hg surface to the solution. The resulting behavior is more reproducible than that with a solid surface, because organic contaminants or adsorbants must undergo reequilibration with each new drop.
2. As mentioned earlier, there is a high overpotential for H^+ ion reduction at mercury. This means that it is possible to analyze many of the metal ions whose standard reduction potentials are more negative than that of the H_2/H^+ ion couple. It is easier, too, to reduce most metals to their mercury amalgam than to a solid deposit. Conversely, however, mercury is easily oxidizable, which severely restricts the use of the DME for the study of oxidation processes.
3. Solid electrodes have surface irregularities because of their crystalline nature. Liquid mercury provides a smooth, reproducible surface which does not depend on any pretreatment (polishing or etching) or on substrate inhomogeneity (epitaxy, grain boundaries, imperfections, etc).

In electroanalyses, diffusion currents are quite small ($< 100\ \mu A$), which means that the aqueous solution *IR* drop between the reference electrode and the DME can be neglected in all but the most accurate work. Electrolytes prepared with organic solvents, however, may have fairly large resistances, and in some instances *IR* corrections must be made.

As each new drop commences to grow and expand in radius, the resulting current is influenced by two important factors. The first is the depletion by electrolysis of the electroactive substance at the mercury drop surface. This gives rise to a diffusion layer in which the concentration of the reactant at the surface is reduced. As one travels radially outward from the drop surface, the concentration increases and reaches that of the bulk homogeneous concentration. The second factor is the outward growth of the drop itself, which tends to counteract the formations of a diffusion layer. The net current wave form for a single drop is illustrated in Fig. 16.22.

A mathematical description of the diffusion current, in which the current i_L is measured at the top of each oscillation just before the drop dislodges, is given by the following equation:

$$i_L = 708nCD^{1/2}m^{2/3}t^{1/6}$$

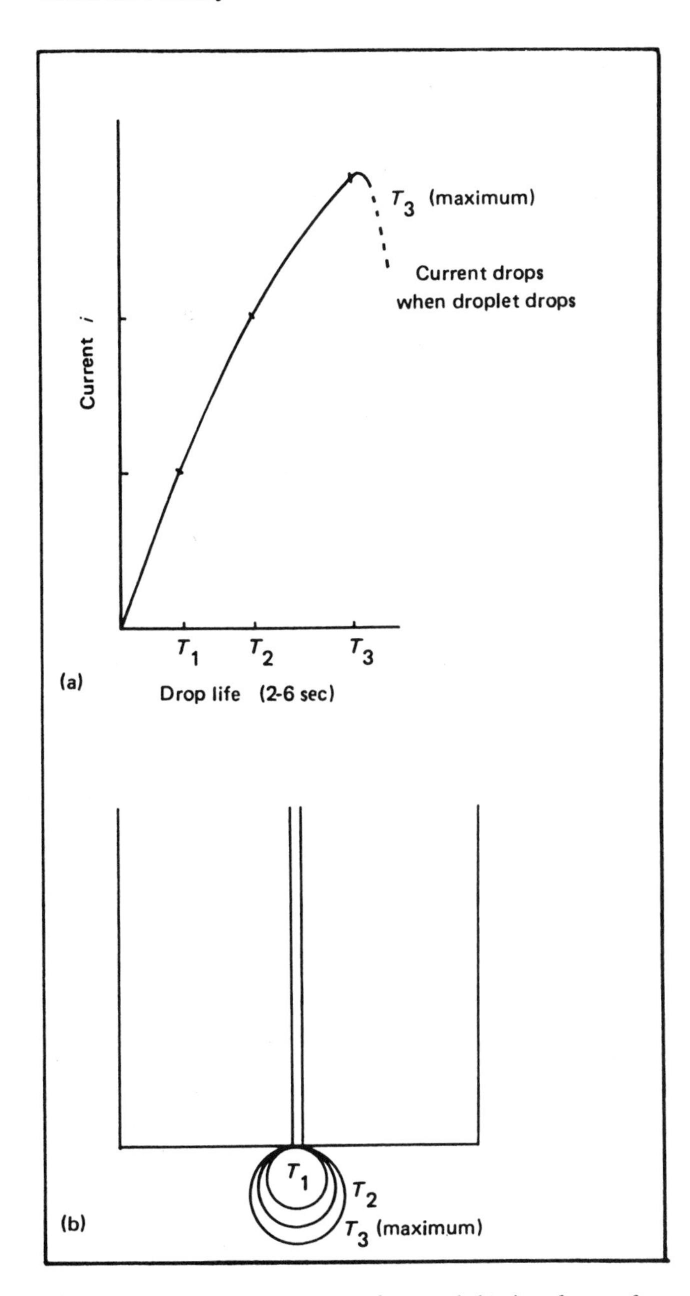

Figure 16.22 (a) Current wave form and (b) time frame of an expanding drop.

where

i_L = maximal current of drop (μA)
n = number of electrons per electroactive species
C = concentration of electroactive species (mM)
D = diffusion coefficient of electroactive species
m = mercury flow rate (mg/sec)
t = drop time (sec)

This is called the *Ilkovic equation*. For a particular capillary and pressure head of mercury, $m^{2/3}t^{1/6}$ is a constant. Also, the value of n and that of the diffusion coefficient for a particular species and solvent conditions are constants. Thus i_L is proportional to the concentration C of the electroactive species, and this is the basis for quantitative analysis. The Ilkovic equation is accurate in practice to within several percent, and routinely $\pm 1\%$ precision is possible. It is commonplace to use standard additions to obtain a calibration curve, or an internal standard. Internal standards are useful when chemical sampling and preparation procedures involve the possibility of losses. The principle is that the ratio of the diffusion currents due to the sample and the added standard should be a constant for a particular electrolyte.

Before a polarogram can be obtained from a sample, two important steps are necessary: (1) The sample must be dissolved in a suitable supporting electrolyte and (2) the solution must be adequately degassed. To understand the purpose of the supporting electrolyte, we should first review the three principal forms of transport that occur in ionic solutions: (1) diffusion, (2) convection, and (3) migration.

Diffusion is the means by which an electroactive species reaches the electrode when a concentration gradient is created by the electrode reaction. The electron transfer process can decrease the concentration of an electroactive species or produce a new species (not originally present in the bulk solution) that diffuses away from the electrode surface. *Convection*, that is, forced motion of the electrolyte, can arise from natural thermal currents always present within solutions or be produced deliberately by stirring the electrolyte or rotating the electrode. In general, natural convection must be minimized by electrode design and by making short time scale measurements (on the order of seconds). The third form of transport is migration of the charged species. *Migration* refers to the motion of ions in an electric field and must be suppressed if the species is to obey diffusion theory. This is done by adding an excess of inert *supporting electrolyte* to the solvent. The role of the supporting electrolyte is twofold: It ensures that the electroactive species reaches the electrode by diffusion and it lowers the resistance of the electrolyte. Typically, supporting electrolyte concentrations are 0.1–1 M. An example is KCl solution. The K^+ ions in this solution are not easily reduced and have the added advantage that the K^+ and Cl^- ions migrate at

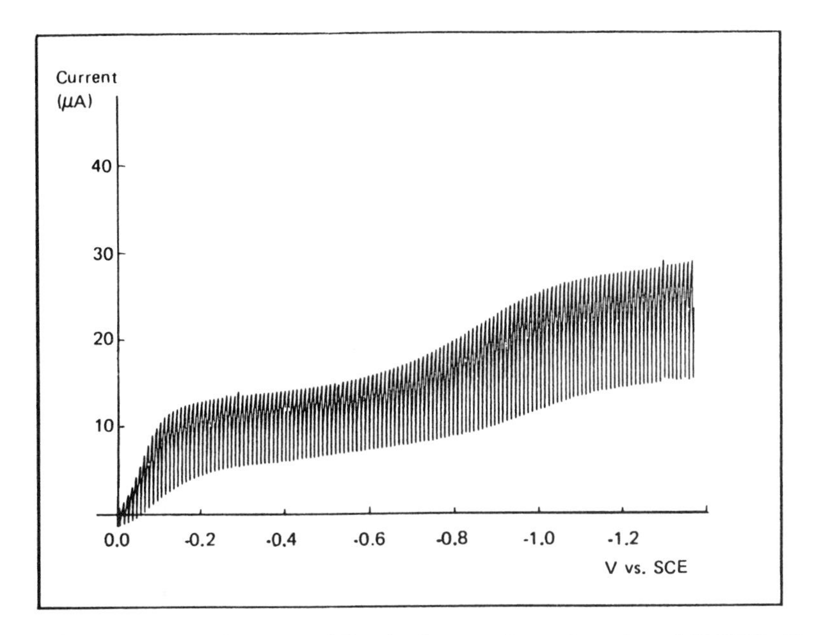

Figure 16.23 Polarogram of dissolved oxygen in an electrolyte, 1 *M* KNO_3 (5 mV/sec, 2-sec drop time, 0.1 mA, *y* scale = 10 V, *x* = 0.2 V/in., *y* = 0.1 V/in.).

about equal velocities in solution. In a solution of $ZnCl_2$ (0.0001 *M*) and KCl (0.1 *M*), the migration current of the zinc is reduced to a negligible amount, and we are therefore able to measure the current response due to zinc ion reduction under diffusion conditions.

Actually, in polarography, even though measurements are made in a quiescent solution (no stirring by either gas bubbling or magnetic bar is permissible), the transient currents arising with each drop depend on both diffusion and the convective motion due to the expanding mercury drop. These effects combine to produce a current that is proportional to $t^{1/6}$.

A further important step is the adequate degassing of the electrolyte by bubbling purified nitrogen or argon. This is necessary because dissolved oxygen from the air is present in the electrolyte and, unless removed, would complicate interpretation of the polarogram. Figure 16.23 shows an actual polarogram of a supporting electrolyte that is air saturated and without added electroactive sample. The abrupt wave at −0.1 V vs. SCE is due to the reduction of molecular oxygen, and the drawn-out wave at −0.8 V vs. SCE is assigned to reduction of a product from the oxygen reduction, namely, hydrogen peroxide. Once the potential exceeds the threshold for peroxide reduction, both of these reactions can occur at each new drop as it grows, and the limiting current is approximately

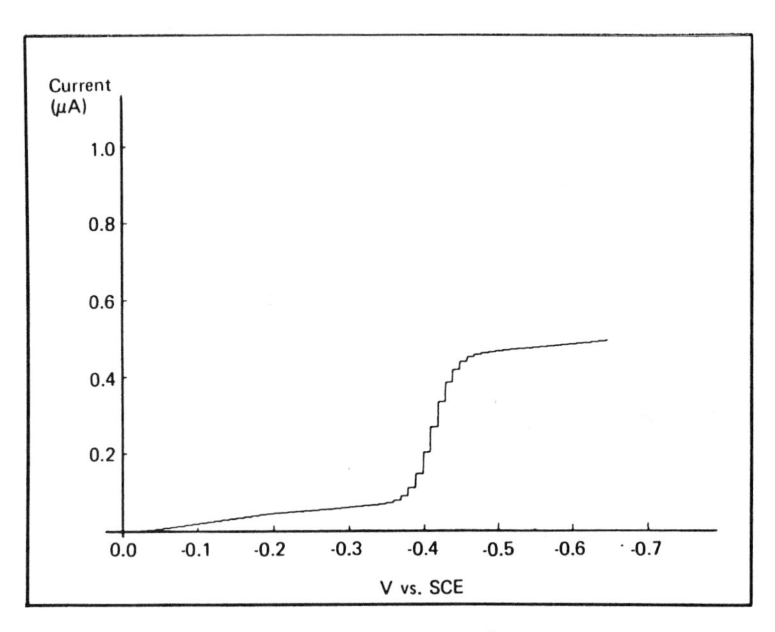

Figure 16.24 Normal pulse polarogram of Pb^{2+} ions (5 ppm) in 1 *M* KNO_3 (scan rate 5 mV/sec, 2-sec drop time).

twice as large. The first reaction is sharp and can be used for the determination of dissolved oxygen in solutions. The addition of a few drops of a dilute Triton X-100 solution is necessary to prevent the formation of maxima in the current at the diffusion plateau. Such maxima distort the wave forms and complicate the measurement of the diffusion currents. Maximum suppressants, such as gelatin and Triton X-100, are capillary-active substances, which presumably damp the streaming currents around the drop that cause the current distortions.

The relationship between current and voltage for a well-degassed solution of $Pb(NO_3)_2$ is shown in Fig. 16.24. Within the voltage range 0 to −0.3 V vs. SCE only a small *residual current* flows. It is conventional in polarography to represent cathodic (reduction) currents as positive and anodic (oxidation) currents as negative. The residual current actually comprises two components; a faradaic contribution and a charging or double-layer current contribution:

$$i_{res} = i_F + i_{ch}$$

The faradaic current i_F arises from any residual electroactive species that can be electrolyzed in this potential range, such as traces of metals or organic contaminants. Such impurities may be introduced by the supporting electrolyte, which will have to be purified in some trace analyses.

The charging current i_{ch} is nonfaradaic; in other words, no electron flow occurs across the metal–solution interface, and neither redox nor permanent chemical changes result from its presence. The mercury–solution interface acts, to a first approximation, as a small capacitor, and charge flows to the interface to create an electrical double layer. At negative potentials this can be thought of as a surplus of electrons at the surface of the metal and a surplus of cations at the electrode surface. In reality, the electrical double layer capacity varies somewhat with the potential; so the fixed capacitor analogy is not strictly accurate. An important consequence of charging currents is that they limit the analytical sensitivity of polarography. This is why pulse polarographic methods, which discriminate against charging currents, can be used to determine much lower concentrations of analyte.

Referring again to Fig. 16.24, we can see that as the potential becomes increasingly negative, it reaches a point where it is sufficient to cause the electroreduction

$$Pb^{2+} + 2e^- \rightarrow Pb(Hg)$$

The Pb^{2+} concentration in the immediate vicinity of the electrode decreases as these ions are reduced to lead amalgam. With a further increase in negative potential, the Pb^{2+} concentration at the surface of the electrode becomes zero, even though the lead ion concentration in the bulk solution remains unchanged. Under these conditions, the electrode is said to be polarized. Notice that the use of an electrode, such as the DME, means that there is no appreciable change of Pb^{2+} ions in the bulk of the solution from a polarographic analysis.

When the electrode has become polarized, a fresh supply of M^+ ions to its surface is controlled by the diffusion of such ions from the bulk of the solution, through the zone depleted of M^+. In other words, the current flowing is dependent on the diffusion of ions from the bulk liquid. This current is called the diffusion current and the plateau region of the curve can be used to measure the limiting diffusion current i_L. It is usual to extrapolate the residual current background and to construct a parallel line through the diffusion current plateau to correct for the residual current contribution to i_L.

Half Wave Potential. An important point on the curve is that at which the diffusion current is equal to one-half of the total diffusion current; the voltage at which this current is reached is the *half-wave potential* $E_{1/2}$. The half-wave potential is used to characterize the current waveforms of particular reactants. Whether a process is termed reversible or not depends on whether equilibrium is reached at the surface of the electrode in the time frame of the measurements. In other words, a process is reversible when the electron transfer reactions are sufficiently fast so that the equilibrium

$$\text{ox} + ne^- \rightleftharpoons \text{red}$$

is established and the Nernst equation describes the ratio [ox]/[red] as a function of potential, that is,

$$E = E_0 + 2.303 \frac{RT}{nF} \log \left(\frac{[\text{ox}]}{[\text{red}]} \right) \tag{16.31}$$

When this is the case, it can be shown that there is a relation between the potential, the current, and the diffusion current i_L, which holds true throughout the polarographic wave:

$$E = E_{1/2} + \frac{0.0591}{n} \log\left(\frac{i_L - i}{i} \right) \text{ (at 25°C)} \tag{16.32}$$

This equation, in fact, may be practically used to test the reversibility (or Nernstian behavior) of an electroactive species. The graph of E versus $\log[(i_L - i)/i]$ will be linear with a slope of $0.0591/n$ and intercept $E_{1/2}$ V. Determination of the slope enables us to determine n, the number of electrons involved in the process. Substitution of $i_L/2$ in Eq. (16.32) reveals that $E = E_{1/2}$ when the surface concentrations [ox] and [red] are equal.

Note, however, that the half-wave potential $E_{1/2}$ is usually similar but not exactly equivalent to the thermodynamic standard potential E_0. First, the product of reduction may be stabilized by amalgam formation in metal ion reductions; second, there will always be a small liquid junction potential in electrochemical cells of this type that should be corrected for; and finally, it can be shown that the potential $E_{1/2}$ is the sum of two terms:

$$E_{1/2} = E_0 + \frac{0.0591}{n} \log\left(\frac{\gamma_{\text{ox}}}{\gamma_{\text{red}}} \right) \left(\frac{D_{\text{red}}}{D_{\text{ox}}} \right)^{1/2}$$

where the γ terms are the activity coefficients and the D terms are the diffusion coefficients of the oxidation reduction species. In most analytical work the activity corrections are ignored and $D_{\text{ox}} \approx D_{\text{red}}$. This is because the size of the electroactive product and reactant is not greatly affected by the gain (or loss) of an electron. The value of the half-wave potential is that it can be used to characterize a particular electroactive species qualitatively. It is not affected by the analyte concentration or by the capillary constant. It can, however, be severely affected by changes in the supporting electrolyte medium. Table 16.4 lists some half-wave potentials of diverse species that may be analyzed by polarography.

b. Normal Pulse Polarography

Unlike classical DC polarography, normal pulse polarography does not use a linear voltage ramp; instead, it synchronizes the application of a square-wave voltage pulse of progressively increasing amplitude with the last 60 msec of the life of each drop. This is shown in Fig. 16.25. It is necessary to use an electronically controlled solenoid to knock the drop from the capillary at a preset time, for

Table 16.4 Half-Wave Potentials of Common Metal Ions[a]

		$E_{1/2}$ vs. SCE
Ag(I)	1 *M* NH_3/0.1 *M* NH_4Cl	−0.24
Cd(II)	1 *M* HCl	−0.64
	0.1 *M* CH_3COONa/0.1 *M* CH_3COOH	−0.65
Cu(II)	1 *M* HCl	−0.22
	0.1 *M* CH_3COONa/0.1 *M* CH_3COOH	−0.07
O_2	0.1 *M* KNO_3	−0.05
Pb(II)	1 *M* HCl	−0.44
	0.1 *M* CH_3COONa/0.1 *M* CH_3COOH	−0.50
Tl(I)	1 *M* HCl	−0.48
Zn(II)	1 *M* NH_3/0.1 *M* NH_4Cl	−1.35
	0.1 *M* CH_3COONa/0.1 *M* CH_3COOH	−1.1

[a]More complete data are available from textbooks (e.g., Bard, A. M., *Modern Polarographic Methods in Analytical Chemistry*, Marcel Dekker, 1980) and from polarographic equipment suppliers (e.g., EG&G Princeton Applied Research Application Briefs and Application Notes).

example, every 2 sec. Each drop has the same lifetime, which is shorter than its natural (undisturbed) span. The initial voltage E_{init} is chosen such that no faradaic reactions occur during most of the growth of the drop. Then, when the rate of change of the drop surface area is less than during the drop's early formation stages, the pulse is applied. As a first approximation the drop surface area can be considered to be constant during the pulse application. If the potential is such that faradaic reaction can take place at the pulse potential, the resultant current decay transient has both a faradaic and a charging current component, as shown in Fig. 16.25. The measurement of current may be made during the last 17 msec of the pulse by a sample-and-hold circuit. This outputs the average current (marked with an X) and holds this value until the next current is ready for output. Thus the output current is a sequence of steady-state values lasting the drop life, for example, 2 sec. The *i–E* curve does not have the large oscillations reminiscent of classical polarography, but may provide a Nernstian wave form directly.

It is readily seen from Fig. 16.25 why the current is sampled in this manner. A faradaic current i_F at an electrode of constant area will decay as a function of $t^{-1/2}$ if the process is controlled by diffusion; on the other hand, the charging current transient i_{ch} will be exponential, with a far more rapid decay. This permits better discrimination of the faradaic current and charging current contributions.

A further important advantage of this technique is that during the first 2–0.060 sec of the life of each drop no current flows. To understand how this is advantageous, we must reconsider what happens in classical DC polarography.

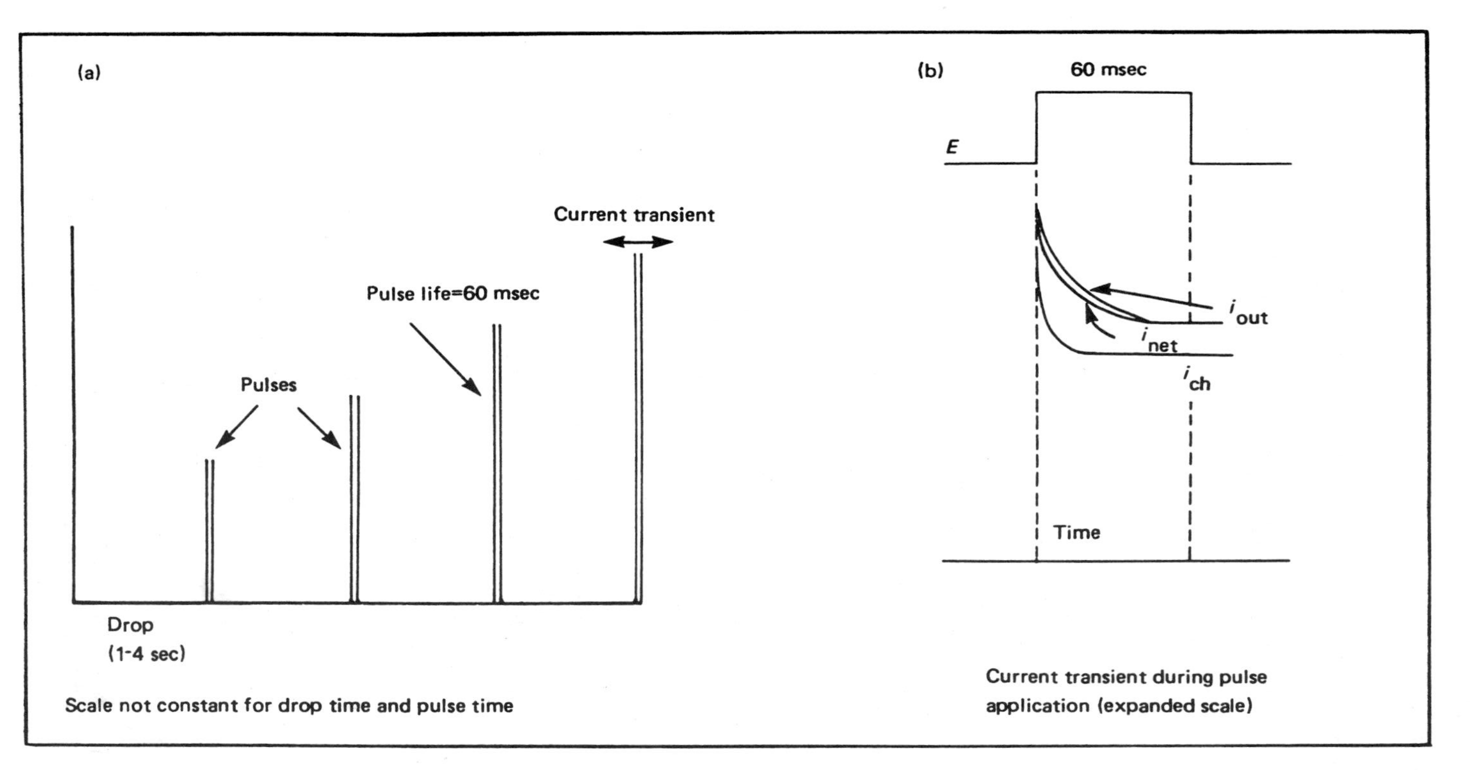

Figure 16.25 (a) Wave form for normal pulse polarography and (b) current transient response (not to scale).

At the start of a drop life, the potential is at a given value and increases slowly as the drop grows. If faradaic reaction can occur, this means that the electroactive species is being depleted during the whole life of the drop, such that when it has grown for about 2–0.060 sec there is a large depletion region extending from the surface. In normal pulse polarography no reaction occurs until the drop area is quite large, and this enhances the current response considerably. The analytical implication is that it is possible to detect concentration levels of about $1 \times 10^{-6} M$ with the normal pulse technique, which is much better than the useful range of about $1 \times 10^{-4} - 1 \times 10^{-2} M$ in classical DC polarography.

c. Differential Pulse Polarography

Differential pulse polarography has the most complex wave forms of the polarographic methods discussed, but it is the easiest to interpret for analytical purposes. The applied voltage is a linear ramp with imposed pulses added during the last 60 msec of the life of each drop (Fig. 16.26). In this technique, however, the pulse height is maintained constant above the ramp and is termed the *modulation amplitude*. The modulation amplitude may be varied between 10 and 100 mV. As in normal pulse polarography, the current is not measured continuously, but in this technique it is sampled twice during the mercury drop lifetime: once just prior to the imposition of a pulse and once just before each drop is mechanically dislodged. As in classical DC polarography the presence of a potential at the start of the life of each drop reduces the analyte concentration by creating a depleted diffusion layer. However, the sampled current output is a differential measurement,

$$\Delta i_{out} = i_{after\ pulse} - i_{before\ pulse}$$

and, as in normal pulse polarography, the time delay permits the charging current to decay to a very small value. As the measured signal is the difference in current, a peak-shaped *i–E* curve is obtained, with the peak maximum close to $E_{1/2}$ if the modulation amplitude is sufficiently small. A differential pulse wave form is shown in Fig. 16.27.

To understand why the wave form is a peak it is necessary to consider what is being measured. The output is a current difference arising from the same potential difference, that is, $\Delta i/\Delta E$. If ΔE becomes very small ($\Delta E \rightarrow dE$), we should obtain the first derivative of the DC polarogram, and the peak and $E_{1/2}$ would coincide. However, because the modulation amplitude is finite and, in fact, has to be reasonably large in order to produce an adequate current signal, the E_{peak} value of the differential pulse polarogram is shifted positive at the half-wave potential for reduction processes:

$$E_{peak} = E_{1/2} - \Delta E/2$$

where ΔE is the modulation amplitude. Increasing ΔE increases the current response but reduces the resolution of the wave form, and experimentally it is best

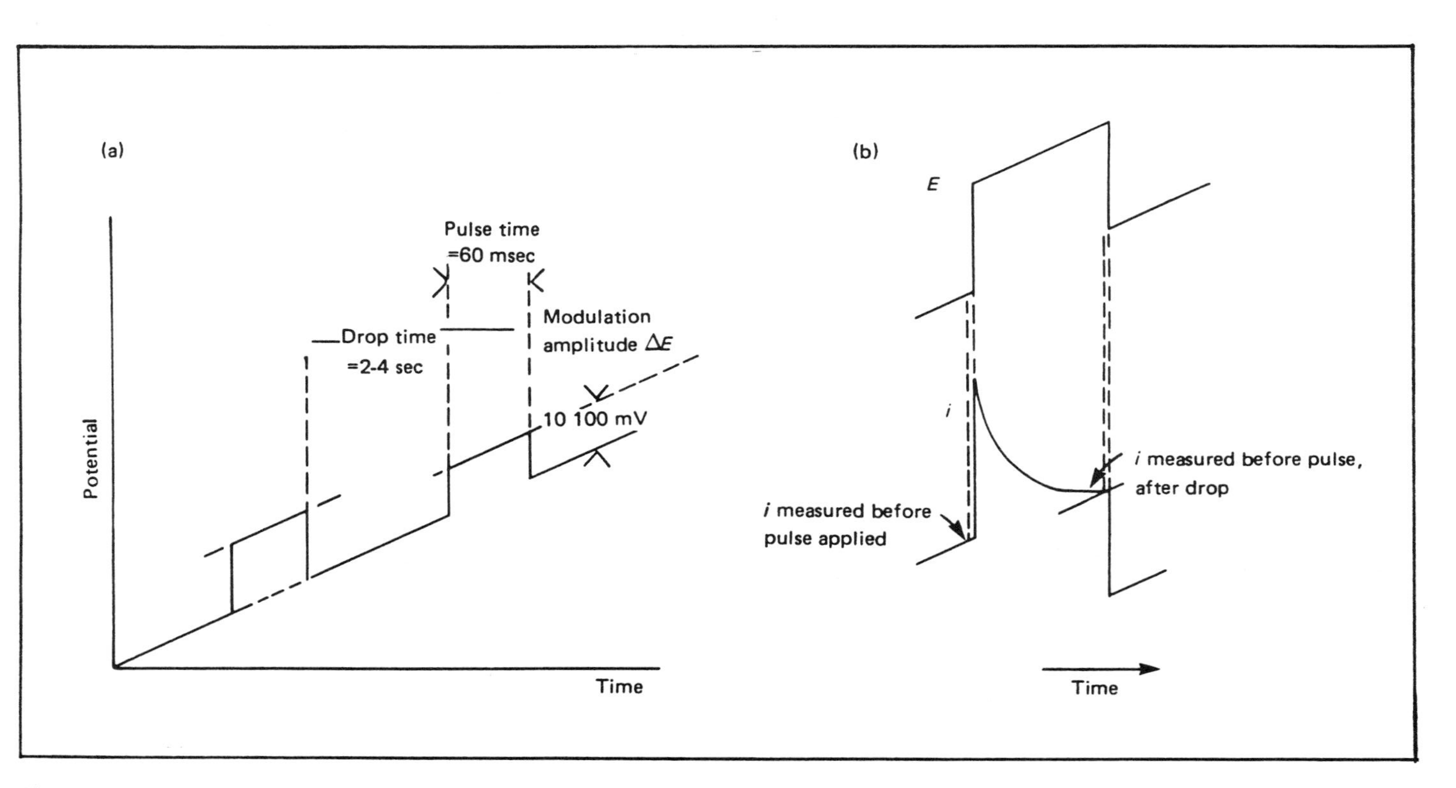

Figure 16.26 Differential pulse polarography (not to scale).

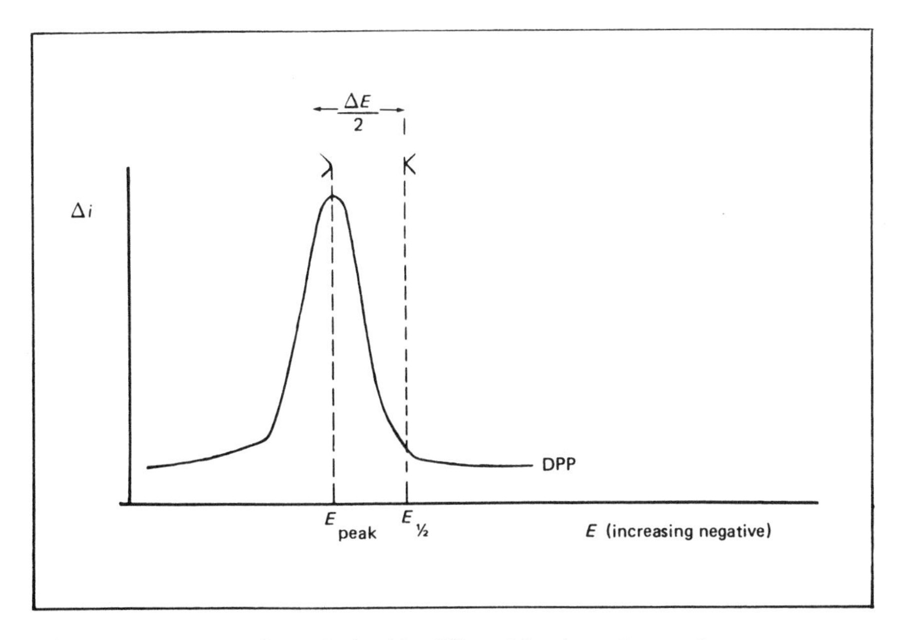

Figure 16.27 Wave form obtained in differential pulse polarography.

to vary the instrument parameters until a suitable combination is found for the ramp scan rate and the modulation amplitude.

The simplicity of the differential pulse polarographic method lies in the fact that the peak height is proportional to the analyte concentration (Fig. 16.28). Note that if adequate resolution exists, it is possible to analyze several ionic species simultaneously with polarography. Differential pulse polarography is especially useful for trace analyses when working close to the electrolyte background reduction or oxidation wave. Its limit of detection is typically $1 \times 10^{-7} M$ or better, depending on the sample and conditions. The addition of surfactants is not always recommended in normal and differential pulse polarography, because their presence may reduce the sensitivity of these methods in certain circumstances. Many other forms of polarography have been suggested and tested. Prominent among these are AC polarographic methods, which use sinusoidal and other periodic wave forms. Most modern instruments offer a range of techniques to the electroanalyst.

The solvent used plays a very important role in polarography. It must be able to dissolve a supporting electrolyte and, if necessary, be buffered. In many cases the negative potential limit involves the reduction of a hydrogen ion or a hydrogen on a molecule. It is therefore vital to control the pH of the solution with a buffer. The buffer normally serves as the supporting electrolyte. Further more, it must conduct electricity. These requirements eliminate the use of many

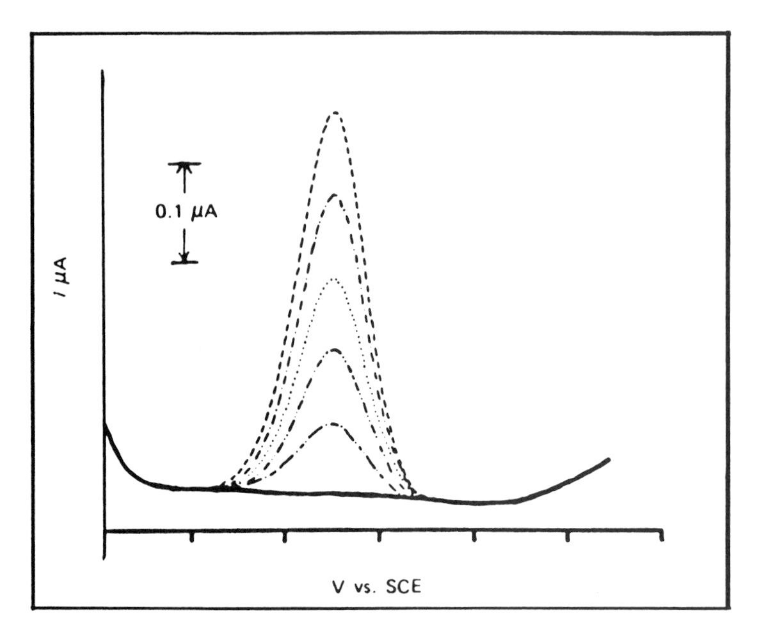

Figure 16.28 Effect of standards addition on a differential pulse polarogram.

organic liquids as such. A polar solvent is necessary, the most popular being water. To dissolve organic compounds in water, a second solvent, such as ethanol, acetone, or dioxane, may first be added to the water. The mixture of solvents dissolves many organic compounds and can be conditioned for polarography. As an alternative to the aqueous/nonaqueous systems, a pure polar solvent may be used, especially if water is to be avoided in the electrolysis. Commonly used solvents are acetonitrile, dimethylformamide, dimethylsulfoxide, and propylene carbonate. Tetralkylammonium perchlorates and tetrafluoroborates are useful supporting electrolytes because their cations are not readily reduced. A list of functional groups that can be determined by polarography is shown in Table 16.5.

In general, simple saturated hydrocarbons, alcohols, and amines are not readily analyzed at the DME. However, aldehydes and quinones are reducible, as well as ketones.

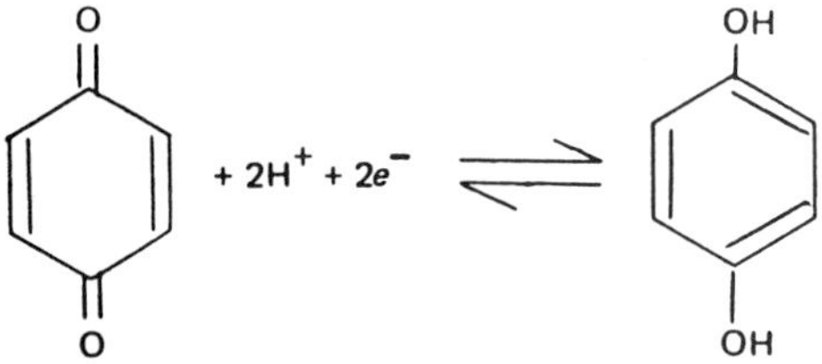

O O $+ 2H^+ + 2e^- \rightleftharpoons$ OH OH

Table 16.5 Typical Functional Groups That Can Be Determined by Polarography[a]

Functional group		Half-wave potential $E_{1/2}$ (V)
R—C(=O)—H	Aldehyde	−1.6
R—C(=O)—OH	Carboxylic acid	−1.8
R—C(=O)—R	Ketone	−2.5
R—O—N═O	Nitrite	−0.9
R—N═O	Nitroso	−0.2
$R—NH_2$	Amine	−0.5
R—SH	Mercaptan	−0.5

[a]Electrochemical data on a large number of organic compounds are compiled in the *CRC Handbook Series in Organic Electrochemistry*, Vols. 1–5 (L. Meites, P. Zuman, and E. B. Rupp, eds.), CRC Press, Inc., Boca Raton, Fla., 1982.

Similarly, olefins (in aqueous solutions) may be reduced according to the equilibrium

$$R—C{=}C—R' + 2e^- + 2H_2O \rightleftarrows R—CH_2CH_2—R' + 2OH^-$$

In short, polarography can be used for the analysis of C—N, C—O, N—O, O—O, S—S, and C—S groups and for the analysis of heterocyclic compounds. Also, many important biochemical species are electroactive, such as vitamin C (ascorbic acid), fumaric acid, vitamin B factors (riboflavin, thiamine, niacin), antioxidants such as tocopherols (vitamin E), *N*-nitrosamines, ketose sugars (fructose and sorbose), and the steroid aldosterone.

d. Voltammetry

We mentioned that polarography is a form of voltammetry in which the electrode area does not remain constant during electrolysis. It is possible, however, to use electrode materials other than mercury for electroanalyses, provided that the potential "window" available is suitable for the analyte in question. Table 16.6 summarizes the accessible potential ranges for liquid mercury and for some solid electrode materials. The precious metals and various forms of carbon are the most common electrodes in use, although a great many materials, both metallic and semiconducting, find use as analytical electrode substrates.

Table 16.6 Potential Windows for Commonly Used Electrodes and Solvents

		Range, V. vs. SCE[a]
Mercury		
Aqueous	1 *M* $HClO_4$	+0.05 to −1.0
	1 *M* NaOH	−0.02 to −2.5
Nonaqueous 0.1 *M* TEAP/CH_3CN[b]		+0.6 to −2.8
Platinum		
Aqueous	1 *M* H_2SO_4	+1.2 to −0.2
	1 *M* NaOH	+0.6 to −0.8
Nonaqueous 0.1 *M* $TBABF_4$/CH_3CN[c]		+2.5 to −2.5

[a]The SCE may only be used in nonaqueous systems if traces of water are acceptable.
[b]TEAP, tetraethylammonium perchlorate.
[c]$TBABF_4$, tetrabutylammonium tetrafluoroborate.

In polarography the DME is renewed regularly during the voltage sweep, with the consequence that the bulk concentration is restored at the electrode surface at the start of each new drop at some slightly higher (or lower) potential. At a solid electrode, the electrolysis process initiated by a voltage sweep proceeds to deplete the bulk concentration of analyte at the surface without interruption. This constitutes a major difference between classical polarography and sweep voltammetry at a solid electrode. Normal pulse polarography is a useful technique at solid electrodes, because the bulk concentration is restored by convection at the surface during the off-pulse, provided that this is at least 10 times longer than the electrolysis period. Note, too, that it is possible to use a mercury electrode in a so-called "static" or "hanging" mode. Such a stationary Hg droplet may be suspended from a micrometer syringe capillary. Alternatively, commercially available electrodes have been developed that inject a Hg drop of varying size to the opening of a capillary. These electrodes find application for stripping voltammetry (see below).

A powerful group of electrochemical methods uses *reversal* techniques. Foremost among these is *cyclic voltammetry*, which is invaluable for diagnosing reaction mechanisms. It is often referred to as the electrochemical equivalent of spectroscopy. A single voltage ramp is reversed at some time after the electroactive species reacts and the reverse sweep is able to detect any electroactive products generated by the forward sweep. This is shown diagrammatically in Fig. 16.29. Usually, an *XY* recorder is used to track the voltage on the time axis so that the reverse current appears below the peak obtained in the forward sweep, but with opposite polarity. The shapes of the waves and their responses at different scan rates are used for diagnostic purposes. Substances are generally examined around the millimolar level, and the electrode potentials at which the species undergo reduction and oxidation may be rapidly determined. Because

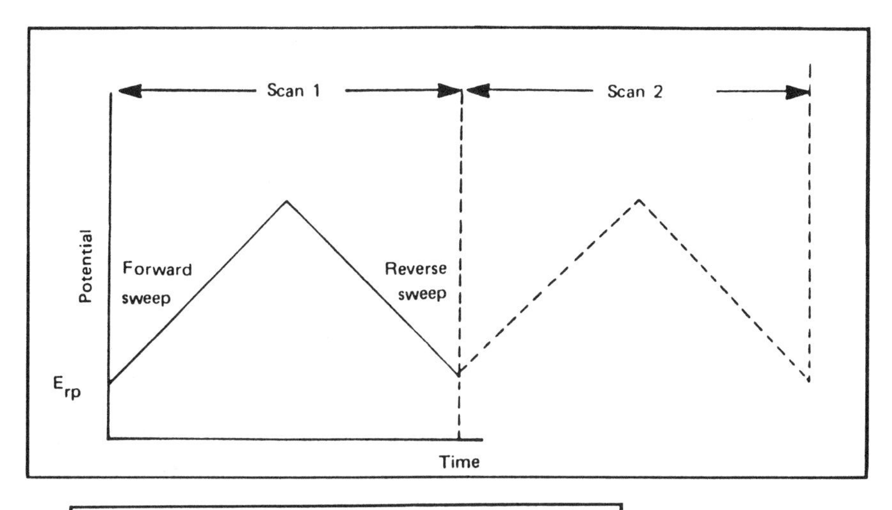

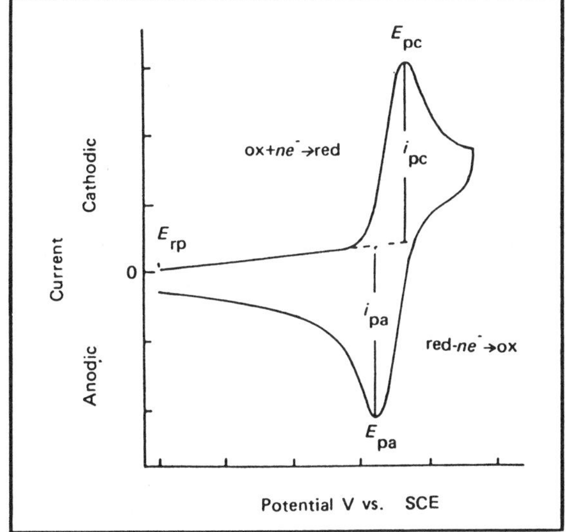

Figure 16.29 (a) Excitation wave form and (b) current response for a reversible couple obtained in cyclic voltammetry.

cyclic voltammetry is not primarily a quantitative analytical technique, it is not described any further here. The references at the end of this chapter provide additional guidance to its applications and interpretation. Its real value lies in the ability to establish the nature of the electron transfer reactions—for example, fast and reversible at one extreme, slow and irreversible at the other—and to explore the subsequent reactivity of unstable products formed by the forward sweep. Suffice it to say that such studies are valuable for learning the fate and degradation of such compounds as drugs insecticides, herbicides, foodstuff contaminants or additives, and pollutants.

Equipment. A cyclic voltammetry mode of operation is featured on many modern polarographs, or a suitable voltage ramp generator may be used in combination with a potentiostat. The three-electrode configuration is required. Pretreatment of the working electrode is necessary for reproducibility (unless a hanging MDE is used), and normally this entails polishing the electrode mechanically with successively finer grades of abrasives.

Some confusion results in choosing an initial voltage at which to start the voltammogram. It is best to measure the open-circuit voltage between the working and reference electrodes with a high-impedance digital voltmeter. This emf is known as the rest potential E_{rp}. Scans may then be made in the negative and/or positive directions starting at the rest potential set on the potentiostat. By this means the potentials for reductions and oxidations, respectively, and the electrolyte limits (the "windows") are established. In other words, it is generally necessary to commence at a potential at which no faradaic reaction is possible, or else voltammograms will be distorted and irreproducible.

Scan rates typically vary from 10 mV/sec to 1 V/sec if an *XY* recorder is used for current output. The lower limit is due to thermal convection, which is always present in an electrolyte. To achieve faster responses an oscilloscope or fast transient recorder is required. The currents in successive scans differ from those recorded in the first scan. For quantitation purposes the first scan should always be used.

e. Stripping Voltammetry

The technique of *stripping voltammetry* may be used in many areas requiring trace analyses to the ppb level. It is especially useful for determining heavy metal contaminants in natural water samples or biochemical studies. Either anodic or cathodic stripping is possible in principle, but analyses by anodic stripping voltammetry are more often used. Stripping analysis is a two-step technique involving (1) the preconcentration of one or several analytes by reduction (in anodic stripping) or oxidation (in cathodic stripping) followed by (2) a rapid oxidation or reduction, respectively, to strip the products back into the electrolyte. Analysis time is on the order of a few minutes. The overall determination involves three phases:

1. Preconcentration
2. Quiescent (or rest) period
3. Stripping process, for example, by sweep voltammetry or differential pulse

It is the preconcentration period that enhances the sensitivity of this technique. In the preconcentration phase precise potential control permits the selection of species whose decomposition potentials are exceeded. The products should form an insoluble solid deposit or an alloy with the substrate. At Hg electrodes the electroreduced metal ions form an amalgam. Usually the potential is set 100–200 mV in excess of the decomposition potential of the analyte of interest. Moreover, electrolysis may be carried out at a sufficiently negative potential to reduce all of the metal ions possible below hydrogen ion reduction at Hg, for example. Concurrent H^+ ion reduction is not a problem, because the objective is to separate the reactants from the bulk electrolyte. In fact, methods have been devised to determine the group I metals and NH_4^+ ion at Hg in neutral or alkaline solutions of the tetraalkylammonium salts. Exhaustive electrolysis is not mandatory and 2-3% removal suffices. Additionally, the processes of interest need not be 100% faradaically efficient, provided that the preconcentration stage is reproducible for calibration purposes, which is usually ensured by standard addition.

Typical solid substrate electrodes are wax-impregnated graphite, glassy (vitreous) carbon, platinum, and gold; however, mercury electrodes are more prevalent in the form of either a hanging mercury drop electrode (HMDE) or thin-film mercury electrode (TFME). Thin-film mercury electrodes may be electro-deposited on glassy carbon electrodes from freshly prepared $Hg(NO_3)_2$ dissolved in an acetate buffer (pH 7). There is an art to obtaining good thin films, and usually some practice is necessary to get uniform, reproducible coverage on the carbon substrates. It is recommended that literature procedures be adhered to carefully. Some procedures recommend simultaneously pre-electrolyzing the analyte and a dilute mercury ion solution ($\sim 10^{-5} M$) so that the amalgam is formed in a single step.

It is important to stir the solution or rotate the electrode during the preconcentration stage. The purpose of this is to increase the analyte mass transport to the electrode by convective means, thereby enhancing preconcentration. In general, in electroanalyses one seeks to obtain proper conditions for diffusion alone to permit mathematical expression of the process rate (the current). In this and controlled flow or rotation cases it is advantageous to purposely increase the quantity of material reaching the electrode surface. Preelectrolysis times are typically 3 min or longer.

In the rest period or quiescent stage, the stirrer is switched off for perhaps 30 sec but the electrolysis potential is held. This permits the concentration gradient of material within the Hg to become more uniform. The rest period is not obligatory for films produced on a solid substrate.

A variety of techniques have been proposed for the stripping stage. Two important methods are discussed here. In the first, in anodic stripping, for exam-

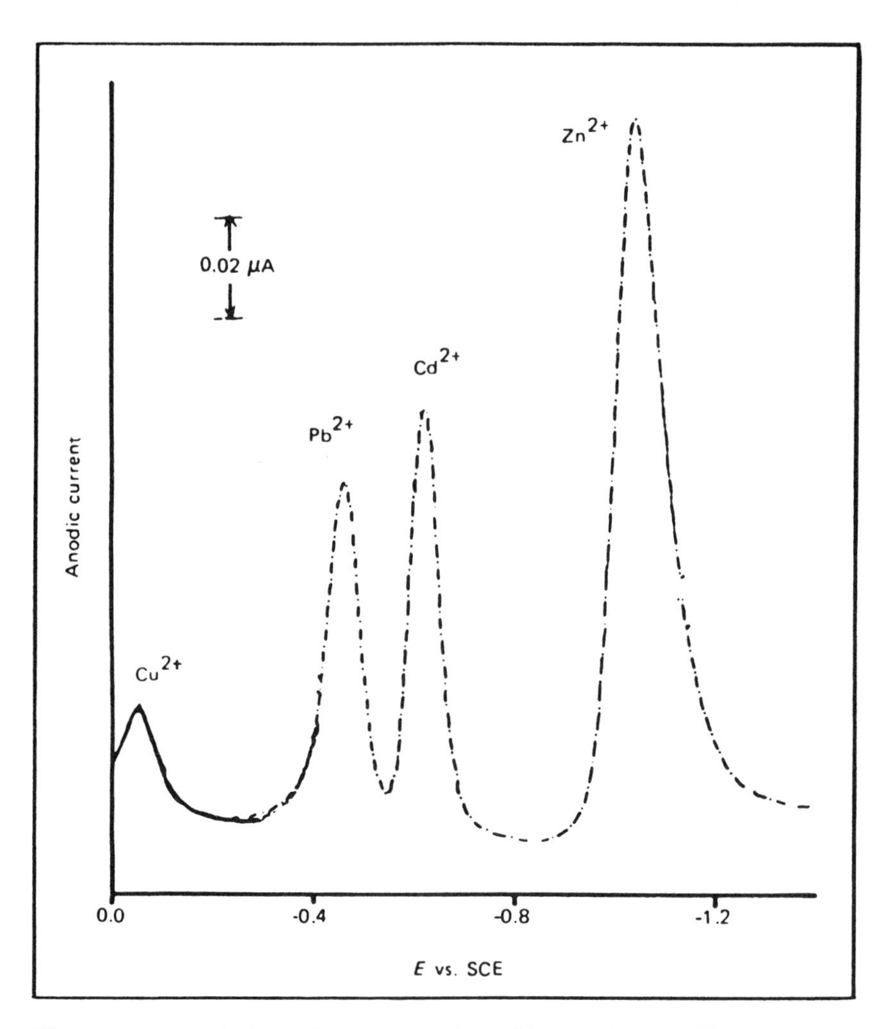

Figure 16.30 Stripping voltammogram of metal ion species at a TFME.

ple, the potential is scanned at a constant rate to more positive values. With this single sweep voltammetry the resolution of a TFME is better than that of a HMDE, because stripping of the former leads to a more complete depletion of the thin film. As illustrated in Fig. 16.30, it is possible to analyze for many metal species simultaneously. The height of the stripping peak is taken to be directly proportional to concentration. Linearity should be established in the working range with a calibration curve. The second method of stripping involves the application of a differential pulse scan to the electrode. As in polarographic methods, an increase in sensitivity is obtained when the differential pulse wave form is used.

Extremely high sensitivities in trace analyses require good analytical practice, especially in the preparation and choice of reagents, solvents, and labware. Glass cells and volumetric glassware must be soaked 24 hr in reagent-purity 6 *M* HNO_3. Plastic electrochemical cells are recommended if loss of sample by adsorption to the vessel walls is a likely problem. The inert gas used to remove dissolved oxygen, N_2, or Ar should be purified so as not to introduce additional contaminants. Solid catalysts and drying agents are recommended for oxygen and water removal. A presaturator is recommended to reduce electrolyte losses by volatilization, especially when low vapor pressure organic solvents are used. A major source of contamination is the supporting electrode itself. This may be purified in part by recrystallization, but the use of sustained preelectrolysis at a mercury pool electrode may ultimately be required. Dilute standards and samples ought to be prepared daily because of the risk of chemical (e.g., hydrolysis) or physical (e.g., adsorption) losses.

Applications. Anodic stripping voltammetry is readily applicable for those metals that form an amalgam with mercury, for example, Ag, As, Au, Bi, Cd, Cu, Ga, In, Mn, Pb, Sb, Sn, Tl, and Zn. One important cause of interferences is intermetallic compound formation of insoluble alloys between the metals within the amalgam, for example, In–Au and Cu–Ni. It is imperative to carefully select the electrolyte such that possibly interfering compounds are complexed and electroinactive. This is also a means of improving resolution when there are two overlapping peaks. Some elements can be analyzed in aqueous electrolytes only with difficulty (groups I and II), but, fortunately, their analysis by flame or atomic absorption methods are sensitive and easy to carry out. Anions of carbon compounds can also be stripped by either (1) anodic preconcentration as sparingly soluble Hg salts, (2) adsorption and decomposition, or (3) indirect methods, such as displacement of a metal complex, for example, thiourea, succinate, and dithiozone. Anions may be determined as mercurous or silver(I) salts if these are sparingly soluble. Based on solubility determinations, it is possible to estimate some theoretical values for the minimum determinable molar concentrations of anions at mercury: Cl^-, 5×10^{-6}; Br^-, 1×10^{-6}; I^-, 5×10^{-8}; S^{2-}, 5×10^{-8}; $CrO_4{}^{2-}$, 3×10^{-9}; $WO_4{}^{2-}$, 4×10^{-7}; $MoO_4{}^{2-}$, 1×10^{-6}; and $C_2O_4{}^{2-}$, 1×10^{-6}. With the exception of Cl^-, the mercurous salts are less soluble than the silver salts.

In conclusion, stripping voltammetry is an inexpensive, highly sensitive analytical tool applicable to multicomponent systems; in fact, it is not recommended for metal ion samples whose concentrations are greater than 1 ppm. Careful selection of operating conditions and especially the electrolyte buffer is necessary. The sensitivity of stripping voltammetry is less for nonmetallic and anionic species. More recently, flow-through systems have been devised for continuous monitoring purposes. Stripping voltammetry has been applied to numerous trace metal analyses and environmental studies, for example, to determine impurities in oceans, rivers, lakes, and effluents; to analyze body fluids, foodstuffs, and soil samples; and to characterize airborne particulates and industrial chemicals.

f. Liquid Chromatography Detectors

Electrochemical detectors for the trace analysis of ionic and molecular components in liquid chromatographic eluents are sensitive, selective, and inexpensive. For many separations they provide the best means of detection, outperforming, for example, UV-visible and fluorescence spectroscopy and refractive index methods. Two major categories of electrochemical detectors will be described below: voltammetric and conductometric. An efficient redox center in the ion or molecule to be detected is necessary for voltammetric methods, but this is not mandatory in the species actually undergoing separation, because many organic compounds can be derivatized with an electroactive constituent before or after column separation. Alternatively, certain classes of organic compounds can be photolytically decomposed by a postcolumn online UV lamp into electroactive products, which can be detected. For example, organic nitrocompounds can be photolyzed to nitrite and nitrate anions, which undergo electrolysis at suitable working electrode potentials. In the case of conductivity detectors the method does not require species that are redox active, or chromophoric; rather, changes in the resistivity of the electrolyte eluent are monitored. Examples of species that may be detected are simple ions such as halides, SO_4^{2-}, PO_4^{3-}, NO_3^-, metal cations, NH_4^+, as well as organic ionic moieties.

Voltammetric Methods. Various wave forms and signal presentation techniques are available, such as steady-state voltammetry with amperometric or coulometric (integrated current) outputs and sweep procedures such as differential pulse. In the simplest mode a sufficient steady-state potential (constant E) is applied to a working electrode to reduce, or oxidize, the component(s) of interest. Commonly the output signal is a transient current peak collected on a Y-time recorder (Fig. 16.31). To obtain reasonable chromatographic resolution, the electrochemical cell, positioned at the outlet of the column, must be carefully designed. The volume of the detector cell must be small to permit maximum band resolution. As the reduction or oxidation is performed under hydrodynamic flow conditions, the kinetic response of the electrode/redox species should be as rapid as possible. Glassy (vitreous) carbon is a popular electrode material, and it is possible to chemically alter the electrode surface to improve the response performance. However, other substrates are available, such as Pt, Hg, and Au; these prove useful in cases where they are not passivated by the adsorption of organic reaction products. Platinum and gold electrodes are particularly prone to blockage from the oxidation products of amino acids, car-

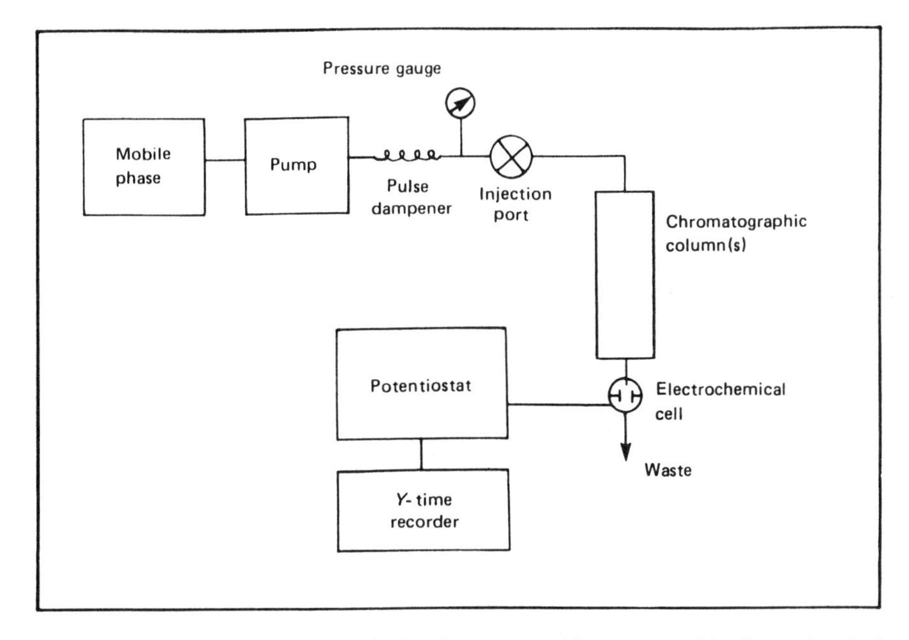

Figure 16.31 Schematic of a liquid chromatographic system with electrochemical detection.

bohydrates, and polyalcohols, for example. Typically, a three-electrode thin-layer cell is used, with the reference and auxiliary electrodes downstream from the working electrode. Dissolved oxygen may interfere in some assays and may therefore have to be removed by prior degassing.

Conductometric Methods. Many important inorganic and organic ions at the trace level are not easily detected by reduction–oxidation or spectroscopic methods. This has led to the development of sophisticated separation and conductometric detectors capable of ultrasensitive analyses. It is difficult to measure small changes in conductivity due to a trace analyte ion when the background level is orders of magnitude larger; consequently procedures have been developed that are capable of suppressing high ionic strength backgrounds, allowing the net signal to be measured more easily. Combined with miniature conductance detectors, these procedures have led to a technique known as *ion chromatography* eluent suppression. Figure 16.32 shows the setup for a typical application: the analysis of the constituents of acid rain samples.

In the first ion exchange column, whose resin exists in its carbonate form ($R^{2+}CO_3^{2-}$), the sample species are selectively separated and eluted as sodium salts in a background of sodium carbonate. Detection at this stage would not be possible because of the large excess of conducting ions Na^+ and CO_3^{2-}. In the

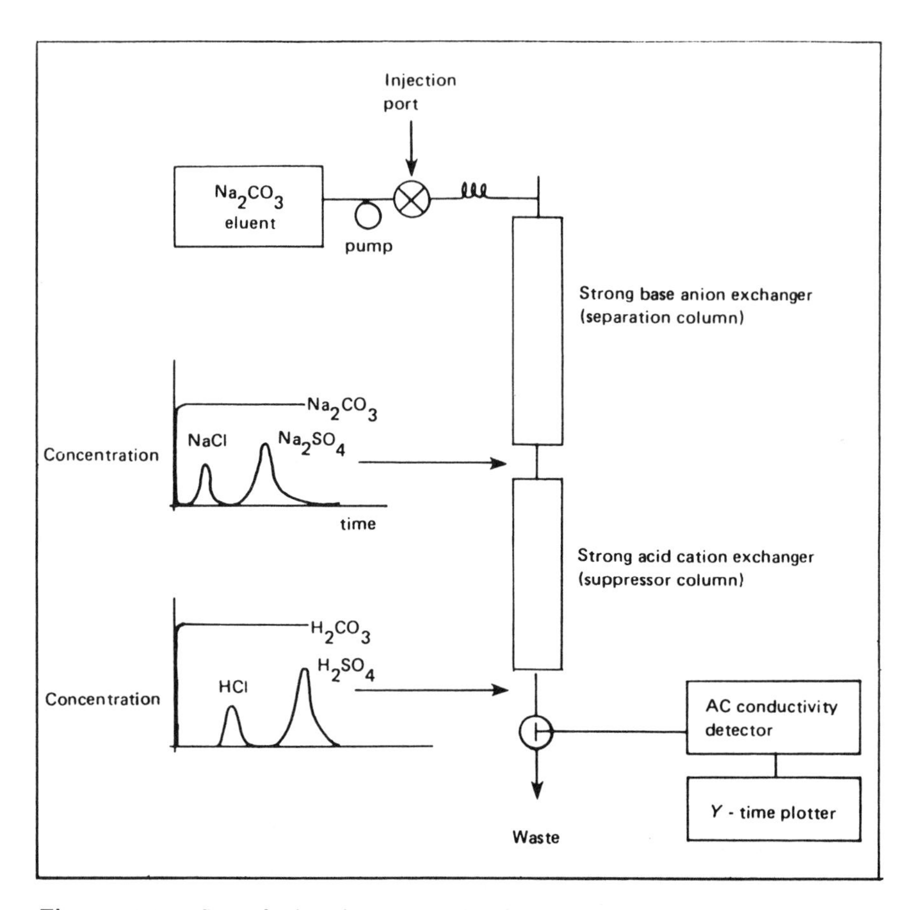

Figure 16.32 Setup for ion chromatography eluent suppression.

second column, the cations undergo ion exchange to produce highly conducting acids (such as the strong acids HCl and H_2SO_4, which are fully ionized) with a background of slightly ionized carbonic acid:

$$H_2CO_3 \rightleftharpoons 2H^+ + CO_3^{2-}$$

A major problem is that the capacity of the second column needs frequent regeneration if large amounts of Na_2CO_3 eluent are used. Various approaches have been tested to improve the practical operation of this form of separation. The first objective is to shorten the time needed to resolve the dilute sample constituents. This is done with a low-capacity separator, which decreases the resolution time and thereby the exchange consumption of the suppressor column. The next objective is to provide continuous regeneration of the suppressor column by automation or by a novel ion exchange membrane methodology that continuously replenishes the column. For cation analyses, lightly sulfonated beads of cross-linked polystyrene have been employed; for anion analysis composite beads are formed by electrostatically coating these beads with anion-exchanging spheres.

Applications. Liquid chromatographic electrochemical detection has been widely used for metabolite studies in complex matrices and has general applicability in many fields, for example, the pharmaceutical industry, forensic science, medicine, the explosives industry, and agriculture.

BIBLIOGRAPHY

Adams, R., *Electrochemistry at Solid Electrodes*, Marcel Dekker, New York, 1969.

Bard, A. J., ed., *Electroanalytical Chemistry*, Marcel Dekker, New York, 1993.

Bard, A. J., and Faulkner, L. R., *Electrochemical Methods—Fundamentals and Applications*, Wiley, New York, 1980.

Baizer, M. M., and Lund, H., *Organic Electrochemistry*, Marcel Dekker, New York, 1983.

Bates, R. G., *Determination of pH*, Wiley, New York, 1973.

Bockris, J. O., *Modern Electrochemistry*, Vols. 1 and 2, Plenum, New York, 1970.

Bond, A. M., *Modern Polarographic Methods in Analytical Chemistry*, Marcel Dekker, New York, 1980.

Dryhurst, G., *Electrochemistry of Biological Molecules*, Academic, New York, 1977.

Ewing, A. G., Mesavres, J. M., Gavin, P. F., Electrochemical detection in microcolumn separations, *Anal. Chem.*, (1994), *9*: 527A.

Fry, A. J., *Synthetic Organic Electrochemistry*, Harper and Row, New York, 1972.

Graaback, A. M., and Jeberg, B., Trace element analysis by computerized stripping potentiometry, *Am, Lab.*, *1*: 27 (1993).

Hart, J. P., *Electroanalysis of Biologically Important Compounds*, Ellis Horwood, New York, 1990.

Ives, D. J. G., and Janz, G. J. (eds.), *Reference Electrodes—Theory and Practice*, Academic, New York, 1961.

Kalvoda, R., *Operational Amplifiers in Chemical Instrumentation*, Ellis Horwood, Chichester, 1975.

Koryta, J., *Ion-Selective Electrodes*, Cambridge University Press, London, 1975.

Sawyer, D. T., and Roberts, Jr., J. L., *Experimental Electrochemistry for Chemists*, Wiley, New York, 1974.

Vydra, F., Stulik, K., and Julakova, E., *Electrochemical Stripping Analysis*, Ellis Horwood, 1976.

Wang, J. *Stripping Analysis*. VCH Publishers, Inc., Deerfield Beach, Fla., 1985.

SUGGESTED EXPERIMENTS

16.1 Run a DC polarogram with a known volume of 1 *M* KNO, electrolyte that has not been deaerated. Observe and identify the two reduction waves of dissolved oxygen.

16.2 Sweep the electrolyte in problem 16.1 with purified nitrogen gas for 20 min or so. Add sufficient lead nitrate solution to make the electrolyte $1.0 \times 10^{-4} M$ in Pb^{2+} ion. Record the DC polarogram and measure the limiting current and the half-wave potential ($E_{1/2}$) for the lead ion reduction wave.

16.3 Measure the mercury flow rate with the column height set as for Problem 16.2. Collect about 20 droplets under the electrolyte (why?) and determine the drop lifetime with a stopwatch. Use the Ilkovic equation to calculate the diffusion equation for Pb^{2+} ion and compare our derived value with a literature value.

16.4 Record a normal pulse polarogram of (a) $1.00 \times 10^{-5} M\ Zn^{2+}$, (b) Cd, (c) Cu ion in a degassed 1 *M* KNO_3 electrolyte. Construct a graph of potential (*E*) versus log $[i_L - i)/i]$. Determine the value of *n* from the slope and comment on the reversibility of this reaction. How does the electrolyte temperature affect the slope of this curve?

16.5 Record a differential pulse polarogram of a background electrolyte without and with added tapwater. Which metal species are present? Use a standard additions calibration to quantify one of the metal impurities. Compare E_{peak} with $E_{1/2i}$.

16.6 Use a F^- ion selective electrode per manufacturer's instructions to find the concentration of F^- ion in your brand of (a) toothpaste and (b) mouthwash. Use a standard additions calibration procedure.

16.7 Collect some samples for pH analyses, for example, rainfall, snow, lake water, pool water, cola. Standardize the pH meter with a standard buffer of pH close to the particular sample of interest. What is the influence of temperature on pH measurements?

16.8 (a) Use a glass pH electrode combination and titrate unknown strength weak acid with 0.2000 *N* NaOH. Stir the electrolyte continuously. (b) Calculate the strengths of the acids and explain the shape of the curve.

16.9 Copper sulfate may be analyzed by gravimetry by exhaustive electrolysis at a weighed platinum electrode. Add 2 ml $C{\cdot}H_2SO_4$, 1 ml $C{\cdot}HNO_3$, and 1 g urea to 25 ml of the copper solution. Suggest a stoichiometry for your copper sulfate based on a variable water of hydration, for example, $CuSO_4{\cdot}CH_2O$.

16.10 Tap waters, as well as a variety of biological fluids and natural waters, can be analyzed for heavy metals by anodic stripping voltammetry at a hanging mercury drop electrode. Clean glassware and sample bottles by leaching for about 5 hours in 6 *N* HNO_3. First, analyze for Cd^{2+} and Pb^{2+} ions at acidic pH (2–4). Deposition is made with gentle stirring for 2 min exactly. After a 15-sec quiescent period, scan from −0.7V SCE in the positive direction. Zn^{2+} and Cu^{2+} ions can be analyzed by raising the pH to 8–9 with 1 M NH_4Cl/1 *M* NH_4OH buffer. Deposit at −1.2V SCE and strip as before. It will be necessary to adjust the deposition time and/or current output sensitivity to achieve best results.

PROBLEMS

16.1 Describe a half-cell. The absolute potential of a half-cell cannot be measured directly. How can the potential be measured?

16.2 (a) What is the Nernst equation? (b) Complete the following table:

pH of solution	Concentration of H^+	E of hydrogen half-cell
1.2		0
3		
	10^{-5}	
	10^{-10}	
7		
12		
14		

16.3 Describe and illustrate a saturated calomel electrode. Write the half-cell reactions.

16.4 How can pH be measured with a glass electrode?

16.5 Describe an op-amp used as a voltage follower. What is its practical use in electrochemistry?

16.6 A salt of two monovalent ions M and A is sparingly soluble. The E_0 for the metal is +0.621 V vs. SHE. The observed emf of a saturated solution of the salt is 0.149 V vs. SHE. What is the solubility product K_{sp} of the salt?

16.7 Describe the process of electrodeposition. What constituents are used in a practical electroplating bath?

16.8 Can two metals such as iron and nickel be separated completely by electrogravimetry? Explain.

16.9 State Faraday's law. A solution of Fe^{3+} has a volume of 1 liter, and 48,246 C of electricity are required to reduce the Fe^{3+} to Fe^{2+}. What was the original molar concentration of iron? What is the molar concentration of Fe^{2+} after the passage of 12,061 C?

16.10 What are the three major forms of polarography? State the reasons why pulse polarographic methods are more sensitive than classical DC polarography.

16.11 What are the advantages of mercury electrodes for electrochemical measurements?

16.12 Describe the method of anodic stripping voltammetry. List the precautions necessary in trace level analyses.

16.13 Describe the principle of ion-selective electrodes and possible interferences.

16.14 How can conductometric measurements be used in analytical measurements? Give examples.

16.15 Briefly outline two types of electrochemical detectors used for chromatography.

Index